W0018234

ICAME 2003

ICAME 2003

*Proceedings of the 27th International Conference
on the Applications of the Mössbauer Effect (ICAME 2003)
held in Muscat, Oman, 21–25 September 2003*

Edited by

M. E. ELZAIN

Sultan Qaboos University, Muscat, Oman

A. A. YOUSIF

Sultan Qaboos University, Muscat, Oman

A. D. AL RAWAS

Sultan Qaboos University, Muscat, Oman

and

A. M. GISMELSEED

Sultan Qaboos University, Muscat, Oman

Reprinted from *Hyperfine Interactions*
Volume 156, Nos. 1–4 (2004)
Volume 157, Nos. 1–4 (2004)

KLUWER ACADEMIC PUBLISHERS
DORDRECHT / BOSTON / LONDON

A C.I.P. Catalogue record for this book is available from the Library of Congress.

ISBN 1-4020-2821-0

Published by Kluwer Academic Publishers,
P.O. Box 17, 3300 AA Dordrecht, The Netherlands

Sold and distributed in North, Central and South America
by Kluwer Academic Publishers,
101 Philip Drive, Norwell, MA 02061, U.S.A.

In all other countries, sold and distributed
by Kluwer Academic Publishers,
P.O. Box 322, 3300 AH Dordrecht, The Netherlands

Printed on acid-free paper

All Rights Reserved
© 2004 Kluwer Academic Publishers
No part of the material protected by this copyright notice
may be reproduced or utilized in any form or by any means,
electronic or mechanical, including photocopying,
recording or by any information storage and retrieval
system, without written permission from the copyright
owner.

Printed in The Netherlands

Table of Contents

Applications in Physics, Including Magnetism and Lattice Dynamics

Biological and Medical Applications

Chemical Applications, Structure and Bonding

Surfaces, Interfaces, Thin Films and Multilayers

Hyperfine Interactions **156/157:** 1–2, 2004.

Preface

These are the proceedings of the 27th International Conference on the Applications of the Mössbauer Effect (ICAME 2003), which was held in Muscat (Oman) during the period 21–25 September 2003. The Iraq war, which took place a few months earlier, shadowed the conference organization during the preparation stages and raised many doubts over its realization. However, the forceful determination and commitment of the faithful participants encouraged the Organizing Committee to carry on. We were pleased of the number of participants that exceeded our expectations. In particular the conference was honored by the participation of Rudolf Mössbauer himself.

The proceedings are divided into nine sections according to the conference topics. All papers were reviewed by at least two referees. Out of the thirteen invited talks presented at the conference, nine were submitted for publication. Each topic section starts with the theme's invited talks wherever available. This is followed by the accepted contributions in alphabetical order of the corresponding author. Contributions, for which the abstracts were received late, are placed towards the end of the relevant section.

A number of people contributed to the realization of these proceedings. Amthauer G., Becker K.D., Bill E., Bonville P., Carbucicchio M., Gancedo R., Greneche J.-M., Genin J.-M., Herber R., Music G., Rueffer R., Pankhurst Q., Sanchez F. and Trautwein A.X., who were selected by the Program Committee as topic coordinators, reviewed and classified the abstracts for presentation at the conference. Following the conference Amthauer G., Bill E., Bonville P., Carbucicchio M., Gancedo R., Greneche J.-M., Genin J.-M., Sanchez F. and Trautwein A.X. helped in the selection of referees to the submitted articles. We very much appreciate the great help extended by the topic coordinators who also tolerated our various and persistent queries and requests. In addition, we would like to thank Guido Langouche, who provided us with additional names whenever we ran short of referees and for his support as the Editor-in-chief of *Hyperfine Interactions*.

The ICAME 2003 was the first major event organized by the Department of Physics. Its staff and students worked in a well-coordinated and cooperative manner, which resulted in a conference that was well praised, in writing, by many participants. The staff of the Public Relation Department at Sultan Qaboos University worked around the clock to ease and facilitate the arrival and departure of all participants. We would like to acknowledge the great contributions of both departments.

Finally we would like to thank Sultan Qaboos University, UNESCO and IS-ESCO for their generous financial contributions.

Mohamed Elzain
Ali Yousif
Ahmed Al Rawas
Abbasher Gismelseed

Hyperfine Interactions **156/157**: 3–8, 2004.
© 2004 *Kluwer Academic Publishers. Printed in the Netherlands.*

Mössbauer Instrument Package MS-2000IP

A. L. KHOLMETSKII[1], V. A. EVDOKIMOV[2], M. MASHLAN[3],
O. V. MISEVICH[2] and A. R. LOPATIK[2]
[1]*Department of Physics, Belarus State University, 4, F. Skorina Ave., 220080 Minsk, Belarus*
[2]*Institute of Nuclear Problems, Belarus State University, 11, Bobruiskaya Str., 220050 Minsk, Belarus*
[3]*Faculty of Experimental Physics, Palacky University, Svobody 26, 771 46 Olomouc, Czech Republic*

Abstract. The paper describes the Instrument Package MS-2000IP, which is based on some new technical ideas of the authors. It allows to increase essentially the productivity of Mössbauer measurements in transmission Mössbauer spectroscopy, in conversion X-ray Mössbauer spectroscopy (XMS), as well as in conversion electron Mössbauer spectroscopy (CEMS).

Key words: transmission Mössbauer spectroscopy, conversion X-ray Mössbauer spectroscopy, conversion electron Mössbauer spectroscopy.

1. General description

The instrument package has been developed on the basis of the personal Mössbauer spectrometer MS-2000 [1]. It contains three spectrometric sections with a common operational module, connected with PC. The first section is based on fast YAP (yttrium aluminum perovskite) scintillation detector in transmission measuring geometry. The second section contains a proportional detector for registration of characteristic iron X-ray radiation in scattering geometry (XMS), while the third section is unitized for CEMS with an air scintillation detector for low-energy electrons. The system of modulation of the energy of resonant gamma-quanta is also common for all sections, and it is based on a mini Doppler modulator [2] with standard feed-back system. Driving system provides an integral non-linearity of the velocity scale less than 0.1%, and the velocity resolution for sodium nitroprusside standard sample is better than 0.24 mm/s. The control system of MS-2000IP allows to choose a velocity form (constant acceleration, constant velocity), velocity range (±100 mm/s), acquisition time, and spectrum name for spectra archiving. The MS-2000IP also contains a section for amplitude analysis on the basis of single channel analyzer with a fixed window and variable position. Data acquisition is realized by PIGGY 32/154/320 microcomputer. Mössbauer spectra of 2048 channels are accumulated in the constant velocity or constant acceleration mode. The main service program is written by the LabVIEW graphical programming language and has a form of a virtual instrument [3].

2. Section for high-performance transmission Mössbauer spectroscopy

In our earlier papers [4, 5] we have shown that the productivity of transmission Mössbauer measurements Q, defined as a ratio of a number of accumulated spectra with a fixed statistic error to the total measuring time, is proportional to

$$Q \sim I_L (S_{s1})^2 / [S_{s1} + 1]^2, \tag{1}$$

where I_L is the limited count-rate of detector, while S_s is the spectrometric selectivity of the detector. The limited count-rate is inversely proportional to a duration of output pulse of the detector. The factor of proportionality is usually taken as $1/10$ for random events [6]. The parameter S_s is defined as a ratio of total count-rate in a selected energy window to the count-rate of resonant events. Equation (1) was used by us in a search of optimal combination of characteristics of detectors in transmission measurements, proceeding from two conclusions:

- if S_s essentially exceeds a unit, than its further increase is not accompanied by essential increase of the productivity Q;
- the productivity Q linearly increases with increase of I_L.

Analysis of conventional detectors for Mössbauer spectroscopy according to Equation (1) reveals incorrectness of traditional approach to a choice of gamma-detectors, when the attention was firstly focused on their energy resolution, without taking into account the value of I_L. In order to increase the productivity of transmission Mössbauer measurements, it is necessary to create such a detector, which has extremely high admissible count-rate I_L and the value of $S_s > 1$. An optimal combination of these requirements was realized in scintillation detector $YAlO_3$:Ce (yttrium aluminum perovskite, YAP). Such a detector has a conversion efficiency about 40% in comparison with NaI(Tl). Therefore, its energy resolution is about 30% worse than for NaI(Tl). It leads to some decrease of S_s. However, this parameter does not play an essential role in productivity of transmission Mössbauer measurements. At the same time, the decay time of YAP is one order of magnitude smaller than for NaI(Tl). This circumstance opens a possibility to enlarger the admissible count-rate of YAP detector. Simultaneously one can choose an optimal thickness of scintillator, which provides almost 100% registration efficiency for 14.4 resonant gamma-quanta with very small registration efficiency for background radiation 122 keV + 136 keV. Under these conditions the fast detector YAP allows to reduce the time of Mössbauer spectra acquisition approximately 6–9 times in comparison with the traditional detectors [4, 5, 7].

The spectrometric section for transmission Mössbauer measurements represents a separate mechanical unit, which includes the Doppler modulator, detector YAP, sample holder and collimating system. The unit contains a double protection from external mechanical vibrations (Figure 1).

Figure 1. Spectrometric section, having a form of tube, is connected with the basic electronic module of MS-2000IP (Mössbauer spectrometer MS-2000).

3. Section for registration of conversion X-ray radiation in back-scattering geometry for XMS

In scattering geometry a detector of radiation is placed outside a direct gamma-beam, that drops a requirement to a high admissible count-rate. In such a case a productivity of measurements is fully determined by the effect-background ratio, which depends on the energy resolution of detector. Due to this reason the fast scintillation detectors with comparably law energy resolution lose their advantages in favor of proportional and semiconductor detectors with high resolution. For registration of characteristic iron X-ray radiation with the energy 6.3 keV we use a xenon proportional counter CI11P, which has a registration efficiency about 100% and the relative energy resolution less than 15%.

Registration section for conversion X-ray radiation represents a separate mechanical unit, which contains a Doppler modulator, proportional counter with shielding, sample holder, collimating system and protective shield. Its view is shown in Figure 2.

4. Section for registration of conversion and Auger electrons for CEMS

It is well known that different kinds of gas detectors with registration of pulses of current are widely-spread detectors for the low energy electrons in CEMS. We suggested and developed a gas detector with registration of pulses of light, accompanying the discharge processes in working gas [8]. Such a method of registration has a number of advantages in comparison with traditional current method. The principal scheme of the developed detector is depicted in Figure 3.

Figure 2. Section for XMS.

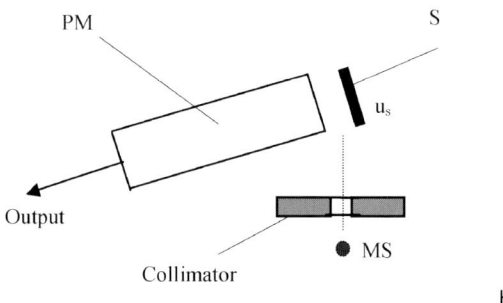

Figure 3. The scheme of air scintillation detector for CEMS.

The sample under investigation (S) is placed near the input window of the photomultiplier (PM). The sample is irradiated by a collimated tangential beam from the Mössbauer source MS. The S, PM and MS are placed in a hermetic chamber HC. The sign of high-voltage on the sample (u_s) is opposite to the sign of high-voltage on the photocathode of PM.

The electrons leave the surface of the sample and cause the micro-discharges in the gap between S and PM. The value of the electric field in the gap is determined by the difference of the electric potentials of the sample and the photocathode of PM. A simplicity of the described construction of the detector is provided by the triple function of the PM: its photocathode is one of the electrodes, its glass bulb plays the role of the isolating film between the electrodes, and the PM properly detects the light pulses. We stress that the isolating film (glass of PM) between the electrodes prevents a development of micro-discharges into self-sustaining discharge in working gas with non-controlled chemical composition. It allows one to use as working gas a natural air. It opens a possibility to conduct measurements with the samples of almost arbitrary form and size. We call the construction in Figure 3 as air scintillation detector (ASD).

Figure 4. Section for CEMS on the basis of air scintillation detector for low-energy electrons.

The selective properties of ASD to low-energy electrons follow from the inverse proportional dependence of the energy loses of electrons on their energy in the range of middle energies. Estimation of S_s for ASD was carried out by method of filter, and in optimal conditions $S_s \approx 2$ under almost 100% registration efficiency for low-energy electrons. In these conditions the value of the resonant effect for a Mössbauer spectrum of a natural sample α-Fe exceeds 10%.

We notice that in case of CEMS the tangential incidence of the gamma-beam on a surface of sample provides an increase of the count-rate by $1/\sin\vartheta$ times in comparison with the case of normal incidence of gamma-beam due to a corresponding increase of the path length of gamma-quanta in the surface layer referring to the maximum escape length of electrons. For chosen value of $\vartheta = 5°$, $1/\sin\vartheta \approx 10$. Hence, the count-rate of the ASD is several times larger compared to normal incidence used in standard CEMS detectors. In addition, the tangential incidence of gamma-beam on a surface of a sample makes the ASD directly sensitive to structural and magnetic anisotropy of the sample, that could be important for practical applications. A general view of the ASD for MS-2000IP is shown in Figure 4.

Currently the developed instrument package MS-2000IP is applied by us in transmission Mössbauer spectroscopy, as well as for investigation of surface layers of materials with involving CEMS and XMS.

References

1. Mashlan, M., Janchik, D., Mulaba, A., Kholmetskii, A. L. and Pollak, H., *Hyp. Interact.* **120–121** (1999), 411. See also the website www.mossp.2000.com.
2. Evdokimov, V. A., Fyodorov, A. A., Misevich, O. V., Mashlan, M., Kholmetskii, A. L. and Zak, D., *Nucl. Instrum. Meth. B* **95** (1995), 278.
3. Kholmetskii, A. L., Mashlan, M., Janchik, D., Zak, D., Dubka, F. and Snasel, V., In: M. Miglierini and D. Petridis (ed.), *Mössbauer Spectroscopy in Material Science*, Kluwer Academic Publisher, Dordrecht, 1999, p. 391.
4. Kholmetskii, A. L., Mashlan, M., Misevich, O. V., Chudakov, V. A., Lopatik, A. R. and Zak, D., *Nucl. Instrum. Meth. B* **124** (1997), 143.

5. Kholmetskii, A. L., Mashlan, M., Nomura, K., Misevich, O. V. and Lopatik, A. R., In: *Current Advances in Materials and Processes*, Vol. 13, The Iron and Steel Institute of Japan, 2000, p. 1417.
6. Lyapidevskii, V. K., *Metody Detektirovaniya Izluchenii*, Atomizdat, Moscow, 1987, 514 p. (in Russian).
7. Mashlan, M., Jancik, D. and Kholmetskii, A. L., *Hyp. Interact.* **139** (2002), 673.
8. Kholmetskii, A. L., Mashlan, M., Misevich, O. V., Anashkevich, A. F., Chudakov, V. A. and Guracevskii, V. L., *Nucl. Instrum. Meth. B* **124** (1997), 110.

Hyperfine Interactions **156/157**: 9–13, 2004.
© 2004 *Kluwer Academic Publishers. Printed in the Netherlands.*

Nuclear Resonant Scattering of Synchrotron Radiation as a Method for Distinction between Covariant Ether Theories and Special Relativity

A. L. KHOLMETSKII[1], W. POTZEL[2], R. RÖHLSBERGER[2], U. VAN BÜRCK[2]
and E. GERDAU[3]
[1]*Department of Physics, Belarus State University, 4, F. Skorina Avenue, 220080 Minsk, Belarus*
[2]*Physik-Department, Technische Universität München, D-85747 Garching, Germany*
[3]*Institut für Experimentalphysik, Universität Hamburg, D-22761 Hamburg, Germany*

Abstract. The paper stresses the importance for basic physics of the new proposed Champeney-like rotor experiment with nuclear resonant scattering of synchrotron radiation. Such an experiment, being sensitive to energy shifts proportional to c^{-3} (c is the light velocity in vacuum), should be able to distinguish between predictions of special relativity theory and covariant ether theories, and thus allow to differentiate between them. The results of computer simulations of experiments with the 14.4 keV resonance in ^{57}Fe show that an energy resolution $\Delta E/E$ at the level of 10^{-16} can be expected which is enough to reveal the third order term.

Key words: Mössbauer effect, special theory of relativity, synchrotron radiation.

Experimental data obtained in high-energy physics and cosmic-ray physics during the past decade again induced an exciting discussion about a possible violation of the Lorentz-invariance in Nature. In this connection some space–time theories with a covariant description of a hypothetical "absolute space" in the Universe (covariant ether theories, CETs) again attract great attention. The ideas of CETs go back to works by Lorentz and Poincaré. However, for a long time, various CETs were considered as physically senseless formal mathematical constructions. The principal possibility of the existence of phenomena, where a hypothetical violation of Einstein's relativity principle might occur within the general relativity principle, was pointed out by Dirac [1]. The possible existence of such phenomena on a laboratory scale was substantiated and predicted in [2].

Let us briefly discuss some important characteristics of the Special Relativity Theory (SRT) and CETs. The SRT is based on two postulates: (a) All inertial reference frames (IRF) have equal rights, they are equivalent. The fundamental physical equations are the same (they are form-invariant) in inertial reference frames. (b) The velocity of light in vacuum c is a constant in all IRF. From these two postulates the Lorentz transformations follow in Minkowski space–time with its Galilean metrics. In particular, an "absolute" inertial frame, distinguished amongst all other inertial frames, does not exist. Lorentz transformations between two IRF are fully determined by their relative velocity. The principal characteristics of CETs are the following: (a) Space–time homogeneity, space–time isotropy, the causality principle as well as the general relativity principles (covariance of fundamental physical equations for admissible space–time transformations) are all valid [2, 3]. (b) An "absolute" inertial frame K_0 is allowed to exist. Therefore the postulates of SRT mentioned above are violated. (c) An "absolute" inertial frame K_0, if it exists, is unique. In K_0 the geometry of space–time is pseudo-Euclidean with Galilean metrics. In any other IRF moving at a constant "absolute" velocity, the metrics of physical space–time is oblique-angled [3]. True (physical) values differ, in general, from their magnitudes measured in experiment.

As a general consequence of these principles, two theorems of CETs follow [3]: (1) A transformation of measured space and time intervals from K_0 to any arbitrary IRF K has a Lorentzian form. (2) Lorentz transformations between two inertial frames $K_1(x_i)$ and $K_2(x_i'')$ always have to proceed via the absolute frame $K_0(x_i')$, where x_i, x_i', and x_i'' are experimentally measured space–time four-vectors. Therefore in CETs, Nature does not "know" a direct relative velocity between two inertial frames K_1 and K_2. Nature only "operates" with absolute velocities, being applied in the Lorentz transformations. A very important consequence of this transformation rule via the absolute frame is the appearance of a frequency (energy) shift between emitter and receiver of electromagnetic radiation, which is proportional to the "absolute" velocity of the Earth [3, 4]. Such a shift appears, e.g., when source and receiver rotate at different distances from a common rotational axis. The shift is proportional to c^{-3}:

$$\Delta E / E = u^2 v / 4c^3 \qquad (1)$$

(u is the linear velocity at the perimeter of the rotor and v is the absolute velocity of the Earth). Although such a possible violation of Einstein's relativity principle represents a tiny effect, it nevertheless can be detected by the modern technique of nuclear resonant scattering of synchrotron radiation. In Ref. [4] we considered a possible experiment with resonant radiation of ^{67}Zn, which, however, faces large experimental difficulties. In this paper we propose an experimental scheme involving the ^{57}Fe resonance, where a high sensitivity is reached due to the application of the recently discovered Nuclear Lighthouse effect [5]. The principal setup, to be realized at an undulator beamline of a third-generation synchrotron radiation source like the ESRF, is shown in Figure 1.

The high-speed rotor carries two targets: the inner target close to the central axis of the rotor and the outer target covering the circumference of the rotor. Both targets are made from metal foils containing the Mössbauer isotope ^{57}Fe with the transition energy of 14.4 keV. After monochromatization to a few meV around the nuclear transition energy achieved by Bragg reflections in Si channel-cut stages, the synchrotron radiation pulse of typically several 100 ps in length excites the nuclei in both targets. This excitation of the nuclei is phased in time by the SR pulse and extends over both spatially separated targets. Such a collective nuclear excitation

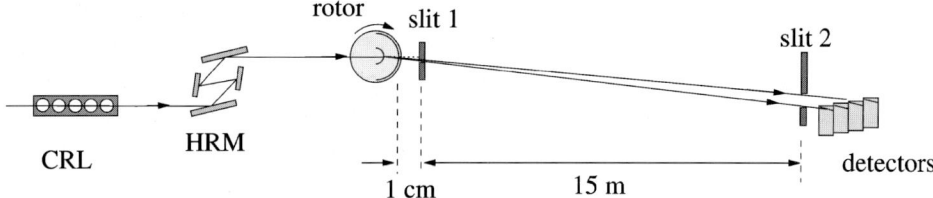

Figure 1. Schematic layout of an experiment at an undulator beamline, e.g., at the ESRF. CRL denotes a compound refractive lens, and HRM a high-resolution monochromator.

(nuclear exciton) follows the rotation of the rotor. This gives rise to the Nuclear Lighthouse Effect because the direction of spatially coherent forward reemission is rotated together with the target. As a result, the time evolution of the nuclear decay is mapped to an angular scale and can be recorded with a position sensitive detector [6]. One can show that background radiation arising from small-angle X-ray scattering (SAXS) from the rotor and the sample itself can be significantly reduced by the use of single-crystalline materials like Al_2O_3 (sapphire) [6]. An additional effect for background reduction relies on the spatial displacement of the nuclear exciton during its lifetime. Due to the motion of the exciton, the radiation sources of the small-angle scattering and the delayed resonant radiation are spatially separated. This allows one to apply a system of slits to almost fully suppress SAXS from the rotor and the sample. In order to avoid SAXS in air, the rotor has to be operated in vacuum.

Due to the energy difference (1), to which the conventional second order Doppler shift (SOD), $\Delta E_{SOD}/E = u^2/2c^2$, has to be added, the radiation from both targets recorded in the detector shows a characteristic Quantum Beat (QB) interference pattern. The QB has the period $T = h/\Delta E$, where h is Planck's constant. In the absence of the effect predicted by CETs, the SOD gives a QB with the period

$$T_{SOD} = \frac{h}{\Delta E_{SOD}}. \tag{2}$$

If the CETs effect is present, as it follows from Equation (1), the T should vary between the extremal values [4]:

$$T_1 = h/(1 + v/2c)\Delta E_{SOD}, \quad \text{and} \quad T_2 = h/(1 - v/2c)\Delta E_{SOD}. \tag{3}$$

Therefore,

$$T_2 - T_1 \approx \frac{v}{c} \cdot \frac{h}{\Delta E_{SOD}} = \frac{v}{c} T_{av}, \tag{4}$$

where the average period T_{av} is given by Equation (2). If the observation window τ_{ob} of the experiment is much larger than T_{av}, the number n of QB maxima within τ_{ob} will be

$$n \approx \frac{\tau_{ob}}{T_{av}}. \tag{5}$$

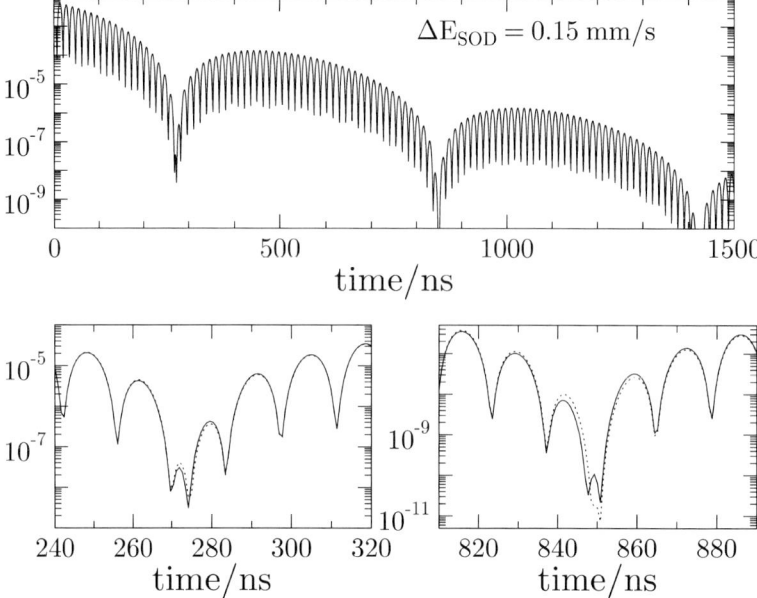

Figure 2. The top graph displays the time spectrum of nuclear resonant forward scattering from two 100 nm thick α-Fe foils with a difference in SOD of 0.15 mm/s. The lower figures display the time ranges around the first two minima of the SOD. It was assumed that CET is valid and introduces the extremal modulation periods of $=0.1499$ mm/s (solid line) and $=0.1501$ mm/s (dashed line).

Then the time difference for n QB maxima between the two extremal periods is given by

$$\Delta T = n(T_2 - T_1) = \frac{\tau_{\mathrm{ob}}}{T_{\mathrm{av}}} \cdot \frac{v}{c} T_{\mathrm{av}} = \frac{v}{c} \tau_{\mathrm{ob}} \approx 10^{-3} \tau_{\mathrm{ob}}. \tag{6}$$

Here and in the further analysis we assume the velocities $u = 300$ m/s and $v = 300$ km/s (a typical value for Galaxy objects relative to the cosmic microwave background radiation). An observation window of $\tau_{\mathrm{ob}} = 1500$ ns has been chosen corresponding to about 10 times the nuclear lifetime τ_s of the Mössbauer level ($\tau_s \approx 141$ ns). Then we obtain $T_{\mathrm{av}} = 575$ ns, i.e. there are about three maxima within τ_{ob}, and $\Delta T \approx 1.5$ ns. Using fast modern electronics such a value for ΔT can be expected to be observable.

A high sensitivity for the measurement of small energy shifts can be achieved when the time response from the foils is additionally modulated by a fast quantum beat pattern. The basic idea is to analyze the structure of the beat pattern in the minima of the SOD oscillations, as well as the shift of the minima. We will explain these features for the case of ferromagnetic Fe metal with an internal hyperfine field of 33.3 T at room temperature. If magnetized perpendicular to the storage ring plane, only the $\Delta m = 0$ transitions are excited, leading to a quantum beat period of 9.5 ns. Figure 2 displays the results of a calculation for an average SOD of 0.15 mm/s ($u = 300$ m/s), that would be modulated between values of 0.1499 mm/s and 0.1501 mm/s if CET were valid. The samples are two iron foils with a thickness of 100 nm each, highly enriched in ^{57}Fe. The calculations show that significant effects can be observed already at early times. Simultaneously it proves the high sensitivity of this method. In particular, the expected relative energy resolution obtained from the calculations of Figure 2 will be better than 10^{-16}. This will be sufficient for a reliable measurement of the effects predicted by CETs.

Concerning the experiment itself, the incident synchrotron radiation beam is focused to the rotor position by a CRL (compound refractive lens) to a vertical beam height of less than 50 μm. The radiation is monochromatized by a HRM (high-resolution monochromator) to a bandwidth of 6.5 meV to reduce the non-resonant background. As indicated in Figure 1, the rotor spins around a horizontal axis with a frequency of 1600 Hz. The detector is located at a distance of approximately 15 m from the rotor, where the resonant radiation is deflected by about 150 mm off the primary beam. The time window of 50 ns around the first QB minimum due to the SOD is selected by a 7.5 mm wide slit. An array of avalanche photodiodes (APDs) covers this time range to monitor the intensity around this minimum. APDs are proposed here because of their very low background noise. However, an ideal detector would be a position sensitive detector with a spatial resolution of about 50 μm and a very low background noise.

Finally, we want to give a rough estimate of the count-rate in such an experiment. The integrated intensity over the time range from 832 ns to 860 ns in Figure 2 amounts to that within an energy range of about 2×10^{-6} Γ_0. This sets a limit for the observable effect. For this reason, the experiment has to be performed at one of the strongest X-ray sources available, like the European Synchrotron Radiation Facility ESRF (Grenoble, France). The best high-resolution monochromator available at beamline ID18 delivers a flux of almost 8×10^9 s^{-1} within a band of 6.4 meV, which corresponds to $6000/(s \cdot \Gamma_0)$ with $\Gamma_0 = 4.7$ neV. With this intensity, one arrives at approximately 130 counts during a 3-hour period falling into the time range mentioned above. Considering the flux available at present third-generation synchrotron radiation facilities, to be sensitive to an effect of $\Delta E / E \approx 3 \times 10^{-16}$ as predicted by CETs, measuring times of several weeks will be required.

Due to the fundamental role of the SRT in modern physics, this new experimental test described here is highly important.

References

1. Dirac, P. A. M., *Nature* **168** (1951), 906.
2. Kholmetskii, A. L., *Physica Scripta* **55** (1997), 18.
3. Kholmetskii, A. L., *Physica Scripta* **67** (2003), 381.
4. Kholmetskii, A. L., *Hyp. Interact.* **126** (2000), 411.
5. Röhlsberger, R., Toellner, T. S., Sturhahn, W., Quast, K. W., Alp, E. E., Bernhard, A., Burkel, E., Leupold, O. and Gerdau, E., *Phys. Rev. Lett.* **84** (2000), 1007.
6. Röhlsberger, R., Quast, K. W., Toellner, T. S., Lee, P., Sturhahn, W., Alp, E. E. and Burkel, E., *Appl. Phys. Lett.* **78** (2001), 2970.

Hyperfine Interactions **156/157**: 15–19, 2004.
© 2004 *Kluwer Academic Publishers. Printed in the Netherlands.*

Mössbauer Spectrometer with Novel Moving System and Resonant Detection of Gamma Rays

MIROSLAV MASHLAN[1], VIKTOR YEVDOKIMOV[2], JIRI PECHOUSEK[1],
RADEK ZBORIL[3] and ALEXANDER KHOLMETSKII[2]
[1]*Department of Experimental Physics, Palacky University, Svobody 26, 771 46 Olomouc,
Czech Republic*
[2]*Department of Physics, Belorussian State University, Skoriny Ave 8, Minsk, Belarus*
[3]*Department of Physical Chemistry, Palacky University, Svobody 8, 771 46 Olomouc,
Czech Republic*

Abstract. A Mössbauer spectrometer with the collective synchronous motion of the radioactive source and resonant detector has been built. The new special transducer with four drive coils and one velocity pickup coil has been developed. The polyamide fibers serve as suspension brackets, barium ferrite magnets are used. The mechanical construction of transducer allows using different cryostats and furnaces, because the sample is immovable. The resonant detector consists of the thin foil of the organic plastic scintillator with the dissolved substance converting the resonant gamma rays to conversion electrons.

1. Introduction

There are two main advantages of the resonant detection of gamma rays in comparison to standard detection in Mössbauer spectroscopy. Firstly, the better signal/noise ratio reduces the time period necessary to the spectrum accumulation [1, 2]. Secondly, the narrower line width allows better to resolve the various Mössbauer subspectra [2–4]. On the other hand, the necessity to use the moving sample restricts the application of both cryostats and furnaces. Principally, the use of the collective synchronous motion of a radioactive source and a resonant detector allows taking advantages of resonant detection of gamma rays. Just one attempt to use the synchronous "source–detector" motion has been made by Maltsev *et al.* [5], but they obtained the satisfactory results only with the harmonic motion.

A Mössbauer spectrometer with resonant detection of gamma rays and with the new special transducer is presented in this paper. Mössbauer spectrum can be accumulated in constant acceleration and constant velocity modes.

2. Moving system

The special transducer of double-loudspeaker type (Figure 1) uses four drive coils (diameters of 24.4 mm), which are made of copper wire (diameter of 0.1 mm). The

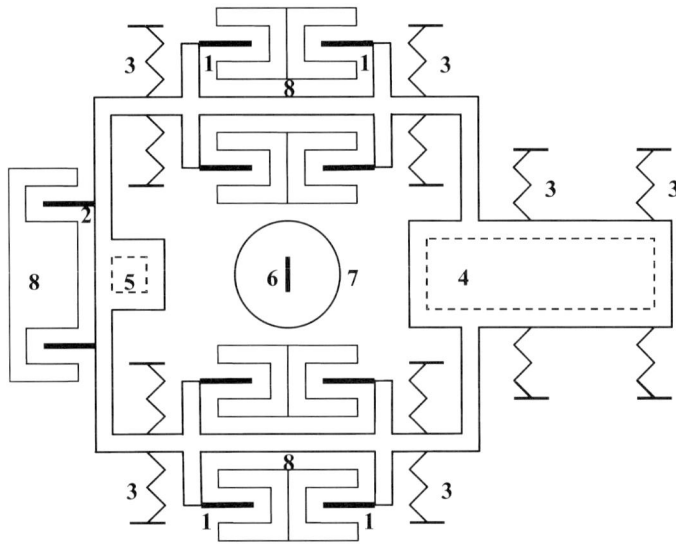

Figure 1. The draft of transducer: 1 – drive coils, 2 – velocity pickup coil, 3 – suspension brackets, 4 – detector unit, 5 – source, 6 – sample, 7 – cryostat, 8 – magnetic systems.

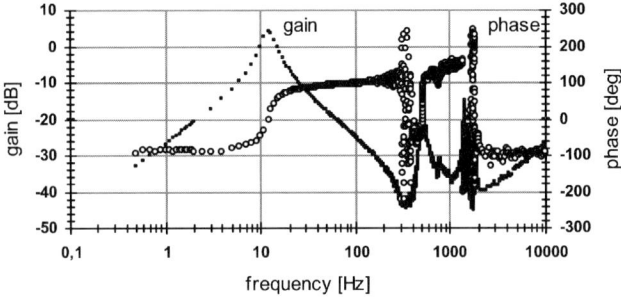

Figure 2. The amplitude and phase frequency characteristics.

resistance of each drive coil is about 16 Ω. The velocity pickup coil (diameter of 27.4 mm) of this transducer is made of 0.07 mm diameter copper wire and its resistance is about 450 Ω. The barium ferrite magnets of the 10 mm high toroid with inner and outer diameters of 32 mm and 72 mm, respectively, are used. Polyamide threads fix the moving part. The transducer amplitude and the phase frequency characteristics are shown in Figure 2. It is obvious that the resonance frequency is about 12 Hz.

Figure 3 shows the schematic circuit diagram of the control drive unit, which consists of the amplifier of the velocity pickup coil signal (IO2), the summator of the reference velocity and the velocity pickup coil signal (IO1), the integrator of the velocity pickup coil signal (IO4A) for correction of error signal, the summator of the error signal with first integral of the velocity pickup coil signal (IO3A), the PID-controller (IO3B), the integrator for correction of dc signal (IO4B) and the power

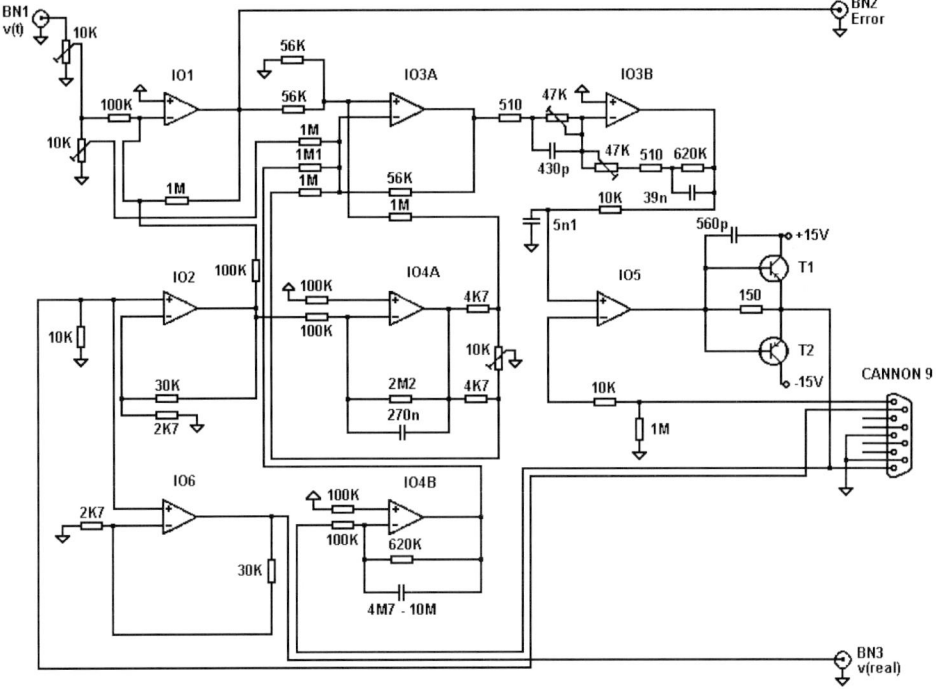

Figure 3. Schematic circuit diagram of the control drive unit.

amplifier with local feedback (IO5, T1, T2). The amplitude and phase frequency characteristics of this control unit are adjusted for the specific transducer [6].

3. Detection system

A thin foil of organic plastic scintillator with dissolved substance of the "resonant gamma rays–electron" convertor (RGEC) constitutes the fundamental element of the detector unit. The principle of operation of resonant scintillation detector is as follows. The resonant gamma photons excite the nuclei of the RGEC grains. In the case the nuclei deexcite by emission of conversion electrons, these electrons will excite along their paths the atoms of the plastic scintillator, which surrounds the RGEC grains. The excited atoms of the scintillator produce photons, which the photoelectronic multiplier tube registers. As the RGEC, $^{119}SnO_2$ and $K_2Mg[^{57}Fe(CN)_6]\cdot H_2O$ are used for ^{119}Sn and ^{57}Fe Mössbauer measurements, respectively. The resonant scintillation detector unit (Figure 4) uses the photomultiplier tube R1924A (Hamamatsu) that is characterized by bialkali photocathode, typical current amplification of 1.1×10^6, and spectral range from 300 to 650 nm (peak wavelength is 420 nm), and low dimension (diameter and length are 25 and 43 mm, respectively). The output signal of the photomultiplier tube is amplified by means of the C6438 (Hamamatsu) fast preamplifier. The fast pulse-height discrimi-

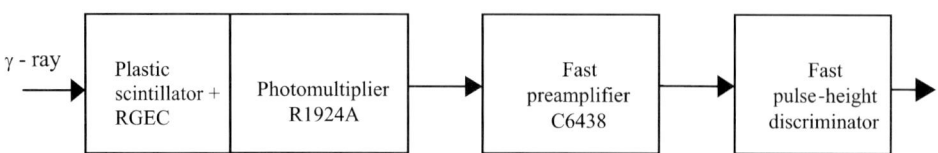

Figure 4. Schematic draft of detection system.

Table I. Results of the nonlinearity measurements

| | | | | | i | | |
		1	2	3	4	5	6
Ascending	$x(i)$ [mm/s]	−8.073	−4.348	−0.826	1.801	5.323	8.622
part	$non(i)$ [%]	0.11	−0.10	−0.03	−0.04	−0.03	0.08
Descending	$x(i)$ [mm/s]	−8.074	−4.348	−0.824	1.803	5.319	8.623
part	$non(i)$ [%]	0.11	−0.10	−0.03	−0.02	−0.05	0.09

nator [7] filters the pulses of the preamplifier output in accordance with Mössbauer resonance gamma rays.

4. Results and discussion

The main parameter characterising the quality of a Mössbauer spectrometer is the nonlinearity of the velocity scale. The following algorithm was used for an estimation of the nonlinearity. The spectral lines of the α-Fe$_2$O$_3$ Mössbauer spectrum, accumulated in 1024 channels, were approximated by Lorentz functions. The nonlinearity for all line positions were calculated by means of fitting of the experimental position of the spectral lines to its theoretical positions by a least square method and by means of the relation

$$non(i) = \frac{x(i) - a \cdot v(i)}{v(6) - v(1)} \quad b,$$

where i ($i = 1$–6), $x(i)$, $v(i)$, a, b are line number, experimental position of the line, theoretical position of the line, parameters obtained from least square method, respectively. The unfolded spectrum was used for the qualitative test of the equipment. The experimental positions of lines and their nonlinearities are shown in Table I.

Two Mössbauer spectra of the BaSnO$_3$ absorber were measured by a YAlO$_3$(Ce) scintillation detector and by a resonance scintillation detector to compare the resonance effect and the line width parameters. The geometries of the experiments were the same. The results of spectra fitting are summarized in Table II. Clearly the Mössbauer effect is significantly higher and the line width narrower with using the resonance detector.

Table II. Fitting results of the BaSnO$_3$ Mössbauer spectra

	YAlO$_3$(Ce) detector	Resonance detector
Resonance effect [%]	7	35
Line width [mm/s]	0.942	0.825

5. Conclusion

The novel equipment significantly improving the efficiency of the Mössbauer measurements and the precision of their results was constructed. New transducer that allows the application of cryostats and furnaces is fully comparable with others double-loudspeaker type transducers. Using such transducer and the resonance scintillation detector the higher signal/noise ratio and narrower line width were obtained in comparison to the conventional equipments.

Acknowledgement

Financial support from The Ministry of Industry and Trade of the Czech Republic under project PROGRES FF-P/108 is gratefully acknowledged.

References

1. Kholmetskii, A. L., Mashlan, M., Misevich, O. V., Chudakov, V. A., Lopatik, A. R. and Zak, D., *Nucl. Instrum. Meth. B* **124** (1997), 143.
2. Maltsev, Y., Mehner, H., Menzel, M. and Rogozev, B., *Hyp. Interact.* **139–140** (2002), 679.
3. Mitrofanov, K. P., Illarionova, N. V. and Shpinel, Y. S., *Prib. Tekhn. Eksp.* **30** (1963), 49.
4. Odeurs, J., Hoy, G. R., L'Abbé, C., Koops, G. E. J., Pattyn, H., Shakhmuratov, R. N., Coussement, R., Chiodini, N. and Paleari, A., *Hyp. Interact.* **139–140** (2002), 685.
5. Maltsev, Y., Mehner, H., Menzel, M. and Rogozev, B., In: *Program and Abstracts, ICAME'99*, Garmisch–Partenkirchen, 29 August–03 September 1999, T9/35.
6. Evdokimov, V. A., Mashlan, M., Zak, D., Fyodorov, A. A., Kholmetskii, A. L. and Misevich, O. V., *Nucl. Instrum. Meth. B* **124** (1995), 287.
7. Mashlan, M., Jancik, D. and Kholmetskii, A. L., In: M. Miglierini and D. Petridis (eds.), *Mössbauer Spectroscopy in Materials Science*, Kluwer Academic Publishers, Dordrecht, Boston, London, 1999, p. 391.

Hyperfine Interactions **156/157**: 21–26, 2004.
© 2004 *Kluwer Academic Publishers. Printed in the Netherlands.*

Angular Distribution of Hyperfine Magnetic Field in Fe_3O_4 and $Fe_{66}Ni_{34}$ from Mössbauer Polarimetry

K. SZYMAŃSKI[1], D. SATUŁA[1] and L. DOBRZYŃSKI[1,2]
[1]*Institute of Experimental Physics, University of Białystok, 15-424 Białystok, Poland*
[2]*The Sołtan Institute for Nuclear Studies, 05-400 Otwock-Świerk, Poland*

Abstract. Experimental determination of some angular averages of hyperfine field is demonstrated. The averages relates to magnetic structure. Exemplary results of the measurements for Fe_3O_4 and $Fe_{66}Ni_{34}$ show that it is possible to obtain valuable information about the field magnitudes and orientations even when distributions of fields are present in the system under study.

In disordered magnetic systems one encounters usually a distribution of both, the intensity and the orientation of hyperfine magnetic field (h.m.f.). Preferred orientation, $P(\Omega)$, of the hyperfine fields is of particular importance in the context of the contribution of selected elements to the magnetic texture [1, 2]. It is usually described in a certain set of basis functions, e.g., spherical harmonics Y_{lm}. Since only M1 dipolar transitions are measured in ^{57}Fe Mössbauer spectroscopy, unpolarized radiation delivers information on Y_{2m} only, while Y_{1m} harmonics can be known when circularly polarized radiation is used [3]. Knowing Y_{1m} and Y_{2m} in the texture function is equivalent to the knowledge of angular averages $\langle \gamma_r \cdot \boldsymbol{m} \rangle$ and $\langle (\gamma_r \cdot \boldsymbol{m})(\gamma_s \cdot \boldsymbol{m}) \rangle$[4], where $\boldsymbol{m}$ is an unit vector parallel to the local hyperfine field $\boldsymbol{B}$, γ_r is a Cartesian versor ($r = x, y, z$) and brackets $\langle \rangle$ denote angular averaging:

$$\langle f(\Omega) \rangle = \int_{4\pi} f(\Omega) P(\Omega) \, d\Omega. \tag{1}$$

In the case of a sample with axial symmetry it is convenient to choose one of the γ_r, denoted by γ, parallel to the $\boldsymbol{k}$ vector of photon. Then the averages $\langle \gamma \cdot \boldsymbol{m} \rangle \equiv c_1$ and $\langle (\gamma \cdot \boldsymbol{m})^2 \rangle \equiv c_2$ can be measured with monochromatic, circularly polarized radiation [4, 5]. In disordered systems one can measure distribution of the intensity of h.m.f., $p(B)$, and for each intensity B, in principle, two averages c_1 and c_2. This paper shows that one can finally get three distributions: $p(B)$, $c_1(B)$ and $c_2(B)$.

Normalized Mössbauer spectrum $S(v)$ consists of a linear combination of N subspectra $s(v, B_i)$:

$$S(v) = \sum_{j=1}^{N} p_j s(v, B_j), \tag{2}$$

where v is Doppler velocity, and p_j is a nonnegative coefficient for a field B_j. Subspectrum $s(v, B)$ is a Zeeman sextet:

$$s(v, B) = \sum_{n=1}^{6} i_n L_n(v, B), \tag{3}$$

where $L_n(v, B)$ describes the shape of the absorption line corresponding to the nth nuclear transition and i_n is the line intensity dependent on the photon polarization and wave vector. For the case of single B and measurements with circularly polarized radiation, the coefficients i_n were given in [3]. One can show [4] that having a distribution of directions of vector B the expressions for i_n should contain already introduced averages, namely:

$$\begin{aligned}
16i_1 &= 48i_4 = 3(1 \pm 2c_1 + c_2), \\
4i_2 &= 4i_5 = 1 - c_2, \\
48i_3 &= 16i_6 = 3(1 \mp 2c_1 + c_2).
\end{aligned} \tag{4}$$

Every subspectrum $s(v, B_j)$ is characterized by its relative area proportional to p_j and two averages c_1 and c_2 (the index j in c_1, c_2 as well as in p coefficients was dropped for simplicity reasons). Using the least squares fitting procedure and varying $3N$ coefficients p, c_1 and c_2, one can find best fit of function $S(v)$, see Equation (2), to the experimental spectrum. Physically possible sets of p, c_1 and c_2 have to be considered only, namely:

$$p \geqslant 0, \quad -1 \leqslant c_1 \leqslant 1, \quad 0 \leqslant c_2 \leqslant 1, \quad c_1^2 \leqslant c_2. \tag{5}$$

The last inequality in (5) is the Buniakovsky–Schwartz relation. In order to make minimisation of χ^2 with conditions (5) effective, we introduce a set of $3N$ new variables, a, b, Δ, related to p, c_1 and c_2 through:

$$\begin{aligned}
p[a, b, \Delta] &= \frac{(a^2 + b^2)^2 + 2\Delta^4}{4}, \\
c_1[a, b, \Delta] &= \frac{a^4 - b^4}{(a^2 + b^2)^2 + 2\Delta^4}, \\
c_2[a, b, \Delta] &= \frac{(a^2 - b^2)^2 + 2\Delta^4}{(a^2 + b^2)^2 + 2\Delta^4}.
\end{aligned} \tag{6}$$

The square brackets were used in transformation functions (6) in order not to confuse them with the distribution functions $p(B), c_1(B)$ and $c_2(B)$. The functions (6) have following properties: (i) they are even, (ii) for a, b, and Δ positive there exist inverse functions $a[p, c_1, c_2], b[p, c_1, c_2], \Delta[p, c_1, c_2]$, (iii) the inequalities (5) hold for any real values of a, b, and Δ, (iv) χ^2 expressed in variables a, b, and Δ is a polynomial of 8th order. The two last properties make numerical process of searching of minimum of χ^2 very effective.

Two different distributions of hyperfine parameters may produce identical spectra. This leads to the ambiguity much discussed in literature [6–8]. One case of

Table I. Fitted probabilities p and magnetic texture coefficients. Two last columns contain average value of the h.m.f. and the width of its Gaussian distribution

	p	c_1	c_2	B [T]	ΔB [T]
Fe_3O_4	0.38 ± 0.01	-0.78 ± 0.02	0.93 ± 0.01	49.92	0.18
	0.62 ± 0.01	0.80 ± 0.02	0.97 ± 0.01	45.13	0.53
$Fe_{0.66}Ni_{0.34}$	0.37 ± 0.02	0.57 ± 0.03	0.60 ± 0.02	28.6	2.04
	0.29 ± 0.02	0.62 ± 0.03	0.62 ± 0.02	25.2	3.36
	0.34 ± 0.02	0.44 ± 0.03	0.63 ± 0.02	17.7	6.45

ambiguity appears when multidimensional distribution is extracted from one dimensional data (i.e. Mössbauer spectrum). Additional independent experimental information usually reduces this ambiguity. We have demonstrated [9], as an example, that two different distributions of h.m.f. reproduce experimental spectra of $Fe_{2.5}Cr_{0.5}Al$ alloy measured with unpolarized beam equally well. The measurements with monochromatic, circularly polarized radiation, showed which of the two is correct one. Continuing this direction we developed an algorithm for simultaneous fitting of the spectra measured with different photon polarization states (on the sample in the same conditions, like external magnetic field, temperature). Three distributions: $p(B)$, $c_1(B)$ and $c_2(B)$ are fitted simultaneously with the help of transformation (6).

In order to apply discussed algorithm to real cases, we have to take into account different isomer shift and quadrupole splitting for every subspectra (the latter is considered as small perturbation of the h.m.f.), and correction for polarization degree [4, 10].

The first example on which the algorithm was tested is Fe_3O_4 powdered absorber, prepared from stoichiometric single crystal of magnetite, with stoichiometry for which Vervey transition of the first kind is observed. The absorbers exposed to external field of 1.1 T were measured at room temperature and the spectra are shown in Figure 1. Fe_3O_4 is ferrimagnet and one expects that in an external magnetic field two hyperfine fields will be oriented antiparallel, resulting in c_1 parameters of opposite signs. This is observed indeed, see Table I. Results of simultaneous fit are shown by solid lines in Figure 1. Majority of the Fe moments, occupying octahedral positions with smaller h.m.f. are oriented parallel to the net magnetization, like in α-Fe (see the inset), and the c_1 parameter is positive. Minority of Fe with larger field (occupying tetrahedral positions) are oriented antiparallel to the net magnetization which is measured quantitatively by negative value of c_1 parameter. Absolute values of c_1 and c_2 parameters are slightly smaller than 1 indicating almost complete saturation of the sample in the applied external magnetic field.

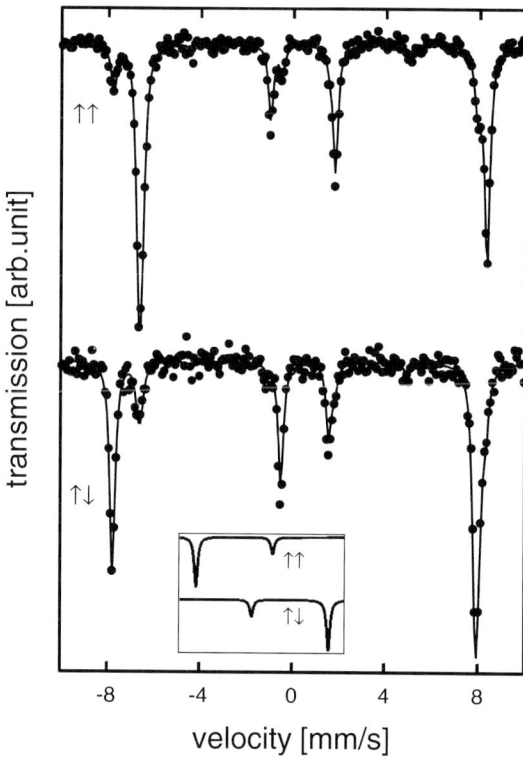

Figure 1. 1 Mössbauer spectra of Fe_3O_4 measured with two opposite circular polarizations of mono-chromatic radiation. Solid lines show the best simultaneous fit obtained from algorithm discussed in the text. The inset shows schematically the shape of α-Fe spectra measured with two opposite circular polarizations abbreviated by ↑↑ and ↑↓ arrows.

The second example is Invar Fe–Ni alloy, whose ground state magnetic structure is still under debate. There are recent experiments performed under high pressure, one of them [11] consistent with 2γ-state model proposed by Weiss [12], and other [13], consistent with low spin non-collinear structure proposed in [14]. Recent polarized neutron diffraction experiments indicated strong coupling of lattice and magnetic degrees of freedom [15]. Circularly polarized polychromatic radiation was used in investigation of Fe–Ni invar alloys [16], the spectra obtained were, however, complicated and difficult for interpretation.

In our experiment, $Fe_{66}Ni_{34}$ was prepared as a foil and measured in the magnetic field perpendicular to the foil at room temperature, see Figure 2. Three Gaussian components, displayed in the inset, describe full spectrum well. Results of the best fit are shown by solid lines in Figure 2; the fitted parameters are listed in the Table I. The most important result is, that the best fit is obtained for almost similar values of c_2 (quite similarly for c_1) for the three components. The smallest value of c_1 is related to the weakest h.m.f. This indicates that this component is more disordered than the remaining two. Our results leave no doubt that the lowest-field component

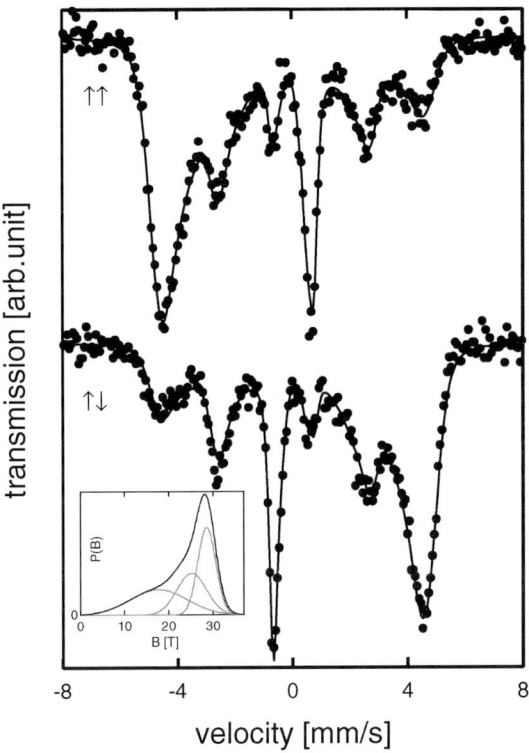

Figure 2. Same as Figure 1 for $Fe_{66}Ni_{34}$. The distribution of h.m.f. resulting from the components used is shown in the inset.

is neither due to the antiferromagnetic ordering postulated in 2γ-state model [12] nor to eventual random disorder. Were any of these two possibilities true, the c_1 should be close to 0.06, i.e. the ratio of the applied field and the mean value of the component (1.1/17.7), while c_2 would be close to 1/3. Moreover our results show that low field component does not have, in average, antiparallel orientation with respect to the net magnetization as interpreted in [16].

References

1. Pfannes, H.-D. and Fisher, H., *Appl. Phys.* **13** (1977), 317.
2. Pfannes, H.-D. and Paniago, R. M., *Hyp. Interact.* **71** (1992), 1499.
3. Frauenfelder, H., Nagle, D. E., Taylor, R. D., Cochran, D. R. F. and Visscher, W. M., *Phys. Rev.* **126** (1962), 1065.
4. Szymanski, K., *NIM B* **134** (1998), 405.
5. Szymanski, K., Dobrzynski, L., Prus, B. and Cooper, M. J., *NIM B* **119** (1996), 438.
6. Le Caer, G., Dubois, J. M., Fischer, H., Gonser, I. U. and Wagner, H. G., *NIM B* **5** (1984), 25.
7. Le Caer, G. and Brand, R. A., *J. Phys.: Condens. Matter* **10** (1998), 10715.
8. Rancourt, D. G., In: G. J. Long and F. Grandjean (eds), *Mössbauer Spectroscopy Applied to Magnetism and Materials Science*, Vol. 5, Plenum, New York, 1996, p. 105.
9. Szymanski, K., Satula, D. and Dobrzynski, L., *J. Phys.: Cond. Matter* **1** (1999), 881.

10. Szymanski, K., *J. Phys.: Cond. Matter* **12** (2000), 7495.
11. Rueff, J. P., Shukla, A., Kaprolat, A., Krisch, M., Lorenzen, M., Sette, F. and Verbeni, R., *Phys. Rev. B* **63** (2001), 132409.
12. Weiss, J., *Proc. R. Soc. London A* **82** (1963), 281.
13. Dubrovinsky, L., Dubrovinska, N., Abrikosov, I. A., Vennström, M., Westman, F., Carlson, S., van Schilfgaarde, M. and Johansson, B., *PRL* **86** (2001), 4851.
14. van Schilfgaarde, M., Abrikosov, I. A. and Johansson, B., *Nature* **400** (1999), 46.
15. Brown, P. J., Kanomata, T., Matsumoto, M., Neumann, K.-U. and Ziebeck, K. R. A., *JMMM* **242–245** (2002), 781.
16. Ulrich, H. and Hesse, J., *JMMM* **45** (1984), 315.

Hyperfine Interactions **156/157**: 27–30, 2004.
© 2004 *Kluwer Academic Publishers. Printed in the Netherlands.*

Two-Dimensional Mössbauer Spectra

YU. MALTSEV[1], S. MALTSEV[2], M. MENZEL[1,*], B. ROGOZEV[2] and
A. SILVESTROV[3]

[1] *Federal Institute for Materials Research and Testing (BAM), Richard-Willstätter-Strasse 11, D-12489 Berlin, Germany; e-mail: Michael.Menzel@BAM.DE*
[2] *Radium Institute, 2nd Murinsky Avenue, 28, 194021 St. Petersburg, Russia*
[3] *RITVERC GmbH, 2nd Murinsky Avenue, 28, 194021 St. Petersburg, Russia*

Abstract. To decrease the spectra measurement time in Mössbauer spectroscopy a new data acquisition system was proposed, which allows to collect data as a two-dimensional distribution.

Key words: data acquisition, signal processor, two-dimensional Mössbauer spectrum.

1. Introduction

In spite of all advances in electronics, the design of Mössbauer spectrometers has not advanced principally. The usual arrangement [1] for collecting a Mössbauer spectrum is shown in Figure 1(a). The main disadvantage of this arrangement is the occurrence of pulse overlapping at high count rates, it is when the next pulse "sits on the tail" of a previous one. In this case a single channel analyzer (SCA) registers the noise pulses and misses the pulses from needed quanta, which will shift out of the working window. Pulse overlapping disturbs the amplitude spectrum, reduces the signal/noise ratio in the Mössbauer spectrum, limits the maximal count rate of the data acquisition system, and, finally, increases the duration of experiment. Another disadvantage of the conventional arrangement is the difficulty in setting the SCA window if the amplitude spectrum has the well-known "exponential decay" shape using CEMS or resonance detectors.

Therefore, a data acquisition system, which is free of these disadvantages, was created.

2. Proposed data acquisition system

The scheme of the proposed data acquisition system is shown in Figure 1(b). It consists of a fast analog-to-digital converter (ADC) AD9224 chip, signal processor (SP) ADSP-21061 chip, and random access memory (RAM).

The ADC digitizes the signals from the detector with a sampling rate of up to 40 million times per second. The signal processor determines the local maximum

---

* Author for correspondence.

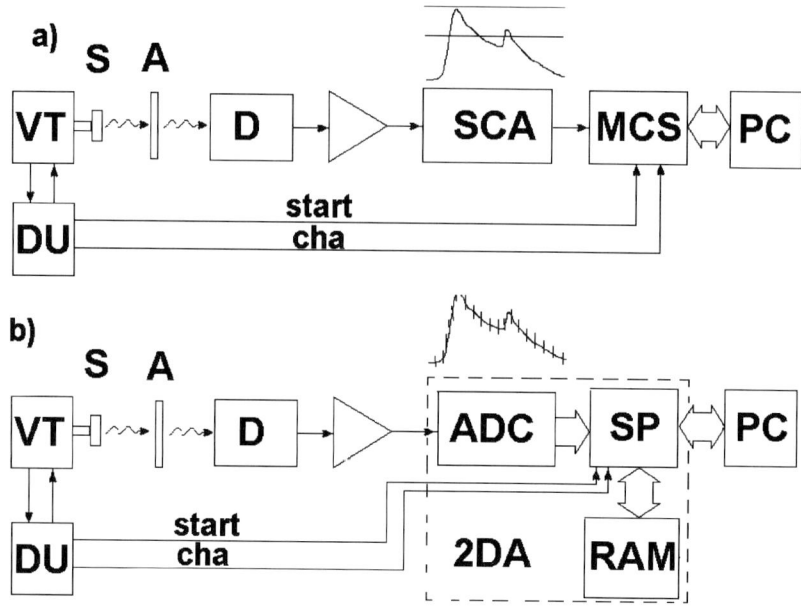

Figure 1. (a) Conventional scheme, where: S – source; A – absorber; VT – velocity transducer; DU – driving unit; D – detector; SCA – single channel analyzer; MCS – multichannel scaler; PC – personal computer. (b) Proposed scheme, where: ADC – fast analog-to-digital converter; SP – signal processor; RAM – random access memory; 2DA – "Two-dimensional analyzer".

and local minimum values, and calculates the correct amplitude for each pulse. Thus, pulse overlapping is eliminated. The operation of the SP is synchronized with the driving system by the signals START and "channel advance" CHA. Using the pulse amplitude in a digital form and the current velocity channel number SP forms a two-dimensional (2D) distribution of pulses in the RAM, where the Y-axis corresponds to the velocity scale, and the X-axis corresponds to the amplitude of pulses from the detector.

3. Experimental results

Examples of two-dimensional Mössbauer spectra are shown in Figures 2 and 3.

Cross-sections parallel to the velocity–counts-plane give Mössbauer spectra, which correspond to different amplitudes of input pulses. Cross-sections parallel to the energy–counts-plane give amplitude spectra, which correspond to different values of the Doppler velocity.

Figure 2 presents a two-dimensional Mössbauer spectrum of an iron foil measured with a proportional counter. There are 6 dips on the 14.4 keV billow and there are 6 small peaks on the 6.3 keV X-rays billow. This example illustrates, that with the new instrumentation absorption and emission spectra are acquired simultaneously.

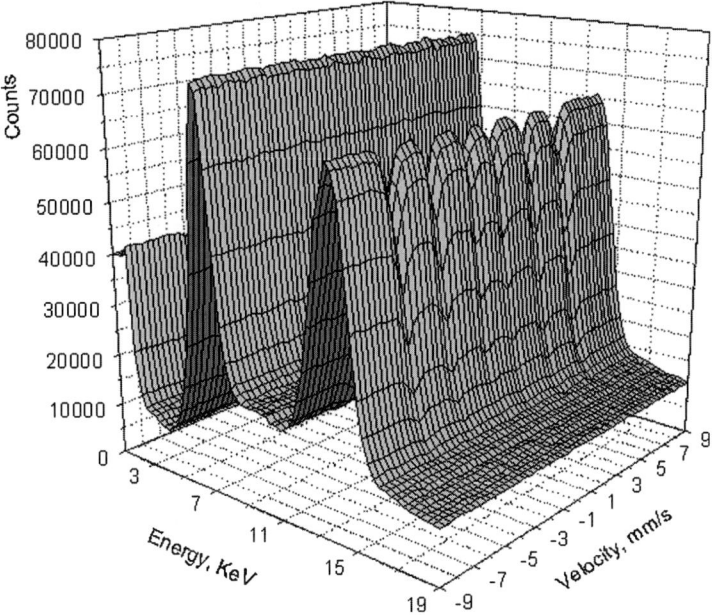

Figure 2. Two-dimensional Mössbauer spectrum of an iron foil, measured with a proportional counter.

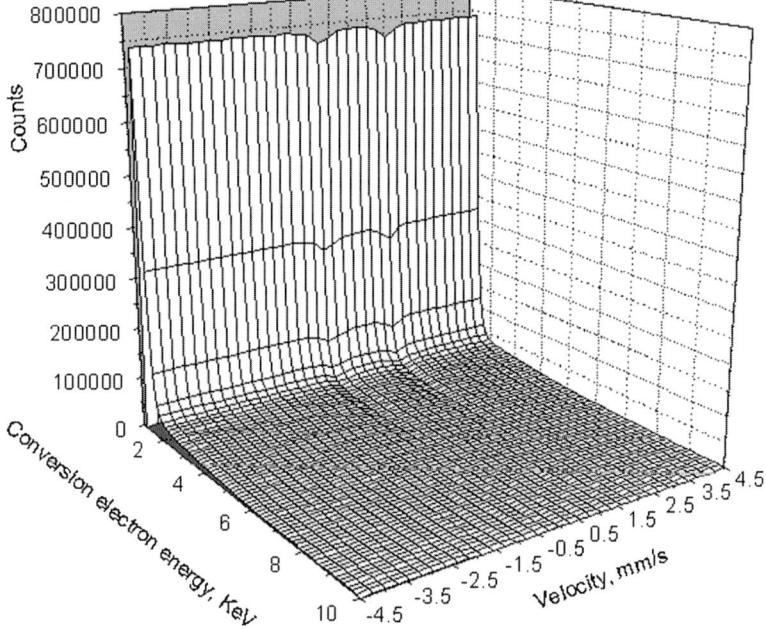

Figure 3. Two-dimensional Mössbauer spectrum of $FeC_2O_4 \cdot 2H_2O$, measured with a resonance scintillation detector.

Figure 3 presents a two-dimensional Mössbauer spectrum of a $FeC_2O_4 \cdot 2H_2O$ sample measured with a resonance scintillation detector [2]. It consists of a set of absorption Mössbauer spectra with different signal/noise ratios. The optimal energy "window" can be chosen *after* the experiment. But from the precision point of view, the best procedure would be to fit each partial Mössbauer spectrum independently, and to calculate the weighted averages of the spectral parameters. This technique reduces the estimated standard deviation by a factor of 1.4 and more, or reduces the data acquisition time by a factor of 2.0 and more.

4. Conclusions

The proposed setup allows to collect more information about the sample in one experiment, and, finally, saves data acquisition time.

The use of a fast ADC together with a modern digital signal processor significantly increases the count rate of Mössbauer spectrometer due to the elimination of pulse overlapping.

The application of a two-dimensional data acquisition system allows to chose the optimal energy "window" in the amplitude spectrum *after* the experiment, and to measure gamma-quanta absorption and X-rays emission spectra simultaneously in the same transmission experiment. In the case of low count rates the proposed scheme also saves data acquisition time, because amplitude and Mössbauer spectra are collected simultaneously.

In the case of CEMS measurements a single experiment gives a number of Mössbauer spectra, which correspond to different surface layers of the sample.

References

1.   Shenoy, G. K. *et al.*, In: *Mössbauer Isomer Shifts*, North-Holland, Amsterdam, 1978, p. 61.
2.   Maltsev, Y., Mehner, H., Menzel, M. and Rogozev, B., *Hyp. Interact.* **139/140** (2002), 679.

Hyperfine Interactions **156/157**: 31–40, 2004.
© 2004 *Kluwer Academic Publishers. Printed in the Netherlands.*

^{57}Fe Mössbauer Study of Magnetic Nanowires

DE-SHENG XUE and FA-SHEN LI
*Key Laboratory for Magnetism and Magnetic Materials of the MOE, Lanzhou University,
Lanzhou 730000, P. R. China*

Abstract. Nanowires of metal, alloy, compound, and ferrite have been electrodeposited in anodic aluminium oxide templates. The structure and magnetic properties of the nanowires are characterized by ^{57}Fe Mössbauer spectroscopy combining with other techniques. It is found that the metal and alloy nanowires have a very strong magnetic anisotropy. The surface distribution of the magnetic moment is different from that of the interior. The Debye temperature of Prussian blue nanowires derived from hyperfine interaction parameters is lower than that of the bulk. The properties of the ferrite nanowires are strongly related to the structure of nanowires.

Key words: nanowires, Mössbauer spectroscopy.

1. Introduction

There has been a rapidly increasing interest in one dimension nanostructure, such as nanotubes, nanowires, nanorods, and nanobelts, because of their potential for fundamental studies of the size effect and for their applications in nanodevices [1, 2]. Since theoretical predictions suggest that one-dimensional Ising model shows no magnetic ordering at nonzero temperature [3, 4], it is an informative way to understand the theoretical result by fabricating and studying the nanowire of molecular-based magnet. From the point of view of applications, the magnetic nanowire arrays are of interest for magnetoresistive devices of very small size [5] and for high-density recording media [6]. For instance, the density in conventional longitudinal recording may be less than 50 Gb/in^2 because of the thermal stability [7]. However, the density of the arrays may potentially be higher than 100 Gb/in^2 [8].

Among known approaches for producing nanostructures, the anodizing anodic aluminum oxide (AAO) template-based method is a popular approach to synthesize a variety of metal and semiconductor nanowires through electrochemical technology [9]. Recently, Fe [10], Fe–Co [11], Fe–Ni [12], Fe$_2$O$_3$ [13], FeOOH [14] and Prussian blue [15, 16] nanowire arrays embedded in AAO templates have been successfully fabricated. The recent development of metal amorphous nanowires arrays such as CoP, FeP and NiP is another highlight [17]. In this paper, combining with other measurement techniques, the information about micro electronic, magnetic and structural properties of the nanowires are studied by the ^{57}Fe Mössbauer spectroscopy.

2. Experimental

The highly ordered porous AAO templates were generated by anodizing aluminum foils (99.999%) in an oxalic acid solution using a two-step anodizing process [15]. Metal, alloy and compound can be directly fabricated by using the AC electrode-position method with a standard double-electrode cell [10, 12, 15]. The ferrite nanowires can be formed by heat-treating FeOOH at different conditions [13]. The images of AAO template, Fe naowires, Prussian blue nanowires and Fe_3O_4 nanowires are shown in Figures 1(a)–(d), respectively.

Structural characterization was performed by means of X-ray diffraction (XRD) using a Rigaku/Max-2400 diffractometer with Cu $K\alpha$ radiation. Transmission electron microscopy (TEM) and selected area electron diffraction (SAED) were performed by using a JEO 2000 microscope, while scanning electron microscopy (SEM) was operated by using JSM-5600 microscope. The Mössbauer spectroscopy (MS) was obtained by using a constant acceleration with a source of ^{57}Co in rhodium. The spectra were fitted with Lorentz lines, and the isomer shifts (*IS*) were referred to that of α-Fe at room temperature (RT).

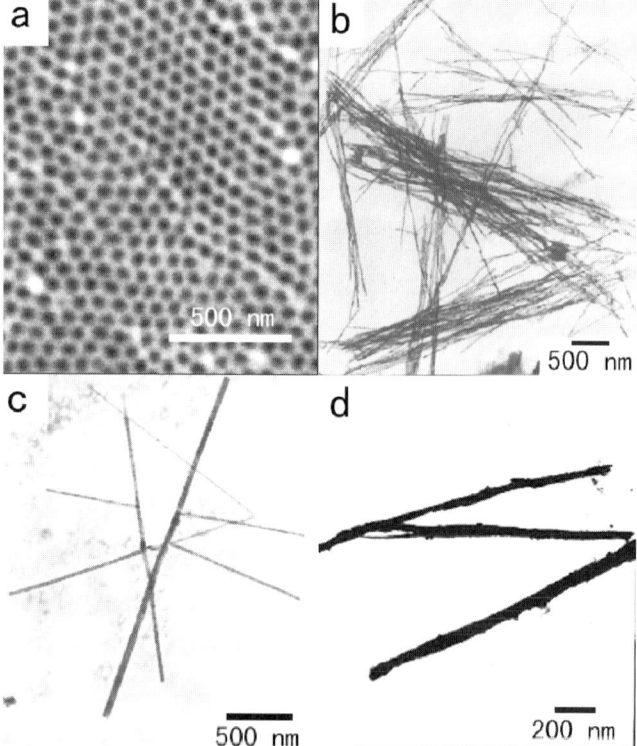

Figure 1. (a) SEM image of porous anodic aluminium oxide template, (b), (c), (d) TEM images of iron, Prussian blue and Fe_3O_4 nanowire, respectively.

3. Results and discussion

3.1. MAGNETIC MOMENT DISTRIBUTION IN ALLOY AND METAL NANOWIRE ARRAYS

3.1.1. $Fe_{1-x}Ni_x$ nanowire arrays

$Fe_{1-x}Ni_x$ $(0 < x \leqslant 0.32)$ nanowire arrays with 16 nm in diameter, 4 μm in length were prepared. The results of XRD showed that the nanowire arrays have a bcc structure with [110] crystallographic orientation along the nanowire axis [12]. The MS obtained for the $Fe_{1-x}Ni_x$ nanowires are shown in Figure 2. Each spectrum consists of a doublet and a sixtet, which are ascribable to a paramagnetic and a magnetic phase, respectively. The vanishing of the second and fifth peaks in the MS indicates that the magnetic moments of the iron atoms in the $Fe_{1-x}Ni_x$ nanowire arrays align on the [110] direction. This means that there is a strong shape anisotropy in [110] direction.

The fitting results of each magnetic spectrum for $Fe_{1-x}Ni_x$ nanowire arrays are listed in Table I. The observed linewidths (*FWHM*) of the MS indicate that the

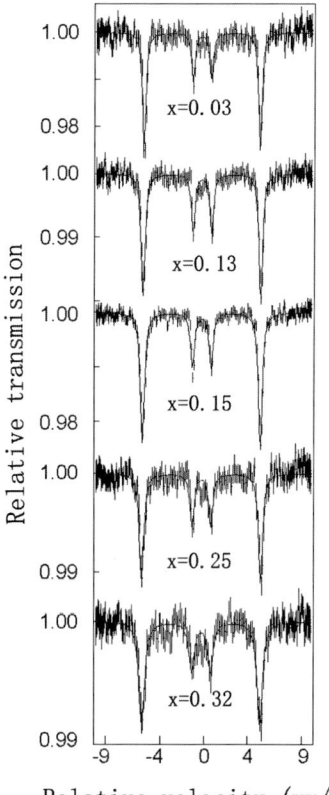

Relative velocity (mm/s)

Figure 2. The room-temperature Mössbauer spectra of $Fe_{1-x}Ni_x$ nanowire arrays.

Table I. Hyperfine parameters of $Fe_{1-x}Ni_x$ nanowire arrays at room temperature: FWHM is the linewidth of the Mössbauer spectrum; δ is the isomer shift; QS is the quadruple splitting; H_{hf} is the hyperfine field

	FWHM $(mm\,s^{-1})$	IS $(mm\,s^{-1})$	QS $(mm\,s^{-1})$	H_{hf} (kOe)
$x = 0.03$	0.37	0.00	0.00	333
$x = 0.13$	0.41	0.02	0.00	338
$x = 0.15$	0.43	0.02	0.00	339
$x = 0.25$	0.46	0.02	0.00	342
$x = 0.32$	0.59	0.01	0.00	338

local environment of iron atoms changes with the substitution of Ni content. The quadruple splittings (QS) of $Fe_{1-x}Ni_x$ nanowires are equal to zero, which suggests that the structure of nanowires still keeps in a cubic symmetry. The IS has little dependence on Ni content. The nickel composition dependence of the hyperfine field (H_{hf}) shows a maximum at $x = 0.25$, which is consistent with the bulk and the fine particles of $Fe_{1-x}Ni_x$ alloy [18].

3.1.2. *Fe nanowire arrays*

In order to understand the distribution of the magnetic moment in metal, the Fe nanowire arrays were fabricated. The conversion electron Mössbauer spectrum (CEMS) was performed on the top of α-Fe nanowire arrays. The CEMS spectra of three samples: (a) α-Fe nanowire arrays with diameters $d = 60$ nm, (b) α-Fe nanowire arrays with $d = 300$ nm, (c) α-Fe foil with 25 μm in thickness, are shown in Figure 3. It is found that the H_{hf} is similar to that in bulk iron, and the $FWHM$ increases with the decreasing diameter of α-Fe nanowires, which indicate that with increasing diameter of nanowires the influence of magnetocrystalline anisotropy becomes more important. The ratio between the second peak and the first peak (I_2/I_1) is 0.215 for $d = 300$ nm while it is 0.089 for $d = 60$ nm. The ratio for the α-Fe foil equals 1.179, which shows the magnetic moment departure from the nanowire axis direction as the diameter of nanowires increase.

Based on the experimental observations, we assume a core–shell structure model, in which the core spins are magnetically coupled, but the surface spins are thermally disordered because of reduction in Fe coordination. The evidence for such a spin configuration comes from the MS measured at RT as shown in Figure 4, the surface spins are thermally activated as paramagnetic contribution, resulting in a central single peak of the Mössbauer spectra, the relative intensity of which is enhanced with decreasing diameter of the Fe nanowire arrays.

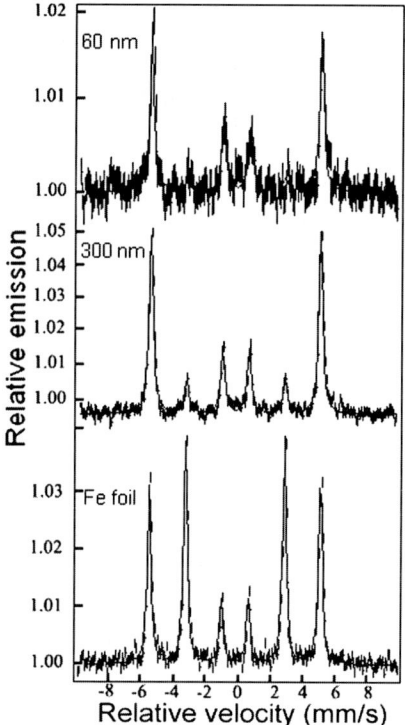

Figure 3. Typical room-temperature conversion electron Mössbauer spectrum of α-Fe nanowires and foils.

3.2. LOW-DIMENSIONAL EFFECT ON PRUSSIAN BLUE NANOWIRES

Prussian blue analogs have played an important role in molecular magnets, and a number of unusual properties were found [19–21]. In our pioneer work, the Curie temperature of highly ordered Prussian blue nanowires embedded in AAO templates was found to been reduced with respect to Prussian blue bulk, resulting from the diminution of the average number of the nearest magnetic interaction neighbors and magnetic exchange interaction constants as the diameters of nanowires decreasing [15]. In order to have a deep look of the chemical binding in the nanowires, the *IS, QS* and the recoil-free fraction (f) were employed.

The highly ordered Prussian blue nanowires with diameter of about 50 nm and length up to 4 μm were electrodeposited into AAO templates. The Mössbauer spectra measured at 15, 77, 150, 230, and 290 K, are shown in Figure 5. Each of the spectra consists of a doublet and a singlet, which are ascribable to the high spin Fe^{3+} ions and low spin Fe^{2+} ions, respectively [16]. The temperature dependence of isomer shift (*IS*) and the spectra area can be fitted by the following Equations (1) and (2), respectively,

$$\delta_{SOD} = -\frac{3k_B T}{2Mc}\left[\frac{3\Theta}{8T} + 3\left(\frac{T}{\Theta}\right)^3 \int_0^{\Theta/T} x^3\left(e^x - 1\right) dx\right], \qquad (1)$$

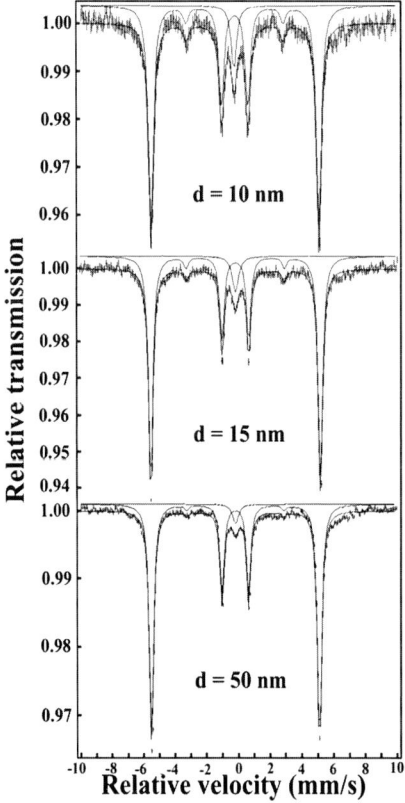

Figure 4. The room-temperature Mössbauer spectrum of α-Fe nanowires with different diameters.

$$f = \exp\left\{-\frac{3}{2}\frac{E_R}{k_B\Theta}\left[1 + 4\left(\frac{T}{\Theta}\right)^2\int_0^{\Theta/T}\frac{x\,dx}{e^x - 1}\right]\right\},\tag{2}$$

where k_B is the Boltzman constant, Θ is the Debye temperature, M is the atomic mass, c is the velocity of light and E_R is the recoil energy. The obtained results show that the Debye temperature (226 ± 5 K) of Prussian blue nanowires decreases with respect to (257 ± 5 K) of Prussian blue bulk, which indicates the strength of the forces binding iron ions in Prussian blue nanowires become smaller, as Prussian blue become a nanowire from a three dimension solid.

3.3. PHASE ANALYSIS OF FERRITES NANOWIRE ARRAYS

The β-FeOOH nanowires with diameter of 120 nm and length about 6 μm embedded in AAO template were prepared. However, it is difficult to check the phase because of its amorphous state. Prior work by Chambaere *et al.* showed that there are two doublets with an intensity ratio of 60 : 40 in MS of β-FeOOH. The *QS* of the doublet with intensity of 60% is 0.51–0.56 mm/s while the *QS* of the doublet with intensity of 40% is 0.92–0.96 mm/s, the *IS* of the two doublets are 0.37 mm/s

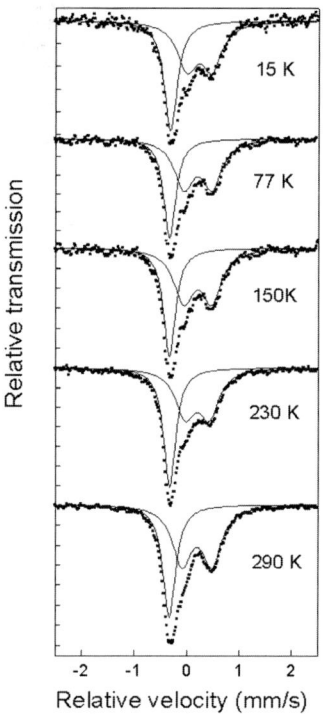

Figure 5. The Mössbauer spectrum of Prussian blue nanowires obtained at different temperatures.

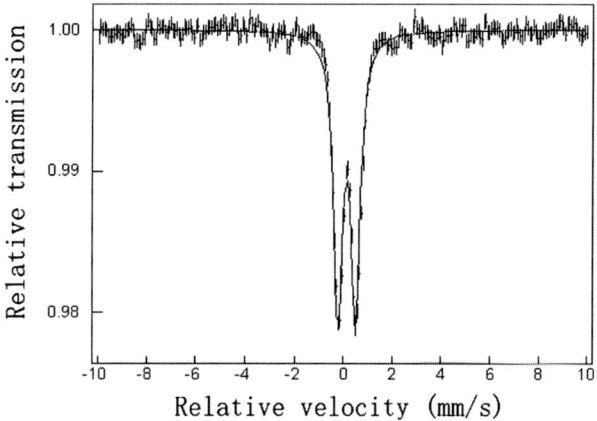

Figure 6. The room-temperature Mössbauer spectrum of β-FeOOH nanowire arrays.

and 0.38 mm/s respectively [22]. The MS of the FeOOH nanowire arrays measured at RT is shown in Figure 6. The average *IS* and *QS* are 0.35 mm/s and 0.74 mm/s, respectively. They are in agreement with the hyperfine parameters of Fe^{3+} in β-FeOOH compounds.

Magnetite nanowire arrays were synthesized in the holes of AAO by heat-treating the precursor β-FeOOH at 600 K for 3 h in H_2. The nanowires have

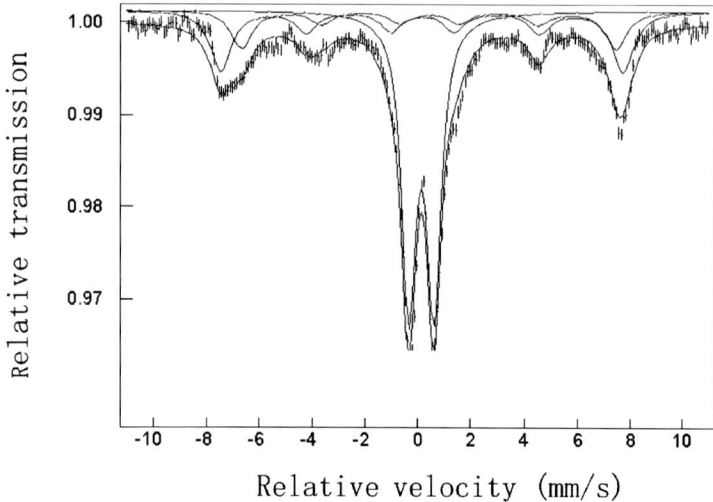

Relative velocity (mm/s)

Figure 7. The room-temperature Mössbauer spectrum of Fe_3O_4 nanowires arrays.

a spinel structure and each nanowire is composed of fine crystallites with size of about 15–40 nm. However, the XRD patterns of Fe_3O_4 are almost similar to that of the γ-Fe_2O_3 due to their similarity in crystal structure. MS was used to verify the phase and study the micro-magnetic properties of Fe_3O_4 nanowires. It is known that the differences of the Mössbauer parameters between Fe_3O_4 and γ-Fe_2O_3 are: (1) the value of QS in Fe_3O_4 is closer to zero than that in γ-Fe_2O_3; (2) the values of the H_{hf} corresponding to A and B sites are different in Fe_3O_4, but almost the same in γ-Fe_2O_3, due to the presence of Fe^{2+} ions at the A sites in Fe_3O_4 [23].

Figure 7 shows the MS of magnetite nanowires in AAO templates with the γ-ray normal to the surface of the membranes at room temperature, which consists of two sub-spectra of sextet and one doublet. The broadening peaks indicate there is a size distribution of the crystallites in the nanowires. In addition, 55% of the doublet is corresponding to a superparamagnetic phase, which is in good agreement with the results of XRD and TEM [14]. The QS is -0.028 mm/s, and the H_{hf} of A and B sites are 47.3 and 43.9 T, respectively.

The larger difference of H_{hf} corresponding to A and B sites as well as the small value of QS indicate that the nanowires are Fe_3O_4 rather than γ-Fe_2O_3. The less value of H_{hf} than that of the bulk materials is due to the existence of collective magnetic excitation caused by the size distribution of the crystallites [24]. The occupation ratio of the cations in A and B sites is 2 : 1, which is different from that of the bulk 1 : 2. It is also found that the intensity ratio of the sextet peaks is nearly 3 : 3 : 1, which suggests that the orientation of magnetic moments of the Fe_3O_4 crystallites is not along or perpendicular to the wire axis. This is also related to the competition of the shape and magnetocrystaline anisotropy.

4. Conclusion

In summary, the metallic and non-metallic nanowires were successfully electrode-posited in AAO templates. Mössbauer spectroscopy is very useful to identify the phase structure, the micro electronic and magnetic properties. It is found that the metal and alloy nanowires have very strong magnetic anisotropy, which make them good candidates for ultrahigh density magnetic recording materials. The surface distribution of the magnetic moment of α-Fe nanowires is different from that of the interior. The Debye temperature derived from hyperfine interaction parameters of Prussian blue nanowire is lower than that of the bulk. Ferrite nanowires can be prepared by heat-treating the precursor FeOOH, and the properties of the ferrite nanowires are strongly related to the structure of nanowires. The preparation of the ferrites and amorphous nanowires remain an open question.

Acknowledgement

This work is supported by the Trans-Century Training Program Foundation for the Talent of MOE, NSFC (Grant No. 10374038, 50171032 and 10274027) and EYTT of China.

References

1. Lieber, C. M., *Solid State Commun.* **107** (1998), 607.
2. Duan, X., Huang, Y., Cui, Y., Wang, J. and Lieber, C. M., *Nature* **409** (2001), 66.
3. Ising, E., *Z. Physics* **31** (1925), 253.
4. Newell, G. F. and Montroll, E. W., *Rev. Mod. Phys.* **25** (1953), 159.
5. Dubois, S., Beuken, J. M., Piraux, L., Duvail, J. L., Fert, A., George, J. M. and Maurice, J. L., *J. Magn. Magn. Mater.* **165** (1997), 30.
6. Chou, S. Y., *Proc. IEEE* **85** (1997), 652.
7. Charap, S., Lu, P. L. and He, Y., *IEEE Trans. Magn.* **33** (1997), 978.
8. Metzger, R. M., Konovalov, V. V., Sun, M., Xu, T., Zangari, G., Xu, B., Benakli, M. and Doyle, W. D., *IEEE Trans. Magn.* **36** (2000), 30.
9. Prieto, A. L., Sander, M. S., Marisol, S. M., Gronsky, R., Sands, T. and Stacy, A. M., *J. Am. Chem. Soc.* **123** (2001), 7160.
10. Li, F. S., Ren, L. Y., Niu, Z. P., Wang, H. X. and Wang, T., *J. Phys.: Condens. Matter.* **14** (2000), 1.
11. Chen, Z. Y., Zhan, Q. F., Xue, D. S., Li, F. S., Zhou, X. Z., Kunkel, H. and Williams, G., *J. Phys.: Condens. Matter.* **14** (2002), 613.
12. Wang, J. B., Liu, Q. F., Xue, D. S., Peng, Y., Cao, X. Z. and Li, F. S., *J. Phys. D.* **34** (2001), 1.
13. Xue, D. S., Gao, C. X., Liu, Q. F. and Zhang, L. Y., *J. Phys.: Condens. Matter.* **15** (2003), 1455.
14. Gao, C. X., Liu, Q. F. and Xue, D. S., *J. Mater. Sci. Lett.* **21** (2002), 1781.
15. Zhou, P. H., Xue, D. S., Luo, H. Q. and Chen, X. G., *Nanolett.* **2** (2002), 845.
16. Zhou, P. H., Xue, D. S., Luo, H. Q. and H. G. Shi, *Hyp. Interact.* **142** (2002), 601.
17. Shima, M., Hwang, M. and Ross, C. A., *J. Appl. Phys.* **93** (2003), 3440.
18. Liu, B., Huang, R. S., Wang, J. H. and Widatallah, H. M. *et al.*, *J. Appl. Phys.* **85** (1999), 1010.
19. Sato, O., Lyoda, T., Fujishima, A. and Hashimoto, K., *Science* **271** (1996), 49.
20. Sato, O., Lyoda, T., Fujishima, A. and Hashimoto, K., *Science* **272** (1996), 704.
21. Ohkoshi, S., Abe, Y., Fujishima, A. and Hashimoto, K., *Phys. Rev. Lett.* **62** (1999), 1285.

22. Chambaere, D. G., Grave, E. D., Vanleerbergher, R. L. and Vandenberche, R. E., *Hyp. Interact.* **20** (1984), 249.
23. Deniels, J. M. and Rosencwaig, A., *J. Phys. Chem. Solids* **30** (1969), 1561.
24. Borzi, R. A., Stewart, S. J., Punte, G., Mercader, R. C., Mansilla, M. V., Zysler, R. D. and Cabanillas, E. D., *J. Magn. Magn. Mater.* **205** (1999), 234.

Hyperfine Interactions **156/157**: 41–46, 2004.
© 2004 *Kluwer Academic Publishers. Printed in the Netherlands.*

Synthesis and Mössbauer Study of Maghemite Nanowire Arrays

DE-SHENG XUE, LI-YING ZHANG and FA-SHEN LI
Key Laboratory for Magnetism and Magnetic Materials of the Ministry of Education,
Lanzhou University, Lanzhou 730000, People's Republic of China

Abstract. Arrays of γ-Fe_2O_3 nanowire were synthesized in anodic aluminum oxide templates. The structure, morphology and magnetic property at room temperature were characterized. Temperature-dependent Mössbauer spectra was collected and the superparamagnetic relaxation was clearly observed. Both hyperfine field and isomer shift increase with decreasing temperature. The anisotropy energy constant is determined from the reduction of the hyperfine field relative to the saturation value caused by the collective magnetic excitations.

Key words: maghemite, nanowires, Mössbauer spectrum.

1. Introduction

Maghemite, γ-Fe_2O_3 has been attached much attentions in the magnetic recording media due to its attractive magnetic properties and chemical stability. In order to improve the magnetic recording density, γ-Fe_2O_3 was diversely prepared. However, with the decreasing of the particle size superparamagnetism of the particle limits further development of γ-Fe_2O_3 on ultrahigh magnetic recording density.

Nanowire arrays are promising candidates to extend this limit [1, 2]. Some metal and alloy nanowires, whose magnetic properties are suitable for ultrahigh-density recording, have been obtained [3–5]. Furthermore, when the dimension of materials reduces, their magnetic properties strongly differ from those of the bulk phases [3–7]. So preparation and study of quasi-one-dimensional γ-Fe_2O_3 are interesting and important not only in fundamental research but also in applications. However, it is difficult to obtain γ-Fe_2O_3 nanowire arrays in anodic aluminum oxide (AAO) templates directly using electrodepositing method or by oxidizing Fe nanowire arrays. In this article, we report the synthesis and Mössbauer study of maghemite nanowire arrays.

2. Experimental

The detailed fabrication of the AAO templates and the precursor FeOOH nanowire arrays can be found elsewhere [8, 9]. In this study, the high pure aluminum foils were anodized at DC 90 V for 3 h in 0.5 M phosphoric acid. FeOOH was electrodeposited into the templates at ac 15 V 70 Hz for 10 minutes in a solution containing $FeCl_3 \cdot 6H_2O$ and $(NH_4)_2C_2O_4 \cdot H_2O$. γ-Fe_2O_3 nanowire arrays can be obtained when the FeOOH nanowires firstly were reduced in H_2 at 325°C for 1 h, and then heat-treated in air at 275°C for 1 h after removing the remaining aluminum layers in a saturated $HgCl_2$ solution.

Structural of the nanowires was characterized by means of X-ray diffraction (XRD) using a Philips X'Pert diffractometer Model with Cu-Kα radiation. The morphology of the nanowires and selected area electronic diffraction (SEAD) were monitored by transmission electron microscopy (TEM) using a JEM-2000EX microscope. The magnetic properties of the nanowire arrays at room temperature (RT) were investigated using a Lake Shore 7304 vibrating sample magnetometry (VSM). Mössbauer spectra (MS) were obtained by using a constant acceleration spectrometer with a source of ^{57}Co in rhodium. The spectra were fitted with lorentzian lines using the least-square method.

3. Results and discussion

The XRD pattern of the AAO templates filled with γ-Fe_2O_3 nanowires are shown in Figure 1. The diffraction patterns are composed of a smoothly varying component and six sharp peaks. The smoothly varying peak intensity results from the amorphous alumina templates. The sharp peaks are due to the diffraction of a polycrystalline spinel structure with $a = 8.3615$ Å, which is corresponding to the γ-Fe_2O_3 standard diffraction pattern of JCPDS (39-1346). The average crystallite size calculated according to the Scherrer formula is about 15 nm from the broadening XRD diffraction peaks.

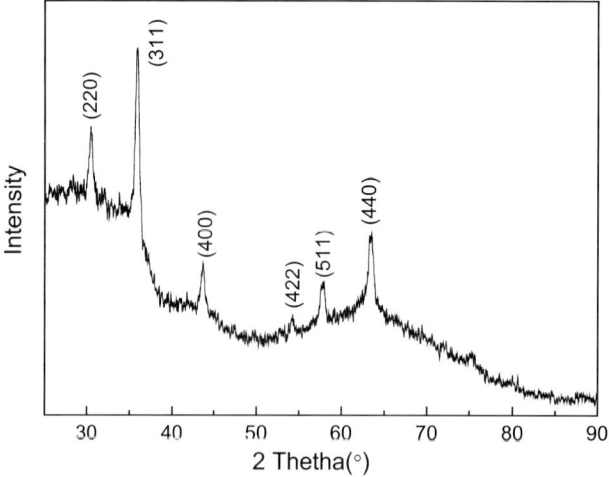

Figure 1. X-ray diffraction pattern of γ-Fe_2O_3 nanowire arrays.

Figure 2 shows a representative TEM image and the SEAD pattern of γ-Fe$_2$O$_3$ nanowires after dissolving the AAO templates. From Figure 2(a) the average diameter of the nanowires was found to be about 120 nm and the length is up to 8 μm. The sharp bright rings in Figure 2(b) indicate a polycrystalline spinel structure which is in agreement with the XRD result.

The RT hysteresis loops of the γ-Fe$_2$O$_3$ nanowires in AAO arrays measured are shown in Figure 3, where $H(//)$ and $H(\perp)$ indicate that the applied field is parallel and perpendicular to the nanowires axis, respectively. The closed hysteresis loops indicate that the nanowires are superparamagnetic. The unsaturated magnetization in the field of 6 kOe also indicates that the size is very small as observed on the fine nanoparticles [10]. The difference of the magnetization may come from the demagnetization contribution due to the shape anisotropy of the arrays.

Figure 2. TEM image (a) and SAED pattern (b) of γ-Fe$_2$O$_3$ nanowire arrays.

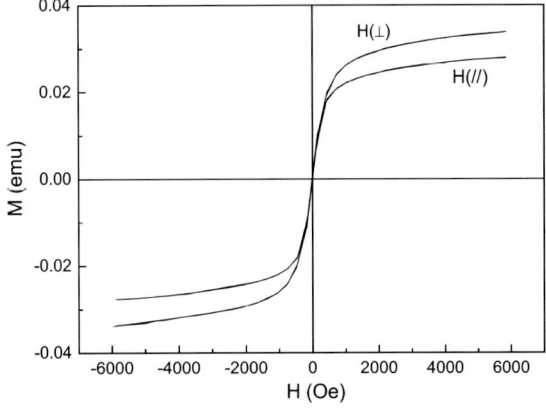

Figure 3. The room temperature hysteresis loops of γ-Fe$_2$O$_3$ nanowire arrays. $H(//)$ and $H(\perp)$ indicate that the applied field is parallel and perpendicular to the nanowire axis, respectively.

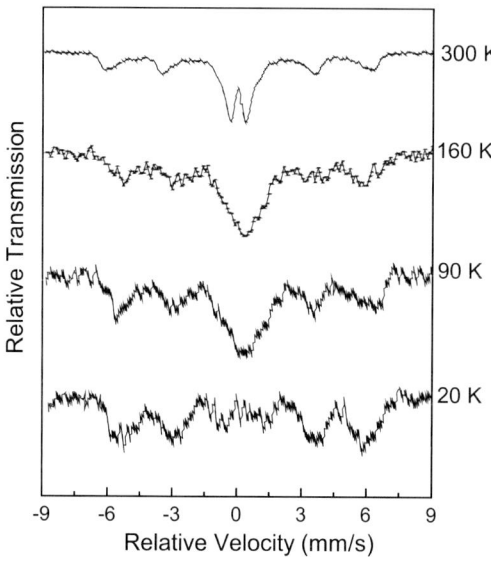

Figure 4. Mössbauer spectra of γ-Fe$_2$O$_3$ nanowire arrays at different temperatures.

Figure 4 shows the temperature-dependent MS of maghemite nanowire arrays in the AAO templates with γ-ray parallel to the wires axis. All MS consist of a broad sextet distribution and a superimposed non-magnetic distribution except for 20 K MS. The vanishing of the non-magnetic phase at 20 K results from the super-paramagnetic fluctuations. The broadening of the peaks results from the relaxation phenomenon like collective magnetic excitations due to the intrinsic finite-size effects. For all temperatures the quadrupole splitting is quite small. The RT isomer shift (*IS*) is 0.33 mm/s, which is almost the same with the bulk phase [11] but little less than the nanoparticles [12]. The *IS* increases with decreasing temperature and up to 0.43 mm/s at 20 K. This differences are due to the thermal variation of the second-order Doppler effect [13]. The relative areas ratio of the six lines in RT MS are about 3 : 4 : 1 : 1 : 4 : 3, suggesting that the spins are almost perpendicular to the wire axis.

Temperature dependence of the hyperfine field is shown in Figure 5. The hyperfine field for the magnetic phase at 300 K is about 46.5 T, which is less than the typical value of 50 T [14], due to the collective magnetic excitations caused by the fine crystallite size. The hyperfine field increases from 46.5 to 50.6 T when the temperature decreased from 300 to 20 K. The reduction of the hyperfine field is due to the collective magnetic excitations. The hyperfine field, $H(V, T)$ relative to the saturation value, $H(\infty, T)$, can be written as [15, 16]

$$H(V, T) = H(\infty, T)(1 - k_B T/2KV),$$

where $H(V, T)$ is the hyperfine field, V is the particle volume at the temperature T, k_B is the Boltzmann constant and K is the anisotropy energy constant. The order of magnitude of the anisotropy constant can be determined from the MS by

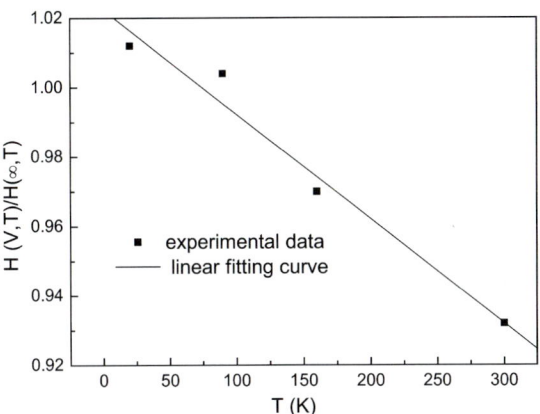

Figure 5. Temperature dependence of normalized hyperfine fields for γ-Fe_2O_3 nanowires in AAO templates, where $H(\infty, T) = 50$ T was used.

comparing the hyperfine field ($H(\infty, T)$) measured in a large magnetic field and that obtained in zero field at the same temperature [15, 16]. The values of $H(\infty, T)$ found from the spectrum obtained at 20 K in zero field (Figure 4) is 50.6 T. Using the size of 15 nm from XRD result, the anisotropy energy constant is found to be $K = 0.69 \times 10^4$ J m^{-3}, which is in good agreement with the result of Hendriksen *et al.* for ultrafine maghemite particles [17].

4. Conclusions

Maghemite nanowire arrays with diameter of about 120 nm and length up to 8 μm were synthesized in AAO templates successfully. The nanowires have a polycrystalline spinel structure with $a = 8.3615$ Å. The closed hysteresis loops and the non-magnetic distribution in MS result from the superparamagnetic fluctuations. From 300 K MS the magnetic moments of the particles are almost perpendicular to the wire axis. The *IS* increases from 0.33 to 0.43 mm/s when the temperature decreased from 300 to 20 K due to the second-order Doppler effect. Whereas, the hyperfine field increases from 46.5 to 50.6 T is due to the collective magnetic excitations. The anisotropy energy constant is found to be 0.69 $\times$ 10^4 J m^{-3} after fitting the temperature dependence of hyperfine field.

Acknowledgement

This work is supported by the NSFC (Grant No 10374038), the Trans-Century Training Program Foundation for the Talent of MOE and EYTT of P.R. China.

References

1. Hwang, M., Abraham, M. C., Savas, T. A., Smith, H. I., Ram, R. J. and Ross, C. A., *J. Appl. Phys.* **87** (2000), 5108.
2. Kaneko, M. P., *IEEE Trans. Magn.* **MAG-17** (1981), 1468.
3. Peng, Y., Zhang, H. L., Pan, S. L. and Li, H. L., *J. Appl. Phys.* **87** (2000), 7405.
4. Zheng, M., Menon, L., Zeng, H., Liu, Y., Bandyopadhyay, S., Kirby, R. D. and Sellmyer, D. J., *Phys. Rev. B* **62** (2000), 12282.
5. Zhan, Q., Chen, Z., Xue, D. and Li, F., *Phys. Rev. B* **66** (2000), 134436.
6. Parker, F. T., Foster, M. W., Margulies, D. T. and Berkowitz, A. E., *Phys. Rev. B* **47** (1993), 7885.
7. Vollath, D., Szabo, D. V., Taylor, R. D. and Willis, J. O., *J. Mater, Res.* **12** (1997), 2175.
8. Masuda, H. and Fukuda, K., *Science* **268** (1995), 1466.
9. Gao, C. X., Liu, Q. F. and Xue, D. S., *J. Mater. Sci. Lett.* **21** (2002), 1781.
10. Pascal, C., Pascal, J. L., Favier, F., Elidrissi Moubtassim, M. L. and Payen, C., *Chem. Mater.* **11** (1999), 141.
11. Zboril, R., Mashlan, M. and Petridis, D., *Chem Mater.* **14** (2002), 969.
12. Zboril, R., Mashlan, M., Barcova, K. and Vujtek, M., *Hyp. Interact.* **139/140** (2002), 597.
13. Chantrell, R. W., Popplewell, J. and Charles, S. W., *Physica* **86–88B** (1997), 1421.
14. Vandenberghe, R. E., de Grave, E., Landuydt, C. and Bowen, L. H., *Hyp. Interact.* **53** (1990), 175.
15. Mørup, S. and Tøpsøe, H., *Appl. Phys.* **11** (1976), 63.
16. Mørup, S., *J. Magn. Magn. Mater.* **37** (1983), 39.
17. Hendriksen, P. V., Bødker, F., Linderoth, S. and Mørup, S., *J. Phys: Condens. Matter.* **6** (1994), 3081.

Hyperfine Interactions **156/157**: 47–50, 2004.
© 2004 *Kluwer Academic Publishers. Printed in the Netherlands.*

Mössbauer Characterization of Iron-Based Nanogranular Films

T. FURUBAYASHI
National Institute for Materials Science, Tsukuba 305-0047, Japan;
e-mail: furubayashi.takao@nims.go.jp

Abstract. Two nanogranular systems, Fe–MgF$_2$ and Fe–SiO, were prepared by co-evaporating the each material in vacuum. Mössbauer studies of the films are described. It was shown that the Fe atoms form nanoparticles of metallic α-Fe in the Fe–MgF$_2$ system. In contrast, the results for Fe–SiO indicate that nanoparticles of a Fe–Si alloy are formed by co-evaporating Fe and SiO.

Key words: iron, nanoparticle, hyperfine field, amorphous alloy.

1. Introduction

Nanogranular films containing magnetic nanoparticles have attracted much attention in view of fundamental studies and applications to recording media, soft magnetic materials, and magnetoresistance materials and so on. Mössbauer spectroscopy is a quite powerful tool for characterizing such nano-structured materials containing iron.

It has been shown that metal particles in insulating materials such as oxides and fluorides can be prepared by co-sputtering or co-evaporation of two materials [1–3]. In this work, two kinds of nanogranular systems, Fe–MgF$_2$ and Fe–SiO, were prepared by co-evaporating two materials in vacuum. The two insulating materials were chosen for the easiness of the evaporation. The obtained films were examined by Mössbauer spectroscopy.

2. Experimental

An evaporating apparatus equipped with two evaporating sources was used for preparing the samples. Iron metal and the other material, MgF$_2$ or SiO, were deposited at the same time in vacuum better than 5×10^{-8} Torr. The films were deposited on Kapton substrates kept at room temperature or cooled by liquid nitrogen. The film thickness was about 0.5 nm. Transmission Mössbauer spectra were recorded at temperatures from 4.2 K to room temperature by using a conventional constant-acceleration spectrometer with a ^{57}Co/Rh source.

3. Results and discussion

Nanogranular structures with the grain size of 2~3 nm were observed by transmission electron microscopy (TEM) for the both systems studied here with the Fe volume fraction less than 30%. X-ray and electron diffraction patterns of the Fe–MgF_2 samples showed reflections from α-Fe and MgF_2. For the Fe–SiO samples, however, halo patterns were observed, suggesting an amorphous structure.

Figure 1(a) shows Mössbauer spectra of the Fe–MgF_2 sample with the Fe volume fraction of 20% prepared on the substrate at 100 K. The spectra show that the Fe particles are superparamagnetic at room temperature and hyperfine splittings appear with decreasing temperature. The spectrum at room temperature is unsymmetrical and indicates the presence of quadrupole splittings QS. The spectra are well reproduced by assuming the components of one singlet and two doublets, as shown by Childress *et al.* in the Fe SiO_2 and the Fe–Al_2O_3 systems [2]. The results of the fittings are summarized in Table I. In fine particles studied here, as small as 3 nm, some tens % of the total atoms are at the interface. Thus, it would be reasonable to attribute two doublets to interfacial Fe atoms, which have no cubic coordination. The presence of two doublets may result from different coordination numbers of atoms in various sites at the interface. The obtained values of IS and QS

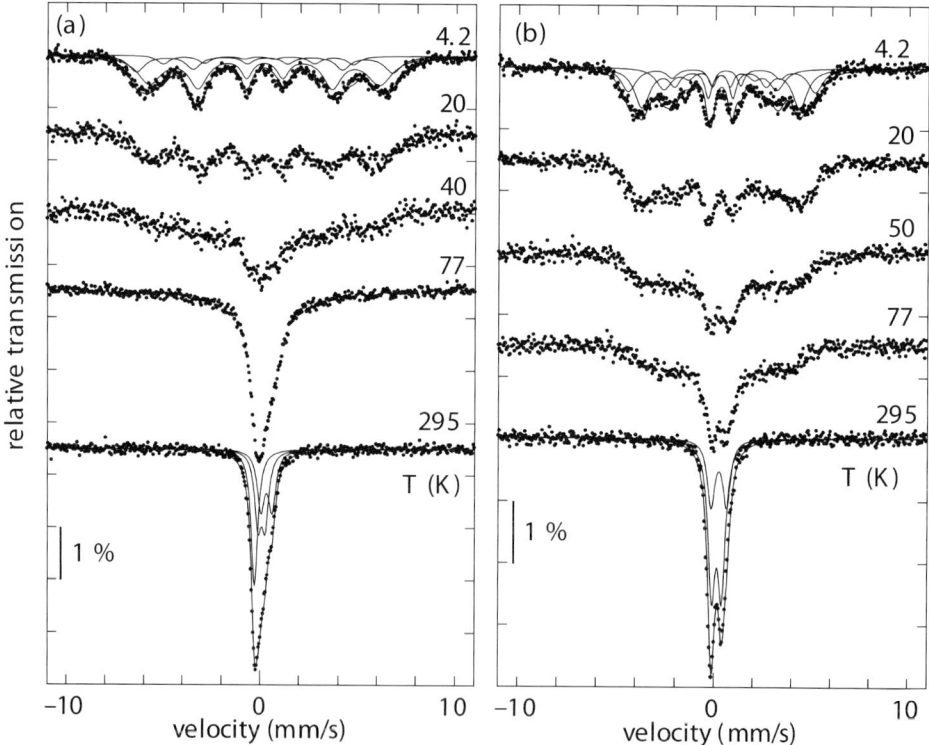

Figure 1. Mössbauer spectra of (a) Fe–MgF_2 (20 vol% Fe) and (b) Fe–SiO (25 vol% Fe) at indicated temperatures.

Table I. Summary of Mössbauer parameters for Fe–MgF$_2$ (20 vol% Fe) and Fe–SiO (25 vol% Fe) at the temperatures of 295 K and 4.2 K. Isomer shifts IS are relative to α-Fe at room temperature

System	T (K)	IS (mm/s)	QS (mm/s)	H_{hf} (kOe)	Rel. area (%)
Fe–MgF$_2$	295	−0.29	–	–	34
		0.08	0.38	–	35
		0.33	0.58	–	31
	4.2	−0.14	−0.06	301	10
		0.15	−0.02	367	65
		0.23	−0.16	409	25
Fe–SiO	295	0.13	0.51	–	71
		0.24	0.80	–	29
	4.2	0.26	−0.04	179	26
		0.26	0.04	250	47
		0.43	−0.18	297	27

are quite similar to those in the Fe–SiO$_2$ and the Fe–Al$_2$O$_3$ systems [2], including the negative IS of the singlet component. Thus, it seems that the electronic state of Fe in the present material is similar to that in such Fe-oxide systems. The negative IS possibly results from enhanced lattice vibrations due to the small size of the particles [2]. The enhanced atomic motion can decrease IS through the second-order Doppler shift. Another contribution would be the modification of the electronic state due to the large surface area, which leads to the change of the electron density at the nucleus.

The spectrum at 4.2 K contains no paramagnetic component and has broad lines indicating the distributed hyperfine field H_{hf}. For evaluating the Mössbauer parameters, we tentatively made the fitting of the spectrum by three components. The average value of H_{hf} is 370 kOe, which is larger than the value of bulk iron, 340 kOe. This can be attributed to the contribution of Fe atoms at the interface. The enhancement has been also observed for interfacial atoms in Fe/MgF$_2$ multi-layered films [4]. No component indicating iron fluorides, which have still much larger H_{hf}, is detected. We find no sign of Fe–Mg alloys, which would lead in turn to smaller H_{hf}. Thus, it is concluded that Fe in the present sample mostly forms metallic iron without mixing with MgF$_2$ at the interface. However, a small part of Fe atoms possibly oxidized during handling in ambient atmosphere and contribute to the component with $H_{hf} = 408$ kOe.

Figure 1(b) shows Mössbauer spectra of Fe–SiO with the Fe volume fraction of 25% prepared on the substrate at 100 K. Similarly with Fe–MgF$_2$, the sample is superparamagnetic around room temperature. Hyperfine splittings appear with decreasing temperature. The spectra in the superparamagnetic state clearly show quadrupole splittings. The spectrum at 295 K is mostly reproduced by superposing two symmetric doublets. The obtained Mössbauer parameters are summarized in Table I. The spectrum at 4.2 K contains no superparamagnetic component and has broad lines indicating the distributed hyperfine field H_{hf}. The average value of H_{hf} at 4.2 K was obtained to be 250 kOe, which is pronouncedly smaller than that of bulk Fe. It is to be noticed that the small value of H_{hf} is not from superparamagnetic fluctuations [5] because the spectra at 4.2 K and 20 K have similar H_{hf}.

It has been reported that, in Fe–Si alloys, Si atoms reduces H_{hf} of Fe atoms in the surrounding sites [6]. In amorphous Fe–Si alloys, H_{hf} is continuously reduced by increasing the concentration of Si [7]. Thus, the small value of H_{hf} in this sample is most reasonably understood by considering that Fe is in an amorphous Fe–Si alloy produced by the chemical reaction of Fe and SiO. Thus, it is reasonable to consider that the film consists of Fe–Si alloy nanoparticles embedded in a SiO_x matrix. The spectra in the superparamagnetic state with quadrupole splittings are also reasonable for an amorphous Fe–Si alloy.

It is to be noticed that the intensities of the 2nd and the 5th lines of the ferromagnetic spectra are weak. The results suggest that the perpendicular anisotropy is enhanced in this material because the γ-ray is in the direction perpendicular to the film. The intensity ratio of the 1st and the 2nd lines was calculated to be 3 : 1.1 at 4.2 K, while 3 : 2 is expected for the case that the direction is random. The tendency to perpendicular magnetization possibly comes from shape or some configuration of the particles. However, the conclusion is not definitive and detailed investigations are needed to clarify the problem.

4. Summary

Nanogranular Fe–MgF$_2$ and Fe–SiO films prepared by co-evaporation in vacuum were examined by Mössbauer spectroscopy at temperatures from room temperature down to 4.2 K. Both films were superparamagnetic at room temperature and showed hyperfine splittings at low temperatures. The Fe–MgF$_2$ film was found to contain α-Fe particles from the analysis of the spectra. In contrast, it was deduced that the Fe–SiO film consists of nano particles of an amorphous Fe–Si alloy embedded in a SiO_x matrix. The intensity ratio of the lines in the spectrum for the Fe–SiO film suggests the enhancement of perpendicular anisotropy.

References

1. Abeles, B., Sheng, P., Coutts, M. D. and Arie, Y., *Adv. Phys.* **24** (1975), 407.
2. Childress, J. R., Chien, C. L., Zhou, M. Y. and Sheng, P., *Phys. Rev. B* **44** (1991), 11689.
3. Furubayashi, T. and Nakatani, I., *J. Appl. Phys.* **79** (1996), 6258.
4. Nishikawa, M., Ono, K., Kita, E., Yanagihara, H., Erata, T. and Tasaki, A., *J. Mag. Mag. Mater.* **238** (2002), 91.
5. Mørup, S., *J. Mag. Mag. Mater.* **37** (1983), 39.
6. Stearns, M. B., *Phys. Rev.* **129** (1963), 1136.
7. Marshal, G., Mangin, P. and Janot, C., *Solid State Commun.* **18** (1976), 739.

Hyperfine Interactions **156/157**: 51–56, 2004.
© 2004 *Kluwer Academic Publishers. Printed in the Netherlands.*

Mössbauer Study of Nanocrystalline ε-Fe₃₋ₓCoₓN System

N. S. GAJBHIYE[1],[*],[**], R. S. NINGTHOUJAM[1] and J. WEISSMÜLLER[2]
[1]*Department of Chemistry, Indian Institute of Technology, Kanpur-208016, India;*
e-mail: nsg@iitk.ac.in
[2]*Institute of Nanotechnology, Forschungszentrum Karlsruhe, 76021 Karlsruhe, Germany*

Abstract. Nanocrystalline ε-Fe₃₋ₓCoₓN system having particle size in the range of 10–15 nm is synthesized by chemical route. For compositions $x = 0.0$–0.2, the system crystallizes in hexagonal structure and for $x = 0.4$–0.8, the system shows a mixture of hexagonal ε-Fe₃₋ₓCoₓN and cubic α-Fe structures. The Mössbauer spectra shows the doublet for the former structure and the sextet for latter when cobalt concentration is in the range $x = 0.4$–0.8 indicating the presence of α-Fe trapped in the nitride matrix. The superparamagnetic nature of ε-Fe₃₋ₓCoₓN phase is confirmed from the magnetization measurements.

1. Introduction

In the phase diagram of iron nitride system [1], the Fe content y varies in the range of $2 \leqslant y \leqslant 3$ for the equilibrium phases of hexagonal ε-Fe$_y$N. When the value of y is 3 and 2, the compounds are ferromagnetic and paramagnetic respectively at room temperature. The Curie temperature decreases to below room temperature with the decrease of x value [2]. However, the Co, Ni substitution can stabilize the giant magnetic phases like Fe₄N and Fe₁₆N₂ in the phase diagram [3, 4]. The FeCo alloys can exit as body centered cubic (bcc) phase and offer potential significance as high-temperature magnets, in applications such as rotors in electrical aircraft engines [5–7]. So, the addition of nitrogen (N) or carbon (C) into the interstitial sites of metal lattice changes the crystal structure and properties because of an increase of metal–metal bond [8, 9]. In this study, the synthesis of the nanocrystalline ε-Fe₃₋ₓCoₓN ($x = 0.0$–0.8) system by oxalate precursor route is carried out and the Mössbauer spectra analyses and magnetic properties are presented.

2. Experimental section

Ferric nitrate and cobalt nitrate are dissolved in water with constant stirring at 30°C in the corresponding proportion of ε-Fe₃₋ₓCoₓN (where $x = 0.0$–0.8). The

---

* Author for correspondence.
** Visiting professor at Institute of Nanotechnology, Forschungszentrum Karlsruhe, 76021 Karlsruhe, Germany.

solution is maintained at 0.1 M and is mixed with excess of oxalic acid, and then it is refluxed for 24 h at 80°C to get homogeneous. It is evaporated slowly to get the mixed metal oxalate gel. This gel is heated at 500–600°C for 3 h to obtain ultrafine Fe–Co-oxide particles. About 2 g of ultrafine Fe–Co-oxide particles are nitrided in an alumina tubular furnace under flowing NH_3 gas at the rate of 120 cm^3/min in the temperature range of 500–550°C for 4 h to give the nanocrystalline ε-$Fe_{3-x}Co_xN$. The chemical analyses of these compositions for Fe, Co are carried out by using standard methods [10]. Nitrogen was estimated by Kjeldahl method using acetanilide as standard for the method [10]. Further, nitrogen content was confirmed using a Perkin–Elmer 240C elemental (C, H, N) analyzer and other elements confirmed from EDAX analysis. Thus, on analysis the stoichiometry of $Fe_{3-x}Co_xN$ compositions is confirmed.

3. Results and discussion

Figure 1 shows the representative X-ray diffraction patterns of ε-$Fe_{3-x}Co_xN$ system; where for $x = 0.0$–0.2 shows the hexagonal structure and for $x = 0.4$–0.8 shows a mixture of phases with hexagonal and bcc structures. The composition ε-Fe_3N, when $x = 0$, crystallizes in the hexagonal phase and the lattice parameters are: $a = 2.75$, $c = 4.40$ Å and unit cell volume of 28.81 Å^3 and corroborate the literature values [11]. The XRD parameters for other compositions are also follow similar trend of ε-Fe_3N phase. The crystallite size is found to be in the range of 10–15 nm. Morphology is studied by SEM and TEM and shows that the shape of particle changes from spherical to oval shape with increase of x.

Figure 2 shows the room-temperature Mössbauer spectra of the ε-$Fe_{3-x}Co_xN$ system for $x = 0.0$–0.4 and 0.8. For $x = 0.0$–0.2, only the superparamagnetic doublet is observed indicating the smaller particle size and for $x = 0.4$–0.8, both superparamagnetic doublet and ferromagnetic sextet are seen. This is further confirmed from the magnetization measurements at room temperature as a function

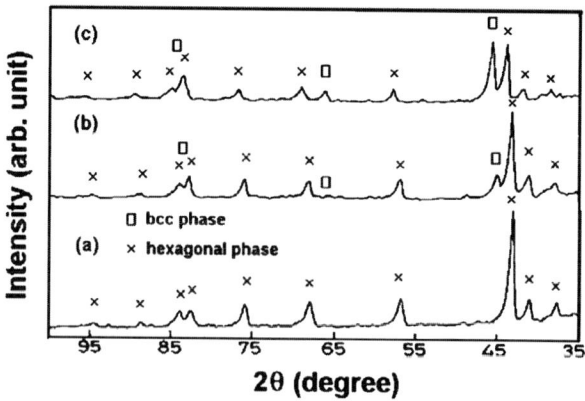

Figure 1. XRD patterns of ε-$Fe_{3-x}Co_xN$: (a) $x = 0.0$, (b) $x = 0.4$ and (c) $x = 0.8$.

of field where no hysteresis is found. Mössbauer parameters – isomer shift (δ), quadrupole splitting (Δ), linewidth (Γ) and magnetic hyperfine field (ΔH_f) of this system are given in Table I. Takahashi et al. [12] reported the presence of both paramagnetic and ferromagnetic phases in (Fe–Co)–N thin film system where Co substitutes Fe site in ε-Fe$_3$N and also in ε-Fe$_3$N phase. But they did not describe the source of paramagnetic doublet. In another study of ultrafine ε-Fe$_3$N particles prepared by precursor route, superparamagnetic phase is observed [11]. In thin film metal nitride system, there is possibility of N deficiency. In their study on ε-Fe$_3$N particles contain non-nitrided Fe atoms in the lattice and give the sextet spectrum. However, the presence of non-nitrided Fe atoms is ruled out in our studies. Also, this is supported by XRD patterns in Figure 1, where there are no peak(s) found corresponding to α-Fe. The crystal structure of ε-Fe$_3$N is described as hexagonal two Fe layers and between these layers N atoms occupy partially one third of the octahedral interstitial positions. Each Fe atom has six nearest neighbor Fe atoms in the same layer (ab-plane) and with the same distance. Similarly, each N has six nearest Fe atoms, above and below the N layer in an octahedral co-ordination.

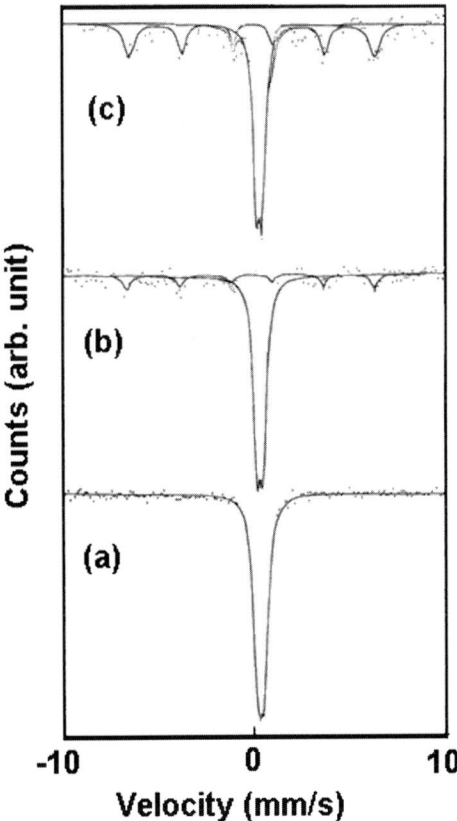

Figure 2. Mössbauer spectra of ε-Fe$_{3-x}$Co$_x$N: (a) $x = 0.0$, (b) $x = 0.4$ and (c) $x = 0.8$.

Table I. Mössbauer resonance parameters of ε-Fe$_{3-x}$Co$_x$N recorded at 300 K

Iron sites in ε-Fe$_{3-x}$Co$_x$N	Hyperfine ΔH_f (T)	Isomer shift δ (mm/s)	Quadrupole splitting Δ (mm/s)	Line widths Γ (mm/s)
$x = 0.0$				
(doublet)		0.44	0.12	0.47
$x = 0.4$				
(sextet)	40.14	0.0	−0.05	0.46
(doublet)		0.45	0.15	0.42
$x = 0.6$				
(sextet)	39.68	0.0	−0.06	0.47
(doublet)		0.46	0.16	0.86
$x = 0.8$				
(sextet)	39.74	0.0	−0.04	0.63
(doublet)		0.45	0.16	0.60

The isomer shift (δ) can give the electron density in Fe nuclei of ε-Fe$_3$N and of ε-Fe$_{3-x}$Co$_x$N systems. So, δ is proportional to the s-electron density at the sample nuclei, i.e. $\delta \propto \psi_s^2(0)_a$. α-Fe is taken as the reference for all Mössbauer measurements. In ε-Fe$_3$N ($x = 0.0$), the isomer shift (δ) value is 0.44 mm/s. This value is slightly higher than the value 0.33 mm/s for Fe occupying cornered or face centered sites in γ'-Fe$_4$N [11]. This indicates that Fe in ε-Fe$_3$N has smaller s-electron density or greater p- and d-electron density as compared to that of γ'-Fe$_4$N. It means that there is an interaction of N and Fe atoms in ε-Fe$_3$N system. In the Fermi-level (E_f), the number of electrons in ε-Fe$_3$N system is due to the mixing of d-electrons for Fe and p-electrons for N. The density of states at the E_f is arose from the unbalance spins between Fe 3d-orbital and N 2p-orbital electrons. This results a partial covalent-character between Fe and N bonds. This isomer shift supports the covalent character [11].

The nanosized magnetic particles are single domain particles and become superparamagnetic below the critical size (d_c) and above a temperature (T_B) called blocking temperature, i.e. magnetization direction fluctuates spontaneously (superparamagnetic relaxation) with a relaxation time (τ) given by Neel expression [13]: $\tau = \tau_0 e^{\Delta E_a/kT}$, where τ_0 is in the order of 10^{-10}–10^{-13} s and ΔE_a, the anisotropic energy barrier equal to the product of an effective anisotropic energy constant (K_{eff}) and the volume (V), k, Boltzmann's constant and T, the temperature. While using Mössbauer spectroscopy, it is found that the superparamagnetic relaxation results in line broadening for the relaxation time of the order of 10^{-8} and τ is less than or equal to 10^{-9} resulting in the disappearance of the magnetic splitting to give the paramagnetic doublet. The single domain ε-Fe$_3$N isolated particles embedded in solidified kerosene shows the blocking temperature in the range of 15–40 K depending on the particle size, but the single domain ε-Fe$_3$N, as prepared particles, having dipolar interactions shows the spin glass cooperative phenomenon at $\sim$70 K [14].

The quadrupole splitting, $\Delta = 0.16$ mm/s with line width, $\Gamma = 0.47$ mm/s indicates that the non-cubic symmetry for ε-Fe$_3$N. With higher Co concentration addition, i.e., $x = 0.4$–0.8, the extra sextet is appeared due to the exaltation of non-nitrided α-Fe (bcc phase) in ε-Fe$_{3-x}$Co$_x$N system. But for $x = 0.0$–0.2 compositions, there are no sextet patterns observed. The magnetic sextet part of ε-Fe$_{3-x}$Co$_x$N system at room temperature Mössbauer spectra corresponds to the formation of α-Fe (bcc structure) in the nitride matrix. The isomer shift value (δ) is almost zero and is consistent. The reported value of δ for sextet of FeCo alloys was 0.10–0.15 [15]. The presence of α-Fe is confirmed from the XRD pattern which gives the characteristic extra peaks at $2\theta = 44.2$, and 66.3 degree for the compositions in the range $0.4 \leqslant x \leqslant 0.8$ (see Figures 1(b) and (c)). The high magnetic hyperfine field value of $\sim$40 T (Table I) observed in these compositions is due to the presence of α-Fe fractions and the enhancement of hyperfine field is attributed to the surface atoms in metallic iron nanoparticles [12, 16].

There is no change of quadrupole splitting in doublet part indicating the random distribution of Co at Fe-site of ε-Fe$_3$N (Table I). The quadrupole splitting for sextet is about 0.05 mm/s, and indicates cubic symmetry of Fe environment. Therefore, it may be concluded that on substitution Co atoms occupy Fe site in ε-Fe$_3$N randomly to give ε-Fe$_{3-x}$Co$_x$N system and simultaneously exalted the extra α-Fe atoms in the nitride matrix. The detailed magnetic properties of the ε-Fe$_{3-x}$Co$_x$N system are under study.

4. Conclusions

The nanocrystalline ε-Fe$_{3-x}$Co$_x$N ($x = 0.0$–0.8) has been prepared in the size range of 10–15 nm by chemical route. The compositions for $x = 0.0$–0.2 show the superparamagnetic nature and for $x = 0.4$–0.8 exhibit the mixed superparamagnetic and ferromagnetic fractions. The Co substitution occupies Fe-sites randomly in ε-Fe$_3$N lattice and for higher concentrations the formation of α-Fe is observed that is trapped in the nitride matrix. The high magnetic hyperfine field of nearly 40 T is found for the ε-Fe$_{3-x}$Co$_x$N ($0.4 \leqslant x \leqslant 0.8$) system and attributed to the presence of α-Fe fraction.

Acknowledgement

We acknowledge DST, New Delhi, India for the financial support during this research work.

References

1. Jack, K. H., *Proc. R. Soc. A* **208** (1951), 200.
2. Chen, G. M., Jaggi, N. K., Butt, J. B., Yeh, E. and Schwartz, L. H., *J. Phys. Chem.* **87** (1983), 5326.
3. Jiang, E. Y., Wang, H. Y. and Ma, Z. W., *Appl. Phys.* **85** (1999), 4488.
4. Panda, R. N. and Gajbhiye, N. S., *J. Magn. Magn. Mater.* **195** (1999), 396.
5. Takahashi, S., Kume, M. and Matsuura, K., *IEEE Trans.* **MAG-26** (1990), 1632.
6. Kim, K. J., Lee, S. J. and Lynch, D. W., *Solid State Comm.* **114** (2000), 457.
7. Bozorth, R. M., In: *Ferromagnetism*, Van Nostrand, New York, 1951.
8. Toth, L. E., In: *Transition Metal Carbides and Nitrides*, Academic Press, New York, 1971.
9. Jhi, S.-H., Louie, S. G., Cohen, M. L. and Ihm, J., *Phys. Rev. Lett.* **86** (2001), 3348.
10. Vogel, A. I., In: *A Text-Book of Quantitative Inorganic Analysis*, Longmans, London, 1960, pp. 295, 461, 248.
11. Panda, R. N. and Gajbhiye, N. S., *IEEE Trans. Magn.* **34** (1998), 542.
12. Takahashi, S., Umeda, K., Kita, E. and Tasaki, A., *IEEE Trans.* **MAG-23** (1987), 3630.
13. Neel, L., *Ann. Geophys.* **5** (1949), 99.
14. Mamiya, H., Nakatami, I. and Furubayashi, T., *Phys. Rev. Lett.* **80** (1998), 177.
15. Gangopadhyay, S., Yang, Y., Hadjipanayis, G. C., Papaefthymiou, V., Sorensen, C. M. and Klabunde, K. J., *J. Appl. Phys.* **76** (1994), 6319.
16. Bødker, F., Mørup, S. and Linderøth, S., *Phys. Rev. Lett.* **72** (1994), 282.

Hyperfine Interactions **156/157**: 57–61, 2004.
© 2004 *Kluwer Academic Publishers. Printed in the Netherlands.*

Mössbauer Studies of Nanosize $CuFe_2O_4$ Particles

N. S. GAJBHIYE[1,*,**], G. BALAJI[1], SAYAN BHATTACHARYYA[1] and
M. GHAFARI[2]

[1]*Department of Chemistry, Indian Institute of Technology, Kanpur-208 016, U.P., India;*
e-mail: nsg@iitk.ac.in
[2]*Materials Science Department, Darmstadt University of Technology, Petersenstr. 23,*
64287 Darmstadt, Germany

Abstract. Nanocrystalline $CuFe_2O_4$ particles are prepared by wet chemical method. The particles of various sizes are obtained by heat treatment in the temperature range 773–1073 K. The room temperature Mössbauer spectrum for all samples shows the presence of both sextet and doublet indicating the presence of superparamagnetic fractions. The isomer shift, quadrupole splitting and hyperfine field values are found to change with particle size. However, these changes in Mössbauer parameters may also be due to the Jahn–Teller effect that essentially arises due to the migration of Cu^{2+} from tetrahedral sites to octahedral sites resulting in crystal structure change from cubic to tetragonal system. These aspects are studied by using Mössbauer spectroscopy and are correlated to the X-ray diffraction data.

Key words: cation distribution, ferrites, nanocrystalline, Jahn–Teller effect.

1. Introduction

Magnetic nanoparticles have generated great interest due to their importance in fundamental understanding of physical processes and technological applications in various areas like magnetic recording media, ferrofluids, magnetic refrigeration, geomagnetism, catalysis, sensor and biomagnetic applications [1]. In the nanoscale regime, the intrinsic magnetic parameters viz. Curie temperature (T_c), saturation magnetization (σ_s), magnetocrystalline anisotropy constant (K), and magnetic hyperfine field (H_f) are found to be dependent on particle size [2]. On similar lines, size dependent cation distribution is also observed in ball-milled nanosize $NiFe_2O_4$, $MnFe_2O_4$ and $CuFe_2O_4$ particles [3–5].

The magnetic structure of $CuFe_2O_4$ system is ferrimagnetic [6]. This can be explained by two sublattices, A and B, where A and B refer to the tetrahedral site and octahedral site, respectively. Thus, in general, the magnetic properties of $CuFe_2O_4$ samples are governed by cation distribution between the two sublattices. The cation distribution can be altered by the synthetic method, heat treatment tem-

---

* Author for correspondence.

** Visiting Professor at Institute of Nanotechnology, Forschungszentrum Karlsruhe, 76021 Karlsruhe, Germany.

perature, heat-duration time and cooling rate. Earlier, it was demonstrated that a crystal distortion can be induced due to the Jahn–Teller effect in partial inverse, bulk $CuFe_2O_4$ samples and this distortion depends on the Cu^{2+} ion distribution in the sublattices [7].

Therefore, the motivation of the present work is to investigate the cation distribution in nanostructured $CuFe_2O_4$ having different particle size and to study the effect of such cation distribution on Mössbauer parameters, magnetic properties and the crystal structure.

2. Experimental section

Nanostructured $CuFe_2O_4$ particles have been synthesized by hydroxide coprecipitation method [8]. 0.1 M solution of cupric chloride and 0.2 M solution of ferric chloride are prepared separately and mixed such that Cu/Fe ratio is maintained as 1 : 2. To this solution, 3 M solution of sodium hydroxide is added until pH of 10 is reached. The resulting solution along with the precipitate is boiled at 373 K for 2 hr. The product is washed with distilled water, filtered and dried. As such prepared samples are amorphous to X-rays. To improve the crystallinity and to prepare the particles of various size, the as-prepared samples are heat-treated in the temperature range 773–1073 K. X-ray diffractograms (Rich Seifert model Isodebyeflex 2002) for these samples are recorded using Cu-Kα radiation to ascertain the formation of mono-phase nanosize $CuFe_2O_4$ particles. The average particle size is calculated from the X-ray line broadening of reflection (311) plane using Scherrer's formula. Particle size and morphology are characterized by TEM (Philips model E-301). Mössbauer spectra of the samples under study are recorded by using conventional spectrometer and ^{57}Co source embedded in Rh matrix. The spectrum is then analyzed by using software PC-MOS supplied by CMTE Electronik, Germany. The values are presented relative to natural iron. The uncertainty in isomer shift and quadrupole splitting is 0.03 and 0.04 mm/s, respectively. The hyperfine field values are correct to 0.1 T.

3. Results and discussion

X-ray diffraction (XRD) patterns indicate the formation of pure and monophasic nanostructured $CuFe_2O_4$ particles. The cubic crystal structure is stabilized in nanosize $CuFe_2O_4$ particles obtained below 973 K whereas tetragonal structure is stable in particles obtained above 973 K. The particle size obtained is in the range 10–40 nm. The tetragonality (or the crystal distortion) is increased with the heat-treatment temperature because of the particle size increase. Electron micrographs (TEM) indicate that particles are spherical in shape.

Figure 1 shows zero-field Mössbauer spectrum of nanostructured $CuFe_2O_4$ particles having size of 10, 16 and 30 nm that is measured at 300 K. The spectrum of 10 nm $CuFe_2O_4$ particles is composed of both doublet and sextet indicating the

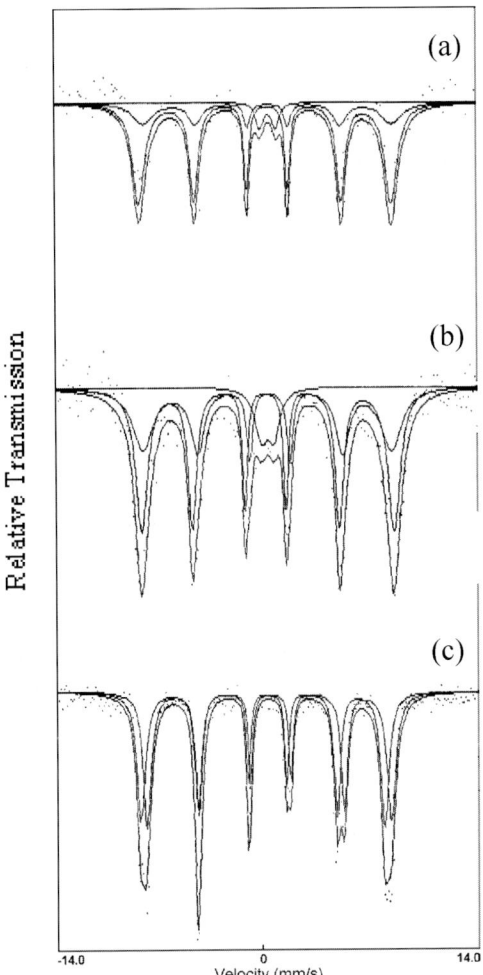

Figure 1. Room-temperature Mössbauer spectrum of CuFe$_2$O$_4$ particles having sizes (a) 10, (b) 16, and (c) 30 nm, respectively.

presence of superparamagnetic fractions. As the size is increased progressively, the intensity of doublet decreases indicating the particle size effect. Finally, the spectrum of 30 nm CuFe$_2$O$_4$ particles is composed of sextet only. The Mössbauer parameters are obtained by fitting the spectra with two sextets and doublet corresponding to A and B sublattices and superparamagnetic fractions respectively and are given in Table I.

The isomer shift (δ) values for both tetrahedral (δ_{tet}) and octahedral (δ_{oct}) sites indicate that the Fe^{3+} is in high spin state. The δ_{tet} and δ_{oct} values for 10 and 16 nm are found to be nearly same (0.28 and 0.40 mm/s, respectively). As the particle size is increased to 30 nm, δ_{tet} value increase from 0.28 to 0.36 mm/s whereas δ_{oct} decreases from 0.40 to 0.25 mm/s. This unequal but yet related change in isomer

Table I. Room-temperature Mössbauer spectral parameters for nanosize $CuFe_2O_4$ particles having different sizes

Particle size (nm)	Isomer shift (δ) (mm/s)			Quadrupole splitting (ΔE_Q) (mm/s)			Hyperfine field (H_f) (T)	
	A	B	D	A	B	D	A	B
10.0	0.29	0.40	0.42	0.085	−0.01	0.28	52.0	51.2
16.0	0.28	0.40	0.36	0.060	−0.10	0.45	51.9	51.2
30.0	0.36	0.25	–	−0.160	−0.12	–	51.2	48.6

shift values is attributed to the change in Fe^{3+}–O^{2-} distance correspondingly at A and B sites. Thus, the change in the bond distances can be envisaged by crystal distortion. The quadrupole splitting (ΔE_Q) values are found to be low at both the sites indicating the overall presence of cubic symmetry at the two occupancy sites. However, ΔE_Q values increase with increasing particle size indicating that the symmetry is distorted progressively.

Interestingly, the hyperfine field (H_{hf}) values for both A and B sites are found to decrease with increasing particle size. This may be due to the nanocrystalline nature and/or crystal distortion. In bulk $CuFe_2O_4$, the cubic to tetragonal crystal structure transformation can be achieved by slow cooling from above 633 K [9]. Earlier, a critical value of 0.25 was deduced for copper ion concentration at the tetrahedral (A) sites to induce crystal distortion from cubic to tetragonal due to Jahn–Teller effect [9]. The Jahn–Teller effect in $CuFe_2O_4$ spinel arises due to the doubly degenerate E_g – type ground state of the Cu^{2+} ion [10]. It is known that the magnetic properties of spinel ferrites are governed by cation distribution. The magnetization decreases with increasing concentration of Cu^{2+} ions at the octahedral (B) sites. Since, the hyperfine field (H_{hf}) is proportional to magnetization, the observed decrease in H_{hf} values is attributed to the decrease in magnetization, and this decrease can only be brought about by the migration of Cu^{2+} to octahedral sites that induce crystal distortion. These observations corroborate earlier results based on the Mössbauer spectroscopy [5, 10] and magnetic measurements [5, 8, 10, 11]. Although, both particle size and cation distribution vary simultaneously with heat-treatment temperature, the observed trend in the present study suggests that cation distribution is dominant factor. It is worth to note that the line width for all the samples is found to be much greater than the natural line width. This is attributed to nanocrystalline nature, cation distribution, and quadrupole and hyperfine field distribution.

4. Conclusion

In the present study, nanostructured $CuFe_2O_4$ particles are synthesized by wet chemical method. The crystal structure of $CuFe_2O_4$ particles remains cubic provided the cation distribution is normal. On increasing the heat-treatment temperature, the cation migration takes place. Subsequently, when the heat treatment temperature is above 973 K, the progressive migration of Cu^{2+} from tetrahedral (A) site to octahedral (B) sites takes place indicating inverse or partially inverse configuration. And this cation migration induces the crystal structure change from cubic to tetragonal. Consequently, a correlated change in isomer shift values at A and B sublattices is observed and attributed to the change in Fe^{3+}–O^{2-} distance correspondingly. The hyperfine field (H_{hf}) values for both A and B sites are found to decrease with increasing particle size. These observations are explained based on the of Jahn–Teller effect in nanostructured CuFe$_2$O$_4$ particles.

Acknowledgement

The authors acknowledge the financial assistance from DST, New Delhi.

References

1. Dormann, J. L. and Fiorani, D. (eds), *Magnetic Properties of Fine Particles*, North-Holland, Amsterdam, 1992.
2. Gajbhiye, N. S. and Balaji, G., In: *Advances in Nanoscience and Nanotechnology*, A. Sharma (ed), NISCAIR, New Delhi, 2003.
3. Chinnasamy, C. N., Narayanasamy, A., Ponpandian, N., Chattopadhyay, K., Shinoda, K., Jeyadevan, B., Tohji, K., Nakatsuka, K., Furubayashi, T. and Naktani, I., *Phys. Rev. B* **63** (2001), 184108.
4. Mahmoud, M. H., Hamdeh, H. H., Ho, J. C., Shea, M. J. and Walker, J. C., *J. Magn. Magn. Mater.* **220** (2000), 139.
5. Goya, G. F. and Rechenberg, H. R., *Nanostruct. Mater.* **10** (1998), 1001.
6. Cullity, B. D., *Introduction to Magnetic Materials*, Addison-Wesley, MA, 1978.
7. Ohnishi, H. and Teranishi, T., *J. Phys. Soc. Jpn.* **16** (1961), 35.
8. Yokoyama, M., Nakamura, A., Sato, T. and Haneda, K., *J. Mag. Soc. Jpn.* **22** (1998), 243.
9. Ohnishi, H., Teranishi, T. and Miyahara, S., *J. Phys. Soc. Jpn.* **14** (1959), 106.
10. Jiang, J. Z., Goya, G. F. and Rechenberg, H. R., *J. Phys.: Condens. Matter* **11** (1999), 4063.
11. Goya, G. F., Rechenberg, H. R. and Jiang, J. Z., *J. Magn. Magn. Mater.* **218** (2000), 221.

Hyperfine Interactions **156/157**: 63–67, 2004.
© 2004 *Kluwer Academic Publishers. Printed in the Netherlands.*

Mössbauer Studies on Nanocrystalline Diol Capped γ-Fe$_2$O$_3$

J. GHOSE[1,*], K. S. K. VARADWAJ[1] and D. DAS[2]
[1]*Department of Chemistry, Indian Institute of Technology, Kharagpur 721302, India*
[2]*Inter-University Consortium for D. A. E. Facilities, Calcutta Centre, 3/LB-8, Salt Lake, Calcutta-700 091, India*

Abstract. Nanocrystalline γ-Fe$_2$O$_3$ was prepared by refluxing ferric nitrate in 1,4-butanediol. Characterisation was done by XRD, FTIR and Raman spectroscopy. TEM showed that the average particle size of the two samples were 9 nm and 6 nm with a small particle size distribution. FTIR and TG showed that the γ-Fe$_2$O$_3$ particles were capped with the diol. Mössbauer spectra of the 6 nm sample showed that the sample is superparamagnetic with a blocking temperature around 225 K. With increase in particle size, blocking temperature increases. 20 K MS of 9 nm sample was fitted to 2 sextets. IS, QS and Hf parameters indicate presence of Fe^{3+} on A and B sites of the spinel lattice. MS of 6 nm sample were fitted to 3 sextets: Fe^{3+} ions on A and B sites and on the surface. As the coating is attached to the surface Fe^{3+} ions, EFG of these ions will be larger than the bulk Fe^{3+} ions. Hence the QS and Hf values are different for the 3rd sextet. These results indicate that although both the samples are nanocrystalline and both are coated, the effect of coating on the MS is only shown by the sample with smaller particle size. Thus when the particles are large the EFG of the Fe^{3+} ions in the bulk and surface are same and hence good fitting of the spectra can be obtained with 2 sextets only. From these results it may be concluded that synthesis of γ-Fe$_2$O$_3$ in diol results in a diol coated sample and the coating can be detected by MS when the particle size is very small.

Key words: Mössbauer spectroscopy, nanocrystalline, capped γ-Fe$_2$O$_3$.

1. Introduction

The exploration of new methods for the synthesis of nanocrystalline magnetic particles and their characterization are being intensively pursued nowadays [1, 2]. It has gained importance not only because of the markedly different novel physical properties exhibited by the magnetic nanoparticles as against their bulk counterparts, but also due to many potential technological applications in magnetic memory devices, colour imaging, magnetic refrigeration, ferrrofluids, medical imaging, control drug delivery, sensors and catalysis [3, 4]. With change in method of synthesis, there is a change in the particle size distribution, degree of crystallinity, surface composition and shape of the nanoparticles. Non-hydrolytic method as against the different aqueous methods for the preparation of metal oxide nanoparticles gives highly crystalline nanoparticles with an active monolayer and it is ex-

---

* Author for correspondence.

pected to show different properties with respect to the defect structure and surface composition [5, 6].

Magnetic nanoparticles below a critical size become single domain and show the unique phenomena of superparamagnetism. Mössbauer spectroscopy provides crucial information for a clear understanding of the magnetic behavior on a macroscopic scale. Moreover these particles also show an abrupt decrease in saturation magnetization and increase in coercive field with respect to that of the bulk material [7].

The polyol process is an extensively studied method for the preparation of different elemental metal as well as bimetallic nanoparticles [8]. In this process, metallic precursor compounds in a polyol medium are refluxed for reduction to get the metallic nanoparticles. In the present work, a non-hydrolytic polyol method has been developed for the synthesis of *in situ* coated γ-Fe_2O_3 nanoparticles, in which the polyol itself acts as a capping agent and the effect of capping has been investigated by Mössbauer spectroscopy.

2. Experimental

Nanocrystalline γ-Fe_2O_3 samples were prepared by dissolving $Fe(NO_3)_3 \cdot 9H_2O$ (10 mmol) in 1,4-butanediol and heating at 413 K for 1 hr. The solution was refluxed at 463 K (sample A) and 483 K (sample C) respectively, for 1 hr and cooled to room temperature. On addition of methanol a precipitate was obtained, which was then separated by centrifugation. The 463 K refluxed sample was further sintered at 573 K in argon atmosphere for 1 hr (sample B).

X-ray diffraction analyses of the samples were carried out with a Phillips X-ray diffraction unit (model PM 1710) using CoK_α ($\lambda = 1.79$ Å) radiation with a Ni filter. The lattice parameter of the samples were calculated by a least square method.

TEM measurements were carried out in a CM12 Phillips microscope operating at 120 kV. The sample was prepared by placing one drop of the well dispersed acetone solution of the powder on a carbon coated copper grid. The average particle size of the samples were determined using TEM. FTIR spectra of all the samples were recorded in KBr medium in the range 400 to 4000 cm^{-1} with a Thermo Nicolet Nexus FTIR (model 870). Thermal analysis (DTA, TG/DTG) of both the samples were carried out with a Thermal Analyser (Pyris Diamond TG/DTA) in the temperature range 323 K to 1273 K, at a rate of 10 deg/min under nitrogen atmosphere. Mössbauer spectroscopy was done with a conventional transmission Mössbauer spectrometer, operating in the constant acceleration mode and a ^{57}Co source in a Rh matrix was used. Spectra were recorded at different temperatures between 20 K and 298 K in a cryostat bath.

3. Results and discussion

Figure 1 shows the XRD patterns of the samples A, B and C. The broad lines in the XRD patterns indicate that all the samples are nanocrystalline in nature.

TEM measurements show that the average particle size of sample B is 6 nm and that of C is 9 nm. The XRD of sample A indicates that it is a ferrihydrite which is also confirmed from FTIR. Sample B shows characteristic peaks of the spinel structure and could be either γ-Fe$_2$O$_3$ or Fe$_3$O$_4$. Figure 1c shows that sample C is also a spinel oxide. From Mössbauer parameters (Table I) sample B and C were identified as γ-Fe$_2$O$_3$ and was further confirmed by Raman spectroscopy. Mössbauer spectra of both the samples were taken in the temperature range 20–298 K (Figures 2, 3). 298 K spectra of the samples consist of paramagnetic doublets with broad linewidth as expected for a superparamagnetic sample [8]. Figures 2 and 3 show that with lowering of temperature the sextet patterns gradually appear in both the samples. The Mössbauer results show that blocking temperature (T_B) increases with increasing particle size. Hence sample B has a blocking temperature of 225 K while T_B for sample C is 250 K. The fitting of the spectra at 20 K show that two hyperfine patterns for the A and B sites of γ-Fe$_2$O$_3$ are present. An additional sextet present in the spectra of sample B probably represent surface Fe^{3+}. A high

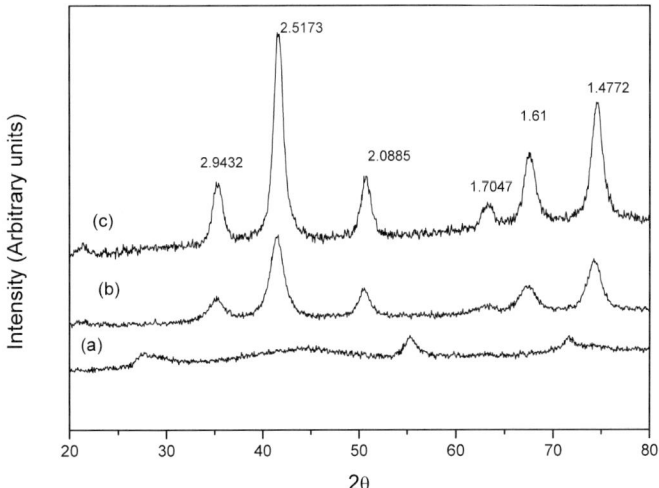

Figure 1. XRD patterns of (a) sample A, (b) sample B, (c) sample C.

Table I. Mössbauer parameters, isomer shift, quadrupole splitting and hyperfine field of sample B and C

Sample	Hyperfine field (kOe)	Isomer shift (mm/s)	Quadrupole splitting (mm/s)	Area (%)
B	534	0.46	−0.07	42
	513	0.45	0.34	41
	475	0.56	0.05	17
C	524	0.47	−0.09	49
	506	0.43	−0.03	51

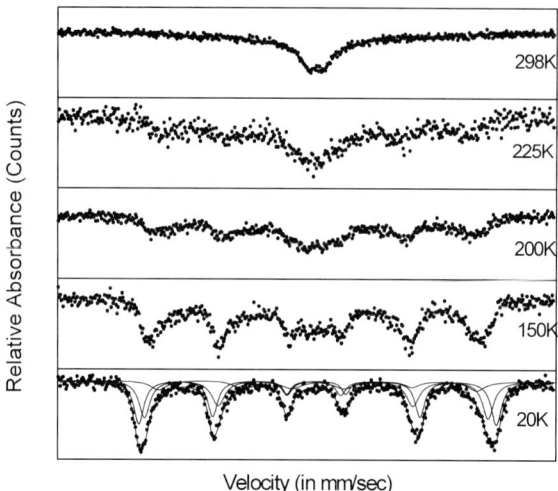

Figure 2. Mössbauer spectra of sample B between 20–298 K.

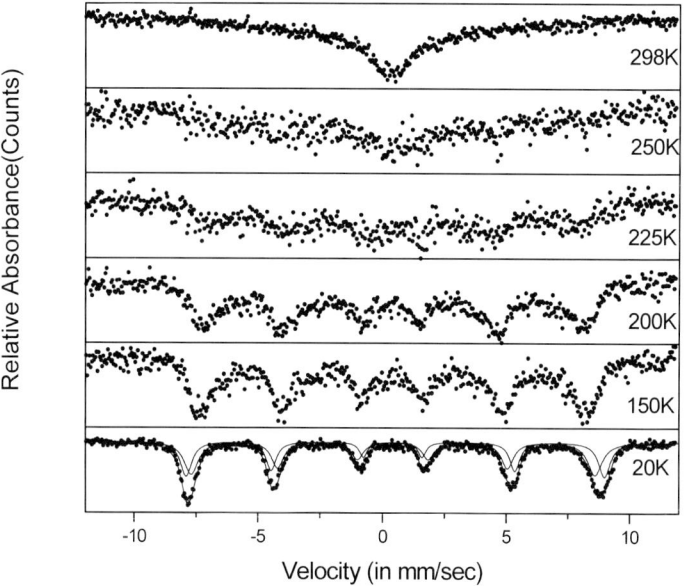

Figure 3. Mössbauer spectra of sample C between 20–298 K.

quadrupole splitting value of this sextet (Table I) also suggests surface Fe^{3+} which is likely to have high EFG. These results suggest that a Fe^{3+} surface state is present in sample B but not in sample C. This could be due to different synthesis processes of the 2 samples.

FTIR studies show that in addition to the absorption band for the spinel phase, some additional bands corresponding to the organic moiety are also present. In sample B the amount of organic moiety appears to be very small. Thermal analysis indicates that on heating the sample, the organic moiety in sample C can be removed only at high temperatures. These results indicate that the γ-Fe$_2$O$_3$ synthesised in 1,4-butanediol is capped with the organic moiety and that the extent of capping is low in sample B. This is perhaps due to sintering of the sample, whereby decomposition of FeO(OH) occurs to form γ-Fe$_2$O$_3$ and consequently the capping is almost removed. Thus the Mössbauer spectra of sample B shows a Fe^{3+} surface state while in sample C, the Fe^{3+} surface state is absent due to the capping. From these results it may be concluded that *in situ* diol capped γ-Fe$_2$O$_3$ can be synthesized by a simple refluxing process and the presence of the coating can be investigated by Mössbauer spectroscopy studies.

References

1. Ziolo, R. F., Giannelis, E. P., Weinstein, B. A., O'Horo, M. P., Ganguly, B. N., Mehrotra, V., Russell, M. W. and Huffman, D. R., *Science* **257** (1992), 219.
2. Trentler, T. J., Denler, T. E., Bertone, J. F., Agrawal, A. and Colvin, V. L., *J. Am. Chem. Soc.* **121** (1999), 1613.
3. Speliotis, D. E., *J. Magn. Magn. Mater.* **193** (1999), 29.
4. Cornell, R. M. and Schwertmann, U. (eds), *The Iron Oxides*, VCH, Weinheim, Germany, 1996, p. 463.
5. Morup, S., *J. Magn. Magn. Mater* **39** (1983), 45.
6. Moser, J., Punchihewa, S., Infelta, P. P. and Grätzel, M., *Langmuir* **7** (1991), 3012.
7. Martinez, B., Roig, A., Obradors, X. and Molins, E., *J. Appl. Phys.* **79** (1996), 2580.
8. Koch, C. C. (ed.), *Nanostructured Materials – Processing, Properties and Potential Applications*, Noyes Publications, William Andrew Publishing, Norwich, New York, USA, 2002, p. 12.

Hyperfine Interactions **156/157**: 69–73, 2004.
© 2004 *Kluwer Academic Publishers. Printed in the Netherlands.*

Hyperfine and Structural Properties of the Mechanically Alloyed (FeMn)$_{30}$Cu$_{70}$ System

J. RESTREPO[1] and J. M. GRENÈCHE[2]
[1]*Grupo de Estado Sólido, Instituto de Física, Universidad de Antioquia, A.A. 1226 Medellín, Colombia; e-mail: jrestre@fisica.udea.edu.co*
[2]*Laboratoire de Physique de l'Etat Condensé, UMR CNRS 6087, Université du Maine, 72085 Le Mans, France*

Abstract. Nanostructured Fe$_{15}$Mn$_{15}$Cu$_{70}$ alloys were prepared by high-energy ball milling. The alloying process spans grinding times from 15 minutes to 114 hours. The Mössbauer isomer shifts are analyzed at 77 K and 4.2 K as a function of the average lattice parameter determined from Rietveld analysis of the X-ray diffraction data. Evidence of two different iron sites at 77 K to account for the asymmetry of the Mössbauer spectra is discussed by correlating hyperfine and structural parameters. This leads to conclude the occurrence of a low-spin to high-spin partial transition of iron at a critical volume of around 50 Å^3 at 77 K, supporting theoretical predictions.

Key words: FeMnCu alloys, spin transition, Mössbauer.

1. Introduction

Both the electronic and the magnetic behaviors of iron in γ-structures have been theoretically and experimentally investigated: it was found two states within a very narrow energy range: one with a low-volume, low-spin, antiferromagnetic or non-magnetic state and the other with a high-volume, high-spin, ferromagnetic state [1–10]. On the other hand, the presence of a copper matrix helps to stabilize the FCC structure despite the low solubility with iron which can be easily overcome through techniques as mechanical alloying [11, 12]. In addition, this technique can induce a wide range of unit cell volumes due to milling-driven strain effects. Specifically, ball-milling gives rise to a lattice expansion as high as 0.6% relative to the Cu lattice parameter as it has been already reported for γ-Fe$_{30}$Cu$_{70}$ [12]. This fact is important as the iron magnetic moment is strongly volume dependent and it can even change discontinuously in the vicinity of a critical volume [6]. Additionally, in γ-MnCu alloys, lattice expansion goes up to 3.7% upon alloying with Mn below the martensitic transformation [13], providing in this way a much wider range of unit cell volumes. These features have motivated us to consider the proposed Cu-rich FeMnCu ball-milled alloy in order to elucidate the magnetic behavior of iron in a γ matrix as a function of the interatomic spacing.

2. Experimental

Samples with nominal stoichiometry $Fe_{15}Mn_{15}Cu_{70}$ were prepared by alloying pure (>99.99%) elemental powders in a planetary ball mill under inert atmosphere. Continuous milling times of 0.25, 0.5, 1, 2, 4, 6, 12, 21, 48, 72 and 114 hours were considered with an average sample-balls weight ratio of 1 : 7. Every sample for every milling time was prepared separately. Hyperfine and structural charac-terization was carried out by means of ^{57}Fe Mössbauer spectrometry and X-ray diffraction (XRD), respectively. Mössbauer samples consisted of 5 mg Fe/cm^2. Zero-field ^{57}Fe Mössbauer spectra were recorded at 77 K and 4.2 K cooling the sample in a bath cryostat, by using $^{57}Co/Rh$ source in transmission geometry. Isomer shift (δ) values are quoted relative to α-Fe at room temperature (~293 K).

3. Results and discussion

Both Mössbauer results and XRD measurements, reveal that most of the α-Fe is dissolved after twelve hours of milling, as evidenced by an increase of the relative area of the paramagnetic component. This component which appears in the first 15 minutes of milling, progressively grows with the milling time as the α-Fe pre-cursor component becomes smaller. This fact reveals effectively the incorporation of iron atoms into the FCC structure. Mössbauer spectra at 77 K were recorded in the velocity range of ±2 mm/s to analyze this paramagnetic component. The observed asymmetry of the doublet in all spectra, as shown, e.g., in Figure 1 for the 48 hours sample, suggests a fitting procedure with at least two components. The same feature was observed for all milled samples. This is because assuming a single asymmetric doublet, the Lorentzian line profile is not well appropriate and the presence of preferential orientation (originating asymmetry) is not verified after rotation of the sample respect to the γ-beam direction. A problem arises when dif-ferent fittings with physical meaning are possible, as can be observed by comparing Figures 1(a) & 1(b). Solutions consist thus of two quadrupolar doublets with either similar quadrupolar splittings (and different isomer shifts) or similar isomer shifts (and different quadrupolar splittings). In order to determine the right model fitting, Mössbauer spectra were recorded at 4.2 K; one example is illustrated in Figure 2, that of the 48 h milled sample. The broadening of the hyperfine field distribution, ranging from around 2 T up to around 30 T, reveals clearly the presence of many Fe environments consistent with atomic disorder. The Fe sites with low hyperfine field are more probable than those with large hyperfine field as can be seen from the probability distribution of the hyperfine field (inset Figure 2). This situation endorses the alloying of the constituent elements at atomic level, and the occur-rence of very diluted Fe sites, i.e. surrounded mainly by copper atoms, as the most probable ones. The presence of Mn atoms in the first Fe coordination shell, is also important since they provoke a decrease in the hyperfine magnetic field [14, 15]. The hyperfine parameters of the observed paramagnetic component with an isomer shift (δ) value of 0.14(1) mm/s, allow choosing the fitting of the 77 K Mössbauer

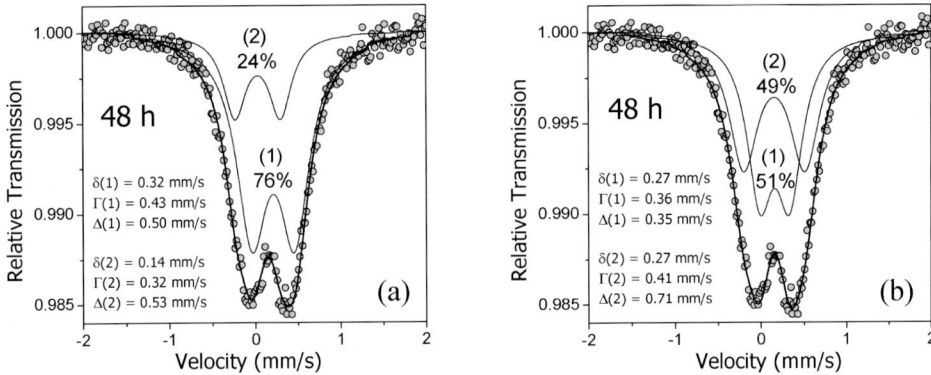

Figure 1. Mössbauer spectrum at 77 K corresponding to the sample Fe$_{15}$Mn$_{15}$Cu$_{70}$ milled for 48 h. Two possible solutions corresponding approximately to (a) same Δ and different δ, and (b) same δ and different Δ, are shown. Γ corresponds to the half-height width.

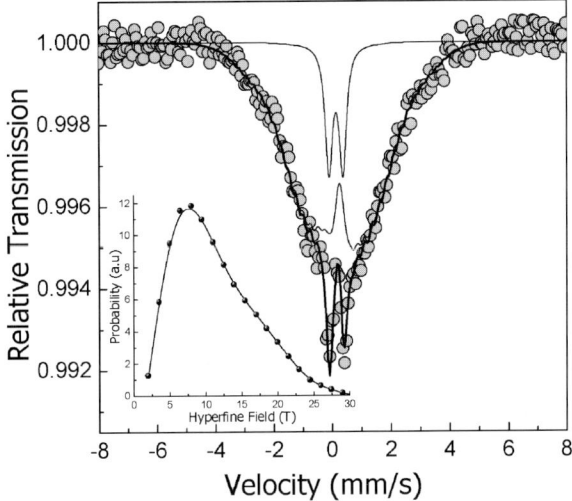

Figure 2. Mössbauer spectrum at 4.2 K corresponding to sample Fe$_{15}$Mn$_{15}$Cu$_{70}$ milled for 48 h. The inset shows the corresponding hyperfine magnetic field distribution.

spectrum, shown in Figure 1(a), as the right one describing our system, since the solution shown in Figure 1(b) exhibits quadrupolar splitting (Δ) values as high as 0.7 mm/s, too large for metallic alloys with cubic symmetry. This well-defined doublet at 4.2 K, is ascribed to the presence of iron magnetic moments in a non-blocked state, whereas the magnetic contribution can be attributed to iron magnetic moments in a blocked state.

On the other hand, for those samples belonging to the first milling stages below 12 h where the lattice parameter increases rapidly, Mössbauer spectra at 77 K exhibit differences relative to those samples with milling times above 12 h where the lattice parameter is stabilized. More accurately, one of the components exhibits δ values much greater than the observed in the respective component of those sam-

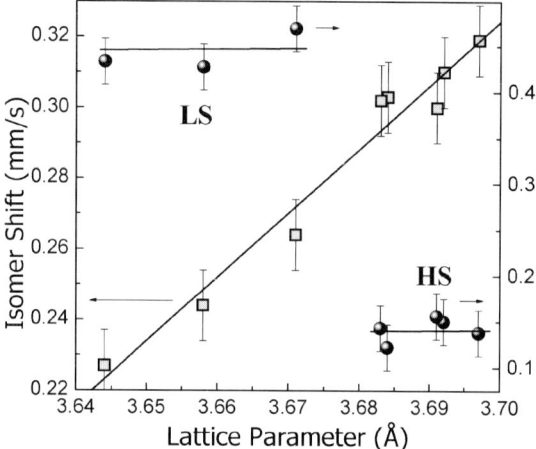

Figure 3. The average lattice parameter dependence of the isomer shift components, as determined from Figure 1.

ples belonging to the steady state. Such difference from around 0.44 mm/s for the early stages, down to 0.14 mm/s for the steady state, resembles a high spin (HS) to low spin (LS) transition, and it is almost two times larger than that observed in γ-FeNi alloys at the Invar composition [16, 17]. As is well established, isomer shifts are very sensitive to interatomic distance, which strongly affects the $4s$-electron charge density. Hence, the onset of this transition is endorsed by the wide range of lattice parameters provided by our alloy. Specifically this range sweeps the interatomic spacing for which the Fe magnetic moments in FCC structures can be found either in a HS or LS state according to theoretical predictions [8–10]. Summarizing the results for all samples, Figure 3 shows the evolution of δ for the two employed components as a function of the lattice parameter. As can be seen from Figure 3, the isomer shift of the second component follows a well-defined linear behavior (with $d(\delta)/d\ln V = 2.2(2)$ mm/s, and $V = a^3$), as already reported to occur in several compounds [18–21]. In addition, the linear dependence is expected when s-electron charge density at the Fe nucleus is controlled by hydrostatic pressure, in sharp contrast with the spin transition predicted by theoretical calculation [8–10]. The origin of this phenomenology and the occurrence of at least two components presumably can be ascribed to the different local symmetries between crystallites and their boundaries for which electronic properties should exhibit differences. Moreover the limitation in the employed methodology when assigning a δ value, which is a local parameter, to an average lattice parameter, makes difficult to clarify this point.

4. Conclusions

The lattice parameter dependence of the isomer shifts suggests the occurrence of a partial LS to HS transition at a critical unit cell volume around 50 Å^3 (3.67 Å $<$

$a < 3.68$ Å), which in turn supports theoretical predictions [8–10]. In contrast, those iron sites following a linear increase of the isomer shift with the average lattice parameter, presumably belong to the core of the grains. Indeed, inside the core the lattice expansion must be ruled out differently (smoothly) from grain boundaries where larger local distortions, a greater density of defects, and consequently greater volume instabilities are expected.

Acknowledgements

This work was supported by Universidad de Antioquia (Project Sostenibilidad 2001–2003, GES) and ECOS-NORD under project C03P01. One of the authors (J.R.) would like to thank the one year postdoctoral fellowship provided by Région Pays de la Loire at the Université du Maine, Le Mans (France), and to the Instituto de Ciencia de Materiales at Madrid where samples were prepared in collaboration with Prof. J. M. González. We are very grateful to A. M. Mercier from Laboratoire des Fluorures of Université du Maine UMR CNRS 6010 for performing XRD measurements.

References

1. Herper, H. C., Hoffmann, E. and Entel, P., *Phys. Rev. B* **60** (1999), 3839.
2. Acet, M., Zähres, H., Wassermann, E. F. and Pepperhoff, W., *Phys. Rev. B* **49** (1994), 6012.
3. Amador, C., Lambrecht, W. R. L. and Segall, B., *Phys. Rev. B* **46** (1992), 1870.
4. Moruzzi, V. L., Marcus, P. M. and Kübler, J., *Phys. Rev. B* **39** (1989), 6957.
5. Krasko, G. L., *Phys. Rev. B* **36** (1987), 8565.
6. Moruzzi, V. L., *Phys. Rev. Lett.* **57** (1986), 2211.
7. Moruzzi, V. L., Marcus, P. M., Schwarz, K. and Mohn, P., *Phys. Rev. B* **34** (1986), 1784.
8. Pinski, F. J., Staunton, J., Gyorffy, B. L., Johnson, D. D. and Stocks, G. M., *Phys. Rev. Lett.* **56** (1986), 2096.
9. Wang, C. S., Klein, B. M. and Krakauer, H., *Phys. Rev. Lett.* **54** (1985), 1852.
10. Bagayoko, D. and Callaway, J., *Phys. Rev. B* **28** (1983), 5419.
11. Bove, L. E., Petrillo, C., Sacchetti, F. and Mazzone, G., *Phys. Rev. B* **61** (2000), 9457.
12. Eckert, J., Holzer, J. C., Krill III, C. E. and Johnson, W. L., *J. Appl. Phys.* **73** (1993), 2794.
13. Cowlam, N., Bacon, G. E. and Gillott, L., *J. Phys. F* **7** (1977), L315.
14. Uhrmacher, M., Kulinska, A., Baldokhin, Yu. V., Tcherdyntsev, V. V., Kaloshkin, S. D., Maddalena, A. and Principi, G., *Intermetallics* **10** (2002), 571.
15. Paduani, C. and Krause, J. C., *Phys. Rev. B* **58** (1998), 175.
16. Lagarec, K., Rancourt, D. G., Bose, S. K. and Dunlap, R. A., *Phase Transitions* **75** (2002), 211.
17. Lagarec, K., Rancourt, D. G., Bose, S. K., Sanyal, B. and Dunlap, R. A., *J. Magn. Magn. Mater.* **236** (2001), 107.
18. Komelj, M., Grotheer, O. and Fähnle, M., *J. Magn. Magn. Mater.* **195** (1999), L275.
19. Kong, Yong, Zhou, Rongjie and Li, Fashen, *J. Phys.: Condens. Matter* **8** (1996), 3829.
20. Fa-Shen Li, Ji-Jun Sun, Chun-Li Yang, De-Sheng Xue, Bao-Gen Shen, Abd-Elmeguid, M. M. and Micklitz, H., *J. Phys.: Condens. Matter* **5** (1993), L577.
21. Brand, R. A., Georges-Gibert, H. and Lelaurain, M., *J. Phys. F* **10** (1980), L257.

Hyperfine Interactions **156/157**: 75–79, 2004.
© 2004 *Kluwer Academic Publishers. Printed in the Netherlands.*

^{197}Au Mössbauer Study of Bimetallic Nanoparticles Prepared by Sonochemical Technique

YASUHIRO KOBAYASHI[1], SHINJI KIAO[1], MAKOTO SETO[1],
HIROE TAKATANI[2], MIOKO NAKANISHI[2] and RYUICHIRO OSHIMA[2]
[1]*Research Reactor Institute, Kyoto University, Kumatori, Osaka 590-0494, Japan*
[2]*Research Institute for Advanced Science and Technology, Osaka Prefecture University, Sakai, Osaka 599-8570, Japan*

Abstract. By ultrasonic wave irradiation, the metallic ions in aqueous solution are reduced to metallic nanoparticles. Using the mixture solution of Au and Pd ions, core(Au)–shell(Pd) structured nanoparticles are prepared. ^{197}Au Mössbauer spectra of core–shell structured nanoparticles contains two components, one is pure Au component due to the inner part of the Au core and the other is the positive isomer shift component due to the alloy layer at the interface of Au core and Pd shell. The thickness of the alloy layer is 0.5∼0.7 nm.

Key words: ^{197}Au Mössbauer spectroscopy, core–shell structure, AuPd nanoparticle, sonochemical technique.

1. Introduction

Metallic nanoparticles are interested for their peculiar properties [1], and many techniques to prepare nanoparticles were developed. The technique using reduction of metallic ions in the solution with surfactant is simple and easy one. While in the sonochemical technique, ultrasonic wave is used for the reduction [2, 3]. By the ultrasonic wave irradiation, cavities are created, expanded and collapsed in the liquid. At the collapse, the active reaction field is made by drastic adiabatic compression. The metallic ions are reduced into the metal by the reaction field, and the nanoparticles are crated. On the irradiation of Au and Pd mixed solution, Au ion is reduced first and Pd ion is reduced latter, so core(Au)–shell(Pd) structured nanoparticles are prepared. On AuPd nanoparticle, an interesting phenomenon was reported on the catalytic activity [4]. On the hydrogenation of cyclohexene, the AuPd core–shell nanoparticle shows higher catalytic activity than Pd nanoparticles. We studied the AuPd nanoparticles to elucidate the structure and electronic state of AuPd nanoparticles using ^{197}Au Mössbauer spectroscopy [5].

2. Experimental procedure

The nanoparticle specimens were prepared by the ultrasonic irradiation. The irradiated aqueous solution was NaAuCl$_4$·H$_2$O/PdCl$_2$·2NaCl·3H$_2$O. The surfactants

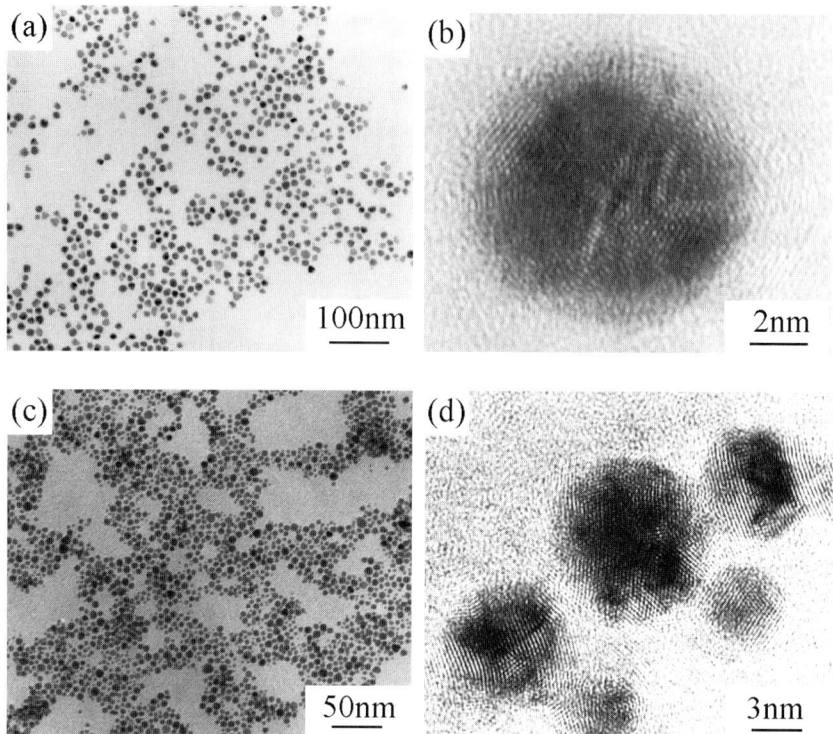

Figure 1. Electron micrographs of AuPd nanoparticles [5]; (a) TEM and (b) HRTEM of AuPd–SDS, (c) TEM and (d) HRTEM of AuPd–PEG.

were sodium dodecyl sulfate (SDS) and polyethylene glycol monostearate (PEG). The concentrations of aqueous solutions are summarized in Table I. Before the irradiation, the aqueous was purged by argon gas and sealed in grass vessels. For the irradiation, a multiwave ultrasonic generator with a barium titanate oscillator of 64 mm diameter was operated for sonication at 200 kHz with an input power of 4.2 W/cm^2.

The electron micrographs of the specimens are shown in Figure 1. In Figures 1(a) and (c), the size distribution of the particles are rather small in AuPd–SDS, and the AuPd-PEG have wide distribution of the size. The mean diameters of the nanoparticles are summarized in Table I. The particles prepared from thin solution of the metallic ions are small, and the particles from thin solution of SDS are large. The particles prepared from the solution of PEG are rather small. In Figure 1(b), the contrasts of AuPd–SDS nanoparticles are different on the center part and rim of the particle. It is due to the core(Au)–shell(Pd) structure, and this structure was confirmed by energy dispersive X-ray spectroscopy (EDX) analyses. In Figure 1(d), the large particles shows core–shell structure's contrasts, and small particles shows uniform contrasts. From EDX analyses, the small particles are not core–shell structure. They are uniform concentration alloy particles. The AuPd–PEG nanoparticles are mixture of core–shell structure and alloy particle [5].

Table I. Summay of specimens

	AuPd–SDS				AuPd–PEG
Au^{3+}	1.0 mM	0.5 mM	0.5 mM	0.25 mM	0.5 mM
Pd^{2+}	1.0 mM	0.5 mM	0.5 mM	0.25 mM	0.5 mM
Surfactant	8 mM	8 mM	12 mM	12 mM	0.4 mM
Mean diameter	11.7 nm	10.2 nm	8.6 nm	7.0 nm	4.9 nm

For [197]Au Mössbauer measurement, the γ-ray source were [197]Pt in Pt metal (98% enriched with [196]Pt) prepared by the neutron irradiation by Kyoto University Reactor, and the absorbers were the nanoparticle specimens. The source and the specimens were cooled by the helium refrigerator. Temperatures of the specimens were 8~11 K. The zero velocity positions of the spectra are peak positions of pure bulk Au.

3. Results and discussion

Figure 2 shows the [197]Au Mössbauer spectra of AuPd nanoparticles. The spectra of AuPd–SDS contain two components, one is the single peak having zero isomer shifts and the other is the single peak having positive isomer shift. The zero isomer shifts suggests the no influence from Pd atoms. On the core(Au)–shell(Pd) structure, the Au atoms without perturbation from Pd atoms are in the middle of the Au cores. The middle of Au cores is pure Au, not alloy. The positive isomer shift shows influence form the Pd atoms. The particle size dependence of the area ratio and the isomer shift of this component were shown in Figure 3. If the interface of the Au core and Pd shell is sharp, only one Au layer on the surface of the Au core is affected by the Pd. The ratio of the Au atoms on surface is estimated only 14~22% from the diameters of the Au cores. However, the area ratio of the positive isomer shift components is 23~31%. This results shows that the interface of the Au core and Pd shell is not sharp, there is alloy layer between the core and the shell. The isomer shifts of this components increase with the particle size. The large isomer shift indicates the increase of Pd content on AuPd alloy. The Pd content of the alloy layer is increase with the particle size. The isomer shifts of the alloy layers correspond to the Pd 40~50% alloy. The particle size dependence of the isomer shift is explained as following. When the AuPd core–shell particles were prepared, the Au ions reduced at first and Pd reduced latter because of the difference of ionization tendency. The alloy layer was prepared by simultaneously reduction of Au and Pd. On the preparation of the large size particle, the reaction was slow, and the overlap of the reduction of Au and Pd is small. Therefore, in the large particle, the Au concentration in the alloy layer is small. The thickness of the alloy layer, which is estimated from the area ratio and isomer shift, is 0.5~0.7 nm.

The [197]Au Mössbauer spectrum of AuPd–PEG nanoparticles is shown in lower part of Figure 2. The spectrum of AuPd–PEG is wider than AuPd–SDS. This spec-

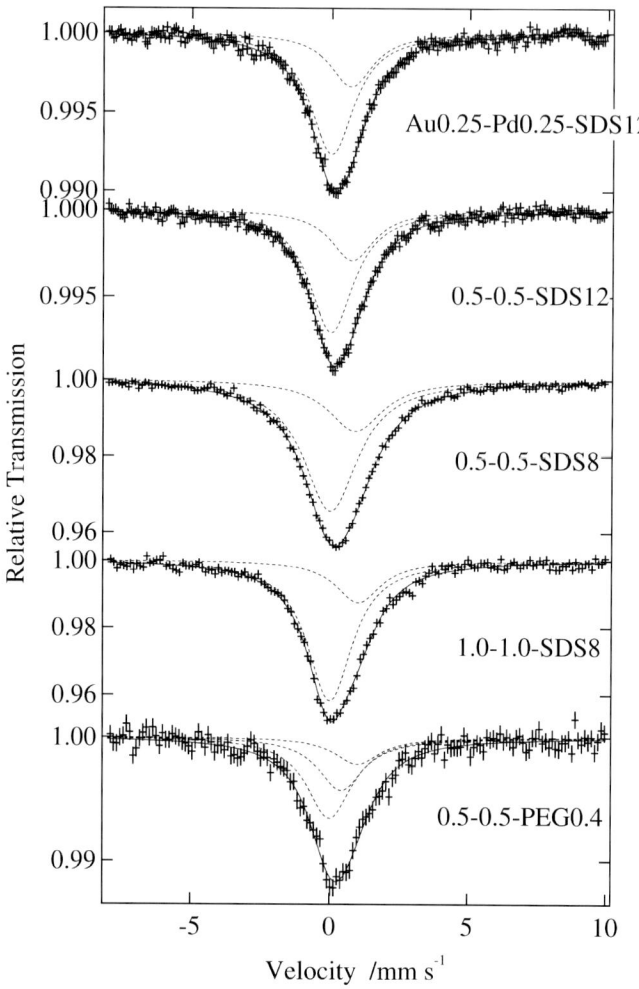

Figure 2. [197]Au Mössbauer spectra of AuPd–SDS and AuPd–PEG nanoparticles.

trum contains three components, zero isomer shift components and two positive isomer shift components. From the area ratio, the larger isomer shift component is estimated the alloy layer of the core–shell particle, like AuPd–SDS nanoparticle. The smaller isomer shift component is estimated AuPd alloy particles component. From the isomer shift, the Pd concentration of the alloy particles was estimated 28%.

4. Conclusion

The [197]Au Mössbauer spectrum of AuPd–SDS nanoparticles consist with two components, one is pure Au components due to the inner part of the Au core, the other is the positive isomer shift component due to the alloy layer at the interface of Au

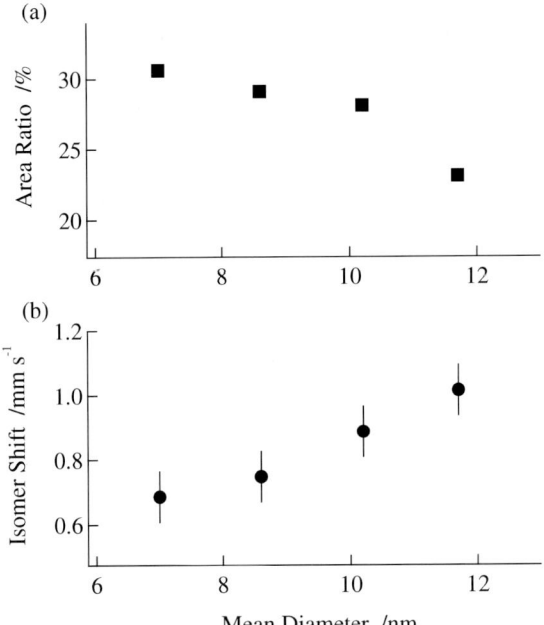

Figure 3. Particle size dependence of Mössbauer parameters of the alloy layer component in spectra of AuPd–SDS nanoparticles, (a) area ratio and (b) isomer shift.

core and Pd shell. The thickness of the alloy layer is 0.5~0.7 nm. The isomer shift of the alloy layer increase with particle size, because the large particles prepared slowly, so the difference of ionization tendency is conspicuous. The spectrum of AuPd–PEG nanoparticles contains three components. The two components are due to the core shell particle as AuPd–SDS particles. The third component is due to the AuPd alloy particles, and its Pd concentration is estimated 28%.

Acknowledgements

This study was partially supported by the Japan Science and Technology Corporation, and the Japan Society for the Promotion of Science (JSPS) Research (Grant-in-Aid for Scientific Research, C, No. 14550658).

References

1. Kubo, R., *J. Phys. Soc. Jpn.* **17** (1962), 975.
2. Flint, E. B. and Suslich, K. S., *Science* **253** (1991), 1397.
3. Okitsu, K., Bandow, H. and Maeda, Y., *Chem. Mater.* **8** (1996), 315.
4. Okitsu, K., Murakami, M., Tanabe, S. and Matsumoto, H., *Chem. Lett.* **29** (2000), 1336.
5. Takatani, H., Kago, H., Nakanishi, M., Kobayashi, Y., Hori, F. and Oshima, R., *Rev. Adv. Mat. Sci.* **5** (2003), 232.

Hyperfine Interactions **156/157**: 81–87, 2004.
© 2004 *Kluwer Academic Publishers. Printed in the Netherlands.*

Phase Composition and Properties of Iron Nanocrystals and Clusters Embedded in MgO Matrix

O. SCHNEEWEISS[1,*], N. PIZÚROVÁ[1], Y. JIRÁSKOVÁ[1], T. ŽÁK[1],
P. BEZDIČKA[2] and H. REUTHER[3]
[1]*Institute of Physics of Materials AS CR, Žižkova 22, 61662 Brno, Czech Republic;*
e-mail: schneew@ipm.cz
[2]*Institute of Inorganic Chemistry AS CR, 25068 Řež, Czech Republic*
[3]*Forschungszentrum Rossendorf e.V., Institut für Ionenstrahlphysik und Materialforschung,*
Postfach 510 119, D-01314 Dresden, Germany

Abstract. Phase composition of the Mg–Fe powder prepared by spark erosion is investigated using Mössbauer spectroscopy and X-ray diffraction. The results are compared with the phase analysis of MgO single crystal wafer implanted by ^{57}Fe.

Key words: nanocrystalline material, implantation, Mössbauer spectroscopy, phase analysis.

1. Introduction

MgO is often used as a substrate for iron thin film deposition as a high-resistivity component for magnetoresistive materials with tunnelling effect conductivity [1]. The nanocrystalline composites based on MgO doped or coated by 3d transition metals have also many applications in the field of catalysis or as gas sensors [2]. Most of these composites have been prepared by physical vapour deposition techniques or by various chemical methods.

    The nanocrystalline and amorphous Mg–Fe materials exhibit high sensitivity to oxidation in the ambient atmosphere at temperatures about 300 K. This means that the Mg–Fe nanopowders transform to MgO–Fe when they are handled in air [3]. Therefore MgO–Fe nanocomposite can be prepared via nanocrystalline Mg–Fe and its subsequent oxidation.

    Magnesium and iron do not form intermetallic compounds and they are almost insoluble elements with each other according to the Mg–Fe equilibrium phase diagram. There has been a lot of effort to extend the negligible equilibrium solubility limit using mechanical alloying [4, 5], coevaporation techniques [6–8], or ion implantation [9].

---

* Author for correspondence.

Stability of an amorphous structure of Fe–Mg thin films prepared by coevaporation technique was investigated by Fnidiki *et al.* [8] using Mössbauer spectroscopy. They found an amorphous phase represented by the doublet with quadrupole splitting $\sigma = 0.52$ mm/s and isomer shift $\delta \sim 0$ mm/s relative to α-Fe in the spectrum of as-prepared samples (mean chemical composition Fe–40 at.%Mg) measured at 293 K. The crystallization into α-Fe was detected in the temperature range of 383–443 K. Curie temperature of the amorphous phase was found to be $T_c = 200 \pm 5$ K and it agrees with the T_c reported in [6]. However, the Mössbauer spectra taken at 77 K did not contain a magnetically split component. Therefore the authors expect superparamagnetic-like behaviour of the phase similar to that reported in [7].

The analysis of the surface layer of Mg implanted by ^{57}Fe using conversion electron Mössbauer spectroscopy and X-ray diffraction (XRD) are described in [9]. The results have shown that paramagnetic doublets appeared in the Mössbauer spectra of the sample after the low doses implantation. The parameters of doublets agreed with those found in [8]. The ferromagnetic α-Fe like phase dominated in the surface after high-dose implantation. The results of Mössbauer phase analysis were in good agreement with XRD where a dilated α-Fe was recognised as well.

A detailed study of states of MgO single crystals implanted by ^{57}Fe ions was carried out by Perez and coworkers [10]. They observed relation between the implantation dose and resulting phase composition. Fe^{3+} has dominated after low-dose implantation, the medium dose led to formation of Fe^{2+}, and metallic Fe clusters were found after high-dose implantation.

Hayashi *et al.* [11] recently investigated embedded iron clusters prepared by ^{57}Fe implantation into MgO. They also found iron as metallic precipitates (superparamagetic and ferromagnetic), Fe^{2+}, and Fe^{3+}. The amount of ferromagnetic precipitates was important after high-dose implantation ($>2.0 \times 10^{17}$ ions/cm^2) and increased after additional annealing at 300°C for 1 hour in Ar.

The aim of this work is an investigation of phase composition of MgO–Fe nanocomposite prepared from the nanocrystalline Mg–Fe powder.

2. Experimental

Mg–Fe nanocrystalline powder was prepared by spark erosion of pure Mg (99.9%) and Fe (99.99%) electrodes carried out in hydrogen atmospheric pressure as dielectric. Ferromagnetic and coarse particles were extracted from the as-prepared powder by magnetic and sedimentation separation, respectively. Controlled heat treatments were carried out in vacuum ($\sim 10^{-3}$ Pa). The samples prepared by ^{57}Fe ion implantation into MgO single crystal were investigated as a model material for the determination of properties of single iron atoms and their clusters in MgO. The MgO (100) single crystal wafers $10 \times 10 \times 0.5$ mm^3 were supplied by SPI Supplies®. The applied dose was 1×10^{16} ^{57}Fe/cm^2 of energy 100 keV and two different current densities: 0.5 μA/cm^2 (sample W1) and 0.3 μA/cm^2 (sample W2).

The sample W1 was annealed at 673 K for 1 hr in vacuum (10^{-3} Pa) to remove defects induced during implantation (sample W1A).

The structure of powder samples was checked using transmission electron microscopy (TEM) and XRD (Cu K_α radiation). ^{57}Fe Mössbauer spectra were measured by the transmission geometry (TMS) in temperature range 20–300 K detecting 14.4 keV gamma radiation and in scattering geometry with integral detection of conversion electrons (CEMS). For calibration an α-iron foil was used. The computer processing of the spectra yielded intensities (i.e. the areas) I of the components, their hyperfine inductions B_{hf}, isomer shifts δ, and quadrupole splittings (shifts) σ. The phase contents are given as the sum of intensities of the corresponding spectra components. Broad Zeeman (magnetic) sextets were described by hyperfine field distribution. Differences in values of Lamb–Mössbauer factors for the phases are not taken into account. Isomer shifts are given relative to α-iron at room temperature. Thermal scan of relative transmissions in selected channels derived from Mössbauer spectra of the as-prepared Mg–Fe powder was used to determine the critical temperatures of a magnetic phase transition. Thermomagnetic curves were measured using Physical Property Measuring System Quantum Design$^\circledR$ magnetometer in 100 mT and 1 T external magnetic fields.

3. Results and discussion

The analysis of the Mössbauer spectra of the ^{57}Fe atoms implanted in the MgO samples (W1, W2) revealed the phases (clusters) similar to those reported in [10, 11]. The results are summarized in Table I. The isomer shifts of the doublets DW1, DW2, and DW3 indicate Fe^{2+}. The single line LW1 represents γ-Fe (fcc).

Table I. Results of the analysis of the Mössbauer spectra of the samples of MgO single crystal with implanted ^{57}Fe. The accuracy of the values (indicated in the first row of the table) was obtained from the statistic errors given by the fitting procedure. (T = temperature of the sample, DW = doublet, LW = single line)

Sample	Method	T [K]	Spectrum components										
			DW1			DW2			DW3			LW1	
			I [%]	δ [mm/s]	σ [mm/s]	I [%]	δ [mm/s]	σ [mm/s]	I [%]	δ [mm/s]	σ [mm/s]	I [%]	δ [mm/s]
W1	CEMS	293	30	1.38	0.60	33	1.03	0.42	17	0.86	0.58	20	−0.08
			±1	±0.01	±0.01	±1	±0.01	±0.01	±1	±0.01	±0.01	±1	±0.01
W2	CEMS	293	32	1.35	0.65	35	1.01	0.43	19	0.73	0.72	14	−0.08
W1A	CEMS	293	20	1.09	0.62	16	1.09	0.97	54	1.07	0.27	10	−0.15
W2	TMS	293	23	1.03	0.60	33	0.72	0.42	16	0.64	0.20	28	−0.14
W2	TMS	22	20	1.31	0.92	15	1.07	0.67	34	0.79	0.39	29	−0.15
W1A	TMS	293	40	0.79	0.34	15	0.70	0.84	25	0.57	0.34	20	−0.15
W1A	TMS	22	14	1.14	1.30	24	1.08	0.67	43	0.78	0.36	19	−0.16

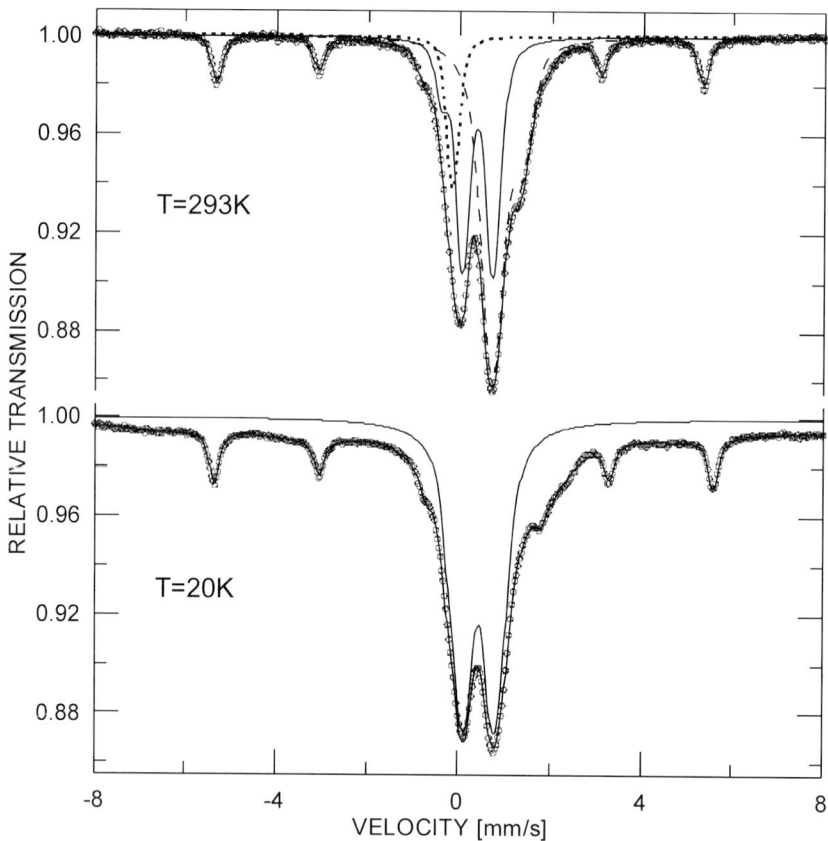

Figure 1. Mössbauer spectra of the as-prepared Mg(O)–Fe sample taken at 293 and 20 K. The circles denote the experimental points, the solid, dashed and doted lines correspond to the fitted components ascribed to Fe–Mg, Fe^{2+} and γ-Fe, respectively.

The presence of this phase is in agreement with the TEM observations on a similar material reported in [12, 13]. A comparison of the W1 and W2 samples shows, that the higher current density produced slight changes in the components DW1, DW2, and DW3 and a decrease in the content of γ-Fe. TMS spectra of W1 and W2 exhibit substantially higher content of γ-Fe than the CEMS ones. This is due to a formation of this phase in deeper layers (>500 nm) of the implanted wafers.

The TMS measurements at 22 and 293 K give similar intensities of the components.

The annealing (W1A sample) caused a decrease in content of γ-Fe due to the formation of MgO–Fe solid solution and redistribution of intensities of DW1, DW2, and DW3 components.

The structure and phase analysis by XRD of the as-prepared spark eroded Fe–Mg powder show Mg, MgO and α-Fe phases and a week amorphous halo. Iron atoms were identified as α-Fe, Fe in Mg, and $Fe_{1-x}O$ phases according to the Mössbauer analysis. An example of the spectra is shown in Figure 1. The results

Table II. Results of the analysis of the Mössbauer spectra of the Mg(O)–Fe samples. (SN = sextet, DN = doublet, LN = single line, HFD = hyperfine field distribution)

Spectrum component	As-prepared, $T = 293$ K				As-prepared, $T = 22$ K				Annealed, $T = 293$ K			
	I [%]	δ [mm/s]	σ [mm/s]	B_{hf} [T]	I [%]	δ [mm/s]	σ [mm/s]	B_{hf} [T]	I [%]	δ [mm/s]	σ [mm/s]	B_{hf} [T]
SN1	3 ±1	0.01 ±0.01	0.01 ±0.01	33.0 ±0.1	5 ±1	0.12 ±0.01	0.01 ±0.01	33.9 ±0.1	41 ±1	0.01 ±0.01	0.01 ±0.01	33.0 ±0.1
SN2					13	0.92	0.42	13.6 (HFD = 7.6T)				
SN3					10	0.83	0.02	42.9 (HFD = 25.0T)				
DN1	22	0.39	0.33		69	0.45	0.33		9	0.35	0.42	
DN2	2	0.15	0.57		3	0.40	0.60					
DN3	4	1.01	0.29						42	1.05	0.28	
DN4	2	1.13	0.41						8	1.02	0.47	
LN1	63	0.72										
LN2	4	−0.17										

of the spectrum analysis are given in Table II. The hyperfine parameters of the sextet SN1 agree with parameters of a pure α-Fe foil. The components DN1 and DN2 are interpreted as Fe in Mg. They occupy 24% of spectrum area at 293 K and 72% at 22 K. This large difference indicates relatively low Debye temperature of this phase. A similar component was not found in the implanted samples (W1, W2). The components DN3, DN4, and LN1 ascribed to the $Fe_{1-x}O$ have the hyperfine parameters, which are in good agreement with the published values, e.g., in [14, 15]. These paramagnetic components split into broad sextets SN2 and SN3 below 180 K as determined from the measurement of the thermal scan of the series of Mössbauer spectra from room temperature down to 22 K (Figure 2). The temperature corresponds well with the Neél temperatures of $Fe_{1-x}O$ [16]. The $Fe_{1-x}O$ was not found in the XRD measurement probably due to its very small coherent volumes that do not yield sharp diffractions. The last component LN2 of the spectra is interpreted as γ-Fe. It also agrees with the component LW1 found in the implanted samples (W1, W2). γ-Fe has magnetic phase transition at temperatures below $\sim$60 K from paramagnetic to antiferromagnetic state [17]. The transition can be observed in the thermal scan of relative transmissions in selected channels derived from Mössbauer spectra and on the thermomagnetic curve drawn in Figure 2. The single line component LN2 split into a sextet which cannot be separated due to strong overlapping with the broad sextets (SN2, SN3) of $Fe_{1-x}O$. The magnetic transitions of $Fe_{1-x}O$ and γ-Fe were not observed in the implanted samples. The $Fe_{1-x}O$ and γ-Fe particles (clusters) in the spark eroded nanocrystalline powder

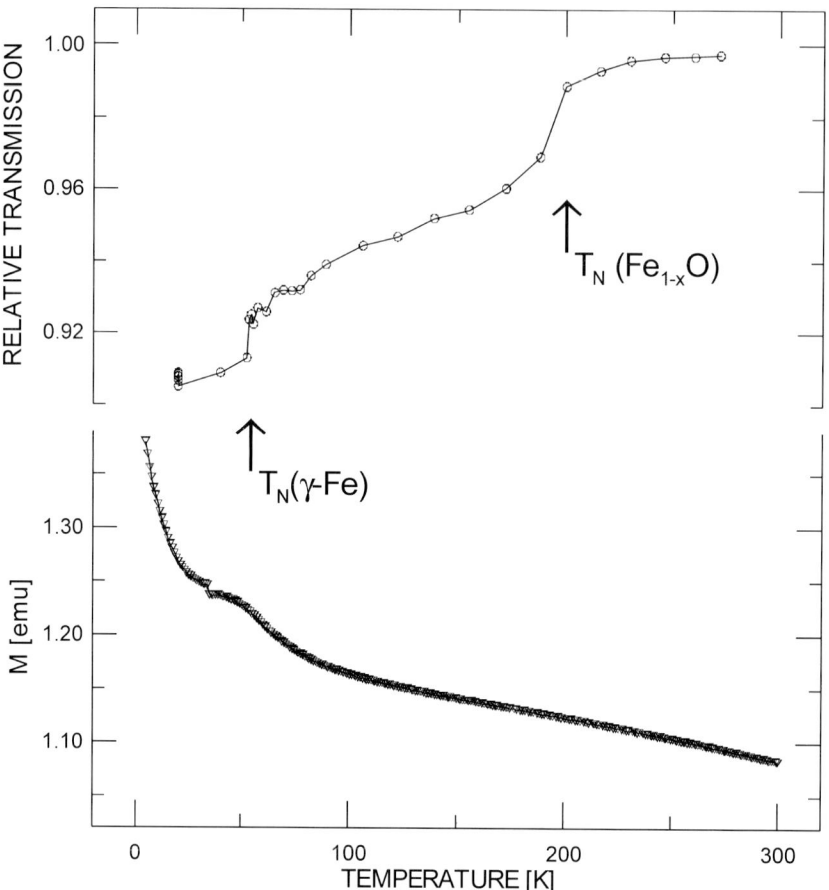

Figure 2. The thermal scan of relative transmissions in selected channels derived from Mössbauer spectra (*top*) and the field cooled (1 T) thermomagnetic curve of the as-prepared Mg–Fe powder (*bottom*). The arrows indicate the Néel temperatures of γ-Fe and $Fe_{1-x}O$.

are coherent with the MgO and therefore not distinguishable in XRD diffractions but they have probably larger sizes and they are not fixed to the MgO as a solid solution.

The annealing of the sample at 973 K for 1 hr in vacuum has caused an increase in α-Fe at the expense of all other phases (Table II) mainly of the component LN1. This is probably the part of $Fe_{1-x}O$ in form of free particles, which can be easily reduced. The component of γ-Fe was not identified there. XRD distinguished MgO and α-Fe phases which are expected as the most stable phases of the composite.

4. Conclusions

The as-prepared spark eroded Fe–Mg powder consists of Mg, MgO, α-Fe, Fe in Mg, $Fe_{1-x}O$, and γ-Fe phases after handling in air at room temperature. The $Fe_{1-x}O$ and γ-Fe phases have similar hyperfine parameters as those in the MgO single crystal wafer implanted by ^{57}Fe. The substantial difference between the spark eroded and implanted samples is in magnetic transitions. In the implanted samples the Fe^{2+} and γ-Fe component remain paramagnetic by cooling down to 22 K in difference to the spark eroded powder where both phases transformed into antiferromagnetic state. Annealing of the spark eroded powder induced changes in the phase composition towards α-Fe particles embedded in MgO phase.

Acknowledgements

This work was supported by the Academy of Sciences of the Czech Republic (K10101040), Grant Agency of the Czech Republic (202/01/0668) and the Czech Ministry of Education, Youth and Sports (OC 523.20).

References

1. Klaus, M., Ullmann, D., Barthel, J., Wulfhekel, W., Kirschner, J., Urban, R., Monchesky, T. L., Enders, A., Cochran, J. F. and Heinrich, B., *Phys. Rev. B* **64** (2001), 134411.
2. Renaud, G., *Surf. Sci. Rep.* **32** (1998), 1.
3. Schneeweiss, O., Jirásková, Y. and Šebek, P., *Phys. Stat. Sol. A* **189** (2002), 725.
4. Konstanchuk, I. G., Ivanov, E. Yu., Pezat, M., Darriet, B., Boldyrev, V. V. and Hagenmüller, P., *J. Less-Common Metals* **131** (1987), 181.
5. Hightower, A., Fultz, B. and Bowman, Jr., R. C., *J. Alloys Comp.* **252** (1997), 238.
6. van der Kraan, A. M. and Buschow, K. H. J., *Phys. Rev. B* **25** (1982), 3311.
7. Shinjo, T., *Hyp. Interact.* **27** (1986), 193.
8. Fnidiki, A., Eymery, J. P. and Denanot, M. F., *Hyp. Interact.* **45** (1989), 295.
9. Reuther, H., Betzl, M., Matz, W. and Richter, E., *Hyp. Interact.* **113** (1998), 391.
10. Perez, A., Marest, G., Sawicka, B. D., Sawicki, J. A. and Tyliszczak, T., *Phys. Rev. B* **28** (1983), 1227.
11. Hayashi, N., Sakamoto, I., Toiyama, T., Wakabayashi, H., Okada, T. and Kuriyama, K., *Surf. Coat. Technol.* **169–170** (2003), 540.
12. Yoshizaki, F., Tanaka, N. and Mihama, K., *J. Electron Microscopy* **39** (1990), 459.
13. Tanaka, N., *J. Mat. Sci. Technol.* **13** (1997), 265.
14. Wilkinson, C., Cheetham, A. K., Long, G. J. and Battle, P. D., *Inorg. Chem.* **23** (1984), 3136.
15. McCammon, C. A. and Price, D. C., *Phys. Chem. Minerals* **11** (1985), 250.
16. Koch, F. and Fine, M. E., *J. Appl. Phys.* **38** (1966), 1470.
17. Tsunoda, Y., Imada, S. and Kunitomi, N., *J. Phys. F* **18** (1988), 1421.

Hyperfine Interactions **156/157**: 89–95, 2004.
© 2004 *Kluwer Academic Publishers. Printed in the Netherlands.*

Shifting the Superparamagnetic Limit of Nanosized Copper Iron Spinel

S. J. STEWART[1,*], R. C. MERCADER[1], G. PUNTE[1], J. DESIMONI[1],
G. CERNICCHIARO[2] and R. B. SCORZELLI[2]
[1]*IFLP, CONICET, Departamento de Física, Facultad de Cs. Exactas, Universidad Nacional de La Plata, C.C.67, 1900 La Plata, Argentina; e-mail: stewart@fisica.unlp.edu.ar*
[2]*Centro Brasileiro de Pesquisas Físicas, Rua X. Sigaud 150, 22290-180 Rio de Janeiro, Brazil*

Abstract. Using a co-precipitation method, a system formed by nanocrystalline Cu–Fe spinel, grain size $D = 6$ nm, and CuO (15 wt.%) was obtained. A Mössbauer blocking temperature (T_B) of the spinel particles that lies between 100 and 200 K was observed. T_B progressively increases as the sample is subjected to high-energy ball milling, being higher than 298 K after milling for 10 hours. The magnetic results also showed the increment of T_B. The maxima of the in-phase component of the AC susceptibility at T_{max} shift toward higher temperatures with the milling. After 4 hours, this shift is ≈ 70 K for all the AC field frequencies. Simultaneously, the increase in the mean grain sizes (up to $D = 13$ nm) and the decrease of the microstrain level indicate that the degree of crystallinity of the spinel phase increases with the milling. After milling for 10 h, the sample is only composed by copper iron spinel with cubic structure.

Key words: ferrites, spinel, ball milling, nanostructures, Mössbauer spectroscopy.

1. Introduction

Assemblies of ferrimagnetic nanoparticles are systems of great complexity due to the simultaneous presence of competing factors, which display different trends in the system's magnetic response. Because of their use as recording media and other technological applications, they have been intensively studied over the last years [1–4]. Notwithstanding, the relation between the surface effects, the interparticle interactions, the surface spin disorder, the degree of inversion or the exchange coupling to the inner ordered grain cores, are still a matter of controversy. This is probably the reason why the precise response of a particular system cannot be assessed *a priori*, and different methods are used semi-empirically to tailor the desired response of a certain material.

High-energy ball milling has been successfully employed to obtain nanostructured ferrites from the material in its bulk state [1–4], which afterwards displays unusual magnetic behavior. In this work, exploring the reverse way, i.e., starting with nanocrystalline copper ferrite and subjecting it to ball milling, we have care-

---

* Author for correspondence.

fully analyzed its structural, hyperfine and magnetic properties to investigate the influence of each parameter on the overall behavior of the system.

2. Experimental

Copper nanoferrite was synthesized by a co-precipitation method [6] and a subsequent annealing at 300°C (AP sample). Afterwards, the AP sample was milled in a vibratory horizontal miller (Retsch) with stainless steel vial and ball for different times up to 12 hours. The mass-to-powder ratio was 10 : 1 and the frequency of operation 40 Hz. The vial was opened after selected times $t_m = 2, 4, 8, 10$ and 12 h, to take out sample for analyses. The XRD diffraction patterns were taken in a Philips PW 1710 diffractometer using $CuK\alpha$ radiation. The Mössbauer spectra in the 25 to 300 K range of temperature were taken in transmission geometry with a nominal 25-mCi ^{57}Co source in a Rh matrix using a Displex closed cycle cryogenic apparatus. Isomer shifts (δ) were calibrated with an α-Fe foil at room temperature. The spectra were fitted using the NORMOS program [5]. The absorbers were prepared with 18 mg/cm^2 of sample. The magnetic measurements were carried out using a commercial SQUID magnetometer and a LakeShore 7130 AC susceptometer.

3. Results and discussion

XRD results showed that the AP sample is composed of copper iron spinel nanograins (mean crystallite sizes $D \approx 6$ nm) and an amount of ≈ 15 wt.% of CuO. The percentage of CuO decreases and, in addition, the spinel line-widths become slightly narrow with the milling. After $t_m = 10$ h, copper iron spinel in its cubic phase is the only phase detected [6]. After $t_m = 12$ h there is a segregation of α-Fe$_2$O$_3$. We observe that D increases with t_m, reaching ≈ 13 nm after milling for 10 h. At the same time, the average microstrain level $\varepsilon = \Delta d/d$ (d is the interplanar distance) decreases. For $t_m = 12$ h, D and ε deviate from the general trend [6].

We observe a broad Fe^{3+} doublet in the room temperature (RT) Mössbauer spectrum of the AP sample (Figure 1). As the measurement temperature is lowered, its relative area decreases while an asymmetric broadened magnetic signal starts to be resolved (Figure 2). At 25 K the magnetic signal accounts for the main part of the Mössbauer spectrum (Figure 3). This thermal evolution is typical of a superparamagnetic (SPM) relaxation behavior, which reveals the small size of the particles. Indeed, at RT the thermal energy is high enough to overcome the anisotropy barriers, and causes the direction of the magnetic moment of the particle to fluctuate between the easy directions of magnetization averaging the hyperfine field to zero within the Mössbauer timescale ($\approx 10^{-8}$ s). As the thermal energy is reduced the relaxation time diminishes and a blocked state of particle magnetic moments is attained, which corresponds to the resolved magnetic signal at 25 K. In this case, the large line-widths as well as the asymmetric line-depths

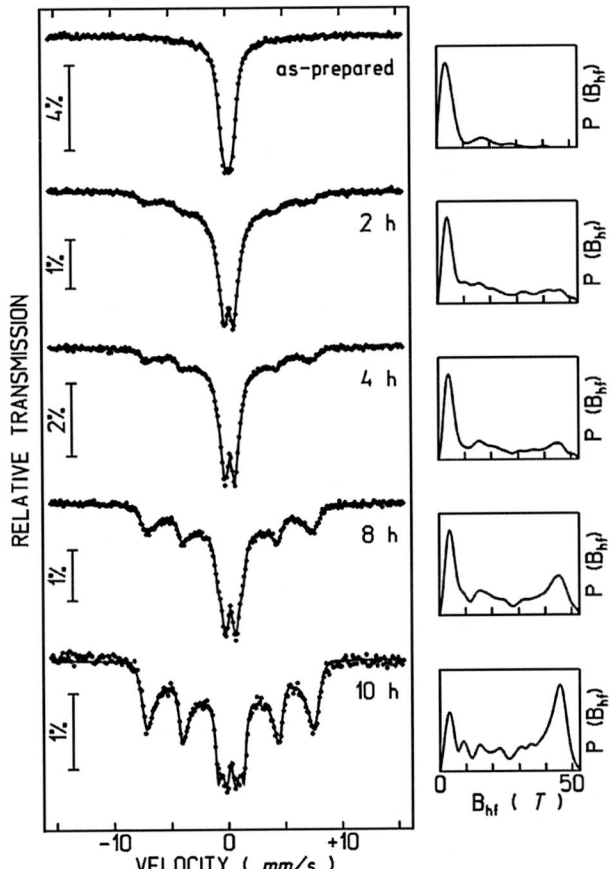

Figure 1. Mössbauer spectra at room temperature for AP sample and after milling during the times indicated (*right*). The solid lines represent the fitting by assuming a distribution of hyperfine fields H_{hf} (*left*).

suggest the presence of at least two different magnetically resolved subspectra I and II. Their hyperfine parameters, assuming Lorentzian line-shapes, are similar to those reported for the B and A spinel sites in cubic $CuFe_2O_4$, respectively ($H_I = 49.6$ T, $\delta_I = 0.41$ mm/s, $2\varepsilon_I = -0.02$ mm/s; $H_{II} = 52.4$ T, $\delta_{II} = 0.48$ mm/s, $2\varepsilon_{II} = -0.02$ mm/s) [7].

After milling the AP sample, a magnetic signal comes out in the RT spectra, which co-exists with the SPM doublet (Figure 1). The relative area of this signal increases with t_m, taking up to $\approx 60\%$ of the total spectrum area for $t_m = 10$ h. For simplicity, we have considered a distribution of static hyperfine fields H_{hf} to fit these spectra. The distribution initially centered at $H_{hf} \approx 10$ T represents the SPM relaxation of the particles with the smaller energy barriers. As the milling advances there is an increasing probability to have contributions from higher H_{hf}, which arise from iron ions in magnetically ordered oxides whose particle magnetic moments

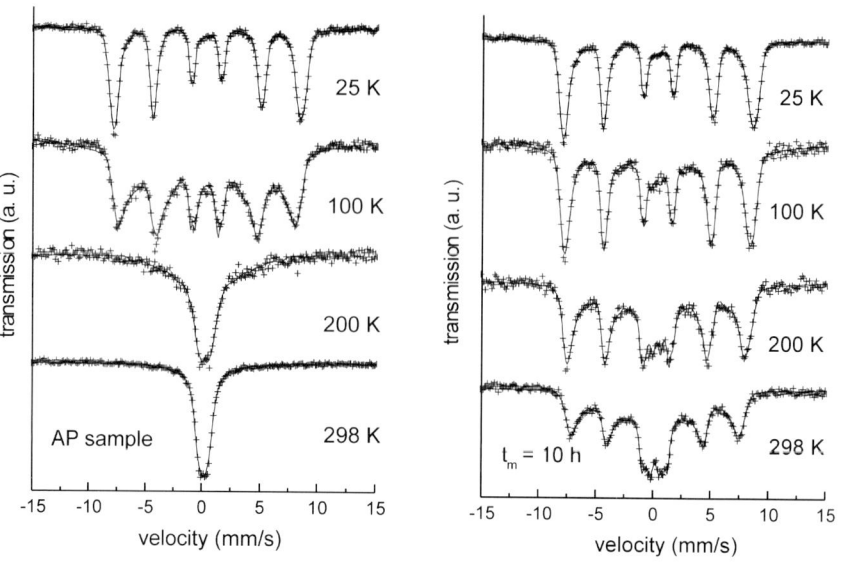

Figure 2. Thermal evolution of the Mössbauer spectra for the copper nanoferrite (AP) and for $t_m = 10$ h. The solid lines correspond to the fitting.

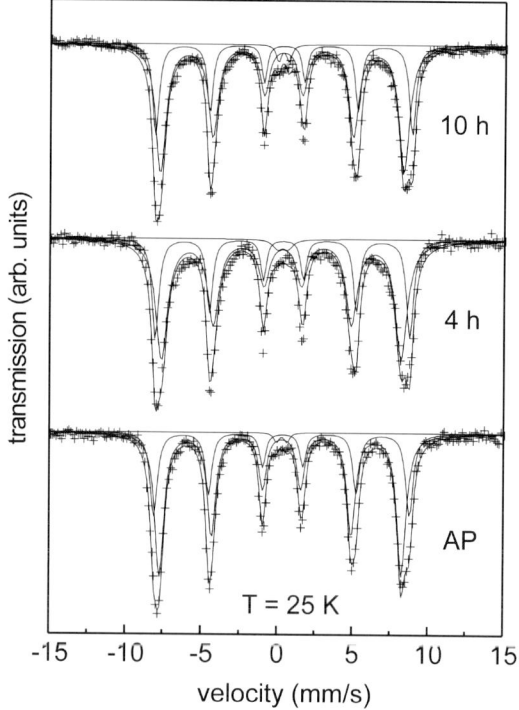

Figure 3. Mössbauer spectra taken at 25 K for the samples shown on the labels. The solid lines are the results of the fittings as described in the text.

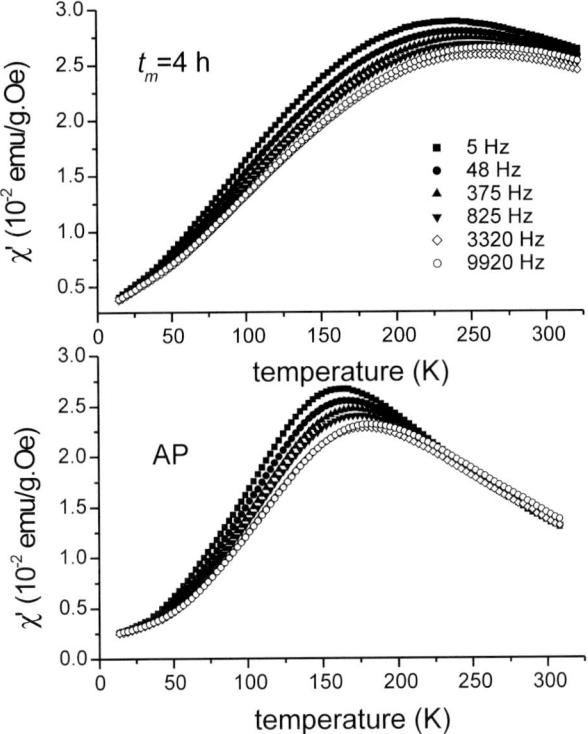

Figure 4. Thermal dependence of the in-phase AC susceptibility, χ', for AP and $t_m = 4$ h samples for different frequencies ν of the AC field H_{AC} of amplitude 1 Oe.

are in a blocked state. The thermal evolution of the spectra of milled samples also displays a SPM behavior (Figure 2). At 25 K we observe that the hyperfine parameters and the relative area ratio of the resolved signals remain almost unchanged with the milling. ($H_I = 49.9$ T, $\delta_I = 0.41$ mm/s, $2\varepsilon_I = -0.02$ mm/s; $H_{II} = 52.6$ T, $\delta_{II} = 0.48$ mm/s, $2\varepsilon_{II} = 0.03$ mm/s.) This would indicate that neither the spinel structure nor the Fe distribution amongst the A and B spinel sites have changed significantly.

The mean Mössbauer blocking temperature T_B, taken as the temperature at which the relative area of the blocked state corresponds to 50% of the total spectrum area, shifts to higher temperatures with t_m. Indeed, T_B is in the 100–200 K interval for AP, while is above 298 K for the 10-h milled sample (Figure 2).

The magnetic results also show an increment of the blocking temperatures. The ZFC maximum of the M vs. T curves (not shown), which is related to the mean T_B, is at 140 K for the AP sample, and shifts to 174, 190, 268 for $t_m = 2$, 4 and 8 h, respectively. The same trend is observed for the temperature at which the ZFC-FC irreversibility starts (T_{irr}), which represents the blocking of particles with the highest energy barriers [6].

The thermal dependence of AC susceptibility χ (Figure 4) shows maxima at T_{max}, which shifts toward higher temperatures when the frequency of the applied

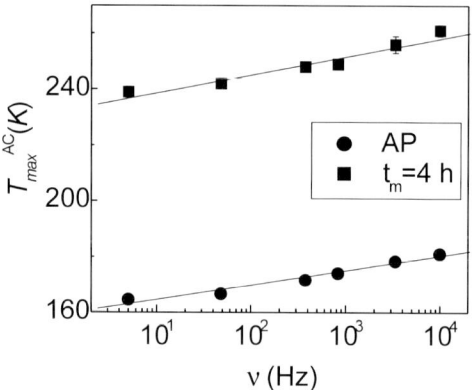

Figure 5. Temperature of the maximum of the in-phase component of the AC susceptibility χ' as a function of the frequency, for AP and 4 h samples.

field, ν, increases. The $T_{\max}$ of the in-phase χ is related to T_B, the latter defined as the temperature at which the relaxation time τ of the nanoparticles equals the measurement time $\tau_m = 1/2\pi\nu$. We observe that the $T_{\max}$ shifts ≈ 70 K for all ν after milling for 4 h. For $t_m = 10$ h, $T_{\max} > 300$ K.

In addition to the growth of T_B there is an increase in the magnetic response with the milling; e.g., we observe that the saturation magnetization M_S at 5 K increases from 17 to 45 emu/g after 10 h.

The present results show that milling nanocrystalline iron–copper spinel induces an important shift in its superparamagnetic limit toward higher temperatures. The poorly crystalline initial state of the AP sample (a mixture of spinel and 15 wt.% CuO) originates a high degree of spin disorder, as evidenced by its relatively low M_S value. The high-energy milling generates an increment of the spinel crystallinity, by increasing D (although remaining in a nanometric scale) and decreasing ε. In addition, the increment of the T_B denotes that the crystallinity improvement causes an increment of the apparent magnetic size. The magnitude of the magnetic response also progressively increases. After $t_m = 10$ h, the Cu–Fe spinel stabilizes in its otherwise metastable cubic phase, which has the larger magnetic moment per formula unit [8].

4. Conclusions

The present results show that the blocking temperatures of nanocrystalline copper ferrite spinel can be shifted to higher temperatures by high-energy ball milling. This process induces an increment of the mean grain size and of the apparent magnetic one. After 10 h of milling the copper nanoferrite stabilizes in the cubic phase. Further milling originates the segregation of hematite.

References

1. Kodama, R. H., Berkowitz, A. E., McNiff, Jr, E. J. and Foner, S., *Phys. Rev. Lett.* **77** (1996), 394.
2. Shi, Y. and Ding, J., *J. Appl. Phys.* **90** (2001), 4078.
3. Chinnasamy, C. N., Narayanasamy, A., Ponpandian, N., Chattopadhyay, K., Shinoda, K., Jeyadevan, B., Tohji, K., Nakatsuka, K., Furubayashi, T. and Nakatani, I., *Phys. Rev. B* **63** (2001), 184108.
4. Goya, G. F., Rechenberg, H. R. and Jiang, J. Z., *J. Appl. Phys.* **84** (1998), 1101.
5. Brand, R. A., Lauer, J. and Herlach, D. M., *J. Phys. F* **18** (1983), 675.
6. Stewart, S. J., Tueros, M. J., Cernicchiaro, G. and Scorzelli, R. B., *Solid State Commun.* **129** (2004), 347.
7. Vandenberghe, R. E. and De Grave, E., In: G. Long and F. Grandjean (eds), *Mössbauer Spectroscopy Applied to Inorganic Chemistry*, Vol. 3, Plenum Press, New York, 1989, p. 59.
8. McCurrie, R. A., In: *Ferromagnetic Materials: Structure and Properties*, Academic, London, 1994, p. 123.

Hyperfine Interactions **156/157**: 97–102, 2004.
© 2004 *Kluwer Academic Publishers. Printed in the Netherlands.*

Synthesis of Nanocrystalline Ni$_{0.5}$Zn$_{0.5}$Fe$_2$O$_4$ by Aerosol Route and Its Characterization

SONAL SINGHAL[1], A. N. GARG[2] and KAILASH CHANDRA[1,*]

[1] *Institute Instrumentation Centre, Indian Institute of Technology-Roorkee, Roorkee 247 667, India;*
e-mail: chandfuc@iitr.ernet.in

[2] *Department of Chemistry, Indian Institute of Technology-Roorkee, Roorkee 247 667, India*

Abstract. Nano-size particles of Ni$_{0.5}$Zn$_{0.5}$Fe$_2$O$_4$ ferrite were prepared through aerosol route. The solutions of iron, nickel and zinc nitrates were mixed in stoichiometric proportion, passed through a pneumatic nebulizer, to get very fine mist (aerosols), and a furnace at $\sim$600°C in air atmosphere. Through various events in succession, metal atoms form ferrite in air. The average particle size was found to be 16±6 nm which increased to 80±8 nm after annealing at 1000°C. The room-temperature magnetic moment of the sample as obtained and after annealing it at various temperatures indicate that the saturation magnetization increases from 1.80 to 72.8 emu/g, while remanent magnetization increases from 0.28 to 25.0 emu/g. Mössbauer spectrum of the sample at room temperature exhibited a doublet with δ(Fe) $= 0.33$ mm s^{-1} and $\Delta E_Q = 0.78$ mm s^{-1} suggesting superparamagnetic nature. However, after annealing at 1000°C this doublet got converted into two magnetic sextets with $B = 52.4$ T and 49.0 T suggesting increase in particle size on annealing. These observations are in conformity with Transmission Electron Microscope (TEM) and X-Ray Diffraction (XRD) results that the particle size increases after annealing the sample at higher temperatures.

1. Introduction

Since the last decade interesting magnetic properties have been reported for nano-crystalline spinel ferrites [1–3]. A reduction in particle size to the nanometer scale results in various special properties such as the quantum-size effects, the high surface area and the lower sintering temperature which is base to obtain fine grain size ceramics with advanced properties, etc. [1, 2]. Ni–Zn ferrites with the spinel crystal structures have been extensively used in many electronic devices because of their high permeability at high frequency, remarkably high electrical resistivity, mechanical hardness, chemical stability and reasonable cost [4]. The electrical and magnetic properties of Ni–Zn ferrites depend strongly on oxidation state and distribution of cations of the tetrahedral (A) and octahedral (B) sites in the lattice [5, 6].

Sedlar *et al.* [7] synthesized nickel zinc ferrite (Ni$_{0.36}$Zn$_{0.64}$Fe$_2$O$_4$) thick films using a dip coating sol–gel process and suggested that the suitable processing temperature is 400°C for the preparation of films with reasonable magnetic prop-

---

* Author for correspondence.

erties. Typical values of magnetization and coercive field were reported as Ms $=$ 110 emu cm^{-3} and Hc $=$ 20 Oe, respectively. Zhiyuan *et al.* [8] proposed that it is possible to obtain a series of good soft materials used for high frequency electronic devices by properly controlling the amount of iron deficiency in Ni–Zn ferrites. Wang and Li [9] prepared $Ni_{1-x}Zn_xFe_2O_4$ ($0.0 \leqslant x \leqslant 1$) by the polyvinyl alcohol sol gel method and reported that the Mössbauer spectra of samples exhibit superparamagnetic relaxation at room temperature and an ordered magnetic structure at 77 K. Recently Sileo *et al.* [10] prepared Ni–Zn ferrites by citrate precursor method and investigated the insertion of small amount of rare earth cations into a nickel zinc ferrites.

Mainly three techniques *viz.* chemical, mechanochemical and thermophysical have been used to prepare nanoparticles. In this paper we report the preparation of nanoparticles of $Ni_{0.5}Zn_{0.5}Fe_2O_4$ through aerosol route and their characterization by transmission electron microscope, X-ray diffraction, magnetic measurements and Mössbauer spectroscopy. Studies were also carried out after annealing the sample at various temperatures for an hour.

2. Experimetal

2.1. PREPARATION OF FERRITES

In order to prepare nano size $Ni_{0.5}Zn_{0.5}Fe_2O_4$ ferrites, through aerosol route desired proportions of iron, nickel and zinc nitrates were weighed in stoichiometric ratio and dissolved in water to prepare 1 mol dm^{-3} solution. This solution was passed through pneumatic nebulizer to get a cloud of fine droplets. These droplets strike on an impact bead in the spray chamber which breaks them into still smaller ones (aerosol). The final aerosols are passed through a furnace at temperatures ~600°C in the presence of air. Through various events in succession (Desolvation, Vapourisation, Atomization and Oxidation) metal atoms form ferrite and nonmetal evaporates in the form of vapours. The ferrite was collected on a Teflon coated pan. The schematic of the system is shown in Figure 1.

2.2. PHYSICAL MEASUREMENTS

The elemental analysis of the ferrite as obtained was carried out using Electron Probe Micro Analyzer (EPMA) (JEOL, 8600M). The particle size was determined by transmission electron microscope (TEM) (Philips, EM400) and X-ray diffractometer (XRD) (Philips, PW 1140/90) using Scherrer's formula. Magnetic measurements were made on a vibrating sample magnetometer (VSM) (Model 155, Princeton Applied Research, USA). Mössbauer spectra were recorded using a constant acceleration transducer driven Mössbauer spectrometer. A 25 mCi ^{57}Co(Rh) source procured from Amersham, UK was used and the spectrometer was calibrated using a natural iron foil as well as recrystallised sodium nitroprusside dihydrate (SNP) as standard.

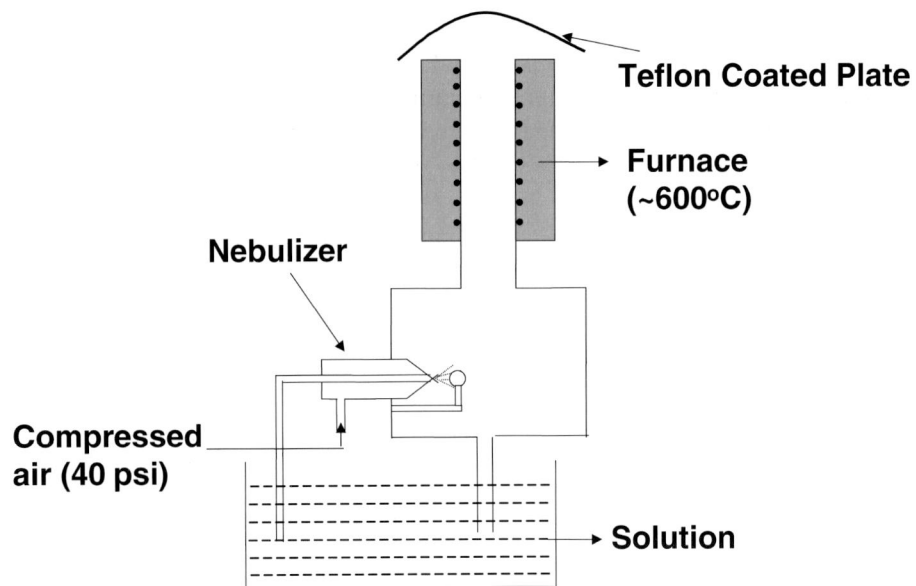

Figure 1. Schematic diagram for the preparation of nanoparticles through aerosol route.

3. Results and discussion

The analysis of Ni, Zn and Fe through EPMA matched well with the formula weight percentage of Ni$_{0.5}$Zn$_{0.5}$Fe$_2$O$_4$. X-ray diffractographs of the sample as obtained and after annealing at various temperatures are shown in Figure 2. Typical transmission electron micrographs and Mössbauer spectra of the ferrite as obtained and after annealing at 1000°C temperature are shown in Figures 3 and 4, respectively. Magnetic data are listed in Table I.

The average particle size as estimated from the micrograph was ~16 nm with nearly spherical in shape (Figure 3(a)). The corresponding selected area electron diffraction pattern confirms the crystallinity of the sample as shown in the inset of Figure 3(a). The average particle size of the sample after heating 1000°C for an hour was ~80 nm and are not spherical now (Figure 3(b)). Thus the particle size of Ni$_{0.5}$Zn$_{0.5}$Fe$_2$O$_4$ varies from ~16–80 nm with the annealing temperature.

The crystallite size of the sample has also been estimated from the broadening of the XRD peaks using the Scherrer equation [11]

$$d = 0.9\lambda/(w - w_1)\cos\theta,$$

where: d is the grain diameter, w and w_1 are the half-intensity widths of the relevant diffraction peak and the instrumental broadening, respectively, λ is the X-ray wavelength and θ is the angle of diffraction. As can be seen in Figure 2, no diffraction pattern was observed for sample as obtained. This can be attributed to very fine size of particles. However, after annealing at various temperatures peaks with varying widths were observed. The particle sizes were calculated using (311) peak and

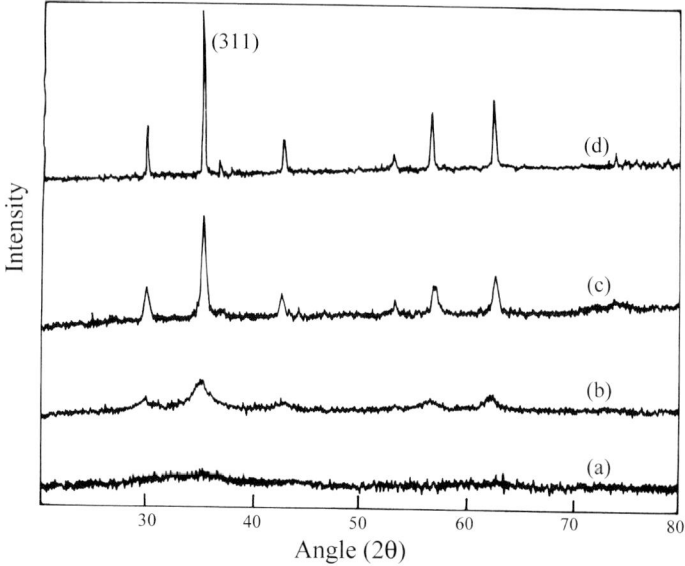

Figure 2. X-ray diffraction pattern for $Ni_{0.5}Zn_{0.5}Fe_2O_4$ at (a) as obtained and after annealing at (b) 400°C, (c) 600°C and (d) 1000°C for 1 h.

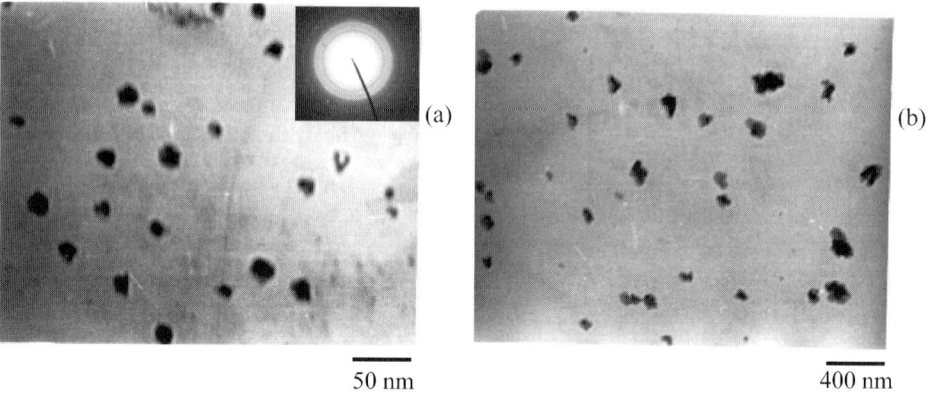

Figure 3. Transmission electron micrographs of the $Ni_{0.5}Zn_{0.5}Fe_2O_4$ (a) as obtained and (b) after annealing at 1000°C. The inset in (a) show the selected area electron diffraction pattern.

found to be 33.4 nm, 55.6 nm and 83.4 nm after annealing the samples at 400°C, 600°C and 1000°C, respectively. These values were in close agreement with the value observed using TEM. XRD patterns of the ferrite sample after annealing at 1000°C showed sharp peaks indicating the formation of well crystalline ferrite phase. It may be noticed that the X-ray line broadening gradually decreases with increasing annealing temperature, attributed to the grain growth of crystallites at higher temperatures. This is in conformity with literature reports [10, 12].

The magnetic data obtained from the hysteresis loop of the sample such as saturation magnetization and remanent magnetization are listed in Table I. From the

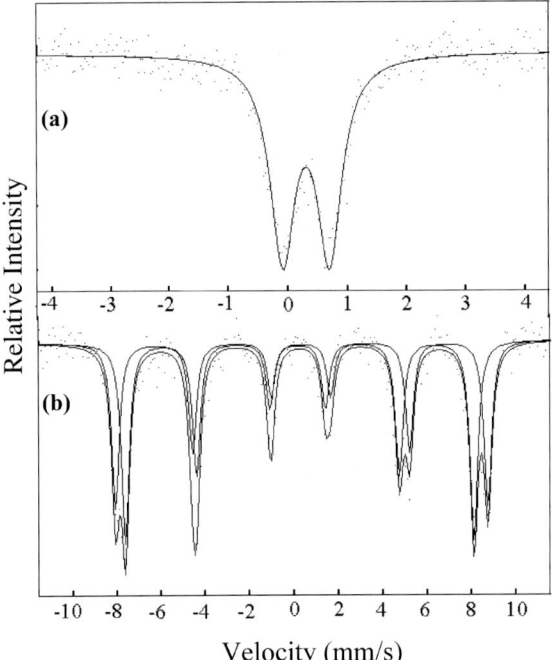

Figure 4. Room-temperature Mössbauer spectra of $Ni_{0.5}Zn_{0.5}Fe_2O_4$ (a) as obtained and (b) after annealing at 1000°C for an hour.

Table I. Magnetic measurements of the Ni–Zn ferrite as obtained and after annealing at various temperatures

Annealing temperatures (°C)	Saturation magnetization (emu/g)	Remanent magnetization (emu/g)
1. RT	1.80	0.28
2. 300	3.07	0.43
3. 400	22.7	4.03
4. 500	39.4	10.65
5. 600	52.6	14.26
6. 700	63.6	21.04
7. 800	69.6	25.13
8. 900	70.9	25.70
9. 1000	72.8	25.03

data it is clear that the saturation and remanent magnetization increases from 1.80 to 72.8 emu/g and 0.28 to 25.03 emu/g, respectively, on increasing the annealing temperature from room temperature to 1000°C, which are within the reported range [13] but less than the bulk materials. The lower value of saturation magnetization with respect to bulk material may be due to their large surface-to-volume ratio, which enhances the spin-canting phenomenon [14]. Both of these parameters become constant after 700°C.

The Mössbauer spectra of $Ni_{0.5}Zn_{0.5}Fe_2O_4$ as obtained and after annealing at 1000°C are shown in Figure 4. The Mössbauer spectra at room temperature exhibit the clear doublet with $\delta(Fe) = 0.33 \pm 0.02$ mm s^{-1} and $\Delta E_Q = 0.78 \pm 0.02$ mm s^{-1} in conformity with the literature [9]. The doublet for the sample reveals the collapse of magnetic ordering due to superparamagnetic relaxation. However, after annealing the sample at 1000°C this doublet gets converted into two magnetic sextets with $B = 52.4$ T and 49.0 T suggesting the increase in particle size. These two sextets are due to the distribution of cations at the tetrahedral (A) and octahedral (B) sites in the lattice [5]. These observations are in conformity with TEM and XRD result that the particle size increases after annealing the sample at higher temperature.

Acknowledgement

Grateful thanks are due to the Council of Scientific Industrial Research, New Delhi for the award of Senior Research Fellowship to S.S.

References

1. Jartych, E., Zurawicz, J. K., Oleszak, D. and Pekala, M., *J. Magn. Mag. Mater.* **208** (2000), 221.
2. Chow, G. M. and Ivanova, N. N., *Nanostructural Materials Science and Technology*, Kluwer Academic Publishers, Dordrecht, 1998.
3. Xiang, L., Deng, X. Y. and Jin, Y., *Scripta Mat.* **47** (2002), 219.
4. Ishino, K. and Narumiya, Y., *Ceramic Bull.* **66** (1987), 1469.
5. Dormann, J. L. and Nogues, M., *J. Phys.: Condens. Matter.* **2** (1990), 1223.
6. Rezlescu, N., Rexiescu, E., Pasnicu, C. and Craus, M. L., *J. Phys.: Condens. Matter.* **6** (1994), 5707.
7. Sedlar, M., Matejee, V., Grygar, T. and Kadlecova, J., *Ceramics Int.* **26** (2000), 507.
8. Zhiyuan, L., Maoren, X. and Qinggiu, Z., *J. Magn. Mag. Mat.* **219** (2000), 9.
9. Wang, L. and Li, F. S., *J. Magn. Mag. Mat.* **223** (2001), 233.
10. Sileo, E. E., Rotelo, R. and Jacoben, F., *Physica B* **320** (2002), 257.
11. Klug, H. P. and Alexander, L. E., *X-Ray Diffraction Procedures for Polycrystalline and Amorphous Materials*, 2nd edn, Wiley, 1974, Chapter 9.
12. Kinemuchi, Y., Ishizaka, K., Suematsu, H., Jiang, W. and Yatsi, K., *Thin Solid Films* **407** (2002), 109.
13. Kim, C. S., Kim, W. C., An, S. Y. and Lee, S. W., *J. Mag. Mag. Mat.* **215–216** (2002), 213.
14. Battle, X., Obradors, X., Medarde, M., Carvajal, J. R., Pernet, M. and Regi, M. V., *J. Mag. Mag. Mater.* **124** (1993), 228.

Hyperfine Interactions **156/157**: 103–111, 2004.
© 2004 *Kluwer Academic Publishers. Printed in the Netherlands.*

Transitions and Spin Dynamics at Very Low Temperature in the Pyrochlores $Yb_2Ti_2O_7$ and $Gd_2Sn_2O_7$

P. BONVILLE[1], J. A. HODGES[1], E. BERTIN[1], J.-PH. BOUCHAUD[1],
P. DALMAS DE RÉOTIER[2], L.-P. REGNAULT[2], H. M. RØNNOW[2],
J.-P. SANCHEZ[2], S. SOSIN[2] and A. YAOUANC[2]
[1]*C.E.A. – Saclay, Service de Physique de l'Etat Condensé, 91191 Gif-sur-Yvette, France*
[2]*C.E.A. – Grenoble, Service de Physique Statistique, de Magnétisme et Supraconductivité,
38054 Grenoble, France*

Abstract. The very low temperature properties of two pyrochlore compounds, $Yb_2Ti_2O_7$ and $Gd_2Sn_2O_7$, were investigated using an ensemble of microscopic and bulk techniques. In both compounds, a first order transition is evidenced, as well as spin dynamics persisting down to the 20 mK range. The transition however has a quite different character in the two materials: whereas that in $Gd_2Sn_2O_7$ (at 1 K) is a magnetic transition towards long range order, that in $Yb_2Ti_2O_7$ (at 0.24 K) is reminiscent of the liquid–gas transition, in the sense that it involves a 4 orders of magnitude drop of the spin fluctuation frequency, with no long range order. We attribute these unusual features to the frustration of the antiferromagnetic exchange interaction in the pyrochlore lattice.

Key words: frustrated magnetic systems, μSR, Mössbauer spectroscopy.

1. Introduction

Frustration of exchange interactions appears in crystallographically ordered materials if the geometry of the lattice is such that it prevents all pairs of exchange bonds from being satisfied throughout the lattice. The simplest example is the bidimensional triangular lattice with isotropic (Heisenberg) antiferromagnetic (AF) nearest neighbour exchange. Villain drew attention to geometrically frustrated systems [1] and showed that no Néel order can occur in a Heisenberg antiferromagnet on a three-dimensional lattice of corner sharing tetrahedra (or pyrochlore lattice). The ground state of such a system was called a "cooperative paramagnet", or a "spin liquid" state, where the spins undergo short range dynamic correlations down to $T = 0$. In the past decade, investigations of geometrically frustrated systems have developed to a large extent [2]. In the case of the Heisenberg antiferromagnet in a pyrochlore lattice, the ground state was shown to have a large degeneracy, where the different configurations are not separated by energy barriers [3]. This has important implications as regards the spin dynamics: zero energy local or extended soft modes are possible, even at very low temperature. Another consequence is

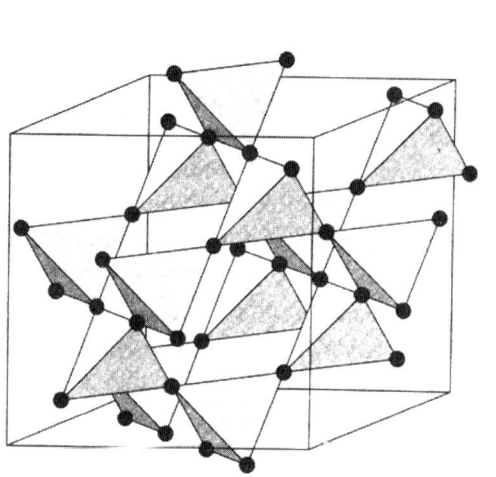
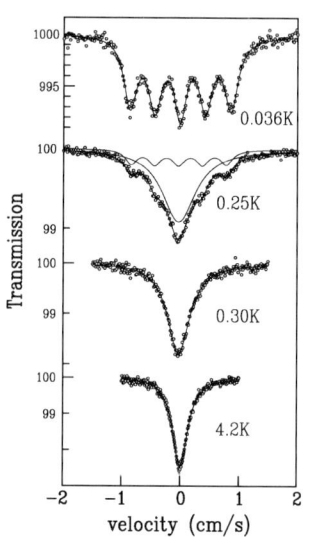

Figure 1. Left: The pyrochlore lattice of $R_2M_2O_7$ materials, where only the R sublattice is shown (black dots); *Right*: selected ^{170}Yb Mössbauer absorption spectra in $Yb_2Ti_2O_7$ between 0.036 K and 4.2 K; the lines are fits as explained in the text.

that any perturbation, like the dipole–dipole interaction, next nearest neighbour exchange, ionic or exchange anisotropy, can select a particular ground state, resulting in a magnetic transition at a finite temperature.

Compounds with formula $R_2M_2O_7$, where R is a rare earth and M a transition or *sp* metal, crystallise in a structure where both R and M ions are located on two interpenetrating pyrochlore lattices (Figure 1 *left*). In the compounds $Yb_2Ti_2O_7$ and $Gd_2Sn_2O_7$, we have discovered novel behaviours at very low temperature. In these materials, the rare earth alone is magnetic and it is located at a site with threefold symmetry (D_{3d}). The paramagnetic Yb^{3+} ($4f^{13}$) and Gd^{3+} ($4f^7$) ions have distinct single ion properties: the crystal electric field splits the ground spin–orbit multiplet $\{J = 7/2\}$ of the Yb^{3+} ion into 4 Kramers doublets, whereas it has practically no influence on the Gd^{3+} ion which has $L = 0$ and $S = 7/2$. Our investigations were mainly carried out using local techniques: Mössbauer spectroscopy on the isotopes ^{170}Yb and ^{155}Gd, Muon Spin Relaxation (μSR) spectroscopy and neutron diffraction. Most of these measurements were performed down to the 30 mK temperature range.

2. $Yb_2Ti_2O_7$

In $Yb_2Ti_2O_7$, a sharp peak in the specific heat had been evidenced near 0.24 K [4], but the nature of the transition had not been further investigated. The magnetic susceptibility follows a Curie–Weiss law below about 50 K, with a small positive paramagnetic Curie temperature $\theta_p \simeq 0.8$ K indicative of a ferromagnetic net exchange interaction [5].

Selected Mössbauer absorption spectra on the isotope ^{170}Yb ($I_g = 0$, $I_e = 2$, $E_0 = 84.3$ keV), represented in Figure 1 *right*, reveal an apparently standard first order magnetic transition at 0.24 K. At 0.036 K, a five-line magnetic hyperfine pattern is observed with a hyperfine field $H_{hf} \simeq 115$ T corresponding to a saturated Yb^{3+} moment of *ca.* 1.15 μ_B. As temperature increases, the spectrum remains unchanged up to 0.22 K; then a broad single line grows superimposed on the five-line spectrum, and at 0.26 K the single line alone is left. These features are characteristic of a first order transition. From these data alone, it cannot be said whether the hyperfine field spectra observed below 0.24 K correspond to a static long-range order (LRO) or to dynamic short range correlations with fluctuation frequencies smaller than the lower limit $\nu_l \simeq 10^8$ s^{-1} of the "Mössbauer relaxation window" for ^{170}Yb [6]. This problem will be dealt with below when describing the neutron diffraction and μSR experiments.

Above 0.24 K, the spectra show a single line whose width decreases as temperature increases. This is interpreted as a dynamic effect in the "extreme narrowing" regime: the spin fluctuation frequency is much larger than the hyperfine coupling and increases as temperature increases. As the neutron diffraction measurements on a single crystal, to be described below, show that spin correlations are present above 0.25 K, we interpret the spectra in the temperature range 0.25 K $\leqslant T \leqslant 0.9$ K in terms of fast hyperfine field fluctuations, where $\mathbf{H}_{hf}$ jumps isotropically at random with a characteristic frequency ν [7]. Then the dynamical broadening of the Mössbauer spectrum is given by: $\Delta\Gamma_R = \nu_{hf}^2/\nu$, where the hyperfine frequency ν_{hf} is $\mu_I H_{hf}$, μ_I being the magnetic moment of the excited ^{170}Yb nuclear state. The fluctuation frequencies obtained thereby are reported in Figure 3 *left* and will be discussed below, together with the μSR data.

An insight into the dynamics of the moments is further provided by μSR measurements (whose characteristic time is about 10^{-6} s), performed at ISIS (Rutherford Appleton Laboratory, Chilton, England) and at PSI (Villigen, Switzerland) [8]. For all temperatures d own to 0.275 K (see Figure 2(a)), the time decay of the muon depolarisation has an exponential form, with a relaxation rate λ_z. In the "extreme narrowing" limit, the following relationship links λ_z and ν, the electronic fluctuation frequency: $\lambda_z = 2(\Delta_p^2/\nu)$, where Δ_p is the root mean square deviation of the distribution of dipolar couplings experienced by the muon spin in its interstitial stopping site. As Δ_p is not known, we extracted ν from the measured λ_z values by adjusting the value of Δ_p so that the frequency values match those measured by Mössbauer spectroscopy (see Figure 3 *left*). We obtain $\Delta_{HT} \simeq 80$ mT, which is a typical order of magnitude for magnetic compounds. Below the temperature of the specific heat peak (0.24 K), the shape of the muon depolarisation changes drastically (see the 0.2 K data in Figure 2): it is no longer an exponential function of time, but it shows a rapid depolarisation within about 0.2 μs followed by a slow quasi-exponential decay. No oscillatory signal is observed, which means that no

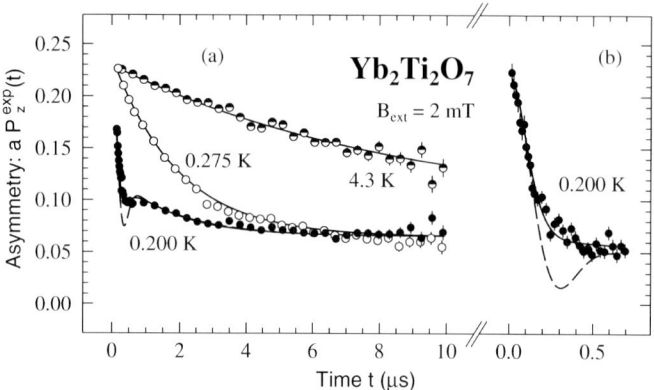

Figure 2. In Yb$_2$Ti$_2$O$_7$: (a) μSR depolarisation spectra, measured at ISIS, on each side of the transition temperature (0.24 K); the constant background asymmetry is $a \simeq 0.065$, and (b) detail of the depolarisation at 0.2 K at short times, measured at PSI. The lines are fits as explained in the text.

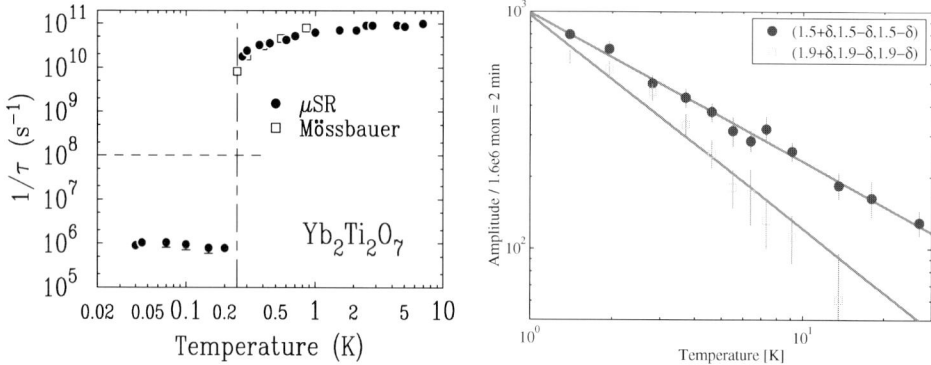

Figure 3. *Left*: Thermal variation of the Yb spin fluctuation frequency as measured by μSR (solid circles) and ^{170}Yb Mössbauer spectroscopy (open squares); the dashed line is the lower limit ν_l of the Mössbauer window for the measurement of fluctuation frequencies; *Right*: thermal variations of the amplitudes of the neutron diffraction peaks near $\mathbf{Q} = (1.5, 1.5, 1.5)$ (black circles) and near $\mathbf{Q} = (1.9, 1.9, 1.9)$ (open squares).

LRO is present, in agreement with the neutron diffraction data described below. The shape of the low temperature depolarisation can actually be accounted for by a dynamic Kubo–Toyabe decay with a rather low fluctuation frequency: $\nu \simeq 10^6$ s^{-1} (dashed line in Figure 2, and see [8] for an explanation of the improved linefit obtained around 0.3 μs).

The thermal variation of the Yb spin fluctuation frequency obtained from the Mössbauer and μSR data is represented in Figure 3 *left*. It is the main result of our study in this compound: the fluctuation frequency undergoes a first order abrupt drop at the temperature of the specific heat anomaly (0.24 K), falling from the 10^4–10^5 MHz range to 1 MHz. So the transition involves the time domain and is reminiscent of the liquid–gas transition, which is first order and involves an abrupt drop of the mean collision frequency between atoms as one enters the liquid phase. The low temperature phase in $Yb_2Ti_2O_7$ is a "spin-liquid" phase [1], where spin dynamics is present down to the lowest temperature.

A neutron diffraction study of a single crystal of $Yb_2Ti_2O_7$ was performed at the Institute Laue Langevin (Grenoble, France), in the temperature range 0.04–30 K. No magnetic Bragg peaks appear as the temperature decreases from 4.2 K to 0.04 K, but mapping of the magnetic elastic scattering in (110) planes reveals the presence of "diffraction rods" along the [111] direction in reciprocal space. This shows that there is no LRO, but that bidimensional antiferromagnetic correlations, probably within planes perpendicular to [111], are present. The width of the peaks obtained from Q-scans across the rod does not vary with temperature and corresponds to a correlation length $\xi \simeq 4$ nm (about 4 times the parameter of the unit cell, which contains 16 Yb ions). No anomaly is found at 0.24 K in the thermal variation of the peak intensities: they increase steadily on cooling from about 25 K down to 0.04 K (see Figure 3 *right*, where the data below 1.4 K are not shown). The short range bidimensional spin correlations in $Yb_2Ti_2O_7$ therefore involve a few hundred Yb ions, and they build up monotonically as temperature decreases. This reinforces the picture sketched above, i.e. that the transition at 0.24 K involves the frequency domain only.

3. $Gd_2Sn_2O_7$

The inverse susceptibility in $Gd_2Sn_2O_7$ follows a Curie–Weiss law with a paramagnetic Curie temperature $\theta_p \simeq -10$ K, indicative of an antiferromagnetic exchange interaction. As the Gd^{3+} ion is isotropic, this compound can be therefore expected to be a good realisation of an AF Heisenberg frustrated system, with no Néel order down to $T = 0$. However, both the magnetic susceptibility and the specific heat, shown in Figure 4 *left*, evidence an anomaly at 1 K. The anomaly of the specific heat $C_p(T)$ reaches the very large value of 120 $J K^{-1}mol.Gd^{-1}$, whereas the expected jump at T_N for a second order magnetic transition is: $\Delta C_p = 20.4$ $J K^{-1}mol.Gd^{-1}$ for Gd^{3+} with $S = 7/2$ [9].

This high ΔC_p value is indicative of a first order transition, which will be confirmed by the ^{155}Gd Mössbauer measurements to be described below. The inset of Figure 4 *left* shows the magnetic entropy variation as temperature increases; at the transition, only 40% of the total entropy $R \ln 8 = 17.3$ $J K^{-1}mol.Gd^{-1}$ has been released, and the full paramagnetic degrees of freedom are recovered only at 8–10 K. This is due to the presence of short range order developing well above the transition temperature, which is a characteristic feature of frustration [2].

Selected Mössbauer absorption spectra on the isotope ^{155}Gd ($I_g = 3/2$, $I_e = 5/2$, $E_0 = 86.5$ keV) in $Gd_2Sn_2O_7$ are shown in Figure 4 *right*. A clear change in the lineshape is observed between 1.05 and 1.1 K: at 1.1 K and above, the spectrum is a pure quadrupolar hyperfine pattern, while at 1.05 K a magnetic hyperfine field

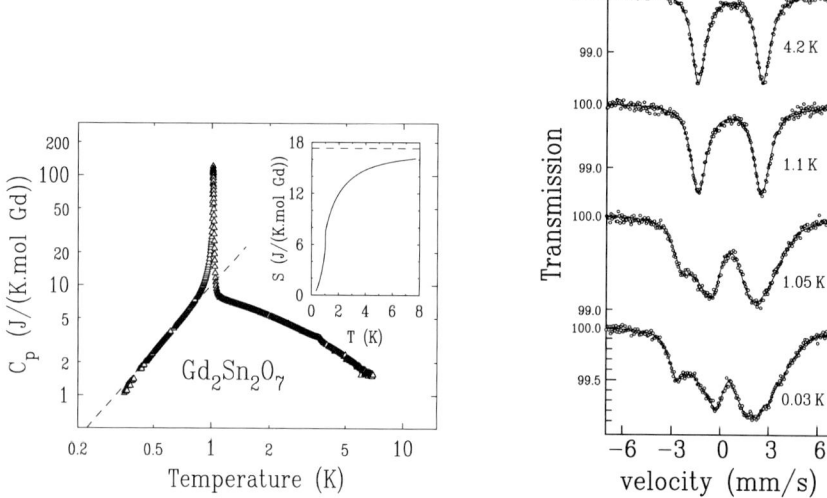

Figure 4. In $Gd_2Sn_2O_7$: *Left*: thermal variation of the specific heat; the dashed line is a T^2 law; inset: thermal variation of the entropy; *Right*: [155]Gd Mössbauer spectra at selected temperatures between 0.027 K and 4.2 K.

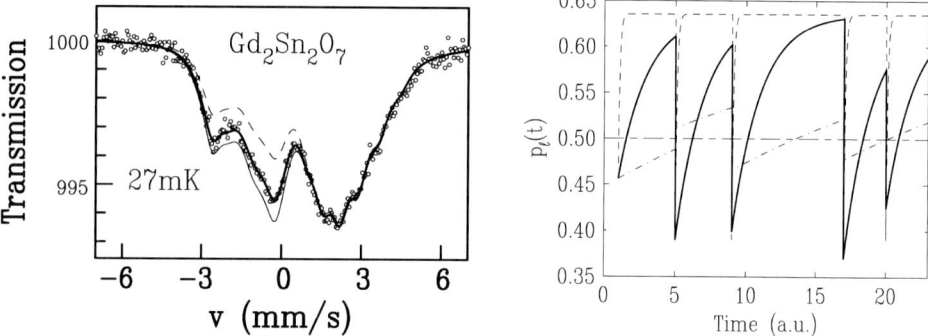

Figure 5. Left: [155]Gd Mössbauer spectrum at 0.027 K; the lines are theoretical curves for various values of the hyperfine temperature T_{hf} (see text); *Right*: Temporal evolution of the ground state population $p_\ell(t)$ of a (nuclear) spin 1/2 doublet, as the (hyperfine) field reverses randomly in time, at arbitrarily chosen instants 5, 9, 17 and 20. Dashed line: $T_1 \ll \tau$: thermalisation is effective; dash-dotted line: $T_1 \gg \tau$: thermalisation is impossible, and solid line: $T_1 \sim \tau$.

has appeared, evidencing short or long range magnetic order. As the temperature is further decreased down to 0.027 K, the hyperfine field increases slightly to reach a saturated value of 30 T. This behaviour is the hallmark of a first order transition, as also inferred from the height of the specific heat peak.

The spectrum at 0.027 K (see Figure 5 *left*) deserves particular attention, because the temperature which can be measured from the spectrum: $\simeq 0.090$ K (referred to hereafter as the hyperfine temperature T_{hf}) is much higher than that of the sample (given by the thermometer): $\simeq 0.03$ K. The possibility of measuring the absolute temperature from the relative line intensities of the Mössbauer spectrum has long been recognised [10]. The condition is that the hyperfine splittings be of the same order of magnitude as $k_B T$. In $Gd_2Sn_2O_7$, the hyperfine splitting in the ground nuclear state is $\Delta_{hf} \simeq 15$ mK. In Figure 5 *left* are shown, together with the experimental spectrum at 27 mK, the theoretical spectra expected for $T_{hf} = 27$ mK (dashed line), for $T_{hf} > 0.2$ K (thin solid line) and for $T_{hf} = 90$ mK (thick solid line). This last line goes perfectly well through the data points, and it is then clear that the hyperfine levels are "hotter" than the lattice, i.e. they are out of thermal equilibrium. The interpretation we give for this phenomenon is the following [11]: the hyperfine levels have no time to reach thermal equilibrium, with a time constant T_1, because the hyperfine field reverses with a time constant τ shorter than or of the same magnitude as T_1. This is illustrated in Figure 5 *right*, which depicts the temporal evolution of the population p_ℓ of the ground state of a fictitious spin 1/2 (modelling the hyperfine levels) as a function of the ratio T_1/τ. The steady state population of the ground state $\langle p_\ell \rangle$ is the Boltzmann population p_ℓ^B (0.63 with our choice of parameters) if hyperfine relaxation occurs very rapidly with respect to τ (dashed line), it is 0.5 (equipopulation) in the reverse case (dash-dotted line), and it should be a function of the ratio $\mu = T_1/\tau$ if the latter is of the order unity (solid line). In order to obtain this function, we devised a model of a spin 1/2 submitted to a magnetic field of constant magnitude, but which undergoes "flips" at random instants of time. The model provided an analytical solution for the steady state population of the ground level, or equivalently for the effective temperature T_{hf} [11]. We obtain, in the temperature range of our experiments: $T_{hf} \simeq T(1 + 2\mu)$. Applying this relation in $Gd_2Sn_2O_7$ at $T = 0.027$ K yields $\mu = T_1/\tau \simeq 0.8$.

The fact that the hyperfine populations are out of equilibrium at 0.027 K reveals therefore the presence of two forms of dynamics of the Gd^{3+} electronic spin. The first one consists in spin flips between one (or more) direction(s) which correspond to flips of the hyperfine field sensed by the ^{155}Gd nuclear spin. Besides, the only plausible mechanism for nuclear relaxation is the coupling to electronic spin-waves [12], which are thus the second form of spin dynamics evidenced at 0.027 K in $Gd_2Sn_2O_7$. The presence of spin-waves at very low temperature is also demonstrated by the μSR experiments, to be described below.

The μSR experiments were performed in zero magnetic field between 20 mK and 100 K at PSI. Above 1 K, the decay has an exponential form, and the relaxation rate λ_z has a temperature independent value of 2 MHz. Below 1 K, the decay signal changes and presents oscillations, indicative of magnetic LRO. The oscillations are clearly visible only at very short times (see Figure 6 *left*); there are in fact two oscillating components, corresponding to dipolar fields of 206 and 441 mT, arising from coupling to the spontaneous Gd moments. Unexpectedly, an exponential damping component is still present in the depolarisation, and is clearly visible at long times after integrating out the oscillations. Its weight is 1/3 of that of the oscillating part, as expected for the μSR spectrum of a polycrystalline sample in the magnetically

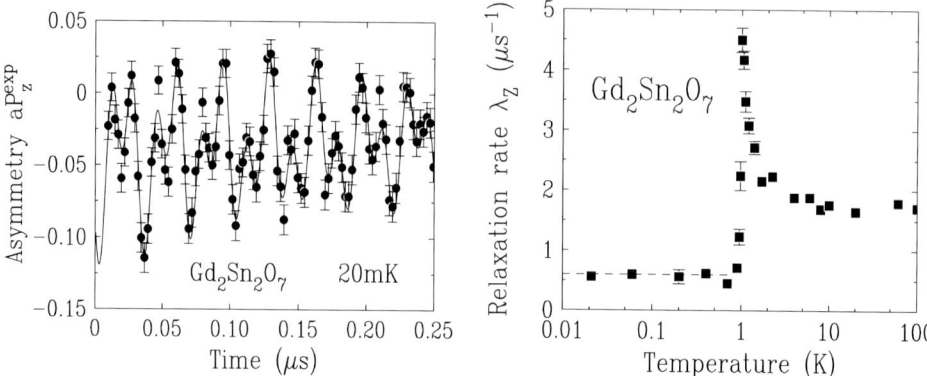

Figure 6. In $Gd_2Sn_2O_7$: *Left*: Short time variation of the μSR depolarisation at 20 mK, measured at PSI; *Right*: Thermal variation of the muon longitudinal relaxation rate λ_z; the dashed line at 0.6 MHz emphasizes the constant relaxation rate below 1 K.

ordered phase [13]. The longitudinal relaxation rate λ_z is then due to the coupling of the muon spin with the spin-waves [14].

The thermal variation of λ_z is shown in Figure 6 *right*. The transition at 1 K is marked by a sharp anomaly, but the remarkable feature is that λ_z below 1 K shows no thermal dependence, remaining at the constant value 0.6 MHz down to 20 mK. This is anomalous, as relaxation of the muon spin by spin-waves should vanish at $T = 0$. For example, it was shown that, for a Heisenberg ferromagnet, λ_z varies as T^2 to a good approximation if one considers two-magnon Raman processes, which are the most likely relaxation mechanism [14]. The explanation for the temperature independence of λ_z must probably be looked for in the unusual spin-wave spectrum of $Gd_2Sn_2O_7$, where soft spin-wave modes with a finite density of states at zero energy could exist.

4. Conclusions

Both studied pyrochlore materials, $Yb_2Ti_2O_7$ and $Gd_2Sn_2O_7$, present a low temperature state where spin fluctuations persist as $T \rightarrow 0$. In $Yb_2Ti_2O_7$, this state seems to be a spin-liquid phase where short range dynamic spin correlations are present, as expected for geometrically frustrated isotropic systems with antiferromagnetic interactions. However, it is reached through a first order transition (at 0.24 K) where the spin fluctuation frequency drops from the GHz to the MHz range, but where the length scale of the spin correlations does not seem to show any change. In $Gd_2Sn_2O_7$, the low temperature phase has long range magnetic order and it is reached through a standard first order transition. However, spin dynamics is present down to 30 mK: it was detected indirectly by [155]Gd Mössbauer spectroscopy through the finding of nuclear levels which are out of thermal equilibrium. This requires the presence of both electronic spin flips and of nuclear relaxation by coupling to spin waves, with similar characteristic time scales. The presence of spin waves down to 20 mK was also evidenced in the μSR measurements by the non-vanishing relaxation rate of the muon spin.

Acknowledgements

We thank A. Forget from SPEC-Saclay for preparing the powder samples and G. Dhalenne, from LPCES-Orsay, for growing the single crystal $Yb_2Ti_2O_7$ sample. The neutron scattering experiments on this single crystal were performed at the D23 diffractometer at the Institute Laue-Langevin, which is a Collaborating Research Group (CRG) instrument operated by the CEA.

References

1. Villain, J., Z. Phys. B **33** (1979), 31.
2. For a recent review, see: Ramirez, A. P., In: K. H. J. Buschow (ed.), *Handbook of Magnetic Materials*, Vol. 13, Elsevier, 2001, p. 423.
3. Moessner, R. and Chalker, J. T., *Phys. Rev. B* **58** (1998), 12049.
4. Blöte, H. W. J., Wielinga, R. F. and Huiskamp, W. J., *Physica* **43** (1969), 549.
5. Hodges, J. A., Bonville, P., Forget, A., Rams, M., Królas, K. and Dhalenne, G., *J. Phys.: Condens. Matter* **13** (2001), 9301.
6. Gonzalez-Jimenez, F., Imbert, P. and Hartmann-Boutron, F., *Phys. Rev. B* **9** (1974), 95.
7. Dattagupta, S., *Hyp. Interact.* **11** (1981), 77.
8. Hodges, J. A., Bonville, P., Forget, A., Yaouanc, A., Dalmas de Réotier, P., André, G., Rams, M., Królas, K., Ritter, C., Gubbens, P. C. M., Kaiser, C. T., King, P. J. C. and Baines, C., *Phys. Rev. Lett.* **88** (2002), 077204.
9. Stanley, H. E., *Introduction to Phase Transitions and Critical Phenomena*, Clarendon Press, Oxford, 1971.
10. Shenoy, G. K. and Maletta, H., *Z. Physik* **269** (1974), 241.
11. Bertin, E., Bonville, P., Bouchaud, J.-Ph., Hodges, J. A., Sanchez, J.-P. and Vulliet, P., *Eur. Phys. J. B* **27** (2002), 347.
12. Beeman, D. and Pincus, P., *Phys. Rev.* **166** (1968), 359.
13. Schenck, A., *Muon Spin Rotation Spectroscopy*, Adam Hilger Ltd, Bristol, 1985.
14. Dalmas de Réotier, P. and Yaouanc, A. *Phys. Rev. B* **52** (1995), 9155.

Hyperfine Interactions **156/157**: 113–122, 2004.
© 2004 *Kluwer Academic Publishers. Printed in the Netherlands.*

Mössbauer and Neutron Diffraction Studies on Co–Al Ferrite

SAM JIN KIM[1], KWANG-DEOG JUNG[2] and CHUL SUNG KIM[1]
[1] *Dept. of Physics, Kookmin University, Seoul 136-702, Korea; e-mail: cskim@phys.kookmin.ac.kr*
[2] *Eco-Nano Center, Korea Institute of Science and Technology, P.O. Box 131, Seoul, Korea*

Abstract. Al substituted $CoAl_xFe_{1-x}O_4$ ($x = 0.1, 0.2, 0.3,$ and 0.5) have been studied with X-ray and neutron diffraction, Mössbauer spectroscopy and magnetization measurements. Neutron diffraction at 10 K for $CoAl_{0.1}Fe_{1.9}O_4$ revealed a cubic spinel structure of ferrimagnetic long range ordering, with magnetic moments of $Fe^{3+}(A)(-4.18\ \mu_B)$, $Fe^{3+}(B)(4.81\ \mu_B)$, $Co^{2+}(B)(2.99\ \mu_B)$, respectively.

The temperature dependence of the magnetic hyperfine field in ^{57}Fe nuclei at the tetrahedral (A) and octahedral (B) sites was analyzed based on the Néel theory of magnetism. In the sample $CoAl_{0.1}$-$Fe_{1.9}O_4$, the intersublattice A–B interaction and intrasublattice A–A superexchange interaction were antiferromagnetic with strengths of $J_{A-B} = -23.3\ k_B$ and $J_{A-A} = -17.6\ k_B$, respectively, while the intrasublattice B–B superexchange interaction was found to be ferromagnetic with a strength of $J_{B-B} = 5.5\ k_B$. With increasing Al substitution the A–B and B–B interaction decreased but the A–A interaction increased. It is interpreted that the reduction of magnetic moment in $Fe^{3+}(A)$ and a noticeable strength of the A–A interaction are closely related to the covalency effects.

Key words: Co–Al ferrite, superexchange interaction, Mössbauer spectroscopy, neutron diffraction, Debye temperature.

1. Introduction

Cobalt spinel ferrites are extensively studied for their promising application as recording media, and for their large magneto-optical effects due to the chemical stability [1]. Recently, various kinds of Co ferrites have been studied for controlling the size of the particles [2, 3]. For understanding of magnetic interaction behavior, nonmagnetic ions (Al^{3+}, Y^{3+}) were employed to dilute the magnetic materials. Specially the Al^{3+} substituted sample has been of great interest for improved Kerr effects and reflectivity [4]. When one tries to obtain small-size particles and to control their magnetic behavior, for an application to nano-scale device, the systematic understanding of superexchange interactions between magnetic ions is essential. Studies on the superexchange interaction of ferrites have been reported by a number of authors [5, 6]. Two sets of data, namely, magnetic moment and paramagnetic susceptibility, can be used to determine accurate superexchange strengths.

When one obtains superexchange parameters, one can only deduce the individual sublattice moments of A and B sites from the net magnetization, because there are no ways to measure the magnetic moments of sublattice separately. However the Mössbauer experiments give the individual magnetic hyperfine fields of A and B sites. Therefore it would be reasonable and interesting that one calculates the exchange parameters from individual sublattice magnetizations that were measured directly by Mössbauer experiments.

In this article, a study of the structural and exchange interactions of Al substituted Co ferrites is presented by using X-ray, neutron diffraction (ND), Mössbauer spectroscopy and magnetization measurements.

2. Experiments

$CoAl_xFe_{2-x}O_4$ ($x = 0.1$, 0.2, 0.3, and 0.5) spinel powders are synthesized by using a sol–gel method. Appropriate portions of $Al(NO_3)_2 \cdot 9H_2O$, $Fe(NO_3)_3 \cdot 9H_2O$, and $Co(CH_3CO_2)_2 \cdot 4H_2O$ were dissolved in 2-methoxyethanol and diethanolamine ($HN(CH_2CH_2OH)_2$). The solution was added to acetic acid and refluxed at $70°C$ for 48 h and then dried at $120°C$ in a dry oven for 24 h. The dried powder was ground and annealed at $1000°C$ for 24 h in air.

The crystal structures of the samples were examined by X-ray diffractometer with Cu Kα radiation and neutron diffractometer at Korea Atomic Energy Research Institute HANARO HRPD (high resolution powder diffractometer, $\lambda = 1.8348$ Å) reactor. The Mössbauer spectra were recorded using a conventional spectrometer of the electromechanical type with a ^{57}Co source in a rhodium matrix. The magnetization and hysteresis curves were obtained using vibrating sample magnetometer (VSM).

3. Results and discussion

The X-ray (Cu-Kα radiation) patterns samples exhibited cubic spinel phase for all our cobalt–aluminum ferrite samples. The lattice parameters of the samples, $x = 0.1$, 0.2, 0.3, and 0.5 at room temperature, were found to be 8.3864(3), 8.3784(3), 8.3670(3), 8.3392(3) Å, respectively. In order to examine the crystallographic and magnetic structures in $CoAl_{0.1}Fe_{1.9}O_4$ and $CoAl_{0.5}Fe_{1.5}O_4$, we obtained neutron diffraction patterns from 10 K to Néel temperatures (T_N), which were 816 and 610 K, respectively. Figure 1(a) shows the neutron diffraction patterns of $CoAl_{0.1}Fe_{1.9}O_4$ at various temperatures from 10 to 816 K. Also, Figure 1(b) shows the neutron diffraction patterns of $CoAl_{0.5}Fe_{1.5}O_4$ at various temperatures from 295 to 610 K.

For all samples the crystal structure was determined to be cubic spinel of $Fd3m$ by Rietveld refinement of the Fullprof program. Figures 2(a) and (b) show examples of the results of refined neutron diffraction refinement patterns for $CoAl_{0.5}$-$Fe_{1.5}O_4$, at 295 and 610 K, respectively. The determined lattice constants, oxygen parameter u, Bragg factor R_B, structure factor R_F, magnetic factor R_M, and magnetic moments of atoms in $CoAl_{0.5}Fe_{1.5}O_4$ are listed in Table I.

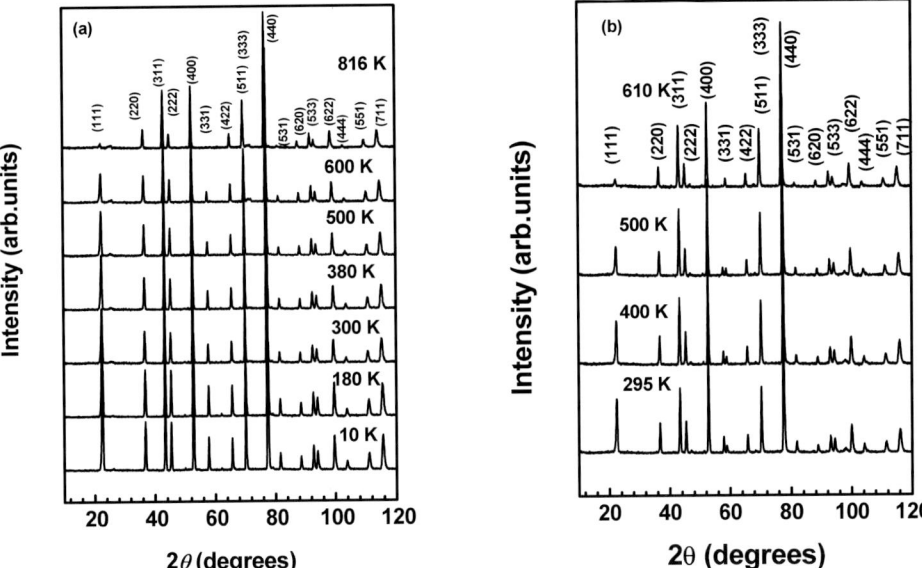

Figure 1. Neutron diffraction patterns of (a) $CoAl_{0.1}Fe_{1.9}O_4$ and (b) $CoAl_{0.5}Fe_{1.5}O_4$ at various temperatures.

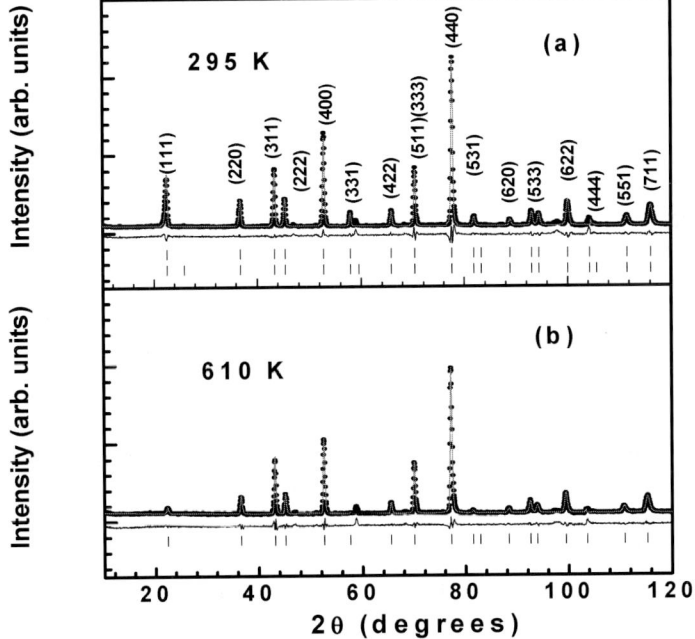

Figure 2. Refined neutron diffraction patterns of $CoAl_{0.5}Fe_{1.5}O_4$ (a) 295 K, (b) 610 K. Tick marks indicate the nuclear (upper) and magnetic (lower) reflections.

Table I. Results of refinement parameters of neutron diffraction on $CoAl_{0.5}Fe_{1.5}O_4$ [$Fd3m$; Fe(8a), Fe, Co, Al (16d), O(32e; (u, u, u))]. Magnetic moment of atoms, Bragg factor R_B, structure factor R_F, magnetic factor R_M are listed

	295 K	610 K
a_0 [Å]	8.3251(3)	8.3480(3)
u (S)	0.7419(3)	0.7420(3)
Fe (A) [μ_B]	−2.29	
Fe (B) [μ_B]	3.81	
Co (B) [μ_B]	2.66	
R_B [%]	2.19	3.30
R_F [%]	3.15	4.31
R_M [%]	3.62	

Figure 2(a) shows the diffraction patterns of nuclei in a magnetic phase, while Figure 2(b) shows the diffraction pattern of paramagnetic lattice of $CoAl_{0.1}Fe_{1.9}O_4$. For all the temperature range below T_N (610 K) as one lower the temperature, it was observed that the intensity of the inner Bragg reflections, namely, (111), (220), (222), and (331), increased significantly, denoting important magnetic contribution to these reflections. The diffraction peaks were found to remain very sharp indicating the presence of magnetic long-range order (LRO) in this sample, as can be seen in Figures 1(a) and (b). The observed magnetic moments of Fe^{3+}(A), Fe^{3+}(B) and Co^{2+}(B) of $CoAl_{0.1}Fe_{1.9}O_4$, have been plotted as a function of temperature in Figure 3. Similar procedures were carried for the sample $CoAl_{0.5}Fe_{1.5}O_4$. Figure 4 shows the temperature dependence magnetic moments of atoms for the sample $CoAl_{0.5}Fe_{1.5}O_4$.

Neutron diffraction for the sample $CoAl_{0.1}Fe_{1.9}O_4$ at 10 K revealed a cubic spinel structure of ferrimagnetic ordering, with magnetic moments of Fe^{3+}(A) (−4.18 μ_B), Fe^{3+}(B)(4.81 μ_B) and Co^{2+}(B)(2.98 μ_B), respectively. It is seen from these values that the magnetic moments of Fe^{3+}(B) and Co^{2+}(B) is close to free ion value, while the magnetic moment of Fe^{3+}(A) is substantially smaller than its estimated free-ion value.

The neutron diffraction results of $CoAl_{0.5}Fe_{1.5}O_4$ at various temperatures from 295 K and 610 K were obtained, too. As can be seen in Table I, the obtained magnetic moments of atoms at 295 K were Fe^{3+}(A)(−2.29 μ_B), Fe^{3+}(B)(3.81 μ_B) and Co^{2+}(B)(2.66 μ_B).

Temperature dependence of magnetization for $CoAl_xFe_{2-x}O_4$ ($x = 0.1, 0.2, 0.3,$ and 0.5) was observed by VSM and gave Néel temperature (T_N) to be 816, 765, 725, 610 K, respectively, in agreement with the following Mössbauer results.

Mössbauer spectra of $CoAl_xFe_{2-x}O_4$ ($x = 0.1, 0.2, 0.3,$ and 0.5) were taken at various absorber temperatures from 4.2 K to the T_N. Figures 5(a) and (b) il-

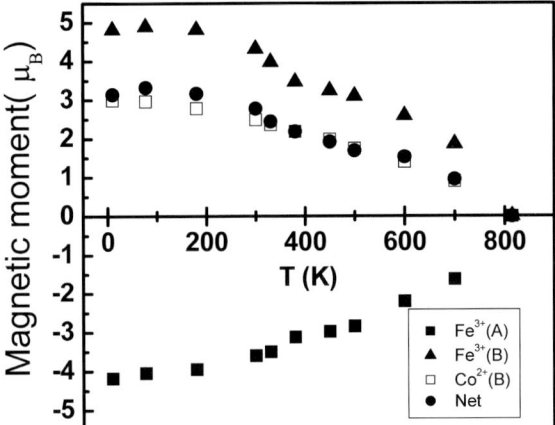

Figure 3. Observed magnetic moments of Fe^{3+}(A) and Fe^{3+}(B), Co^{2+}(B) and the net magnetic moments for $CoAl_{0.1}Fe_{1.9}O_4$.

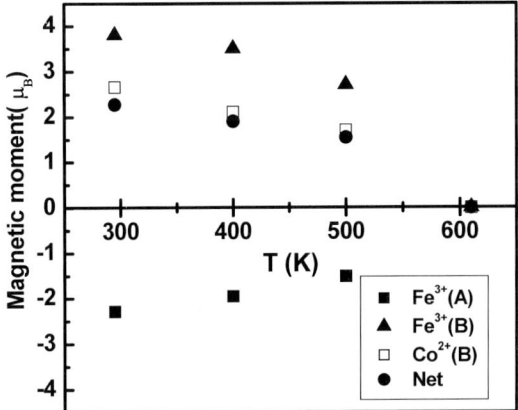

Figure 4. Observed magnetic moments of Fe^{3+}(A) and Fe^{3+}(B), Co^{2+}(B) and the net magnetic moments for $CoAl_{0.5}Fe_{1.5}O_4$.

lustrate some of the spectra of the samples $CoAl_{0.1}Fe_{1.9}O_4$ and $CoAl_{0.3}Fe_{1.7}O_4$, respectively, which were composed of two six-line hyperfine patterns A and B. The respective magnetic hyperfine fields of A and B at 4.2 K are 513 ± 2 and 546 ± 2 kOe in sample $x = 0.1$. The isomer shifts at room temperature for the A and B patterns were found to be 0.26 and 0.37 mm/s relative to the metallic iron, which were consistent with either highly covalent high spin Fe^{3+} or low spin Fe^{+3} [7]. But the magnitude of magnetic hyperfine fields at 4.2 K mentioned above exclude the second alternative. In addition to magnetic hyperfine field, one can consider that the electrons are in low spin state, on which the crystal filed is larger than the Hund coupling energy. Also, large crystal filed may distort the charge distributions; finally, one gets a large quadruple splitting. However, the obtained

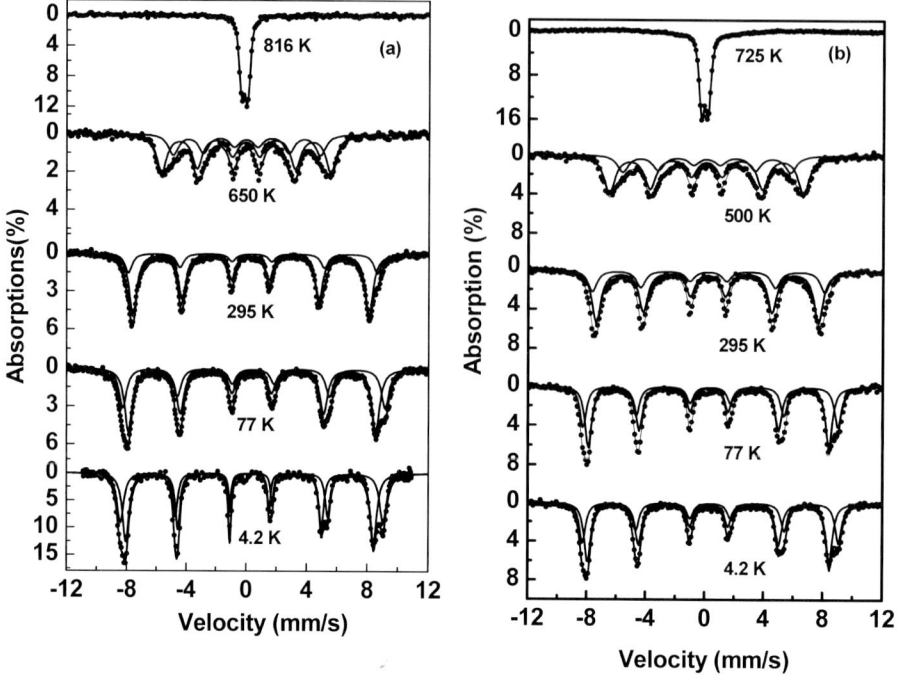

Figure 5. Mössbauer spectra of the (a) $CoAl_{0.1}Fe_{1.9}O_4$ and (b) $CoAl_{0.3}Fe_{1.7}O_4$ at various temperatures.

quadrupole shifts are nearly 0 mm/s, which means spherical or symmetrical charge distribution. Therefore, the possibility of low spin state is removed.

The possibility of the Fe^{2+} has been examined, too. If the iron ion is in Fe^{2+}, then the valence state of cobalt ion will be Co^{3+}, from the charge neutrality. However, generally, the hyperfine fields of most ferrous ion are known to have values less than 300 kOe. Furthermore, the electric quadrupole splitting of the ferrous (Fe^{2+}) ions are quite larger than the ferric (Fe^{3+}) ions due to orbital contribution. The electric quadrupole splitting are nearly zero in the whole temperature range in this sample. Hence, the possibility of Fe^{2+} is excluded, simultaneously, the possibility of Co^{3+} is discarded from the charge neutrality. The smaller value of isomer shift of the A site was due to a larger covalence at the A site. The ferric character of the Fe ions was also manifested by the magnitudes of magnetic hyperfine fields. The hyperfine field values of A and B patterns at room temperature in the $CoAl_{0.1}Fe_{1.9}O_4$ are found to be 487 ± 2 and 514 ± 2 kOe, respectively, which are typical values for Fe^{3+} ions. These assignments agree with De Grave group's results [8], which were carried under external magnetic field on samples of Ga doped Co ferrites. From the isomer shifts value, electric quadrupole splitting, and magnetic hyperfine fields we conclude that the iron ions in $CoAl_{0.1}Fe_{1.9}O_4$ are ferric.

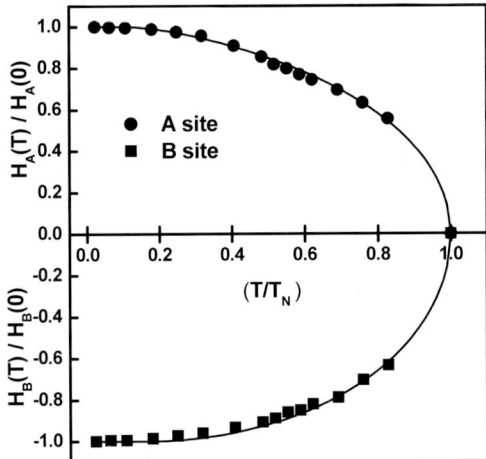

Figure 6. Reduced magnetic hyperfine fields $H(T)/H(0)$ for tetrahedral (A) and octahedral (B) sites for $CoAl_{0.3}Fe_{1.7}O_4$ as a function of reduced temperature T/T_N. The solid line represents calculated reduced magnetization.

Figure 6 shows the reduced magnetic hyperfine fields $H(T)/H(0)$ for the A and B sites of the sample $CoAl_{0.3}Fe_{1.7}O_4$ as a function of the reduced temperature $\tau = T/T_N$ ($T_N = 725$ K). Since both magnetic moment and magnetic hyperfine field of the ferric ion are proportional to its spin, the reduced magnetic hyperfine fields shown in Figure 6 should be equal to the reduced sublattice magnetization σ_A and σ_B. To extract information on the superexchange interactions from Figure 6, we applied the Néel theory of ferrites to the two sublattices of $CoAl_{0.3}Fe_{1.7}O_4$. The detailed theoretical procedure was published in elsewhere [9]. A good agreement between experimental data and theoretical values of σ_A and σ_B was obtained, and finally exchange parameters J_{A-B} and J_{A-A}, were calculated to be antiferromagnetic with the strengths of -20.2 ± 0.2 and -20.6 ± 0.2 k_B, respectively, J_{B-B} was ferromagnetic with the strength of 4.5 ± 0.2 k_B. Here k_B is the Boltzman constant. According to the previous Anderson and Kanamori approaches, the B–B interaction has two terms, one ferromagnetic and the other antiferromagnetic. Positive value of B–B interaction is due to the prevailing ferromagnetic contribution [10]. We notice that the strength of A–A superexchange interaction parameter is comparable to the strong A–B interaction. Also, increase of A–A interaction correlates well with related to the magnetic moments of individual atoms, which are $Fe^{3+}(A)(-2.29 \mu_B)$, $Fe^{3+}(B)(3.81 \mu_B)$, $Co^{2+}(B)(2.66 \mu_B)$, at 295 K, respectively. It is noticeable that magnetic moment of $Fe^{3+}(A)$ is fairly low compared with purely ionic moment of 5.0 μ_B. There are two important factors, bond length and angle, affecting to the magnetic moments. From Table I, the atomic bond lengths corresponding to $Fe^{3+}(A)-O^{2-}$ and $Fe^{3+}(B)-O^{2-}$ were calculated to be 1.91 and 2.03 Å, respectively. Therefore one can easily expect the larger covalence effects of the A site than that of the B sites. Simultaneously electron density of the A site

Table II. Values of superexchange interactions and Debye temperature for $CoAl_xFe_{2-x}O_4$ ($x = 0.1, 0.2, 0.3,$ and 0.5)

x	a_0 (Å)	T_N (K)	J_{A-B} (k_B)	J_{A-A} (k_B)	J_{B-B} (k_B)	Θ_A (K)	Θ_B (K)
0.1	8.3865	816	−23.3	−17.6	5.53	746	204
0.2	8.3784	758	−21.3	−19.6	4.8	709	197
0.3	8.3670	725	−20.2	−20.6	4.5	442	220
0.5	8.3392	610	−19.3	−21.6	3.8	421	207
	±0.0003	±2	±0.1	±0.1	±0.1	±5	±5

would be reduced compared to that of the B sites due to larger covalence and results in reduction of localized magnetic moment of the A site [12, 13].

Similar procedures were carried out for the other samples $CoAl_xFe_{2-x}O_4$ ($x = 0.1, 0.2, 0.3,$ and 0.5). Table II presents the calculated superexchange strength parameters, Néel temperatures, lattice constants, and Debye temperatures. The strength of superexcahge interactions non Al doped $CoFe_2O_4$ were published elsewhere [14]. It is noticeable that the A–B superexchange interaction is antiferromagnetic, and its strength is larger than that of A–A and B–B interaction in the sample $x = 0.1$. However the A–A interaction is comparable to the A–B interaction. We note that the A–B and B–B interaction decrease while A–A interaction increases, with increase of aluminum substitution. This is related to the cation distribution.

According to the results of the X-ray and neutron diffraction refinement, the final cation distribution of $CoAl_xFe_{2-x}O_4$ ($x = 0.1, 0.2, 0.3,$ and 0.5) is $[Fe^{3+}]_A[Co^{2+}Al_x^{3+}Fe_{1-x}^{3+}]_BO_4$. In the refinement process, the neutrality of total charge, including oxygen, has been assumed. The best fitting is obtained, when Al^{3+} and Co^{2+} ions enter into the B sites. Also, if we consider the both iron ions in A and B sites are in Fe^{3+}, one leads to the conclusion that the possibility of Co^{3+} is ruled out. This cation distribution accords with the area ratio of the Mössbauer absorption spectra, too.

Now, we interpret that, implicitly, Al ions weaken A–B intersublattice interaction and this corresponds to the decreasing of magnetic ordering temperature T_N as shown in Table II. Since Al ions are substituted in B sites, a decreasing tendency of B–B interactions is easily understood. Now, how can we interpret increasing tendency of A–A interaction? According to our results, the main interaction of $Fe^{3+}(A)–O^{-2}–(Co^{2+}, Fe^{3+})(B)$ is decreased with increasing Al concentrations and finally it results in a decrease of covalence between A and B sites of iron ions via oxygen $2p$ orbital. Then the electrons of A site, which are participating in A–B interaction, would decrease. Simultaneously, these electrons at A site, which are participating in A–B interaction, would have more chances to participate in A–A interaction. Therefore an increasing tendency of A–A interaction is interpreted well.

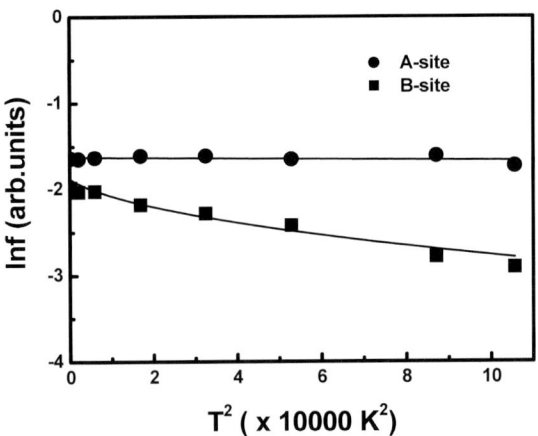

Figure 7. Natural logarithm of the absorption area, f, vs T for the A and B subspectra of $CoAl_{0.1}Fe_{1.9}O_4$. The solid lines were calculated from Equation (2).

Figure 7 shows the temperature dependence of the Mössbauer absorption area for the $CoAl_{0.1}Fe_{1.9}O_4$. The Debye model gives the following expression for the recoil-free fraction [15]:

$$f = \exp\left[-\frac{3E_R}{2k_B\Theta}\left(1 + \frac{4T^2}{\Theta^2}\int_0^{\Theta/T}\frac{x\,dx}{e^x - 1}\right)\right], \tag{1}$$

where E_R is the recoil energy of ^{57}Fe for the 14.4 keV gamma ray, and Θ represents Debye temperatures. Taking the logarithm of both sides of Equation (1), one obtains

$$\ln f = -\frac{3E_R}{2k_B\Theta}\left(1 + \frac{4T^2}{\Theta^2}\int_0^{\Theta/T}\frac{x\,dx}{e^x - 1}\right). \tag{2}$$

When $\ln f$ is plotted as a function of T^2, one obtains a curve, which becomes almost linear at low temperatures. Equation (2) with a proper additive constant was fitted to the data in Figure 2 using a least-squares computer program to determine a Debye temperature. The Debye temperature can be calculated from the temperature dependence of the resonant absorption area at low temperatures. The calculated Debye temperatures A and B sites are $\Theta_A = 746 \pm 5$ K and $\Theta_B = 204 \pm 5$ K, respectively. With the similar procedure for the other samples, Debye temperatures Θ_A and Θ_B were also calculated and are listed in Table II.

In summary, we have studied magnetic properties and interaction mechanism in $CoAl_xFe_{2-x}O_4$ ($x = 0.1, 0.2, 0.3$, and 0.5). Neutron diffraction for $CoAl_{0.1}Fe_{1.9}O_4$ and $CoAl_{0.5}Fe_{1.5}O_4$ revealed a cubic spinel structure of ferrimagnetic long range ordering. The weakening of magnetic moment in A site compared to B site is closely related to covalence of A site. The temperature dependence of the magnetic hyperfine field in ^{57}Fe nuclei at the tetrahedral (A) and octahedral (B) sites was analyzed based on the Néel theory of magnetism.

Acknowledgement

This Research was performed for the Hydrogen Energy R&D Center, one of the 21st Century Frontier R&D Program, funded by the Ministry of Science and Technology of Korea.

References

1. Bouet, L., Tailhades, P. and Rousset, A., *J. Magn. Magn. Mater.* **153** (1996), 389.
2. Hochpied, J. F. and Pieleni, M. P., *J. Appl. Phys.* **87** (2000), 2472.
3. Kim, W. C., Kim, S. J., Uhm, U. R. and Kim, C. S., *IEEE. Trans. Mag.* **37** (2001), 2362.
4. Zhou, B., Zhang, Y. W., Liao, C. S., Yu, Y. J., Yan, C. H., Chen, L. Y. and Wang, S. Y., *Appl. Phys. Lett.* **82** (2003), 1188.
5. Srivastava, C. M., Sirinivasan, G. and Nanadikar, N. G., *Phys. Rev. B* **19** (1979), 499.
6. Dionne, G. F., *J. Appl. Phys.* **63** (1988), 3777.
7. Baek, K. S., Hahn, E. J. and Ok, H. N., *Phys. Rev. B* **36** (1987), 763.
8. de Bakker, P. M. A., Vandenberghe, R. E. and De Grave, E., *Hyp. Interact.* **94** (1994), 2023.
9. Kim, S. J., Lee, S. H. and Kim, C. S., *Jpn. J. Appl. Phys.* **40** (2001), 4897.
10. Srivastava, C. M., Sirinivasan, G. and Nanadikar, N. G., *Phys. Rev. B* **19** (1979), 499.
11. Chakravarthy, R., Madhav Rao, L., Paranjpe, S. K., Kulshrestha, S. K. and Roy, S. B., *Phys. Rev. B* **43** (1991), 6031.
12. Palmer, H. M. and Greaves, C., *Physica B* **276–278** (2000), 568.
13. Bronger, W. and Muller, P., *J. Alloys. Comp.* **246** (1997), 27.
14. Kim, S. J., Lee, S. W., An, S. Y. and Kim, C. S., *J. Magn. Magn. Mater.* **215–216** (2000), 210.
15. Mössbauer, R. L. and Widermann, W. H., *Z. Phys.* **159** (1962), 33.

Hyperfine Interactions **156/157**: 123–127, 2004.
© 2004 *Kluwer Academic Publishers. Printed in the Netherlands.*

Mössbauer Studies of $Fe_{0.7-x}Si_{0.3}Mn_x$ Alloys

I. A. AL-OMARI*, A. GISMELSEED, A. RAIS, H. M. WIDATALLAH,
A. AL RAWAS, M. ELZAIN and A. A. YOUSIF
*Department of Physics, P.O. Box 36, Sultan Qaboos University, PC 123, Muscat,
Sultanate of Oman; e-mail: ialomari@yahoo.com*

Abstract. In this work we present Fe^{57} Mössbauer study for the alloy system $Fe_{0.7-x}Si_{0.3}Mn_x$, where $0 \leqslant x \leqslant 0.3$. Mössbauer spectroscopic results show that all the samples studied are magnetically ordered at 77 K, and at room temperature, except for $x = 0.3$ at 300 K where it shows paramagnetic behavior. The average magnetic hyperfine field is found to decrease with increasing the manganese concentration at 77 K and 300 K. The average magnetic hyperfine field is found to increase with decreasing the temperature from 300 K to 77 K for all samples under investigation. The average isomer shift is found to decrease with increasing the manganese concentration.

Key words: Mössbauer effect, hyperfine field, isomer shift.

1. Introduction

Materials based on Fe_3Si and Fe_3Al alloy systems are of great interest, because of their high temperature strength, excellent oxidation and corrosion resistance. Substitution of Fe by transition metal element affect the magnetic properties, the lattice parameter, and the structural ordering of these compounds [1–3]. Neutron diffraction study for $Fe_{3-x}Mn_xSi$ by Yoon and Booth [2] showed that Mn preferentially occupies the B site of the DO_3-type structure for $x < 0.75$, and then starts to occupy the A and C sites for higher Mn concentrations. They also found that this system exhibits ferromagnetic behavior for $x < 0.75$ and a complex magnetic behavior evolves for $x > 0.75$. Substituting Fe by Mn was found to decrease the magnetic moment, interaction energy, and Curie temperature with increasing the Mn concentration. Waliszewski *et al.* [3] studied the $Fe_{3-x}Cr_xSi$ system and found that Curie temperature decreases from 840 K for $x = 0$ to 712 K for $x = 0.4$, and the magnetic moment of iron at the B site to be 2.44 μ_B, and about 1.18 μ_B for (A, C) sites, while the magnetic moments of Cr were determined to be (2.03 ± 1.3) μ_B for Cr at the B site and (0.41 ± 0.63) μ_B for the (A, C) sites with orientation antiparallel to the magnetic moments of Fe.

The structural and magnetic properties of the intermetallic Fe–Si alloys depend on the Si concentration. $FeSi_2$ was found to form the tetragonal type structure, FeSi forms the B_2O type structure, while Fe_3Si forms the DO_3 type cubic structure.

---

* Author for correspondence.

Fe_3Si is a well ordered ferromagnetic alloy with four sites; 2-equivanlt sites (A, C) occupied by Fe and the other two sites are occupied by Fe (B site) and Si (D site) [4, 5]. Mössbauer and NMR [6, 7] investigations for this system have yielded information about the hyperfine field and Si site occupation. These studies showed that the hyperfine field and the magnetic moment at the Fe sites strongly depend on the Si occupancy and the number of Fe nearest neighbors and they decrease with increasing the Si concentration. The aim of this work is to study the effects of substituting Fe by Mn on the magnetic and structural properties of the alloy system $Fe_{0.7-x}Si_{0.3}Mn_x$.

2. Experimental methods

The bulk samples of $Fe_{0.7-x}Si_{0.3}Mn_x$ ($x = 0$, 0.05, 0.10, 0.15, 0.2 and 0.3) were prepared from at least 99.99% pure elements by arc melting under a flowing argon atmosphere. Each sample was melted four to five times to insure homogeneity. Mössbauer spectra were collected using a standard constant acceleration Mössbauer spectrometer over 1024 channels. The samples for Mössbauer studies were circular disks of diameter 1.3 cm prepared by sprinkling a thin layer of the finely powdered alloys on a piece of scotch tape. The γ-ray source was a 50 mCi Co^{57} in a palladium matrix. Isomer shifts were measured relative to the centroid of the α-iron spectrum at room temperature, and α-iron spectrum was also used for calibration.

3. Results and discussion

Figures 1 and 2 show the, room temperature and 77 K, Mössbauer spectra (dots) for the alloy system $Fe_{0.7-x}Si_{0.3}Mn_x$, and the fitting is represented by the solid curves. It is clear from these figures that the spectra of all the samples show a magnetically split component, and a central paramagnetic line that starts to appear at $x = 0.15$ at room temperature and at $x = 0.3$ at 77 K. As x increases, the paramagnetic line becomes more intense. The spectra are fitted with a distribution of magnetic hyperfine fields $P(H)$, in which the fields are linearly correlated with the isomer shift, and the results are also shown in Figures 1 and 2. The distribution of the magnetic hyperfine filed suggests a disorder alloys, which indicates the presence of iron atoms with different configurations for the surrounding environment. The different broad peaks in the hyperfine filed distribution are associated with the Fe sites with different numbers of Si and/or Mn nearest environment, where the hyperfine field decreases with the increase of number of nonmagnetic atoms surrounding the iron atoms. The average hyperfine field is found to decrease with increasing the Mn concentration at room temperature and at 77 K, as shown in Figure 3. This decrease is due to the replacement of magnetic Fe by non-magnetic Mn and can be attributed to the reduction of the interatomic exchange interaction, and hence that this reduction drops the Curie temperature and has the effect of depressing the saturation magnetization as well as the magnetic hyperfine field. Another reason

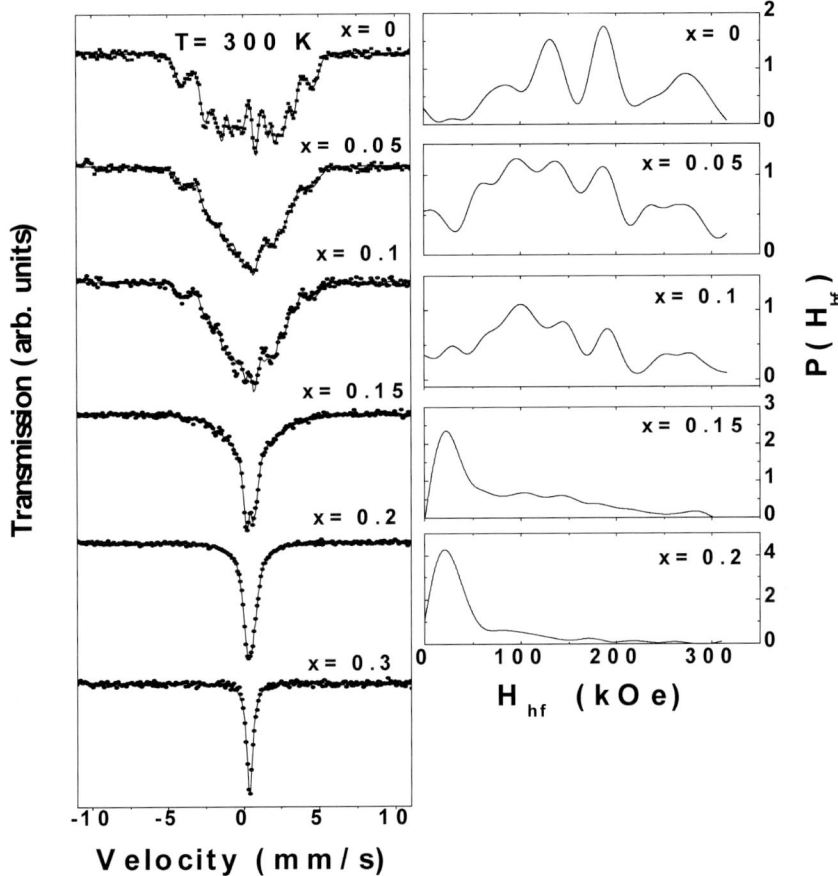

Figure 1. Mössbauer spectra, at room temperature, of $Fe_{0.7-x}Si_{0.3}Mn_x$ alloys (the solid curves represent the fitting), and the probability hyperfine field distribution for the different alloys.

for the decrease is when Fe is replaced by Mn, the manganese surrounding the iron will hybridize the iron's 3d, 4s, and 4p atomic orbits and broaden their energy levels into energy bands. The increase in the average hyperfine field with decreasing the temperature can be attributed to the increase in the interatomic exchange interaction and the enhancement of the magnetic moments with decreasing the temperature below Curie temperature.

The average isomer shift is found to be positive and to decrease with increasing the manganese concentration from 0.30 mm/s for $Fe_{0.7}Si_{0.3}$ to 0.19 mm/s for $Fe_{0.4}Si_{0.3}Mn_{0.3}$ at room temperature, and from 0.38 mm/s for $Fe_{0.7}Si_{0.3}$ to 0.27 mm/s for $Fe_{0.4}Si_{0.3}Mn_{0.3}$ at 77 K. Using the average value of the magnetic hyperfine field for $Fe_{0.7}Si_{0.3}$ and the correlation parameters between the isomer shift and the hyperfine field from Ref. [8] for $Fe_{0.76}Si_{0.24}$, we calculated a value of 0.31 mm/s for the average isomer shift at room temperature for $Fe_{0.7}Si_{0.3}$. This indicates that our value of the average isomer shift for $Fe_{0.7}Si_{0.3}$ is in good agree-

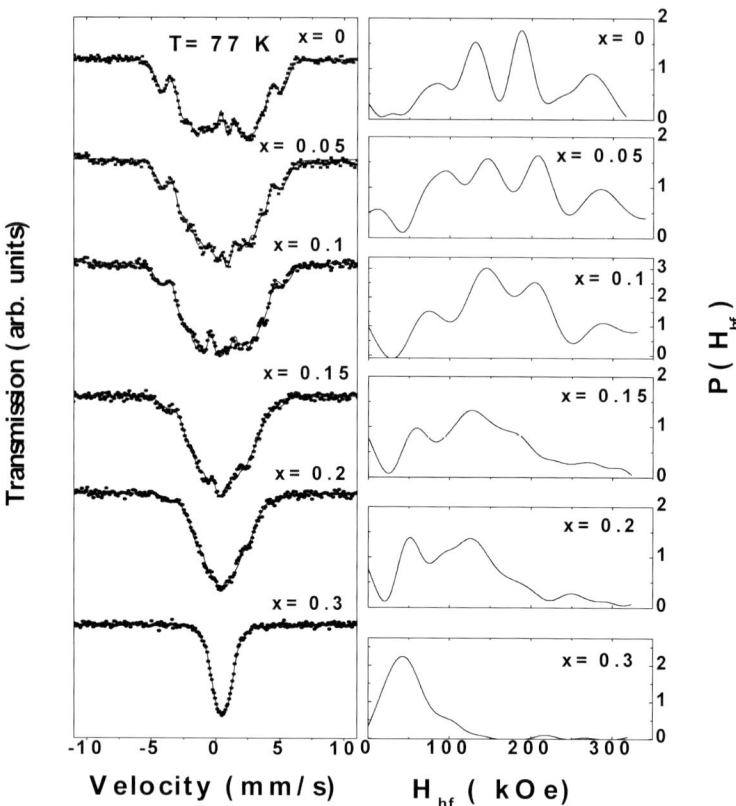

Figure 2. Mössbauer spectra, at 77 K, of $Fe_{0.7-x}Si_{0.3}Mn_x$ alloys (the solid curves represent the fitting), and the probability hyperfine field distribution for the different alloys.

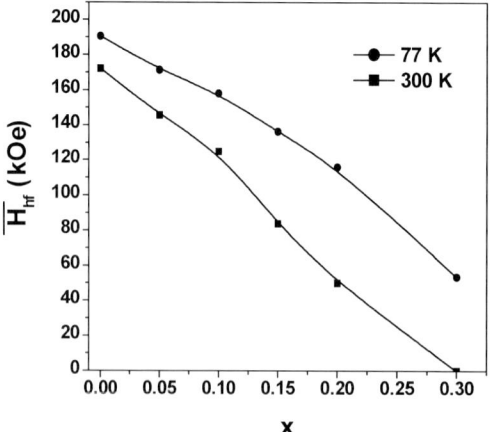

Figure 3. Dependence of the average hyperfine field on the Mn concentration (x) for $Fe_{0.7-x}Si_{0.3}Mn_x$ alloys at room temperature and 77 K.

ment with the expected value. The positive isomer shift in Fe–Si alloys is due to the screening of the 4s electrons in Fe as some of the s, p electrons partially transfer from Si to the iron's 3d band. The decrease of the average isomer shift with increasing the Mn concentration can be attributed to the transfer of electrons from the Mn atoms to the Fe atoms during the substitution. This behavior is in agreement with previous observations for similar compounds [9, 10]. The isomer shift for all samples at 77 K is larger than that at room temperature by 0.08 mm/s. This difference is equal to the second order Doppler shift (δ_R). This value of 0.08 mm/s is close to the value of 0.11 mm/s for $(Fe_{0.67}V_{0.33})_3Al$ by Bara et al. [11]. The difference between these two values can be attributed to the difference in Debye temperature (θ_D) for Al, Mn, and Si.

4. Conclusions

Mössbauer spectroscopic results of the alloy system $Fe_{0.7-x}Si_{0.3}Mn_x$ show that all the samples studied are magnetically ordered at 77 K, and at room temperature, except for $x = 0.3$ at 300 K where it shows paramagnetic behavior. The average magnetic hyperfine field is found to decrease with increasing the Manganese concentration at 77 K and 300 K. The average magnetic hyperfine field is found to increase with decreasing the temperature from 300 K to 77 K for all samples under investigation. The average isomer shift is found to decrease with increasing the manganese concentration. The above results indicate that Mn dissolves in the bcc lattice in this system resulting in the collapse of the hyperfine field with increasing the Manganese concentration.

References

1. Ishikawa, K., Kainuma, R., Ohnuma, I., Aoki, K. and Ishida, K., *Acta Materialia* **50** (2002), 2233.
2. Yoon, S. and Booth, J., *J. Phys. F* **7** (1997), 1079.
3. Waliszewski, J., Dobrzynski, L., Malinowski, A., Satula, D., Szmanski, K., Prandl, W., Brückel, Th. and Schärpf, O., *J. Magn. Magn. Mater.* **132** (1994), 349.
4. Niculescu, V. A., Burch, T. J. and Budnick, J. I., *J. Magn. Magn. Mater.* **39** (1983), 223.
5. Hansen, M., *Constitution of Binary Alloys*, McGraw-Hill, New York, 1958.
6. Stearns, M. B., *Phys. Rev.* **129** (1963), 1336.
7. Budnick, J. J., Skalski, S., Burch, T. J. and Wernick, J. H., *J. Appl. Phys.* **38** (1967), 1137.
8. Flutz, B., Gao, Z.-Q., Hamdeh, H. H. and Oliver, S. A., *Phys. Rev. B* **49** (1994), 6312.
9. Shobaki, J., Al-Omari, I. A., Hassan, M. K., Albiss, B. A., Azez, K. A., Al-Akhras, M.-Ali, Hamdeh, H. H. and Mahmood, S. H., *J. Magn. Magn. Mater.* **213** (2000), 51.
10. Dubiel, S. M. and Zinn, W., *J. Magn. Magn. Mater.* **45** (1984), 298.
11. Bara, J. J. et al., *J. Magn. Magn. Mater.* **59** (1986), 208.

Hyperfine Interactions **156/157**: 129–135, 2004.
© 2004 *Kluwer Academic Publishers. Printed in the Netherlands.*

Observation of Substitutional Fe in CEMS Measurements on Synthetic CVD Diamond

K. BHARUTH-RAM[1], J. E. BUTLER[2], D. NAIDOO[1,3] and
G. KLINGELHOFER[4]
[1] *School of Physics, University of KwaZulu-Natal, Durban, 4041, South Africa*
[2] *Naval Research Laboratory, Washington, DC 20375, USA*
[3] *Physics Department, University of the Witwatersrand, WITS 2050, Johannesburg, South Africa*
[4] *Institut für Anorganische Chemie, Universität Mainz, D-55099 Mainz, Germany*

Abstract. Conversion electron Mössbauer spectroscopy measurements have been made on a diamond sample synthesized by chemical vapour deposition. The sample was implanted with 70 keV
^{57}Fe to a dose of 5×10^{14} cm^{-2} and Mössbauer measurements were made on the as-implanted
sample and after annealing at temperatures of 600 K, 950 K and 1470 K. The spectra at the lower
temperatures were characterized by broad doublets, but the annealing at 1470 K resulted in dramatic
decrease in the intensity of the doublet components, and the appearance of a strong single line with
an isomer shift of $\delta = -0.90(5)$ mm/s and areal intensity of 30%, and a weaker line (5%) with
$\delta = +0.07(4)$. Arguments are presented to attribute the single lines to substitutional and interstitial
Fe, respectively. In contrast, identical measurements made on diamonds synthesized in the high
temperature high pressure process showed little evidence of any strong single line component.

Key words: CVD diamond, ion implantation, Mössbauer spectroscopy.

1. Introduction

Properties of diamond such as its wide band gap (5.4 eV), high thermal conductivity, high mobility for both p- and n-type carriers, and extreme hardness, make it an
ideal base for semiconducting devices for specific applications. This has prompted
much research on the production of n- and p-type semiconducting diamond layers
with incorporation of foreign atoms. While p-type semiconducting diamonds exist
in nature and have been achieved in the laboratory with boron incorporation during
growth, the production of n-type diamond layers has proved more problematic.
The high formation energy of most of the potential dopants in diamond rules out
thermal diffusion and favours ion implantation as a means of incorporating these
atoms in diamond. Ion implantation allows the introduction of individual dopant
atoms in the lattice in a very controlled dilute way, with implantation profiles
tailored to particular needs. However, the energy spikes that accompany the penetration of each ion in the lattice invariably result in lattice damage around the
implanted ions and the formation of implanted ion–defect complexes. Thermal
treatment of the implanted samples is required to heal the lattice and reduce the

concentration of such defects. Mössbauer spectroscopy (MS) offers a very sensitive probe with which to study the immediate neighbourhood of the implanted atoms and the probe–defect complexes that may be formed, and follow their behaviour with thermal treatment.

The most significant early Mössbauer studies on diamond were those of Sawicka et al., who implanted 50 keV ^{57}Co into natural diamond crystals [1–4]. Measurements made at temperatures between 4.2 K and 1100 K showed strong annealing effects at the higher temperatures where the resonance spectra were resolved into a broad symmetric doublet and a single line. The single line had an isomer shift of $-0.89(5)$ mm/s and a fractional intensity, after annealing above 800 K, of $f = 20\%$. These authors also found that implantation at $T \geqslant 800$ K or implantation at low temperature followed by annealing above 800 K gave similar results. The investigations of de Potter and Langouche, on lattice damage brought on by the implantation process, found that transformation from highly damaged, but recrystallizable, to amorphous diamond occurs suddenly at an implantation dose between 10^{14} and 10^{15} atoms/cm^2 [5, 6].

In recent years, MS studies on Fe in diamond have been pursued using a variety of techniques. These include in-beam Mössbauer Spectroscopy (IBMS) [7], Conversion Electron Mössbauer Spectroscopy (CEMS) on synthetic high temperature high pressure synthesized diamonds [8], and investigations following ^{57}Mn implantation [9, 10]. These measurements, as well as those of Sawicka et al., have shed significant light on the components seen in Mössbauer spectra of ^{57}Fe in diamond. The single line with $\delta = -0.92(5)$ mm/s, clearly due to Fe at a high symmetry site in the diamond lattice, has been identified with substitutional Fe and one with $\delta = +0.15(5)$ mm/s with interstitial Fe.

Recent studies report of considerable progress towards the achievement of diamond based electronic devices with appropriately doped synthetic diamonds [11–14]. The present contribution seeks to add to this endeavour, by presenting the results of CEMS measurements made on a diamond sample synthesized by chemical vapour deposition (CVD) and ion implanted with ^{57}Fe. The results are compared with those of identical measurements made on a high pressure high temperature (HTHP) synthesized diamond [8].

2. Experimental details

The CVD diamond used in these measurements was produced by high power microwave deposition on a silicon substrate. The substrate was removed after growth. The nitrogen concentration in the diamond was <5 ppm. The average crystallite size in the polycrystalline sample was ≈100 μm, which is several orders of magnitude larger than the estimated mean implantation range (310 Å) and straggle (90 Å) of the ^{57}Fe ions. The ^{57}Fe probes, except for those implanted at the grain boundaries, were therefore sampling essentially single crystal environments.

The CVD diamond was implanted with 70 keV [57]Fe ions to a dose of 5×10^{14} cm^{-2}. The CEMS measurements were made on the as-implanted sample and after annealing the diamond for 10 min in vacuum at temperatures, T_A, of 600 K, 950 K and 1470 K. The measurements were made at room temperature using a parallel plate avalanche detector.

3. Results and analysis

Figure 1(a) show the CEMS spectra obtained from the as-implanted CVD diamond and after annealing the diamond at the temperatures indicated. The spectra at the lower temperatures are characterized by broad doublets, but displaying structure evident of the presence of several components. The broad spectral components could be fitted with combinations of different quadrupole doublets. The doublet components, labelled D1 and D2, shown in Figure 1(a) gave consistent fits to the CVD data as well as to data acquired in similar measurements on an HTHP synthesized diamond [8]. Two single lines with isomer shifts close to those observed in earlier measurements and attributed to interstitial and substitutional Fe were included in the analysis. The spectral parameters of the different components are listed in Table I, and compared with those of components observed in identical measurements on HTHP synthesized diamonds [8]. In the CVD diamond, annealing at 1470 K resulted in dramatic changes in the spectrum: the intensity of doublet D2 decreased markedly to reveal a strong single line with isomer shift $\delta = -0.90(5)$ mm/s, and a second single line with $\delta = 0.07(4)$ mm/s also became evident. A puzzling observation was the large reduction in the resonance absorption effect, indicating that a significant fraction of the implanted Fe, presumably in graphitized complexes close to the grain surface, had diffused out of the sample.

The Mössbauer parameters of the different components, determined from fits o the data, are listed in Table I, and the areal intensities are presented in Figure 2 as a function of annealing temperature. The isomer shifts are expressed as absorber shifts relative to α-Fe. The spectral parameters of the single lines compare well with previous measurements. The isomer shift of S1 ($\delta = -0.90(5)$ mm/s) is in very good agreement with that of the singlet component first observed by De Waard and his co-workers [1–4], and also observed in recent MS measurements following [57]Mn implantation in natural type IIa diamond [9, 10], while that of S2 is in good agreement with the single line observed in IBMS measurements [7].

4. Discussion

The discussion focuses on the two singlet components, S1 and S2, observed in the CEMS measurements on the CVD diamond, and a comparison of the CVD and HTHP diamond data.

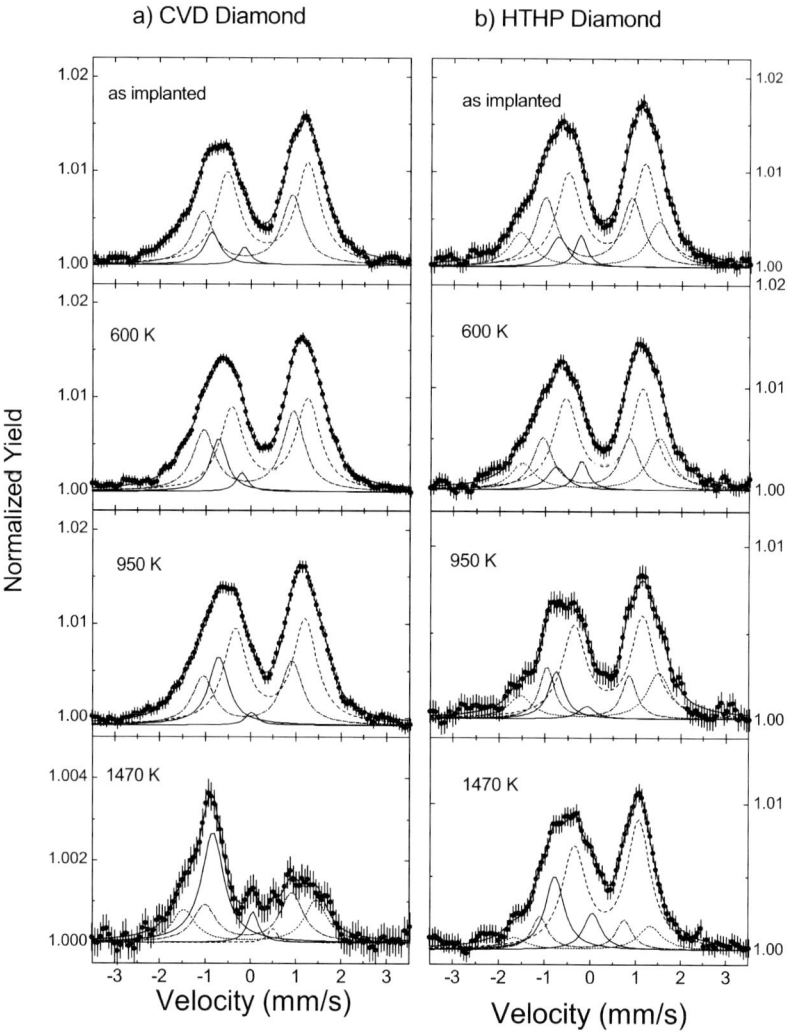

Figure 1. CEMS spectra of (a) CVD diamond and (b) HTHP diamond, implanted with ^{57}Fe and annealed at temperatures indicated.

The isomer shifts and the symmetry of the two lines allow some deductions about the sites occupied by the Fe probe nuclei. The isomer shift is directly related to the total s-electron density at the ^{57}Fe nucleus. A large negative isomer shift therefore implies an increased s-electron density at the Fe nucleus, and therefore a large compression of the s-electron orbitals. The absence of any quadrupole splitting in S1 and S2 shows that these spectral components are due to Fe at sites of tetrahedral symmetry. There are two such sites in the diamond lattice, substitutional or tetrahedral interstitial (T_d). Fe has eight electrons outside its closed $3s$ and $3p$ sub-shells. At a substitutional site, the Fe atom forms four covalent bonds with the neighbouring carbon atoms, leaving the unpaired electrons in configuration $3d^4$,

Table I. Isomer shifts (δ), quadrupole splittings (ΔE_Q), and linewidths (Γ) of the components observed in CEMS measurements on a CVD diamond; for comparison, δ and ΔE_Q values observed in CEMS measurements on synthetic HTHP diamonds are also listed

| | CEMS: CVD diamond | | | CEMS: HTHP diamond | |
| | δ | ΔE_Q | Γ (FWHM) | δ | ΔE_Q |
	(mm/s)	(mm/s)	(mm/s)	(mm/s)	(mm/s)
S1	$-0.90(5)$	–	0.60	$-0.81(5)$	–
S2	$+0.07(5)$	–	0.40	$+0.03(5)$	–
D1	$-0.05(5)$	2.05(5)	0.70	$-0.04(5)$	1.92(5)
D2	$+0.40(5)$	1.79(5)	0.70	$+0.37(5)$	1.68(5)
D3	–			$-0.03(5)$	3.15(5)

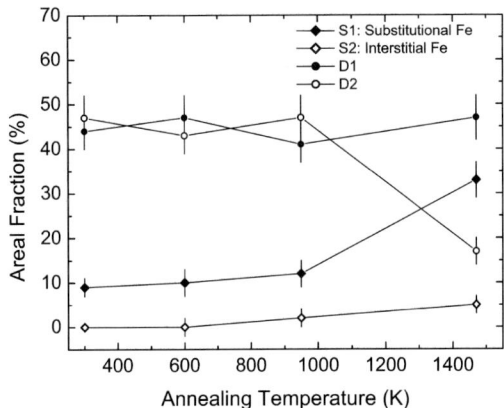

Figure 2. Areal fractions of components required to fit the CEMS spectra of the CVD diamond, as a function of annealing temperature.

while at the T_d site no covalent bonds are formed and the configuration of the valence electrons would be $3d^8$. The small volume available to the Fe atoms in the tight diamond lattice is expected to lead to a large compression of the s-electron shells of the Fe atom. At the T_d site the eight $3d$ electrons offer a strong measure of shielding to the inner s electrons, and should lead to a smaller compressions of the inner s orbitals and to a correspondingly smaller isomer shift. For Fe atoms at substitutional sites the screening by the $3d$ electrons is less effective (since there are only four $3d$ electrons), and the strain fields are expected to lead to a large compression of the s-electron shells, and a correspondingly large negative isomer shift. Our results are hence consistent with attributing component S1, with $\delta = -0.90(5)$ mm/s, to substitutional Fe, and component S2 ($\delta = 0.07(4)$ mm/s) with interstitial Fe.

We now compare our results for the CVD diamond with identical measurements on HTHP synthesized diamonds. The spectra for the HTHP diamond, displayed in Figure 1(b), show little change with annealing temperature, and no clear single line component is evident even after the annealing at 1470 K. The HTHP synthesized diamonds are of type Ib, containing typically 100 ppm nitrogen at single substitutional sites in the lattice. The substitutional N could present trapping centres for the Fe ions and result in the formation of Fe–N complexes. The extra doublet component required to fit the wings of the resonance structure in Figure 1(b) may be due to such complexes.

5. Conclusions

The lattice sites taken up by ion-implanted Fe in CVD diamond has been investigated in CEMS measurements following ^{57}Fe implantation and annealing up to 1470 K. Our results show that the incorporation of significant fractions of Fe at high symmetry substitutional sites in CVD synthesized diamond can be achieved with implantation at room temperature and annealing at $T \geqslant 1470$ K. These results are in agreement with emission channelling measurements in natural type IIa diamond, which show the incorporation of 45(5)% Fe atoms close to substitutional sites [15]. In contrast, little evidence of a high symmetry component is seen in CEMS measurements on HTHP synthesized diamonds. Here nitrogen in single substitutional sites presumably provide trapping centres for the Fe and lead to the formation of Fe–N complexes.

Acknowledgement

Support of the National Research Foundation (South Africa) under grant GUN No. 2064730 is acknowledged.

References

1. Sawicka, B. D., Sawicki, J. A. and De Waard, H., *Phys. Lett.* **85A** (1981), 303.
2. Sawicka, J. A. and Sawicka, B. D., *Nucl. Instrum. Meth.* **194** (1982), 465.
3. Sawicka, J. A., Sawicki, B. D. and De Waard, H., *Hyp. Interact.* **15/16** (1983), 483.
4. Sawicka, J. A. and Sawicki, B. D., *Nucl. Instrum. Meth. B* **46** (1990), 38.
5. de Potter, M. and Langouche, G., *Hyp. Interact.* **15/16** (1983), 479.
6. de Potter, M. and Langouche, G., *Z. Phys. B* **53** (1983), 89.
7. Bharuth-Ram, K., Hartick, M., Kankeleit, E., Dorn, C., Held, P., Sielemann, R., Wende, L. and Sellschop, J. P. F., *Phys. Rev. B* **58** (1998), 8955.
8. Bharuth-Ram, K., Naidoo, D., Klingelhofer, G. and Butler, J. E., *Hyp. Interact. C* **5** (2002), 111.
9. Weyer, G., Gunnlaugsson, H. P., Dietrich, M., Fynbo, H. and Bharuth-Ram, K., *Proc. 10th International Conference on Defect Recognition, Imaging and Physics of Semiconductors (DRIPX)*, Loire Atlantique, France, Sept. 3, 2003.
10. Bharuth-Ram, K., Gunnlaugsson, H. P., Weyer, G., Dietrich, M., Naidoo, D., Mantova, R. and Sielemann, R., *J. Appl. Phys.* (to be published).

11. Aleksov, A., Denisenko, A., Kunze, M., Vescan, A., Bergmeier, A., Dollinger, C., Ebert, W. and Kohn, E., *Semiconductor Sci. Tech.* **18** (2003), S59.
12. Koizumi, S., Watanabe, K., Hasegawa, M. and Kanda, H., *Science* **292** (2001), 1899.
13. Butler, J. E., Gries, W., Krohn, K. E., Lawless, J., Lyszczarc, T. M., Fletchner, D. and Wright, R., *Semiconductor Sci. Tech.* **18** (2003), S67.
14. Isberg, J., Hammersberg, J., Johanesson, E., Wikström, T., Twitchen, D. J., Whitehead, A. J., Coe, S. E. and Scarsbrook, G. A., *Science* **297** (2003), 1670.
15. Bharuth-Ram, K., Wahl, U. and Correia, G., *Nucl. Instrum. Meth. B* **206** (2003), 941.

Hyperfine Interactions **156/157**: 137–142, 2004.
© 2004 *Kluwer Academic Publishers. Printed in the Netherlands.*

Interaction between Interstitial Hydrogen and Fe Atoms within the β-Hydride Phase in $\text{Nb}_{1-y}\text{Fe}_y\text{H}_x$ Alloys

B. BRZESKA-MICHALAK and A. OSTRASZ

Institute of Experimental Physics, University of Wrocław, Maksa Borna Sq. 9, PL-50-204 Wrocław, Poland

Abstract. The influence of hydrogen on the electronic and structural properties of $\text{Nb}_{1-y}\text{Fe}_y$ alloys has been studied with ^{57}Fe Mössbauer spectroscopy and X-ray diffraction. The results clearly indicate the β-NbH_x hydride formation in place of α-Nb phase. The quadrupole splitting and the isomer shift of β-hydride subspectrum have been considered in terms of interaction between the interstitials and the Mössbauer atoms within the hydride phase. We show that in disordered non-stoichiometric β-hydride formed in $\text{Nb}_{1-y}\text{Fe}_y\text{H}_x$ alloy the Fe atoms can be really treated as probes that are sensitive to the change of the hydride sublattice configuration. The results indicate that the hydrogen-induced isomer shift relative to the unloaded alloys can be used as a measure of the mean number of hydrogen atoms next to the Fe probes. Both, the isomer shift and the quadrupole splitting indicate a configuration with three hydrogen atoms as nearest neighbours of Fe captured in β-hydride sublattice in $\text{Nb}_{(0.985 \div 0.90)}\text{Fe}_{(0.015 \div 0.10)}\text{H}_x$ alloys.

Key words: iron–niobium alloys, niobium hydrides, Mössbauer spectroscopy of ^{57}Fe.

1. Introduction

In contrast to niobium that is easy to charge with hydrogen up to the stoichiometric composition of NbH_2 [1], even a small amount of hydrogen is difficult to pump into iron unless high pressure of H_2 gas is used ($p_{\min} \geqslant 7$ GPa [2]). In alloys however, absorption of reasonable amounts of hydrogen is possible straight through the high temperature synthesis (ca. 1300 K) at hydrogen pressure below 10^5 Pa. A significant lattice expansion and the lattice distortions near interstitial hydrogen atoms are observed after such reaction [3]. The lattice distortion can fluctuate in connection with the diffusion of the interstitials. The influence of interstitial diffusion on the Mössbauer spectra of iron solutes in $\text{NbH}_{0.84}$ and $\text{NbH}_{0.78}$ has been studied by Wagner *et al.* [4] and Wordel *et al.* [5], respectively. They emphasised sensitivity of the intensity and the shape of the Mössbauer pattern to the magnitude of the displacements of the probe atoms by neighbouring interstitials. In disordered, non-stoichiometric hydrides, the probes have different hydrogen environments. When the diffusion of hydrogen causes the hyperfine interactions to become time independent, as in the fast diffusion case, the resulting hyperfine patterns, as well as the motional averaging of the isomer shift and electric quadrupole splitting reflect the structure of different hydride phases. In this work we want to show that the Fe atoms can use as such probes that are sensitive to the configuration of non-stoichiometric hydride formed in $Nb_{1-y}Fe_yH_x$ alloy.

2. Experimental and results

The alloys used in the experiment were prepared from fine polycrystalline powders of niobium and iron with a purity of 99.9+% and 99.999%, respectively. To allow the β-NbH_x hydride formation the alloys were previously heated up and homogenised near 1300 K, and than exposed to hydrogen gas during slow cooling down to room temperature. The amount of hydrogen absorbed was determined with an accuracy of about 1% from the change of the hydrogen pressure in a standard glass apparatus of known volume. The detailed description of the similar apparatus for hydriding with hydrogen gas can be found in [6]. In our case the TiH_2 metal hydride has been used as a pure hydrogen gas source. The concentration dependence of the Mössbauer spectra were measured at room temperature for several samples of $Nb_{1-y}Fe_yH_x$ as well as for unloaded $Nb_{1-y}Fe_y$ absorbers ($y = 0.015 \div 0.8$). The reduced scale of velocity ± 2 mm s^{-1} was applied for lack of magnetic splitting or satellite lines outside of the main peak. Some of the Mössbauer spectra in the range of iron concentrations from 1.5 at.% Fe to 30 at.% Fe are shown in Figure 1. This is the range of phase transition from the single α-Nb phase to nearly pure η-Fe_2Nb_3 phase, i.e. the range where the hydrogen can interfere in the phase structure of the alloy. The iron-rich phases appear to take up no hydrogen.

The plots of the central parts of the spectra reveal a mixed pattern composed of a singlet and/or a symmetric doublet in the case of hydrogen-free samples, and two symmetric doublets in the case of hydrogenated alloys. Some of the hyperfine parameters of the individual components fitted in compliance with suitable phases that may be formed before and after hydrogenation are shown in Table I. In order to verify the phase composition, the X-ray diffraction method was employed. The results obtained confirm in general the outcome of the analysis of the Mössbauer spectra. As can be seen from Figure 1(*right*), going from the lowest iron concentration to the highest one, the intensity of the β-hydride reflection (the right-hand side doublet) decreases, while that of the Fe_2Nb_3 phase (the left-hand side doublet) increases. This is probably due to the less negative value of the enthalpy of formation of niobium hydrides NbH_x (-59 kJ/mol H_2) in comparison with the one for almost all compounds of the Nb_mFe_n type ($-66 \div -71$ kJ/mol formula unit) [1].

The α-Nb phase is not detected in the hydrogenated system and the formation of the niobium β-hydride in place of the bcc solid solution of Fe in Nb seems to be the main effect of hydrogen on in the phase structure of the alloy. Furthermore, the hyperfine parameters of the right-hand side doublets in the spectra of

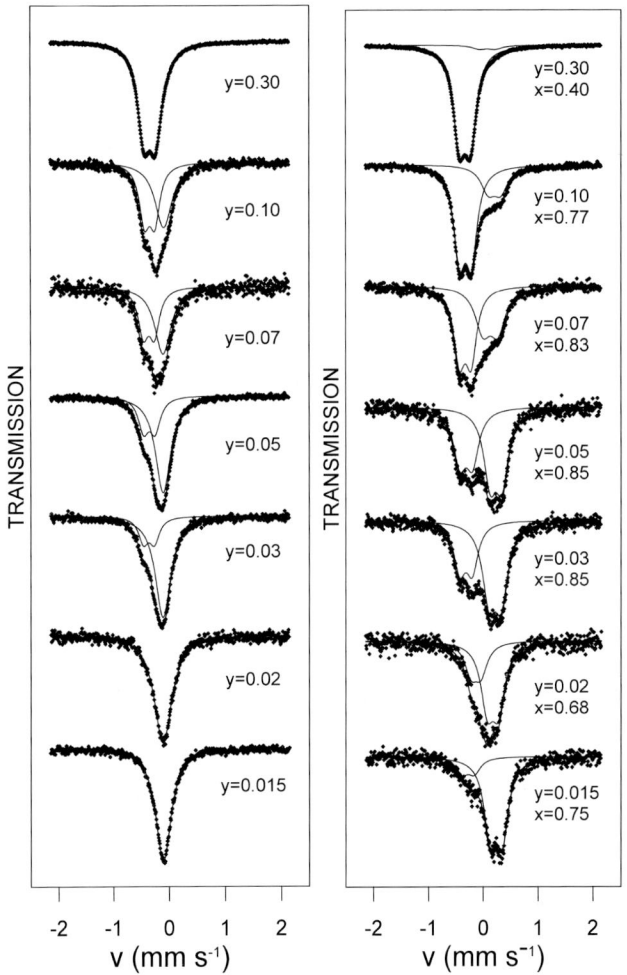

Figure 1. Mössbauer spectra of (*left*) Nb$_{1-y}$Fe$_y$ and (*right*) Nb$_{1-y}$Fe$_y$H$_x$ alloys in the range of iron concentration of $y = 0.015 \div 0.30$.

Nb$_{(0.985 \div 0.90)}$Fe$_{(0.015 \div 0.10)}$H$_x$ are, within experimental error, the same as those in the spectrum of NbH$_{0.84}$ ($\sim$300 K) [4]. However, in [4] the alloys were made from enriched ^{57}Fe and the absorbers consisted of 0.3 at.% Fe in Nb. It seems that in our systems ($y \leqslant 0.3$) the Fe atoms simultaneously play the role of lattice atoms in the intermetallic compounds and of substitutional probes in the hydride sublattice. The resulting hyperfine patterns reflect therefore the structural properties of the hydride as well. At higher concentration of iron, the Fe atoms affect the structure of the alloy to form intermetallic compound with niobium. Thus the hydride fraction formed from α-Nb phase decreases and finally becomes unobservable. In other words, an iron atom may reflect the hydride structure so long as the hydride fraction is dominant.

Table I. The isomer shift *IS* (with respect to α-Fe), and the quadrupole splitting *QS* of the components fitted in the ^{57}Fe Mössbauer spectra of $Nb_{1-y}Fe_y$ and $Nb_{1-y}Fe_yH_x$ alloys

Phase	$Nb_{1-y}Fe_y$		$Nb_{1-y}Fe_yH_x$	
	IS (mm s^{-1})	*QS* (mm s^{-1})	*IS* (mm s^{-1})	*QS* (mm s^{-1})
α-Nb (bcc) $y = 0.015 \div 0.10$	0.003(5)$\div$0.023(5)	–	–	–
β-NbH$_{0.68 \div 0.85}$	–	–	0.326(5)[a]	0.22(2)[a]
β-NbH$_{0.40}$ orthorhombic; tetrahedral H-site	–	–	0.186(5)	0.29(3)
η-Fe$_2$Nb$_3$ (bcc)	−0.224(4)	0.21(3)	−0.220(2)	0.21(2)

[a] On average in β-NbH$_{0.68 \div 0.85}$ phase ($y = 0.015 \div 0.10$).

3. Discussion

The quadrupole splitting (QS) of the hydride subspectrum varies insignificantly with iron concentration of $y \leqslant 0.10$ (see inserted graph in Figure 2). This shows that the Fe atoms interact alike with surrounding interstitials, as long as the hydride stoichiometry is not changed significantly. The explanation of such an interaction and hence of the hyperfine parameters seems possible taking into account the diffusion of hydrogen within the hydride sublattice. In the β-NbH$_x$ the structurally allowed interstitial sites form a plane rectangle around each metal atom [4, 5] (see insertion in Figure 2). In stoichiometric β-NbH$_{1.0}$ each metal atom is surrounded by four hydrogen atoms [7]. In the sub-stoichiometric hydride, some of these nearest neighbours will be missing and the diffusion of hydrogen is supposed to take place mainly between sites allowed by the structure [8]. At a hydrogen-to-metal ratio of $x \approx 0.75$, three of these sites should be occupied on the average, unless the environment of the impurities is modified by a repulsive or attractive interaction with the interstitials, but such interactions have been found to be small in the hydrides of niobium [4].

Similarly as for NbH$_{0.84}$ hydride [4], we assume that in our case three of these four sites are always occupied, and that the hydrogen diffusion can be described as hopping of the vacancy on the fourth free corner of the rectangle around the Fe probe. Configurations with zero or four nearest neighbours will not disturb the central metal atom. For ^{57}Fe in palladium hydride it was found [9] that the hydrogen-induced isomer shift is, with good approximation, proportional to the number of nearest hydrogen neighbours around the probe atom and can be used as a measure of the mean number of hydrogen atoms next to the Fe probes. In our case the stoichiometry of the hydride is as 3 : 4 on average, and the isomer shift is about of +0.3 mm s^{-1} with respect to the pure α-Nb. Therefore, a configuration

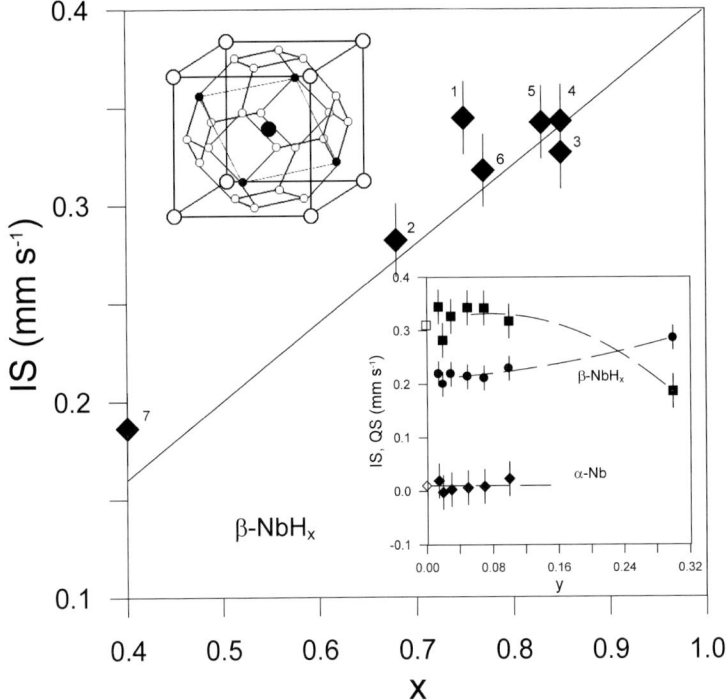

Figure 2. Hydrogen concentration (x) dependence of the isomer shift of β-hydride fraction represented by the right doublets in the Mössbauer patterns of Nb$_{1-y}$Fe$_y$H$_x$ alloys (1 – Nb$_{0.985}$Fe$_{0.015}$H$_{0.75}$, 2 – Nb$_{0.98}$Fe$_{0.02}$H$_{0.68}$, 3 – Nb$_{0.97}$Fe$_{0.03}$H$_{0.85}$, 4 – Nb$_{0.95}$Fe$_{0.05}$H$_{0.85}$, 5 – Nb$_{0.93}$Fe$_{0.07}$H$_{0.83}$, 6 – Nb$_{0.90}$Fe$_{0.10}$H$_{0.77}$, 7 – Nb$_{0.70}$Fe$_{0.30}$H$_{0.40}$). The solid line shows the dependence calculated on the assumption that hydrogen-induced isomer shift is proportional to the number of nearest hydrogen neighbours around the Fe atoms. In the inserted graph is shown the isomer shift (■) and the quadrupole splitting (●) of β-NbH$_{0.85 \div 0.77}$ doublets in Mössbauer patterns of Nb$_{1-y}$Fe$_y$H$_x$ ($y \leqslant 0.3$) alloys. The isomer shift of the single line representing the α-Nb phase in uncharged Nb$_{1-y}$Fe$_y$ alloys is shown by (♦). Also the values of the isomer shift of 0.3 at.% ^{57}Fe in NbH$_{0.84}$ and in pure Nb, taken from [4], are indicated on the left-side axis by (□) and (◇) symbols, respectively. The isomer shift values are given with respect to the α-Fe. The upper insertion shows the lattice structure of the β-NbH$_x$ hydride, with large circles for the metal atoms and small ones for 24 nearest interstitial sites surrounding a Mössbauer atom in the body-centred position. Full circles indicate the four hydrogen positions allowed in the β phase of NbH$_x$.

with three hydrogen neighbours occurs. Thus, the mean isomer shift per nearest hydrogen neighbour is about +0.1 mm s^{-1}, and is exactly the same as in the case of NbH$_{0.84}$ hydride [4]. Moreover, the planar arrangement of the nearest hydrogen neighbours around the metal atoms in the β-NbH$_x$ phase explains why the electric field gradient at the iron nuclei does not disappear when the diffusion becomes sufficiently fast for motional averaging, that is certainly above 300 K [5].

According to the conclusion drawn from the isomer shift, the configuration of the hydride in the Nb$_{0.70}$Fe$_{0.30}$H$_{0.40}$ sample should have two hydrogen atoms missed on the four nearest interstitials that are allowed by the structure or in the other words, two hydrogen atoms as the nearest neighbours of the Fe atom treated as the body-centred Mössbauer probe in the hydride cage. In this case the *IS* parameter is about of $+0.2$ mm s^{-1} (Table I) and the hydrogen-to-metal ratio is approximately close to the hydride fraction stoichiometry of $1 : 2$.

4. Conclusion

We have found that the Fe atoms alloying with Nb, which constitute lattice atoms in the intermetallic compounds, are also sensitive probe for estimation of configuration of hydrogen occupation in β-NbH$_x$ hydride. The shapes and centre shifts of Mössbauer patterns reflect therefore the structure of the hydride as well. Regarding the results obtained by Wagner *et al.* [4], we can conclude that the hydrogen-induced isomer shift, with a good approximation, is equally good measure of the number of nearest hydrogen neighbours around the Fe probe atoms in the hydride fraction of the Nb$_{1-y}$Fe$_y$H$_x$ system, as in the case of Fe solutes straight in the hydride of Nb.

Acknowledgement

This work was supported by the University of Wrocław under the Grant number 2016/W/IFD/2003.

References

1. Buschow, K. H. J., Bouten, P. C. and Miedema, A. R., *Rep. Prog. Phys.* **45** (1982), 937.
2. Antonov, V. E., Belash, I. T., Ponyatovskii, E. G., Thiessen, V. G. and Shiryaev, V. I., *Phys. Stat. Sol. A* **65** (1981), K43.
3. Peisl, H.: In: G. Alefeld and J. Völkl (eds), *Hydrogen in Metals I*, Springer Verlag, Berlin, 1978, p. 53.
4. Wagner, F. E., Wordel, R. and Zegler, M., *J. Phys. F* **14** (1984), 535.
5. Wordel, R., Litterst, F. J. and Wagner, F. E., *J. Phys. F* **15** (1985), 2525.
6. Schober, T. and Wenzl, H., In: G. Alefeld and J. Völkl (eds), *Hydrogen in Metals II*, Springer Verlag, Berlin, 1978, p. 60.
7. Schober, T. and Wenzl, H., In: G. Alefeld and J. Völkl (eds), *Hydrogen in Metals II*, Springer Verlag, Berlin, 1978, p. 11.
8. Alefeld, B., Bohn, H. G. and Stump, N., In: *Jül-Bericht Jül-Conf-6*, Vol. 1, KFA, Jülicht, 1972, p. 286.
9. Pröbst, F., Wagner, F. E. and Karger, M., *J. Less-Common Met.* **88** (1982), 201.

Hyperfine Interactions **156/157**: 143–149, 2004.
© 2004 *Kluwer Academic Publishers. Printed in the Netherlands.*

First Observation of Photoinduced Magnetization for the Cyano-Bridged 3d–4f Heterobimetallic Assembly Nd(DMF)$_4$(H$_2$O)$_3$(μ-CN)Fe(CN)$_5$·H$_2$O (DMF = N,N-Dimethylformamide)

GUANGMING LI[1,3], TAKASHIRO AKITSU[1], OSAMU SATO[2] and YASUAKI EINAGA[1]

[1] *Department of Chemistry, Faculty of Science and Technology, Keio University, 3-14-1 Hiyoshi, Yokohama 223-8522, Japan; e-mail: einaga@chem.keio.ac.jp*
[2] *Special Research Laboratory for Optical Science, Kanagawa Academy of Science and Technology, KSP Building East 412, 3-2-1 Sakado, Takatsu-ku, Kawasaki 213-0012, Japan*
[3] *Faculty of Chemistry and Chemical Engineering, Heilongjiang University, Harbin 150080, P. R. China*

Abstract. Photoinduced magnetization of the cyano-bridged 3d–4f hetero-bimetallic assembly Nd (DMF)$_4$(H$_2$O)$_3$(μ-CN)Fe(CN)$_5$·H$_2$O (**1**) (DMF = *N,N*-dimethylformamide) is described in this paper. The $\chi_M T$ values are enhanced by about 45% after UV light illumination in the temperature range of 5–50 K. We propose that UV light illumination induces a structural distortion in **1**. This small structural change is propagated by molecular interactions in the inorganic network. Furthermore, the cooperativity resulting from the molecular interaction functions to increase the activation energy of the relaxation processes, which makes observation of the photoexcited state possible. The flexible network structure through the hydrogen bonds in **1** plays an essential role for the photoinduced phenomenon. This finding may open up a new domain for developing molecule-based magnetic materials.

Key words: molecule-based magnet, photoinduced magnetization, 3d–4f assembly, Mössbauer spectroscopy.

1. Introduction

Photoinduced magnetic and optical properties have attracted much attention [1–3], because photocontrol of these properties is a challenging topic of interest in material science in view of the possible implementation in optical and memory devices [1b]. It is noted that almost all previous reports have focused on Prussian blue analogues or transition metal complexes [2, 3]. However, crystallization of Prussian blue analogues is difficult, making verification of the proposed structures by X-ray crystallographic analysis impractical [2, 3]. In contrast, the synthesis and structural characterization of various cyano-bridged 3d–4f hetero-bimetallic assemblies have been reported driven by the interest in the molecular magnetism [4]. Keeping the

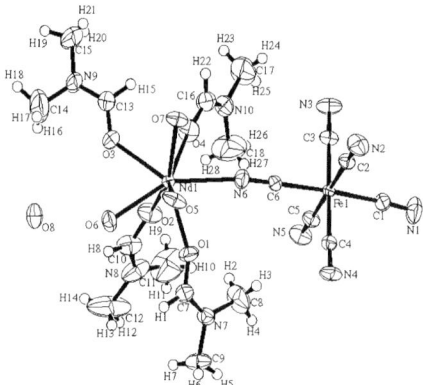

Figure 1. A view of the molecular structure for **1**. Hydrogen atoms are omitted for clarity. Possible hydrogen bond distances: O(8)–O(6) 2.662(7) Å, O(8)–N(1) 2.77(1) Å, O(8)–N(2) 2.85(1) Å.

similarities of the two types of complex in mind, we studied the effects on the magnetic properties of the cyano-bridged 3d–4f hetero-bimetallic assemblies by light stimulation. As a result, a significant photoinduced magnetization was observed for $Nd(DMF)_4(H_2O)_3(\mu\text{-}CN)Fe(CN)_5 \cdot H_2O$ (**1**) (DMF = N,N-dimethylformamide) in a given range of temperature. Herein, we describe the photoinduced magnetization for **1**.

2. Experimental

A UV lamp was used as the light source in the investigation of photoinduced magnetization changes. The UV light was introduced through an optical fiber into the SQUID magnetometer for illumination of the sample. Upon irradiation at 5 K, the magnetization was slowly increased and gradually saturated after several tens of minutes. The plots of $\chi_M T$ versus temperature for **1** (Figure 3) show that the $\chi_M T$ values are enhanced by about 45% as compare to those before irradiation in the temperature range of 5–50 K. However, the UV light induced conversion rate depends on the sample geometry and the light transmission. The photoexcited state could remain for at least several hours at 5 K. When the temperature was increased to 50 K, the photoinduced magnetization was erased. This indicates that the photoexcited state had returned to the ground state.

3. Results and discussion

The crystal structure* for **1** (Figure 1) reveals that the Nd^{3+} ions are eight-coordinated in a square-antiprism arrangement, while the Fe^{3+} ions are six-coordinated

---

* Crystallographic data for **1** ($C_{18}H_{36}FeN_{10}O_8Nd$): FW = 720.63, yellow prism (0.40 × 0.30 × 0.30 mm), monoclinic, space group $P2_1/n$ (No.14), a = 19.930(9) Å, b = 8.914(5) Å, c = 17.642(7) Å, β = 95.87(3)°, V = 3117(2) Å^3, Z = 4, D_c = 1.535 g cm^{-3}, R_1 = 0.077, R_w = 0.112.

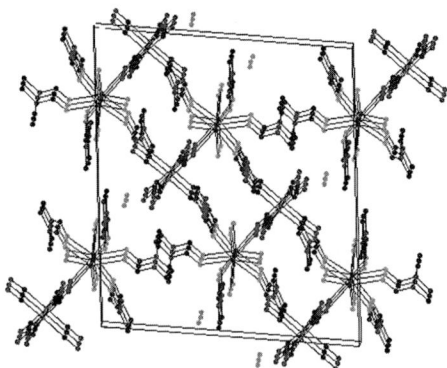

Figure 2. A projection of the three-dimensional crystal packing for **1** formed by intermolecular hydrogen bondings and van der Waals cohesive forces.

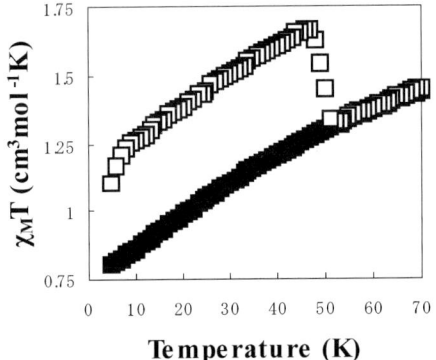

Figure 3. Plot of magnetic susceptibilities ($\chi_M T$) vs temperature (T) for **1** at $H = 5000$ G: before UV light illumination (■); after UV light illumination (□).

in an octahedral environment. A cyanide bridge links a Nd^{3+} ion to an Fe^{3+} ion. In addition, a crystalline water molecule is involved in a neutral dinuclear unit. The possible intermolecular bond distances, which are shorter than the sum (3.2 Å) of van der Waals radii [5], reveal the presence of hydrogen bonds. Thus, a three-dimensional flexible network of the crystal lattices in **1** (Figure 2) was essentially formed by intermolecular hydrogen bonds and van der Waals cohesive forces. It is this flexible network structure that plays a dominant role in theformation of a photoexcited state during light illumination [2d].

To explain this interesting photoinduced magnetic phenomenon, IR, UV, XRD, and Mössbauer spectra for **1** at lower temperature were performed before and after irradiation. Meanwhile, consideration was given to the well-known metal to metal charge transfer mechanism (redox reaction) which was used to explain photoinduced magnetization for Prussian blue analogues [2b, 3a]. However, Mössbauer spectra of ^{57}Fe for **1** at 10 K (Figure 4) show that the doublet absorption peak (isomer shift $= -0.18$ mm/s) remains after UV light illumination. These reveal that the oxidation state for iron ions remains unchanged [3c]. In addition,

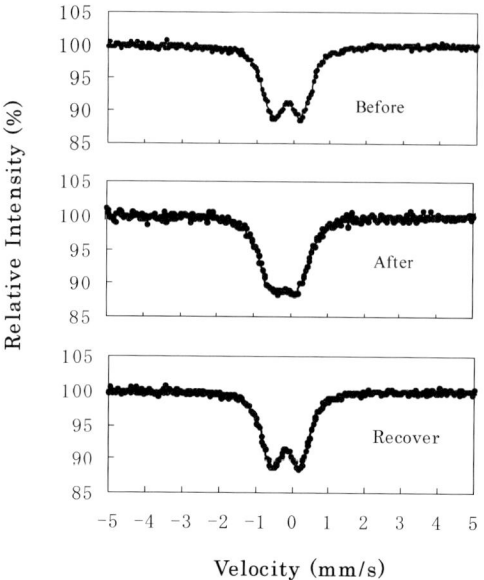

Figure 4. Mössbauer spectra for **1** at 10 K: before UV light illumination (*top*), after UV illumination (*middle*), and annealing (*bottom*).

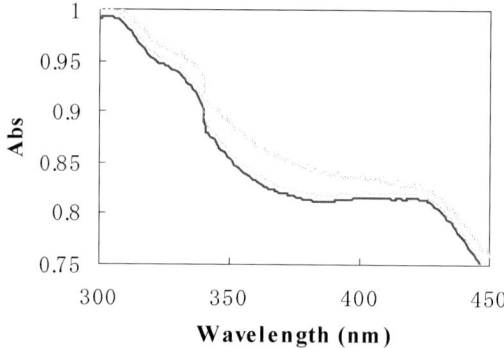

Figure 5. UV–visible spectra for **1** at 10 K: before (solid line), after (dotted line) UV light illumination, and annealing test (dashed line).

charge transfer involving neodymium ions is unlikely because the valence would remain at +3. Thus, the charge transfer mechanism for interpreting the photoinduced magnetic phenomenon of **1** was disproved. Nevertheless, the quadrupole splitting shrinks from 0.74 to 0.65 mm/s suggesting that the ligand–metal bond distance and symmetry of the $Fe^{III}(CN)_6$ moiety may distort [6].

Therefore, another mechanism, proposed to explain the photoexcited state for $Fe(phen)_2(NCS)_2$, is contemplated as being applicable in the present case. This has been proven by single crystallographic analysis of both the irradiated and the non-irradiated states of the photoinduced compound [7]. It demonstrates that the major differences between the ground and the photoexcited states are in the bond lengths

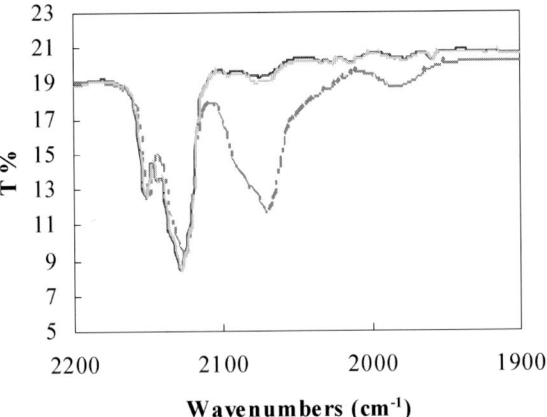

Figure 6. IR spectra (KBr pellet) for **1** at 10 K: before (solid line), after (dotted line) UV light illumination, and annealing test (dashed line).

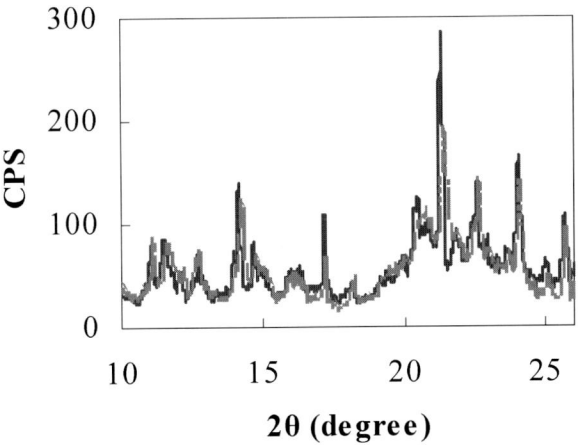

Figure 7. X-ray powder diffraction spectra for **1** at 25 K: before (solid line) and after (dotted line) UV light illumination.

and the crystal cell volumes; however, these differences are very small. Thus, on the base of previous work and our experimental results, we propose that UV illumination of complex **1** induces a LMCT transition. This is supported by the UV spectra* (Figure 5), and the accompanying increase in the electron density on Fe^{3+} revealed by the IR spectra** (Figure 6). It is followed by an increase in the metal–ligand

---

  * The absorbance band at 420 nm, which is assigned to the LMCT of $Fe(CN)_6{}^{3-}$, is enhanced after UV light illumination. It suggests that the UV light irradiation induces the LMCT for **1**. See [8].
  ** The ν(CN) of IR spectra for **1**, after irradiation, reveals small red shifts (averaging about 7 cm^{-1}) from 2152, 2144 (sh), 2129 and 2075 (w) cm^{-1} to 2146, 2123 and 2067 (s) cm^{-1}. It is indicative of the increment of electron density for the iron ions. However, it is too small to conclude that the change of oxidation state from $[Fe^{III}(CN)_6]^{3-}$ to $[Fe^{II}(CN)_6]^{4-}$ occurrs. Because the ν(CN) are 2135, 2130 and 2118 cm^{-1} for $K_3Fe^{III}(CN)_6$ whilst 2098, 2062 and 2044 cm^{-1} for $K_4Fe^{II}(CN)_6$

bond length and a small structural change as indicated by IR spectra[‡] and X-ray powder diffraction[‡‡] (Figure 7) for **1**. This small structural change is propagated by molecular interactions in the inorganic network. Furthermore, the cooperativity resulting from the molecular interaction functions to increase the activation energy of the relaxation processes, which makes observation of the photoexcited state possible. It is the cooperativity arising from the flexible network that leads to the photoinduced magnetization changes. This is in accordance with the photoinduced memory effect that can be triggered by electron excitation in a given material inducing a rearrangement of the lattice and the electronic configuration [10].

4. Conclusions

The above-mentioned results clearly show that the magnetic properties of cyano-bridged 3d–4f hetero-bimetallic compounds can be influenced by external stimuli in a given temperature range. It should be noted that the size of effect on **1** in terms of the photoinduced magnetization changes is very much smaller than that for Prussian blue's. In the former, it is still in a range of para-magnetization although the $\chi_M T$ values are increased by about 45% after light irradiation. However, the para-magnetization turns to ferri-magnetization after light irradiation in the latter. Our strategy to realize a photoexcited state in **1** is the introduction of a flexible network structure by means of the hydrogen bonds. These findings may open up a new domain for developing molecule-based magnetic materials. Because a number of cyano-bridged 3d–4f assemblies, which are of facile synthesis and simple composition referred to Prussian blue analogues, are known, novel molecule-based magnetic materials with superior properties may be developed and discovered soon. Finally, this will promote applications in the fields of photocontrolled switches and information storage.

References

1. (a) Gutlich, P., Garcia, Y. and Woike, T., *Coord. Chem. Rev.* **219–221** (2001), 838; (b) Gutlich, P., Hauser, A. and Spiering, H., *Angew. Chem., Int. Ed. Engl.* **33** (1994), 2024.
2. (a) Bleuzen, A., Lomenech, C., Escax, V., Villain, F., Varret, F., Moulin, C. C. D. and Verdaguer, M., *J. Am. Chem. Soc.* **122** (2000), 6648; (b) Cartier, C., Villain, F., Bleuzen, A., Arrio, M.-A., Sainctavit, P., Lomenech, C., Escax, V., Baudelet, F., Dartyge, E., Gallet, J.-J.

---

respectively, the red shifts from $K_3Fe^{III}(CN)_6$ to $K_4Fe^{II}(CN)_6$ are in a range of 37 to 74 cm^{-1}. See [9].

[‡] The weak peak at 2075 cm^{-1} for **1** is significantly enhanced which results in the change of peak pattern. It proposes that the symmetry of the $Fe(CN)_6^{3-}$ moiety changes, which may attribute to the small structure distortion of the $Fe(CN)_6^{3-}$ moiety.

[‡‡] X-ray powder diffraction spectra for **1** reveal a slight change of the diffraction pattern between 20 and 30°(2θ). The characteristic difference was that the peak at 20.5° was broadened and the intensity of the peak at 21.3° was reduced, indicating a small structure change induced by the light illumination.

and Verdaguer, M., *J. Am. Chem. Soc.* **122** (2000), 6653; (d) Hayami, S., Gu, Z.-Z., Shiro, M., Einaga, Y., Fujishima, A. and Sato, O., *J. Am. Chem. Soc.* **122** (2000), 7126.

3. (a) Sato, O., Iyoda, T., Fujishima, A. and Hashimato, K., *Science* **272** (1996), 704; (b) Sato, O., Einaga, Y., Fujishima, A. and Hashimato, K., *Inorg. Chem.* **38** (1999), 4405; (c) Sato, O., Hayami, S., Einaga Y. and Gu, Z.-Z., *Bull. Chem. Soc. Jpn.* **76** (2003) 443.

4. (a) Figuerola, A., Diaz, C., Ribas, J., Tangoulis, V., Granell, J., Lloret, F., Mahia, J. and Maestro, M., *Inorg. Chem.* **42** (2003), 641; (b) Yang, B. and Chen, Z., *Prog. Nat. Sci.* **11** (2001), 401 and references therein.

5. Desiraju, G. R. and Steiner, T., *The Weak Hydrogen Bond*, Oxford University Press, New York, 1999.

6. Gibb, T. C., *Principle of Mössbauer Spectroscopy*, Chapman and Hall, London, 1976.

7. Marchivie, M., Guionneau, P., Howard, J. A. K., Chastanet, G., Letard, J.-F., Goeta, A. E. and Chasseau, D., *J. Am. Chem. Soc.* **124** (2002), 194.

8. Alexander, J. J. and Gray, H. B., *J. Am. Chem. Soc.* **90**, (1968), 4260.

9. Griffith, W. P. and Turner, G. T., *J. Chem. Soc. A* (1970), 858.

10. Wojtowicz, T., Kolesnik, S., Miotkowski, I. and Furdyna, J. K., *Phys. Rev. Lett.* **70** (1993), 2317.

Hyperfine Interactions **156/157**: 151–155, 2004.
© 2004 *Kluwer Academic Publishers. Printed in the Netherlands.*

Non-Magnetic Stainless Steels Reinvestigated – a Small Effective Field Component in External Magnetic Fields

T. ERICSSON[1], Y. A. ABDU[1], H. ANNERSTEN[1] and P. NORDBLAD[2]

[1] *Uppsala University, Department of Earth Sciences, Villavägen 16, SE-75236 Uppsala, Sweden;*
e-mail: Tore.Ericsson@geo.uu.se
[2] *Uppsala University, Department of Materials Science, Box 534, SE-75121 Uppsala, Sweden*

Abstract. Three standard non-magnetic stainless steels of composition (wt%) $Fe_{70}Cr_{19}Ni_{11}$, $Fe_{70}Cr_{17}Ni_{13}$ and $Fe_{69}Cr_{18.5}Ni_{10.3}Mn_{1.8}Ti_{0.4}$ have been investigated by Mössbauer spectroscopy (5–295 K and in external fields $\leqslant$7 T at room temperature) and magnetization measurements (10–300 K) using a SQUID magnetometer. There are indications of a field induced ferromagnetic interaction in the samples at room temperature.

Key words: stainless steel, external field, field induced ferromagnetism.

1. Introduction

Non-magnetic stainless steels, containing normally Fe, Cr and Ni as major elements and sometimes Ti, Mn as minor elements have the fcc structure, the same as γ-Fe, being thermodynamically stable above $\approx$900°C. The γ-phase is stable above 400°C in $Fe_{60}Ni_{40}$ [1]. Introduction of Cr stabilizes the γ-phase to even lower temperatures [2]. The γ-phase may also be retained in the miscibility region at room and lower temperatures in quenched samples, e.g., in small grains [3] and after mechanical alloying producing great amounts of lattice defects, but frequently a transformation to martensite structure occurs. The magnetic phase diagram of fcc FeCrNi stainless steels is very complicated. Increased Ni contents leads to ferromagnetic (FM) phases, decreased to paramagnetic (PM) or antiferromagnetic (AFM) phases. Other phases also occur, e.g., $Fe_{80-x}Ni_xCr_{20}$ is FM for $x = 30$, FM and spin glass (SG) for $x = 26$ and 23, field induced ferromagnets (FIFM) and SG for 21 and 19, AFM for $x = 17$ and 14 [2]. However, the Neél- and Curie temperatures, T_N and T_C, are low, normally <50 K [2]. The complicated magnetic properties of fcc FeCrNi alloys can, at least partly, be understood introducing mixed exchange interactions. Kondorsky and Sedov [4] introduced the phrase "latent antiferromagnetism" when studying Fe–Ni invar alloys: an AFM coupling between Fe–Fe nearest neighbour pairs, but FM exchanges between Fe–Ni and Ni–Ni pairs. However, there are details still not well understood and we have here studied three FeCrNi stainless steels, and focused on Mössbauer spectroscopy (MS) in external fields at room temperature.

2. Experimental

Three non-magnetic stainless steels of standard commercial type have been investigated using Mössbauer spectroscopy at room and low temperatures and in external fields up to 7 T at room temperature. Magnetization measurements, using a SQUID magnetometer, have been performed as M versus temperature, $M(T)$, from 10 K to 300 K and as M versus magnetizing field H ($\leqslant$50 kOe), $M(H)$, at 10 K and 300 K. Two powder samples ("Ni11", "Ni13") from Alfa Aesar, $Fe_{70}Ni_{11}Cr_{19}$ (in wt%, type 304-L, -100 mesh) and $Fe_{70}Ni_{13}Cr_{17}$ (type 303-L, -140 mesh) and one foil (25 μm of a type often delivered with ^{57}Co-sources) of composition $Fe_{69.0}Ni_{10.3}Cr_{18.5}Mn_{1.8}Ti_{0.4}$ (determined using a Cameca electron-microprobe) were used in the investigation. A superconducting magnet up to 7 T from Cryogenic Ltd., having a 10 cm open bore hole was used for the Mössbauer meaurements at room temperature. The Mössbauer velocity scale and center shifts CS are given with α-Fe at room temperature as reference.

3. Results and discussion

Figure 1 shows $M(T)$-curves at increasing temperatures for the three samples in an external field of 1 kOe (=0.1 T). The weak temperature dependence of the curves may look typical to metallic PM samples. Knowing about the existence of a possible low temperature AFM transition, indications of such a transition are seen at low temperatures for the powder samples. The magnetization of the foil only shows a continuous decrease, but this sample had a remanent magnetization of 4.3 emu/g when the external field was switched off. We attribute this remanence to martensite or a ferrite impurity of $\approx$2 wt%, not characteristic

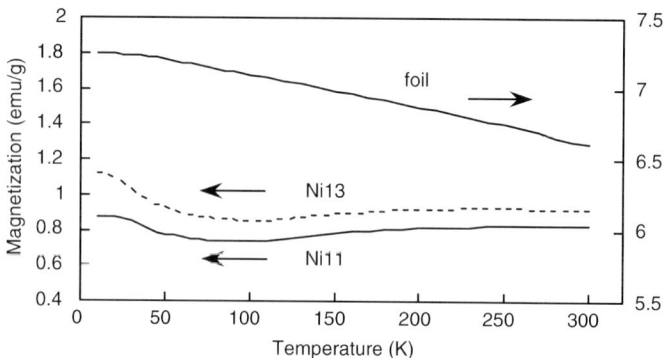

Figure 1. $M(T)$ curves measured at increasing temperature for zero field cooled (ZFC) samples in an external field of 1 kOe.

for the fcc stainless steel, and that the temperature dependence of the magnetization is influenced by the decrease of the remanence with increasing temperature.

Figure 2 shows magnetization versus applied field curves at 300 K for the three samples. The $M(H)$-curves measured at 10 K did have a rather similar shape as those in Figure 2: a technical saturation well below 10 kOe and then concave curves not at all saturated even at 50 kOe. However, the magnetization at higher fields (and slope of the curves) was about 2–3 times stronger than at 300 K, albeit with an initial slope at lower fields (<2 kOe) that remained quite weakly temperature dependent (cf. Figure 1). The rapid increase of the magnetization at low fields might be assigned to nano sized regions (few % in volume) of martensitic or ferritic phases. Somewhat larger regions of martensite or ferrite in the foil could then also cause the remanence.

Mössbauer spectra (5–300 K) recorded in zero external field showed a singlet that broadened below 50 K. The broadening (in FWHM) was $\approx$0.5 mm/s, indicating a PM–AFM transition in agreement with the SQUID measurements. However, the saturated moments seem to be very small, $\approx$0.1μ_B (as 2.2 μ_B corresponds to 10.62 mm/s in α-Fe), but in line with results for the low moment AFM phase in the Fe–Ni Invar region [3]. The $CS(T)$-curves for the three samples were also quite "normal", -0.1 mm/s at 295 K, then a slope in agreement with the second order Doppler-shift (SODS) for a Debye-temperature of $\approx$400 K. However, at the PM–AFM transition there seems to be a very small extra shift of ≈-0.005 mm/s.

The in-field spectra for the Ni11 and Ni13 samples are similar to those obtained for the foil, shown in Figure 3. Due to the high absorption in the middle of the Mössbauer pattern, it seems necessary to use a low field component in fitting the spectra, much too strong to be related to the earlier mentioned martensitic or ferritic phases. Accordingly, the samples cannot be in a pure PM state at room temperature. A possible interpretation could be to introduce a field induced fer-

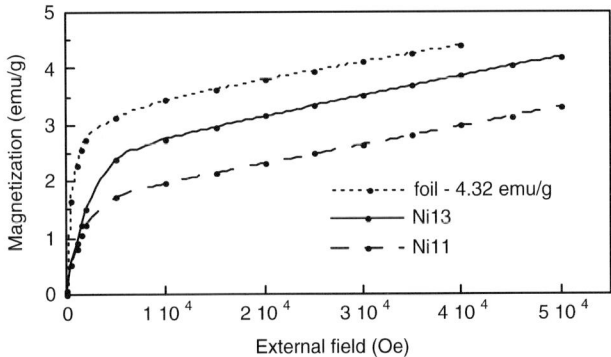

Figure 2. $M(H)$ curves for the three samples measured at 300 K. The remanence detected in the foil is subtracted.

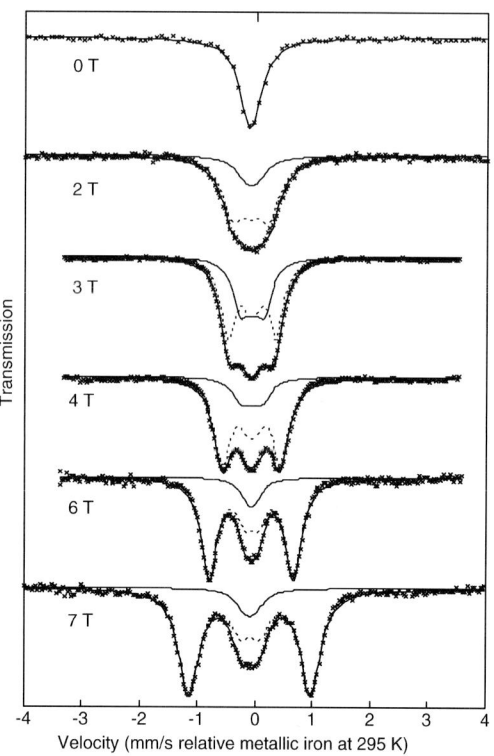

Figure 3. MS-spectra for the foil recorded at room temperature in longitudinal external fields 0–7 T. The spectra are fitted using two quadruplets (lines 2 and 5 missing in an ordinary sextet). The high field (dashed line) and low field (full line) components are also shown.

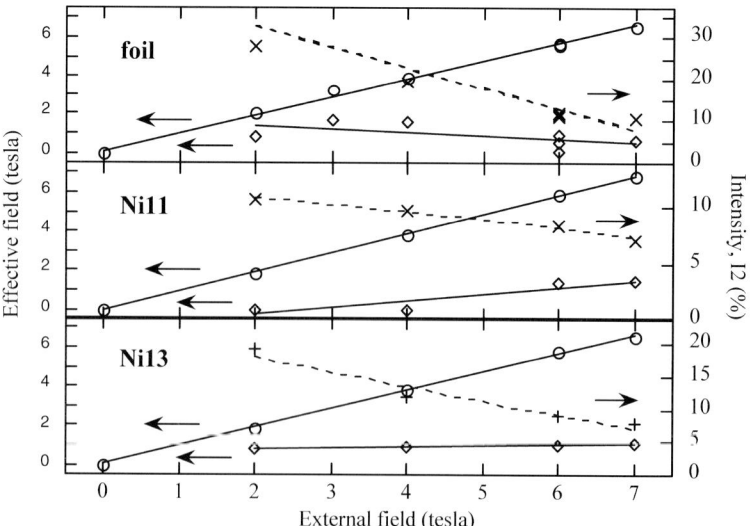

Figure 4. Measured effective fields and intensity of low field component (I2, right scale) versus external field.

romagnetic state (FIFM). The measured effective field will then roughly be the magnitude difference between the external field and the induced hyperfine field. Our results above then indicate a heterogeneous situation. A fraction of the sample is PM (high field component) and another part, having an intensity decreasing with increasing field (Figure 4), is FIFM with a magnetic moment component increasing with ≈ 0.07 μ_B/T external field. Interestingly, we have found the same behavior in Fe–Ni (21–27%) fcc alloys and also in the anti-taenite Fe–Ni fraction in two chondrites (not published yet). FIFM states have been proposed in fcc dilute FeCu [5] and FeCrNi stainless steels [2] (magnetization study). Metamagnetism has been proposed at low temperature in Fe–Ni Invar alloys [6], also from magnetization measurements.

References

1. Reuter, K. B., Williams, D. B. and Goldstein, J. I., *Metallurgical Transactions A* **20** (1989), 719.
2. Majumdar, A. K. and Blanckenhagen, P. V., *Phys. Rev. B* **29** (1984), 4079.
3. Asano, H. J., *Phys. Soc. Japan* **27** (1969), 542.
4. Kondorsky, E. I. and Sedov, V. L., *J. Appl. Phys.* **31** (Suppl.) (1960), 331S.
5. Frankel, R. B., Blum, N. A., Schwartz, B. B. and Kim, D. J., *Phys. Rev. Lett.* **18** (1967), 1051.
6. Pauthenet, R., Maruryama, H., and Yamada, O., *J. Magn. Magn. Mater.* **31–34** (1983), 835.

Hyperfine Interactions **156/157**: 157–163, 2004.
© 2004 *Kluwer Academic Publishers. Printed in the Netherlands.*

Magnetic Response and Hyperfine Magnetic Fields at Fe Sites of $Sr_3Fe_2MO_9$ (M = Mo, Te, W, U) Double-Perovskites

L. A. BAUM[1], S. J. STEWART[1], R. C. MERCADER[1] and J. M. GRENÈCHE[2]

[1]*Departamento de Física, IFLP, Facultad de Ciencias Exactas, Universidad Nacional de La Plata, C.C.67, 1900 La Plata, Argentina*
[2]*Laboratoire de Physique de L'Etat Condensé UMR CNRS 6087, Université du Maine, 72085 Le Mans, France; e-mail: greneche@univ-lemans.fr*

Abstract. We have studied the isostructural series of double-perovskites $Sr_3Fe_2MO_9$ (M = Mo, Te, W, U) by Mössbauer spectrometry and AC susceptibility measurements. The hyperfine structure of Mössbauer spectra at room temperature is attributed to the presence of high spin state Fe^{3+} ions sensing both static and fluctuating magnetic hyperfine fields with different relative areas that depend on M. The magnetically split signal – indistinguishable from the background in the Mo compound spectrum – increases with the Fe-site disorder in the sequence Mo < U < Te < W. The spectra at 77 K demonstrate that the W-perovskite sample is already magnetically fully ordered, while the other three cations suggest ordering temperatures that increase from Te to U to Mo. At 4.2 K all the spectra are completely magnetically split and display hyperfine fields that range from 49 up to 53 T. Coincident with the X-ray and neutron diffraction results, the hyperfine parameters are consistent with Fe atoms centered in oxygen octahedral units, coordinated to different numbers of M-centered octahedra. The AC susceptibility response is $\chi'_{max} \approx 3.5 \times 10^{-5}$ emu/g.Oe for the Mo compound and increases for the W, Te and U compounds with values of $\chi'_{max} 1.6 \times 10^{-4}$, 3.0×10^{-4}, and 9.4×10^{-3} emu/g.Oe, respectively. The out-of-phase component, χ'', could only be detected for the U compound. Its frequency dependence displays a shift that denotes a spin-glass-like state arising from the chemical disorder.

Key words: [57]Fe Mössbauer spectroscopy, magnetic measurements, double-perovskites, magnetic ordering transition.

1. Introduction

After the discovery of tunneling and colossal magnetoresistances (TMR and CMR) in Sr_2FeMoO_6 [1] and other members of the series of double perovskites $AB'_{1/2}B1_{1/2}O_3$ [2, 3] many investigations have followed aiming at the understanding of the magnetic properties of these compounds. In particular, a special type of magnetic coupling has been proposed for Sr_2FeMoO_6, where it has been claimed that the double exchange or superexchange mechanisms cannot account for its magnetic properties [4]. A new type of magnetic ordering mechanism based on sizeable hopping interaction strengths that couple Fe d states to Mo d states via the oxygen p orbitals has been suggested [4] and found appropriate to reproduce EPR experiments [5]. However, in the same system a different model based on a ferromagnetic stabilization mechanism [6] has also been proposed to explain the relatively much higher ordering temperatures of these perovskites as compared to the manganites. In the series of perovskites A_2FeMoO_6 (A = Ca, Sr, Ba), however, a model based on single-site disorder and antiphase boundaries was necessary to match the seemingly differences found between the magnetization and Mössbauer data [7]. Other authors [8] have found that the saturation magnetization can be effectively controlled by the synthesis conditions that lead to different concentration of antisite defects in the ferrimagnetic lattice.

Because of the diverse interpretations of the origin of the magnetism in these perovskites, we have considered important to study the structural and magnetic properties of the double perovskites, which have different stoichiometry, namely, $Sr_3Fe_2MO_9$ (M = Mo, U, Te, and W) but that have very similar environments for the magnetic ions. Recently, a study has been reported [9] on the relation between the magnetic behavior and the chemical disorder of the oxygen octahedra that surround the Fe centered octahedra. In this work, we report low-temperature and AC susceptibility results, that give additional evidence towards the understanding of the magnetic behavior.

2. Experimental section

The synthesis conditions and crystallographic data have already been reported elsewhere [9]. The transmission Mössbauer spectra were taken in a conventional constant acceleration spectrometer with a $^{57}CoRh$ source in transmission geometry. The absorbers were powdered samples of about 36, 22, 30, and 17 mg/cm^2 for the compounds of Mo, U, Te and W, respectively, which yield the optimum absorber thickness. The low-temperature measurements were taken in a liquid helium bath cryostat. The hyperfine parameters were refined by means of a least-squares program using discrete distributions of hyperfine field, keeping isomer shift and quadrupole shifts common to the different sextets. Isomer shifts are quoted to that of α-Fe foil at room temperature.

AC susceptibility measurements were carried out in a LakeShore 7130 susceptibility at temperatures increasing from 13 to 325 K and different frequencies ranged from 5 to 9.92 kHz. The measuring AC exciting field was 10 Oe for frequencies up to 1 kHz and 1 Oe for the higher ones. The frequencies were set to 5, 40, 375, 825, 3320 and 9920 Hz. Each point was taken at a 3 K interval after the temperature had been stabilized.

3. Results

Figure 1 shows the Mössbauer spectra of the compounds of Mo (*left*) and W (*right*), at 295, 77 and 4.2 K. The spectra of U (*left*) and Te (*right*) are shown in Figure 2 taken at the temperatures indicated. The room temperature hyperfine parameters

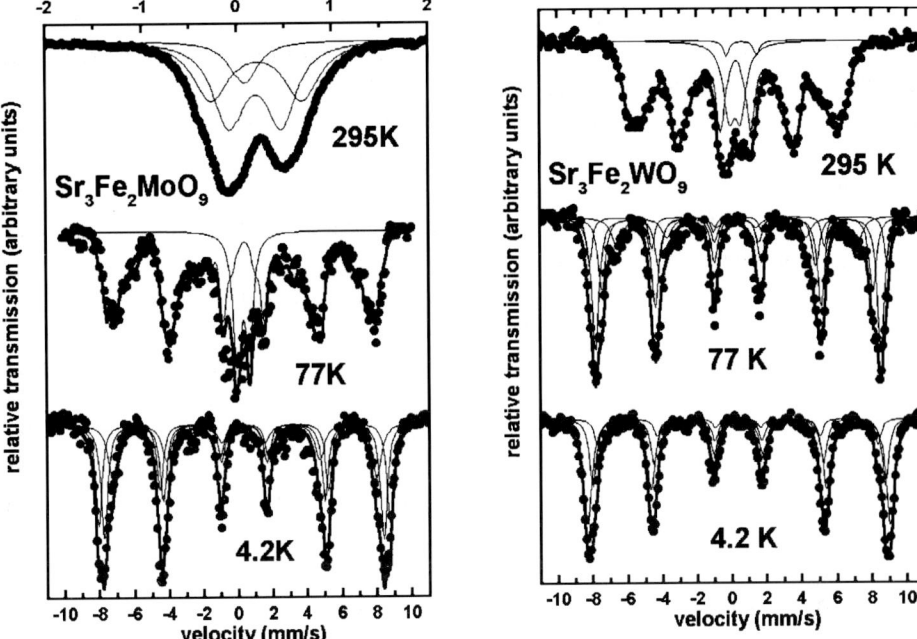

Figure 1. Mössbauer spectra at the temperatures indicated of $Sr_3Fe_2MoO_9$ (*left*) and $Sr_3Fe_2WO_9$ (*right*). The spectrum of $Sr_3Fe_2MoO_9$ at 295 K was taken within the lower velocity range indicated at the top. The solid line is the result of the fit as described in the text. The thinner solid lines are the magnetic and quadrupole components that make up the spectra.

for the four compounds ($IS = 0.33$–0.36 mm/s) are consistent with the presence of at least two Fe^{3+} ions with high spin state located at lattice positions with different degrees of distortion and/or disorder. Within the statistical uncertainty of the spectra, no Fe^{2+} phase exists in the samples. At 295 K the spectrum of the Mo compound exhibits a paramagnetic state while the spectra of the other three compounds display both paramagnetic and magnetic hyperfine structure. The magnetic absorption areas increase in approximately the same way as the crystallographic disorder: Mo < U < Te < W [9]. At 77 K the spectrum of $Sr_3Fe_2WO_9$ (Figure 1, *right*) is already magnetically split, while the other three spectra suggest Fe probes sensing different relaxation regimes, with a magnetic ordering increasing from U to Mo to Te.

At 4.2 K all compounds have a fully magnetically ordered state. Their average hyperfine fields are ranged from 49 to 53 T (Mo: 49.8 T; U: 51.6 T; W: 52.6 T; U: 54.6 T) while their isomer shifts vary from 0.45 to 0.50 mm/s, confirming thus high spin state Fe^{3+} ions. The spectra have been decomposed into different components under the same assumptions described in [7]. They are assigned to Fe atoms whose first neighbors have well-ordered M environments, disordered M environments, and Fe ions in M sites with six Fe neighbors. Details of the description of the hyperfine structure will be published elsewhere.

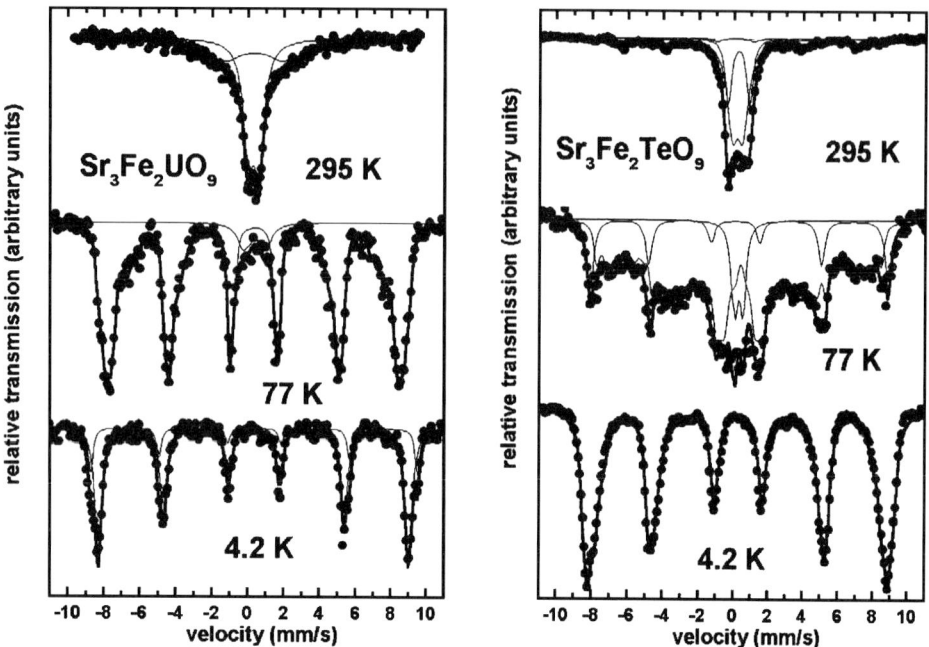

Figure 2. Mössbauer spectra at the temperatures indicated of Sr$_3$Fe$_2$UO$_9$ (*left*) and Sr$_3$Fe$_2$TeO$_9$ (*right*). The solid line is the result of the fit as described in the text. The thinner solid lines are the magnetic and quadrupole components that make up the spectra.

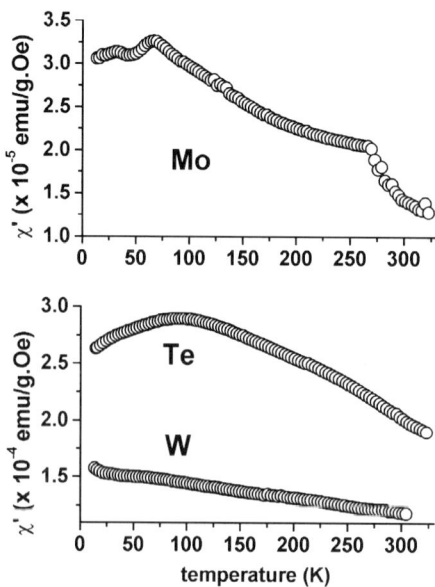

Figure 3. AC susceptibility in-phase component for the compounds labeled with their respective cations at 825 Hz.

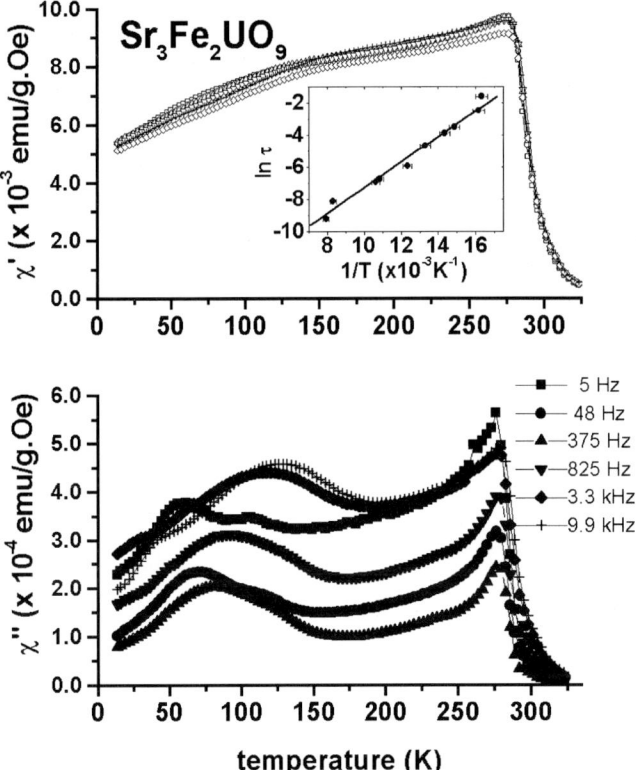

Figure 4. In-phase, χ', and out-of-phase, χ'', components of the AC susceptibility for $Sr_3Fe_2UO_9$ at the frequencies indicated. The inset shows the dependence of $\ln \tau$ *versus* T^{-1} and the resulting linear fit.

The AC in-phase component of the magnetic susceptibility is shown in Figure 3. The most striking feature is the drastically different response yielded by the four compounds of the series. The U compound has a magnetization (Figure 4) which is at all temperatures and frequencies one order of magnitude higher than those of Te and W, which, on their turn, are one of order higher than the Mo one. The out-of-phase component of $Sr_2Fe_3UO_9$ was the only one detectable and is also shown in Figure 4. The frequencies *versus* the inverse temperatures belonging to the secondary maxima of the $\chi''(T)$ curves seen at around 100 K are shown in the inset.

4. Discussion

The structure of the compounds, which crystallize in the tetragonal crystal system (space group I 4/m, $Z = 4$) is made up of Fe and B$'$ centered-oxygen octahedra that form a three-dimensional arrangement sharing all the corner-located oxygen atoms with the six neighbors.

It has to be noticed that the prevailing component at 295 K is due to the broadened magnetic sextet, in the case of $Sr_3Fe_2WO_9$. This result is consistent with recent results reported using neutron powder diffraction [10] that showed that the compound acquires a ferrimagnetic ordering below 373 K.

The behavior of the hyperfine fields thermal dependence, i.e., a distribution of hyperfine fields at all temperatures and probes sensing different magnetic regimes, is likely related to inhomogeneous exchange pathways across the samples. This occurs for the four compounds studied in this work and seems to reveal similar disorder features as those observed in Sr_2FeMoO_6. The main cause for the observed lack of saturation of the magnetic moment and behavior of the hyperfine fields in Sr_2FeMoO_6 was assigned to the existence of antisite defects and antiphase boundaries brought about by the chemical disorder [7]. A similar situation for $Sr_3Fe_2MoO_9$ has been reported with respect to the occupation sites refined by Rietveld analysis on neutron diffraction data [9].

At both 77 K and 300 K, the evolution of the hyperfine fields observed for each of the four different members of double perovskites studied here, is consistent with different Mössbauer magnetic ordering temperatures. On the other hand, the very similar average values of the hyperfine fields at 4.2 K reveal that the Fe ions are not so sensitive to the type of cation but the different width of the hyperfine field distribution suggest that the cationic disorder is dependent on the composition of the double perovskite.

The AC susceptibility response is very weak for the Mo compound (χ'_{max} 3.5 $\times$ 10^{-5} emu/g.Oe) and increases for the W, Te and U compounds with values of χ'_{max} 1.6 $\times$ 10^{-4}, 3.0 $\times$ 10^{-4}, and 9.4 $\times$ 10^{-3} emu/g.Oe, respectively. However, the temperature dependence of the in-phase component of the susceptibility of the Mo, Te and W compounds, is quite complex and requires a better knowledge of other magnetic parameters in order to suggest a model that may explain their behaviors.

The out-of-phase component could only be detected for the U compound (Figure 4). The curves display a principal maximum at around 280 K that is almost the same for all the measuring frequencies. This is probably related to the magnetic long range ordering of the U sublattice. However, they also exhibit a secondary maximum, at about 100 K that shifts with the frequency, $\omega = \tau^{-1}$, of the exciting AC field. The shift of the temperatures at which it displays the maxima for the different frequencies exhibits an Arrhenius dependence on the temperature, i.e., almost a linear dependence of the $\ln \tau$ versus T^{-1} (shown in the inset of Figure 4). The thermal shift of the peak of the susceptibility per decade of frequency variation, 0.09, allows establishing that the behavior can be related to a spin-glass-like state, i.e. neither a canonical spin glass nor a superparamagnetic regime [11]. This is likely related to the ordering of the Fe ions and so may be yet one more indication of the spin-glass-like state produced by the chemical disorder. This behavior may be common to the other three cations, but their χ'' values are too small to be detected.

5. Conclusions

From this preliminary study on the series of isostructural double perovskites $SrFe_{2/3}B1_{1/3}O_3$ (B = Mo, U, Te, and W), the present hyperfine and AC susceptibility results show that the overall magnetic response of the compounds changes drastically from one cation to the other. However, the low temperature hyperfine fields, although displaying slight alterations caused by the cation disorder, does not depend much on the type of cations occupying the centers of the octahedra neighbors to the Fe probes.

The frequency dependence of the maximum temperatures of the peaks of the out-of-phase component χ'' of $Sr_3Fe_2UO_9$ is indicative of a spin-glass-like state created by the chemical disorder.

References

1. Kobayashi, K. L., Kimura, T., Sawada, H., Terakura, K. and Tokura, Y., *Nature* **395** (1998), 677.
2. Kobayashi, K. L., Kimura, T., Tomioka, Y., Sawada, H., Terakura, K. and Tokura, Y., *Phys. Rev. B.* **59** (1999), 11159.
3. Maignan, A., Raveau, B., Martin, C. and Hervieu, M. J., *J. Solid State Chem.* **144** (1999), 224.
4. Ray, S., Kumar, A., Sarma, D. D., Cimino, R., Turchini, S., Zennaro, S. and Zema, N., *Phys. Rev. Lett.* **87** (2001), 097204.
5. Tovar, M., Causa, M. T., Butera, A., Navarro, J., Martínez, B., Fontcuberta, J. and Passeggi, M. C. G., *Phys. Rev. B* **66** (2002), 024409.
6. Fang, Z., Terakura, K. and Kanamori, J., *Phys. Rev. B* **63** (2001), 180407.
7. Greneche, J. M., Venkatesan, M., Suryanarayanan, R. and Coey, J. M. D., *Phys. Rev. B* **63** (2001), 174403.
8. Balcells, L., Navarro, J., Bibes, M., Roig, A., Martínez, B. and Fontcuberta, J., *Appl. Phys. Lett.* **78** (2001), 781.
9. Viola, M. C., Augsburger, M. S., Pinacca, R. M., Pedregosa, J. C., Carbonio, R. E. and Mercader, R. C., *J. Solid State Chem.* **175** (2003), 252.
10. Ivanov, S. A., Eriksson, S.-G., Tellgren, R. and Rundlof, H., *Mat. Res. Bull.* **36** (2001), 2585.
11. Mydosh, J. A., *Spin Glasses: an Experimental Introduction*, Taylor and Francis, London, 1993.

Hyperfine Interactions **156/157**: 165–168, 2004.
© 2004 *Kluwer Academic Publishers. Printed in the Netherlands.*

Modification of Nuclear Decay Constant in the Finite Space

M. T. JEONG
Department of Physics, Dongshin University, Naju 520-714, Korea

Abstract. The decay constants of nuclei are calculated based on the Cheon's idea, namely, the population of nuclear excited state can be modified by the mechanism that the gamma-ray emitted from the radioactive nuclei in a solid is reabsorbed by the source after reflected on a metal surface. Our results show that the nuclear lifetime is enhanced inside a lead cylinder.

Key words: nuclear lifetime, γ-emission, Debye–Waller factor.

1. Introduction

One of the challenging problems in low energy nuclear physics is to change nuclear decay constants. Then, various attempts were made so far, i.e. altering the chemical state of radioactive nuclei [1, 2] and applying high pressure [3] or low temperature [4], but most of them found only negligible variations.

On the other hand, the fact that the nuclear lifetime could be modified by presence of two parallel plates was first discovered by Cheon [5]. His interpretation of this phenomenon was that it would be induced by modulation of the population of the nucleus staying in the excited state through reabsorption of the emitted γ-ray by the source [6, 7]. Of course, in a free space the γ-ray emitted from the nucleus never come back to it. However, if the γ-ray is reflected by some sorts of mirrors, it can be return to the original source nucleus and be absorbed again.

2. Modified nuclear decay constant

Let us present here the formulation given by Cheon [7, 8]. The γ-ray absorption cross section is given by [9]

$$\hat{\sigma} = 2\pi \lambda_\gamma^2 \frac{2J_f + 1}{2J_i + 1} \frac{1}{1 + \alpha}, \tag{1}$$

where $\lambda_\gamma = \hbar c / E_\gamma$ with the γ-ray energy E_γ and α is the internal conversion coefficient. J_i and J_f are spins of initial and final nuclear states. The emission cross section is also given by the detail balanced relation. The returning γ-rays are absorbed by the number of source nuclei per cm^2. This number $\hat{n}$ can be estimated as follows: the mass of the molecule forming the solid in which the radioactive nuclei are planted is given by the mole number divided by the Avogadro's number. The number of molecules in cm^3 is obtained by dividing the density by its mass. Let this number and the source thickness be n and d, then we have $\hat{n} = nd$ in unit of cm^{-2} when d is measured in cm.

Now, the γ-ray scattering on the metallic surface occurs mostly through the Compton scattering which is actually inelastic process. The elastic scattering with no energy loss is the Rayleigh scattering which takes place in wavelength of visible light. In comparatively high energy, it is the Thomson scattering. The differential cross section of the Thomson scattering is given by the electron classical radius $r_0 = 2.8$ fm as $d\sigma(\theta)/d\Omega = r_0^2 \frac{1}{2}(1 + \cos^2\theta)$. The scattering angle is $\theta = \pi$ in our case, because the emitted photon returns back to the source. Therefore, $d\sigma(\theta)/d\Omega|_{\theta=\pi} = \sigma_\pi = r_0^2 = 7.8 \times 10^{-26}$ cm^2. The number of atoms per cm^2 in metal of thickness d_0 is given by its density ρ as $n_0 = \rho d_0 \times (Avogadro's\ number)/(gram\ molecule)$. d_0 must be replaced by $d_1 = d_0/\cos\theta_1$ in the direction of θ_1 against the normal direction. Therefore, the probability of the photon returning to the source is obtained as $Z n_0 \sigma_\pi$, where Z is the atomic number. However, these atoms as scattering bodies are bound in the solid and, thereby effect of lattice vibration should be taken into account. This effect depends on temperature, and is well known as the Debye–Waller factor, which is expressed as

$$f = \exp\left\{ \frac{-3E_0^2}{Mc^2 k_B \theta_D} \left[\frac{1}{4} + \left(\frac{T}{\theta_D}\right) \int_0^{\theta_D/T} \frac{x\,dx}{e^x - 1} \right] \right\}, \tag{2}$$

where E_0 is the photons energy, M the mass of the atom, k_B the Boltzmann constant, θ_D the Debye temperature of the metal and T the temperature. The number of photons surviving in passing through the source by a depth d is given by

$$I = I_0 \exp(-f_a d n \hat{\sigma}) = I_0 \exp(-f_a \hat{n} \hat{\sigma}), \tag{3}$$

where I_0 is the number of the incident photons and f_a is the DW factor of the source. Therefore, the fraction of incident photons which are absorbed is

$$\zeta = \frac{I_0 - I}{I_0} = 1 - \exp(-f_a \hat{n} \hat{\sigma}). \tag{4}$$

Notice that ζ never be larger than unity.

Generally, the radioactive nuclei emit γ-ray isotropically and, then, the number of photons emitted in one direction during the time dt is given by

$$\frac{d}{d\Omega}(dN) = \frac{\lambda N}{4\pi} dt, \tag{5}$$

where λ is the decay constant. Therefore, the number of photons hitting the reflector is obtained by integrating over the solid angle which the source looks the whole reflector.

Thus, if emitted photons return once after being reflected coherently by the metallic surface, the equation of decay is given by the decay constant λ as

$$dN = -\lambda N \, dt + \zeta \frac{1}{4\pi} \int (f Z n_1 \sigma_\pi) \lambda N \, dt \, d\Omega \equiv -\lambda' N \, dt, \tag{6}$$

where $\lambda' = a\lambda$ with $a = 1 - \zeta \frac{1}{4\pi} \int (f Z n_1 \sigma_\pi) d\Omega$. such a process repeats m times during the lifetime, the decay constant is found in the form $\tilde{\lambda} = a^m \lambda$. m is actually dependent on the distance between the source nucleus and the point of reflection. Let this distance be u, then the time which the photon takes in a round trip is $t_r = 2u/c$. When the photon made its round trips m times, the amount of radioactive nuclei would become a half, i.e., $\frac{1}{2} = \exp[-\lambda^{(m)} m t_r] = \exp[-\lambda^{(m)} m \frac{2u}{c}]$. By this equation, we can evaluate the value of m as $m = c \ln 2/[2u\lambda^{(m)}] = c\tau_{1/2}/[2ua^m]$.

Assuming a cylinder with a radius R and length L, one can obtain $u(z) = (R^2 + z^2)^{1/2}$, when the source is located at the center of the cylinder. z is the position of the scattering body measured from the middle of the cylinder.

For this case, we find

$$a = 1 - 0.727\zeta f Z n_0 \sigma_\pi. \tag{7}$$

3. Results and discussion

To examine the theory described above, let us take the case of the first excited state $\frac{3}{2}^+$ of ^{57}Fe which is 14.4 keV level with the lifetime 97.81 ns. The source was placed at the center of a lead cylinder of 1.4 cm inner diameter and 5 cm length. Thickness of the lead cylinder was 5 mm, and the absorber thickness was 11 mg/cm^2. When the thickness of the source is 11 mg/cm^2, we find $\hat{n} = 1.120 \times 10^{20}$ molecules/cm^2. Since the spin of the ground state is $\frac{1}{2}^+$ and the internal conversion coefficient α is 9.0, the absorption cross section can be obtained as $\hat{\sigma} = 2.359 \times 10^{-16}$ cm^2. The thickness $D = 0.5$ mm of a lead cylinder gives the value of $n_1 = 1.642 \times 10^{22}$ cm^{-2}. The Thomson cross section may approximately be used, because the electron binding energy in Pb is still small compared with 14.4 keV photon energy. The Debye temperature of lead is 88 degrees.

Result of our numerical calculations yields $\tilde{\tau}_{1/2} = 2.53\tau_{1/2}$ at 293 K. Namely, the half-life of ^{57}Fe (14.4 keV) is prolonged in a cylinder by 153%. This surprising large value may be due to neglect of screening effect on Z which would be significant for large photon wave length. If the effective atomic number $Z' = 14.6$ is used, it turns out to be 8.3% prolongation.

Recently, Russian experiment [10] was performed with the gamma-source ^{57}Co(^{57}Fe*) and absorber ^{57}Fe. The result obtained in this measurement by time delayed coincidence method yields 10% increase of the lifetime of 14.4 keV excited level. Another measurement of ^{133}Cs (81 keV) level lifetime showed 8~26% enhancement when the gamma-ray source was placed between two parallel silicon flat plates [8].

It should be stressed that the recent experiment [11] contains not only radiation trapping but also alteration of nuclear lifetime through the mechanism explained above. Another case is that the γ-ray emitted from the nucleus excited in the matter might be delayed to reach the detector [12]. This case is related to the collision time and not directly to the lifetime.

Acknowledgements

The author would like to thank a professor Il-T. Cheon for providing most of the original works presented here. This work was supported by the Korea Research Foundation Grant (KRF-2003-015-C00124).

References

1. Segrè, E., *Phys. Rev.* **71** (1947), 274.
2. Bainbridge, K. T., Goldhaber, M. and Wilson, E., *Phys. Rev.* **90** (1953), 430.
3. Bainbridge, K. T., *Chem. Eng. News* **30** (1952), 654.
4. Byers, D. H. and Stump, R., *Phys. Rev.* **112** (1958), 77.
5. Cheon, Il-T., Presented at the Workshop on "Application of Field Theory"; the 1999 Meeting of Phys. Soc. Japan.
6. Cheon, Il-T., *J. Phys. Soc. Japan* **70** (2001), 3193.
7. Cheon, Il-T., to be published in *Prog. Theor. Phys.* (2004).
8. Cheon, Il-T. and Jeong, M. T., T3/5 in this volume, ICAME03.
9. Condon, E. U. and Odishaw, H., *Handbook of Physics*, 2nd edn, McGraw-Hill, 1967, p. 9.
10. Vysotskii, V. I., *et al.*, In: *Proc. 8th Russian Conf. on Cold Nuclear Transmutation of Chemical Elements*, Dagomys, Oct. 4, 2000.
11. Chumakov, A. I., *et al.*, *Phys. Rev. B* **56** (1997), R8455.
12. Smirnov, G. V. and Shvyd'ko, Yu. V., *JETP* **68** (1989), 444.

Hyperfine Interactions **156/157**: 169–174, 2004.
© 2004 *Kluwer Academic Publishers. Printed in the Netherlands.*

Mössbauer Studies and Magnetic Properties of $Y_{3-x}Ce_xFe_5O_{12}$

JUN SIG KUM, SAM JIN KIM, IN BO SHIM and CHUL SUNG KIM*

Dept. of Physics, Kookmin University, Seoul 136-702, Korea; e-mail: cskim@phys.kookmin.ac.kr

Abstract. Magnetic and crystallographic properites of $Y_{3-x}Ce_xFe_5O_{12}$ ($x = 0.0$, 0.1, and 0.3) have been studied with X-ray diffraction, vibrating sample magnetometer (VSM), and Mössbauer spectroscopy. A small coercivity ($H_c = 5.8$ Oe), was obtained for the sample $Y_{2.9}Ce_{0.1}Fe_5O_{12}$, which is comparable to that of an undoped sample $Y_3Fe_5O_{12}$ ($H_c = 54.1$ Oe). Mössbauer spectra of $Y_{3-x}Ce_xFe_5O_{12}$ were measured at various absorber temperatures from 4.2 K to Néel temperature. It is found that Debye temperatures of octahedral ($16a$) and tetrahedral ($24d$) site for $Y_{2.9}Ce_{0.1}Fe_5O_{12}$ are $\Theta_a = 353$, $\Theta_d = 464$ K, respectively, and for $Y_{2.7}Ce_{0.3}Fe_5O_{12}$, $\Theta_a = 380$, $\Theta_d = 444$ K, respectively. The intersublattice a–d superexchange interaction was found to be antiferromagnetic with the strength of $J_{a-d} = -21.42$ k_B, while the intrasublattice interactions a–a, d–d were found to be ferromagnetic with strengths of $J_{a-a} = 4.50$ k_B and $J_{d-d} = 0.02k_B$, respectively, in the sample $Y_{2.9}Ce_{0.1}Fe_5O_{12}$.

Key words: superexchange interaction, Mössbauer spectroscopy, Debye temperature, coercivity, Ce doped garnet.

1. Introduction

A low-coercivity, high-remanence, soft magnetic material, having a square hysteresis loop, is required for microwave operation. For a magnetic material to be applied in microwave devices, the most important static magnetic properties are the saturation magnetization (M_s), anisotropy constants, Néel temperature (T_N), remanent magnetization (M_{rr}), coercivity (H_c), and temperature derivative of these quantities. In general, high T_N and low H_c are required. Also, low anisotropy is one of the conditions for low coercivity [1].

Great deals of works have been done on the substitution of magnetic ions to vary magnetic coercivity of yttrium iron garnet (YIG) systems [2]. Cerium-substituted YIG (Ce:YIG) have been found to exhibit a large magneto-optic (MO) effect and low propagation loss, which will be good candidate materials for devices with higher quality [3]. In YIG ($Y_3Fe_5O_{12}$) system, both octahedral (a) and tetrahedral (d) sites are occupied by Fe^{3+} ions. The molecular fields act on each Fe ion in a site and d site [4]. The strongest magnetic interactions are the intersublattice exchange interaction between the Fe^{3+} ions in the a and d sublattices, while intrasublattice exchange interactions (a–a and d–d) are known to be small [5]. In order to control

---

* Author for correspondence.

magnetic behaviors, the thorough understanding of interaction mechanism between magnetic ions is very important. In this paper, we have studied exchange integrals, J_{a-d}, J_{a-a}, and J_{d-d} of superexchange interactions of the garnet-related structures of composition $Y_{3-x}Ce_xFe_5O_{12}$ ($x = 0.0, 0.1$, and 0.3) and their examinations by Mössbauer spectroscopy.

2. Experiments

Compounds of composition $Y_{3-x}Ce_xFe_5O_{12}$ ($x = 0.0, 0.1$, and 0.3) were prepared using the sol–gel method. Weighted amounts of $Y(NO_3)_3 \cdot 5H_2O$, $Ce(NO_3)_3 \cdot 6H_2O$, and $Fe(NO_3)_3 \cdot 9H_2O$ were first dissolved in 2-methoxyethanal (2-MOE) and acetic acid.

The solution was refluxed at 80°C for 12 h to allow gel formation, and then dried at 120°C for 24 h. The dried powder were ground and annealed at a temperature 1400°C for 3 h in air. The compositions of the sintered samples were identified by an X-ray diffractometer. Magnetization was measured with a vibrating sample magnetometer. Mössbauer spectra were recorded at temperatures ranging from 4.2 K to T_N using a constant acceleration Mössbauer spectrometer with a ^{57}Co in Rh matrix source.

3. Results and discussion

It is shown that powders have only a single phase of the garnet structure regardless of the amount of Ce substitution according to X-ray diffraction patterns as shown in Figure 1. The lattice constants of $Y_3Fe_5O_{12}$ and $Y_{2.7}Ce_{0.3}Fe_5O_{12}$ were found to be

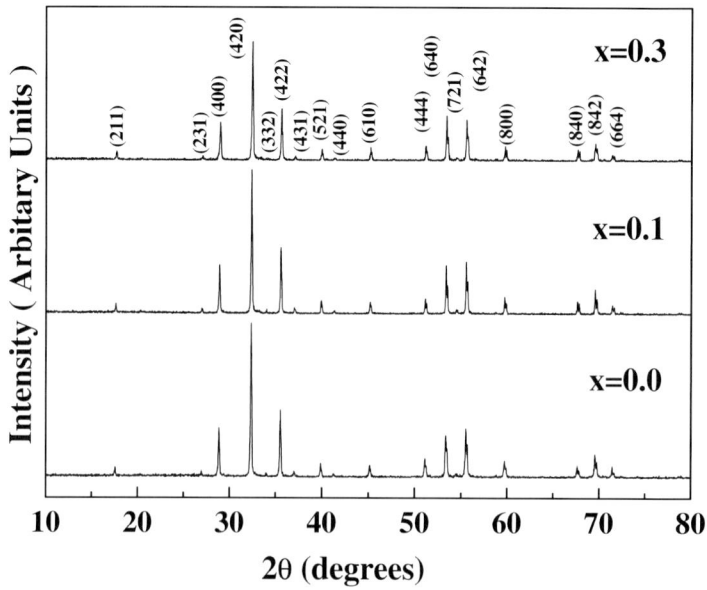

Figure 1. XRD patterns for $Y_{3-x}Ce_xFe_5O_{12}$ ($x = 0.0, 0.1$, and 0.3).

Table I. Lattice parameter (a_0), saturated magnetization (M_s), coercivity (H_c), Néel temperature (T_N), magnetic hyperfine field (H_{hf}), quadrupole splitting (ΔE_Q), and isomer shifts (δ), at room temperature for $Y_{3-x}Ce_xFe_5O_{12}$ ($x = 0.0, 0.1,$ and 0.3)

x	a_0 (Å)	M_s (emu/g)	H_c (Oe)	T_N (K)	H_{hf} (kOe)		ΔE_Q (mm/s)		δ (mm/s)	
					16a	24d	16a	24d	16a	24d
0.0	12.381	26.5	54.1	590	486	394	0.01	0.03	0.26	0.02
0.1	12.399	27.5	5.8	585	491	399	0.01	0.01	0.36	0.14
0.3	12.408	26.7	12.4	580	492	398	0.02	0.01	0.27	0.04

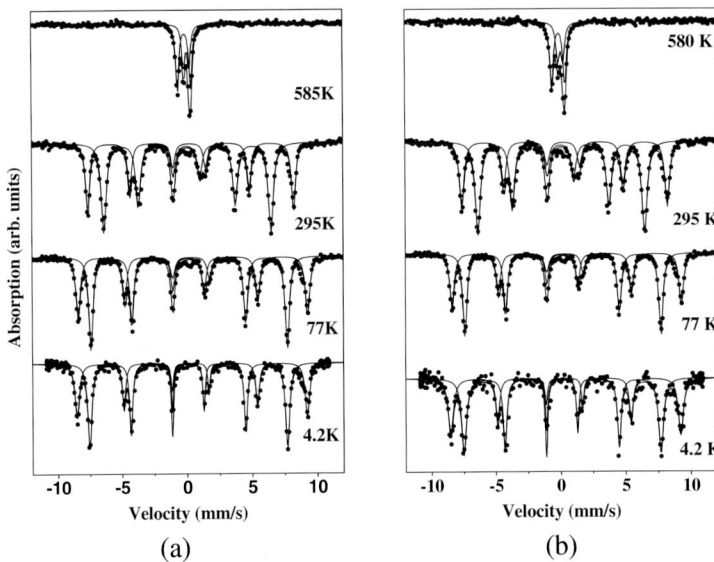

Figure 2. Mössbauer spectra for $Y_{3-x}Ce_xFe_5O_{12}$ at various temperature ranging from 4.2 K to Néel temperatures: (a) $Y_{2.9}Ce_{0.1}Fe_5O_{12}$, (b) $Y_{2.7}Ce_{0.3}Fe_5O_{12}$.

$a_0 = 12.381$ and 12.408 Å, respectively. The lattice constants increase linearly with increasing Ce concentration. This results accords with Vegard's law.

The saturation magnetization and coercivities of sintered powders were measured using a VSM up to a maximum applied field of 5000 Oe at room temperature. The saturation magnetization was not changed fairly but coercivity was minimum at $Y_{2.9}Ce_{0.1}Fe_5O_{12}$, with increasing Ce concentration. The minimum coercivity of $Y_{2.9}Ce_{0.1}Fe_5O_{12}$ powder with applied field up to 5000 Oe is about 5.8 Oe. The lattice parameter saturated magnetization, and coercivities for the samples of $Y_{3-x}Ce_xFe_5O_{12}$ are listed in Table I.

Mössbauer absorption spectra of $Y_{3-x}Ce_xFe_5O_{12}$ were taken at various temperatures. Figures 2(a) and (b) illustrate the representative spectra for the $Y_{2.9}Ce_{0.1}$-Fe_5O_{12} and $Y_{2.7}Ce_{0.3}Fe_5O_{12}$ samples, respectively, which were composed of two six-line hyperfine patterns d (inner sextet) and a (outer sextet). The Mössbauer parameters, magnetic hyperfine filed, quadrupole splitting, and isomer shifts at

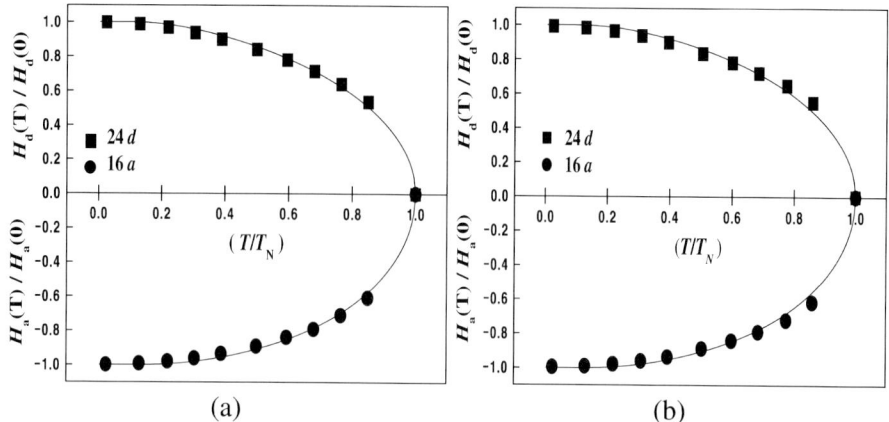

Figure 3. Reduced magnetic hyperfine fields $H(T)/H(0)$ for $16a$ and $24d$ sites in $Y_{3-x}Ce_xFe_5O_{12}$ as a function of reduced temperature T/T_N. The solid lines represent calculated reduced magnetizations. (a) $Y_{2.9}Ce_{0.1}Fe_5O_{12}$, (b) $Y_{2.7}Ce_{0.3}Fe_5O_{12}$.

Table II. Strength of superexchange parameters and Debye temperatures for $Y_{3-x}Ce_xFe_5O_{12}$ ($x = 0.0, 0.1,$ and 0.3)

x	J_{a-d} (k_B)	J_{a-a} (k_B)	J_{d-d} (k_B)	Θ_a (K)	Θ_d (K)
0.0	−24.71	0.17	0.02	385	472
0.1	−21.42	4.50	0.02	353	464
0.3	−21.75	3.87	0.02	342	417

room temperature for $Y_{3-x}Ce_xFe_5O_{12}$ are presented in Table I. Figure 3 shows the reduced magnetic hyperfine field $H(T)/H(0)$ for the $24d$ and $16a$ sites of $Y_{3-x}Ce_xFe_5O_{12}$ as a function of the reduced temperature $\tau = T/T_N$. The temperature dependence of the magnetic hyperfine field in ^{57}Fe nuclei at the d and a sites were analyzed based on the Néel theory of ferrimagnetism. The detailed theoretical procedures are published in [6]. The superexchange parameters and Debye temperatures of $16a$ and $24d$ sites for $Y_{2.9}Ce_{0.1}Fe_5O_{12}$ are listed Table II. The intersublattice a–d superexchange interaction, for $Y_{2.9}Ce_{0.1}Fe_5O_{12}$ was found to be antiferromagnetic with a strength of $J_{a-d} = -21.42\ k_B$, while the intra-sublattice interactions a–a, d–d were found to be ferromagnetic with a strength of $J_{a-a} = 4.50\ k_B$ and $J_{d-d} = 0.02\ k_B$, respectively. For $Y_{2.7}Ce_{0.3}Fe_5O_{12}$, the strengths of superexchange integrals were $J_{a-d} = -21.77\ k_B$, $J_{a-a} = 3.86\ k_B$ and $J_{d-d} = 0.00\ k_B$. Here, k_B is the Boltzmann constant. The strongest a–d interaction does not changed fairly, finally it showed no considerable changes in Néel temperatures and saturation magnetization. However, coercivities of samples were changed sensitively with Ce contents.

As the Ce concentration x increases, the d–d interaction increases form $0.76\ k_B$ to $4.50\ k_B$ when x increases form 0.0 to 0.1, and took maximum at $x = 0.1$ before

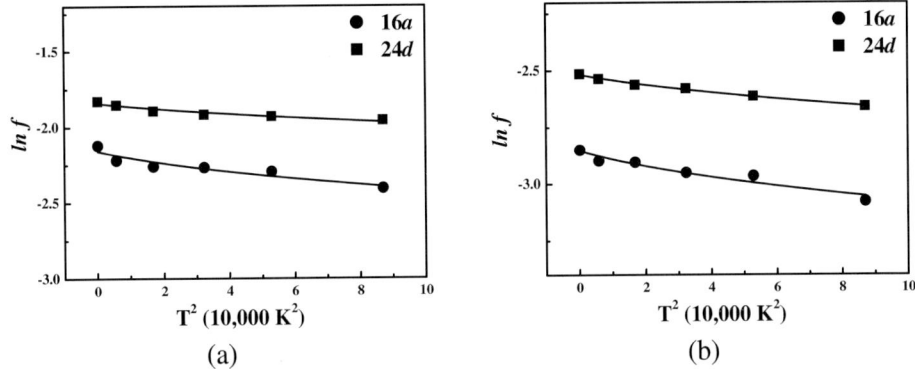

Figure 4. Natural logarithm of the absorption area, f, vs T^2 for the $16a$ and $24d$ subspectra of (a) $Y_{2.9}Ce_{0.1}Fe_5O_{12}$ and (b) $Y_{2.7}Ce_{0.3}Fe_5O_{12}$. The solid lines were calculated from Equation (2).

decreasing again and then decreased to 3.86 k_B for $x = 0.3$. We note that there is an opposite between the H_c and the strength of J_{d-d}.

Implicitly, it can be concluded that d–d interaction weakens magnetic anisotropy and leads to lower coercivity in Ce doped Garnet materials [7, 8].

Figure 4 shows the temperature dependence of the resonant absorption area of the $Y_{3-x}Ce_xFe_5O_{12}$ samples. The Debye model gives the following expression for the recoil-free fraction [9],

$$f = \exp\left[-\frac{3E_R}{2k_B\Theta}\left(1 + \frac{4T^2}{\Theta^2}\int_0^{\Theta/T}\frac{x\,dx}{e^x - 1}\right)\right]. \tag{1}$$

Here E_R is the recoil energy of ^{57}Fe for the 14.4 keV gamma ray, and Θ represents Debye temperatures. Taking the logarithm of both sides in Equation (1), one can get

$$\ln f = -\frac{3E_R}{2k_B\Theta}\left(1 + \frac{4T^2}{\Theta^2}\int_0^{\Theta/T}\frac{x\,dx}{e^x - 1}\right). \tag{2}$$

When $\ln f$ is plotted as a function of T^2, one will get a curve, which becomes almost a straight line at low temperatures. Equation (2), with a proper additive constant, was fitted to the data in Figure 4 using a least-squares computer program to get a Debye temperature [10, 11]. The estimated Debye temperatures of octahedral ($16a$) and tetrahedral ($24d$) site for $Y_{2.9}Ce_{0.1}Fe_5O_{12}$ are $\Theta_a = 353$, $\Theta_d = 464$ K, respectively, and for $Y_{2.7}Ce_{0.3}Fe_5O_{12}$, they are $\Theta_a = 380$, $\Theta_d = 444$ K, respectively. This suggests a larger inter-atomic binding force for the d-site than for the a-site.

In summary, we have studied the magnetic properties of $Y_{3-x}Ce_xFe_5O_{12}$ ($x = 0.0$, 0.1, and 0.3). The T_N was not changed fairly from 590 K ($x = 0.0$), with increasing Ce contents, while coercivity of the sample drastically reduced from 54.1 Oe ($x = 0.1$) to 5.8 Oe ($x = 0.1$). For an application to the microwave devices, samples should have a lower coercivity. Therefore, it is concluded that Ce doped $Y_{2.9}Ce_{0.1}Fe_5O_{12}$ sample is the pronounced candidate for microwave devices material. In addition, the temperature dependence of magnetic hyperfine fields was analyzed by Néel theory of ferrimagnetism.

Acknowledgement

This work was supported by the KOSEF (R02-2003-000-10046-0).

References

1. Pardavi-Horvath, M., *J. Mag. Magn. Mater.* **215–216** (2000), 171.
2. Kim, H., Grishin, A. and Rao, K. V., *J. Appl. Phys.* **89** (2001), 4380.
3. Shintaku, T., Tate, A. and Mino, S., *Appl. Phys. Lett.* **71** (1997), 1640.
4. Kim, C. S., Uhm, Y. R., Kim, S. B. and Lee, J. G., *J. Magn. Magn. Mater.* **215–216** (2000), 551.
5. Vaquerio, P., Crosner-Lopez, M. P. and Lopez-Quintela, M. A., *J. Solid State Chem.* **126** (1987), 161.
6. Kim, S. J., Lee, S. W. and Kim, C. S., *Jpn. J. Appl. Phys.* **40** (2001), 4897.
7. Yunus, S. M., Ahmed, F. U. and Asgar, M. A., *J. Alloys Compd.* **315** (2001), 90.
8. Chakravarthy, R., Madhav Rao, L., Paranjpe, S. K., Kulshrestha, S. K. and Roy, S. B., *Phys. Rev. B* **43** (1991), 6031.
9. Mössbauer, R. L. and Widermann, W. H., *Z. Phys.* **159** (1962), 33.
10. Kim, C. S., Shim, I. B., Ha, M. Y., Choi, H. and Sur, J. C., *J. Kor. Phys.* **23** (1990), 166.
11. Ok, H. N. and Kim, Y. K., *Phys. Rev. B* **36** (1987), 5120.

Hyperfine Interactions **156/157**: 175–179, 2004.
© 2004 *Kluwer Academic Publishers. Printed in the Netherlands.*

Control of Charge Transfer Phase Transition in Iron Mixed-Valence System $(n\text{-}C_nH_{2n+1})_4N[Fe^{II}Fe^{III}(dto)_3]$ $(n = 3\text{–}6; dto = C_2O_2S_2)$

N. KOJIMA[1], Y. ONO[1], Y. KOBAYASHI[2] and M. SETO[2]
[1] *Graduate School of Arts and Sciences, The University of Tokyo, Tokyo 153-8902, Japan*
[2] *Research Reactor Institute, Kyoto University, Osaka 590-0494, Japan*

Abstract. Iron mixed-valence complex, $(n\text{-}C_3H_7)_4N[Fe^{II}Fe^{III}(dto)_3]$ (dto $= C_2O_2S_2$), shows a charge transfer phase transition around 120 K and a ferromagnetic transition at 6.5 K. In order to control the charge transfer phase transition by changing the cation size, we synthesized $(n\text{-}C_nH_{2n+1})_4$-$N[Fe^{II}Fe^{III}(dto)_3]$ $(n = 3\text{–}6)$ and investigated the charge transfer phase transition by means of ^{57}Fe Mössbauer spectroscopy. The charge transfer phase transition takes place around 120 K for $n = 3$ and 4, while it does not take place between 300 K and 4 K for $n = 5$ and 6.

Key words: mixed valence, charge transfer, ^{57}Fe Mössbauer spectroscopy, ferromagnetism.

1. Introduction

Octahedral transition metal complexes with d^4–d^7 configuration have a possibility of spin transition between a low-spin state and a high spin state. In the case of mixed-valence complexes whose spin states are situated in the spin-crossover region, it is expected that new types of conjugated phenomena coupled with spin and charge take place between neighboring metal ions in order to minimize the free energy in the whole system [1].

Recently, we have discovered a new type of first order phase transition around 120 K for $(n\text{-}C_3H_7)_4N[Fe^{II} Fe^{III}(dto)_3]$ (dto $= C_2O_2S_2$), where the thermally induced charge transfer between Fe^{II} and Fe^{III} occurs reversibly, which is schematically shown in Figure 1 [2]. Moreover, we have found a ferromagnetic transition at 6.5 K [2]. $(n\text{-}C_3H_7)_4N[Fe^{II} Fe^{III}(dto)_3]$ has a two-dimensional honeycomb network structure with an alternating array of Fe^{II} and Fe^{III} atoms through dto bridges, which was determined by the X-ray structural analysis for the single crystal [3]. In this paper, we report the charge transfer phase transition for $(n\text{-}C_nH_{2n+1})_4N$-$[Fe^{II}Fe^{III}(dto)_3]$ $(n = 3\text{–}6)$ and the cation size effect on the appearance of the phase transition by means of ^{57}Fe Mössbauer spectroscopy.

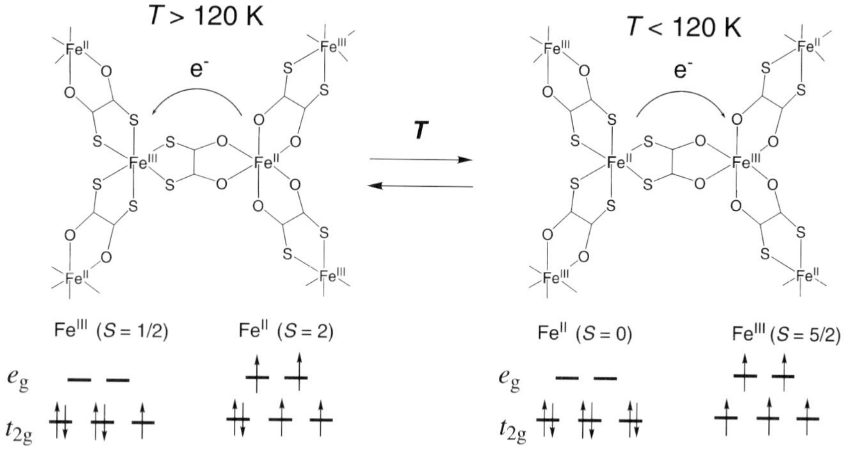

Figure 1. Charge transfer phase transition for $(n\text{-}C_3H_7)_4N[Fe^{II}Fe^{III}(dto)_3]$.

2. Experimental

$(n\text{-}C_nH_{2n+1})_4N[Fe^{II}Fe^{III}(dto)_3]$ ($n = 3$–6) were prepared in the similar way as reported in [2]. For ^{57}Fe Mössbauer spectroscopic measurement, ^{57}Co in Rh was used as a Mössbauer source. The spectra were calibrated by using the six lines of a body-centered cubic iron foil (α-Fe), the center of which was taken as zero isomer shift.

3. Results and discussion

Figure 2 and Table I show the ^{57}Fe Mössbauer spectra and the Mössbauer parameters for $(n\text{-}C_nH_{2n+1})_4N[Fe^{II}Fe^{III}(dto)_3]$ ($n = 3$–6), respectively. The assignment of the spectra A, B, C and D were confirmed by the ^{57}Fe Mössbauer spectra of $(n\text{-}C_3H_7)_4N[^{57}Fe^{II}Fe^{III}(dto)_3]$ and $(n\text{-}C_3H_7)_4N[Fe^{II\,57}Fe^{III}(dto)_3]$.[*] At 200 K, the line profiles of all the complexes are quite similar to each other. As shown in Table I, the isomer shift (*IS*) and the quadrupole splitting (*QS*) of the spectrum A at 200 K for $(n\text{-}C_nH_{2n+1})_4N[Fe^{II}Fe^{III}(dto)_3]$ ($n = 3$–6) are quite similar to those (*IS* = 0.33 mm/s, *QS* = 0.35 mm/s at 196 K) of the ^{57}Fe Mössbauer spectrum for the Fe^{III} ($S = 1/2$) site in $KBa[Fe^{III}(dto)_3]\cdot 3H_2O$ [4], where the Fe^{III} site is coordinated by six S atoms. On the other hand, the *IS* and *QS* of the spectrum B at 200 K are quite similar to those (*IS* = 1.235 mm/s, *QS* = 1.42 mm/s at 190 K) of the ^{57}Fe Mössbauer spectrum for the Fe^{II} ($S = 2$) site in $(n\text{-}C_4H_9)_4N[Fe^{II}Fe^{III}(ox)_3]$ (ox = oxalato) [5], where the Fe^{II} site is coordinated by six O atoms. Therefore,

---

[*] In our previous assignment for $(n\text{-}C_3H_7)_4N[Fe^{II}Fe^{III}(dto)_3]$ [2], the single peak of Fe^{III} ($S = 1/2$) is inside the doublet of Fe^{II} ($S = 2$) at 200 K, and the single peak of Fe^{II} ($S = 0$) is outside the doublet of Fe^{III} ($S = 5/2$) at 77 K. In our present assignment, however, *IS* and *QS* of Fe^{II} ($S = 0, 2$) and Fe^{III} ($S = 1/2, 5/2$) for $(n\text{-}C_3H_7)_4N[Fe^{II}Fe^{III}(dto)_3]$ are quite similar to those of Fe^{II} ($S = 0, 2$) and Fe^{III} ($S = 1/2, 5/2$) for typical iron complexes.

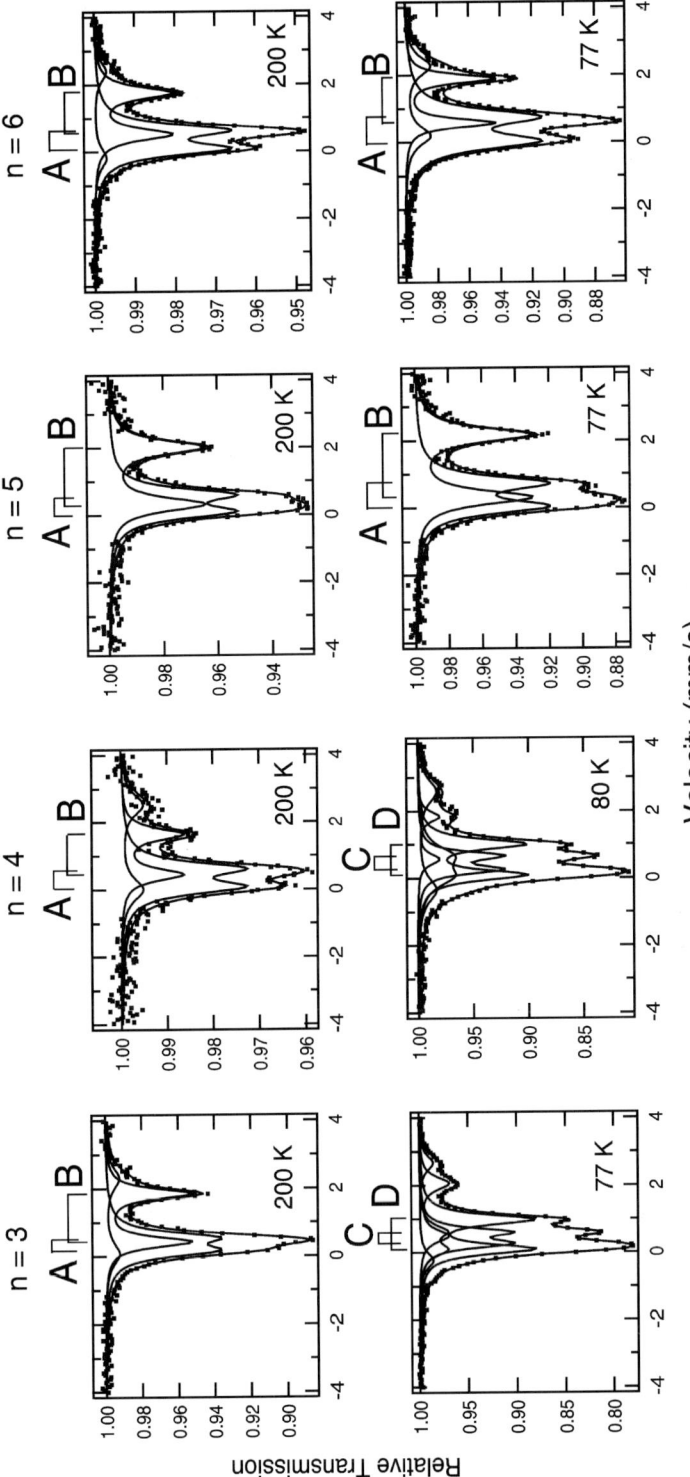

Figure 2. ^{57}Fe Mössbauer spectra of $(n\text{-}C_nH_{2n+1})_4N[Fe^{II}Fe^{III}(dto)_3]$ $(n = 3–6)$. A: Fe^{III} $(S = 1/2)$, B: Fe^{II} $(S = 2)$, C: Fe^{II} $(S = 0)$, D: Fe^{III} $(S = 5/2)$.

Table I. Mössbauer parameters for $(n\text{-}C_nH_{2n+1})_4N[Fe^{II}Fe^{III}(dto)_3]$ $(n = 3\text{–}6)$

Spectrum	Assignment	T (K)	IS (mm/s)				QS (mm/s)			
			3	4	5	6	3	4	5	6
A	Fe^{III} $(S = 1/2)$	200	0.35	0.29	0.35	0.31	0.38	0.55	0.51	0.56
		77(80)	0.38	0.36	0.39	0.38	0.55	0.60	0.68	0.74
B	Fe^{II} $(S = 2)$	200	1.14	1.05	1.17	1.14	1.43	1.20	1.69	1.24
		77(80)	1.28	1.20	1.24	1.23	1.57	1.27	1.84	1.31
C	Fe^{II} $(S = 0)$	77(80)	0.43	0.46	–	–	0.35	0.41	–	–
D	Fe^{III} $(S = 5/2)$	77(80)	0.54	0.55	–	–	0.89	0.92	–	–

it is concluded that the Fe^{II} $(S = 2)$ and Fe^{III} $(S = 1/2)$ sites in $(n\text{-}C_nH_{2n+1})_4$-$N[Fe^{II}Fe^{III}(dto)_3]$ $(n = 3\text{–}6)$ are coordinated by six O atoms and six S atoms, respectively. In the cases of $n = 3, 4$ and 6, besides the ^{57}Fe Mössbauer spectra due to the $Fe^{III}S_6$ and $Fe^{II}O_6$ octahedra, a weak spectrum with two branches at about 0.0 mm/s and 2.5 mm/s is observed, which is attributed to the high-spin state $(S = 2)$ of $Fe^{II}O_4S_2$ octahedron caused by the linkage isomerization on precipitation process.

In the cases of $n = 3$ and 4, at 77 K (80 K for $n = 4$), the spectra A and B decrease by about 80%. Instead of these spectra, the spectra C and D appear. The IS and QS of the spectrum C for $n = 3$ and 4 are similar to those ($IS = 0.325$ mm/s, $QS = 0.39$ mm/s) of the Fe^{II} $(S = 0)$ site in $[Fe^{II}(bipy)_3](ClO_4)_2$ (bipy $= 2,2'$-bipyridine) [6]. On the other hand, the IS and QS of the spectrum D for $n = 3$ and 4 are similar to those ($IS = 0.486$, $QS = 0.64$ at 90 K) of the ^{57}Fe Mössbauer spectrum for the Fe^{III} $(S = 5/2)$ site in $(n\text{-}C_4H_9)_4N[Fe^{II}Fe^{III}(ox)_3]$ [5].

From the ^{57}Fe Mössbauer spectra for $(n\text{-}C_nH_{2n+1})_4N[Fe^{II}Fe^{III}(dto)_3]$ $(n = 3$ and 4), it is obvious that the charge transfer phase transition takes place between 200 K and 77 K for $n = 3$ and 4. The coexistence of the higher and lower temperature phases at 77 K is typical of first order phase transition, which reflects on the thermal hysteresis in magnetic susceptibility [2]. From the analysis of heat capacity, the critical temperatures were determined to be 122.4 K and 141.1 K for $n = 3$ and 4, respectively [1, 7].

In the cases of $n = 5$ and 6, the line profile of the ^{57}Fe Mössbauer spectra remains unchanged between 200 K and 77 K, which implies that the charge transfer phase transition does not take place for $n = 5$ and 6. In fact, the higher temperature phase exists between 300 K and 4 K for $n = 5$ and 6. The charge transfer phase transition is sensitive to the size of 2D honeycomb network structure. The increase of cation size expands the honeycomb ring, which presumably stabilizes the higher temperature phase. Consequently, the increase of cation size suppresses the charge transfer phase transition, which is supported by pressure effect on the charge transfer phase transition [8].

Acknowledgement

This work was supported by a Grant-in-Aid for Scientific Research and a Grant-in-Aid from 21st Century COE (Center of Excellence) Program (Research Center for Integrated Science) from the Ministry of Education, Culture, Sports, Science and Technology, Japan.

References

1. Nakamoto, T., Miyazaki, Y., Itoi, M., Ono, Y., Kojima, N. and Sorai, M., *Angew. Chem., Int. Ed.* **40** (2001), 4716.
2. Kojima, N., Aoki, W., Itoi, M., Ono, Y., Seto, M., Kobayashi, Y. and Maeda, Yu., *Solid State Commun.* **120** (2001), 165.
3. Itoi, M., Taira, A., Enomoto, M., Matsushita, N., Kojima, N., Kobayashi, Y., Asai, K., Koyama, K., Nakano, T., Uwatoko, Y. and Yamaura, J., *Solid State Commun.* **130** (2004), 415.
4. Birchall, T. and Tun, K. M., *Inorg. Chem.* **15** (1976), 376.
5. Carling, S. G., Visser, D., Hautot, D., Watts, I. D., Day, P., Ensling, J., Gütlich, P., Long, G. J. and Grandjean, F., *Phys. Rev. B* **66** (2002), 104407.
6. Collins, R. L., Pettit, R. and Baker, Jr., W. A., *J. Inorg. Nuclear Chem.* **28** (1966), 1001.
7. Miyazaki, Y., private communication.
8. Kobayashi, Y., Itoi, M., Kojima, N. and Asai, K., *J. Magn. Magn. Mater.* **272–276** (2004), 1093.

Hyperfine Interactions **156/157**: 181–185, 2004.
© 2004 *Kluwer Academic Publishers. Printed in the Netherlands.*

Mössbauer Spectroscopy Study on the Effect of Al–Cr Co-Substitution in Yttrium and Yttrium–Gadolinium Iron Garnets

A.-F. LEHLOOH[1,*], S. MAHMOOD[2], M. MOZAFFARI[3] and J. AMIGHIAN[3]

[1] *Nizwa College for Education, Nizwa, Oman. On leave from Physics Department, Yarmouk University, Irbid 211-68, Jordan; e-mail: aflehlooh@yu.edu.jo, alehlooh@yahoo.com*
[2] *Physics Department, Yarmouk University, Irbid 211-68, Jordan*
[3] *Physics Department, University of Isfahan, 817-44 Isfahan, Iran*

Abstract. Room-temperature Mössbauer spectra were recorded for two series of Al–Cr co-substituted rare-earth iron garnets: YIG ($Y_3Al_xCr_xFe_{5-2x}O_{12}$) and Y–GdIG ($Y_{1.5}Gd_{1.5}Al_xCr_xFe_{5-2x}O_{12}$). All the spectra were fitted with two magnetic sextets, one sextet corresponds to the (**a**) octahedral site and the other corresponds to the (**d**) tetrahedral site. The hyperfine fields for all the samples show a reduction of ($\sim$5 Tesla) per substituted atom with increasing x. The values of the hyperfine fields of the Y–GdIG samples are slightly higher than those of the YIG samples. The relative intensities of the two sextets show that upon substitution, the Al–Cr occupy the (**a**) site rather than the (**d**) site. Also, the line widths of the two sextets were found to increase systematically upon substitution, due to the increasing atomic disorder as the Al–Cr contents increase.

Key words: Mössbauer spectroscopy, garnet, hyperfine parameters, substituted YIG.

1. Introduction

Interests in the yttrium and rare earth (RE) iron garnets $R_3Fe_5O_{12}$ and substituted garnets continue due to their wide range of applications [1–2]. Systems of aluminum substituted YIG have been studied by many investigators [2–5].

The rare earth (RE) iron garnet (RIG) have the formula unit $R_3Fe_5O_{12}$, when the RE ions are paramagnetic, there are three magnetic sublattices, one containing 24 trivalent R^{3+} ions occupying the dodecahedral (**c**) sites, another containing 16 iron (Fe^{3+}) ions occupying the octahedral (**a**) sites and the third containing 24 iron (Fe^{3+}) ions occupying the tetrahedral (**d**) sites. The magnetic moments are antiferromagnetically coupled because of the superexchange angle.

Muramkar *et al.* [6] have studied the magnetic and electrical properties of aluminum and chromium co-substituted YIG ($Y_3Al_xCr_xFe_{5-2x}O_{12}$). They found that the saturation magnetization of these systems decreases linearly with increasing x indicating a reduction in ferrimagnetic behaviour.

---

* Author for correspondence.

Previously, studies of Al substituted YIG [2, 4], Al substituted RIG [7], and on Al–Cr co-substituted YIG [6] have shown that the Al and Cr ions preferentially occupy the tetrahedral (**d**) sites.

In this work we use Mössbauer spectroscopy to investigate the hyperfine parameters of the Al–Cr co-substituted YIG and Y–GdIG. This study is conducted in order to obtain more information about the distribution of aluminium and chromium atoms in the various sites of the garnets under investigation. Also, the effects of these substitutions on the hyperfine interactions are investigated.

2. Experimental

Systems of Al–Cr co-substituted YIG ($Y_3Al_xCr_xFe_{5-2x}O_{12}$) and Y–GdIG ($Y_{1.5}Gd_{1.5}Al_xCr_xFe_{5-2x}O_{12}$) with $x = 0, 0.2, 0.4$ and 0.6 were prepared as follows:

The Fe, Y, Gd and Al oxides powders were all from Bayer company (min purity of 99.9%), while Cr oxide powder was from a Hungarian company named Chemolab (min purity of 99.9%). The elemental powders weighed in proper ratio to get en homogeneous and wet mixed for 1 hour in acetone. These were then dried in air, the dried powders calcined for 10 hours at 1200°C in air. The calcined powders were milled for 1 hour to get a homogenous submicron powders. The powders were pressed in blocks with 2 tons/cm^2 using PVA as binder. The blocks were then sintered in oxygen atmosphere for 10 hours. The sintering temperature was 1360°C for the aluminum free samples and the samples with 0.2 mole ratio while for the rest it was 1480°C. The sintered blocks were then crushed in to powders and a final annealing at 700°C for 2 hours was performed to relieve possible strain defects from crushing.

Mössbauer absorber was prepared as a thin layer of the sample powder on a transparent tape. The Mössbauer spectra were collected at room temperature (RT) using a constant acceleration Mössbauer spectrometer with ^{57}Co/Pd source. The isomer shifts were measured in mm/s with respect to α-iron at RT. The spectra were then fitted with Lorentzian lines using fitting routines based on least square analysis.

3. Results and discussion

Mössbauer spectra of the two Al–Cr co-substited systems, YIG ($Y_3Al_xCr_x$-$Fe_{5-2x}O_{12}$) and Y–GdIG ($Y_{1.5}Gd_{1.5}Al_xCr_xFe_{5-2x}O_{12}$) with $x = 0, 0.2, 0.4$ and 0.6, were recorded at RT and plotted in Figure 1. The spectra show a well-defined hyperfine magnetic splitting, each spectrum is fitted with two magnetic sextets with variable width and free line widths ratios. The fitting parameters are listed in Table I. The first magnetic sextet corresponds to iron at the octahedral (**a**) site while the second sextet corresponds to the iron at tetrahedral (**d**) site.

The values of the hyperfine magnetic fields and isomer shifts are found to be similar to those reported in other studies [4, 5, 8].

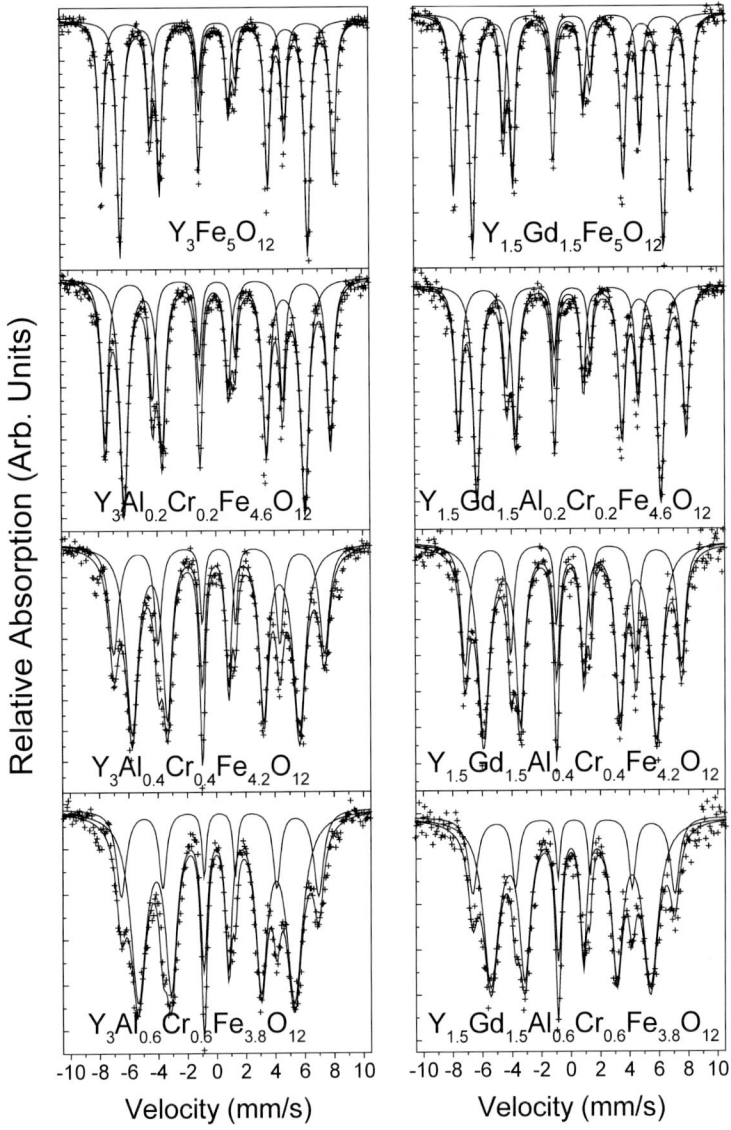

Figure 1. Room-temperature Mössbauer spectra of Al–Cr co-substituted YIG system (Y_3Al_x-$Cr_xFe_{5-2x}O_{12}$ with $x = 0$, 0.2, 0.4 and 0.6) and Al–Cr co-substituted Y–GdIG system ($Y_{1.5}Gd_{1.5}Al_xCr_xFe_{5-2x}O_{12}$ with $x = 0, 0.2, 0.4$ and 0.6).

The data in Table I and Figure 2 show that the values of the hyperfine magnetic field (B_{hf}) at the iron nucleus in both (**a**) and (**d**) sites of both systems are found to decrease as the Al–Cr contents (x) increase. The reduction in B_{hf} is ($\sim$5 Tesla) per substituted atom. This reduction agrees with the reduction in the saturation magnetization of $Y_3Al_xCr_xFe_{5-2x}O_{12}$ with increasing x [6].

As seen from Table I, the values of B_{hf} corresponding to both (**a**) and (**d**) sites for the YIG system is ($\sim$0.5 Tesla) higher than those for the Y–GdIG. This could

Table I. Mössbauer hyperfine parameters for Al–Cr substituted YIG ($Y_3Al_xCr_xFe_{5-2x}O_{12}$) and Al–Cr substituted Y–GdIG ($Y_3Al_xCr_xFe_{5-2x}O_{12}$) with $x = 0.0$, 0.2, 0.4 and 0.6; *IS*: isomer shift in mm/s relative to α-iron, B_{hf}: hyperfine magnetic field in Tesla, *W*: line width in mm/s and *I%*: the relative intensity

	Al–Cr co-substituted YIG ($Y_3Al_xCr_xFe_{5-2x}O_{12}$)				Al–Cr co-substituted Y–GdIG ($Y_{1.5}Gd_{1.5}Al_xCr_xFe_{5-2x}O_{12}$)			
x	0.0	0.2	0.4	0.6	0.0	0.2	0.4	0.6
IS (1)	0.37	0.34	0.38	0.38	0.36	0.36	0.38	0.37
IS (2)	0.13	0.11	0.14	0.14	0.11	0.11	0.15	0.17
B_{hf} (1)	49.6	48.2	45.0	42.2	50.1	48.4	46.0	43.0
B_{hf} (2)	40.0	38.6	35.8	33.4	40.4	39.0	36.7	34.0
*W*11	0.43	0.54	0.79	0.80	0.41	0.52	0.66	0.76
*W*12	0.40	0.49	0.59	0.60	0.40	0.50	0.48	0.55
*W*13	0.37	0.38	0.37	0.35	0.39	0.37	0.33	0.32
*W*21	0.44	0.66	0.99	1.28	0.46	0.56	0.95	1.24
*W*22	0.41	0.59	0.73	0.95	0.45	0.53	0.70	0.90
*W*23	0.38	0.47	0.46	0.55	0.43	0.40	0.48	0.52
I% (1)	40	37	32	22	40	39	31	22
I% (2)	60	63	68	78	60	61	69	78

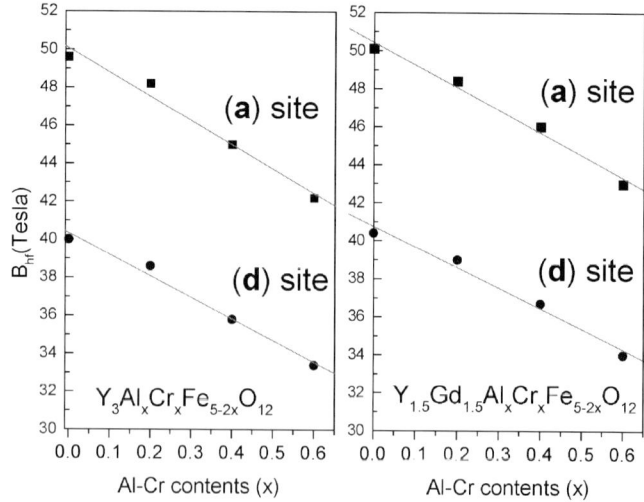

Figure 2. The hyperfine magnetic fields (B_{hf}) in Tesla versus the Al–Cr contents (x) for $Y_3Al_xCr_xFe_{5-2x}O_{12}$ and $Y_{1.5}Gd_{1.5}Al_xCr_xFe_{5-2x}O_{12}$ with $x = 0, 0.2, 0.4$ and 0.6. fitted by linear regression lines.

be due to the dipolar effect of the magnetic Gd^{3+} ions having a magnetic moment of $7\,\mu_B$.

The rate at which B_{hf} decreases with increasing x for the $Y_3Al_xCr_xFe_{5-2x}O_{12}$ is faster than that for $Y_3Al_xFe_{5-x}O_{12}$ and is consistent with previously reported results [4]. Also, the estimated rate of decrease in B_{hf} with increasing x for the substituted YIG is found to be slightly larger than that for the substituted Y–GdIG. The relative intensity I (1) of the (a) octahedral site clearly decreases with increasing x, as Table I shows. This indicates that the Al and Cr ions prefer to occupy the (a) octahedral sites. This preference is different than that found in previous studies [5–8], where the tetrahedral (d) sites were preferred.

The Mössbauer line widths of the two magnetic sextets are clearly increasing (for both systems) as the Al and Cr contents increase. The widths of the outer (1, 6) lines are show in Table I. The rate of broadening is higher in the sextet of the tetrahedral (d) sites, which could be explained as a result of local fluctuations of the chemical environments of the iron atoms at these sites arising from the substitution of Al and Cr atoms at the (a) sites. This is further supporting evidence of the preferential occupation of the octahedral (a) sites by the Al and Cr atoms.

4. Conclusion

We conclude from this study that the Al–Cr ions in both garnet systems prefer to replace the Fe ions in the octahedral (a) sites rather than the Fe atoms in the tetrahedral (d) sites. Substituting Al–Cr in the YIG and Y–GdIG systems causes a broadening in the Mössbauer lines of the two iron sites. This broadening is greater in the lines corresponding to the tetrahedral (d) sites. Also, a reduction in the B_{hf} at both sites is observed as the Al–Cr contents increase in both systems.

Acknowledgements

The authors wish to thank the University of Isfahan for financially supporting this project and Yarmouk University for using the XRD and Mössbauer facilities. One of the authors (A.-F. Lehlooh) acknowledges Yarmouk University for partially supporting him during his Sabbatical.

References

1. Dionne, G. F., *J. Appl. Phys.* **81** (1997), 5064.
2. Srinivasan, T. T., Prakash, O. and Patni, M. J., *Trans. Ind. Ceramic Soc.* **40**(1) (1981), 1.
3. Suresh, K. and Patil, K. C., *J. Alloys Compounds* **209** (1994), 203.
4. Thongmee, S., Winotai, P. and Tang, I. M., *Solid State Commun.* **109** (1999), 471.
5. Lataifeh, M. S. and Lehlooh, A.-F. D., *Solid State Commun.* **97**(9) (1996), 805.
6. Muramkar, V. D., Modi, K. B., Jad Hav, K. M., Bichile, G. K. and Kulkarni, R. G., *Mater. Lett.* **32** (1997), 281.
7. Lataifeh, M. S., Lehlooh, A.-F. D. and Mahmood, S. H., *Hyp. Interact.* **122** (1999), 253.
8. Vandormael, D., Grandjean, F., Hautot, D. and Long, G., *J. Phys.: Condens. Matter.* **13** (2001), 1759.

Hyperfine Interactions **156/157**: 187–194, 2004.
© 2004 *Kluwer Academic Publishers. Printed in the Netherlands.*

Mössbauer Study of Microstructure and Magnetic Properties (Co, Ni)–Zr Substituted Ba Ferrite Particles

A. GRUSKOVA[1], J. LIPKA[2], M. PAPANOVA[1], D. KEVICKA[1],
A. GONZALEZ[3], G. MENDOZA[3], I. TOTH[2] and J. SLAMA[4]
[1] *Department of Electrotechnology, Faculty of Electrical Engineering and Information Technology, Slovak University of Technology, Ilkovičova 3, 812 19 Bratislava, Slovak Republic*
[2] *Department of Nuclear Physics and Technology, Faculty of Electrical Engineering and Information Technology, Slovak University of Technology, Ilkovičova 3, 812 19 Bratislava, Slovak Republic*
[3] *Cinvestav-Saltillo, Carr. Saltillo-Mty. Km. 13, P.O. Box 663, 25000 Saltillo, Coahuila, Mexico*
[4] *Department of Electromagnetic Theory, Faculty of Electrical Engineering and Information Technology, Slovak University of Technology, Ilkovičova 3, 812 19 Bratislava, Slovak Republic*

Abstract. The barium hexaferrites of M-type $BaFe_{12-2x}(Co, Ni)_x Zr_x O_{19}$ were synthesized by citrate precursor method in $0.0 \leqslant x \leqslant 0.6$ range. For comparison, Ni–Zr compounds were also prepared by mechanical milling. Mössbauer measurements show that of Zr^{4+} ions have preference for $2b$ and $2a + 4f_1$ sites whereas Co^{2+} ions prefer $4f_2$ and $12k$ positions. Different occupations were observed on Ni–Zr substitutions. The temperature coefficient of the coercivity, dH_C/dT, was nearby to zero (0.01 kA/m°C) for the Co–Zr substitutes and positive for the rest of the samples. A fast reduction of coercivity H_c with the increase in substitution level (370 to 150 kA/m) were underwent for all (Co, Ni)–Zr mixtures.

Key words: Mössbauer spectroscopy, hexagonal ferrite, magnetic properties, coercivity, substitution.

1. Introduction

Substituted barium ferrites (BaM) have been intensely investigated due to their use as high density magneto-optical recording [1], for the microwave applications [2] and as filler in the magnetocomposites [3]. The Co^{2+}–Ti^{4+} ions substituted BaM ferrites exhibit good magnetic properties and the coercivity decreases rapidly (380 kA/m to 80 kA/m) as x increases [4]. However, these ferrites have a drawback, namely high-temperature coefficient of the coercivity. The cation site preferences depend on their nature but they may also depend on the prepared procedure. Quantitative analysis have shown that both Co^{2+} and Ti^{4+} ions are found into $4f_2$ and $12k$ sublattices, respectively [5]. Both Fe^{3+} ions on $2b$ site and Co^{2+} ions on $4f_1$, $4f_2$ and $12k$ sites should have strong influence on the exchange interactions among other sublattices. The investigation by Mössbauer spectroscopy of Ba-ferrite prepared by modified ceramic method shows that Co^{2+} and Ti^{4+} cations prefer to occupy the $4f_2$ and $2b$ sites at $x \leqslant 0.6$ [6]. The Mössbauer results of Co–Zr substituted BaM prepared by spray-drying in NaCl flux [7] and citrate gel [8] indicate some differences in cations substitution compared with Co–Ti substitutions [4]. The dH_c/dT can be modified by adding Co–Zr ions in BaM [9]. Also, It is known that the presence of Ni^{2+} ions reduces the temperature coefficient of coercivity. On the other hand, it has been reported that Zr^{4+} ions replace Fe^{3+} ions at the $2b$ site and Ni^{2+} ion occupies the octahedral $4f_2$ and $12k$ sites [10].

The present investigation reports the magnetic properties and site preferences of the nonmagnetic Zr^{4+} ions and magnetic Co^{2+} or Ni^{2+} ions in hexagonal magnetic structure. Also, the influence of processing parameters such as the heat treatment and the variation of the initial ratio Fe/Ba on powder samples prepared by two methods are described.

2. Experimental

$BaFe_{12-2x}(Co,Ni)_x Zr_x O_{19}$ hexaferrites with $0.0 \leqslant x \leqslant 0.6$ called as Sk were prepared by citrate precursor method with Fe/Ba initial ratio of 10.8 [11]. The samples named as Mx were prepared by mechanical milling with initial ratio 10.0 [12]. The (Sk) samples with Co–Zr ions were annealed at 975°C and 1070°C with a soaking time of 2 hours. The Ni–Zr substituted BaM (Sk) samples were heat treated at 1070°C for 2 hours and (Mx) ones were annealed at 1050°C/1.5 hours. The specific magnetic polarization $J_{s\text{-}m}$ and $J_{s\text{-}r}$, the coercivity H_c, and the temperature coefficient of the coercivity dH_c/dT were determined using a vibrating sample magnetometer (VSM) with an external magnetic field of 540 kA/m. The Mössbauer spectra were performed using the conventional constant acceleration mode with γ-ray source of ^{57}Co in Rh matrix at the room temperature. The temperature dependencies of magnetic susceptibility $\chi(\vartheta)$ and ordering temperature, T_c were performed by the bridge method at constant heat of 4°C/min.

3. Results and discussion

The room temperature Mössbauer spectra of (Sk) samples for Co–Zr mixture with limit values of substitution ($x = 0.0–0.6$) and annealed at 1070°C/2 h are shown in Figure 1. The pure BaM ferrite was fitted with four sextets corresponding to $4f_2$, $2a + 4f_1$, $12k$ and $2b$ sites.

It can be notice that the Co–Zr ($x = 0.6$) spectrum has the linewidths broaden, especially for the $12k$ site. As x increases, the $12k$ site can be fitted well by two sublattices $12k$ and $12k'$. This may be the result of changes in the environment of the $12k$ site when the substitutions take place in R block. The intensity of each sextet is directly proportional to the number of iron ions in each site, so that, it gives us on estimate of the occupancy rate of substitution elements. It is known that less electronegative ions prefer tetrahedral coordination [8]. The electronegativity of Ni^{2+}, Co^{2+} and Zr^{4+} ions are 1.91, 1.88 and 1.33, respectively, therefore Zr^{4+} ions

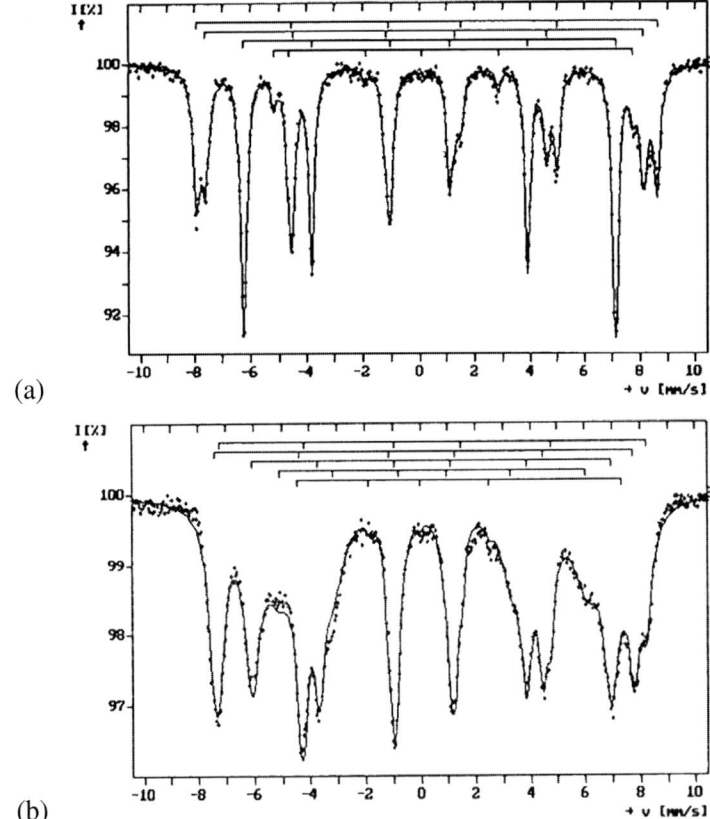

(a)

(b)

Figure 1. The room-temperature Mössbauer spectra Co–Zr substituted BaM for (a) $x = 0.0$ and
(b) $x = 0.6$.

should prefer tetrahedral sites. On the other hand, it has been shown that Zr^{4+} ions
have preference for the bipyramidal $2b$ site due to its five-fold coordination [13].
The changes of hyperfine field $\Delta B_{hf}/\Delta x$ of the Co–Zr and Ni–Zr samples are
shown in Table I. The ratio $\Delta B_{hf}/\Delta x$ of Co–Zr (Sk) samples was negative in
all sites and at both temperatures. The difference $\Delta B_{hf}/\Delta x$ of samples treated at
975°C/2 h is higher than that for samples annealed at 1070°C/2 h especially at
substitution ($x > 0.4$) where $2b$ site can be replaced, likely by Zr^{4+} ions. It was
possible to see that $\Delta B_{hf}/\Delta x$ for Ni–Zr substitution in (Mx) and (Sk) samples vs.
x was negative for all sites when x changes from 0.0 to 0.6. However at higher
(Mx) Ni–Zr substitutions least occupied were $2b$ and $4f_2$ sites presumably owing
to the apparition of a secondary phase.

Table II shows the results of the occupancy calculation of the fractions $F_{Co–Zr}$.
It can be observe that the Co–Zr ions preferentially occupy the $2b$, $4f_2$ and $12k$
sites, also the $2a + 4f_1$ sites for samples treated at 975°C and 1070°C. The occupancy fractions for both Ni–Zr substituted samples were calculated (Figure 2).
Our results for Ni–Zr ions show that the Zr^{4+} ions replaced iron ions on $4f_1$ site

Table I. The $\Delta B_{hf}/\Delta x$ values for Co–Zr, Ni–Zr substituted BaM

	x (−)	$4f_2$	$2a + 4f_1$	$12k$	$12k'$	$2b$
Co–Zr						
(Sk)	0.0	0.00	0.00	0.00		0.00
975°C	0.2	−0.68	−0.84	−0.04	0.00	−0.17
	0.4	−2.03	−1.7	−0.63	−1.88	−0.24
	0.6	−3.87	−2.36	−1.18	−2.71	−18.62
(Sk)	0.0	0.00	0.00	0.00		0.00
1070°C	0.2	−0.81	−0.75	−0.23	0.00	−1.44
	0.4	−1.45	−1.22	−0.32	−0.18	−1.96
	0.6	−3.46	−1.79	−1.04	−0.95	−3.34
Ni–Zr						
(Mx)	0.0	0.00	0.00	0.00		0.00
1050°C	0.2	−0.25	−0.25	−0.09	0.00	−0.2
	0.4	−0.26	−0.24	−0.04	−1.37	−0.05
	0.6	−0.26	−0.27	−0.05	−3.46	0.04
(Sk)	0.0	0.00	0.00	0.00		0.00
1070°C	0.2	−0.41	−0.42	−0.48	0.00	−0.76
	0.4	−1.15	−1.16	−0.74	−1.01	−0.73
	0.6	−1.65	−1.45	−0.64	−0.63	−1.10

Table II. The values of $F_{(Co–Zr)}$ for Co–Zr substituted BaM

(Sk)	x (−)	$4f_2$	$2a + 4f_1$	$12k$	$12k'$	$2b$	$12k + 12k'$
975°C/2 h	0.0	0.00	0.00	0.00	0.00	0.00	0.00
	0.2	20.04	19.05	21.33	0.00	53.03	−17.93
	0.4	40.82	19.43	35.52	0.00	86.1	−21.88
	0.6	34.37	42.9	37.91	0.00	63.31	−23.61
1070°C/2 h	0.0	0.00	0.00	0.00	0.00	0.00	0.00
	0.2	22.22	13.09	11.06	0.00	27.05	−12.98
	0.4	27.24	16.44	23.65	0.00	65.93	−13.22
	0.6	39.28	28.04	35.91	0.00	70.95	−24.67

and at higher values ($x > 0.4$) on $2b$ sites which are in agreement with Rane *et al.* [10]. While Ni^{2+} ions occupy the $4f_2$ site and the $12k$ site at higher substitutions. The substitution Ni^{2+} in these sites is expected from the crystal stabilization energy [10].

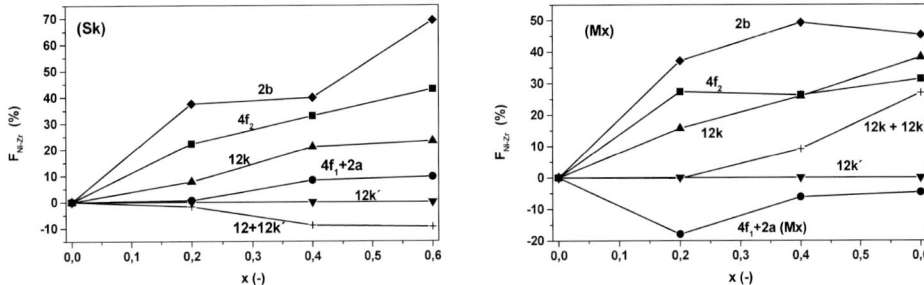

Figure 2. The occupancy fractions $F_{(Ni–Zr)}$ of Ni–Zr substituted BaM. *Left*: (Sk). *Right*: (Mx) samples.

The relative areas $S(i)$ (%) were obtained from Mössbauer spectra where $i = 1$–4, which correspond to the number (N) of iron ions at the respective sites and $N(i)$ is the occupation number for the ith site. According to the formulas described in [13] all separate values were calculated

$$N_{Fe}(i) = C_{Fe} \frac{S(i)}{\sum_{i=1}^{5} S(i)}, \tag{1}$$

$$N_{(Co,Ni)–Zr} = N(i) - N_{Fe}(i), \tag{2}$$

$$F_{(Co,Ni)–Zr} = \frac{N_{(Co,Ni)–Zr}(i)}{N(i)} \times 100\%. \tag{3}$$

Where $N_{Fe}(i)$ and $N_{(Co,Ni)–Zr}(i)$ are the occupation numbers in the ith sites for Fe and (Co, Ni)–Zr ions, respectively. The C_{Fe} refers to the composition of Fe ions and $S(i)$ denotes the relative area at the ith site. $F_{(Co,Ni)–Zr}$ is occupancy fractions.

The Mössbauer spectra are more complex with raising substitutions Ni–Zr and therefore the formulas (1 to 3) cannot be used for the calculation of the presence of secondary phase. On the spectra for BaM samples recorded at 460°C and 520°C only paramagnetic Fe^{3+} doublets were showed and magnetic measurements confirmed that they are magnetic. Our fitting model do not consider different values of f-factor for the crystallographic sites of iron in the BaM structure.

The behaviour of the magnetic parameters of the Co–Zr substituted Ba hexaferrites are shown in Table III measured according to [14]. The saturated magnetic polarization J_{s-m} for (Sk) decreased up to 0.6 while the remanent magnetic polarization J_{s-r} stayed almost constant at both treated temperatures. For Ni–Zr (Mx) samples J_{s-m}, J_{s-r} and H_c magnetic parameters decreased. The temperature coefficient of the coercivity dH_c/dT was defined according to [15] as $dH_c/dT = [H_c(120°C) - H_c(20°C)]/100°C$ and the coercivity was measured at 120°C and 20°C, respectively. The (Co, Ni)–Zr BaM ferrite particles have positive temperature coefficient of H_c, which decrease as x increased.

The temperature dependencies of the magnetic susceptibility χ versus temperature ϑ for Co–Zr substituted BaM samples are shown in Figure 3. The principle of the method is described in [16] and the measures showed that the room temperature

Table III. The magnetic properties of (Co, Ni)–Zr substituted BaM

	x (−)		$J_{\text{s-m}}$ ($10^{-6}\,\text{T m}^3\,\text{kg}^{-1}$)	$J_{\text{s-r}}$ ($10^{-6}\,\text{T m}^3\,\text{kg}^{-1}$)	H_c (kA m^{-1})	dH_c/dT (kA/m °C)	T_c (°C)
Co–Zr							
975°C	0.0	(Sk)	65.10	34.23	320	0.55	450
	0.2	(Sk)	76.16	39.22	253	0.26	403
	0.4	(Sk)	78.33	38.92	220	0.05	343
	0.6	(Sk)	80.51	36.46	160	0.01	315
1070°C	0.0	(Sk)	72.19	35.57	376	0.34	450
	0.2	(Sk)	75.78	38.90	295	0.2	403
	0.4	(Sk)	78.06	38.19	205	0.05	347
	0.6	(Sk)	82.70	37.99	130	0.01	333
Ni Zr							
1050°C	0.0	(Mx)	80.77	42.91	350	0.5	443
	0.2	(Mx)	75.06	38.47	215	0.22	439
	0.4	(Mx)	72.46	35.37	152	0.08	443
	0.6	(Mx)	69.95	31.78	127	0.08	430
1070°C	0.0	(Sk)	72.19	35.57	376	0.34	450
	0.2	(Sk)	82.63	42.76	254	0.44	398
	0.4	(Sk)	85.94	42.28	207	0.37	388
	0.6	(Sk)	79.99	38.24	190	0.25	310

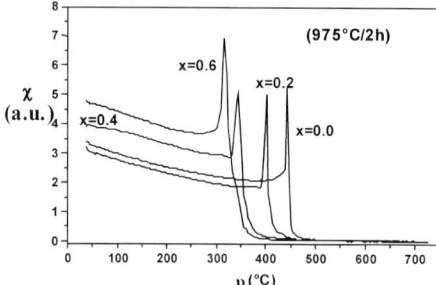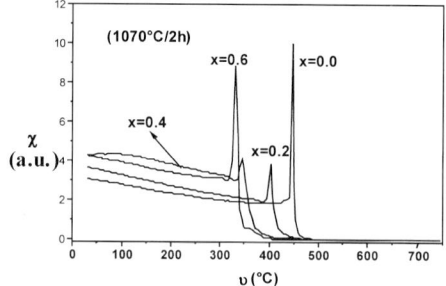

Figure 3. The temperature dependence of the magnetic susceptibility $\chi(\vartheta)$ for Co–Zr ions $(0.0 \leqslant x \leqslant 0.6)$ substituted BaM, treated at 975°C/2 h and 1070°C/2 h.

χ increase with substitution level. The initial susceptibility is given in arbitrary units and is related to the same amount of the sample for all cases. The results show, in the vicinity of the Curie temperature, a sharp Hopkinson peak for monophasic M-hexaferrite.

The $\chi(\vartheta)$ dependencies for the pure $x = 0.0$ and 0.6 for Ni–Zr substituted (Sk) and (Mx) samples are shown in Figure 4. From the $\chi(\vartheta)$ dependencies, it can see that Ni–Zr (Sk) samples are single phase whilst in the (Mx) samples appeared two phases. The linear decrease of the Curie temperature with x for all samples

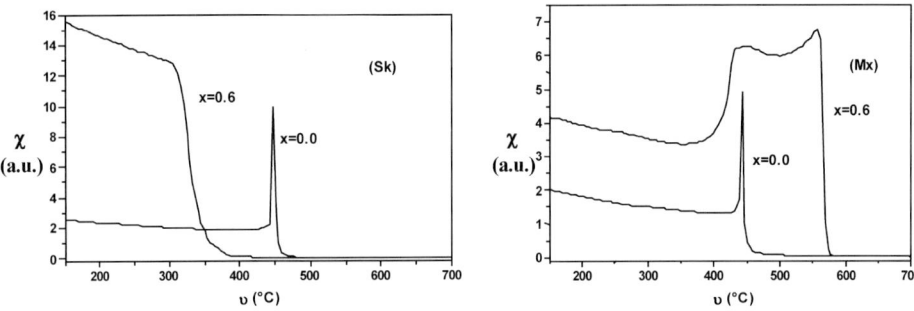

Figure 4. The temperature dependence of the magnetic susceptibility $\chi(\vartheta)$ for Ni–Zr substituted BaM with $x = 0.0$ and 0.6. *Left*: (Sk) samples. *Right*: (Mx) samples.

was caused by weakening of the superexchange interactions between Fe^{3+} ions via O^{2-}. In the (Mx) samples the lower diminution in the T_c due to the presence of the secondary magnetic phase were observed [17].

4. Conclusions

The magnetocrystalline structure of $BaFe_{12-2x}(Co$ or $Ni)_x Zr_x O_{19}$ ferrites prepared by two synthesised methods and their effects on the magnetic parameters were analyzed by Mössbauer spectroscopy. According to Mössbauer studies is believed that iron ions were substituted on bipyramidal ($2b$) and tetrahedral ($4f_1$) sites by Zr^{4+} ions and on octahedral $4f_2$ and $12k$ sites by the Ni^{2+} ions. The difference of $\Delta B_{hf}/\Delta x$ showed that the Co–Zr ions occupy the spinel blocks $2a + 4f_1$. The calculations of the occupancy fractions $F_{(Ni-Zr)}$ confirmed that Ni–Zr ions have preference for $2b$, $4f_2$, $12k$ and slightly at $2a + 4f_1$ sites. The (Sk) samples show a slightly occurence substituted ions in $12k$ sites. However at higher (Mx) Ni–Zr substitutions the $2b$ an $4f_2$ sites were least substituted probably due to the presence of the secondary phase. It should be notted that by X-ray diffraction measurements this phase was not verified. We can conclude that probably the presence of nonhomogeneous grain size likewise diminish the magnetic parameters.

The coercivity H_c was fast reduced (376 kA/m to 130 kA/m) for all samples. The J_{s-m} was increased up to $x = 0.6$ in (Sk) samples and decreased for (Mx) samples in the same interval. The Curie temperature decrease faster for (Sk) samples with substitution x compared with the (Mx) samples. The dH_c/dT is expressive fall nearby zero at Co–Zr samples.

Acknowledgements

The work has been supported by VEGA Scientific Agency of Ministry of Education of the Slovak Republic under project No. G-1/0163/03. Authors would like to thanks CONACyT - Mexico under project J28283U for the support given to carry out this work.

References

1. Teh, G. B. *et al.*, *J. Sol. Stat. Chem.* **167** (2002), 254.
2. Kreisel, J. *et al.*, *J. Mag. Mag. Mat.* **213** (2000), 262.
3. Singh, P. *et al.*, *Mat. Scien. Eng. B* **78** (2000), 70.
4. Kubo, O. *et al.*, *IEEE Trans. Magn.* **MAG-18** (1982), 1122.
5. Williams, J. M. *et al.*, *J. Mag. Mag. Mat.* **220** (2000), 124.
6. Zhou, X. Z. *et al.*, *IEEE Trans. Magn.* **MAG-27** (1991), 4654.
7. Chin, T. S. *et al.*, *J. Mag. Mag. Mat.* **120** (1993), 64.
8. Rane, M. V. *et al.*, *J. Mag. Mag. Mat.* **153** (1996), L1.
9. Yang, Z. *et al.*, *J. Mag. Mag. Mat.* **115** (1992), 77.
10. Rane, M. V. *et al.*, *J. Mag. Mag. Mat.* **192** (1999), 288.
11. Gruskova, A. *et al.*, *Czech. J. Phys.* **52**(2) (2001), 135.
12. Gonzalez, A. *et al.*, *J. El. Eng.* **53**(10/S) (2002), 127.
13. Li, Z. W. *et al.*, *Phys. Rev. B* **62**(10) (2000), 6530.
14. Dosoudil, R. *et al.*, *J. El. Eng.* **53**(10/S) (2002), 135.
15. Yang, Z. *et al.*, *Mat. Scien. Eng. B* **90** (2002), 142.
16. Jancarik, V. *et al.*, *J. El. Eng.* **50** (1999), 63.
17. Slama, J. *et al.*, *J. Mag. Mag. Mat.*, will be published.

Hyperfine Interactions **156/157**: 195–200, 2004.
© 2004 *Kluwer Academic Publishers. Printed in the Netherlands.*

Mössbauer Studies of Dilute ^{119}Sn and ^{57}Fe in SrRuO$_3$ and Sr$_2$FeRuO$_6$

I. NOWIK and I. FELNER
Racah Institute of Physics, The Hebrew University, Jerusalem 91904, Israel

Abstract. In both compounds the ^{57}Fe^{3+} reveals no magnetic hyperfine structure at 90 K, yet large hyperfine fields at 4.2 K. ^{119}Sn^{4+} in SrRuO$_3$ reveals hyperfine fields starting from $T_c = 165$ K. In Sr$_2$Fe^{3+}Ru^{5+}O$_6$ the iron and tin exhibit at 4.2 K wide distributions of hyperfine interactions, typical to Fe/Ru disorder and spin glass structure.

Key words: Mössbauer spectroscopy, ternary ruthenates, magnetic order, spin glass.

1. Introduction

Ternary ruthenates exhibit a wide range of electronic and magnetic properties, ranging from superconductivity to ferromagnetism. One class of oxides that has attracted renewed interest is the family of orthorhombic perovskite SrRuO$_3$ and CaRuO$_3$ compounds, due to their unusual magnetic properties. It is well established that the perovskite SrRuO$_3$ with an orthorhombic distortion, is a ferromagnet (FM, $T_c = 165$ K). The iso-structural CaRuO$_3$ shows the characteristics of spin-glass, $T_{sg} = 87$ K [1] while SrFeO$_3$ is a helical antiferromagnet, $T_N = 134$ K [2]. Research has recently been extended to compounds with ordered double-perovskite structure Sr$_2$FeMO$_6$ (M = Mo, W and Re). We present here the study of magnetization and dilute ^{57}Fe and ^{119}Sn Mössbauer spectra of Sr$_2$FeRuO$_6$ and compare them to those of SrRuO$_3$. Our results are consistent with [3, 4], who demonstrated that Sr$_2$FeRuO$_6$ exhibits Fe/Ru disorder and a spin-glass behavior below ~50 K.

In a previous publication [5], the conditions under which the magnetic hyperfine interactions of dilute Mössbauer probes (^{57}Fe^{3+}, ^{151}Eu^{3+}), represent faithfully the magnetic order of the Ru sublattice, in magnetic ruthenium compounds, have been investigated. ^{119}Sn^{4+} as a probe in the same compounds will behave quite differently. In Ru^{4+} compounds the magnetic transferred hyperfine field will follow faithfully the Ru magnetization temperature dependence, while in Ru^{5+} compounds local distortions will destroy the relation with the Ru magnetization. These phenomena are here demonstrated.

2. Experimental details

Polycrystalline samples of pure and doped $SrRuO_3$ and Sr_2FeRuO_6 materials, were prepared by solid-state reaction from the appropriate stoichiometric mixtures of $SrCO_3$, Fe (all 99.99%) and Ru (99.9%). Pressed pellets were preheated at 1000°C for 24 h, and then sintered at 1300°C for 72 h in air, with two intermediate grindings. Powder X-ray diffraction (XRD) measurements confirmed the purity of the compounds. Magnetic dc measurements were performed in a SQUID magnetometer. The Mössbauer studies were performed using a standard constant velocity drive spectrometer. A ^{57}Co:Rh source (50 mCi) was used for iron studies and a $Ba^{119m}SnO_3$ source (5 mCi) for the tin studies. The obtained spectra were compared to theoretical models by a least-square fit procedure. The iron isomer shift (IS) is relative to iron metal at room temperature.

3. Experimental results and discussion

3.1. X-RAY DIFFRACTION STUDIES

The XRD studies confirm the orthorhombic structure (space-group Pnma) for both Sr_2FeRuO_6 and $SrRuO_3$ compounds. The lattice parameters for Sr_2FeRuO_6 and $SrRuO_3$ are: $a = 5.522(1)$ and $5.545(2)$ Å, $b = 5.522(1)$ and $5.609(2)$ Å and $c = 7.808(1)$ and $7.847(4)$ Å, respectively. Since the crystal structure of both compounds is so similar, from the XRD studies alone it is not possible to be completely certain that our Sr_2RuFeO_6 sample does not contain a small fraction of $SrRuO_3$. In such a case the Sr_2RuFeO_6 sample is actually $Sr_2Ru_{1-x}Fe_{1+x}O_6$. In fact this what occurred to our sample.

3.2. MÖSSBAUER STUDIES

3.2.1. *SrRuO₃*

This compound has been studied in the past by many experimental methods and is known to order at 165 K. Dilute ^{57}Fe Mössbauer studies [5] show that the iron probe (Fe^{3+}) hyperfine field does not follow the magnetic order of the Ru. Here we report dilute ^{119}Sn Mössbauer studies which show that the transferred hyperfine field at the Sn^{4+} nucleus does follow the Ru host magnetization. The observed experimental spectra, Figure 1, were analyzed in terms of two subspectra, one corresponding to Sn substituting Ru in $SrRuO_3$, the other (~40%) composed of a single line at all temperatures, probably corresponds to $SrSnO_3$ clusters. The dilute tin nuclei in $SrRuO_3$ experience a magnetic hyperfine field below 165 K, consistent with the temperature dependence of the Ru magnetization as measured by us and [4], Figure 2.

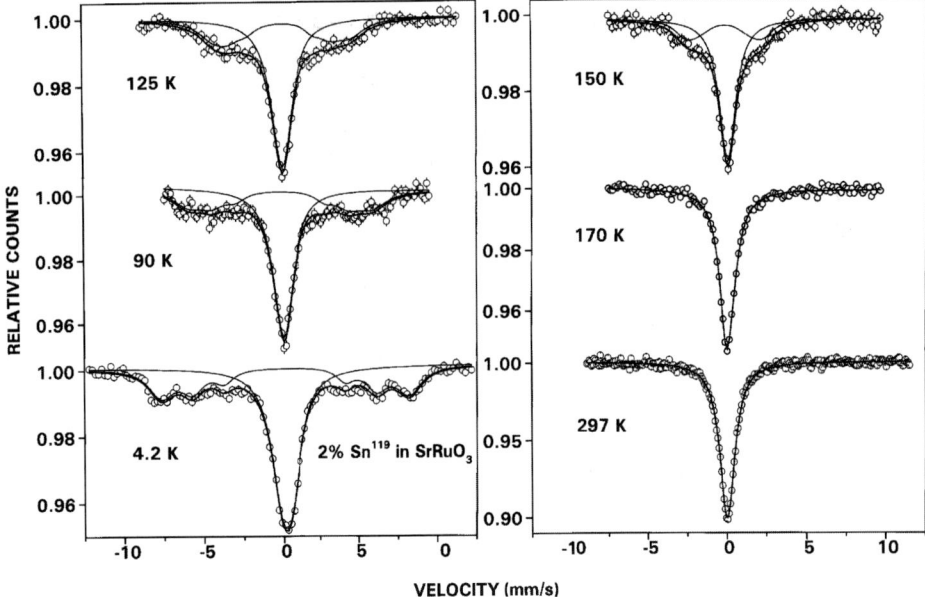

Figure 1. Mössbauer spectra of ^{119}Sn in SrRuO$_3$ at various temperatures, stressing the difference with those spectra of ^{57}Fe in SrRuO$_3$ [5].

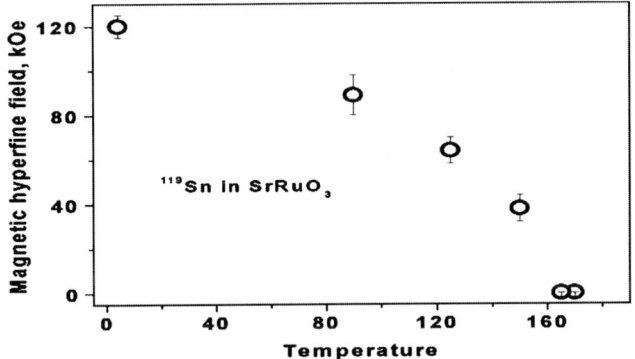

Figure 2. Temperature dependence of the magnetic hyperfine field acting on dilute Sn nuclei in SrRuO$_3$.

3.2.2. Sr$_2$FeRuO$_6$

This compound was studied by Mössbauer spectroscopy using both ^{57}Fe and ^{119}Sn isotopes. Some of the experimental spectra are shown in Figures 3 and 4. The ^{57}Fe spectra above 90 K, Figure 3, were analyzed as due to two subspectra, the major quadrupole doublet ($\sim$90%, $IS = 0.36$ mm/s, 1/2eq $Q = 0.40$ mm/s) is certainly of Fe^{3+} and is composed of a wide asymmetric distribution of quadrupole interactions, the minor subspectrum ($IS = -0.13$ mm/s, 1/2eq $Q = 0.22$ mm/s) is probably due to the appearance of some Fe^{4+}. This is consistent with the 4.2 K

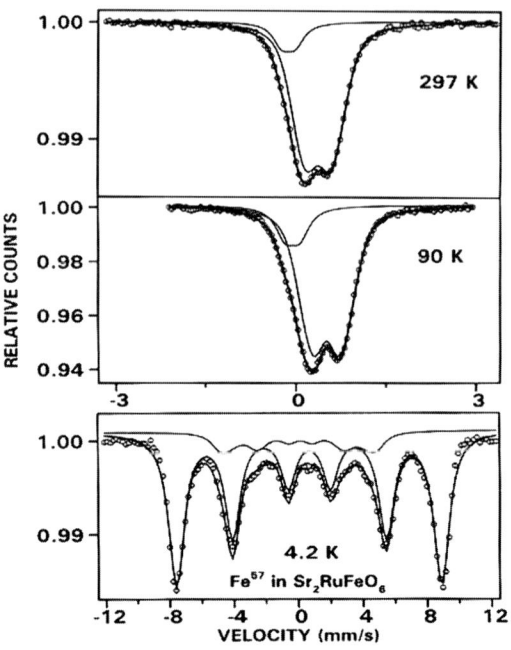

Figure 3. Mössbauer spectra of ^{57}Fe in Sr_2RuFeO_6 at various temperatures.

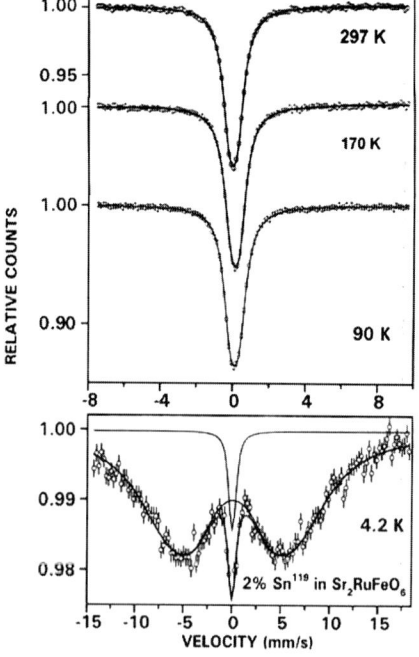

Figure 4. Mössbauer spectra of dilute ^{119}Sn in Sr_2RuFeO_6 at various temperatures.

spectrum which is composed of two magnetic subspectra. The intensities, isomer shifts (0.67 and 0.07 mm/s) and hyperfine fields (516 kOe and 314 kOe, respectively) of these two subspectra are in agreement with the assumption that the minor subspectrum is due to Fe^{4+} [6]. The observed distribution in quadrupole interactions proves that the Fe^{3+} (Fe^{4+}) and Ru^{5+} are randomly distributed in their identical site. For the same reason also the magnetic hyperfine fields observed at 4.2 K have a wide distribution. The average effective quadrupole interaction (the component along the magnetic field) is zero, indicating the random spin orientation of the Fe ions, consistent with the claim that they order as a spin glass [3, 4]. The presence of Fe^{4+} in our sample may be the result of some excess Fe relative to Ru in our Sr$_2$RuFeO$_6$ sample (in the extreme case of SrFeO$_3$ all the Fe ions are Fe^{4+}) and the presence of pure SrRuO$_3$. This is also observed in our raw magnetization results, where the presence of a small amount of SrRuO$_3$ ($\sim$8%) leads to a fictitious 165 K ordering temperature. The ^{119}Sn spectrum at 4.2 K, Figure 4, exhibits a very wide distribution of hyperfine fields, the average of which is 115 kOe.

3.3. MAGNETIZATION STUDIES

ZFC and FC magnetic susceptibility curves for SrRuO$_3$ and our sample of Sr$_2$FeRuO$_6$ have been measured. For both samples a FM transition around $T_C = 165$ K is observed. This result is in obvious contradiction to the past reported results [3, 4]. We tried to solve this discrepency, by considering the fact that our XRD measurements indicated the possible presence of a small amount of SrRuO$_3$, the Mössbauer studies indicated the presence of excess Fe in our Sr$_2$FeRuO$_6$ sample, and hence again the possible presence of SrRuO$_3$. Thus we tried to plot the magnetic susceptibility of our sample "Sr$_2$FeRuO$_6$" less various amounts of the magnetic susceptibility of pure SrRuO$_3$. Such a trial (with the effort to cancel the phase transition at 165 K) leads to a peak in the susceptibility curve around 55 K and is consistent with the spin glass structure reported in [3, 4]. Thus, we concluded that our sample "Sr$_2$FeRuO$_6$" is actually Sr$_2$Fe$_{1.04}$Ru$_{0.96}$O$_6$.

4. Conclusions

The results of the XRD, magnetization and in particular the Mössbauer studies lead to the following conclusions; In pure SrRuO$_3$ (Ru^{4+}), a ferromagnet with $T_C = 165$ K, the ^{57}Fe Mössbauer probe (Fe^{3+}) does not follow the host magnetization [5], whereas the ^{119}Sn probe (Sn^{4+}) does. The transferred hyperfine field acting on the tin nucleus is observable up to T_C (Figures 1 and 2). The difference in the behaviour of the two probes is a result of the fact that Fe^{3+} disturbs its local environment of Ru^{4+} ions, while Sn^{4+} does not. Our expected sample of Sr$_2$FeRuO$_6$ turned out to be Sr$_2$Fe$_{1.04}$Ru$_{0.96}$O$_6$ which is also a spin glass ($T_{sg} \sim 55$ K). Both probes, ^{57}Fe and ^{119}Sn exhibit a wide spread of hyperfine interaction parameters, which proves the Fe/Ru crystallographic disorder. The excess iron in the Sr$_2$Fe$_{1.04}$Ru$_{0.96}$O$_6$ sample causes the presence of Fe^{4+} ions in the sample. Samples with excess Ru exhibit the presence of only Fe^{3+} ions [4].

Acknowledgements

We acknowledge E. Galstyan for his help. This research was supported by the Klachky Foundation.

References

1. Felner, I., Nowik, I., Bradaric, I. M. and Gospodinov, M., *Phys. Rev. B* **62** (2000), 11332; Felner, I., Asaf, U., Nowik, I. and Bradaric, I., *Phys. Rev. B* **66** (2002), 054418.
2. Fang, Z., Terakura, K. and Kamamori, J., *Phys. Rev. B* **63** (2001), 180407(R).
3. Battle, P. D., Gibb, T. C., Jones, C. W. and Studer, F., *J. Solid. State Chem.* **78** (1989), 281.
4. Gibb, T. C., Greatrex, R., Greenwood, N. N. and Snowdon, K. G., *J. Sol. State Chem.* **14** (1975), 193.
5. Nowik, I., Felner, I. and Asaf, U., *Hyp. Interact.* **141** (2002), 213.
6. Fawcett, I. D., Veith, G. M., Greenblatt, M., Croft, M. and Nowik, I., *Sol. State Sci.* **2** (2000), 821.

Hyperfine Interactions **156/157**: 201–204, 2004.
© 2004 *Kluwer Academic Publishers. Printed in the Netherlands.*

Mössbauer Studies on the Quasibinary System FeTe$_{1.45}$–TiTe$_{1.45}$

O. YU. PANKRATOVA, A. V. ZABOLOTNAYA, K. A. HISTIAEV,
V. V. PANCHUCK, V. G. SEMENOV, R. A. ZVINCHUK and A. V. SUVOROV
St. Petersburg State University, 198504 Russia

Abstract. In this work the synthesis, powder XRD and Mössbauer studies of the quasibinary system FeTe$_{1.45}$–TiTe$_{1.45}$ were carried out. A series of nonstoichiometric tellurides Fe$_x$Ti$_{1-x}$Te$_{1.45}$ ($x = 0.1, 0.2, \ldots, 1.0$) was synthesized by the direct reaction from elements at 850°C. The Mössbauer results are in the good accordance with the XRD data. Behaviors lattice parameters and ^{57}Fe Mössbauer parameters from composition of the tellurides allow to make some conclusions about their compensation mechanism for nonstoichiometry (CMN). For both Fe$_x$Ti$_{1-x}$Te$_{1.45}$, and nonstoichiometric Ti$_3$Te$_4$ CMN comes from the formation of Te–Te bonds at the constant metal valence. On the whole, in system titanium-tellurium CMN could be different and depends on the interval of composition.

Key words: nonstoichiometric tellurides, powder XRD, ^{57}Fe Mössbauer spectroscopy.

1. Introduction

The compensation mechanism for nonstoichiometry (CMN) in chalcogenides of 3d-transition metals could be different and depends on both nature metal, and chalcogen. On the one hand, in many nonstoichiometric compounds cations are found in different oxidation states. In these chalcogenides CMN is provided by valence change of a part of metal atoms (mixed-valence) at the constant chalcogen valence, e.g., in the interval from sesqui- to ditellurides of titanium, where Ti(II) and Ti(IV) coexist [1]. On the other hand a new point of view on the chalcogenides arise from the Mössbauer data on Fe$_7$S$_8$ and Fe$_7$Se$_8$ which revealed that all cations are in the ferrous state [2, 3]. In this case CMN comes from the formation of Ch–Ch bonds, at the constant metal valence. A set of the investigation was designed to examine CMN in the quasibinary system FeTe$_{1.45}$–TiTe$_{1.45}$. In both Ti–Te and Fe–Te systems there are phases Me$_3$Ch$_4$ with NiAs-like structures. The range of homogeneity of Ti$_3$Te$_4$ (TiTe$_y$) is $1.15 < y < 1.45$ [4] and Fe$_3$Te$_4$ (FeTe$_y$) is $1.45 < y < 1.65$ [5]. The aim of this work was synthesis and Mössbauer studies of the quasibinary system FeTe$_{1.45}$–TiTe$_{1.45}$.

2. Experimental

A series of nonstoichiometric tellurides Fe$_x$Ti$_{1-x}$Te$_{1.45}$ ($x = 0.1, 0.2, \ldots, 1.0$) was synthesized by the direct reaction from elements. Iron powder was obtained from carbonyl iron (99.9%), metallic titanium was obtained from TiH_2 by de-hydrogenation in a vacuum at 850°C, and tellurium was "special purity" grade. The stoichiometric amounts of initial elements were put into compressed quartz ampoules, which were evacuated to a residual pressure of 0.13 Pa, and sealed. The synthesis was carried out in a furnace using step-like heating: at 550°C for 72 h and at 850°C 1 week. After the synthesis ampoules were quenched by to put into cold water. The phase composition and lattice parameters were determined by use DRON-3 (with $CuK\alpha$, external standard α-Al_2O_3) diffractometer. The samples were also studied by Mössbauer spectroscopy on the ^{57}Fe at room temperature.

3. Results and discussion

X-ray diffraction data showed that in $Fe_xTi_{1-x}Te_{1.45}$ the mutual solubility is lim-ited. In Figure 1 is shown the dependencies of the hexagonal lattice parameters a, c from composition. It should be noted that the solid solution ranges take place only at $0.8 \leqslant x < 1.0$ (hexagonal Phase I) and $0 \leqslant x < 0.4$ (hexagonal Phase IV). At an intermediate x-value some superstructures of the stacked layers mainly occupied by different atoms appear (Phases II and III) [6]. Mössbauer spectra at room temperature have been taken for all the samples. Isomer shift (δ), quadrupole splitted (ΔE) and hyperfine field (H) spectra of the tellurides are collected in Table I. It is worth noting that the values of isomer shift correlate with the iron

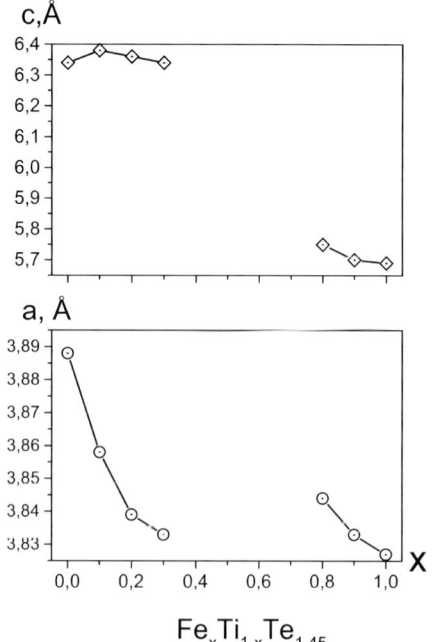

Figure 1. The dependencies of the hexagonal lattice parameters a, c from composition.

Table I. The Mössbauer spectra parameters isomer shift (δ), quadrupole splitting (ΔE) and hyperfine field (H) for Fe$_x$Ti$_{1-x}$Te$_{1.45}$

Fe$_x$Ti$_{1-x}$Te$_{1.45}$ where x	Subspectra	δ, mm/s	ΔE, mm/s	H, kOe	%, Fe[a]
0.1	singlet	0.75 ± 0.01	–	–	59.20
	doublet	0.33 ± 0.03	0.50 ± 0.06	–	18.91
	sextet	0.06 ± 0.02	0.10 ± 0.04	316.6 ± 0.2	21.90
0.2	singlet	0.78 ± 0.01	–	–	51.57
	doublet	0.41 ± 0.05	0.40 ± 0.08	–	15.70
	sextet	-0.02 ± 0.03	-0.15 ± 0.06	319.9 ± 0.3	32.73
0.3	singlet	0.76 ± 0.01	–	–	58.66
	doublet	0.37 ± 0.05	0.36 ± 0.09	–	3.59
	sextet	0.01 ± 0.01	0 ± 0.02	328.4 ± 0.1	37.75
0.4	doublet	0.77 ± 0.01	0.22 ± 0.01	–	55.64
	doublet	0.33 ± 0.01	0.29 ± 0.01	–	44.36
0.5	singlet	0.83 ± 0.01	–	–	25.21
	doublet	0.54 ± 0.01	0.25 ± 0.01	–	74.79
0.6	doublet	0.82 ± 0.01	–	–	22.57
	doublet	0.54 ± 0.01	0.24 ± 0.01	–	77.43
0.7	singlet	0.81 ± 0.01	–	–	21.66
	doublet	0.52 ± 0.01	0.26 ± 0.01	–	78.34
0.8	singlet	0.86 ± 0	–	–	8.64
	doublet	0.57 ± 0	0.29 ± 0.01	–	91.36
0.9	singlet	0.86 ± 0	–	–	6.92
	doublet	0.56 ± 0	0.30 ± 0.01	–	93.08
1.0	doublet	0.531	0.33 ± 0.01	–	100

[a] The content of Fe atoms for different spectra are given with the precision up to F-factor.

content. The Mössbauer parameters for chalcogenides of iron previously reported in literature [2, 3, 7]. The special feature of chalcogenides is their isomer shifts that much different from oxide ones: they are lowered.

The distribution of iron atoms for Phases I, II, III, IV is shown in Figure 2. In Phase IV there are three different states of iron atoms: Fe(II)$_{-sym.}$ in the symmetric environment, Fe(II)$_{-asym.}$ in the asymmetric environment and diluted iron atoms Fe(0) that, evidently, must stabilize the solid solution. When the iron concentration enlarges $x \geqslant 0.4$ diluted iron atoms are not observed, Fe(II)$_{-sym.}$ decreases and Fe(II)$_{-asym.}$ enlarges. At FeTe$_{1.45}$ ($x = 1.0$) there is only Fe(II)$_{-asym.}$. FeTe$_{1.45}$ gives only a quadrupole doublet that points at the absent of magnetic order state of this phase at room temperature. In the range of the superstructures $0.5 \leqslant x \leqslant 0.7$ (Phases II and III) the ratio of Fe(II)$_{-sym.}$ and Fe(II)$_{-asym.}$ are nearly constant.

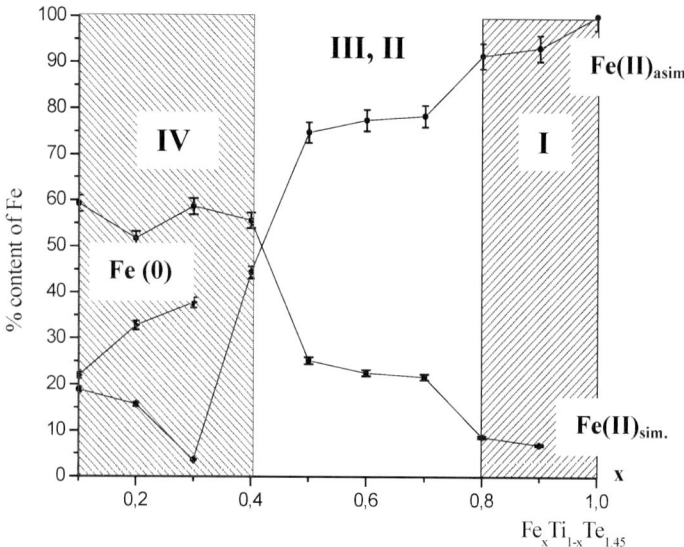

Figure 2. The distribution of iron atoms for $Fe_xTi_{1-x}Te_{1.45}$ depends on the composition.

This fact that in the series of nonstoichiometric tellurides $Fe_xTi_{1-x}Te_{1.45}$ only Fe(II) takes place can be explained by the fact that in this case CMN comes from the formation of Te–Te bonds at the constant metal valence Fe(II) as well as it takes place in Fe–Te system. As far as Ti–Te system CMN could be different and depends on the interval of composition, namely for Ti_2Te_3–$TiTe_2$ the CMN is provided a valence change of a part of titanium atoms at the constant chalcogen valence and for nonstoichiometric compound Ti_3Te_4 CMN comes from the formation of Te–Te bonds at the constant metal valence.

Acknowledgement

This work is supported by the grant RF Ministry of Education No. E02-5.0-226.

References

1. Pankratova, O. Yu., Grigor'eva, L. I., Zvinchuk, R. A. and Suvorov, A. V., *Russian J. Inorg. Chem.* **38** (1993), 383.
2. Reddy, K. V. and Chetty, S. C., *Phys. Status Solidy A* **32** (1975), 585.
3. Boumford, C. and Morrish, A. H., *Phys. Status Solidy A* **22** (1974), 435.
4. Pankratova, O. Yu., Novozhilova, S. A., Grigor'eva, L. I. and Zvinchuk, R. A., *Russian J. Inorg. Chem.* **39** (1994), 1539.
5. Vladimirova, V. A., Zvinchuk, R. A., Morozova, M. P. and Pankratova, O. Yu., *Zhurnal Fizicheskoi Khimii* **56** (1982), 565.
6. Pankratova, O. Yu., Zabolotnaya, A. V., Histiaev, K. A. *et al.*, In: *Proc. XV Int. Conf. XRD & Crystal Chem. of Minerals*, St. Petersburg, Russia, 2003, p. 49.
7. Tsuji, T., Howe, A. T. and Greenwood, N. N., *J. Solid State Chem.* **17** (1976), 157.

Hyperfine Interactions **156/157**: 205–212, 2004.
© 2004 *Kluwer Academic Publishers. Printed in the Netherlands.*

Magnetic Properties of Iron Clusters in Silver

M. ELZAIN[1,*], A. AL RAWAS[1], A. YOUSIF[1], A. GISMELSEED[1], A. RAIS[1],
I. AL-OMARI[1], K. BOUZIANE[1] and H. WIDATALLAH[2]
[1]*Department of Physics, College of Science, Box 36, Alkhod 123, Oman;
e-mail: elzain@squ.edu.om*
[2]*Department of Physics, Faculty of Science, Khartoum University, Sudan*

Abstract. The discrete variational method is used to study the effect of interactions of iron impurities on the magnetic moments, hyperfine fields and isomer shifts at iron sites in silver. We study small clusters of iron atoms as they grow to form FCC phase that is coherent with the silver lattice. The effects of the lattice relaxation and the ferromagnetic and antiferromagnetic couplings are also considered. When Fe atoms congregate around a central Fe atom in an FCC arrangement under ferromagnetic coupling, the local magnetic moment and the contact charge density at the central atom hardly change as the cluster builds up, whereas the hyperfine field increases asymptotically as the number of Fe nearest neighbors increases. Introduction of antiferromagnetic coupling has minor effect on the local magnetic moments and isomer shifts, however it produces large reduction in the hyperfine field. The lattice relaxation of the surrounding Fe atoms towards a BCC phase around a central Fe atom leads to reduction in the magnetic moment accompanied by increase in the magnetic hyperfine field.

Key words: cluster size, magnetic moment, magnetic hyperfine field, isomer shift.

1. Introduction

How metallic magnetism develops in a material has remained an intriguing question. The simplest system of a magnetic impurity in a nonmagnetic metallic matrix has received considerable attention. The conduction electrons of the host material added more complexity to the problem. The direct interaction between the magnetic and conduction electrons led to the Kondo effect, whereas the indirect interaction between the magnetic atoms through the conduction electrons resulted in spin glasses. Metallic magnetism, particularly in 3d transition metals, arises from the direct interaction between magnetic electrons.

Magnetic nanoparticles are of great importance because of their practical applications in various areas in addition to their fundamental theoretical interest. In the majority of applications, magnetic nanoparticles are usually embedded in non-magnetic matrices. The relevant questions in the field of nanoparticles address the nature of inter-particle interactions and the internal structure of individual

---

* Author for correspondence.

nanoparticles. We focus here on the latter, where we study the magnetic structures of embedded iron clusters as they buildup in size in a silver matrix.

Silver and iron are completely immiscible. Hence, supersaturated solutions of iron in silver are usually prepared using deposition techniques, mechanical alloying or implantation. Kahl and Krebs [1], using pulsed laser deposition technique, synthesized single-phase BCC FeAg alloy films with iron concentration greater than 60%. This nonequilibrium phase was found to be stable for temperatures up to 450 K. Silver–iron films having dilute Fe concentrations were prepared by Morales *et al.* [2] using vapor quenching. *In situ* and *ex situ* Mössbauer spectroscopy was used to study the films [2]. The Mössbauer spectra were analyzed in terms of three Fe components: monomers, dimers and clusters. The *in situ* spectra at 10 K show that in the as-deposited film a large percentage of Fe atoms enter the Ag matrix as monomers. On heating the film, monomers and dimers diffuse forming FCC clusters. Since very small magnetic hyperfine fields were obtained, it was considered that the blocking temperature is less than 10 K. The *ex situ* measurements were performed at low temperatures. Larger magnetic hyperfine fields were observed indicating that the blocking temperature is around 4.2 K. The annealed samples exhibit the presence of BCC Fe clusters with their characteristic small isomer shift and associated large magnetic hyperfine field.

The *ab initio* calculation of the electronic and magnetic structures of iron impurities in silver could reasonably be compared to sd-model calculations with degenerate d-bands. This is because the d-band of silver lies at low energy below the Fermi level, whereas the conduction band is formed from s-electrons (see [3] and references therein). Oswald *et al.* [4], using KKR Green function method, calculated the magnetic properties of Fe monomers and dimers in Ag. Their emphasis was on the density of states, magnetic moment and exchange energy. The local magnetic moment per Fe impurity was found to be around 3 μ_B for monomers and ferromagnetically and antiferromagnetically coupled dimers. Nogueira and Petrilli [5] used the first principle real-space linear muffin-tin orbital-atomic-sphere approximation method to study the magnetic moment and hyperfine field of iron monomers, ferromagnetic dimers and FCC clusters of 13, 19 and 43 atoms in silver. The magnetic moments were found to change slightly with increasing number of atoms, whereas the contact hyperfine fields were found to depend on the cluster size and in general decrease in magnitude from the center of the cluster outward towards the surface.

In this study we used the first principle discrete variational method (DVM) to calculate the local magnetic moments, contact hyperfine fields and charge densities at Fe sites in FCC Ag. The spin polarised Kohn–Sham equation in the local density approximation with von Barth and Hedin exchange-correlation potential was solved using linear combinations of atomic orbitals of Fe and Ag. The equations were integrated employing the diophantine sampling and Gaussian quadrature in different regions of space [6]. The contact hyperfine field was obtained as a sum of core and valence terms. The core term was estimated from its linear relation to the local d magnetic moment with a proportionality constant of 11 T/μ_B, whereas the valence term was directly calculated [7]. The sign and magnitude of the valence contribution are primarily determined by sd-antiferromagnetic interaction between the s-electrons at the central atom and the d-electrons at the neighboring sites [7]. In a recent calculation on Fe in FCC 4d metals we have found that the results obtained using the DVM and the full-potential linear-augmented-plane-waves formalism (FP-LAPW, utilizing WIEN2k code) are comparable [8]. The calculated values of the hyperfine fields are in general larger than the experimental results. Our emphasis here is on the trends exhibited by the hyperfine fields rather than on their absolute values. The Fe–Ag systems are represented by, 55 atom clusters. The Fe particles are built around the central Fe atom in the cluster.

In the following section we present and discuss results of the DVM calculation of local magnetic moments, contact magnetic hyperfine fields and contact charge density at Fe sites. Concluding remarks are given in the last section.

2. Results and discussions

The FCC iron has complex magnetic structure, where ferromagnetism, various forms of antiferromagnetism, ferrimagnetism and modulated spin-density waves were all reported ([9] and references therein). Coherent FCC iron clusters in Ag lattice are expected to be ferromagnetic because of the large Ag lattice constant. At reduced lattice constant the antiferromagnetic phases are more stable and with further reduction stability shifts towards the ferromagnetic BCC phase [9].

We have calculated the electronic and magnetic properties for ferromagnetic (FM) and antiferromagnetic (AFM) FCC and ferromagnetic BCC iron clusters. Results are presented below for local moments, contact hyperfine fields and charge densities. In particular, we give here the results for a single Fe impurity in Ag matrix, which will be used as a reference to results of other configurations. The local magnetic moments for the Fe impurity in Ag with no lattice relaxation are 2.93 μ_B and 3.07 μ_B when employing the DVM and FP-LAPW, respectively. The respective hyperfine fields are −16 T and −11 T, while the contact charge density using DVM is 3.76 au and that using FP-LAPW is 5.29 au. These contact charge densities are to be compared to the corresponding results for α-Fe, which are 5.75 au and 6.97 au, respectively. The calculated isomer shifts for Fe impurity in Ag, using the DVM and FP-LAPW are then 0.50 mm/s and 0.42 mm/s, respectively. An isomer shift of the same value was reported by Morales et al. [2].

2.1. FERROMAGNETIC FCC CLUSTERS

When iron grows coherently within the silver FCC lattice it forms a FM phase. With Fe atoms in this phase, we have calculated the local properties at the central atom for Fe clusters extending from one Fe atom to 55 Fe atoms. The contact charge density and local magnetic moment at central Fe atom remain almost con-

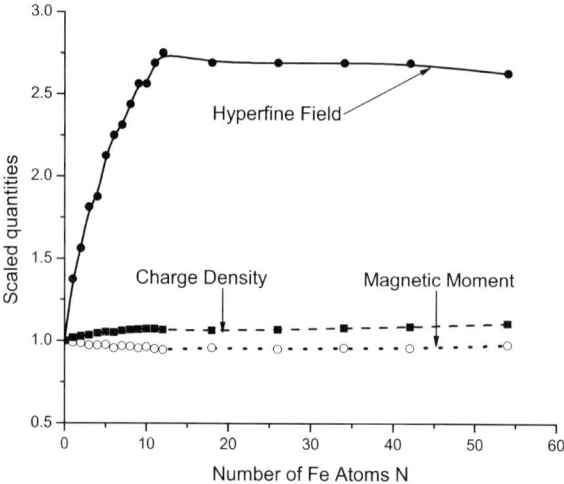

Figure 1. The local magnetic moment (open circles), the contact charge density (full boxes) and the hyperfine field (full circles) at the central Fe site, scaled by the corresponding single Fe impurity in Ag quantities, versus the iron cluster size N. The iron clusters are coherent with the underlying silver FCC lattice. Lines are for eye guidance.

stant with increasing number of Fe atoms, while the local magnetic hyperfine field increases steadily with increasing number of surrounding Fe atoms, reaching saturation at 12 Fe atoms (i.e. at filled nearest neighbor shell). These trends are illustrated in Figure 1, where the three quantities are scaled by the corresponding single Fe atom values. Consequently, it is expected that in a system of FM iron nanoparticles in Ag, the magnetization per Fe atom and the Mössbauer isomer shift remain constant as Fe content or Fe particle size change. On the other hand, the average and the distributions of the contact hyperfine field vary with particle size since the Fe atoms in the nanoparticle have different local environment. However, the observation that magnetic fields saturate for filled nearest neighbor shell configuration, simplify the deduction of fields in larger clusters. For a cluster of N iron atoms with shells filled consecutively from the center, we calculated the number of Fe atoms, n_j, with j Fe atoms in its neighboring shell. The hyperfine field distribution, $P_N(H)$, is then determined using

$$P_N(H) = \sum_{j=0}^{12} \frac{n_j}{N} \delta(H - B_j),$$

where B_j, is the calculated field at the central Fe atom with j Fe neighbors. For example, we show the in Figure 2 the field distribution for clusters with 13, 141 and 16757 Fe atoms. In the 13 Fe atom clusters the distribution is peaked around a value corresponding to the surface atoms with tails extending into values corresponding to the bulk. On the other hand, the distribution for the 16757 atoms cluster is peaked around the bulk hyperfine field with tails extending into the lower fields region. The field distribution for the 141 atoms cluster spreads over a wide range indicating

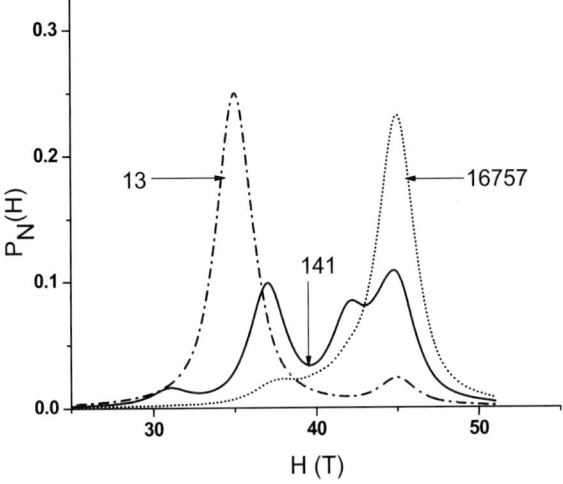

Figure 2. The contact hyperfine field distribution $P_N(H)$ for 13 (dash-dotted), 141 (continuous) and 16757 (dotted) iron atom clusters with FCC crystal structure coherent with that of silver.

the presence of various local environments. Of course, some Ag atoms may be distributed inside the Fe clusters and this will lead to additional structures in the field distribution. We note that Morales *et al.* [2] reported a distribution of hyperfine fields at 4.2 K with 3 peaks.

2.2. ANTIFERROMAGNETIC FCC CLUSTERS

We have considered the AFM coupling between the central Fe atom and the Fe atoms in the nearest neighbor (NN) shell in the equilibrium Ag lattice. The local magnetic moment at the Fe central site barely changes with introduction of this AFM coupling as the number of AFM iron NN atoms (N_{AFM}) increases. Clusters of more than 8 Fe atoms exhibit a dip in magnetic moment when four of the NN iron atoms couple AFM to the central atom. With further increase in N_{AFM} the moment remains constant. This is shown in Figure 3 for scaled moments in clusters of 12 Fe atoms or more.

The magnitude of contact hyperfine field is observed to decrease linearly with increasing N_{AFM}. For clusters with 12 Fe atoms or more, the contact field saturates for $N_{AFM} \geqslant 12$. This is also shown in Figure 3 for scaled contact fields.

The trends observed for AFM coupled atoms agree with those obtained for FM coupled atoms in the sense that the local properties at a Fe atom are determined mainly by its immediate neighborhood. The local magnetic moment is not sensitive enough to describe changes in the immediate neighborhood, while the contact field varies with changes in the type of the surrounding atoms as well as with changes in the nature of the magnetic coupling.

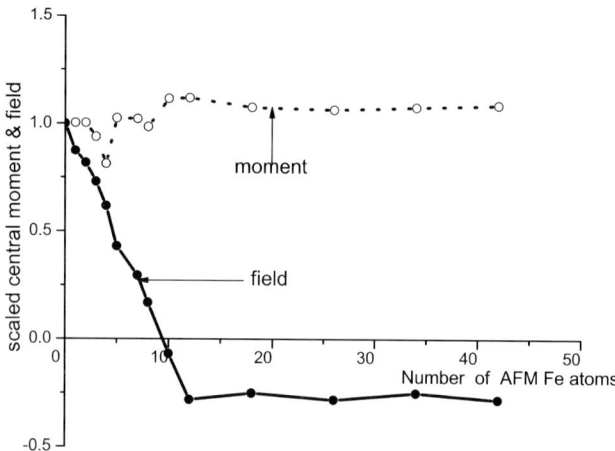

Figure 3. The local magnetic moment (open circles) and the hyperfine field (full circles) at the central Fe site, scaled by the corresponding single Fe impurity in Ag quantities, versus the number of surrounding AFM coupled Fe atoms. The iron clusters are coherent with the underlying silver FCC lattice. Lines are for eye guidance.

2.3. RELAXED CLUSTERS

The expanded FCC clusters of iron atoms in silver may relax maintaining the FCC structure or transform to the BCC phase. When the lattice constant of the FCC iron clusters is reduced from Ag lattice constant of 4.09 Å to 3.60 Å the trends of the local properties undergo various changes. In Figure 4 we show the trends of the local magnetic moment, the contact magnetic hyperfine field and the contact charge density at the central Fe site versus the number of iron atoms N. The values are scaled relative the corresponding single Fe impurity quantities. The local moment decreases with increasing N and for $N > 23$, the central Fe atoms couples AFM to its neighbors. The magnetic hyperfine field increases initially reaching saturation in a similar manner to the expanded lattice. However for $N > 23$, where the central Fe atoms couple AFM to its surroundings, the hyperfine field drops to small values. The contact charge density increases initially with increasing N and then remains constant.

In Table I we show the local magnetic moments, the contact charge densities and hyperfine fields for a relaxed FCC cluster of 13 Fe atoms in Ag versus N_{AFM}. The magnetic structure of iron clusters with a relaxed lattice and in presence of AFM coupling is more complex and does not exhibit regular trends as observed in the case of the expanded lattice. However, we found that in general the local moment at the central site decreases with increasing N_{AFM} resulting in small moment when the number of NN atoms is greater or equal to 12. In some cases this local magnetic moment changes sign and align parallel to the AFM neighbors. The trends of the contact hyperfine field follows suit, while the contact charge density hardly changes

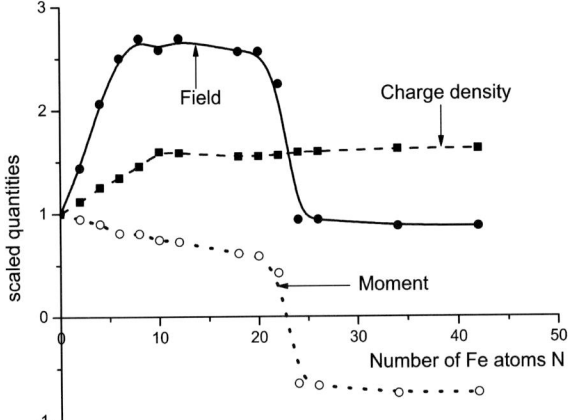

Figure 4. The local magnetic moment (open circles), the contact charge density (full boxes) and the hyperfine field (full circles) at the central Fe site, scaled by the corresponding single Fe impurity in Ag quantities, versus the iron cluster size N. The iron clusters form a relaxed FCC structure in the underlying silver lattice. Lines are for eye guidance.

Table I. The local magnetic moment μ in Bohr magnetons, the valence contact charge density ρ in atomic units and the contact hyperfine field B_{hf} in Tesla at the central Fe site in a relaxed 13 Fe atoms clusters embedded in Ag versus the number of antiferromagnetically coupled Fe atoms in NN shell N_{AFM}

	N_{AFM}							
	0	1	2	3	6	8	10	12
μ (μ_B)	2.12	1.82	1.71	−0.19	0.88	0.13	−1.67	2.06
ρ (au)	5.96	6.00	5.98	6.02	6.05	6.00	5.98	6.00
B_{hf} (T)	−43	−36	−31	−12	−4	10	31	16

with N_{AFM}. A negative isomer shift of order −0.1 mm/s will be observed in this case.

A BCC environment is established around the central Fe atom in the Fe cluster when 14 of the surrounding Fe atoms relax to form the NN (8) and next-nearest neighbors (6) BCC shells, respectively. For this cluster we found that the local moment at the central site is 2.23 μ_B, while the contact field is −39 T and the contact charge density is 5.76 au. These are to be compared with the calculated respective values for α-Fe, which are 2.30 μ_B, −41 T and 5.76 au. For environments with less Fe atoms larger magnetic moments and smaller hyperfine fields are obtained.

3. Conclusion

The magnetic properties of Fe clusters embedded in Ag exhibit various trends depending on their local crystal structure. For Fe clusters forming FCC structure coherent with the underlying Ag lattice, the magnetic properties at the central Fe site exhibit simple features. When the cluster is FM, the local magnetic moment and contact charge density remain constant as the Fe cluster builds up, whereas the hyperfine field increases steadily and saturates when the NN shell is full. This observation helps in predicting the hyperfine field distributions in a cluster of any size. In presence of AFM coupling between the central atom and its neighbors, the magnetic moment does not experience large variation, while the hyperfine field decreases in magnitude with increasing number of AFM Fe neighbors until it saturates when the Fe atom at the NN shell all couple antiparallel to the central atom.

When the Fe clusters relax maintaining the FCC structure, FM solutions at the central site are obtained for clusters with less than 23 Fe atoms. For larger clusters the AFM solution are obtained. Antiferromagnetic coupling between the central Fe atom and its Fe neighbors leads to small magnetic moments and hyperfine fields. Relaxation into BCC structure results in magnetic moments and hyperfine fields comparable to those of α-Fe. To make useful connection to experimental observations, the information about the lattice structure of the embedded clusters is needed.

Acknowledgement

The financial support of Sultan Qaboos University under grant IG/SCI/PHYS/02/01 is acknowledged.

References

1. Kahl, S. and Krebs, H., *Phys. Rev. B* **63** (2001), 172103.
2. Morales, M. A., Passamani, E. C. and Baggio-Saitovitch, E., *Phys. Rev. B* **66** (2002), 144422.
3. Gubanov, V. A., Liechtenstein, A. I. and Postnikov, A. V., *Magnetism and Electronic Structure of Crystals*, Springer Series in *Solid-State Sciences*, Vol. 98, Springer-Verlag, 1992.
4. Oswald, A., Zeller, R., Braspenning, B. J. and Dederichs, P. H., *J. Phys. F (Met. Phys.)* **15** (1985), 193.
5. Nogueira, R. and Petrilli, H., *Phys. Rev. B* **60** (1999), 4120.
6. Averil, F. W. and Ellis, D. E., *J. Chem. Phys.* **59** (1973), 6412.
7. Elzain, M. E., Ellis, D. E. and Guenzberger, D., *Phys. Rev. B* **34** (1986), 1430.
8. Elzain, M., Al Rawas, A., Yousif, A., Gismelseed, A., Rais, A., Al Omari, I. and Widatallah, H., to be published.
9. Herper, H. C., Hoffmann, E. and Entel, P., *Phys. Rev. B* **60** (1999), 3839.

Hyperfine Interactions **156/157**: 213–221, 2004.
© 2004 *Kluwer Academic Publishers. Printed in the Netherlands.*

Magnetic Properties of Nanocrystalline Fe_xCu_{1-x} Alloys Prepared by Ball Milling

A. YOUSIF[1], K. BOUZIANE[1,*], M. E. ELZAIN[1], X. REN[2], F. J. BERRY[2],
H. M. WIDATALLAH[3], A. AL RAWAS[1], A. GISMELSEED[1] and
I. A. AL-OMARI[1]

[1] *Physics Department, College of Science, Sultan Qaboos University, P.O. Box 36, PC 123, Muscat,
Sultanate of Oman; e-mail: bouzi@squ.edu.om*
[2] *Department of Chemistry, The Open University, Milton Keynes, MK7 6AA, UK*
[3] *Institute of Nuclear Research, Sudan Atomic Energy Commission, Khartoum 11115, Sudan*

Abstract. X-ray diffraction, Mössbauer and magnetization measurements were used to study Fe_xCu_{1-x} alloys prepared by ball-milling. The X-ray data show the formation of a nanocrystalline Fe–Cu solid solution. The samples with $x \geqslant 0.8$ and $x \leqslant 0.5$ exhibit bcc or fcc phase, respectively. Both the bcc and fcc phases are principally ferromagnetic for $x \geqslant 0.2$, but the sample with $x = 0.1$ remains paramagnetic down to 78 K. The influence of the local environment on the hyperfine parameters and the local magnetic moment are discussed using calculations based on the discrete-variational method in the local density approximation.

Key words: Fe–Cu alloys, nanocrystalline, bcc phase, fcc phase, magnetism.

1. Introduction

A lot of attention has been paid recently on mechanical alloying of Fe and Cu. Several non-equilibrium nano-crystalline phases of the Fe–Cu alloy system may be synthesized by the ball-milling technique, although Fe and Cu do not form solid solutions under equilibrium conditions. The miscibility between Fe and Cu has been studied as function of ball-milling time and annealing temperature using techniques such as Mössbauer spectroscopy and X-ray diffraction (XRD) [1–4]. For Fe_xCu_{1-x} system, single-phase bcc solid solutions are formed for $x \geqslant 0.8$, and fcc solid solutions are obtained for $x \leqslant 0.6$, while the samples with intermediate concentration contain both structures [5–7]. The existence of multi-phases (bcc-Fe, fcc-Cu and fcc-FeCu) was reported in the preparation of samples with $x = 0.5$ for milling times that extended from 0.5 to 100 hours, although the development of the solid solution Fe–Cu increases with milling time [8]. Similarly, the structure of the samples with $x = 0.8, 0.6$, or 0.5 milled for 100 hours have been interpreted according to the hyperfine field values as having three phases [9]. On the other hand, single- fcc phase was formed for samples with $x = 0.6$ after milling for 80

---

* Author for correspondence.

hours [10]. Thermal treatment of the solid solution Fe–Cu would develop back the immiscible phases [4, 10].

The magnetic structure of Fe–Cu system may be used as a fingerprint to study the miscibility alloying Fe and Cu by ball milling. While the Fe–Cu solid solution, with either bcc or fcc structure experiences ferromagnetic ordering, bcc-Fe in Fe rich environment is ferromagnetic but fcc Fe formed as precipitates in a Cu matrix is antiferromagnetic. We have managed to synthesize a wide range of concentrations of Fe–Cu solid solution by milling elemental Fe and Cu powders for 52 hours. Within the detection limits of Mössbauer spectroscopy and XRD, all samples have single-phase structures. The measured local magnetic moments deduced from magnetization measurements and the distribution of hyperfine fields are correlated with those obtained theoretically for different Fe environments using the discrete-variational method in the local density approximation [11].

2. Methodology

$Fe_x Cu_{1-x}$ ($x = 0.1, 0.2, 0.5, 0.8, 0.9$) alloys were prepared from crystalline Fe and Cu powders with purity of better than 99.99%. The powders were mixed in the desired compositional ratio. The mixtures were milled for 52 hours in argon atmosphere at room temperature in a Retsch PM 400 planetary ball-mill with stainless steel vial of volume 250 ml and balls of radius 20 mm at a speed of 200 rpm. The balls-to-powder mass ratio was 20 : 1. XRD measurements were carried out on a Philips PW 1700 diffractometer with CuK_α source. The average crystallite size was inferred from XRD patterns using the Scherrer relation. The magnetization was measured with a DMS 1660 vibrating sample magnetometer (VSM) in a magnetic field up to 13 kOe, and the temperature range 120–1010 K. The VSM was calibrated using pure nickel ($M_s = 54.9$ emu/g). The Mössbauer spectra were recorded at 300 and 78 K using $^{57}Co/Rh$ source with the spectrometer in the transmission mode. The spectra were fitted with a distribution of hyperfine fields using NORMOS program.

The discrete variational method was employed to solve the Kohn–Sham equation in the local density spin polarized approximation. The $Fe_x Cu_{1-x}$ alloys were represented by clusters of 51 and 55 atoms in bcc and fcc configurations, respectively. It is assumed that the plastic deformations resulting from the ball milling do not significantly change the local environments in bcc and fcc lattices. The local properties at the central site were calculated for different cluster configurations with a varying number of nearest and next-nearest neighbors of Cu atoms. The average properties were obtained using binomial distribution. In particular the probability distribution of the hyperfine field was calculated using the relation

$$P(H) = \sum \delta\big(H - B_{hf}(n, m)\big) p(n, m),$$

where $p(n, m)$ is the binomial probability distribution for the configuration with n and m atoms at nearest and next-nearest neighboring shells for bcc or fcc structures.

The magnetic hyperfine field was assumed to result from the Fermi contact term and was split into two terms. The core term was assumed to be directly proportional to the local 3d moment with a constant of proportionality of -10 T/μ_B whereas the valence term was directly calculated from the contact valence spin density [12].

3. Results and discussion

The XRD patterns of the Fe_xCu_{1-x} samples milled for 52 hours are presented in Figure 1. For samples with $x \leqslant 0.5$, the bcc Fe peaks have fully disappeared, and only the broad peaks of the fcc phase can be observed. The corresponding position of the peaks shift to lower angles relative to that of the peaks of pure fcc Cu. Besides, this shift to lower angles increases when Fe content increases. Moreover, the analysis of the intensity ratio (I_{200}/I_{111}) of the (200) reflection to (111) reflection (Table I) shows an opposed expected tendency; i.e. the ratio I_{200}/I_{111} slightly increases as the Fe content decreases, with however lower ratio values than that of pure Cu. These unexpected results may be explained by the formation of Fe–Cu solid solution with fcc phase. For samples with $x \geqslant 0.8$, only bcc peaks are observed. The intensity of (110) reflection is found to increase in detriment of (200) reflection as the Fe content increases, as expected. Similarly, the slight shift of all peaks to higher angles suggests alloying of Fe with Cu leading to the formation of solid solution with bcc phase. The angular position shifts of the peaks reflect a change of the lattice constant due to the Cu and Fe incorporation in the bcc and fcc structure, respectively. The fcc phase lattice constant changed from 3.621 Å for $x = 0.1$ to 3.645 Å for $x = 0.5$ and in bcc phase from 2.925 Å for $x = 0.8$

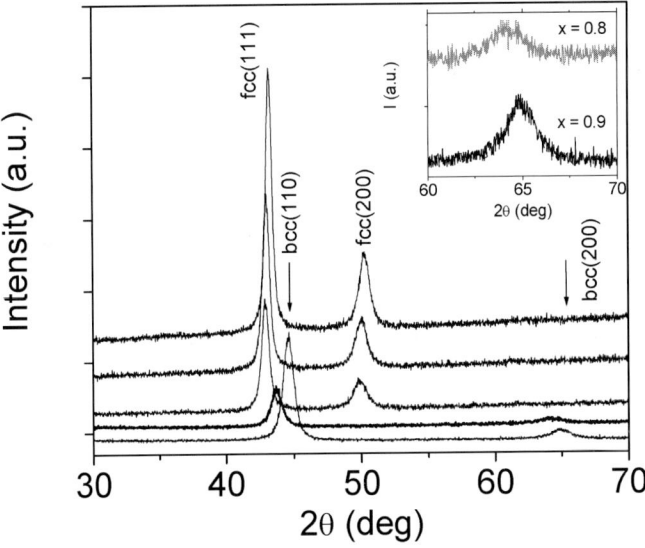

Figure 1. XRD patterns for the Fe_xCu_{1-x} samples. The top-to-bottom order is: $x = 0.1, 0.2, 0.5$, 0.8 and 0.9. Inset is an enlargement for $x = 0.8$ and 0.9.

Table I. Fitted XRD structural parameters

Fe content x	Grain size (nm)	Lattice constant a (Å)	I_{200}/I_{111} fcc	I_{200}/I_{110} bcc
0.1	32.6	3.621	0.29	–
0.2	29.5	3.639	0.27	–
0.5	28.3	3.645	0.26	–
0.8	20.1	2.925	–	0.14
0.9	20.1	2.870	–	0.08

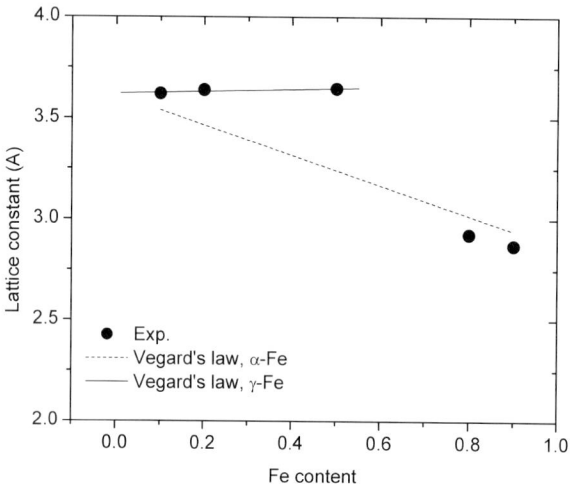

Figure 2. Lattice constant vs. Fe content. Solid and dashed lines correspond to Vegard's law based on fcc-Fe and bcc-Fe, respectively.

to 2.870 Å for $x = 0.9$. Figure 2 displays lattice constant of Fe$_x$Cu$_{1-x}$ alloys as function of Fe content. As one can notice, while the variation of the lattice constant is fundamentally in line with Vegard's law (dashed line in Figure 2) for $x \geqslant 0.8$; there is a strong deviation for $x \leqslant 0.5$. This may not be surprising since for high Fe content, Fe has essentially bcc phase (the dashed line corresponds to Vegard's law based on the bcc-Fe lattice constant). However, this deviation is corrected when considering Vegard's law based on a theoretically predicted lattice constant for fcc-Fe [13] (solid line in Figure 2). The best corresponding fit for $x \leqslant 0.5$ gives lattice constant values of 3.672 Å and 3.621 Å for Fe and Cu, respectively. This is another indication of the formation of supersaturated solid solution in the range of 0.1–0.5 Fe content. It should be pointed out that XRD peaks of oxides were not detected.

The average crystallites size is about 30 nm for the samples with $x \leqslant 0.5$ and decreases to around 20 nm for the samples with $x \geqslant 0.8$, which are slightly small

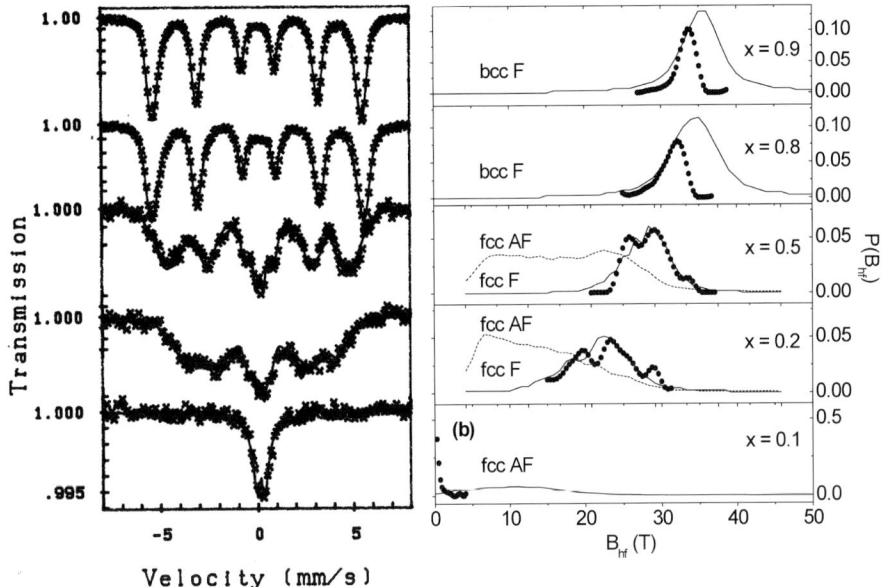

Figure 3. Experimental and fitted Mössbauer spectra for Fe$_x$Cu$_{1-x}$ at 78 K (*left*), and their hyperfine field distributions $P(B_{hf})$ (*right*): $P(B_{hf})$ derived from the Mössbauer spectra (•), calculated $P(B_{hf})$ (solid and dashed lines), F and AF denote ferromagnetic and antiferromagnetic phase, respectively.

relative to those reported in the literature [1–3]. We can reasonably admit that our samples present a nanocrystalline structure.

The ^{57}Fe Mössbauer spectra and the corresponding hyperfine field distributions $P(B_{hf})$ for Fe$_x$Cu$_{1-x}$ (with $x = 0.1, 0.2, 0.5, 0.8$ and 0.9) at 78 K are shown in Figure 3 (left) and (rigth), respectively. The corresponding theoretical distributions $P(B_{hf})$, associated with various possible structural and magnetic phases, are also plotted in Figure 3 (solid and dashed lines) for comparison. The fitted and theoretical average hyperfine field and local magnetic moment are reported in Table II. The Mössbauer spectra of the samples with $x = 0.9$ and 0.8 exhibit broadening and an average hyperfine field that decreases with decreasing Fe content, but both are smaller than that of pure α-Fe. This feature is attributed to Fe and Cu alloying since the average hyperfine fields ($\overline{B}_{hf}$) of 33.63 and 31.87 T for $x = 0.9$ and 0.8, respectively, are also typical of bcc-FeCu solid solution [4]. The corresponding experimental and theoretical hyperfine distributions $P(B_{hf})$, based on ferromagnetic bcc-FeCu phase, overlap but with slightly different $\overline{B}_{hf}$ values (see Table I). The reduction in $\overline{B}_{hf}$ is related to the decrease in the exchange interaction between the ferromagnetic atoms/entities present due to the introduction of Cu atoms. As the Fe content decreases to $x = 0.5$, the Mössbauer spectrum shows more pronounced broadening and a singlet at the center of the spectrum evolves (in fact, this singlet has been neglected for $x = 0.8$). The singlet could be attributed to Fe atoms within a predominantly Cu environment and persists as a paramagnetic component down to 78 K. The relatively wide $P(B_{hf})$ of Fe$_{0.5}$Cu$_{0.5}$ and $\overline{B}_{hf}$ of 28.88 T is consistent

Table II. Fitted (μ_{exp}) and calculated (μ_{cal}) local magnetic moments, and average hyperfine field ($\overline{B}_{hf}$) for samples with $Fe_x Cu_{1-x}$

x	Experimental $\overline{B}_{hf}$ (T) 78 K	Calculated $\overline{B}_{hf}$ (T)			μ_{exp} (μ_B) 0 K	μ_{cal} (μ_B) 0 K
		fcc-AF	fcc-F	bcc-F		
0.9	33.63	18.01	–	35.43	2.13	2.20
0.8	31.87	–	–	34.46	2.25	~2.26
0.5	28.88	14.27	28.90	29.78	2.34	2.34
0.2	22.05	7.98	21.46	–	3.05	2.65
0.1	0.00	10.05	16.92	–	0.67**	

* Magnetic moment of $Fe_x Cu_{1-x}$ system ($0.2 \leqslant x \leqslant 0.9$) calculated based on ferromagnetic phase.

** Effective paramagnetic moment μ_{eff}.

with an fcc Fe–Cu phase [4]. The closest corresponding calculated $\overline{B}_{hf}$ of 28.90 T was based on ferromagnetic-fcc Fe–Cu phase. It should be emphasized that there is striking disagreement between the experimental and theoretical $P(B_{hf})$ when the calculations were based on antiferromagnetic-fcc Fe–Cu phase which would lead to a $\overline{B}_{hf}$ of 14.27 T.

Similar features were observed for samples with $x = 0.2$ but with broader $P(B_{hf})$ and a large reduction in $\overline{B}_{hf}$ ($= 22.05$ T). This latter value agrees well with the calculated one ($\overline{B}_{hf} = 21.46$ T) based on ferromagnetic-fcc Fe–Cu phase but strongly disagrees with that based on antiferromagnetic-fcc Fe–Cu phase ($\overline{B}_{hf} = 7.98$ T).

Figure 4 shows the magnetization curve (up to 13.3 kOe) at room temperature of Fe–Cu alloys with metastable fcc and bcc phases. The magnetization curves of samples with $x \geqslant 0.2$ show hysteresis loops, which indicates that they are ferromagnetic. On the other hand, the sample with $x = 0.1$ show almost a straight line with small magnetization under a field of 13.3 kOe, which means that it is paramagnetic.

The existence of well-defined ferromagnetic state for $x \geqslant 0.2$ could be related to the volume expansion of our nanocrystalline Fe–Cu alloys indicated by the introduction of Cu into the Fe structure. The ferromagnetic ordering is also supported by magnetic measurements since the Bloch's $T^{3/2}$ law based on spin–wave excitations of ferromagnets fits quite well the magnetization $M(T)$ of samples with $x \geqslant 0.2$ at low temperatures (Figure 5). The magnetic moments of $Fe_x Cu_{1-x}$ with $x \geqslant 0.2$ were obtained from $M(0)$ by extrapolating Bloch's law [$M(T) = M(0)(1 - BT^{3/2})$] to $T = 0$ K. This plot shows that the spin waves excitation extends over a large temperature range, which increases when increasing Fe content (up to 200 K for $x = 0.2$ and up to 400 K for $x = 0.9$). This feature enables us to

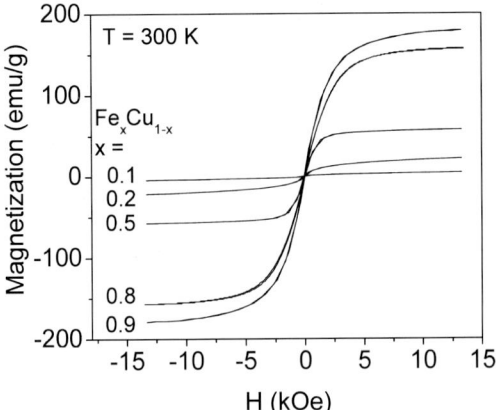

Figure 4. Magnetization curves up to 13.3 kOe at room temperature of the Fe$_x$Cu$_{1-x}$ ($x = 0.1, 0.2$, 0.5, 0.8 and 0.9) metastable bulk alloys.

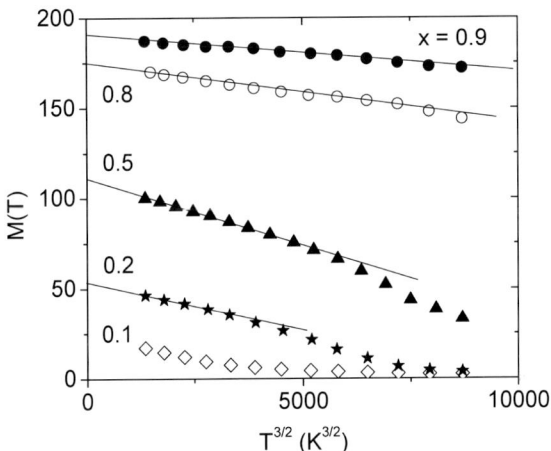

Figure 5. The magnetization at 13 kOe as function of $T^{3/2}$, based on Bloch $T^{3/2}$ law, for Fe$_x$Cu$_{1-x}$ samples.

use reasonably the Bloch's law for relatively high temperature. The magnetic moment μ_{cal} was calculated using the formula: $n(\mu_B) = $ [molar mass $\times M_s(0)$]/5586. The effective moment μ_{eff} was determined from the slope of the linear portion of the reciprocal of the susceptibility (χ^{-1}) vs. the temperature (T). These features were, also, well reflected in the experimental values of the local magnetic moment (μ_{exp}) which increases with decreasing Fe content from 2.13 μ_B for $x = 0.9$ to 3.05 μ_B for $x = 0.2$ (see Table II). This may also explain the unexpected increase of μ_{exp} accompanying the slight decrease in the atomic volume in the fcc Fe–Cu phase; corresponding to a $= 3.645$ Å ($x = 0.5$) to $a = 3.639$ Å ($x = 0.2$).

The theoretical and experimental average magnetic moments are reported in Figure 6. The theoretical moment was calculated by taking the average of magnetic

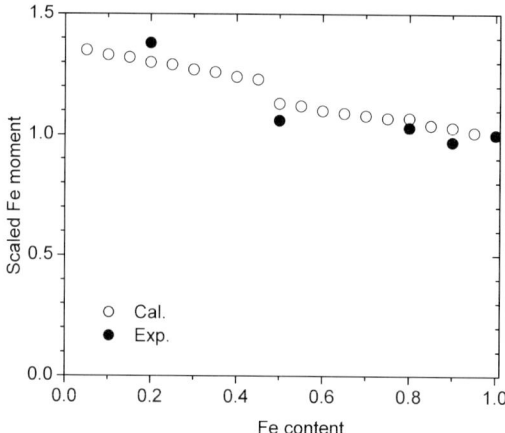

Figure 6. Average local magnetic moment ($\bullet$) experimental ($\circ$) theoretical. The plots are scaled by the bcc α-Fe moment ($2.2\,\mu_B$).

moments of different configurations at the first and second shell to central Fe atom. The plots are scaled by the bcc α-Fe moment ($2.2\,\mu_B$). The average magnetic moment is found to increase when Fe content decreases in agreement with previous results [11], that show the local magnetic moment at the Fe site increases when Fe is progressively substituted by Cu at least in the first and second shells to central Fe atom. The lattice constants obtained from XRD data supports this result since for $x \leqslant 0.5$, the lattice constant is greater than 3.6 Å consistent with high-moment ferromagnetic state [13]. For samples with $x \geqslant 0.8$, the slight but noticeable decrease in the magnetic moment with Fe content may be due principally to the increase of atomic volume of Fe [1].

The Mössbauer spectrum of sample with $x = 0.1$ is composed of a singlet characteristic of a nonmagnetic fcc phase down to 78 K. This MS singlet show no broadening when measured at different temperatures, supporting the paramagnetic state of this sample (not shown here). This is also reflected by the linear behavior of $M(T)$ vs. T/H (where H is the applied field). We can assume for the sample with $x = 0.1$ that the Fe atoms are well diffused into the Cu matrix. Besides, the experimental hyperfine field distribution totally disagrees with the theoretical one $P(B_{hf})$ that is based on antiferromagnetic-fcc of $Fe_{0.1}Cu_{0.9}$ with $\overline{B}_{hf}$ of 10.05 T.

4. Conclusion

XRD and Mössbauer spectra show that mechanically alloyed Fe_xCu_{1-x} milled for 52 hours form a nanocrystalline solid solution. Samples with $x \geqslant 0.8$ adopt ferromagnetic bcc Fe_xCu_{1-x} phase while those with $0.2 \leqslant x \leqslant 0.5$ adopt a ferromagnetic fcc phase with a possibility of nonmagnetic Fe phase. The sample with $x = 0.1$ shows a paramagnetic state crystallizing in fcc-structure. The local magnetic moment as well as the distribution of hyperfine fields were found to be very sensitive to the local environment of Fe atoms.

References

1. Jiang, J. Z., Pankhurst, Q. A., Johnson, C. E., Gente, C. and Bormann, R., *J. Phys.: Condensed Matter* **6** (1994), L227.
2. Hernando, A., Gómez-Polo, C., El Ghannami, M. and García Escorial, A., *J. Magn. Magn. Mater.* **173** (1997), 275.
3. Yavari, A. R., Desre, P. J. and Benamour, T., *Phys. Rev. Lett.* **68** (1992), 2235.
4. Yu-zhi, L., Tie, L., Yu-heng, Z., Li-wen, W., Chen, G. and Wen-han, L., *J. Phys.: Condensed Matter* **8** (1996), 7191.
5. Eckert, J., Holzer, J. C. and Johnson, W. L., *J. Appl. Phys.* **73** (1993), 131.
6. Eckert, J., Holzer, J. C., Krill, C. E. and Johnson, W. L., *J. Appl. Phys.* **73** (1993), 2794.
7. Ma, E., Atzmon, M. and Pinkerton, F. E., *J. Appl. Phys.* **74** (1993), 955.
8. Jiang, J. Z., Gonser, U., Gente, C. and Bormann, R., *Appl. Phys. Lett.* **63** (1963), 1056.
9. Li, T., Li, Y.-Z., Zhang, Yu.-H., Gao, C., Wei, S.-Q. and Liu, W.-H., *Phys. Rev. B* **52** (1995), 1120.
10. Yang, Y., Zhu, Y., Li, Q., Ma, X., Dong, Y., Wang, G. and Wei, S., *Physica B* **293** (2001), 249.
11. Elzain, M. E., Al Rawas, A., Gismelseed, A., Yousif, A., Ren, X., Berry, F. J. and Widatallah, H. M., *Hyp. Interact. (C)* **5** (2001), 535.
12. Elzain, M. E., Ellis, D. E. and Guenzberger, D., *Phys. Rev. B* **34** (1986), 1430.
13. Del Bianco, L., Ballesteros, C., Rojo, J. M. and Hernando, A., *Phys. Rev. Lett.* **81** (2004), 4500 and references herein.

Hyperfine Interactions **156/157**: 223–228, 2004.
© 2004 *Kluwer Academic Publishers. Printed in the Netherlands.*

The Formation of Lithiated Ti-Doped α-Fe$_2$O$_3$ Nanocrystalline Particles by Mechanical Milling of Ti-Doped Lithium Spinel Ferrite

H. M. WIDATALLAH[1,2,*], A. M. GISMELSEED[2], K. BOUZIANE[2],
F. J. BERRY[3], A. D. AL RAWAS[2], I. A. AL-OMARI[2], A. A. YOUSIF[2] and
M. E. ELZAIN[2]
[1]*Department of Physics, Khartoum University, P. O. Box 123, Khartoum 11115, Sudan;*
e-mail: hisham@ictp.trieste.it
[2]*Department of Physics, Sultan Qaboos University, P. O. Box 36; Muscat, Oman*
[3]*Department of Chemistry, The Open University, Milton Keynes, MK7 6AA, UK*

Abstract. The milling of spinel-related Ti-doped Li$_{0.5}$Fe$_{2.5}$O$_4$ for different times is studied with XRD, Mössbauer spectroscopy and magnetic measurements. Milling converts the material to Li–Ti-doped α-Fe$_2$O$_3$ nanocrystalline particles via an intermediate γ-LiFeO$_2$-related phase. The role played by the dopant Ti-ion in the process is emphasized.

Key words: spinel ferrites, hematite, hyperfine field.

1. Introduction

Mechanical milling has become an attractive technique for synthesizing inorganic solids since in addition to producing nanometer grain size particles, it can induce chemical and structural transformations; see, e.g., [1]. Here, we show that milling Ti-doped Li$_{0.5}$Fe$_{2.5}$O$_4$ results in the formation of lithiated Ti-doped α-Fe$_2$O$_3$ nanoparticles. The phases that develop in this formation process are investigated structurally and magnetically.

2. Experimental

Ti-doped Li$_{0.5}$Fe$_{2.5}$O$_4$ of composition Li$_{0.61}$Ti$_{0.29}$Fe$_{2.07}$O$_4$ was prepared as described previously [2]. The as-prepared material was milled in air using a Retsch PM 400 planetary ball mill with stainless steel vials (250 ml) and balls (20 mm) at 200 rpm. The powder-to-balls mass ratio was 1 : 20. Samples were labeled S0, S10, S20, S30 and S50, where the number indicates milling time in hours. X-ray powder diffraction (XRD) patterns were collected with a Philips PW1820 diffractometer using Cu$K\alpha$ radiation. Mössbauer measurements were recorded at 298 K

---

* Author for correspondence.

and 77 K using a 25 mCi ^{57}Co/Rh source. Isomer shifts are relative to that of α-Fe at 298 K. The magnetization was measured with a vibrating sample magnetometer (Model DMS-1660-VSM) in a magnetic field up to 1.3 T, and the temperature range 120–1010 K. The VSM was calibrated using pure nickel ($M_s = 54.9$ emu/g).

3. Results and discussion

The XRD patterns recorded from S0 and the milled samples are given in Figure 1. The Mössbauer spectra of the samples at 298 K and 77 K are given in Figure 2(a) and 2(b) and their fitted hyperfine parameters are presented in Table I. The Mössbauer spectra of S0 at 298 K and 77 K, were fitted with two sextets for Fe^{3+} at tetrahedral A and octahedral B sites. The fitting parameters presented in Table I that are in good agreement with the previous results [3]. The XRD pattern of sample S10 (Figure 1, S10) shows that while the spinel phase is largely preserved, a weak γ-LiFeO$_2$ related phase starts to develop. The average crystallite size of the dominant spinel phase after 10 h of milling, as determined by the Scherrer method, was found to be *ca.* 100 nm. The Mössbauer spectra of S10 (Figure 2) both at 298 K and 77 K were fitted with two spinel A and B sextets with higher isomer shift (δ) and lower hyperfine field (H_{eff}) values relative to those of S0. This feature can be attributed to a decrease in the particle size [4]. The Mössbauer spectrum at 300 K for γ-LiFeO$_2$ is an unresolved magnetic sextet, whose Mössbauer parameters are close to those of Li$_{0.5}$Fe$_{2.5}$O$_4$, superimposed on a paramagnetic doublet [6]. Hence, it was difficult to resolve the component corresponding to this weak phase from

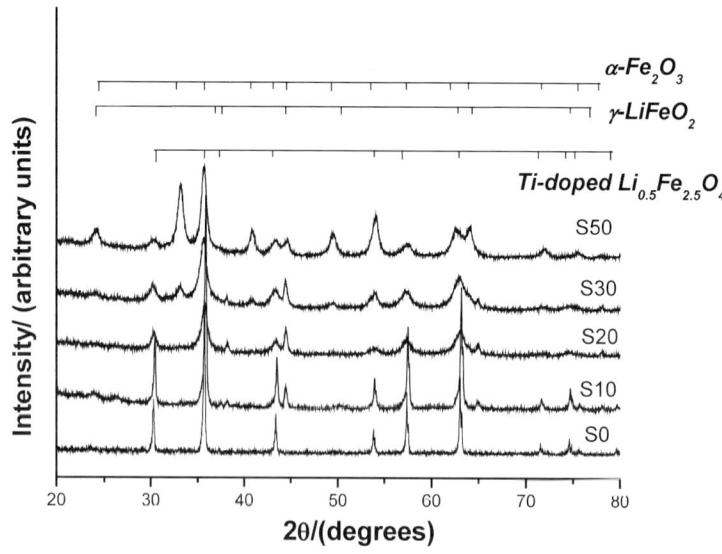

Figure 1. The XRD patterns of the materials resulting from milling Ti-doped Li$_{0.5}$Fe$_{2.5}$O$_4$ for different times (see text). The bars show the reflection peak positions of the different phases.

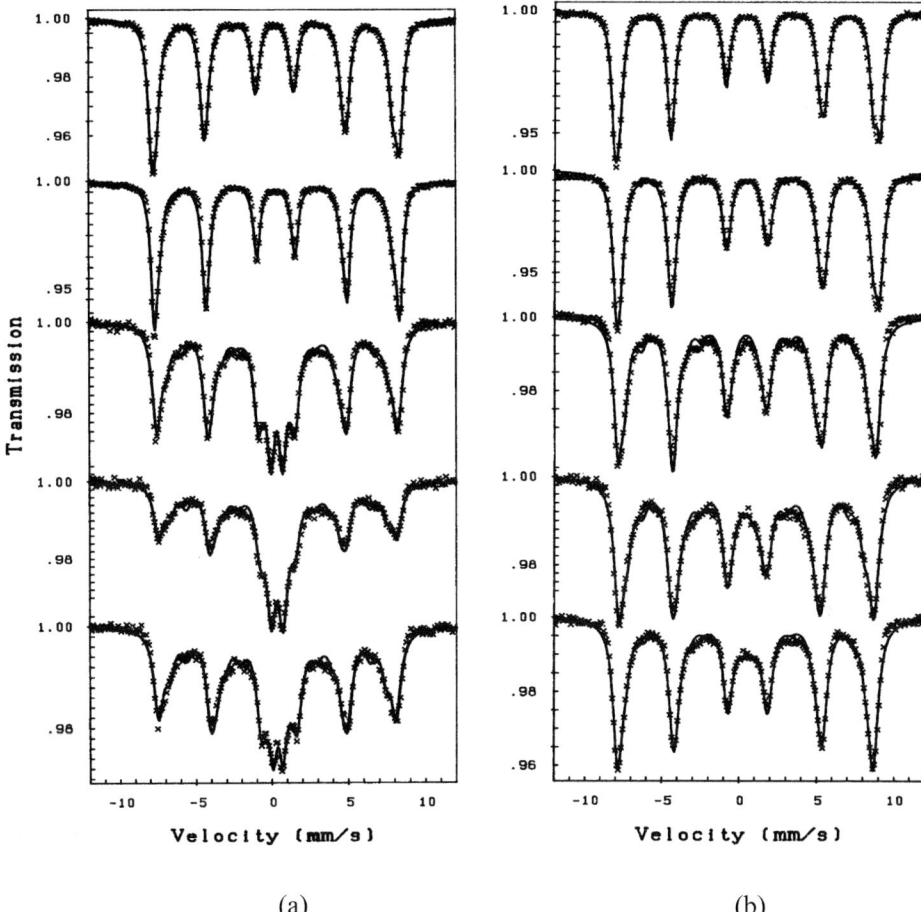

(a)   (b)

Figure 2. The Mössbauer spectra of the milled Ti-doped Li$_{0.5}$Fe$_{2.5}$O$_4$ samples at 298 K (a) and 77 K (b). The top-to-bottom order of the samples is: S0, S10, S20, S30 and S50.

the Mössbauer spectrum of S10, even though it was clear in the XRD pattern of the sample.

After 20 h of milling (Figure 1, S20), the XRD peaks of the Ti-doped Li$_{0.5}$Fe$_{2.5}$O$_4$ phase decreased in intensity and broadened substantially giving a crystallite size of 15 nm while the intensity of the reflection peaks of the γ-LiFeO$_2$ related phase increased. The Mössbauer spectrum of S20 at 298 K (Figure 2(a)) which shows a broad six-line pattern superimposed on a central doublet was fitted with two sets of sextets (A,B) and (A′,B′). Sextets A and B correspond to larger particles in the sample while A′ and B′, with lower H_{eff} values can be associated with smaller particles whose size is just above the 'superparamagnetic' volume [4]. The doublet is attributed to Ti-doped Li$_{0.5}$Fe$_{2.5}$O$_4$ superparamagnetic particles having effective volume with blocking temperatures $T_B < 298$ K. This superparamagnetic relaxation is suppressed at 77 K. Milling for 30 h (Figure 1, S30) leads, in addition

Table I. The Mössbauer parameters of the milled samples at 298 K. The corresponding parameters for the samples at 77 K are given between parentheses

Sample	Component	δ (mm/s)		Δ (mm/s)		H_{eff} (T)		Area (%)	
S0	A	0.23	(0.32)	0.02	(0.02)	48.09	(50.95)	42.2	(48.7)
	B	0.28	(0.41)	0.00	(0.00)	50.31	(53.2)	57.8	(51.3)
S10	A	0.27	(0.36)	0.03	(0.00)	47.73	(50.67)	41.2	(44.1)
	B	0.33	(0.47)	0.00	(0.00)	49.87	(52.69)	58.8	(55.9)
	A	0.31	(0.35)	0.06	(0.13)	45.01	(49.00)	15.0	(27.4)
S20	B	0.33	(0.47)	−0.01	(−0.05)	48.93	(51.78)	47.9	(54.5)
	A$'$	0.28	(0.27)	−0.11	(−0.12)	27.88	(30.23)	8.2	(8.6)
	B$'$	0.33	(0.49)	0.02	(0.06)	39.40	(42.79)	9.8	(9.5)
	Doublet	0.33		0.78				19.0	
	A	0.25	(0.40)	0.01	(0.03)	27.00	(44.00)	15.1	(27.4)
S30	B	0.35	(0.46)	−0.04	(−0.05)	48.42	(51.78)	29.9	(54.5)
	A$'$	0.34	(0.28)	0.07	(0.05)	37.44	(30.23)	11.6	(8.6)
	B$'$	0.37	(0.39)	−0.01	(−0.10)	44.03	(49.00)	15.7	(27.4)
	Doublet	0.33	(0.41)	0.74	(0.58)			27.7	(15.1)
	A	0.40	(0.49)	0.01	(0.01)	37.40	(48.42)	15.5	(3.4)
S50	B	0.41	(0.44)	−0.03	(−0.05)	26.95	(30.10)	10.4	(17.2)
	Hem.1	0.39	(0.48)	−0.13	(−0.15)	44.60	(42.11)	18.0	(6.0)
	Hem.2	0.36	(0.64)	−0.15	(−0.13)	48.30	(51.31)	40.7	(70.0)
	Doublet	0.37	(0.57)	0.61	(0.58)			15.4	(3.4)

δ = isomer shift; Δ = quadrupole splitting; H_{eff} = hyperfine magnetic field.

to the γ-LiFeO$_2$ phase, to the formation a small α-Fe$_2$O$_3$ (hematite) related phase. It was not possible to resolve the spectral component of the emerging α-Fe$_2$O$_3$ related phase by fitting the Mössbauer spectra of S30 at 298 K and 77 K which were fitted in a similar way as that of S20 (at 298 K).

The XRD pattern of S50 (Figure 1) shows that the resulting material is dominantly a α-Fe$_2$O$_3$ related phase. The peaks of the γ-LiFeO$_2$ related phase almost disappear after 50 h of milling, showing it to be an intermediate phase in the conversion process. The small amount of the initial spinel phase that exists after 50 h indicates that the complete conversion to the α-Fe$_2$O$_3$ related phase requires additional milling time. The crystallite size of the α-Fe$_2$O$_3$ related phase in S50 remained unchanged relative to that of S30. This is supported by the Mössbauer spectra of S50 at 298 K and 77 K showing the α-Fe$_2$O$_3$ related phase to be dominant. The spectrum of the α-Fe$_2$O$_3$ related phase was best fitted to two-sextets, *Hem.1* and *Hem.2*, with values for δ and quadrupole splitting (Δ) similar to those of lithiated Ti-doped α-Fe$_2$O$_3$ [6]. The broader sextet *Hem. 1* with low intensity results from Fe^{3+} with Ti^{4+} and/or Li$^+$ nearest neighbors. The decrease in H_{eff}

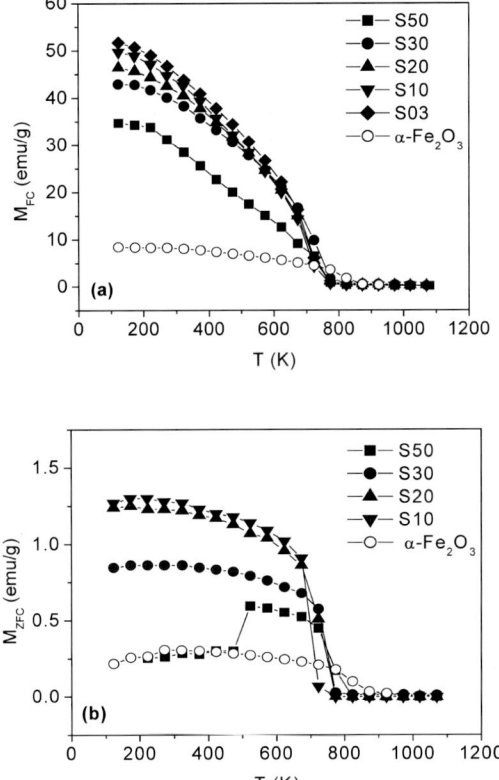

Figure 3. Temperature dependence of magnetization for the milled Ti-doped Li$_{0.5}$Fe$_{2.5}$O$_4$ samples: S10, S20, S30 and S50, when cooling at (a) 1.3 T (M_{FC}) and (b) 0.05 T (M_{ZFC}). The corresponding curves for α-Fe$_2$O$_3$ are also shown for comparison.

relative to the pure α-Fe$_2$O$_3$ is due to the locations of the non-magnetic ions as well as the small particle size. The narrow and more intense sextet *Hem. 2*, is attributed to Fe^{3+} with no Ti^{4+} and/or Li$^+$ ions in the nearby lattice sites. The values of H_{eff} for this sextet are comparable to those of α-Fe$_2$O$_3$ with similar particle size [7]. It is seen from the Mössbauer results (Table I) that after 50 h of milling, 76% of the Ti-doped Li$_{0.5}$Fe$_{2.5}$O$_4$ initial sample converts to lithiated titanium-doped α-Fe$_2$O$_3$.

Figure 3 shows the magnetization behavior when cooling from 1100 K down to 77 K at a field of 1.3 T (a) and <0.05 T (b) (hereafter FC and ZFC, respectively) for the different samples. The temperature-dependence of the magnetization at 1.3 T (Figure 3(a)) reveals a similar trend for S10, S20 and S30, which all have a Curie temperature of about 770 K. While, the magnetization at zero-temperature M(0) (determined by extrapolating the Bloch's law M_{FC} vs. $T^{3/2}$ to $T = 0$ K) decreases with increasing the milling time from 52.63 emu/g for S10 to 44.4 emu/g for S30. This could be attributed to the reduction of crystallites size in the dominant Ti-doped Li$_{0.5}$Fe$_{2.5}$O$_4$ phase as revealed by XRD and Mössbauer data. However, the substantial decrease in M_{ZFC} for S30 relative to those of S10 and S20 (Figure 3(b)) cannot be explained only in terms of crystallites becoming progressively nano-sized. The partial change of phase to a weak ferromagnetic α-Fe$_2$O$_3$ and antiferromagnetic γ-LiFeO$_2$, revealed by XRD and Mössbauer techniques, may be the origin of M$_{ZFC}$ drop after 30 h of milling. The relatively smaller value of M(0) of S50 is consistent with a dominant α-Fe$_2$O$_3$ phase with smaller magnetic moment at zero-temperature. Moreover, M_{ZFC} of S50 is similar to that of α-Fe$_2$O$_3$ for $T < 500$ K, showing that the material has almost transformed to an α-Fe$_2$O$_3$ related phase as inferred from both XRD and Mössbauer data. The decrease in the Curie temperature of S50 (820 K) relative to that of bulk α-Fe$_2$O$_3$ (870 K) [8] can be associated with both the small particle size and the presence of the non-magnetic Li$^+$ and Ti^{4+} corundum-related structure. The sharp decrease in M_{ZFC} of S50 at 500 K with decreasing temperature may be associated to the remaining superparamagnetic Ti-doped Li$_{0.5}$Fe$_{2.5}$O$_4$.

The conversion of Ti-doped Li$_{0.5}$Fe$_{2.5}$O$_4$ discussed in the present study to lithiated Ti-doped α-Fe$_2$O$_3$ is to be contrasted with the stability of both pure [9] and Mg-doped Li$_{0.5}$Fe$_{2.5}$O$_4$[4] where the spinel structure was preserved under a similar milling regime. The role of Ti^{4+} ions in inducing the observed conversion is not clear and awaits for more investigation.

References

1. Begin-Collin, S., Girot, J., Caer, G. L. and Macellin, A., *J. Solid State Chem.* **149** (2000), 41.
2. Widatallah, H. M., Berry, F. J., Johnson, C., Moore, E. A., Jartych, E., Pekala, M. and Grabski, J., ICTP Preprint IC/2002/158.
3. Dormann, J. L., Tomas, A. and Nogues, M., *Phys. Stat. Sol. A* **77** (1983), 611.
4. Widatallah, H. M., Berry, F., Gismelseed, A., Johnson, C. and Pekala, M., *Hyp. Interact. C* **5** (2002), 87.
5. Tabuchi, M., Tasutusi, S., Masqulier, R., Kanno, C., Ado, K., Matsubara, I., Nasu, S. and Kageyama, H., *J. Solid State Chem.* **140** (1998), 159.
6. Berry, F. J., Marco, J. F., Stewart, S. J. and Widatallah, H. M., *Solid State Commun.* **117** (2001), 235.
7. Vasquez-Mansilla, M., Zysler, R. D., Arciprete, C., Dimitrjewits, M., Rodriguez-Sirra, D. and Saragovi, C., *J. Mag. Mag. Mater.* **226–230** (2001), 1907.
8. Stewart, S. J., Borzi, R. A., Cabanillas, E. D., Punte, G. and Mercader, R. C., *J. Magn. Magn. Mater.* **260** (2003), 447–454.
9. Widatallah, H. M. and Berry, F., to be published.

Hyperfine Interactions **156/157**: 229–234, 2004.
© 2004 *Kluwer Academic Publishers. Printed in the Netherlands.*

Effect of Mg^{2+} on the Magnetic Compensation of Lithium–Chromium Ferrite

A. RAIS*, A. A. YOUSIF, A. GISMELSEED, M. E. ELZAIN, A. AL RAWAS
and I. A. AL-OMARI
Department of Physics, Sultan Qaboos University, P.O. Box 36, Muscat 123, Sultanate of Oman;
e-mail: amrais@yahoo.com

Abstract. Mg-substituted ferrite $Li_{0.5}Mg_xFe_{1.25-(2/3)x}Cr_{1.25}O_4$ ($0 \leqslant x \leqslant 0.3$) was studied using X-ray diffraction, Mössbauer spectroscopy and magnetic measurements. X-ray diffraction patterns show that all samples have cubic spinel structure. The temperature-dependent magnetic measurements revealed that the compensation point T_K of $Li_{0.5}Fe_{1.25}Cr_{1.25}O_4$ starts to approach the Neel temperature T_N as Mg^{2+} substitution of Fe^{3+} increases, until the compensation disappears at $x = 0.3$. This effect is investigated in relation to the cation distribution established using the Mössbauer study of this system.

Key words: magnetic compensation, X-ray diffraction, Mössbauer spectroscopy.

1. Introduction

The ferrite $Li_{0.5}Fe_{2.5}O_4$ has been the subject of extensive technical and fundamental studies both in its pure form as well as its substituted form [1–3]. Various researchers have reported the effect of additions of divalent, trivalent and tetravalent ions in lithium ferrites and the different parameters have been measured depending on the desired application [4–6].

The phenomenon of magnetic compensation is the disappearance of the magnetization at a temperature other than the Neel temperature and Gorter *et al.* [7] were the first to observe this phenomenon in $Li_{0.5}Fe_{1.25}Cr_{1.25}O_4$. The chromium-doped lithium ferrite $Li_{0.5}Fe_{2.5-x}Cr_xO_4$ is among the few systems exhibiting this phenomenon. In this work, we report the effect of Mg^{2+} substitution for Fe^{3+} on the magnetic compensation of lithium–chromium ferrite. Moreover, using a Mössbauer study of this system, we propose a cation distribution for $Li_{0.5}Mg_xCr_{1.25}$ $Fe_{1.25-(2/3)x}O_4$ and investigate its relationship with the magnetization measurements.

---

* Author for correspondence.

2. Experimental

Five samples of the ferrite system $Li_{0.5}Mg_xCr_{1.25}Fe_{1.25-(2/3)x}O_4$ ($0 \leqslant x \leqslant 0.3$) were prepared by the conventional double-sintering ceramic technique [8]. The X-ray data were collected using a Philips PW1820 vertical goniometer with mono-chromator attached to a CuK_α PW1700 generator operating at a voltage of 40 kV. The Mössbauer spectra were collected in the transmission mode with α-Fe for cal-ibration. The magnetization measurements were performed on a vibrating sample magnetometer (VSM) of 10^{-5} emu sensitivity in the magnetic field range of 0 kOe to 13.5 kOe and in the temperature range of 77 K to 600 K.

3. Results and discussion

X-ray diffraction spectrum showed the formation of cubic spinel structure for all five samples. A representative diffraction pattern of $Li_{0.5}Cr_{1.25}Fe_{1.25}O_4$ is shown in Figure 1. The major peaks are indexed while the minor peaks indicate the presence of superlattice structures which suggest an ordered arrangement of Li^{2+} and Fe^{3+} cations on the octahedral sublattice.

The lattice parameter (a) was obtained by extrapolation to $\theta = 90°$ of a for different indexed planes against the Nelson–Riley function. The lattice parameter value of $Li_{0.5}Cr_{1.25}Fe_{1.25}O_4$ agrees well with the literature taking into account our preparation technique. As can be seen in the inset of Figure 1, within error bars the lattice parameter increases slowly with magnesium content. This variation reflects the larger radius of Mg^{2+} as compared to Fe^{3+}.

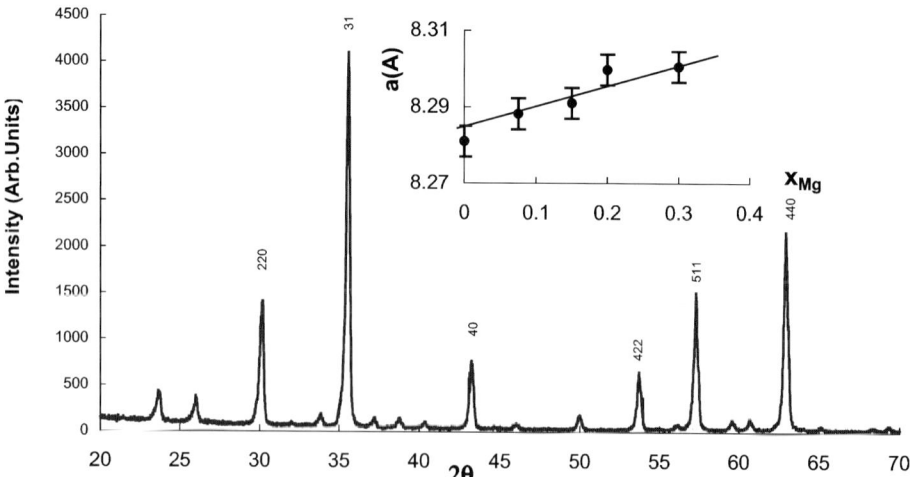

Figure 1. X-ray representative spectrum of $Li_{0.5}Fe_{1.25}Cr_{1.25}O_4$. The inset shows the lattice param-eter a (Angstroms) versus magnesium content.

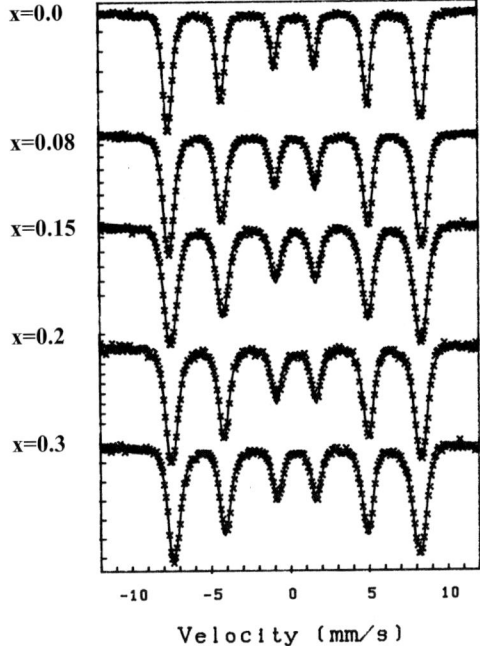

Figure 2. Mössbauer spectra of Li$_{0.5}$Mg$_x$Cr$_{1.25}$Fe$_{1.25-(2/3)x}$O$_4$ at 77 K.

Figure 2 shows the Mössbauer spectra of Li$_{0.5}$Mg$_x$Cr$_{1.25}$Fe$_{1.25-(2/3)x}$O$_4$ (0 $\leqslant$ $x \leqslant 0.3$) at 77 K. The spectra were fitted with two magnetic sextets assigned to the octahedral (B) and tetrahedral (A) sites.

The linewidth of the outer absorption line of each spectrum shows a broadening and increases with magnesium content. This could give an estimate of the distribution of the hyperfine fields at each site. The 77 K spectra shows patterns similar to room temperature, but with less broadened outer lines. All the hyperfine interaction parameters, given in Table I, were allowed to vary freely in the fitting process. Note that all the magnetic hyperfine fields are decreasing slowly with increasing magnesium content.

Taking the fitted absorption areas as proportional to iron site occupancies, Table II shows the proposed cation distributions. It is well known from previous studies that Mg^{2+} occupies A-sites [9] and Cr^{3+} occupies B-sites [10]. On the other hand from earlier reports, lithium–chromium ferrites [11] have been interpreted with Li$^+$ occupying preferentially the octahedral sites. The rest of the tetrahedral and octahedral sites have been filled with Fe^{3+} cations in accordance with our Mössbauer absorption areas. On the basis of Neel's molecular field model [12] and the cation distributions given in Table II, the magnetic moment in ferrites is mainly from the parallel-uncompensated electron spin of the individual ions, and the spin alignments in the two sub-lattices are arranged antiparallel. Also, the A–B exchange interaction is predominant over the A–A and B–B interactions. Hence the

Table I. Mössbauer parameters of $Li_{0.5}Mg_xCr_{1.25}Fe_{1.25-(2/3)x}O_4$ at 77 K. Isomer shift (IS); quadrupole splitting (QS); line width (LW); hyperfine field (H); relative peak areas (A)

x	Tetrahedral A-site					Octahedral B-site				
	IS mm/s	QS mm/s	LW mm/s	H (T)	$A\%$	IS mm/s	QS mm/s	LW mm/s	H (T)	$A\%$
0.0	0.38	0.02	0.56	48.5	79.8	0.39	0.04	0.46	50.3	20.2
0.08	0.41	0.03	0.92	48.8	77.3	0.44	0.0	0.46	50.3	22.9
0.15	0.42	0.01	0.92	48.0	73.9	0.42	0.04	0.5	50.4	26.1
0.2	0.42	0.01	0.81	48.2	71.4	0.44	0.03	0.73	50.2	28.6
0.3	0.42	0.01	0.81	47.1	66.7	0.43	0.05	0.54	49.6	33.3

Table II. Cation distribution of $Li_{0.5}Mg_xCr_{1.25}Fe_{1.25-(2/3)x}O_4$ and calculated magnetic moments M_A, M_B and M_0 in Bohr magnetons. Round brackets denote A-sites and square brackets denote B-sites

x	Cation distribution	M_B (μ_B)	M_A (μ_B)	M_0 (μ_B) = $M_B - M_A$
0	$(Fe) [Li_{0.5}Cr_{1.25}Fe_{0.25}]$	5	5	0
0.08	$(Fe_{0.925}Mg_{0.08}) [Li_{0.5}Cr_{1.25}Fe_{0.275}]$	5.1	4.6	0.5
0.15	$(Fe_{0.85}Mg_{0.15}) [Li_{0.5}Cr_{1.25}Fe_{0.3}]$	5.3	4.3	1.0
0.2	$(Fe_{0.8} Mg_{0.2}) [Li_{0.5}Cr_{1.25}Fe_{0.32}]$	5.4	4.0	1.4
0.3	$(Fe_{0.7} Mg_{0.3}) [Li_{0.5}Cr_{1.25}Fe_{0.35}]$	5.5	3.5	2.0

net magnetic moment of the lattice is given by the algebraic sum of the magnetic moments of A and B sub-lattices, i.e. $M_0 = M_B - M_A$. Given the electronic configurations of the B- and A-site cations [Fe^{3+} ($3d^5$) and Cr^{3+} ($3d^3$)], we can estimate the sublattice magnetizations (Fe^{3+} has no orbital momentum, whereas in Cr^{3+} it is crystal-field frozen). The calculated spin magnetic moments M_0 at 0 K are shown in Table II and appear to increase with magnesium content.

Measured saturation magnetization M_S of all compositions against temperature are shown in Figure 3. At $x = 0$, M_S values are in good agreement with those of Gorter *et al.* [7] as well as the compensation temperature T_K (310 K) and the Neel temperature T_N (500 K). The Neel temperatures were obtained using the method of the "intersecting tangents" to M_S against T curves. The inset shows T_N and T_K versus magnesium content.

It is interesting to emphasize the gradual increase of M_S with the addition of magnesium at temperatures below T_K. This trend is consistent with the calculated M_0 values (Table II) obtained using the cation distribution deduced from our Mössbauer analysis. We could not extract the experimental values of M_S at 0 K from our

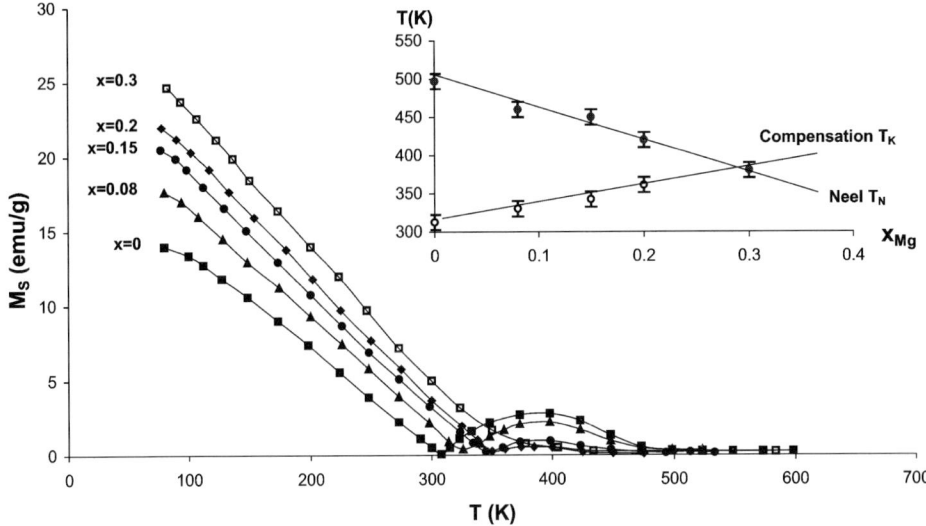

Figure 3. Saturation magnetizations versus temperature of Li$_{0.5}$Mg$_x$Cr$_{1.25}$Fe$_{1.25-(2/3)x}$O$_4$. The inset shows the Neel and compensation temperatures T_N and T_K versus magnesium content.

data because the long extrapolation from 77 K to 0 K would make the comparison with the calculated M_0 unreliable.

As expected, the inset shows a decreasing T_N as more diamagnetic Mg^{2+} replaces Fe^{3+}. Moreover, Figure 3 shows that M_S decreases with magnesium content at temperatures between T_K and T_N, this is also expected since T_K gradually approaches T_N as the inset shows. Indeed, on extrapolating the compensation line, the intersection with the Neel line represents the point where compensation disappears, i.e. $x \approx 0.3$.

This value is confirmed by our direct observation of the magnetization behavior at this concentration. This effect of the disappearance of magnetic compensation can be interpreted on the basis of Neel's molecular field model [12] and the concept of 'weak' magnetic sublattice as introduced by Belov [13]. In the case of Li$_{0.5}$Mg$_x$Cr$_{1.25}$Fe$_{1.25-(2/3)x}$O$_4$, we may consider A-sites as the magnetically strong sub-lattice because of the relatively higher content of Fe^{3+} cations and the B-sites as the weak sub-lattice. As diamagnetic Mg^{2+} substitutes Fe^{3+} in A-sites only, A–A exchange interactions are modified in such a way that the magnetization M_A of A sub-lattice A becomes weaker. This means that the temperature dependence of M_A with temperature is such that the compensation occurs at a point T_K approaching gradually T_N. The compensation disappears at a magnesium content where the A sub-lattice magnetization has weakened sufficiently so that compensation of B sub-lattice over the whole temperature range up to T_N does not occur. Note that as Mg^{2+} substitutes Fe^{3+} in A-sites, the net moment $M = M_B - M_A$ at $T < T_K$ increases because M_A decreases while M_B stays almost constant. This is due to the relatively unchanged content of the magnetic cations Fe^{3+} and Cr^{3+} in B-sites.

4. Conclusions

We conclude from the study of the effect of Mg^{2+} on the magnetic compensation of lithium–chromium ferrite that:

1. The magnetic compensation in $Li_{0.5}Fe_{1.25}Cr_{1.25}O_4$ disappears when Fe^{3+} in A-site is partially replaced by Mg^{2+}.
2. Below the compensation temperature, the observed magnetic moment of these ferrites increases with magnesium content.
3. The magnetic moments calculated using a cation distribution consistent with the Mössbauer study of these ferrites show an increasing trend with increasing magnesium content. This agrees with the magnetic measurements.

References

1. Krishnan, R., *Physica B* **86–88** (1977), 1457.
2. Dormann, J. L., *Rev. Phys. Appl.* **15** (1980), 1113.
3. Gill, N. K. and Puri, R. K., *Spectrochimica Acta A* **41** (1985), 1005.
4. Deepika, K., Sumitra, P. and Baijal, J. S., *J. Mater. Sci.* **25** (1990), 5142.
5. Song, J. M. and Koh, J. G., *IEEE Trans. Magn.* **32** (1996), 411.
6. Mazen, S. A., Metawe, F. and Mansour, S. F., *J. Phys. D* **30** (1997), 1799.
7. Gorter, E. W. and Schulkes, J. A., *Phys. Rev.* **89** (1953), 487.
8. Ahmed, M. A., Darwish, K. A., Mikhail, H., Mounir, M. and El-Khawas, E. H., *Phys. Scripta* **55** (1997), 750.
9. Fatseas, G. A., Dorman, J. L. and Blanchard, H., *J. Phys.* **12** (1976), 787.
10. Chen, Y. L., Xu, B. B. and Chen, J. G., *Hyp. Interact.* **70** (1976), 1029.
11. Kuznetsov, M., Pankhurst, Q. A. and Parkin, I. P., *J. Phys. D* **31** (1998), 2886.
12. Neel, L., *Ann. Phys.* **3** (1948), 137.
13. Belov, K. P., *Phys. Usp.* **39** (1996), 623.

Hyperfine Interactions **156/157**: 235–240, 2004.
© 2004 *Kluwer Academic Publishers. Printed in the Netherlands.*

High-Pressure ^{57}Fe Mössbauer Spectroscopy of Octamethyl-Ethynyl-Ferrocene

T. SUENAGA[1], S. NASU[1], T. KAWAKAMI[2] and R. H. HERBER[3]

[1]*Department of Materials Engineering Science, Graduate School of Engineering Science, Osaka University, Toyonaka, Osaka 560-8531, Japan*
[2]*Institute of Quantum Science, Nihon University, Tokyo 101-8308, Japan*
[3]*Racah Inst. of Physics, Hebrew University of Jerusalem, 91904 Jerusalem, Israel*

Abstract. High-pressure ^{57}Fe Mössbauer measurements of octamethyl-ethynyl-ferrocene (OMFA) were performed using a diamond anvil cell (DAC) in order to clarify the pressure effects on the phase transition occurring at ∼248 K. A sharp resonance can be seen at 300 K and 12.6 GPa, suggesting a pressure-induced recovery to a rather large Lamb–Mössbauer factor. The resonance effect gradually decreases with a decrease of the external pressure.

Key words: octamethyl-ethynyl-ferrocene, DAC, phase transition, Lamb–Mössbauer factor.

1. Introduction

Octamethyl-ethynyl-ferrocene (OMFA) exhibits a sharp decrease of the Lamb–Mössbauer factor at a temperature $T_c \approx 248$ K, well below the melting point $T_m = 436$ K [1, 2], which points to a strong increase in the mean square displacement of the iron atoms. It has been suggested that this phenomenon is of intramolecular origin, e.g., due to the onset of oscillations due to the rotation of the two cyclopentadienyl rings [1, 3]. We have performed a high-pressure Mössbauer spectroscopic study in order to understand this phase transition from the low temperature phase to the high temperature phase.

In this investigation, we performed high-pressure ^{57}Fe Mössbauer measurements of OMFA using a diamond anvil cell (DAC) in order to show the pressure effects on the phase transition occurring at 248 K. A sharp resonance was observed at 300 K and 12.6 GPa, suggesting a pressure-induced recovery to a rather large Lamb–Mössbauer factor. The resonance absorption area gradually decreases with a decrease of the external pressure. A small resonance absorption is still visible at 0.1 MPa after pressurization to 12.6 GPa, indicating that the external pressure at 300 K induces a first-order phase transition from the high temperature phase to the low temperature phase. However, we repeated the Mössbauer measurements and found recently small resonance absorption is not always visible at 0.1 MPa after pressurization above 10 GPa.

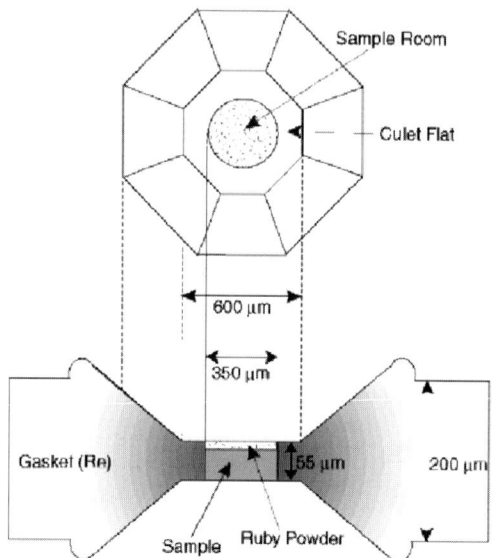

Figure 1. A schematic drawing of the arrangement of the diamond anvil, Re gasket, ruby powder as a manometer, and the specimen. The Re gasket, of 200 μm thickness, is pressed by two opposing diamond anvils. At the center of the Re gasket there is a hole for the specimen chamber of 350 μm diameter and 55 μm thickness.

2. Experimental

A sample of octamethyl-ethinyl-ferrocene (OMFA) enriched to 95% in ^{57}Fe was used. We used Basset type DAC and ruby fluorescence manometry to measure the distribution of pressure in the specimen chamber of the DAC. For the pressure-transmitting medium, we used Daphne oil 7373 (IDEMITSU Co. Japan). A rhenium gasket was used to collimate the Mössbauer γ-rays. A high-density ^{57}Co source (1 mm × 1 mm, 370 MBq) was used. A standard spectrometer (WISSEL Germany) and data acquisition system with a PC were used for the Mössbauer measurements. The velocity scale used was relative to α-Fe at 300 K.

Figure 1 shows a typical arrangement of the opposed two diamond anvil, Re gasket, ruby powder and specimen in the pressure transmitting medium (not shown).

3. Results and discussion

In order to study the pressure effect, we subjected the OMFA specimens to pressures up to 12.6 GPa, from an ambient pressure of 0.1 MPa, and measured the Mössbauer spectrum at 300 K. We then decreased the pressure gradually from 12.6 GPa to an ambient pressure of 0.1 MPa. Figure 2 shows typical ^{57}Fe Mössbauer spectra of OMFA obtained at 300 K and at different pressures, during the process of decreasing the pressure from 12.6 GPa to 0.1 MPa. All the spectra consist of a large quadrupole-split doublet, with the magnitude of the resonance

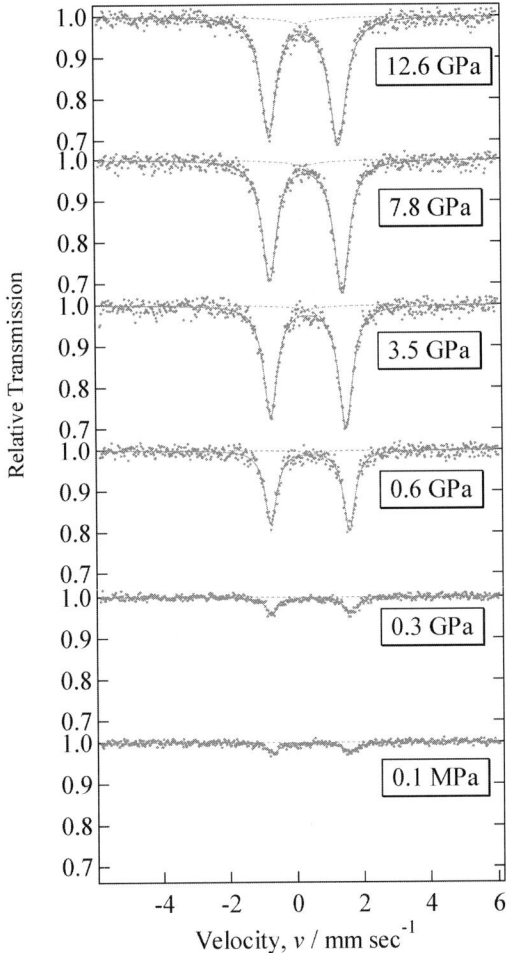

Figure 2. Typical ^{57}Fe Mössbauer spectra obtained from octamethyl-ethynyl-ferrocene (OMFA) at 300 K at pressures decreasing from 12.6 GPa to 0.1 MPa. The vertical axis indicates the relative transmission.

absorption area decreasing as the pressure decreases. It is well known for typical solids that the magnitude of the Lamb–Mössbauer factor, reflected in the area under the resonance absorption curve and its temperature dependence, follow a simple exponential behavior in the high temperature limit [4]. Resonance absorption curves can be extracted from precision data up to a temperature within a few degrees (or less) of the melting point. However, the ^{57}Fe Mössbauer spectra of OMFA measured in the temperature range between 90 K and 350 K at 0.1 MPa pressure exhibit a sharp decrease of the resonance absorption (related directly to the Lamb–Mössbauer factor) at 248 K, well below the melting point of 436 K, which suggests the existence of a phase transition from a low temperature phase to a high temperature phase. Figure 3 shows a sketch of OMFA [3] and an Fe atom

Figure 3. A sketch of octamethyl-ethynyl-ferrocene (OMFA). The Fe atom (largest sphere) is located between the upper cyclopentadienyl ring and the lower ring [3].

(large sphere) located between the upper cyclopentadienyl ring and the lower ring. The Fe atom motion in the low temperature phase is nearly static and the magnitude of the Lamb–Mössbauer factor in this phase is rather large, showing a clear resonance absorption curve. On the other hand, in the high temperature phase the Fe atom motion, that is, the mean square displacement of the atom, is anomalously large and the resonance absorption in this phase nearly vanishes. The origin of this anomalously large atomic displacement might be due to ring rotation or oscillations of the molecule coupled with rotation of the substituted cyclopentadienyl ring. Figure 4 shows the resultant Mössbauer parameters at 300 K as a function of pressure, decreasing from 12.6 GPa. As shown in Figure 4, the full widths at half maximum and intensity ratio of the 1st and 2nd absorption curves do not show any appreciable change on pressure reduction from 12.6 GPa to 0.1 MPa. The isomer shift value and the magnitude of the quadrupole splitting increase slightly with a reduction of the pressure from 12.6 GPa, due to the pressure effects on the charge density at the ^{57}Fe nucleus and the unit cell volume of OMFA. These Mössbauer parameter results show that there is no large structural change, such as a complete collapse of the structural relationship between the upper and lower ring. A clear enhancement of the area in the resonance absorption curve has been observed at 12.6 GPa and 300 K, as shown in Figure 2, which suggests that an external pressure of 12.6 GPa induces a phase transition from a high temperature phase to a low temperature phase. As the pressure decreases from 12.6 GPa, the area of the resonance absorption curve decreases. However, at an ambient pressure of 0.1 MPa, reduced from 12.6 GPa, a resonance absorption curve is still visible in

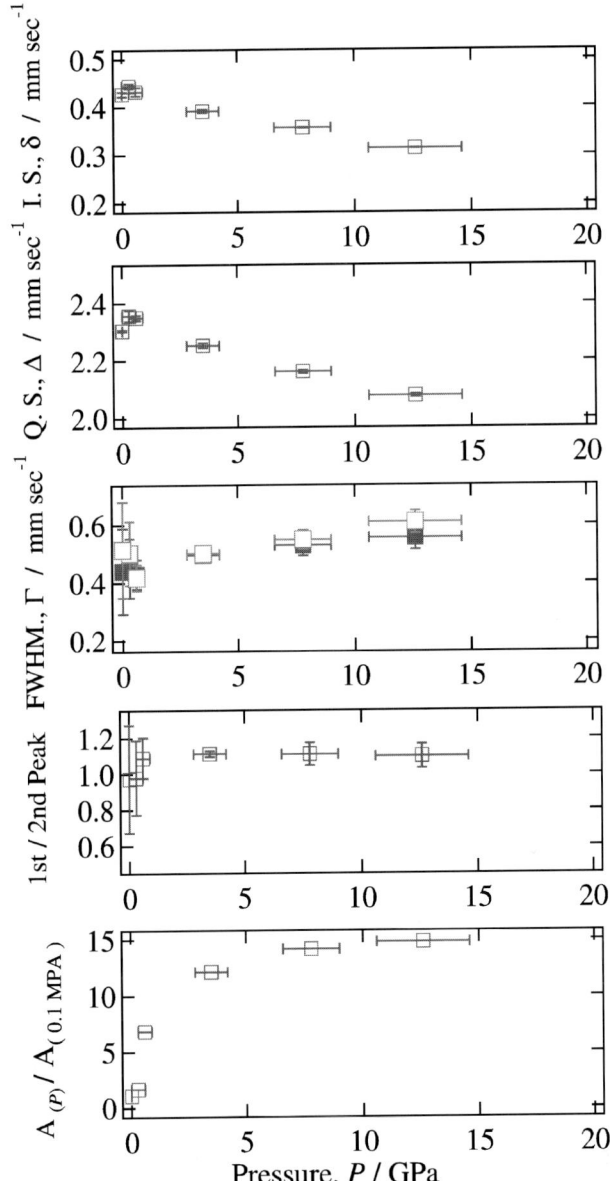

Figure 4. Mössbauer parameters obtained for octamethyl-ethynyl-ferrocene (OMFA) at 300 K, where I.S. (δ) is the isomer shift value, Q.S. (Δ) is the quadrupole splitting, FWHM (Γ) is the full width at half maximum, ($\square$ is obtained from a singlet at positive velocity side, $\blacksquare$ is from a singlet at negative velocity side), 1st/2nd is the intensity ratio of the first absorption line to the second absorption line, and $A_{(P)}/A_{(0.1\ \text{MPa})}$ is the ratio between the resonance absorption area under pressure P and the resonance absorption obtained under 0.1 MPa, that is, after complete pressure release from 12.6 GPa.

Figure 2. The fact that the pressure induced low temperature phase survives after the release of the pressure from 12.6 GPa, implies that this pressure induced phase transition is a first order transition.

4. Summary

We performed high-pressure ^{57}Fe Mössbauer spectroscopy of octamethyl-ethynyl-ferrocene (OMFA) using a diamond-anvil-cell (DAC). A clearly enhanced resonance absorption spectrum was observed at 12.6 GPa at 300 K, suggesting a pressure-induced phase transition from a high temperature phase to a low temperature phase.

Ackowlegement

The authors wish to thank Prof. H. Schottenberger and Dr. E. Reichel of the University of Innsbruck, Austria for supplying the OMFA sample.

References

1. Nowik, I. and Herber, R. H., *Inorg. Chim Acta* **310** (2000), 191; Scottenberger, H., Wurst, K. and Herber, R. H., *J. Organomet. Chem.* **625** (2001), 200.
2. Herber, R. H. and Nowik, I., *Hyp. Interact.* **126** (2000), 127.
3. Asthalter, T., Franz, H., van Bürck, U., Messel, K., Schreier, E. and Dinnebier, R., *J. Phys. Chem. Solids* **64** (2003), 677.
4. Herber, R. H., *Chemical Mössbauer Spectroscopy*, Plenum Press, New York, 1985, p. 199.

Hyperfine Interactions **156/157**: 241–245, 2004.
© 2004 *Kluwer Academic Publishers. Printed in the Netherlands.*

^{57}Fe Mössbauer Spectroscopic Study of Fe–B Compounds

TAKASHI SEGI[1,*], SUBURO NASU[1], SHOTARO MORIMOTO[1] and HISATO TOKORO[2]

[1]*Department of Materials Engineering Science, Graduate School of Engineering Science, Osaka University, Toyonaka, Osaka 560-8531, Japan; e-mail: segi@moss.mp.es.osaka-u.ac.jp*
[2]*Advanced Electronics Research Laboratory, Hitachi Metal Co. Ltd., Kumagaya, Saitama 360-0843, Japan*

Abstract. ^{57}Fe Mössbauer measurements of Fe doped β-rhombohedral boron (β-B) were performed in the temperature range between 300 K and 4.5 K. The spectra of Fe doped β-B consist of three paramagnetic doublets at room temperature and show a magnetic ordered state below 11.8 K, the temperature of which was determined by the ^{57}Fe Mössbauer thermal scanning method at zero Doppler-velocity. The magnitude of the hyperfine magnetic field of Fe atoms in β-B at 4.5 K is about 30 T or more according to hyperfine distribution analysis.

Key words: beta-Boron, Fe–B, icosahedral-Boron.

1. Introduction

The lattice of B-rich solids is formed by periodically arranged B-12 icosahedra. The crystal of β-rhombohedral B (β-B) is an icosahedral substance and shows p-type semiconductor properties with an energy gap of about 1.2 to 1.5 eV [1]. The icosahedral crystal of β-B belongs to the rhombohedral space group $R\bar{3}m$ with lattice parameters $a = 1.09251$ and $c = 2.38143$ nm [2, 3]. This rhombohedral unit cell contains 105 B atoms. It is known that Fe atoms partially occupy two crystallographic doping sites, A_1 and D. The other site, the E site, remains empty, as revealed by Rietveld analysis of powder XRD measurements for Fe-doped β-B specimens with varied Fe concentration [4]. Hence, the β-B crystal can contain as many as six Fe atoms in A_1 sites and eighteen Fe atoms in D sites in the unit cell. In ^{57}Fe Mössbauer measurements of Fe doped β-B (Fe$_1$B$_{105}$) [5], the line shape observed at liquid helium temperature suggests a magnetic ordered state, however the magnetic nature for this substance is not yet clear.

In this work, ^{57}Fe Mössbauer measurements of Fe doped β-B were performed in the temperature range between 300 K and 4.5 K in order to clarify the magnetic properties of Fe in β-B.

---

* Author for correspondence.

2. Experimental method

A specimen of Fe doped β-B sample was prepared by arc-melting with a powder mix of 2 at.% Fe (Toho zinc Co. Ltd., purity 99.9%) and 98 at.% B (H. C. Starck Inc., purity 99.5%). ^{57}Fe Mössbauer measurements of Fe doped β-B were performed in the transmission geometry in the temperature range between 300 to 4.5 K using a ^{57}Co in Rh source. To find the magnetic transition temperature, the ^{57}Fe Mössbauer thermal scanning method was carried out at zero velocity between 20 K and 5.3 K. The velocity scale is relative to α-Fe at 300 K. The resultant spectra were analyzed by conventional least square fit with a superposition of various Lorentz functions and a continuous distribution of hyperfine fields using the program package NORMOS-DIST developed by Brand [6].

3. Results and discussion

^{57}Fe Mössbauer spectra of Fe doped β-B obtained at room temperature are shown in Figure 1. There are two crystallographic positions of Fe sites in a unit cell of β-rhombohedral B, the A_1 and D sites, however analysis with a superposition of two

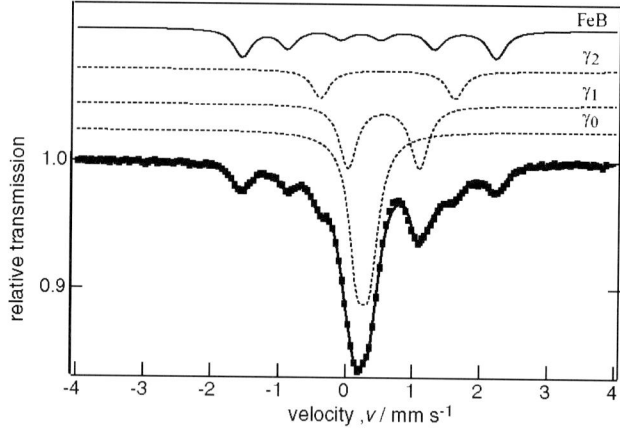

Figure 1. ^{57}Fe Mössbauer spectra of Fe doped β-B and FeB obtained at room temperature.

Table I. Hyperfine parameters for the specimen prepared by arc melting method

	$B_{\text{int.}}$ (T)	δ (mm s^{-1})	Δ (mm s^{-1})	I (%)
FeB	11.9	+0.27	+0.13	22.50
γ_0	–	+0.28	+0.19	43.52
γ_1	–	+0.58	+1.07	23.70
γ_2	–	+0.63	+2.03	10.28

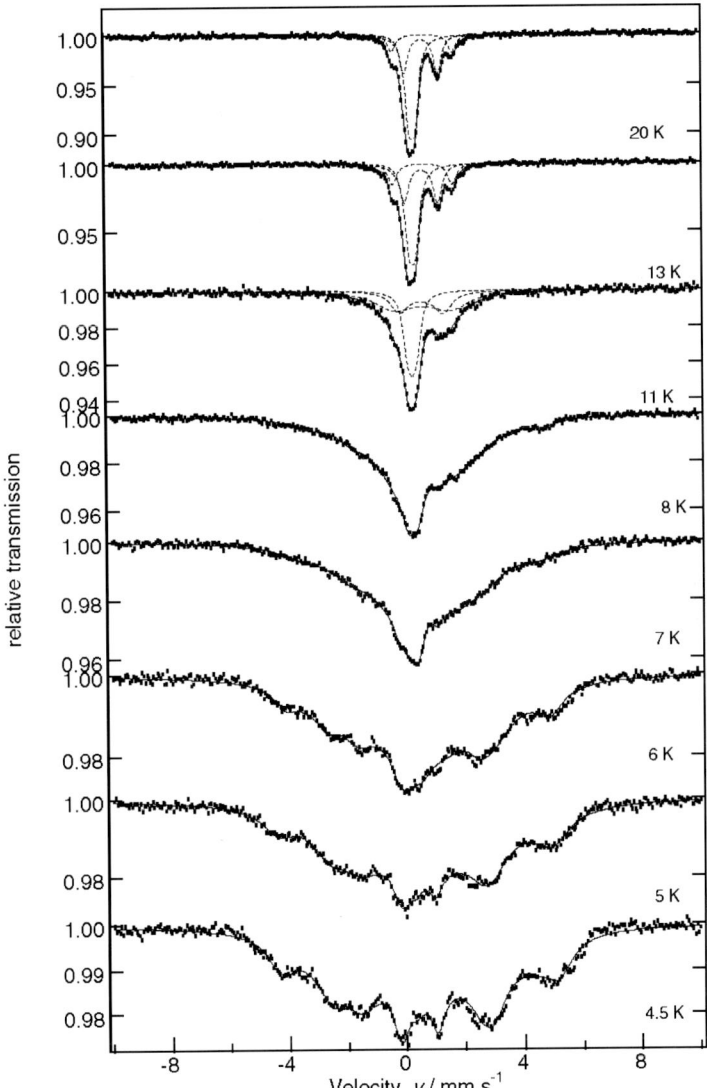

Figure 2. ^{57}Fe Mössbauer spectra of Fe doped β-B obtained in the temperature range between 20 and 4.5 K.

doublets gives a poor fit to the data; decomposition by three paramagnetic doublets, γ_0, γ_1 and γ_2 in Figure 1, improves the value of χ^2. This indication shows that the interstitial sites of the specimen of β-B are not occupied completely.

The spectrum can be decomposed into three paramagnetic doublets, γ_0, γ_1 and γ_2, and a magnetically ordered sextet. The magnetically ordered sextet of magnitude 11.8 T is assigned to FeB, as it agrees with the reported hyperfine field of FeB [7]. Fe atoms do not exist within 0.3 nm of the A_1 site, while Fe atoms in the D site can adjoin other Fe atoms. Since the A_1 site has a higher symmetry than

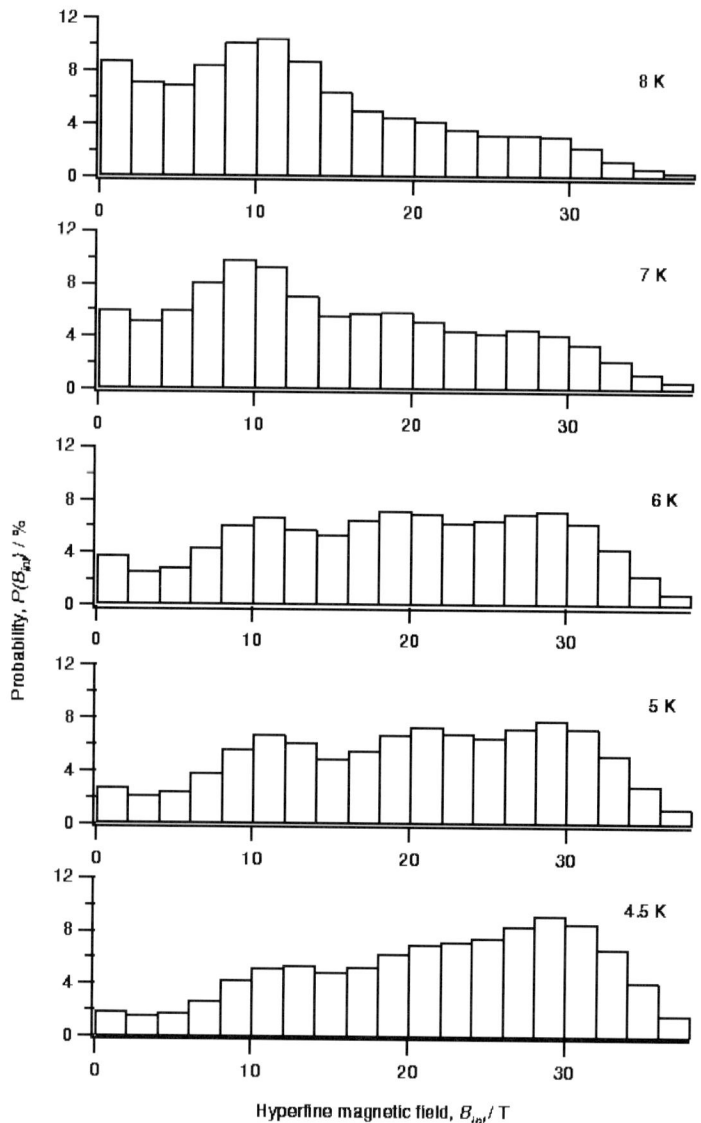

Figure 3. Distribution of the hyperfine field of Fe doped β-B spectra obtained at various temperatures.

the D site, the magnitude of the quadrupole splitting of the doublet γ_0, with the smallest value, is attributed to the A_1 site. On the other hand, the γ_1 and γ_2 components originate from the D site. The hyperfine parameters of the paramagnetic components are listed in Table I and agree with those reported in the literature for Fe doped β-B [4].

Figure 2 shows the [57]Fe Mössbauer spectra, for which the contribution from the FeB component has been subtracted from the experimental line shape, at low temperature in the range between 20 K and 4.5 K. In the temperature range between 20 K and 13 K, the spectral shape does not show any appreciable change except for a shift of the center position due to the second order Doppler shift. Below 13 K, magnetic ordered states appear. The magnetic transition temperature was found to be 11.8 K, determined by [57]Fe Mössbauer thermal scanning method at zero velocity.

Because the Mössbauer spectra obtained below 8 K shows a very broad distribution of the hyperfine field, it is impossible for the experimental resultant to be decomposed into three sub-spectra. Figure 3 shows the distribution of the hyperfine magnetic field at each temperature determined by the NORMOS-DIST program [6]. The magnitude of the hyperfine field increases with decreasing temperature and reaches a value of about 30 T, which is much larger than that of FeB.

4. Conclusion

[57]Fe Mössbauer measurements of Fe doped β-B were performed for the first time in the temperature range between 300 K and 4.5 K. Below 11.8 K, the spectra of Fe doped β-B show a magnetic ordered state suggesting a broad distribution of the hyperfine magnetic fields. From the hyperfine distribution analysis, the magnitude of the hyperfine magnetic fields at Fe nucleus in β-B at 4.5 K is about 30 T or more.

References

1. Werheit, H., Schmechel, R., Kueffel, V. and Lundström, T., *J. Alloys Compounds* **262–263** (1997), 372.
2. Callmer, B., *Acta Cryst. B* **33** (1977), 1951.
3. Sands, D. E. and Hoard, J. L., *J. Am. Chem. Soc.* **79** (1957), 5582.
4. Takizawa, H., Haze, N., Okamoto, K., Uheda, K. and Endo, T., *Mater. Res. Bull.* **37** (2002), 113.
5. Nakayama, T., Matuda, H., Kimura, K. and Ino, H., *J. Solid State Chem.* **133** (1997), 342.
6. Brand, R. A., *Nucl. Instrum. Methods Phys. Res. B* **28** (1987), 398.
7. Shinjo, T., Itoh, F. and Takaki, H., *J. Phys. Soc. Jpn.* **19** (1964), 1252.

Hyperfine Interactions **156/157**: 247–256, 2004.
© 2004 *Kluwer Academic Publishers. Printed in the Netherlands.*

Mössbauer- and EPR-Snapshots of an Enzymatic Reaction: The Cytochrome P450 Reaction Cycle

V. SCHÜNEMANN[1], C. JUNG[2], F. LENDZIAN[3], A.-L. BARRA[4],
T. TESCHNER[1] and A. X. TRAUTWEIN[1]
[1]*Institute of Physics, University of Lübeck, D-23538 Lübeck, Germany*
[2]*Max-Delbrück-Center for Molecular Medicine, D-13125 Berlin, Germany*
[3]*Max-Volmer Laboratory for Biophysical Chemistry, Technical University, PC 14, D-10623 Berlin, Germany*
[4]*Grenoble High Magnetic Field Laboratory, CNRS, B.P. 166, F-38042 Grenoble Cedex, France*

Abstract. In this communication we present a complimentary Mössbauer- and EPR-study of the time dependance of the reaction of substrate free P450cam with peracetic acid within a time region ranging from 8 ms up to 5 min. An Fe(IV) species as well as a tyrosyl radical residing on the amino acid residue Tyr96 have been identified as reaction intermediates. These species possibly are formed by the reduction of compound I by means of transferring an electron from Tyr 96 to the heme moiety.

Key words: cytochrome P450, peracetic acid, Mössbauer spectroscopy, high-field EPR.

1. Introduction

The physiologically important enzyme superfamily cytochrome P450 catalyzes a variety of reactions, such as aliphatic and aromatic hydroxylations, epoxidations, heteroatom oxidation, and N- and O-dealkylation, by transfer of an active oxygen atom from its heme unit to the substrates. All enzymes of the cytochrome P450 family have a hydrophobic binding pocket, in which resides a protoporphyrine IX with its iron center being coordinated to an axial cysteine ligand [1].

The enzyme cytochrome P450cam from *Pseudomonas putida* hydroxylases (1R)-camphor as natural substrate and is regarded to be a representative enzyme for the whole P450 family. Mössbauer spectroscopic studies on this enzyme have been performed by Debrunner and coworkers in the seventies [2] of the last century. The results of these studies provided insight into the catalytic mechanism of the enzyme even before the crystal structure of the resting state of P450cam was published [3].

The postulated enzymatic reaction cycle is shown in Figure 1. In the resting state of the enzyme the catalytically active heme iron center acquires the ferric low-spin state ($S = 1/2$). After binding of the substrate camphor to the amino acid residue Tyr96 inside the heme pocket, the iron changes from the ferric low-spin to the ferric high-spin state ($S = 5/2$). The transfer of the first electron originating from NAD(P)H via redox proteins (flavin and iron–sulfur proteins) reduces the iron

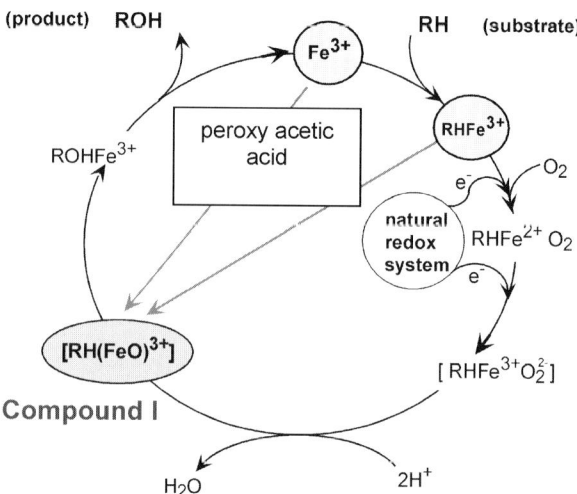

Figure 1. Reaction cycle of cytochrome P450 indicating the shunt pathway using peroxy acetic acid as oxidant.

to the ferrous high-spin state ($S = 2$). Subsequent binding of molecular oxygen to the iron forms a diamagnetic FeO_2 center, similar to the oxygenated state of myoglobin. The transfer of a second electron initiates catalytic steps which lead to an iron-oxo intermediate called compound I (cpd I). It is this intermediate which inserts the active oxygen atom into the substrate camphor.

Neither the electronic structure of cpd I of P450cam nor of any other P450 has been unambiguously determined up to now, but it is generally assumed that the putative cpd I in cytochrome P450 should contain the same electron and spin distribution as is observed for cpd I of peroxidases [4], catalases and many synthetic cpd I analogues [5]. In these systems one oxidation equivalent resides on the Fe(IV)=O unit (d^4, $S = 1$) and one is located on the porphyrin ($S' = 1/2$), constituting a magnetically coupled ferryl iron-oxo porphyrin π-cation radical system. We have recently reported the detection of a ferryl iron ($S = 1$) and a radical ($S' = 1/2$), via Mössbauer and EPR studies of a 8 ms-reaction intermediate of substrate-free P450cam from *Pseudomonas putida*, prepared by a freeze-quench method using peroxy acetic acid as the oxidizing agent [6]. This species, however, is different from the above-mentioned iron-oxo porphyrin radical system.

Under the same reaction conditions, but in the presence of the substrate camphor, only trace amounts of the tyrosine radical are formed and no Fe(IV) is detectable by Mössbauer spectroscopy [7]. Also in this case a porphyrin-π-cation radical could not be observed and only small amounts of hydroxylation product was found by gas chromatography and mass spectrometry. From studies with variation of oxidant to P450 ratios, with substrates which are more mobile in the active site, and by studies on the stability against oxidative porphyrin-ring cleavage we conclude that the access of the oxidant to the heme pocket is blocked in the presence of the substrate.

Within this work we report the identification of the observed radical as Tyr96*
as well as the time-dependent formation of the Fe(IV) and the Tyr96* intermediate.
Furthermore the formation of a ferric high-spin reaction intermediate is identified,
which most likely represents a 5-coordinated heme center. We believe that it has
been formed after the reduction of the Fe(IV) species and that it converts back to
the ferric resting state within a time scale of minutes.

2. Material and methods

The plasmids carrying the wild type P450 (CYP101) and Y96F mutant gene are
described elsewhere [8, 9]. The proteins were expressed in *E. coli* TB1, purified
and removed from camphor as described previously [10]. Rapid freeze-quench experiments were performed with a System 1000 apparatus from Update Instruments
by mixing a 0.93 mM P450cam solution and a five-fold excess of peroxy acetic
acid (PA) solution in 100 mM potassium phosphate buffer, pH 7 (volume mixing
ratio of 1 : 1) [6, 7]. A Delrin cup with 4 mm inner diameter and a volume of
50 μl was attached to a quartz tube, which was connected to a funnel. The funnel
was completely immersed into an isopentane bath at $T = -100°C$. The reaction
mixtures were sprayed into the cold isopentane; the so obtained frozen protein
material was packed at the bottom of the Delrin cup using a packing rod made of
Teflon. This procedure made it possible to record Mössbauer as well as EPR spectra
from the same sample. EPR spectra at 9.6 GHz were recorded with a conventional
X-band spectrometer (Bruker 200D SRC) equipped with a He-flow cryostat (ESR
910, Oxford Instruments) at 20 K. In order to obtain relative EPR intensities the
spectra have been simulated according to the procedure described by Beinert and
Albracht [11] using orientation-dependent Gaussian line shapes. Absolute signal
quantifications have been obtained by assuming a packing factor of 1 and taking the
concentration of the ferric low-spin heme obtained from Mössbauer spectroscopy
as an internal standard.

For the radical signal continuous-wave high-field EPR measurements at 94 GHz
were performed on a Bruker Elexsys 680 spectrometer equipped with a fundamental mode microwave resonator. EPR at 190 and 285 GHz was performed at the
Grenoble High Magnetic Field Laboratory with a set-up described in [12].

Mössbauer spectra were recorded using a conventional spectrometer in the constant-acceleration mode. Isomer shifts are given relative to α-Fe at room temperature. The spectra were measured in a He-bath cryostat (Oxford Instruments),
equipped with permanent magnets. Typical measuring time was three weeks per
spectrum. The magnetically split spectra were simulated within the spin-Hamiltonian formalism [13].

3. Results and discussion

Figure 2a shows the Mössbauer spectrum of the starting material obtained at 4.2 K
in a small field of 20 mT applied perpendicular to the γ-beam. The magneti-

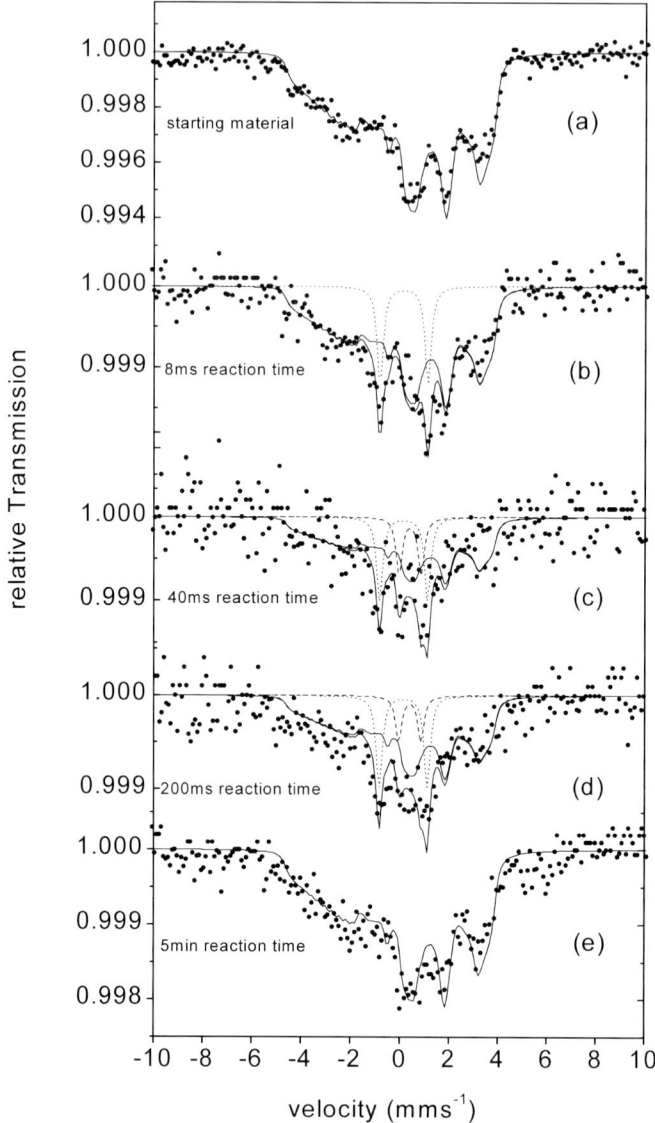

Figure 2. Mössbauer spectra of freeze-quenched ^{57}Fe P450cam after addition of peroxy acetic acid for reaction times as indicated. (a) Starting material, (b)–(e) 8 ms, 40 ms, 200 ms, and 5 min reaction time, respectively. The solid lines are simulations with parameters given in Table I. For the spin-Hamiltonian simulation of the native P450cam the g-tensor $g = (1.91, 2.26, 2.45)$, the magnetic hyperfine coupling tensor $A/\mu_n g_n = (-45, 10, 19)$ T and the asymmetry parameter $\eta = -1.8$ were used [2, 17]. All spectra have been obtained at 4.2 K in a field of 20 mT applied perpendicular to the γ-beam.

cally split pattern is characteristic for an anisotropic ferric low-spin system with anisotropic magnetic hyperfine coupling constants and has been successfully simulated with the Mössbauer parameters given by Sharrock *et al.* [14]. The EPR trace

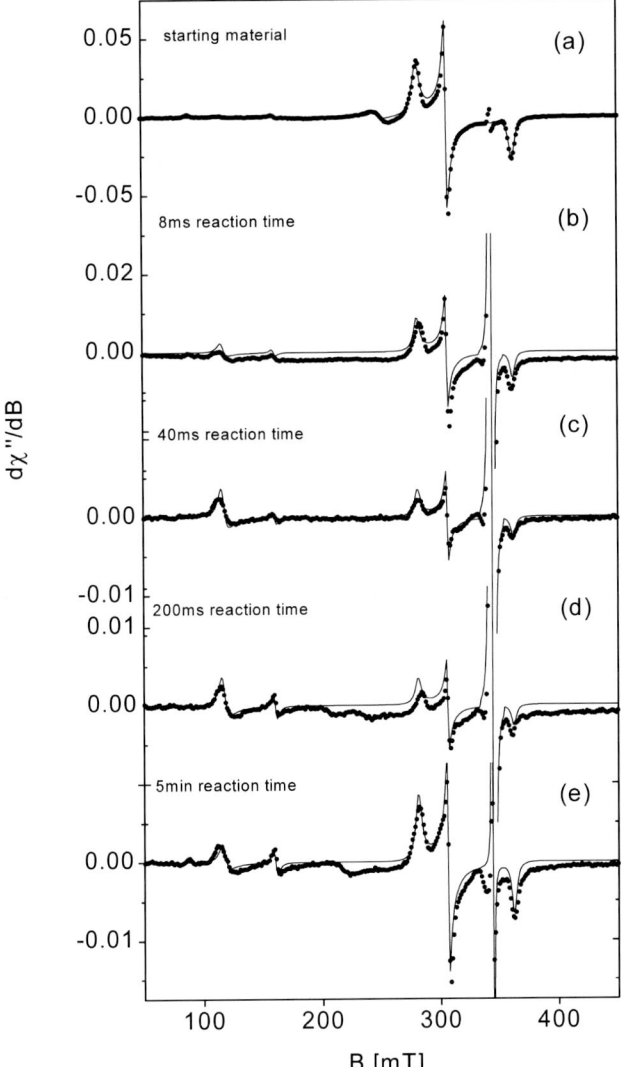

Figure 3. EPR spectra of freeze-quenched ^{57}Fe P450cam samples after addition of peroxy acetic acid for reaction times as indicated. (a) Starting material, (b)–(e) 8 ms, 40 ms, 200 ms, and 5 min reaction time, respectively. The strong radical signal at $g = 2.00$ in (b)–(d) is off scale. The solid lines are simulations with parameters given in Table II. The spectra show a minor rhombic signal at $g = 4.3$, which accounts for less than 1% and which is not included in Table II. Experimental conditions: $T = 20$ K; $P = 80$ μW; $\nu = 9.64$ GHz; modulation amplitude 0.5 mT; modulation frequency 100 kHz.

of the same sample (Figure 3a) shows an anisotropic EPR signal within the $g \sim 2$ region which corroborates the $S = 1/2$ state of the enzyme.

The Mössbauer spectrum obtained after reaction of the substrate-free P450cam with peroxy acetic acid for 8 ms is shown in Figure 2b. While approximately

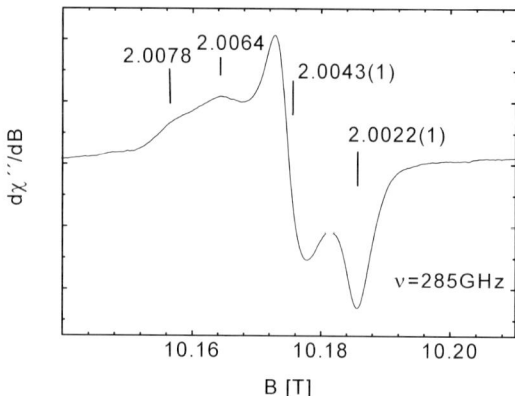

Figure 4. High field EPR-spectrum of the radical observed after 40 ms reaction time of ^{56}Fe P450cam and peroxy acetic acid. Experimental conditions: $T = 5$ K; $\nu = 285$ GHz; modulation amplitude 2.4 mT; modulation frequency 0.8 kHz.

~87% of the spectral area corresponds to the ferric low-spin iron with g-factors and hyperfine parameters of the starting material, a new component (~13%) with $\delta = 0.13$ mm s^{-1} and $\Delta E_Q = 1.94$ mm s^{-1} appears. This doublet has an isomer shift which is in the range of ferryl iron with $S = 1$. Correspondingly the EPR data shown in Figure 3b do not show any signal of a magnetically coupled cpd I species. Instead, the pattern of the starting material is accompanied by a strong radical signal at $g = 2.00$ (~15%).

High-field EPR at 285 GHz (Figure 4) allows to resolve the g-tensor components of the radical (stretched g_x-values from 2.0078–2.0064, $g_y = 2.0043$, and $g_z = 2.0022$). These are fingerprints for tyrosine radicals in a heterogeneous polar environment. The measurements at 94 GHz (not shown) using a fundamental mode microwave resonator setup confirm the 285 GHz study. By simulating the hyperfine structure of the 94 GHz EPR signal of the wild-type protein and of a Tyr96-deficient Y96F mutant of P450cam we have recently shown that the radical in the wild-type protein is located at Tyr96 [15]. In fact, Tyr96 is close to the heme on the distal side. We believe that the tyrosine radical in the protein is due to an intramolecular electron transfer from Tyr96 to the porphyrin which leads to the reduction of the porphyrin-π-radical within a time period of much shorter than 8 ms.

In order to fully simulate the experimental low-field Mössbauer spectra obtained after 40 and 200 ms reaction time a second doublet in addition to the doublet representing the ferryl iron has been introduced (Figures 2c, d). This doublet exhibits $\delta = 0.4$ mm s^{-1} and $\Delta E_Q = 0.9$ mm s^{-1}, parameters which are comparable to those found for 5-coordinated ferric high-spin heme centers [16]. The spin relaxation time at 4.2 K of isolated ferric high-spin centers is generally larger than the Mössbauer time window (nuclear precession time) of ~10^{-7} s. This causes a magnetically split six-line pattern in the Mössbauer spectrum even at small applied

Table I. Parameters derived from the Mössbauer spectra shown in Figure 2

Reaction time	Species	δ (mm s^{-1})	ΔE_Q (mm s^{-1})	Γ (mm s^{-1})	Rel. area (%)
start	native P450cam	0.38	2.85	0.30	100
8 ms	native P450cam	0.38	2.85	0.30	87 ± 3
	Fe(IV) $S = 1$	0.13	1.94	0.30	13 ± 3
	Fe(III) $S = 5/2$	–	–	–	–
40 ms	native P450cam	0.38	2.85	0.30	71 ± 4
	Fe(IV) $S = 1$	0.13	1.94	0.30	17 ± 4
	Fe(III) $S = 5/2$	0.40	0.90	0.30	12 ± 4
200 ms	native P450cam	0.38	2.85	0.30	76 ± 4
	Fe(IV) $S = 1$	0.13	1.94	0.30	16 ± 4
	Fe(III) $S = 5/2$	0.40	0.90	0.30	8 ± 4
5 min	native P450cam	0.38	2.85	0.30	100
	Fe(IV) $S = 1$	–	–	–	–
	Fe(III) $S = 5/2$	–	–	–	–

fields of 20 mT. In this study, however, we observe a doublet (~12%) which is characteristic for a fast relaxing ferric high-spin species with a relaxation time of $\ll 10^{-7}$ s. This behavior is most likely due to the presence of spin–spin relaxation between the ferric high-spin site ($S = 5/2$) and the spin ($S' = 1/2$) on the tyrosine radical. The doublet is not seen when a field of 1 T is applied to the sample [17], likely because the application of that field lowers the spin relaxation rate and gives rise to a magnetically split pattern which disappears in the spectral background. Details of the Mössbauer analysis are summarized in Table I.

Corresponding EPR spectra contain also a (so far overlooked) ferric high-spin signal: After 8 ms reaction time a weak resonance (~6%) at $g = (5.9, 5.9, 2.0)$ characteristic for a ferric, slightly spin admixed, high-spin species ($S = 5/2$) in an axial ligand field (rhombicity parameter $E/D = 0$) has been detected which is too weak to be visible in the Mössbauer spectrum (Figure 3b). After 40 ms reaction time an increase of the radical signal (~36%) is accompanied by an increase (~13%) of the ferric high-spin species ($S = 5/2$) (Figure 3c). The EPR spectrum obtained after 200 ms (Figure 3d) resembles that of the pattern observed for 40 ms reaction time, i.e. a dominating radical signal (~39%), the ferric high-spin species (~14%) and the ferric low-spin signal (~47%) of the resting state. After 5 min reaction time the EPR spectrum of the ferric low-spin heme center of the enzyme is almost restored (~91%). The signal intensity at $g = (5.9, 5.9, 2.0)$ has decreased (~6%) and only trace amounts of radical (~3%) have been detected (Figure 3e). A summary is provided in Table II.

Figure 5 shows the time dependence of the formation of the tyrosine radical, of the ferryl iron and of the axial ferric high-spin intermediate. After 8 ms of reac-

Table II. Parameters derived from the EPR spectra shown in Figure 3

Reaction time		g	Rel. contribution (%)
start	native P450cam	(1.91, 2.26, 2.45)	96.5 ± 1.0
	impurity	(2.0, 2.75, 2.80)	3.5 ± 1.0
8 ms	native P450cam	(1.91, 2.26, 2.45)	79 ± 8
	axial Fe(III) $S = 5/2$	(5.9, 5.9, 2.0)	6 ± 2
	Tyr*	(2.0, 2.0, 2.0)	15 ± 3
40 ms	native P450cam	(1.91, 2.26, 2.45)	51 ± 5
	axial Fe(III) $S = 5/2$	(5.9, 5.9, 2.0)	13 ± 3
	Tyr*	(2.0, 2.0, 2.0)	36 ± 6
200 ms	native P450cam	(1.91, 2.26, 2.45)	47 ± 5
	axial Fe(III) $S = 5/2$	(5.9, 5.9, 2.0)	14 ± 3
	Tyr*	(2.0, 2.0, 2.0)	39 ± 6
5 min	native P450cam	(1.91, 2.26, 2.45)	91 ± 6
	axial Fe(III) $S = 5/2$	(5.9, 5.9, 2.0)	6 ± 2
	Tyr*	(2.0, 2.0, 2.0)	3 ± 1

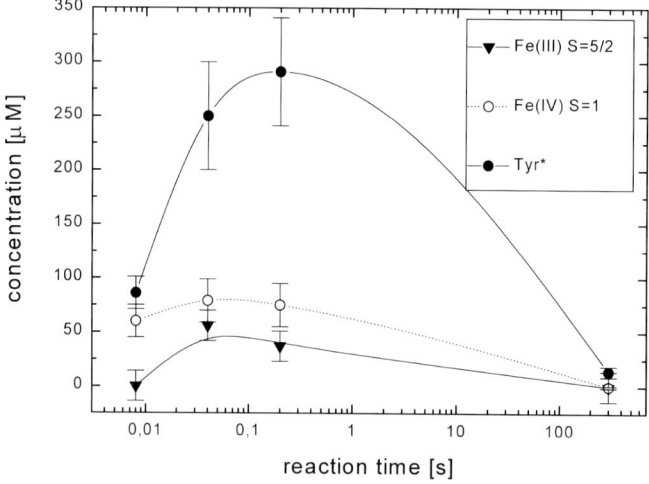

Figure 5. Time dependence of Tyr96 radical formation (as determined from EPR) and formation of Fe(IV) and 5-coordinated ferric heme (as determined from Mössbauer spectroscopy).

tion time the amount of Tyr96* scales approximately with the amount of Fe(IV) observed. This correlation is lost already after 40 ms reaction time. From the concentrations shown in Figure 5 a correlation between the amount of radical and the sum of the amounts of ferryl and axial ferric high-spin species seems likely. After a reaction time of 5 min only traces of intermediates have been detected.

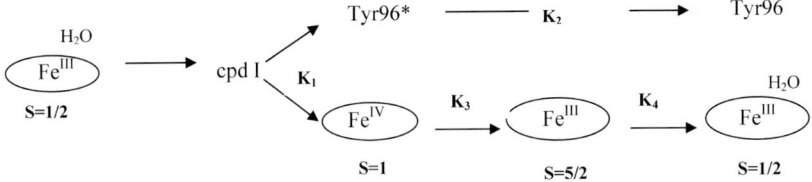

Figure 6. Proposed mechanism for the decay of compound I (cpd I) in P450cam.

The time dependences can be rationalized by the reaction scheme displayed in Figure 6. All spectroscopic results presented here are devoid of any evidence for a cpd I-like species (i.e. a ferryl iron-oxo porphyrin-π-cation radical system). Yet we believe that cpd I has been created, which however, reacts very fast ($\ll 8$ ms) with an electron from Tyr96 forming the neutral porphyrin with a ferryl iron center.

Since both routes of the decay reaction in Figure 6 start from cpd I the constant of formation K_1 is equal for both species. The ferryl intermediate decays with a reaction constant K_3 to the axial ferric high-spin heme (most probably by releasing the oxygen in the iron-oxo-species as a neutral water molecule or as an OH^- molecule). The axial ferric high-spin state finally converts back to the initial state possibly by re-arranging and re-binding of the water cluster within the heme pocket and by subsequent binding of a water molecule to the heme iron (reaction constant K_4). The tyrosine radical seems to be relatively stable (K_2) on this time scale.

4. Conclusion

We have presented a complimentary EPR- and Mössbauer-analysis of the formation and decay of reaction intermediates within the P450 reaction cycle using the shunt reaction with peroxy acetic acid as oxidant and with substrate-free P450cam as enzyme. With the freeze-quench reaction we were able to cover reaction times from milliseconds to minutes. Although cpd I has not been identified directly, we have been able to trap and identify its decomposition products, i.e. a ferryl iron and a tyrosine radical (Tyr96*) as well as an axial ferric high-spin heme which most likely represents the state of the active site of P450 before being converted back into the resting ferric low-spin state.

Acknowledgements

This work has been supported by the German Research Foundation (DFG, TR26-1,2,3; Ju229/4-1,2,3) and by the European Community "Access to Research Infrastructure action of the Improving Human Potential Programme".

References

1. Lewis, D. F. V., *Cytochromes P450 – Structure, Function and Mechanism*, Taylor & Francis Ltd, London, 1996.
2. Sharrock, M., Debrunner, P. G., Schulz, C., Lipscomb, J. D., Marshall, V. and Gunsalus, I. C., *Biochim. Biophys. Acta* **420** (1976), 8.
3. Poulos, T. L., Finzel, B. C. and Howard, A. J., *Biochemistry* **25** (1986), 5314.
4. Rutter, R., Hager, L. P., Dhonau, H., Hendrich, M., Valentine, M. and Debrunner, P., *Biochemistry* **23** (1984), 6809.
5. Bill, E., *Hyp. Interact.* **90** (1994), 143.
6. Schünemann, V., Jung, C., Trautwein, A. X., Mandon, D. and Weiss, R., *FEBS Lett.* **479** (2000), 149.
7. Schünemann, V., Jung, C., Terner, J., Trautwein, A. X. and Weiss, R., *J. Inorg. Biochem.* **91** (2002), 586.
8. Unger, B. P., Gunsalus, I. C. and Sligar, S. G., *J. Biol. Chem.* **262** (1986), 1158.
9. Atkins, W. M. and Sligar, S. G., *J. Biol. Chem.* **263** (1988), 18842.
10. Jung, C., Hui Bon Hoa, G., Schröder, K.-L., Simon, M. and Doucet, J. P., *Biochemistry* **35** (1992), 12855.
11. Beinert, H. and Albracht, S., *Biochim. Biophys. Acta* **683** (1982), 245.
12. Barra, A. L., *Appl. Magn. Res.* **21** (2001), 619.
13. Trautwein, A. X., Bill, E., Bominaar, E. L. and Winkler, H., *Structure and Bonding* **78** (1991), 1.
14. Sharrock, M., Debrunner, P. G., Schulz, C., Lipscomb, J. D., Marshall, V. and Gunsalus, I. C., *Biochim. Biophys. Acta* **420** (1976), 8.
15. Jung, C., Schünemann, V., Lendzian, F., Contzen, J., Barra, A.-L. and Trautwein, A. X., In: P. Anzenbacher and J. Hudecek (eds), *Cytochromes P450, Biochemistry, Biophysics and Drug Metabolism, 13th Internat. Conf. on Cytochromes P450*, Pague, 2003, Monduzzi Editore Bologna, p. 201.
16. Debrunner, P. G., In: A. B. P. Lever and H. B. Gray (eds), *Iron Porphyrins*, Part III, VCH Publishing, Heidelberg, 1989, p. 139.
17. Schünemann, V., Trautwein, A. X., Jung, C. and Terner, J., *Hyp. Interact.* **141/142** (2002), 279.

Hyperfine Interactions **156/157**: 257–263, 2004.
© 2004 *Kluwer Academic Publishers. Printed in the Netherlands.*

The Effect of Biocompatible Coating Layers on Magnetic Properties of Superparamagnetic Iron Oxide Nanoparticles

M. MIKHAYLOVA[1], Y. S. JO[1], D. K. KIM[1], N. BOBRYSHEVA[2],
Y. ANDERSSON[3], T. ERIKSSON[3], M. OSMOLOWSKY[2], V. SEMENOV[2] and
M. MUHAMMED[1]

[1] *Dept. of Materials Science and Engineering, Materials Chemistry Division,
Royal Institute of Technology, SE 100 44, Stockholm, Sweden*
[2] *Dept. of Chemistry, Saint Petersburg State University, 198904, Saint Petersburg, Russia*
[3] *Dept. of Materials Chemistry, Ångström Laboratory, Uppsala University,
SE 751 21 Uppsala, Sweden*

Abstract. The effect of the surface coating on the magnetic properties of superparamagnetic iron oxide nanoparticles (SPION) with 8 nm in size has been studied. Four different biocompatible coating layers are considered: poly L,L-lactic acid (PLLA), poly ε-caprolactone (PCL), bovine serum albumin (BSA) and gold. The presence of coating layer on the surface of SPION is confirmed by FT-IR spectroscopy. Mössbauer spectroscopy and magnetic susceptibility measurements show that for uncoated SPION and Au@SPION the superparamagnetic fraction is retained. The formation of clusters in the case of BSA@SPION and chain-like structure for PCL@SPION and PLLA@SPION increase the inter-particle interactions resulting in hyperfine magnetic structure observed in the Mössbauer spectra at ambient temperature.

Key words: magnetic nanoparticles, magnetite, superparamagnetism, Mössbauer spectroscopy, biocompatible coating.

1. Introduction

Superparamagnetic iron oxide nanoparticles (SPION) with particle size less than 10 nm and a strong magnetic response under external magnetic field have received considerable attention for different biomedical applications [1]. Although magnetite nanoparticles have shown a low toxicity in biological systems when encapsulated in a protein cage [2], some studies of biocompatibility demonstrated the toxicity of magnetic fluid consisted of $MnFe_2O_4$ [3]. Therefore, several studies have been performed to modify the surface of magnetic particles with biocompatible materials in the past decade, which also results in improving the biocompatibility, colloidal stability, dispersibility, and selective uptakes of the magnetic particles by the target cells [4]. The choice of the coating layer is critical since the coating process may change the particle size and result is the changing of superparamag-

netic (SPM) properties, which is considered to be most important physical characteristic for the *in vivo* and *in vitro* application of SPION. Mössbauer spectroscopy is
a convenient technique to study the magnetic and electronic properties of systems
containing iron atom. A number of studies have been performed on the iron oxide
nanoparticles coated with different materials by using this technique [5, 6]. However, the influence of organic and inorganic sheaths on the magnetic properties for
further medical application has not been fully investigated. Therefore, the present
research is focused on the study of the effects of different biocompatible organic
and inorganic coating layers on magnetic properties of SPION.

2. Experimental methods

The phase composition of all samples was determined by using a Siemens D500
XRD diffractometer with Cu K_α radiation. The size and morphology of particles
were analyzed by a JEOL-2000EX transmission electron microscope. FT-IR spectra were recorded by using Nicolet, Avatar 360, FT-IR spectrometer with the
Smart DuraSamplIR Diamond ATR Accessory. Mössbauer spectra were recorded
with a conventional transmission Mössbauer spectrometer operating at a constant
acceleration mode ($T = 300$ K). A $\sim$100 mCi of ^{57}Co(Rh) source was used and the
spectrometer was calibrated using α-iron at room temperature. The magnetization
of the uncoated SPION and Au@SPION was measured with a 7 T Quantum Design
SQUID magnetometer in the temperature range of 5–300 K. For zero-field cooled
(ZFC) experiments, the sample was cooled down and a constant field was applied
during the warm up scan.

3. Description of synthesis

Uncoated SPION have been prepared by chemical coprecipitation of appropriate commercial chlorides ($FeCl_3 \cdot 6H_2O$ and $FeCl_2 \cdot 4H_2O > 99\%$ obtained from
Sigma) in an alkali solution of NaOH (0.5 M in water) as reported earlier [7].
The gold coating process is performed by *in situ* reduction of gold ions on the
surface of SPION and a homogeneous gold layer is obtained by very slow reduction of gold from 2 mM chloroauric acid ($HAuCl_4 \cdot 4H_2O$) using citric acid
(1%, $C_6H_8O_7$) [8]. The immobilization of BSA on the surface of SPION was
achieved by a preliminary surface modification with amino-group followed by
carbodiimide activation [9]. In case of PCL and PLLA coatings, *in situ* poly-
merization process is undertaken. For this purpose, the desired amount of corresponding monomer and Sn(II) 2-ethylhexanoate, as an initiator, are added to a
previously prepared and redispersed uncoated SPION in anhydrous toluene. The
surface modification-polymerization process is performed under nitrogen flow at
110°C for 12 hrs.

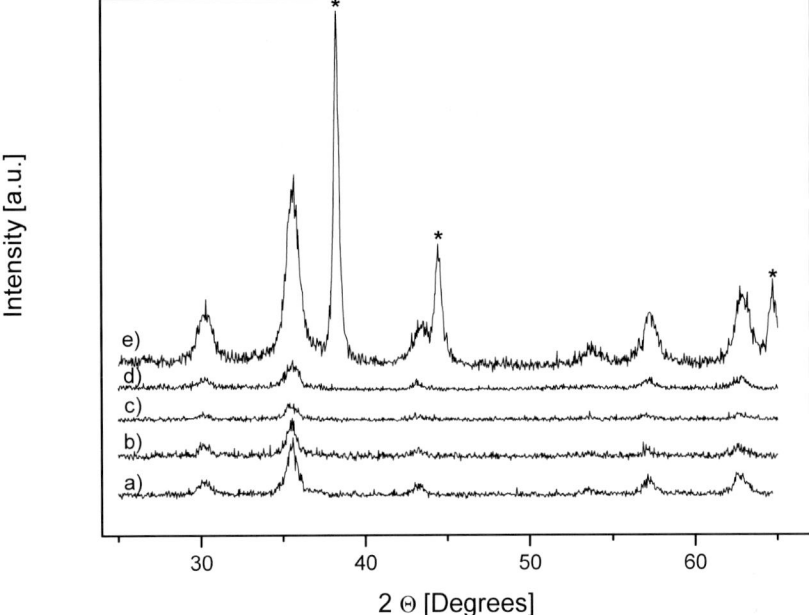

Figure 1. X-ray powder diffraction patterns of (a) uncoated SPION, (b) BSA@SPION, (c) PLLA@ SPION, (d) PCL@SPION and (e) Au@SPION. (*: Au peaks.)

4. Results and discussion

TEM analysis of the SPION, prepared by chemical co-precipitation, showed particles with almost spherical morphology and average size of 8 nm. SPION with these characteristics exhibit a high saturation magnetization ($M_s = 89$ emu/g) and is used for further surface modification. XRD patterns (Figure 1) demonstrate pure magnetite phase for all samples, apart from Au@SPION sample, where the characteristics gold peaks are also observed. The peak broadening is due to the small size of the crystallites. The particle size estimated from TEM micrographs are in an agreement with the values calculated from XRD by using Debye–Scherrer equation [10]. The existence of organic coating layers on the surface of SPION is established by FT-IR analysis (data not shown) and characteristic absorption bands of PCL, PLLA and BSA are observed for coated samples. The amount of coating, estimated from magnetic measurements and thermogravimetric analysis is ~40% of the total particle weight for polymer-coated SPION and ~60% for BSA@SPION. The difference between these values may be due to the nature of the interactions with the surface of SPION since polymers are adsorbed on the SPION surface, but BSA is covalently bound to the surface-attached amino groups, which results in the forming several coating layers. The amount of gold formed on the surface of particles, in case of Au@SPION, is estimated form the saturation magnetization measurements and showed that the particles consist of 58% Au and 42% of magnetite. The ZFC and FC magnetization measurements for Au@SPION

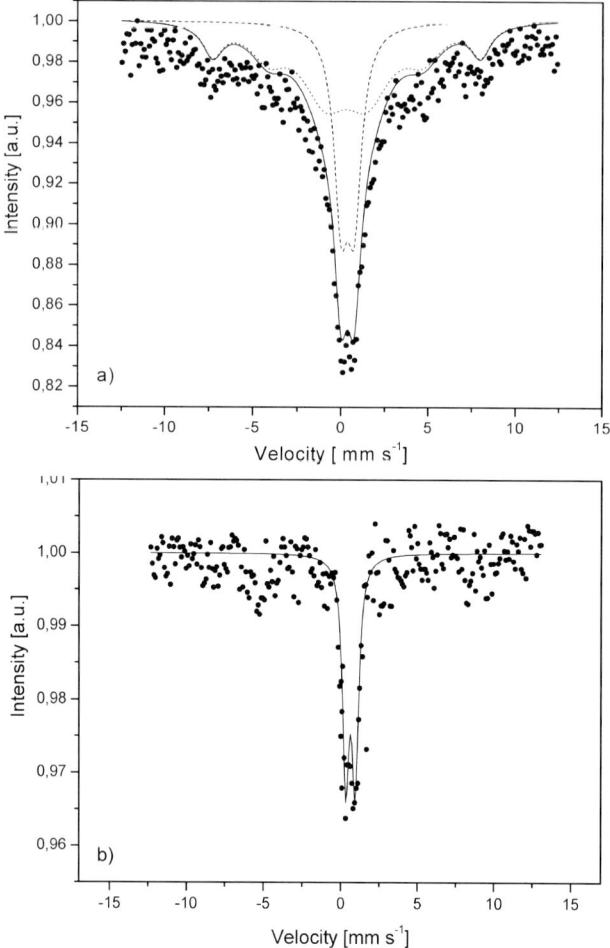

Figure 2. Mössbauer spectra of (a) uncoated SPION and (b) Au@SPION.

show a maximum value at 80 K corresponding to the blocking temperature. The saturation magnetization decreases for all samples as a result of the presence of nonmagnetic sheath on the surface of particles.

Mössbauer spectra of (a) uncoated and (b) Au@SPION are shown in Figure 2. The spectrum for uncoated SPION recorded at ambient temperature represents a broad line consisting of two components in the absence of external magnetic filed. The hyperfine Mössbauer parameters are given in Table I. The spectrum consists of two components: doublet with isomer shift of $\delta = 0.38$ mm s^{-1} and sextet with isomer shift of $\delta = 0.32$ mm s^{-1}, respectively. The presence of two different components for uncoated SPION can be explained by variation in relaxation time of particles. For the particles with mean diameter of 8 nm the relaxation time of magnetic moments are in the range of $\sim 10^{-8}$ s. Therefore, both constituents of spectrum can be assigned to superparamagnetic component. The line

Table I. Hyperfine parameters of Mössbauer spectra of prepared samples

Sample	HF		δ [mm/s]	QS [mm/s]	H_{eff} [T]	S [%]	Site
SPION	doublet	1	0.38	0.76		34	A
	sextet	1	0.32	~0	47.4	66	A
BSA@SPION	sextet	1	0.30	~0	47.2	13	A
		2	0.46	~0	43.4	87	B
Au@SPION	doublet	1	0.63	0.62		100	A
PLLA@SPION	sextet	1	0.31	~0	47.1	22	A
		2	0.48	~0	43.1	78	B
PCL@SPION	sextet	1	0.32	~ 0	46.8	25	A
		2	0.49	~0	42.7	75	B

widening is due to the broad size distribution and the various types of crystallographic surrounding of iron atoms. Mössbauer spectrum of bulk magnetite particles consists of two sextets, corresponding to tetrahedral A-site (Fe^{3+}) and octahedral B-site (Fe^{2+} and Fe^{3+}) [5]. When a H_{ext} (0.17 T) is applied the doublet is disappeared due to the induced magnetization of the sample (data not shown). The Mössbauer spectrum of Au@SPION is attributed to one doublet with a high value of isomer shift ($\delta = 0.63$ mm s^{-1}) due to the transfer of electron density from Au to the surface of SPION [11]. Figure 3 displays the spectra for BSA@SPION, PCL@SPION and PLLA@SPION. All samples show two magnetic hyperfine field distributions with defined A- and B-sites and close values of isomer shifts (Table I). The SPION, when sheathed by protein have a tendency to form clusters as indicated by TEM micrographs (data not shown). The PCL@SPION and PLLA@SPION samples have the formation of islands, which is typical for polymer coating where the inter-particle interactions are observed. There is also the possibility of additional interactions between SPION through polymer matrix which takes place due to the presence of excess of polymer. Therefore, we observe the decrease of the SPM fraction in Mössbauer spectra while the inter-particle interactions and the size of agglomerates increase. These results are in agreement with data reported earlier. The magnetite nanoparticles coated with polymers as polyvinyl alcohol [6] and styrene-divinylbenzene [12] form chain-like structure and as a result the blocking temperature increases with the appearance of hyperfine magnetic structure according to the Mössbauer spectroscopy measurements at ambient temperature. However, polysaccharide, such as dextran, creates clusters of magnetite nanoparticles according to Pardoe *et al.* [13] and the Mössbauer spectrum recorded at ambient temperature consists of significantly low fraction of doublet and dominated fraction of sextet. More detailed investigations are currently in progress to analyze the effect of the amount of polymer on magnetic

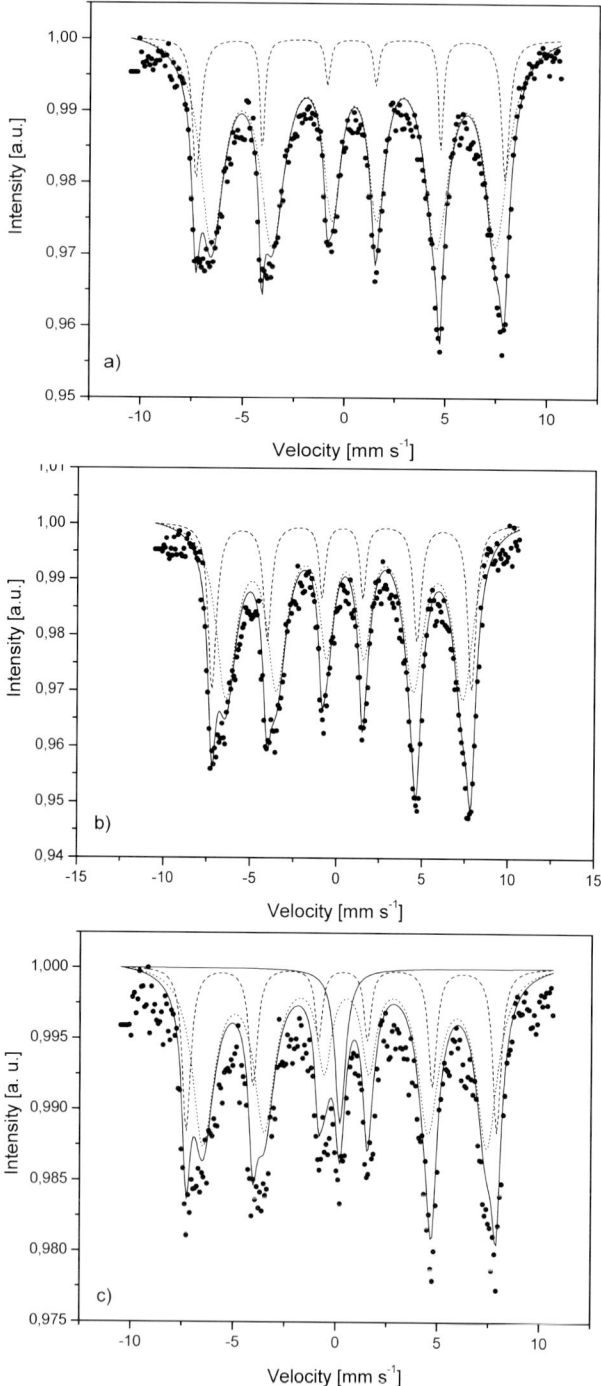

Figure 3. Mössbauer spectra of (a) BSA@SPION, (b) PCL@SPION, and (c) PLLA@SPION.

properties of SPION. Low temperature Mössbauer spectroscopy measurements are required to clarify the statement about the superparamagnetic properties for coated samples. Au@SPION and BSA@SPION have been tested *in vitro* and proven useful for medical application [8].

Acknowledgements

This work was supported by Swedish Foundation for Strategic Research (SSF). The authors would like to thank Dr. Sang Ho Yun and Prof. Judy Wu from the University of Kansas for SQUID measurements.

References

1. Shinkai, M., *J. Biosci. Bioeng.* **94** (2002), 606.
2. Dickson, D. P. E., Walton, S. A., Mann, S. and Wong, K., *Nanostruct. Mater.* **9** (1997), 595.
3. Lacava, Z. G. M., Azevedo, R. B., Martins, E. V., Lacava, L. M., Freitas, M. L. L. Garcia, V. A. P., Rebula, C. A., Lemos, A. P. C., Sousa, M. H., Tourinho, F. A., Da Silva, M. F. and Morais, P. C., *J. Magn. Magn. Mater.* **201** (1999), 435.
4. Berry, C. and Curtis, A., *J. Phys. D* **35** (2002), R1.
5. Kilcoyne, S. H. and Gorisek, A., *J. Magn. Magn. Mater.* **177–181** (1998), 1457.
6. Novakova, A. A., Lanchinskaya, V. Yu., Volkov, A. V., Gendler, T. S. Kiseleva, T. Yu., Moskvina, M. A. and Zezin, S. B., *J. Magn. Magn. Mater.* **258–259** (2003), 354.
7. Kim, D. K., Mikhaylova, M., Zhang, Yu. and Muhammed, M., *Chem. Mater.* **15** (2003), 1617.
8. Mikhaylova, M., Berry, C. C., Kim, D. K., Jo, Y. S., Curtis, A. S. G. and Muhammed, M., *Ann. Transplant.*, Submitted for publication.
9. Koneracka, M., Kopcansky, P., Antalik, M., Timko, M., Ramchand, C., Lobo, D., Mehta, R. and Upadhyay, R., *J. Magn. Magn. Mater.* **201** (1999), 427.
10. Blanco-Mantecon, M. and O'Grady, K., *J. Magn. Magn. Mater.* **203** (1999), 50.
11. Mikhaylova, M., Kim, D. K., Tsakalakos, T. and Muhammed, M., *Langmuir* **20** (2004), 2472.
12. Rodriguez, A. F. R., Oliveira, A. C., Rabelo, D., Lima, E. C. D. and Morais, P. C., *J. Magn. Magn. Mater.* **252** (2002), 77.
13. Pardoe, H., Chua-Anusorn, W., Pierre, T. G. St. and Dobson, J., *J. Magn. Magn. Mater.* **225** (2001), 41.

Hyperfine Interactions **156/157**: 265–272, 2004.
© 2004 *Kluwer Academic Publishers. Printed in the Netherlands.*

Spin States of Iron(III) in Highly Saddled Dodecaphenylporphyrin Complexes

T. OHYA, J. TAKEDA and M. SATO
Laboratory of Biophysics, Faculty of Pharmaceutical Sciences, Teikyo University, Sagamiko, Kanagawa 199-0195, Japan; e-mail: ohyat@pharm.teikyo-u.ac.jp

Abstract. Iron(III) complexes of highly saddled dodecaphenylporphyrin, Fe(DPP)X (X = Cl, Br or I) have been prepared and characterized by Mössbauer, UV–Vis and magnetic measurements. The Mössbauer spectra, recorded at temperatures from 5 to 300 K, contain two components A and B. Component A is attributed to iron(III) in a spin-admixed ($S = 3/2, 5/2$) state. The UV–Vis spectra of solution samples of these complexes exhibit broad and red-shifted absorption bands. The effective magnetic moments derived from the molar magnetic susceptibilities measured by modified Gouy method at 298 K for X = Cl, Br and I are 5.52, 5.10 and 4.28 μ_B, respectively.

Key words: intermediate-spin iron(III), spin admixture, dodecaphenylporphyrin, iron porphyrin complexes, Mössbauer spectra.

1. Introduction

Systematic studies of porphyrin derivatives have resulted in the discovery of a number of spin-state/stereochemical relationships. In particular, sterically crowded and hence distorted porphyrins have been investigated [1] for their relevance to the functional consequences of similar distortions observed for porphyrin derivatives in various proteins. We have prepared iron(III) complexes of highly saddled dode-caphenylporphyrin, Fe(DPP)X (X = Cl, Br or I) (Figure 1), and characterized them by Mössbauer, UV–Vis and magnetic measurements. Spin-admixed ($S = 3/2, 5/2$) states of iron(III) have been observed in the complexes.

2. Materials and methods

2,3,5,7,8,10,12,13,15,17,18,20-dodecaphenylporphyrin, H_2DPP, was prepared as described in the literature [2]. Insertion of iron was carried out in refluxing ben-zonitrile with $FeBr_2$. The crude iron complex was dissolved in dichloromethane and passed through a short alumina column. The reddish band was collected and treated with a saturated aqueous solution of NaX (X = Cl, Br or I) made strongly acidic with corresponding hydrohalic acid. The organic layer was dried over corre-sponding sodium salt, and evaporated to dryness. Recrystallization from CH_2Cl_2-heptane gave purplish crystals of Fe(DPP)X. ^{57}Fe Mössbauer measurements were

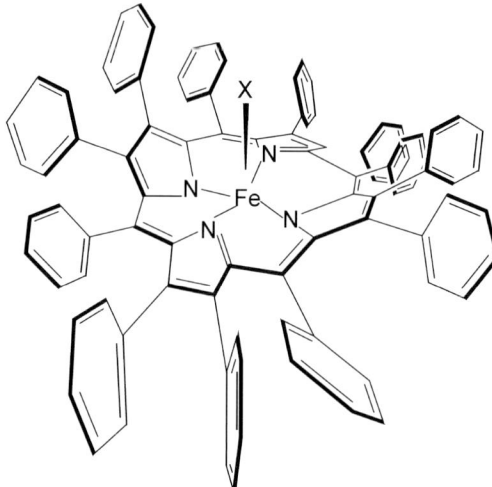

Figure 1. Fe(DPP)X. X = Cl, Br or I.

performed with a Topologic Systems MA-100 Mössbauer spectrometer in transmission geometry using a ^{57}Co/Rh source. The spectra were fitted with appropriate
combinations of Lorenzian lines. The isomer shift values are given relative to α-
iron at room temperature. UV–Vis spectra were measured for solution samples in
toluene with a Hitachi U-3310 spectrophotometer. The molar magnetic susceptibilities were measured at room temperature with a modified Gouy balance (Sherwood
MSB-MK I).

3. Results and discussion

Figures 2, 3 and 4 show the zero-field Mössbauer spectra of Fe(DPP)X (X = Cl,
Br or I) at various temperatures. Each spectrum at a lower temperature has been
analyzed with two doublets A and B, the quadrupole splitting ΔE_Q of A being
larger than that of B (Table I), whereas some spectra at higher temperatures have
not satisfactorily been resolved into components. The first component A of each of
these pairs of components is characterized by an isomer shift δ around 0.4 mm s^{-1}
and ΔE_Q in the range 1.9–3.1 mm s^{-1} at lower temperatures. The splitting is
temperature independent up to about 80 K. Such Mössbauer data have previously
been found for complexes that were either 'pure' $S = 3/2$ iron(III) porphyrin
species or spin admixed 3/2, 5/2 iron(III) porphyrin materials [3, 4]. The area
fractions of component A at lower temperatures are estimated to be about 40, 50
and 80% for X = Cl, Br and I, respectively. The second component B is typical of
an $S = 5/2$ high-spin iron(III) complex with δ around 0.5 mm s^{-1} and ΔE_Q in the
range 0.9–1.2 mm s^{-1} at lower temperatures. Quadrupole splittings of this size in
$S = 5/2$ complexes are rather large compared to the values 0.5 to 0.9 mm s^{-1} for
usual porphyrin complexes. The temperature dependence of the asymmetry of the

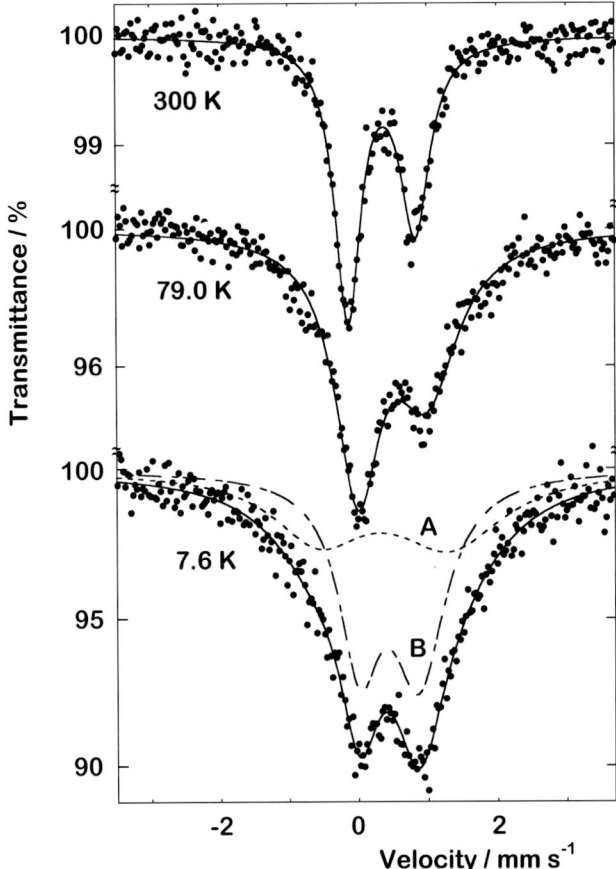

Figure 2. Mössbauer spectra of Fe(DPP)Cl. The component with larger quadrupole splitting is denoted as A, and the one with smaller value as B.

quadrupole-split pattern of component B in Figure 3, results from the temperature-dependent spin–spin relaxation of high-spin iron(III) ion and the off-diagonal terms of the hyperfine operator [5, 6].

The spin states of ferric hemes are governed by the d-orbital splittings. Figure 5 gives the schematic representation of the ground-state spin of iron(III) in a tetragonally distorted field as a function of the orbital-energy separations [7, 8]; ε_1 is the separation of the d_{xy} and $d_{x^2-y^2}$ orbitals, while ε_2 is the separation of the d_{yz}/d_{zx} and d_{z^2} orbitals. The intermediate-spin $S = 3/2$ state arises with a large ε_1 and a small ε_2, that is, when the equatorial ligand field is strong and the axial ligand field is weak, causing the $d_{x^2-y^2}$ orbital to be considerably higher than the other four. Near the border line $S = 3/2, 5/2$ of Figure 5, spin–orbit coupling leads to a strong spin mixture. The arrow indicates the proposed [8, 9] behavior of iron(III) porphyrin complexes under weakening of the axial ligand field: for a given equatorial (porphyrin) ligand, decrease in the axial field leads to a compensating increase in the equatorial field.

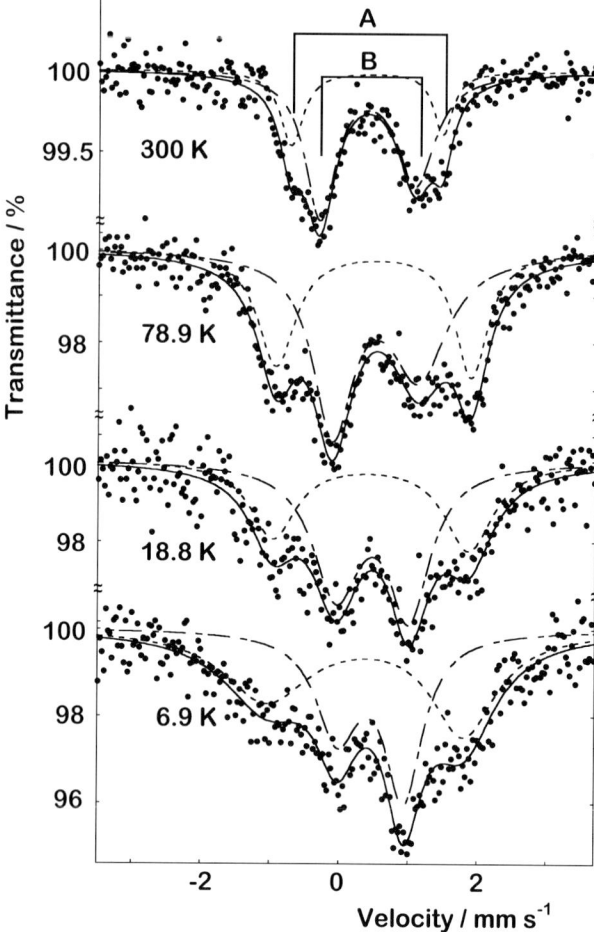

Figure 3. Mössbauer spectra of Fe(DPP)Br. The component with larger quadrupole splitting is denoted as A, and the one with smaller value as B.

Halides are relatively strong axial ligands in 'magnetochemical' series [10], and form high-spin iron(III) complexes with usual porphyrins, while extremely weak ligands such as ClO_4^- and SbF_6^- give spin-admixed $3/2$, $5/2$ states. Sterically crowded DPP and its Ni complex have been reported to be of highly saddled conformation [11] and exhibit unusual physicochemical properties. The $d_{x^2-y^2}$ orbital of an iron in a deformed porphyrin complex is destabilized in consequence of core contraction [11], resulting a large ε_1 in Figure 5. A substantial $S = 3/2$ spin admixture responsible for component A of Fe(DPP)X (X = Cl, Br or I) (Figures 2 to 4 and Table I) is not very common in complexes with halide ligands [1, 12, 13]. Cheng *et al.* [12] showed that the chloroiron(III) derivative of the sterically crowded octaethyltetraphenylporphyrin Fe(OETPP)Cl has spin-admixed state with approximately 40% contribution from $S = 3/2$. Weiss and coworkers [13],

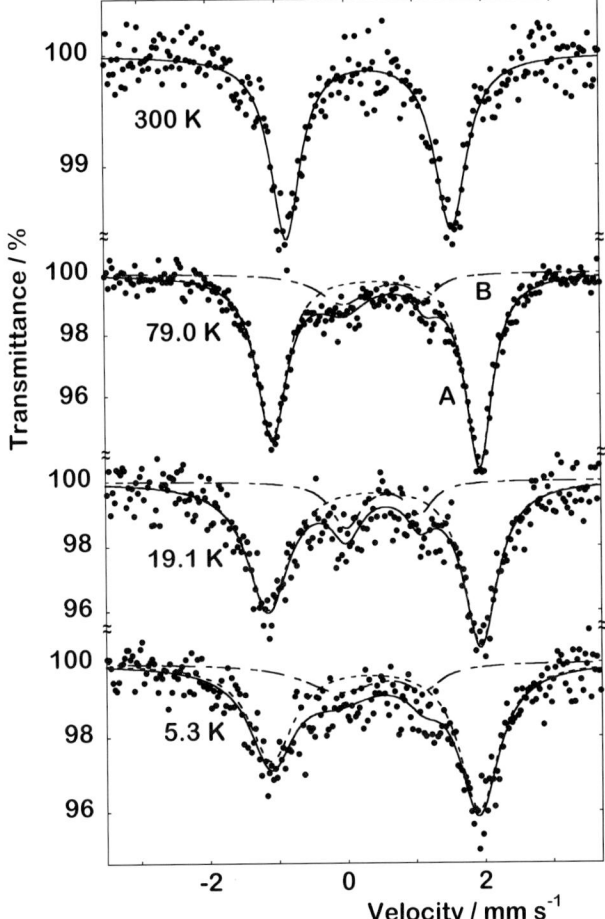

Figure 4. Mössbauer spectra of Fe(DPP)I. The component with larger quadrupole splitting is denoted as A, and the one with smaller value as B.

however, pointed out subsequently that a significantly smaller (4–10%) $S = 3/2$ admixture is present in the complex. The large ΔE_Q values for component A in Fe(DPP)I indicate that the complex is in an essentially pure intermediate-spin state. The component B of each of the Mössbauer spectra has rather large ΔE_Q for $S = 5/2$ iron(III), and is considered to be in a slightly admixed 5/2, 3/2 state. The value of ΔE_Q for Fe(OETPP)Cl is reported [13] to be 0.95 mm s^{-1} at 280 K and is not larger than the values for components B of the present complexes.

   Although it is difficult at present to explain why two components are observed in each of the complexes studied, we believe on the basis of some observed facts that the phenomenon is inherent in the complexes, not attributed to coexistence of any impurities such as the hydroxo complex. The values of ΔE_Q for both components at low temperatures depend on X in the order Cl < Br < I, and also area fractions of component A at low temperatures change in the same order. Furthermore, we have

Table I. Mössbauer data for Fe(DPP)X (X = Cl, Br or I). The isomer shift δ is given relative to α-iron at room temperature

T [K]	Component	δ [mm s^{-1}]	ΔE_Q [mm s^{-1}]
	Fe(DPP)Cl		
300		0.35(1)	0.96(1)
79.0		0.49(1)	0.99(1)
7.6	A	0.37(1)	1.88(11)
	B	0.44(2)	0.85(3)
	Fe(DPP)Br		
300	A	0.36(2)	2.18(4)
	B	0.38(2)	1.38(4)
78.9	A	0.48(1)	2.82(1)
	B	0.50(2)	1.23(3)
18.8	A	0.45(4)	2.83(7)
	B	0.48(2)	1.05(4)
6.9	A	0.38(3)	2.85(6)
	B	0.46(2)	0.96(2)
	Fe(DPP)I		
300		0.35(1)	2.41(1)
79.0	A	0.44(1)	3.02(1)
	B	0.57(3)	1.21(5)
19.1	A	0.40(1)	3.10(2)
	B	0.53(4)	1.07(7)
5.3	A	0.40(2)	3.03(3)
	B	0.49(10)	1.23(14)

prepared the hydroxo complex Fe(DPP)OH and measured the Mössbauer spectra. The spectrum at 9.5 K of this complex itself has two components A and B, the area fraction of A being 20%. The values of δ (mm s^{-1}) and ΔE_Q (mm s^{-1}) are 0.43(5) and 2.43(5) for component A, and 0.45(3) and 0.97(5) for component B. It is likely that the two components, that is, the two spin states for each of the complexes are correlated to slowly relaxing two species with different geometrical configurations. The relaxation probably becomes fast at high temperatures making it impossible for the spectra to be resolved into components. The broad band widths of the UV–Vis absorption spectra mentioned in the following section also support this view.

UV–Vis spectra of the solutions of the present complexes in toluene have been found to be characterized by red-shifted broad absorption bands compared with common planar porphyrin complexes such as Fe(TPP)X (TPP = 5,10,15,20-tetra-phenylporphyrin) [14]. The values of λ_{max} and ε of B, Q_v and Q_0 bands are given in Table II. Red-shifted electronic spectra have been observed for the majority of nonplanar porphyrins studied [15].

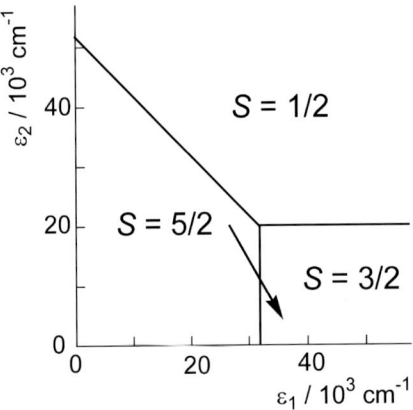

Figure 5. Schematic representation of the ground-state spin of iron(III) in a tetragonally distorted field as a function of the orbital-energy separations [7, 8]. ε_1 is the separation of the d_{xy} and $d_{x^2-y^2}$ orbitals, while ε_2 is the separation of the d_{yz}/d_{zx} and d_{z^2} orbitals. The Racah parameters B and C are taken as 1100 and 3500 cm^{-1}, respectively. The arrow indicates the proposed [8, 9] behavior of iron(III) porphyrin complexes under weakening of the axial ligand field.

Table II. Absorption spectral data for Fe(DPP)X[a] and Fe(TPP)X[b] (X = Cl, Br or I) at 300 K

Complex	λ_{max} [nm] (ε [l mmol^{-1} cm^{-1}])		
	B	Q_v	Q_0
Fe(DPP)Cl	452.2 (88.09)	533.2 (17.29)	580.0 (12.56)
Fe(DPP)Br	452.8 (83.34)	540.0 (16.22)	585.0 (12.26)
Fe(DPP)I	436.8 (74.98)	547.0 (13.55)	587.0 (11.24)
Fe(TPP)Cl	418.8 (121.4)	507.6 (14.26)	573.0 (4.07)
Fe(TPP)Br	420.2 (89.38)	511.2 (15.19)	578.0 (3.59)
Fe(TPP)I	414.0 (90.79)	516.0 (14.89)	572.4 (5.22)

[a] Solution in toluene.
[b] Solution in benzene [14].

The molar magnetic susceptibilities of the complexes were measured for powder samples at 298 K with a modified Gouy balance. The effective magnetic moments per molecule are derived with corrections of diamagnetic contributions to be 5.52, 5.10 and 4.28 μ_B for X = Cl, Br and I, respectively. The values are considerably lower than the spin-only value 5.78 μ_B for a pure $S = 5/2$ state; and the value for the iodide is close to 3.87 μ_B expected for a pure $S = 3/2$ state.

4. Conclusions

The Mössbauer spectra of Fe(DPP)Cl, Fe(DPP)Br and Fe(DPP)I contain two components A and B, the former corresponding to strongly spin-admixed iron(III). The effective magnetic moments of the iron(III) ions in these complexes are lower than the value for a pure high-spin state, and the value for the iodide is close to the one for a pure intermediate-spin state. The solution samples of the present complexes exhibit red-shifted optical spectra as is often the case with nonplanar porphyrins.

References

1. Nakamura, M., Ikeue, T., Ohgo, Y., Takahashi, M. and Takeda, M., *Chem. Commun.* (2002), 1198.
2. Takeda, J., Ohya, T. and Sato, M., *Chem. Phys. Lett.* **183** (1991), 384; *Chem. Pharm. Bull.* **38** (1990), 264.
3. Spartalian, K., Lang, G. and Reed, C. A., *J. Chem. Phys.* **71** (1979), 1832.
4. Abu-Soud, H. and Silver, J., *Inorg. Chim. Acta* **32** (1988), 61.
5. Dattagupta, S., *Phys. Rev. B* **12** (1975), 3584.
6. Ohya, T. and Sato, M., *J. Chem. Soc., Dalton Trans.* (1996), 1519.
7. Rispin, A. S., Sato, M. and Kon, H., *Theoret. Chim. Acta (Berl.)* **50** (1978), 95.
8. Gismelseed, A., Bominaar, E. L., Bill, E., Trautwein, A. X., Winkler, H., Narsi, H., Doppelt, P., Mandon, D., Fischer, J. and Weiss, R., *Inorg. Chem.* **29** (1990), 2741.
9. Reed, C. A., Mashiko, T., Bentley, S. P., Kastner, M. E., Scheit, W. R. Spartalian, K. and Lang, G., *J. Am. Chem. Soc.* **101** (1979), 2948.
10. Reed, C. A. and Guiset, F., *J. Am. Chem. Soc.* **118** (1996), 3281.
11. Medforth, C. J., Senge, M. O., Smith, K. M., Sparks, L. D. and Shelnutt, J. A., *J. Am. Chem. Soc.* **114** (1992), 9859.
12. Cheng, R.-J., Chen, P.-Y., Gau, P.-R., Chen, C.-C. and Peng, S.-M., *J. Am. Chem. Soc.* **119** (1997), 2563.
13. Schünemann, V., Gerdan, M., Trautwein, A. X., Haoudi, N., Mandon, D., Fischer, J., Weiss, R., Tabard, A. and Guilard, R., *Angew. Chem. Int. Ed.* **38** (1999), 3181.
14. Ohya, T. and Sato, M., *Bull. Chem. Soc. Jpn.* **69** (1996), 3201.
15. Wasbotten, I. H., Conradie, J. and Ghosh, A., *J. Phys. Chem. B* **107** (2003), 3613.

Hyperfine Interactions **156/157**: 273–277, 2004.
© 2004 *Kluwer Academic Publishers. Printed in the Netherlands.*

Mössbauer Spectroscopy of Iron Containing Vitamins and Dietary Supplements

M. I. OSHTRAKH[1], O. B. MILDER[2] and V. A. SEMIONKIN[2]

[1] *Division of Applied Biophysics, Faculty of Physical Techniques and Devices for Quality Control, Ural State Technical University – UPI, Ekaterinburg 620002, Russian Federation*
[2] *Faculty of Experimental Physics, Ural State Technical University – UPI, Ekaterinburg 620002, Russian Federation*

Abstract. Mössbauer spectroscopy was used to study various industrial samples of vitamins containing ferrous fumarate and ferrous bisglycinate chelate (Ferrochel®) and dietary supplements containing ferrous sulfate. The presence of small quantities of various ferric impurities was found. Two vitamins contained major iron compounds that did not correspond to ferrous fumarate and ferrous bisglycinate chelate.

Key words: Mössbauer spectroscopy, iron containing vitamins, iron containing dietary supplements.

1. Introduction

Iron is one of the vitally important metals deficiency of which causes anemia and other pathological changes in the body. To prevent and treat iron deficiency various injectable and oral iron containing pharmaceuticals such as iron–dextran and iron–polysaccharide complexes, iron containing vitamins and dietary supplements are used. It is well known that free iron ions are very toxic. Therefore, iron containing pharmaceuticals must contain iron compounds in non-toxic and bioavailable forms. For instance, ferrous iron compounds were more readily bioavailable than ferric one for oral iron containing supplements [1, 2]. The toxic side effects were more severe with ferric salts because absorption of ferric iron was relatively slower than ferrous one after administration [3]. The U.S. Food and Drug Administration (FDA) requires that ferrous fumarate should not contain more than 2% of ferric iron (Food and Drugs, Sec. 172.350). Thus, the knowledge of the iron state in these subjects is very important for quality control because the valence/spin state of iron may be related to the effect and toxicity of pharmaceutical product. Previously we studied various injectable ferric iron–dextran complexes and found small differences of the values of quadrupole splitting for some complexes and the presence of 4–7% of ferrous impurity [4, 5]. In this work we have studied ferrous iron containing vitamins and dietary supplements with the aim of analyzing of the iron state and quality of these products.

2. Materials and methods

Multiple vitamins Multiple Vitamins With Iron and Therapeutic M (Walgreen Co., USA), My Favorite® Take One™ (Natrol, Inc., USA), Essential Balance® (Nature Made Nutritional Products, USA) contained iron in the form of ferrous fumarate ($FeC_4H_2O_4$). Multiple vitamin My Favorite® Multiple (Natrol, Inc., USA) contained iron in the form of ferrous bisglycinate chelate (Ferrochel® from Albion Laboratories, USA). Dietary supplements Your Life® Maximum Pak® (P. Leiner Nutritional Products, Inc., USA) and Feosol® (SmithKline Beecham Corporation, USA) contained ferrous sulfate ($FeSO_4$). All studied vitamins contained 18 mg of iron per tablet while dietary supplements contained 8.3 mg of iron (Your Life® Maximum Pak®) and 50 mg of iron (Feosol®) per tablet. Samples were used as powder prepared from about 1/2 of each product tablet. The effective thickness of these samples varied from 2 to 8 mg Fe/cm^2. Sample of My Favorite® Multiple contained 1000 mg of Ca that led to the decrease of the absorption effect.

Mössbauer spectra were measured with the constant acceleration computer-ized precision spectrometer that was a part of multi-dimension parametric Möss-bauer spectrometer SM-2201 [6]. The noise of velocity signal of spectrometer was 1.5×10^{-3} mm/s, the drift of zero point velocity was $\pm 2.6 \times 10^{-3}$ mm/s, the non-linearity of velocity signal was 0.01%, the harmonic distortion factor was 0.005% for the frequency band in the range of 0–1120 Hz. The 0.5×10^9 Bq ^{57}Co(Cr) source was used at room temperature. Mössbauer spectra of vitamins and dietary supplements were measured at room temperature in transmission geometry with moving absorber. Mössbauer spectra of all samples were computer fitted with the least squares procedure using Lorentzian line shape. Mössbauer hyperfine parameters (isomer shift δ and quadrupole splitting ΔE_Q) as well as line width Γ and subspectrum area S were determined. The values of isomer shift are given relative to α-Fe at 295 K.

3. Results and discussion

Mössbauer spectra of vitamins containing ferrous fumarate and ferrous bisglyci-nate chelate as well as dietary supplements containing ferrous sulfate are shown in Figures 1–3, respectively. Mössbauer parameters are given in Table I. Two vitamin samples with ferrous fumarate contained two components: the main component 1 related to $FeC_4H_2O_4$ and minor component 2 (~4–5%) related to the high spin ferric compound. One vitamin sample contained the main component 1 related to $FeC_4H_2O_4$ and additional component 3 (3%) related to another high spin ferrous compound and component 2 (6%) related to the high spin ferric compound. An-other vitamin sample also contained 3 components: high spin ferrous compounds 5 (84%) and 4 (10%) with parameters that were different from those of $FeC_4H_2O_4$, and high spin ferric compound 6 (6%) with different parameters than that of other vitamins. The high spin ferric compound 2 may be a result of ferrous fumarate

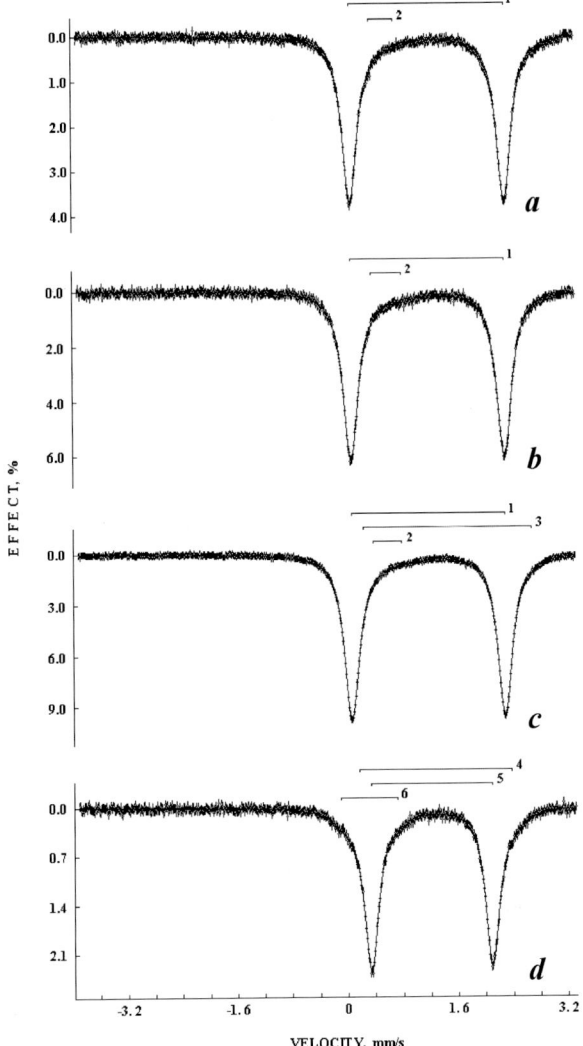

Figure 1. Mössbauer spectra of vitamins containing ferrous fumarate: a – Essential Balance®; b – Therapeutic M; c – Multiple Vitamins With Iron; d – My Favorite® Take One™. Components 1, 3, 4 and 5 are ferrous compounds; components 2 and 6 are ferric compounds. $T = 295$ K.

oxidation and formation of ferric fumarate as it was supposed in [3] or any other ferric salt during vitamin production.

Vitamin sample with ferrous bisglycinate chelate announced as ferrous compound contained two different high spin ferrous compounds 1 (28%) and 2 (13.5%) and the high spin ferric compound 3 (58.5%). This large amount of ferric compound in the vitamin may be dangerous in case of administration. We should point out that the values of isomer shift for the components 3 (Figure 1c) and 2 (Figure 2) appeared to be unusually large.

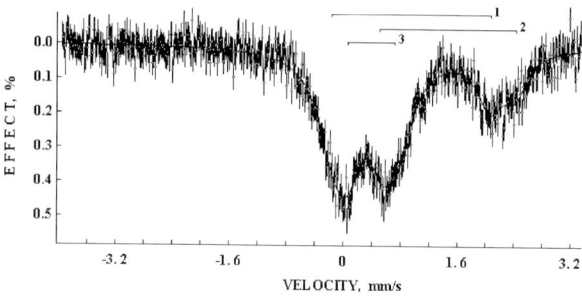

Figure 2. Mössbauer spectrum of vitamin My Favorite® Multiple containing ferrous bisglycinate chelate (Ferrochel®). Components 1 and 2 are ferrous compounds; component 3 is ferric compound. $T = 295$ K.

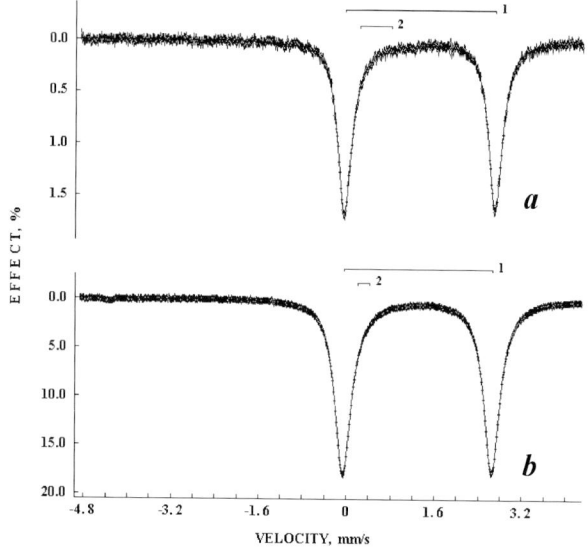

Figure 3. Mössbauer spectra of dietary supplements containing ferrous sulfate: *a* – Your Life® Maximum Pak®; *b* – Feosol®. Component 1 is ferrous compound; component 2 is ferric compound. $T = 295$ K.

Samples of dietary supplements with ferrous sulfate contained major compound of $FeSO_4$ and additional minor component (~2–4%) of the high spin ferric compound. The content of ferric impurities in these samples appeared to be lower than that of observed previously in dietary supplements with ferrous sulfate (8–13%) [3].

4. Conclusion

This study showed that all ferrous pharmaceutical samples contained small amount of ferric impurity. The content of ferric impurity in vitamins containing ferrous fumarate was larger than FDA limitation for these products. Two vitamins con-

Table I. Mössbauer parameters of vitamins and dietary supplements measured at 295 K

Sample	Γ^a (mm/s)	δ^b (mm/s)	ΔE_Q^b (mm/s)	S (%)	Compound^c
Essential Balance®	0.258	1.185	2.210	96.0	Ferrous fumarate (1)
	0.297	0.533	0.351	4.0	Ferric high spin (2)
Therapeutic M	0.264	1.183	2.219	95.3	Ferrous fumarate (1)
	0.405	0.586	0.445	4.7	Ferric high spin (2)
Multiple Vitamins	0.259	1.179	2.218	91.1	Ferrous fumarate (1)
With Iron	0.298	1.457	2.435	2.9	Ferrous high spin (3)
	0.581	0.585	0.409	6.0	Ferric high spin (2)
My Favorite® Take	0.409	1.274	2.194	9.9	Ferrous high spin (4)
One™	0.248	1.215	1.751	84.1	Ferrous high spin (5)
	0.336	0.317	0.818	6.0	Ferric high spin (6)
My Favorite®	0.493	0.957	2.234	28.0	Ferrous high spin (1)
Multiple	0.371	1.473	1.920	13.5	Ferrous high spin (2)
	0.559	0.387	0.656	58.5	Ferric high spin (3)
Your Life®	0.307	1.309	2.749	96.0	Ferrous high spin (1)
Maximum Pak®	0.459	0.500	0.577	4.0	Ferric high spin (2)
Feosol®	0.355	1.288	2.712	98.3	Ferrous high spin (1)
	0.289	0.302	0.212	1.7	Ferric high spin (2)

^{a} Experimental error is ±0.0286 mm/s.
^{b} Experimental error is ±0.0143 mm/s.
^{c} The numbers of components in Figures 1–3 are indicated in parentheses.

tained major iron compounds that were different from ferrous fumarate and ferrous bisglycinate chelate, respectively, which were claimed by the manufacturer (Natrol, Inc., USA). In this case Mössbauer spectroscopy may be a useful tool for the control of the quality of iron containing vitamins and dietary supplements production.

References

1. Jeppsen, R. B. and Borzelleca, J. F., *Food Chem. Toxicol.* **37** (1999), 723.
2. Davidsson, L., Kastenmayer, P., Szajewska, H., Hurrell, R. F. and Barclay, D., *Am. J. Clin. Nutr.* **71** (2000), 1597.
3. Sud, M., Bala, N., Liveleen, Vishwamittar and Puri, S. P., *Indian J. Pure Appl. Phys.* **26** (1988), 701.
4. Oshtrakh, M. I., Semionkin, V. A., Milder, O. B., Livshits, A. B. and Kozlov, A. A., *Z. Naturforsch.* **55a** (2000), 186.
5. Oshtrakh, M. I., Semionkin, V. A., Prokopenko, P. G., Milder, O. B., Livshits, A. B. and Kozlov, A. A., *Int. J. Biol. Macromol.* **29** (2001), 303.
6. Irkaev, S. M., Kupriyanov, V. V. and Semionkin, V. A., British Patent No 10745, 7 May, 1987.

Hyperfine Interactions **156/157**: 279–284, 2004.
© 2004 *Kluwer Academic Publishers. Printed in the Netherlands.*

Comparative Study of Human Liver Ferritin and Chicken Liver by Mössbauer Spectroscopy. Preliminary Results

M. I. OSHTRAKH[1], O. B. MILDER[2], V. A. SEMIONKIN[2],
P. G. PROKOPENKO[3] and L. I. MALAKHEEVA[4]
[1]*Division of Applied Biophysics, Faculty of Physical Techniques and Devices for Quality Control, Ural State Technical University – UPI, Ekaterinburg 620002, Russian Federation*
[2]*Faculty of Experimental Physics, Ural State Technical University – UPI, Ekaterinburg 620002, Russian Federation*
[3]*Faculty of Biochemistry, Russian State Medical University, Moscow 117869, Russian Federation*
[4]*Science Consultation Department, Simbio Holding, Ekaterinburg 620044, Russian Federation*

Abstract. A comparative study of normal human liver ferritin and livers from normal chicken and chicken with Marek disease was made by Mössbauer spectroscopy. Small differences of quadrupole splitting and isomer shift were found for human liver ferritin and chicken liver. Mössbauer parameters for liver from normal chicken and chicken with Marek disease were the same.

Key words: Mössbauer spectroscopy, human liver ferritin, chicken liver.

1. Introduction

Mössbauer study of iron storage protein ferritin from various sources (human, limpet and bacterial cells) demonstrated different magnetic ordering temperature (above 50, about 30 and 3 K, respectively), differences of magnetic hyperfine field at 1.3 K (494 ± 5, 477 ± 5 and 420 ± 5 kOe, respectively) and small differences of quadrupole splitting at 87 K (0.69 ± 0.02, 0.75 ± 0.02 and 0.70 ± 0.02 mm/s, respectively) [1]. Several tissues such as liver and spleen contain large amount of iron storage protein to measure its Mössbauer spectra [2, 3]. Small differences of quadrupole splitting were found for spleen samples from normal people and patients with β-thalassemia (0.51 ± 0.10 and 0.68 ± 0.03 mm/s, respectively) [4]. In this work we present preliminary results of comparative study of the extracted normal human liver ferritin and liver tissues from normal chicken and chicken with Marek disease, a malignant tumor with transformation of lymphoid cells, by Mössbauer spectroscopy.

2. Materials and methods

Preparation of lyophilized normal human liver ferritin was described earlier [5]. Liver tissues from normal chicken and chicken with Marek disease were washed from the blood and lyophilized. Mössbauer spectra of the sample powders were measured at room temperature with the constant acceleration computerized high precision and stable spectrometer that was a part of multi-dimension parametric Mössbauer spectrometer SM-2201 [6]. The noise of velocity signal of spectrometer was 1.5×10^{-3} mm/s, the drift of zero point velocity was $\pm 2.6 \times 10^{-3}$ mm/s, the nonlinearity of velocity signal was 0.01%, the harmonic distortion factor was 0.005% for the frequency band in the range of 0–1120 Hz. The 0.5×10^9 Bq ^{57}Co(Cr) source was used at room temperature. Spectra of liver samples were measured up to 10 days with statistical rate from 10×10^6 to 26×10^6 counts per channel. Mössbauer spectra were measured in transmission geometry with moving absorber to exclude parabolic distortion of the spectrum and contribution of the ^{57}Fe in the beryllium window of scintillator detector to the spectrum in case of samples with low iron content. Mössbauer spectra were computer fitted with the least squares procedure using Lorentzian line shape. Mössbauer parameters isomer shift δ, quadrupole splitting ΔE_Q, line width Γ and subspectrum area S were determined. Additionally Mössbauer spectra were fitted with a program of quadrupole splitting distribution. The values of isomer shift are given relative to α-Fe at 295 K.

3. Results and discussion

Mössbauer spectra of ferritin and chicken livers are shown in Figure 1. Spectra of chicken livers demonstrate the presence of ferritin-like iron in tissues. It is well known that paramagnetic Mössbauer spectra of ferritin, like any ferric hydrous oxides, have non-Lorentzian line shape and, therefore, are better fitted using more than one quadrupole doublet or using distribution of quadrupole splitting [7]. The distributions of quadrupole splitting for these spectra are shown in Figure 2. The values of average δ and ΔE_Q and ΔE_Q with maximal probability obtained for this fitting are given in Table I. The average values of ΔE_Q for human liver ferritin was higher than that of chicken liver while ΔE_Q with maximal probability for human liver ferritin was lower than that of chicken liver. However, the fitting of Mössbauer spectra of ferric hydrous oxides by the distribution of quadrupole splitting should not be considered as final. Recently we showed that there was no agreement in quantity of peaks in distribution and the maximal quantity of quadrupole doublets required for better fitting [8]. Therefore, these spectra were fitted using one quadrupole doublet and superposition of two quadrupole doublets. The results of one quadrupole doublet fit of Mössbauer spectra of human liver ferritin and chicken liver are shown in Figure 3 as a plot of ΔE_Q and δ. The difference between Mössbauer parameters of human liver ferritin and chicken liver is clearly seen. Mössbauer parameters obtained by the spectra fitting using two quadrupole doublets are given in Table II. The differences of ΔE_Q and δ values for the first

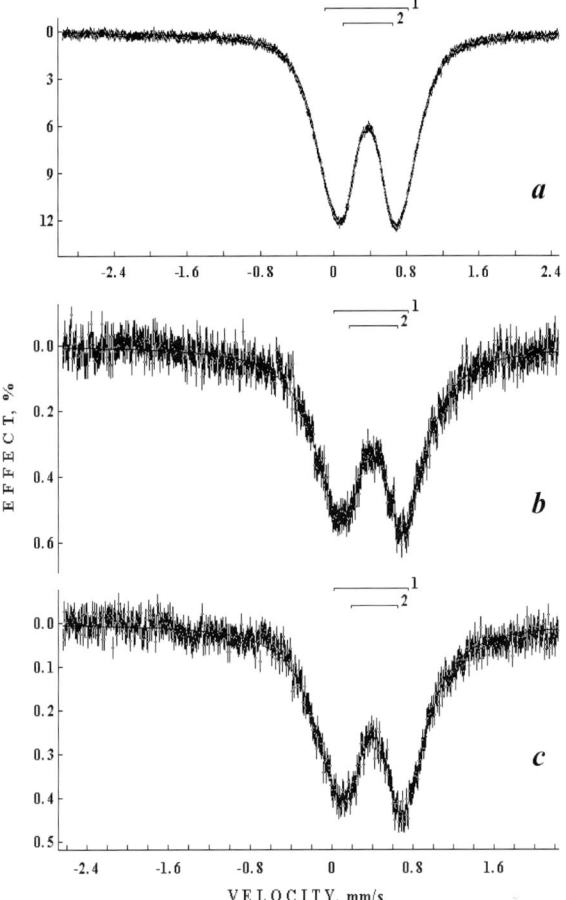

Figure 1. Mössbauer spectra of human liver ferritin (a) and chicken liver: normal (b) and with Marek disease (c) in lyophilized form. 1 and 2 are the result of two quadrupole doublets fit.

and the second doublets were found for human liver ferritin and chicken liver. The relative areas of the first and the second doublets were also different. These results indicate structural differences of the iron core structure in human liver ferritin and ferritin-like iron in chicken liver. The differences between normal chicken liver and liver from chicken with Marek disease were not found. Earlier the differences between normal human liver and liver from patients with β-thalassemia/HbE were also not found in contrast to difference that was found for normal and patient's spleen [4].

4. Conclusion

The results obtained showed some differences between normal human ferritin and chicken liver in the shape of the distributions of quadrupole splitting, in the values

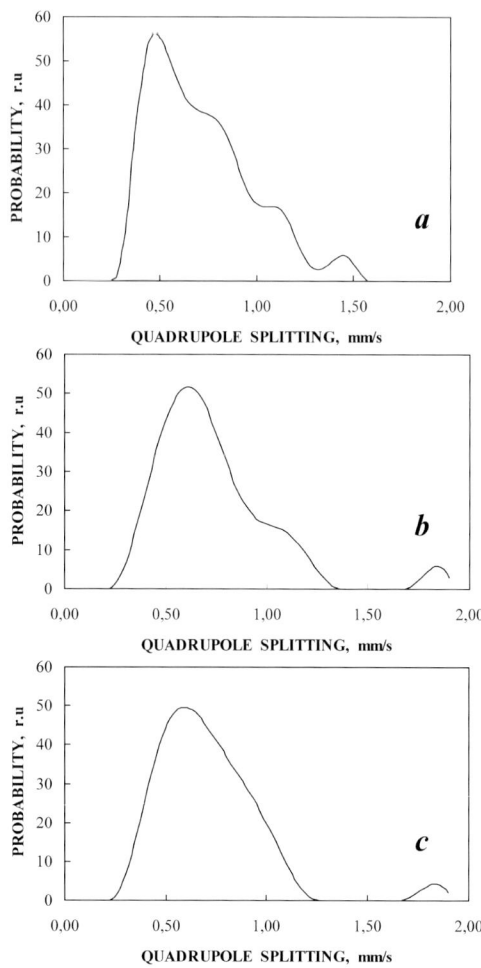

Figure 2. Distributions of quadrupole splitting for human liver ferritin (a) and chicken liver: normal (b) and with Marek disease (c) in lyophilized form.

Table I. Average δ and ΔE_Q values and values of ΔE_Q with maximal probability obtained for the distributions of quadrupole splitting for Mössbauer spectra of normal human liver ferritin and chicken liver

Sample	δ average, mm/s	ΔE_Q average, mm/s	ΔE_Q max, mm/s
Normal human liver ferritin	0.366 ± 0.002	0.710 ± 0.017	0.482 ± 0.013
Normal chicken liver	0.390 ± 0.015	0.696 ± 0.012	0.584 ± 0.016
Liver from chicken with Marek disease	0.390 ± 0.015	0.680 ± 0.008	0.576 ± 0.012

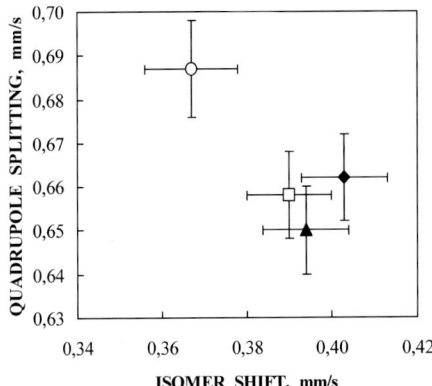

Figure 3. Plot of Mössbauer hyperfine parameters of normal human liver ferritin (○), normal chicken liver (□) and liver from chickens with Marek disease (▲, ◆) obtained by one quadrupole doublet fit. $T = 295$ K.

Table II. Mössbauer parameters of normal human ferritin and chicken liver obtained by two quadrupole doublets fit

Sample	Γ, mm/s	δ, mm/s	ΔE_Q, mm/s	S, %
Normal human liver	0.333 ± 0.022	0.371 ± 0.011	0.541 ± 0.011	46
ferritin	0.402 ± 0.022	0.360 ± 0.011	0.917 ± 0.011	54
Normal chicken liver	0.250 ± 0.020	0.410 ± 0.010	0.476 ± 0.010	15
	0.531 ± 0.020	0.385 ± 0.010	0.726 ± 0.010	85
Liver from chicken with	0.264 ± 0.020	0.418 ± 0.010	0.463 ± 0.010	17
Marek disease	0.527 ± 0.020	0.385 ± 0.010	0.729 ± 0.010	83

of ΔE_Q and δ in case of one quadrupole doublet fit, in the values of ΔE_Q, δ and S in case of two quadrupole doublets fit. These differences may be a result of small variations of the iron core structure in human liver ferritin and ferritin-like iron in chicken liver. Mössbauer parameters for livers from normal chicken and chicken with Marek disease were the same. In the latter case further studies are required for detailed analysis of ferritin-like iron in various tissues.

References

1. St. Pierre, T. G., Bell, S. H., Dickson, D. P. E., Mann, S., Webb, J., Moore, G. R. and Williams, R. J. P., *Biochim. Biophys. Acta* **870** (1986), 127.
2. Chua-Anusorn, W., St. Pierre, T. G., Black, G., Webb, J., Macey, D. J. and Parry, D., *Hyp. Interact.* **91** (1994), 899.
3. Chua-Anusorn, W., St. Pierre, T. G., Webb, J., Macey, D. J., Yansukon, P. and Pootrakul, P., *Hyp. Interact.* **91** (1994), 905.

4. St. Pierre, T. G., Chua-Anusorn, W., Webb, J., Macey, D. J. and Pootrakul, P., *Biochim. Biophys. Acta* **1407** (1998), 51.

5. Oshtrakh, M. I., Semionkin, V. A., Prokopenko, P. G., Milder, O. B., Livshits, A. B. and Kozlov, A. A., *Int. J. Biol. Macromol.* **29** (2001), 303.

6. Irkaev, S. M., Kupriyanov, V. V. and Semionkin, V. A., British Patent No 10745, 7 May, 1987.

7. Murad, E., Bowen, L. H., Long, G. J. and Quin, T. G., *Clay Minerals* **23** (1988), 161.

8. Oshtrakh, M. I., Milder, O. B., Semionkin, V. A., Prokopenko, P. G., Livshits, A. B., Kozlov, A. A. and Pikulev, A. I., *Z. Naturforsch.* **57a** (2002), 566.

Hyperfine Interactions **156/157**: 285–291, 2004.
© 2004 *Kluwer Academic Publishers. Printed in the Netherlands.*

Low-Spin Ferriheme Models of the Cytochromes: Correlation of Molecular Structure with EPR and Mössbauer Spectral Parameters

T. TESCHNER[1], A. X. TRAUTWEIN[1,*], V. SCHÜNEMANN[1],
L. A. YATSUNYK[2] and F. A. WALKER[2]
[1]*Institute of Physics, University of Lübeck, 23538 Lübeck, Germany*
[2]*Department of Chemistry, University of Arizona, Tucson, AZ 85721-0041, USA*

Abstract. The magnetic Mössbauer spectra of a series of low-spin ferriheme complexes have been investigated and compared with their EPR spectral parameters and molecular structures. To date there has been little systematic analysis of either estimated or fitted values of the hyperfine coupling constants for low-spin ferriheme centers and no meaningful correlation has been established between the Mössbauer parameters and the axial ligands of such species. With the results of the present study, we have been able to find correlations of molecular structures with iron-orbital splittings, g-tensor values derived from EPR signals and magnetic hyperfine interaction components A_{zz} obtained from magnetic Mössbauer spectra. These correlations should be useful to future workers in the field of heme-containing enzymes.

Key words: heme model compounds, cytochromes, molecular structure, magnetic hyperfine interaction, g-tensor, iron d-orbital splitting.

1. Introduction

The biological function of six-coordinated heme centers ranges from participation in electron transfer processes of cytochrome-containing systems in inner mitochondrial membranes (e.g. in cytochrome c552 of *Nitrosomonas europaea*) to enzymatic catalysis like in the case of the cytochromes P450, which play important roles in detoxification in a number of organisms. Bis-histidine-coordinated heme centers are involved in electron transfer in photosynthesis and cell respiration. EPR data for the cytochrome bc_1 complex of mitochondria and the related cytochrome $b_6 f$ complex of chloroplasts show that both of the b hemes of each protein exhibit EPR signals known as the "large g_{max}" or "HALS" type (highly anisotropic low-spin, with the g_{max} value $\geqslant 3.2$). Model systems have greatly aided the correlation of the structure of heme centers with their spectroscopic properties. Walker, Scheidt and their coworkers have shown that the "large g_{max}" signal occurs for ferriheme complexes with $(d_{xy})^2(d_{xz}, d_{yz})^3$ electronic ground state when

---

* Author for correspondence.

the planar axial ligands are aligned in mutually perpendicular planes [1, 2], and likewise a "normal rhombic" signal occurs when the ligands are aligned in parallel planes. The arrangement of the planar axial ligands is of particular importance in defining the spectroscopic properties and may be also in estimating the reduction potentials of heme proteins. Several crystal structures of these protein complexes have such a low resolution that determination of the orientation of the axial ligands is not possible. Therefore low-spin ferriheme model complexes with different types and orientations of axial ligands are useful in correlating structural and spectroscopic parameters in heme proteins such as the electron-transferring cytochromes. In this work, three structurally-characterized model heme complexes [3] with 1-methylimidazole axial ligands have been studied by EPR and Mössbauer spectroscopy.

2. Materials and methods

Crystals of the heme model complexes were grown by liquid diffusion methods as reported previously; [OMTPPFe(1-MeIm)$_2$]$^+$ crystallized in two forms, one with 90° (perpendicular) and the other with 19.5° (nearly parallel) axial ligand dihedral angles [3]. EPR spectra were recorded on a Bruker ESP-300E EPR spectrometer (operating at 9.4 GHz with 100 kHz field modulation) equipped with an Oxford Instruments ESR 900 continuous flow helium cryostat. Spectra were obtained for crystalline samples at 4.2 K. Microwave frequencies were measured using a Systron-Donner frequency counter. Mössbauer spectra were recorded using a conventional spectrometer in the constant-acceleration mode. Isomer shifts are given relative to α-Fe at room temperature. The spectra obtained at 20 mT were measured in a He bath cryostat (Oxford MD 306) equipped with a pair of permanent magnets. For the high-field spectra a cryostat equipped with a superconducting magnet was used (Oxford Instruments). Magnetically split spectra of paramagnetic samples were simulated in the spin-Hamiltonian approximation, otherwise spectra were analysed by least-squares fits using Lorentzian line shapes.

3. Results and discussion

In Figure 1 are shown field-dependent 4.2 K Mössbauer spectra of [OMTPPFe(1-MeIm)$_2$]$^+$, the planar axial ligand planes of which are oriented perpendicular to each other [3]. Based upon the "large g_{max}" ($= g_z$) value of 3.61, as obtained from EPR measurements, the spread of the magnetically split Mössbauer lines can be simulated with the following magnetic hyperfine coupling tensor elements: $A_{zz} = +90.1$ T, $A_{yy} = +25.2$ T, $A_{xx} = -16.9$ T. For the magnetic splitting the effective hyperfine field, which is the sum of the external field and the internal field, must be considered. Nevertheless, the internal field dominates strongly, and therefore the magnetic splitting of this ferric low-spin species is effectively determined by the large and positive A_{zz} component. The physical basis of both the large g_z

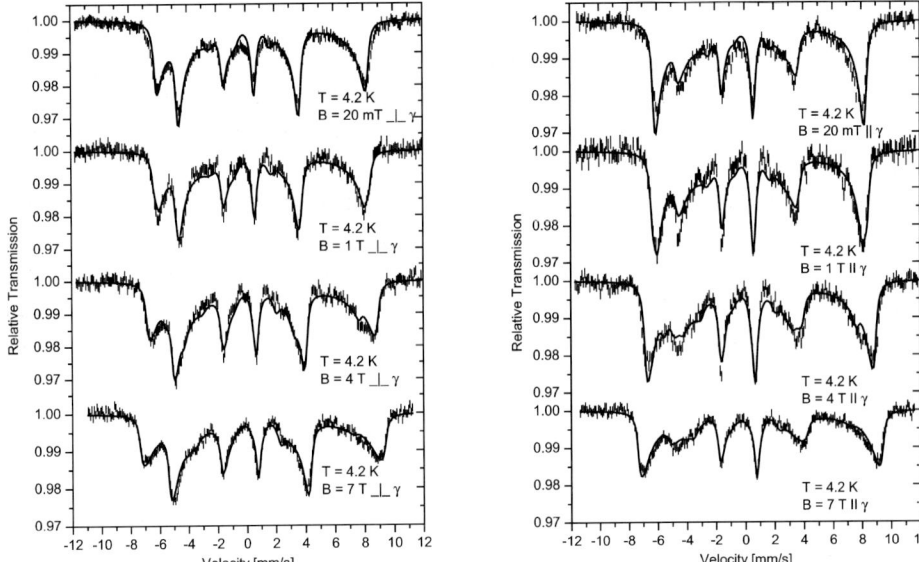

Figure 1. Mössbauer spectra of [OMTPPFe(1-MeIm)$_2$]$^+$ with axial ligand planes rotated by 90° relative to each other [3]. The spectra are simulated in the slow relaxation limit with parameters given in Table I (*g*-values) and in Table II.

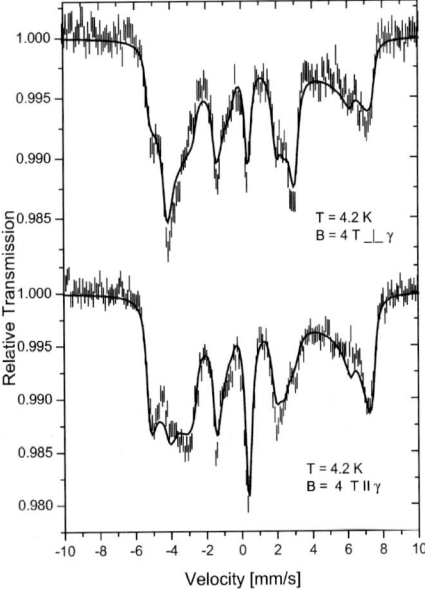

Figure 2. Mössbauer spectra of [OETPPFe(1-MeIm)$_2$]$^+$ with axial ligand planes rotated by 73° relative to each other [3]. The spectra are simulated in the slow relaxation limit with parameters given in Table I (*g*-values) and in Table II.

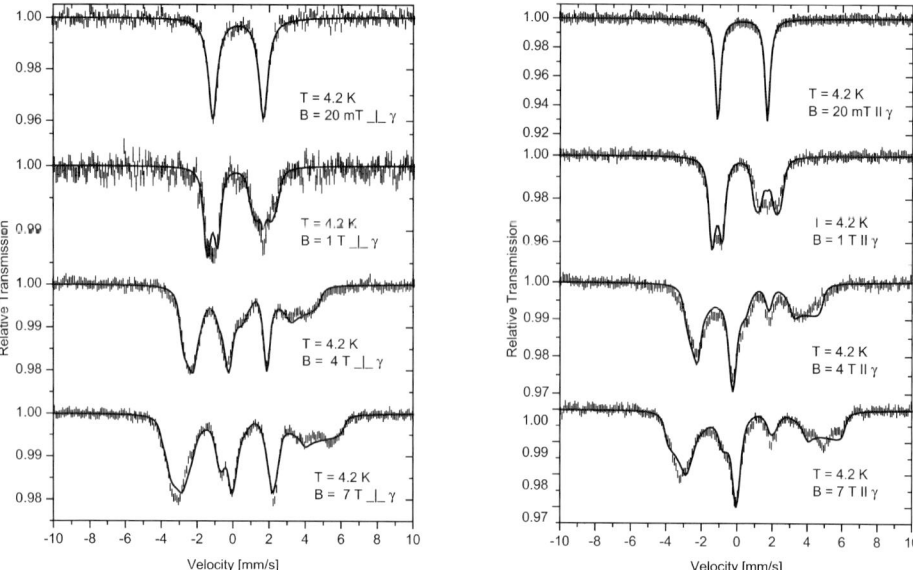

Figure 3. Mössbauer spectra of [OMTPPFe(1-MeIm)$_2$]$^+$ with axial ligand planes rotated by 19.5°
relative to each other [3]. The spectra are simulated in the fast relaxation limit with parameters given
in Table I (g-values) and in Table II.

value and the large and positive A_{zz} component is the near energy degeneracy of
iron d_{xz} and d_{yz} orbitals ($V/\lambda = 0.64$), which in turn is caused by the perpendicular arrangement of the axial ligand planes in [OMTPPFe(1-MeIm)$_2$]$^+$ [3]. As a
consequence of the small orbital splitting, $V = E(d_{yz}) - E(d_{xz})$, strong spin–orbit
coupling within the ferric low-spin $(d_{xy})^2(d_{xz}, d_{yz})^3$ configuration exists, which
yields strong orbital contributions to g_z and A_{zz}. Opposite to the negative Fermi-
contact contribution to the internal field, the orbital contribution to A_{zz} is positive
which, in the present case, acquires the large value of $\sim+90$ T. This qualitative line
of arguments is supported by quantitative calculations on the basis of Oosterhuis
and Lang's [4] crystal-field treatment which correlates g-values, quadrupole splitting ΔE_Q, orbital splittings V/λ and Δ/λ and A-components and which yields the
theoretical value $A_{zz} = +91.9$ T (Table II).

Figure 2 presents field-dependent 4.2 K Mössbauer spectra of [OETPPFe(1-
MeIm)$_2$]$^+$, the planar axial ligand planes of which are oriented nearly perpendicular (73°) to each other [3]. This species falls in the same category as the one
described above, with the only difference being that its orbital contributions to g_z
and A_{zz} are slightly smaller because of its somewhat larger rhombic orbital splitting
V/λ (Table I).

According to the arguments described above, ferric low-spin heme species with
planar axial ligand planes being oriented parallel (0°) or nearly parallel (19.5°) to
each other are expected to exhibit larger rhombic orbital splitting V/λ compared
to species with perpendicular (90°) or nearly perpendicular (73°) orientation of

Table I. g-values and crystal-field parameters for the complexes of this study

Complex	(i)	g-tensor[ii]	V/λ[iii]	Δ/λ[iii]	V/Δ
parallel $[OEPFe(4\text{-}NMe_2Py)_2]^+$	0° [5]	$g_{xx} = 1.58$ $g_{yy} = 2.30$ $g_{zz} = 2.83$	2.15	3.17	0.68
nearly parallel $[OMTPPFe(1\text{-}MeIm)_2]^+$	19.5° [3]	$g_{xx} = 1.54$ $g_{yy} = 2.51$ $g_{zz} = 2.71$	2.44	1.87	1.31
nearly perpendicular $[OETPPFe(1\text{-}MeIm)_2]^+$	73° [3]	$(g_{xx} = 1.14)$ $(g_{yy} = 2.00)$ $g_{zz} = 3.27$	1.16	3.44	0.36
perpendicular $[OMTPPFe(1\text{-}MeIm)_2]^+$	90° [3]	$(g_{xx} = 0.63)$ $(g_{yy} = 1.53)$ $g_{zz} = 3.61$	0.64	3.82	0.17

[i]Dihedral angle between the planes of the axial ligands. Reference to structure given in brackets.
[ii]The g-values obtained result from EPR measurements for the largest g-value and from Mössbauer spectral fits for the two smaller g-values.
[iii]$V/\lambda = E(d_{yz}) - E(d_{xz}) = g_{xx}/(g_{zz} + g_{yy}) + g_{yy}/(g_{zz} - g_{xx})$; $\Delta/\lambda = E(d_{yz}) - E(d_{xy}) - 1/2V/\lambda = g_{xx}/(g_{zz} + g_{yy}) + g_{zz}/(g_{yy} - g_{xx}) - 1/2V/\lambda$.

axial ligand planes. This is indeed the case, as shown in Table I. Consequently, larger V/λ values correlate with smaller orbital contributions to g_z and A_{zz} (Tables I and II) and therefore also to considerably smaller overall magnetic hyperfine splitting. The field-dependent 4.2 K Mössbauer spectra (Figure 3) of $[OMTPPFe(1\text{-}MeIm)_2]^+$, which has axial ligand planes being rotated by 19.5° to each other, indeed proves that this is the case.

4. Conclusions

We have shown that investigations of the molecular structures, EPR, and Mössbauer spectra of well-defined low-spin ferriheme model compounds have provided conclusive proof that perpendicular or near perpendicular alignment of planar axial ligands correlates with near-degeneracy of iron d_{xz} and d_{yz} orbitals and with large orbital contributions to g_z and A_{zz}. Hence, these species are characterized by $g_z \geqslant 3.2$ and $A_{zz} \sim 70\text{--}100$ T. Likewise, species with parallel or near parallel orientation of planar axial ligands are characterized by considerably smaller orbital contributions to g_z and A_{zz}. The sequence of increasing dihedral angles (0°, 19.5°, 73°, 90°) of axial ligand planes (1-methylimidazole), including literature data for $[OEPFe(4\text{-}NMe_2Py)_2]^+$, which has different planar axial ligands (4-NMe_2Py) [5],

Table II. Mössbauer data obtained from the simulation of the Mössbauer spectra for the complexes of this study

Complex	(i)	δ [mm/s]	ΔE_Q [mm/s]	(ii)	η	$A/g_N\mu_N$ [T]
parallel [OEPFe(4-NMe$_2$Py)$_2$]$^+$	0° [5]	–	2.15	–	−1.83	−41.6 17.7 44.6
nearly parallel [OMTPPFe(1-MeIm)$_2$]$^+$	19.5° [3]	0.28 ± 0.01	2.80 ± 0.01 (2.37)[iii]	–	−2.03 ± 0.10 (−1.46)[iii]	−42.2 ± 1.5 (−42.0)[iii] 23.6 ± 0.5 (28.4)[iii] 50.2 ± 0.5 (38.2)[iii]
nearly perpendicular [OETPPFe(1-MeIm)$_2$]$^+$	73° [3]	0.26 ± 0.01	1.94 ± 0.01 (2.19)[iii]	–	−0.94 ± 0.05 (−1.07)[iii]	−37.6 ± 2.0 (−35.8)[iii] 2.4 ± 1.0 (10.0)[iii] 71.4 ± 0.5 (71.4)[iii]
perpendicular [OMTPPFe(1-MeIm)$_2$]$^+$	90° [3]	0.27 ± 0.02	1.70 ± 0.02 (1.98)[iii]	$\alpha = 43°$	−0.62 ± 0.10 (−0.62)[iii]	−17.0 ± 75.0 (−29.5)[iii] 25.2 ± 15.0 (14.9)[iii] 90.1 ± 0.5 (91.9)[iii]

(i) Dihedral angle between the planes of the axial ligands. Reference to structure given in brackets.
(ii) Euler angles between the principle axes systems of the g- and efg-tensor.
(iii) Values in brackets are theoretical values obtained from the crystal-field model of Oosterhuis und Lang [4].

is directly reflected by the sequence of iron orbital splitting values, V/λ (2.15, 2.44, 1.16, 0.64), g_z-values (2.83, 2.71, 3.27, 3.61) and A_{zz}-values (+44.6 T, +50.2 T, +71.4 T, +90.1 T). Thus, we are able to correlate for the first time a series of ferriheme complexes of known molecular structures [3, 5] simultaneously with both EPR and Mössbauer data.

Acknowledgement

FAW acknowledges the support by the Alexander von Humboldt-Stiftung.

References

1. Walker, F. A., Huynh, B. H., Scheidt, W. R. and Osvath, S. R., *J. Am. Chem. Soc.* **108** (1986), 5288.
2. Scheidt, W. R., Kirner, J. F., Hoard, J. L. and Reed, C. A., *J. Am. Chem. Soc.* **109** (1987), 1963.
3. Yatsunyk, L., Carducci, M. D. Walker, F. A., *J. Am. Chem. Soc.* **125** (2003), 15986.
4. Oosterhuis, W. T. and Lang, G., *Phys. Rev.* **178**(2) (1969), 439.
5. Safo, M. K., Gupta, G. P., Walker, F. A. and Scheidt, W. R., *J. Am. Chem. Soc.* **113** (1991), 5497.

Hyperfine Interactions **156/157**: 293–298, 2004.
© 2004 *Kluwer Academic Publishers. Printed in the Netherlands.*

Iron–Sulfur Proteins Investigated by EPR-, Mössbauer- and EXAFS-Spectroscopy

P. WEGNER[1], M. BEVER[1], V. SCHÜNEMANN[1], A. X. TRAUTWEIN[1], C. SCHMIDT[2], H. BÖNISCH[3], M. GNIDA[4] and W. MEYER-KLAUCKE[4]
[1]*Institute of Physics, University of Lübeck, 23538 Lübeck, Germany*
[2]*Institute of Biochemistry, University of Lübeck, 23538 Lübeck, Germany*
[3]*Karolinska Institutet, Dept. of Biosciences at NOVUM, Center for Structural Biochemistry, S-14157 Huddige, Sweden*
[4]*EMBL Outstation Hamburg, DESY, 22603 Hamburg, Germany*

Abstract. The structural and spectroscopic properties of the biologically active [Fe–4S] site of three different mutants of the wild-type rubredoxin from the archaeon *Pyrococcus abyssi* were investigated and compared with each other and additionally with those of the rubredoxin from the bacterium *Clostridium pasteurianum*.

Key words: iron–sulfur-proteins, rubredoxins, Mössbauer-spectroscopy, EXAFS, DET-calculations.

1. Introduction

Rubredoxins are small proteins ($\sim$6 kDa) that contain a mononuclear [Fe–4S] site. Their biochemical function is the electron transfer; hence the iron can acquire two different oxidation states: the oxidized (Fe(III) $S = 5/2$) and the reduced state (Fe(II) $S = 2$) [1]. The aim of this investigation was to study three different mutants of the wild-type rubredoxin from the archaeon *Pyrococcus abyssi* and to correlate spectroscopic and structural information about their biologically active [Fe–4S] sites with those of the rubredoxin from the bacterium *Clostridium pasteurianum*.

2. Materials and methods

The proteins studied here were expressed in *E. coli* cells. *Pyrococcus abyssi* occurs in hyperthermal vents (black smokers) [2]. Its optimal growth temperature is $\sim$97°C at pH 5.5 under anaerobic conditions [3]. Three different rubredoxin mutants (Rm) have been prepared:

Protein Rm 2-4 is a mutant in which two amino acids (Trp3 and Arg4) have been replaced by Leu3 and Ser4 as compared to the actual wild-type found in *P. abyssi*. The exchange of the two amino acids does not affect the geometry of the iron sphere at the active site. In the mutant Rm A44S the amino acid Ala44 has been exchanged by Ser44. This mutation causes an additional OH-bond between

Ser44 and the iron-coordinated sulfur of Cys42. In the mutant Rm 2-13 the amino acid Lys7 which is located next to the iron-coordinated Cys6 has been deleted.

Mössbauer spectra were recorded using a conventional spectrometer in the constant-acceleration mode. Isomer shifts are given relative to α-Fe at room temperature. The spectra obtained at 20 mT (perpendicular to the γ-beam) were measured in a He-bath cryostat (Oxford MD 306), equipped with a pair of permanent magnets. For the high field spectra up to 7 T a cryostat equipped with a superconducting magnet was used (Oxford Instruments). The obtained spectra were analyzed by least-square fits using Lorentzian line shapes, while the magnetically split spectra of the paramagnetic samples were simulated within the nuclear Hamilton formalism [4]:

$$\hat{H} = \frac{eQV_{zz}}{4I(2I-1)}[3\hat{I}_z^2 - I(I+1) + \eta(\hat{I}_x^2 - \hat{I}_y^2)] - g_N\mu_N\vec{I}\cdot\vec{B} + \langle\vec{S}\rangle\cdot\vec{A}\cdot\vec{I}.$$

I denotes the spin of the nuclear ground or excited state, Q the nuclear quadrupole moment of the nuclear excited state, V_{zz} the z-component of the electric-field-gradient (efg), η the asymmetry parameter of the efg, g_N the nuclear g factor and μ_N the nuclear magneton.

EPR spectra were recorded on a Bruker ER200D-SRC spectrometer equipped with an Oxford Instruments liquid-helium cryostat (ESR 910).

EXAFS spectra were recorded at the EMBL beamline D2 (DESY, Hamburg) in fluorescence mode.

3. Results

For the set of oxidized proteins the parameters given in Table I were extracted from field-dependent Mössbauer spectra recorded at $T = 4.2$ K by means of the spin-Hamiltonian formalism for a single iron center in the Fe(III) $S = 5/2$ state. The variation of parameters was carried out by simultaneous simulation of the spectra shown in Figure 1 and of spectra recorded at 4.2 K in a field of 20 mT (data not shown). The corresponding procedure was applied to the spectra obtained for oxidized Rm 2-13 and Rm A44S (not shown) and also to the spectra of the reduced rubredoxins, i.e. Rm 2-4 (Figure 2) as well as Rm 2-13 and Rm A44S (not shown). The obtained parameter sets are summarized and compared with the parameter set for rubredoxin from *Clostridium pasteurianum* [5] for the oxidized and reduced proteins in Table I and Table II, respectively.

EPR spectra from oxidized proteins Rm 2-4 and Rm 2-13 (not shown) have been recorded at 13 K and 8 K, respectively. The spectrum of Rm 2-4 exhibits a broad distribution of effective g-values around 4.3, arising from the $S_z = \pm 3/2$ Kramers doublet of the Fe(III) $S = 5/2$ state ($g_x = 4.28$, $g_y = 4.01$, $g_z = 4.53$). This spectral pattern yields the rhombicity parameter $E/D = 0.28$ which is in reasonable agreement with the corresponding parameter from the Mössbauer results of this species. The spectrum recorded for Rm 2-13 exhibits a sharp peak at the

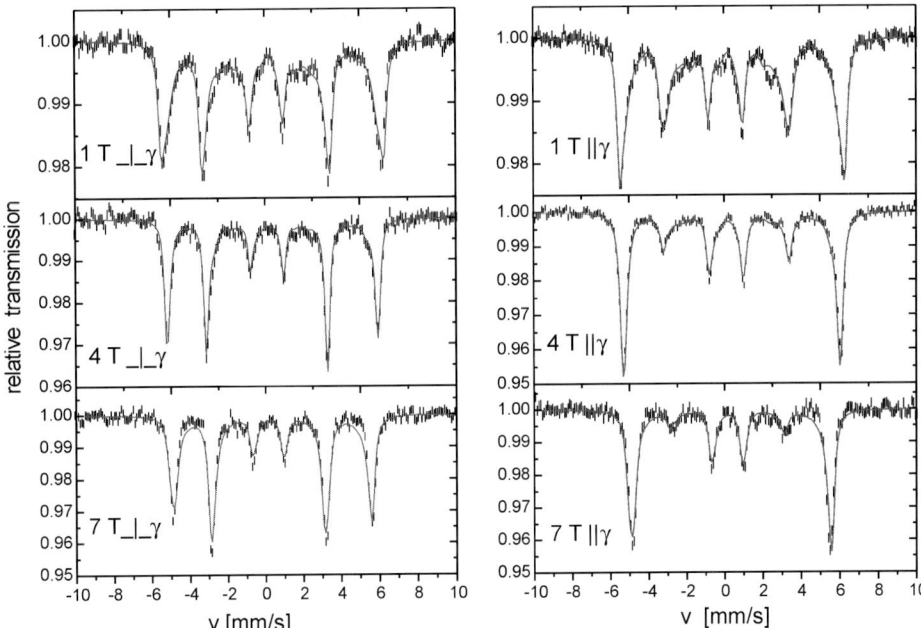

Figure 1. Mössbauer spectra obtained for oxidized Rm 2-4 at 4.2 K in applied fields as indicated. The spectra are simulated in the slow relaxation limit with parameters given in Table I.

Table I. Mössbauer parameters obtained from the simulations for the oxidized set of rubredoxins from the archeon *P. abyssi* compared to the parameters for rubredoxin from *C. pasteurianum* [5]. The value for the isomer-shift has recently been remeasured by Yoo et al. [9]

	Rm 2-4	Rm 2-13	Rm A44S	Ref. [5, 9]
δ [mm/s]	0.24 ± 0.01	0.23 ± 0.01	0.24 ± 0.01	0.24 ± 0.01
ΔE_Q [mm/s]	-0.36 ± 0.04	-0.49 ± 0.04	-0.43 ± 0.04	-0.50 ± 0.05
Γ [mm/s]	0.33 ± 0.02	$0.31 + 0.02$	0.35 ± 0.02	0.35
$g_x = g_x = g_y$	2	2	2	2
$A_{xx}/g_N \mu_N$ [T]	-17 ± 2	-16 ± 1	-16 ± 1	-16 ± 1
$A_{yy}/g_N \mu_N$ [T]	-15.7 ± 0.7	-15.7 ± 0.4	-15.7 ± 0.1	-15.9 ± 0.3
$A_{zz}/g_N \mu_N$ [T]	-16 ± 4	-17 ± 2	-17 ± 2	-16.9 ± 0.3
η	0.8 ± 0.2	0.6 ± 0.2	0.5 ± 0.2	0.2 ± 0.1
D [cm^{-1}]	2.13 ± 0.60	1.55 ± 0.50	1.4 ± 0.6	1.9 ± 0.3
E/D	0.23 ± 0.03	0.33 ± 0.01	0.26 ± 0.02	0.23 ± 0.02

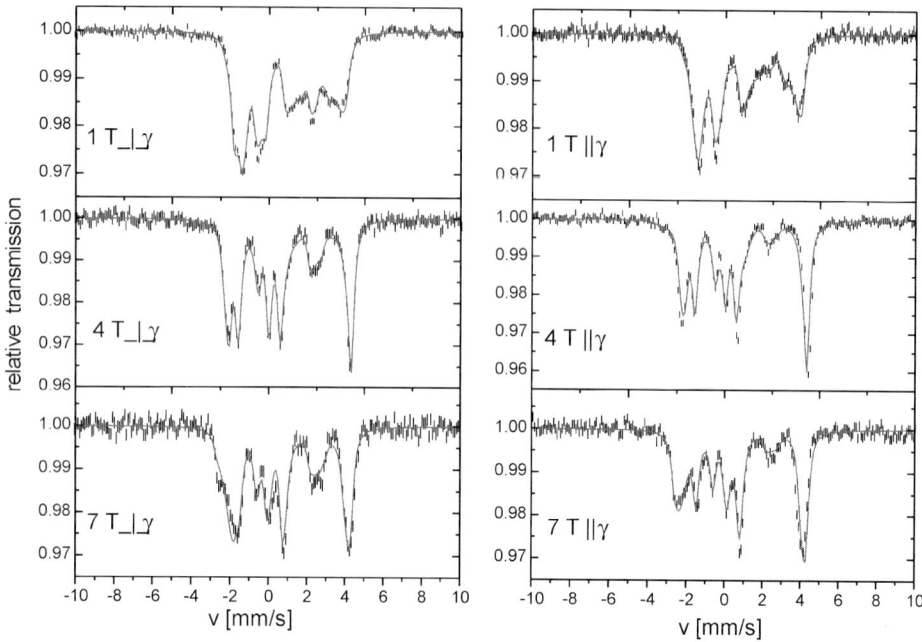

Figure 2. Mössbauer spectra obtained for reduced Rm 2-4 at 4.2 K in applied fields as indicated. The spectra are simulated in the slow relaxation limit with parameters given in Table II.

Table II. Mössbauer parameters obtained from the simulations for the reduced set of rubredoxins from the archeon *P. abyssi* compared to the parameters for rubredoxin from *C. pasteurianum* [5]. In this early work by Schulz and Debrunner only uncertainties for δ, ΔE_Q and η are given

	Rm 2-4	Rm 2-13	Rm A44S	Ref. [5]
δ [mm/s]	0.70 ± 0.01	0.69 ± 0.01	0.70 ± 0.01	0.7 ± 0.02
ΔE_Q [mm/s]	-3.25 ± 0.01	-3.14 ± 0.01	-3.23 ± 0.01	-3.25 ± 0.01
Γ [mm/s]	0.32 ± 0.01	0.33 ± 0.01	0.33 ± 0.01	0.35
g_x, g_y, g_z	2.11, 2.19, 2	2.11, 2.19, 2	2.11, 2.19, 2	2.11, 2.19, 2
$A_{xx}/g_N\mu_N$ [T]	-14.5 ± 0.5	-12.6 ± 0.5	-14.5 ± 0.5	-20.1
$A_{yy}/g_N\mu_N$ [T]	-9.2 ± 0.3	-9.7 ± 0.4	-8.5 ± 0.3	-8.3
$A_{zz}/g_N\mu_N$ [T]	-28 ± 5	-25 ± 4	-26 ± 4	-30.1
η	0.74 ± 0.10	0.40 ± 0.10	0.68 ± 0.10	0.65 ± 0.1
D [cm^{-1}]	7.2 ± 0.5	5.9 ± 0.5	5.6 ± 0.5	7.6
E/D	0.16 ± 0.02	0.09 ± 0.02	0.23 ± 0.02	0.28

Table III. EXAFS-parameters extracted from the simulations of the EXAFS-spectra. N indicates the number of backscattering sulfur atoms per Fe-atom, R is the averaged Fe–S distance, $2\sigma^2$ is the Debye–Waller parameter and E_0 is the energy correction. (Due to limited beamtime the oxidized mutant Rm A44S could not be investigated)

	Rm 2-4 red.	Rm 2-13 red.	Rm A44S red.	Rm 2-4 ox.	Rm 2-13 ox.
N	4	4	4	4	4
$2\sigma^2$ [Å^2]	0.007	0.006	0.007	0.008	0.007
E_0 [eV]	−14.1	−13.1	−14.2	−13.4	−14.4
R [Å]	2.33	2.31	2.32	2.28	2.28

effective g-value of $g_x = g_y = g_z = 4.3$ yielding $E/D = 0.33$ which is in exact correspondence with E/D as derived from the Mössbauer spectrum.

The aim of the EXAFS investigation was to determine and compare Fe–S distances. Data reduction such as normalization of the measured spectra and extraction of the EXAFS spectra followed procedures described previously [6]. The simulation with EXCURV98 [7] yielded the average Fe–S distances as summarized in Table III. Within the experimental error the reduced rubredoxins show the same average Fe–S distance of 2.32 Å and the two oxidized rubredoxins have the same average Fe–S distance of 2.28 Å.

4. Discussion

The average Fe–S distances of the investigated mutants of rubredoxin from the archaeon *P. abyssi* are found to be practically the same, i.e. 2.28 Å for the oxidized form and 2.32 Å for the reduced form, though at least two of the mutations (Rm 2-13 and Rm A44S) cause structural changes in close vicinity of the [Fe–4S] site of the proteins. From this finding it is expected that the gross pattern of Mössbauer- and EPR-spectral properties is the same for the three mutants. The differences of the rhombicity parameters E/D of Rm 2-13 as compared to Rm 2-4 and Rm A44S (both in the oxidized and reduced form) could be attributed to small changes in bond angles $S_{i-1,2,3}$–Fe–S_{Cys6} in Rm 2-13, the mutation of which is located next to the iron-coordinated Cys6. Preliminary X-ray structures (1.6 Å resolution, H. Bönisch *et al.*, unpublished) confirm that the oxidized forms of the mutants Rm 2-4 and Rm A44S indeed do not exhibit any structural geometric variation of the [Fe–4S] site with respect to each other. Corresponding X-ray studies of Rm 2-13 are in progress.

Comparing the Mössbauer parameters of the three rubredoxins under study here with those of the rubredoxin from *C. pasteurianum* [5, 9] we notice that the oxidized forms of the former exhibit similar isomer shifts and quadrupole splittings compared to the latter. This finding is in agreement with the result that the average

Fe–S distances of both the oxidized rubredoxins studied here and the oxidized rubredoxin from *C. pasteurianum* [8] are practically the same.

However, preliminary X-ray structures of oxidized Rm 2-4 and Rm A44S indicate that three Fe–S bonds, out of four, are significantly shorter while the fourth is considerably longer than the corresponding bonds in oxidized rubredoxin from *C. pasteurianum* [8]. Our plan for the near future is to perform DFT calculations on the basis of resolved X-ray structures (rather than on the basis of averaged Fe–S bonds) of the [Fe–4S] moiety of the various rubredoxins to quantitatively correlate structural and spectroscopic properties of these proteins in their oxidized and reduced state.

References

1. Meyer, J. and Moulis, J., In: A. Messerschmidt, R. Huber, T. Poulos and K. Wieghardt (eds), *Handbook of Metalloproteins*, John Wiley and Sons, Chichester, 2001, p. 505.
2. Godfroy, A., Raven, N. and Sharp, R. J., *FEMS Microbiol. Lett.* **186** (2000), 127.
3. Vielle, C. and Zeikus, G. J., *Microbiol. Mol. Biol. Rev.* **65** (2001), 1.
4. Schünemann, V. and Winkler, H., *Rep. Prog. Phys.* **63** (2000), 263.
5. Schulz, C. E. and Debrunner, P. G., *J. Physique Colloq.* **37** (1976), 153.
6. Meyer-Klaucke, W., Winkler, H., Schünemann, V., Trautwein, A. X., Nolting, H. F. and Haavik, J., *Eur. J. Biochem.* **241** (1996), 432.
7. Binsted, N., Strange, R. W. and Hasnain, S. S., *Biochemistry* **31** (1992), 12117.
8. Dauter, Z., Wilson, K. S., Sieker, L., Moulis, J. and Meyer, J., *Proc. Natl. Acad. Sci.* **93** (1996), 8836.
9. Yoo, S. J., Meyer, J., Achim, C., Petersen, J., Hendrich, M. P. and Münck, E., *JBIC* **5** (2000), 476.

Hyperfine Interactions **156/157**: 299–303, 2004.
© 2004 *Kluwer Academic Publishers. Printed in the Netherlands.*

Mössbauer Study of Lanthanum–Strontium Ferromanganite Oxides

M. ABDELMOULA[1], M. PETITJEAN[2], G. CABOCHE[2], J.-M. GÉNIN[1] and L. C. DUFOUR[2]

[1]*Laboratoire de Chimie Physique et Microbiologique pour l'Environnement (LCPME),
UMR 7564 CNRS UHP Nancy I, 405 rue de Vandœuvre, 54600 Villers-les-Nancy, France;
e-mail: abdelmou@lcpe.cnrs-nancy.fr*
[2]*Laboratoire de Recherches sur la Réactivité des Solides (LRRS), UMR 5613 CNRS-Université de
Bourgogne, 9, avenue Alain Savary, BP47870 21078 Dijon, France*

Abstract. The $La_{0.8}Sr_{0.2}Mn_{(1-y)}Fe_yO_{(3\pm\delta)}$ (LSMF with $y = 0, 0.2, 0.5, 0.8, 1$) compounds are prospective cathode materials for advanced solid oxide fuel cells (SOFC) application operating at $700°C$. Usual analysis methods like thermogravimetric analysis or redox titration enable to determine the average oxidation state of both manganese and iron cations. The comparative role of iron and manganese in B-site was evaluated by ^{57}Fe Mössbauer spectroscopy. Spectra revealed that the complete substitution of iron for manganese induces the formation of Fe^{5+} for the compound with $y = 1$. However, no tetravalent iron cation was observed in air for the LSMF compounds with $y = 0.2, 0.5$ and 0.8. This means that only manganese cations are electronically active in the bulk with valence states Mn^{3+} and Mn^{4+}. Then, when the iron content increases, the concentration in oxygen vacancies increases also facilitating anionic diffusion. The charge disproportionation Fe^{3+}/Fe^{5+} can improve the electrical properties for the compound with $y = 1$.

Key words: perovskite, lanthanum–strontium ferromanganites, Mössbauer spectroscopy.

1. Introduction and experimental

$[La_{0.8}Sr_{0.2}][Mn_{(1-y)}Fe_y]O_{3\pm\delta}$ perovskites are of great interest as cathode materials for developing advanced solid oxide fuel cells (SOFCs) operating at $700°C$ [1]. A previous study showed a thermomechanical and structural consistency of LSMF with electrolyte YSZ having a good electrical and ionic conductivity [2]. Several outstanding bulk and surface properties (metal–insulator transition, high magneto-resistance or catalytic activity) are obtained by forming high valence B-cation within the perovskite ABO_3 structure of both Mn- and Co-based oxides, i.e., Mn^{4+} and Co^{4+} [3, 4]. For instance, the presence of $Mn^{3+}-O-Mn^{4+}$ units gives rise to ferromagnetism and semiconductor–metal transition due to a double-exchange mechanism [5]. Thus, one excepts that the substitution of other transition metal ions, such as Fe for the Mn, can lead to the formation of Fe^{4+} in the solid phase leading to important modifications in the magnetic, transport and electrocatalytic properties. For instance, Fe is in Fe^{4+} state, more precisely a mixed Fe^{3+}/Fe^{5+} state in $CaFeO_3$ [6] and $SrFeO_3$ [7].

This study aims at assessing the role of Mn and Fe in the working conditions of SOFC. The Fe state is studied by Mössbauer spectroscopy while the structural properties and off-stoichiometry oxygen is checked by XRD and thermogravimetric analysis, respectively. The Fe state determined at low temperature is compared with that at high temperature. Mössbauer data were also collected to correlate magnetic behaviour and structural changes. Spectra were recorded between 15 and 150 K with a constant acceleration Mössbauer spectrometer using a $^{57}Co/Rh$ source and fitted with Lorentzian-shaped lines. $(La_{0.8}Sr_{0.2})(Mn_{1-y}Fe_y)O_{3\pm\delta}$ samples ($y = 0.2, 0.5, 0.8, 1$) were synthesised by the glycine-nitrate process equilibrated towards oxygen by successive powder grindings and annealing [8]. Some were dry-pressed and sintered into disc-shaped specimens at $1400°C$ for 4 hours and pulverised and sintered compounds were heated at $800°C$ for 60 hours before quenching in the air.

2. Structural properties and off-stoichiometry oxygen in SOFC working conditions

Structural properties were established by XRD from powdered samples analysis in the range RT–700°C in the air. Rhombohedral symmetry is kept in the whole range for $y = 0.2$. Two symmetries, rhombohedral and orthorhombic, are present at ambient that becomes only rhombohedral above 80 and 160°C for $y = 0.5$ and 0.8, respectively. An orthorhombic-rhombohedral transition occurs approximately 300°C for $y = 1$. Thus, adding iron shifts the structural transition temperature to higher temperatures. No measurement was achieved below ambient but an extrapolation is correlated to Mössbauer analysis for $y = 0.2$. Oxygen composition changes were determined from pulverised samples by thermogravimetry under 200 mbars of oxygen between ambient and 700°C. The results show that the oxygen stoichiometry is very little modified in air between RT and 700°C for all compounds. That means that the state of oxidation of any cation is independent of temperature in these conditions. Thus, the valence state of Fe at 700°C is the same as that determined at low temperature by Mössbauer spectroscopy if no charge transfer between Mn and Fe is assumed.

3. Oxidation state of active cations and magnetic behaviour

The chemical formula of the compounds was checked by ICP-AES analysis of the metallic elements. The amount of B-site cation, $Mn^{n+}+Fe^{n+}$ with $n > 3$, and, therefore, the oxygen content were determined by Mohr salt titration. The results imply the existence of vacancies which are rather cationic for iron-poor compounds and anionic for iron-rich ones.

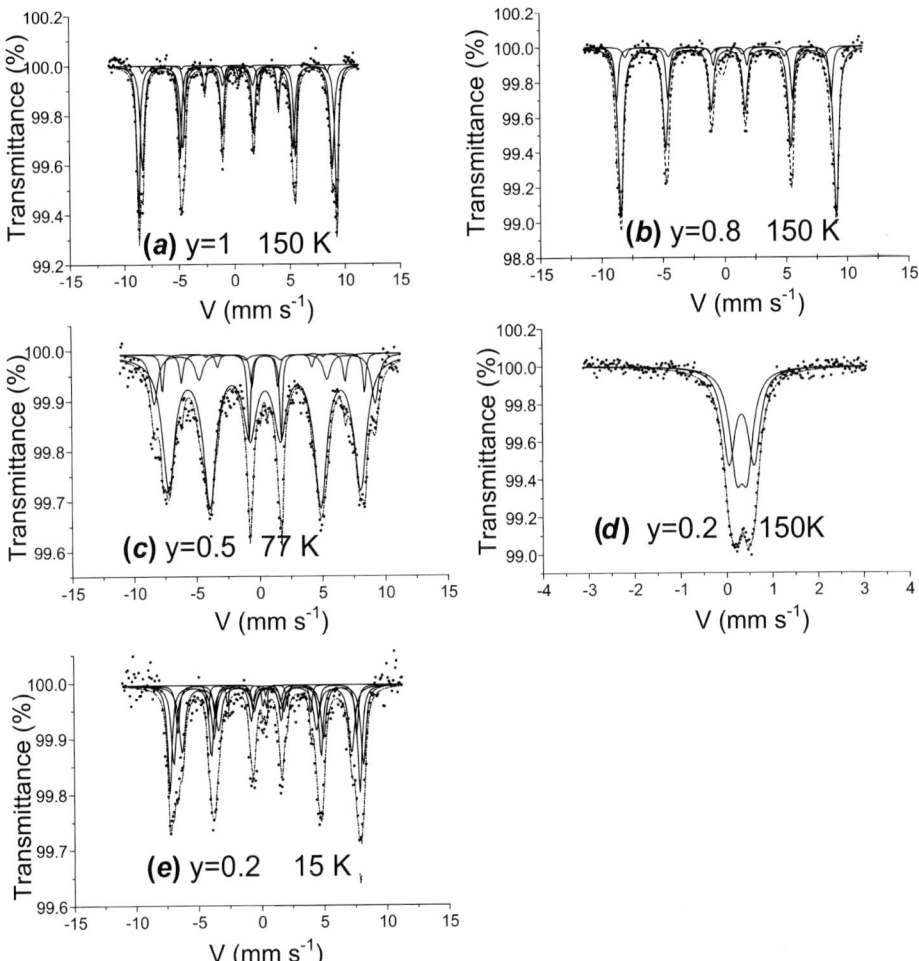

Figure 1. Mössbauer spectra of LSMF at various compositions with (a) $y = 1$, (b) 0.8, (c) 0.5 and (d)–(e) 0.2.

The study was completed by Mössbauer analysis yielding information about the Fe oxidation state. In both samples $y = 0.8$ and $y = 1$, the spectra measured at 150 K, gave rise to a magnetic ordering and exhibited a set of hyperfine split sextets. For $y = 0.8$, the fitting can be described as a superimposition of three sextets. The isomer shifts *IS* vary between 0.26 and 0.4 mm/s while the magnetic fields vary between 505 and 545 kOe. These values are ascribable to Fe^{3+}. Furthermore, according to these *IS* values, Fe^{3+} cations are located in distorted octahedral sites. This arises from the anisotropic deformation of the environment of the Fe^{3+} ions due to oxygen vacancies [9] and correlates with the chemical titration.

For the sample with $y = 1$, the same components were observed with an additional sextet whose hyperfine parameters correspond to Fe^{5+} ($IS = -0.08$ mm/s

Table I. Mössbauer parameters corresponding to spectra of Figure 1

y (Fe)	T (K)		IS (mm s^{-1})	H (kOe)	Δ or ε (mm s^{-1})	Abundance (%)
0.2	15		0.27		0.29	1.5
		Fe^{3+}	0.48	441	0.02	11
		Fe^{3+}	0.49	418	−0.02	26.5
		Fe^{3+}	0.61	468	−0.004	26
		Fe^{3+}	0.35	469	−0.05	30
		Fe^{5+}	−0.20	254	0.09	4
0.2	150	Fe^{3+}	0.45		0.54	51
		Fe^{3+}	0.45		0.23	49
0.5	77	Fe^{3+}	0.45	478	−0.02	79
		Fe^{3+}	0.44	498	−0.05	6
		Fe^{3+}	0.41	545	0.05	10
		Fe^{3+}	0.44	403	−0.05	5
0.8	150		−0.05		0.26	1
		Fe^{3+}	0.45	545	0.02	74
		Fe^{3+}	0.26	505	0.05	6
		Fe^{3+}	0.36	542	−0.5	19
1	150	Fe^{3+}	0.46	558	−0.05	48
		Fe^{3+}	0.44	535	−0.016	40
		Fe^{3+}	0.26	524	−0.05	4
		Fe^{5+}	−0.08	261	0	8

and $H = 261$ kOe). In fact, there is a low Fe^{3+}/Fe^{5+} charge disproportionation, as already reported in [10, 11].

For $y = 0.5$, the spectrum at 150 K evidenced the coexistence of paramagnetic and magnetically ordered phases. The spectrum revealed at 77 K signals only due to magnetic phases with H ranging between 403 and 498 kOe and $IS = 0.4$ mm s^{-1} characterising Fe^{3+}.

For the iron poor composition, i.e., $y = 0.2$, the spectrum at 150 K consisted of a paramagnetic phase. It was analysed by two quadrupole doublets with IS of ≈ 0.45 mm s^{-1} (Fe^{3+}) having a large quadrupole splitting of $\varepsilon \approx 0.54$ mm s^{-1} and a small one ≈ 0.23 mm s^{-1}. The latter doublet corresponds most probably to Fe^{3+} in an octahedral oxygen lattice. Another spectrum for this composition ($y = 0.2$) at 15 K was collected to promote magnetic ordering: it reveals four sextets with the values of H within the range 418–469 kOe such as those found for $y = 0.5$ at 77 K and typical for high-spin Fe^{3+}. In addition, a fifth component was observed at low intensity characterised by IS and H values (respectively –0.2 mm/s and 254 kOe) lower than Fe^{3+}. This sextet corresponds probably to Fe^{5+}. Furthermore, these Mössbauer data seem to correlate the magnetic order change between 150 and 15 K with the structural changes observed by XRD. Indeed the change in structural symmetry is probably due to this magnetic-paramagnetic transition [12]. The Mössbauer parameters are listed in Table I.

4. Conclusion

The Mössbauer spectra for all $[La_{0.8}Sr_{0.2}][Mn_{(1-y)}Fe_y]O_{(3\pm\delta)}$ compounds exhibit a doublet or a sextet resulting from Fe^{3+} ions. For iron rich composition, $y = 0.8$ and 1, Fe^{3+} cations are located in distorted octahedral sites as a result of the anisotropic distorsion due to oxygen vacancies [10]. Mössbauer spectroscopy reveals also that the complete substitution of Mn^{3+} by Fe^{3+} induces the formation of Fe^{5+} for the compound with $y = 1$. However, for the LSMF compounds with $y = 0.2, 0.5$ and 0.8, no tetravalent iron cation exists in the air and only the Mn^{3+}/Mn^{4+} couples are electronically active in the bulk. In other words, the redox stability of Fe^{3+} is higher than that Mn^{3+}. As a consequence, doping with Fe reduces the Mn^{3+}/Mn^{4+} ratio and decreases the number of available hopping sites. The double exchange is suppressed, limiting both ferromagnetic exchange and metallic conduction. Thus, when considering the bulk properties, the LSMF compounds with $y \neq 1$ does not seem to be a good material for electrochemical applications such as SOFC electrodes. However, a comparison with surface properties determined by XPS has shown that iron plays a key role in the oxygen surface reduction [13].

References

1. Simner, S. P., Bonnett, J. F., Canfield, N. L., Meinhardt, K. D., Shelton, J. P., Sprenkle, V. L. and Stevenson, J. W., *J. Power Sources* **113** (2003), 1.
2. Steele, B. C. H., Carter, S., Kajda, Kontoulis, I. and Kilner, J. A., In: F. Grosz, P. Zeglers, C. C. Singhal and O. Yamamoto (eds.), *Proc. 2nd Int. Symp. on SOFC*, 1991, pp. 517.
3. Meadowcroft, D. B., *Nature* **226** (1970), 847.
4. Seiyama, T., Yamasoe, N. and Eguchi, K., *Ind. Eng. Chem.* **24** (1985), 19.
5. Zener, C., *Phys. Rev.* **82** (1951), 403.
6. Kuzushita, K., Morimoto, S., Nasu, S. and Nakamura, S., *J. Phys. Soc. Japan* **69** (2000), 2767.
7. Kawasaki, S., Takano, M. and Takeda, Y., *J. Solid State Chem.* **121** (1996), 174.
8. Morin, F., In: P. Stevens (ed.), *Proceedings of the 3rd European SOFC Forum*, 1998, p. 193.
9. Figueiredo, F. M., Waerenborgh, J., Kharton, V. V. and Näfe, H., *Solid State Ionics* **156** (2003), 371.
10. Dann, S. E., Currie, D. B., Weller, M. T., Thomas, M. F. and Al-Rawwas, A. D., *J. Solid State Chem.* **109** (1994), 134.
11. Leontiou, A. A., Ladavos, A. K., Bakas, T. V., Vaimakis, T. C. and Pomonis, P. J., *Appl. Catal. A-Gen.* **241** (2003), 143.
12. Abdelmoula, N., Guidera, K., Dhahri, E., and Joubert, J. C., *J. Solid State Chem.* **151** (2000), 139.
13. Petitjean, M., Caboche, G. and Dufour, L.-C., *Ann. Chim. – Sci. Mat.*, accepted.

Hyperfine Interactions **156/157**: 305–309, 2004.
© 2004 *Kluwer Academic Publishers. Printed in the Netherlands.*

Iron-57 Mössbauer Spectroscopic Investigation of Manganese-Doped γ-Fe$_2$O$_3$

FRANK J. BERRY[1], ÖRN HELGASON[2] and J. W. FRED MOSSELMANS[3]

[1] *Department of Chemistry, The Open University, Walton Hall, Milton Keynes MK7 6AA, United Kingdom*
[2] *Science Institute, University of Iceland, IS-107 Reykjavik, Iceland*
[3] *Daresbury Laboratory, Daresbury, Warrington WA4 4AD, United Kingdom*

Abstract. Manganese-doped γ-Fe$_2$O$_3$ has been prepared by precipitation techniques and shown by Mn K-edge XANES and EXAFS to contain Mn^{3+} in the octahedral sites of the spinel-related structure. The ^{57}Fe Mössbauer spectra recorded *in situ* between 295 and 750 K show that conversion of the spinel-related γ-Fe$_2$O$_3$ -to the corundum-related α-Fe$_2$O$_3$-structure occurs at *ca.* 700 and 730 K in the samples containing *ca.* 1.5 and 4.1% manganese, respectively. The presence of manganese therefore stabilises the γ-Fe$_2$O$_3$-related structure relative to conversion to the α-Fe$_2$O$_3$ phase. The temperature dependence of the spectra recorded from the manganese-doped α-Fe$_2$O$_3$ is similar to that of pure α-Fe$_2$O$_3$ but with a smaller hyperfine magnetic field reflecting the presence of the manganese dopant.

Key words: Manganese-doped γ-Fe$_2$O$_3$, Mössbauer spectroscopy.

1. Introduction

The doping of cations into the iron oxides Fe$_3$O$_4$, γ-Fe$_2$O$_3$ and α-Fe$_2$O$_3$ has attracted attention because of the magnetic, electrical, gas sensing and catalytic properties of the materials [1–5]. The effects of the dopants on reduction–reoxidation processes and on the phase transitions between the oxides have also been examined [6, 7]. We report here on the investigation by ^{57}Fe Mössbauer spectroscopy, supported by Mn K-edge XANES and EXAFS data, on the effect of manganese on the γ-Fe$_2$O$_3$ to α-Fe$_2$O$_3$ phase transition and magnetic properties of these materials.

2. Experimental details

Materials of the type γ-Fe$_{2-x}$Mn$_x$O$_3$ ($x = ca.$ 0.03, 0.08 corresponding to *ca.* 1.5 and 4.1 at% Mn) were prepared by adding aqueous ammonia to a 2 : 1 mixture of aqueous solutions of iron(III) chloride hexahydrate and iron(II) chloride tetrahydrate with an aqueous solution of manganese(II) chloride tetrahydrate. The mixtures were boiled under reflux (3 h). The precipitates were removed by filtration, washed with 95% ethanol until no chloride ions could be detected in the washings by silver nitrate solution, and heated in air at 250°C (12 h).

The ^{57}Fe Mössbauer spectra were recorded from the powdered samples with a constant acceleration spectrometer and a *ca.* 400 MBq ^{57}Co/Rh source and using a furnace described in detail elsewhere [8]. The sample thickness was 50–80 mg/cm^2. The linewidth (FWHM) of the calibration spectrum was 0.24 mm s^{-1}. The isomer shifts are given relative to that of metallic iron at 298 K.

Manganese K-edge X-ray absorption near edge structure (XANES) and extended X-ray absorption fine structure (EXAFS) were recorded on Station 8.1 at the Synchrotron Radiation Source at Daresbury Laboratory with an average current of 200 mA at 2 GeV. The Si(III) monochromator was calibrated with a manganese foil the X-ray absorption edge of which was taken to be at 6539.0 eV. The data were collected from ground samples pressed between adhesive tape in transmission (γ-Fe$_{1.92}$Mn$_{0.08}$O$_3$)- and fluorescence (γ-Fe$_{1.97}$Mn$_{0.03}$O$_3$)-mode. XANES data were also collected from MnO, Mn$_2$O$_3$ and MnO$_2$ standards. The X-ray absorption edge positions were identified at the points of greatest slope on the edge. The EXAFS data were reduced with the programs Excalib and Exspline and fitted with the program Excurv 98 [9].

3. Results and discussion

The X-ray absorption edge positions as determined by XANES from γ-Fe$_{1.97}$Mn$_{0.03}$O$_3$ and γ-Fe$_{1.92}$Mn$_{0.08}$O$_3$ together with those determined from MnO, Mn$_2$O$_3$ and MnO$_2$ are collected in Table I and show that, in both samples, manganese is present as Mn^{3+}.

The Mn K-edge EXAFS were best fitted to manganese occupying an octahedral site in the spinel-related γ-Fe$_2$O$_3$ structure (Table II) although the large Debye–Waller factors for the first oxygen shell indicate that the site distribution may be complicated.

The ^{57}Fe Mössbauer spectra recorded *in situ* from γ-Fe$_{1.97}$Mn$_{0.03}$O$_3$ and γ-Fe$_{1.92}$Mn$_{0.08}$O$_3$ as a function of increasing temperature are shown in Figure 1. The spectra recorded from both samples at 285 K could be satisfactorily fitted with either a main sextet characteristic of γ-Fe$_2$O$_3$ together with two discrete sextets of rather broad line widths as shown in Figure 1 or with a distribution of sextets

Table I. X-ray absorption edge data

Sample	Edge position (eV $\pm$ 0.5)
MnO	6545.0
Mn$_2$O$_3$	6548.0
MnO$_2$	6558.5
γ-Fe$_{1.97}$Mn$_{0.03}$O	6548.1
γ-Fe$_{1.92}$Mn$_{0.08}$O$_3$	6548.0

Table II. Best fit parameters to Mn K-edge EXAFS data

Sample	Atom type	Coordination number	Distance (Å) ±0.03	$2\sigma^2$ (Å^2)
γ-Fe$_{1.97}$Mn$_{0.03}$O$_3$	O	6	1.95	0.03
	Fe	4.67	2.97	0.03
	Fe	6	3.48	0.03
γ-Fe$_{1.92}$Mn$_{0.08}$O$_3$	O	6	1.97	0.04
	Fe	4.67	2.98	0.03
	Fe	6	3.52	0.03

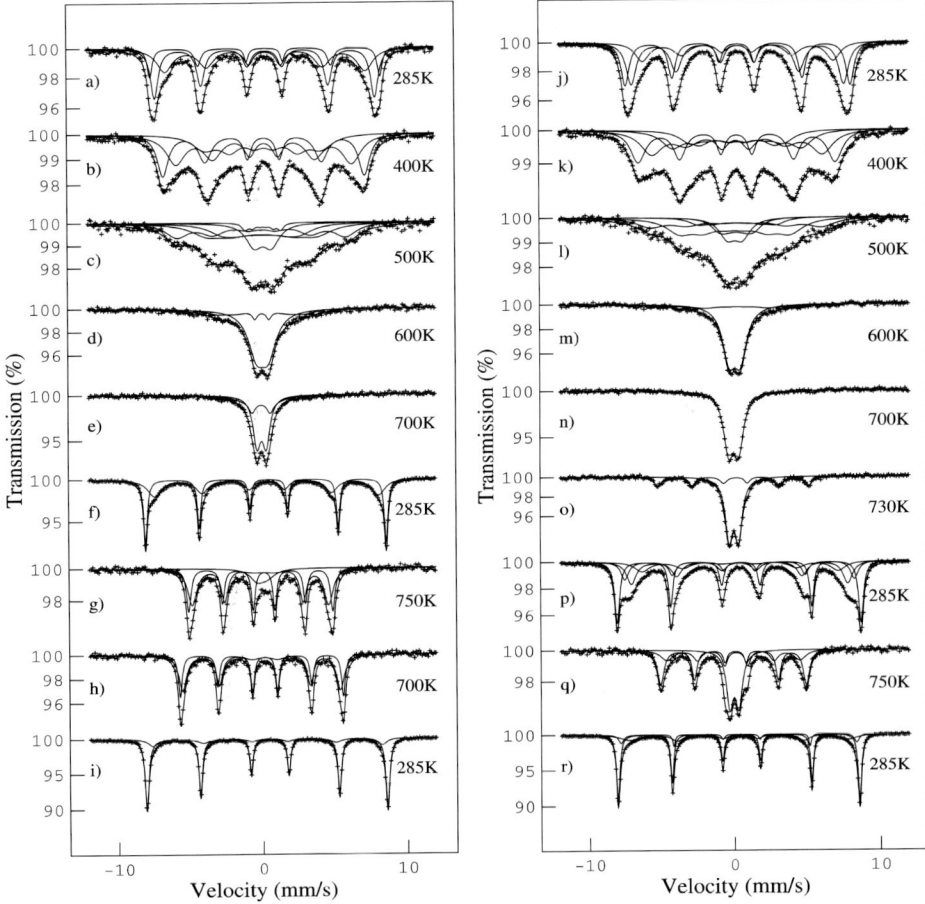

Figure 1. ^{57}Fe Mössbauer spectra recorded from (*left*) γ-Fe$_{1.97}$Mn$_{0.03}$O$_3$ and (*right*) γ-Fe$_{1.92}$Mn$_{0.08}$O$_3$.

with magnetic hyperfine fields in the range 40–46 T. Both fits to the data gave chemical isomer shifts and quadrupole shifts identical to those of γ-Fe$_2$O$_3$. The results, together with the spectra recorded at 400, 500 and 600 K, which show superparamagnetic and relaxation effects, are indicative of a distribution of particle sizes down to *ca.* 10 nm. The detailed fitting of the spectrum recorded from γ-Fe$_{1.97}$Mn$_{0.03}$O$_3$ at 700 K showed the emergence of a weak magnetically split contribution in the spectral wings. The spectrum recorded at 285 K (Figure 1f) shows that *ca.* 50% of the spectral area can be assigned to a sextet with hyperfine parameters characteristic of the corundum-related structure of α-Fe$_2$O$_3$.

The same effect is seen in the spectrum recorded from γ-Fe$_{1.92}$Mn$_{0.08}$O$_3$ at 730 K (Figure 1o) and in the spectrum recorded at 285 K (Figure 1p). Further heating for 24 hours at 750 K results in *ca.* 90% conversion of the spinel-related structure to the corundum-related structure characteristic of α-Fe$_2$O$_3$ (Figures 1i and 1r). X-ray powder diffraction showed that manganese did not segregate from the structure during the phase transformation. The spectra indicate a lower degree of relaxation consistent with the sintering of the particles at these temperatures. The results also show that the conversion of γ-Fe$_2$O$_3$ to α-Fe$_2$O$_3$, which we have previously shown [10] to occur between *ca.* 650–700 K in similarly prepared pure oxides, is increased by the incorporation of manganese.

The main magnetic component in the spectra recorded from both samples following conversion to the α-Fe$_2$O$_3$ structure shows a temperature dependence (Figure 2) which is very similar to that of pure α-Fe$_2$O$_3$ but with a slightly lower hyperfine magnetic field due to the incorporation of manganese within the corundum-related structure. These data were extrapolated from 750 K according to the relationship, $B(T) = B_0(1 - T/T_N)^\beta$ where T_N is the Néel temperature and $\beta = 0.3$. Hence a magnetic transition temperature for the Mn-doped samples of 855–870 K, compared to $T_N = 955$ K for undoped α-Fe$_2$O$_3$, has been deduced. The figure also indicates the maximum magnetic hyperfine field of γ-Fe$_{1.97}$Mn$_{0.03}$O$_3$ before the phase transition takes place.

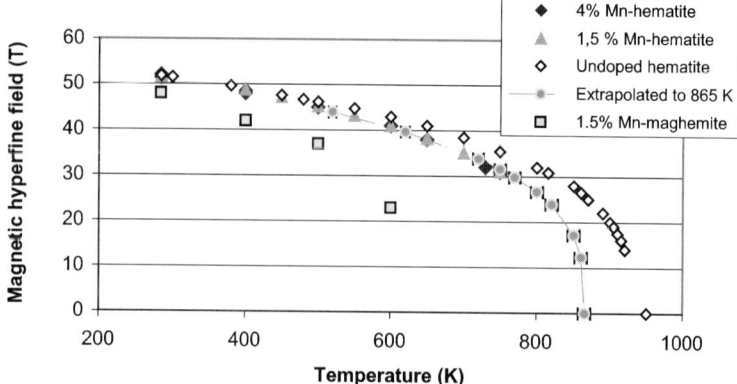

Figure 2. Temperature dependence of hyperfine magnetic field for manganese-doped α-Fe$_2$O$_3$.

Acknowledgement

We thank Dr. I. Ayub for preparing the two samples.

References

1. Morin, F. J., *Phys. Rev.* **83** (1951), 1005.
2. Kanai, H., Mizutani, H., Tanaka, T., Funabiki, T., Yoshida, S. and Takano, M. J., *Mater. Chem.* **2** (1992), 703.
3. Bonzi, P., Depero, L. E., Parmigiani, F., Perego, C., Sberveglieri, G. and Quattroni, G. J., *Mater Res.* **9** (1994), 1250.
4. Uekawa, N., Watanabe, M., Maneko, K. and Mizukami, F., *J. Chem. Soc. Faraday Trans.* **91** (1995), 2161.
5. Berry, F. J., Borhorquez, A., Helgason, Ö., Jiang, J., McManus, J., Moore, E., Mortimer, M., Mosselmans, F. and Mørup, S., *J. Phys.: Condens. Matter* **12** (2000), 4043.
6. Berry, F. J., Helgason, Ö., Jonsson, K. and Skinner, S. J., *J. Solid State Chem.* **122** (1996), 353.
7. Berry, F. J. and Helgason, Ö., *Hyp. Interact.* **126** (2000), 269.
8. Helgason, Ö., Gunnlaugsson, H. P., Jonsson, K. and Steinthorsson, S., *Hyp. Interact.* **91** (1994), 595.
9. Binsted, N., Campbell, J. W., Gurman, S. J., Ross, I. and Stephenson, P. C., Excurv 98, CCLRC Daresbury Laboratory Computer Program, 1998.
10. Berry, F. J., Greaves, C., Helgason, Ö. and McManus, J., *J. Mater. Chem.* **9** (1999), 223.

Hyperfine Interactions **156/157**: 311–314, 2004.
© 2004 *Kluwer Academic Publishers. Printed in the Netherlands.*

Study on Chemical Bond and Electronic State of New Gold Mixed Valence Complexes $Cs_2[Au^IX_2][Au^{III}Y_4]$ (X, Y = Cl, Br, I) by Means of ^{197}Au Mössbauer Spectroscopy

K. IKEDA[1], N. KOJIMA[1], Y. ONO[1], Y. KOBAYASHI[2], M. SETO[2], X. J. LIU[3] and Y. MORITOMO[3]
[1]*Graduate School of Arts and Sciences, The University of Tokyo, Tokyo 153-8902, Japan*
[2]*Research Reactor Institute, Kyoto University, Osaka 590-0494, Japan*
[3]*CIRSE, Nagoya University, Nagoya 464-8601, Japan*

Abstract. $Cs_2[Au^IX_2][Au^{III}X_4]$ (X = Cl, Br, I) is known for a perovskite-type gold mixed-valence system. We have synthesized new gold mixed valence complexes, $Cs_2[Au^IX_2][Au^{III}Y_4]$ (X, Y = Cl, Br, I), which was proved by means of Raman spectroscopy. From the analysis of ^{197}Au Mössbauer spectra, it was elucidated that the charge transfer between $Au^I(5d_{x^2-y^2})$ and $Au^{III}(5d_{x^2-y^2})$ becomes to be predominant in the order of Y = Cl < Br < I for $Cs_2[Au^IX_2][Au^{III}Y_4]$ (X, Y = Cl, Br, I).

Key words: ^{197}Au Mössbauer spectroscopy, charge transfer, mixed-valence, gold complex.

1. Introduction

Gold mixed-valence complexes, $M_2[Au^IX_2][Au^{III}X_4]$ (M = alkali metal; X = halogen) are well known for halogen-bridged gold mixed valence complexes, which have a distorted perovskite-type structure [1, 2].

These complexes undergo the pressure-induced Au valence transition from the mixed valence state of $Au^{I,III}$ to the single valence state of Au^{II} [2, 3], which is coupled with a structural phase transition. In certain pressure region, a dynamic two-electron exchange between the Au^I and Au^{III} states reflects the good conductivity [4]. The metallic phase appearing commonly for these complexes under high pressure and high temperature exists as metastable state at ambient pressure and room temperature. Recently, a photo-induced Au valence transition has been discovered for $Cs_2[Au^IBr_2][Au^{III}Br_4]$ [5].

For the sake of controlling its chemical bond and electronic state, we have controlled their bridging halogens.

In this paper, we report on the synthesis of hetero-halogen bridged gold mixed-valence complexes, $Cs_2[Au^IX_2][Au^{III}Y_4]$ (X, Y = Cl, Br, I) and their Mössbauer spectra and discuss the chemical bond and the charge transfer between Au^I and Au^{III}.

2. Experimental

$Cs_2[Au^IX_2][Au^{III}Y_4]$ (X, Y = Cl, Br, I) was prepared in the following manner. Solutions of $(n\text{-}Bu)_4N[Au^IX_2]$ and $(n\text{-}Bu)_4N[Au^{III}Y_4]$ in 1,1,2-trichloroethane were stirred at $-20°C$, individually. A solution of $C_6H_5SO_3Cs$ in methanol was stirred at $-20°C$. All of these solutions were mixed and stirred at $-20°C$ for an hour. Thus, $Cs_2[Au^IX_2][Au^{III}Y_4]$ (X, Y = Cl, Br, I) was obtained as black colored precipitate. This method has a possibility of synthesizing pseudo-halogen bridged gold mixed valence complexes.

Raman spectroscopy was performed using an argon ion laser (514.5 nm) in a backward configuration.

Mössbauer spectroscopy of the 77.34 eV transition in ^{197}Au was employed with both source and absorber cooled down to 16 K. The γ-ray source (^{197}Pt) was obtained by neutron irradiation for 98% enriched ^{196}Pt metal with the nuclear reaction $^{196}Pt(n, \gamma)^{197}Pt$ in the Kyoto University Reactor.

3. Results

In order to confirm the synthesis of hetero-halogen bridged gold mixed-valence complex, we have investigated Raman spectra for $Cs_2[Au^II_2][Au^{III}Br_4]$, $Cs_2[Au^IBr_2][Au^{III}Br_4]$, and $(n\text{-}Bu)_4N[Au^II_2]$. In the case of $Cs_2[Au^IBr_2][Au^{III}Br_4]$, two A_{1g} modes are observed in the $c(a + b, a + b)c$ configuration, while one B_{1g} mode is observed in the $c(a + b, a - b)c$ configuration. The lower-lying two modes, i.e. B_{1g} (178 cm^{-1}) and A_{1g} (179 cm^{-1}) modes correspond to the stretching modes of the $[Au^{III}Br_4]^-$ molecule, while the higher-lying A_{1g} (220 cm^{-1}) mode corresponds to that of the $[Au^IBr_2]^-$ molecule. For $Cs_2[Au^II_2][Au^{III}Br_4]$, the A_{1g} (220 cm^{-1}) mode corresponding to the $[Au^IBr_2]^-$ molecule disappears. Instead, the Raman spectrum corresponding to the $[Au^II_2]^-$ molecule appears at about 150 cm^{-1}. In this way, it is proved that the hetero-halogen bridged gold mixed-valence complex, $Cs_2[Au^II_2][Au^{III}Br_4]$, is formed.

In Table I, the Mössbauer parameters of $Cs_2[Au^IBr_2][Au^{III}Y_4]$ and $Cs_2[Au^IX_2][Au^{III}Br_4]$ (X, Y = Cl, Br, I) are listed. Figure 1 shows the isomer shifts of Au^I and Au^{III} for these complexes.

4. Discussion

In $Cs_2[Au^IX_2][Au^{III}Y_4]$, there are two relative orientation between $[Au^IX_2]^-$ and $[Au^{III}X_4]^-$. In one, the $\cdots X\text{--}Au^I\text{--}X\cdots Au^{III}\cdots X\text{--}Au^I\text{--}X\cdots$ network lies along the c-axis, while in the other, the $\cdots Y\text{--}Au^{III}\text{--}Y\cdots Au^I\cdots Y\text{--}Au^{III}\text{--}Y\cdots$ network lies in the ab plane. In both cases, there are two types of charge transfer between Au^I and

Table I. Mössbauer parameters of $Cs_2[Au^IBr_2][Au^{III}Y_4]$ (Y = Cl, Br, I) and $Cs_2[Au^IX_2][Au^{III}Br_4]$ (X = Cl, Br, I). *IS*: -1.276 mm/s shift relative to α-Fe metal

	Site	IS	QS	FWHM	$I(Au^I)/I(Au^{III})$
$Cs_2[Au^IBr_2][Au^{III}Cl_4]$	Au^I	0.25	5.04	2.09	0.802
	Au^{III}	1.54	0.92	2.14	
$Cs_2[Au^IBr_2][Au^{III}Br_4]$	Au^I	0.40	4.73	1.87	0.771
	Au^{III}	1.41	1.10	1.94	
$Cs_2[Au^IBr_2][Au^{III}I_4]$	Au^I	0.92	4.60	1.87	0.758
	Au^{III}	1.74	1.42	2.04	
$Cs_2[Au^ICl_2][Au^{III}Br_4]$	Au^I	0.47	4.89	2.33	0.757
	Au^{III}	1.39	1.08	2.33	
$Cs_2[Au^II_2][Au^{III}Br_4]$	Au^I	0.39	4.48	1.91	0.793
	Au^{III}	1.07	1.31	2.29	

IS (mm/s): isomer shift, *QS* (mm/s): quadrupole splitting, *FWHM* (mm/s): full width at half maximum, $I(Au^I)/I(Au^{III})$: intensity ratio.

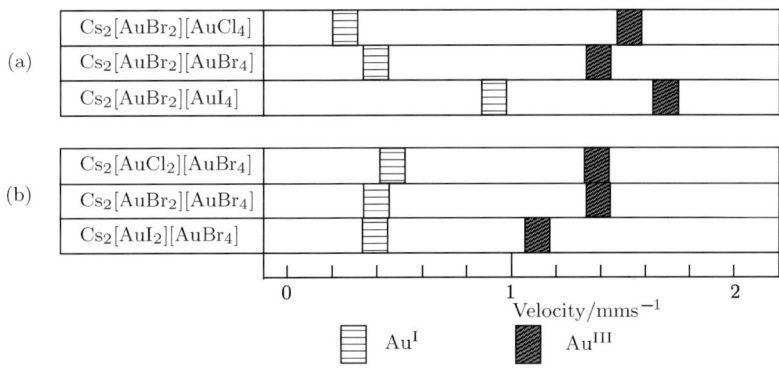

Figure 1. Isomer shifts of Au^I and Au^{III} in $Cs_2[Au^IBr_2][Au^{III}Y_4]$ (Y = Cl, Br, I) and $Cs_2[Au^IX_2]$ $[Au^{III}Br_4]$ (X = Cl, Br, I). *IS*: -1.276 mm/s shift relative to α-Fe metal.

Au^{III}. One is the charge transfer between $Au^I(5d_{z^2})$ and $Au^{III}(5d_{x^2-y^2})$. The other is that between $Au^I(5d_{x^2-y^2})$ and $Au^{III}(5d_{x^2-y^2})$. According to the MO calculation for $Cs_2[Au^IX_2][Au^{III}X_4]$ [4], the $Au^I(5d_{z^2})$ orbital in $[Au^IX_2]^-$ consists of $6s$ component because of ds hybridization. However, the $Au^I(5d_{x^2-y^2})$ orbital of $[Au^IX_2]^-$ does not consist of $6s$ component because of the restriction of symmetry. Therefore, if the charge transfer between $Au^I(5d_{z^2})$ and $Au^{III}(5d_{x^2-y^2})$ is predominant, the isomer shift of Au^I becomes to be small, since the s-electron population at the Au^I site decreases due to the charge transfer from Au^I to Au^{III}. On the other hand, if the charge transfer between $Au^I(5d_{x^2-y^2})$ and $Au^{III}(5d_{x^2-y^2})$ is predominant, the isomer shift of Au^I becomes to be large because of the decrease in the screening. In the case of $Cs_2[Au^IBr_2][Au^{III}Y_4]$, in the order of Y = Cl < Br < I, the isomer shift of Au^I remarkably increases which implies that the charge transfer between $Au^I(5d_{x^2-y^2})$ and $Au^{III}(5d_{x^2-y^2})$ becomes to be strong in the order of $Y = Cl < Br < I$. In contrast, the isomer shift of Au^I in $Cs_2[Au^IX_2][Au^{III}Br_4]$ does not depend on X because of the competition between two types of charge transfer mentioned above. On the other hand, the remarkable decrease in the isomer shift of Au^{III} for $X = I$ is due to the increase in screening effect. These facts describe that the charge transfer strongly depends on $[Au^{III}Y_4]^-$, which strongly supports that the charge transfer interaction between Au^I and Au^{III} in $Cs_2[Au^IX_2][Au^{III}Y_4]$ is two-dimentional.

Acknowledgements

This work has partly been supported by a Grant-in-Aid for Science Research and a Grant-in-Aid from The 21st Century COE (Center of Excellence) program (Research Center for Integrated Science) from the Ministry of Education, Science, Sports and Culture, Japan.

References

1. Raubenheimer, H. G. and Cronje, S., In: H. Schmidbaur (ed.), *Gold*, John Wiley & Sons Ltd., Chichester, 1999, p. 568.
2. Kojima, N., Hasegawa, M., Kitagawa, H., Kikegawa, T. and Shimomura, O., *J. Am. Chem. Soc.* **116** (1994), 11368.
3. Liu, X. J., Moritomo, Y., Nakamura, A. and Kojima, N., *J. Chem. Phys.* **110** (1999), 9174.
4. Kojima, N., *Bull. Chem. Soc. Jpn.* **73** (2000), 1445.
5. Liu, X. J., Moritomo, Y., Ichida, M., Nakamura, A. and Kojima, N., *Phys. Rev. B* **61** (2000), 20.

Hyperfine Interactions **156/157**: 315–319, 2004.
© 2004 *Kluwer Academic Publishers. Printed in the Netherlands.*

Magnetic Properties of $TlCo_2Se_2$ Studied by Mössbauer Spectroscopy

SAEED KAMALI[1], LENNART HÄGGSTRÖM[1], SABINA RONNETEG[2] and ROLF BERGER[2]
[1]*Department of Physics, Uppsala University, Box 530, SE-751 21 Uppsala, Sweden*
[2]*Department of Materials Chemistry, Uppsala University, Box 538, SE-751 21 Uppsala, Sweden*

Abstract. The quasi two-dimensional antiferromagnetic compound $TlCo_2Se_2$ has been studied by [57]Fe Mössbauer spectroscopy. Small single crystals have been made with 2% Fe doping. The Mössbauer spectrum, developed below the Néel temperature of 85 K, reveals a distribution of magnetic hyperfine fields (MHF). The fit is reasonably good if we suppose a connection between the values and mutual orientation of the EFG and MHF. That allows us to suggest the existence of a modulation in the helical magnetic structure, discovered earlier by neutron diffraction.

Key words: Mössbauer spectroscopy, magnetic moment, magnetic hyperfine field, electric quadrupole splitting, helical magnetic structure, two-dimensional magnetism.

1. Introduction

One interesting feature of $TlCo_2Se_2$ ($ThCr_2Si_2$ structure type, I4/mmm [1]) is its magnetic structure. The Co atoms are situated in sheets and arranged in a simple square structure ($a' \approx 2.7$ Å) with the sheets separated by half the c-axis, i.e. about 7 Å. Below 85 K the magnetic structure has been characterized as a helix running along the c-axis. The magnetic moments lie in the basal tetragonal plane with a helical turn angle of about 121° between two adjacent Co layers [2] as shown in Figure 1. In order to study this helix in a more local way we decided to perform a Mössbauer study.

2. Experimental

Powder of $TlCo_{1.96}Fe_{0.04}Se_2$ was synthesized by mixing stoichiometric amounts of TlSe, Co, [57]Fe and Se. The materials were brought to reaction in an evacuated silica-tube, homogenised twice and heat-treated at 773 K. The single crystals were prepared by bringing $TlCo_{1.96}Fe_{0.04}Se_2$ powder to melt, also in an evacuated silica-tube, and then slowly cooled (1 K/h) to room temperature. The crystal structure was investigated by X-ray diffraction, using a Guinier–Hägg camera with $CuK\alpha_1$ radiation and germanium as an internal standard. No significant difference caused by the [57]Fe doping was observed on the cell parameters. The

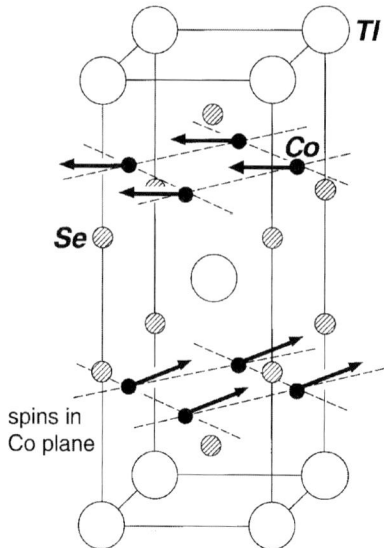

Figure 1. The magnetic structure [2].

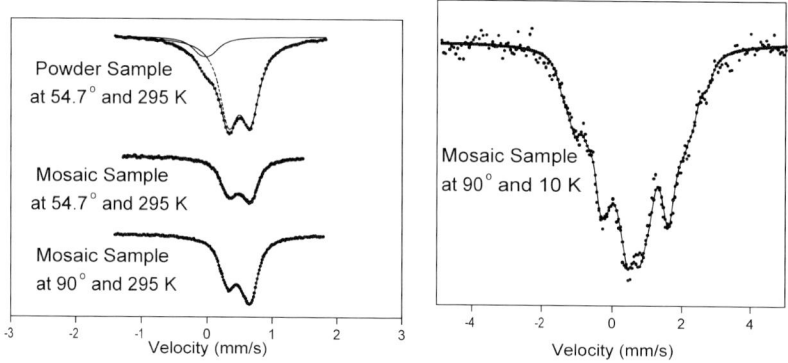

Figure 2. The Mössbauer spectra at room temperature (*left*) and at 10 K (*right*).

single crystals were about 1 mm² in size and flat. Many flat crystals were placed on a tape, like a mosaic with the [001] orientation perpendicular to the tape. For ^{57}Fe Mössbauer spectroscopic measurements, ^{57}Co in Rh was used at room temperature as a Mössbauer source. The spectra were calibrated by using the six lines of α-Fe, the center of which was taken as zero isomer shift. The mosaic sample consisting of single crystals as well as a powdered sample mixed with boron nitride were used as absorber.

3. Result and discussion

Typical Mössbauer spectra are shown in Figure 2. At room temperature the spectra for the mosaic sample reveal an asymmetric doublet both when measured at an angle of incidence of 54.7° (the magic angle) and of 90° with respect to the sample plane. The spectrum for the powder sample, measured at the magic angle, also shows an asymmetric doublet in addition to a marked shoulder at the low velocity side.

3.1. ROOM-TEMPERATURE SPECTRA

The fact that the two spectra measured at the magic angle show asymmetry might be due to a Goldanskii–Karyagin effect in the structure. This effect could however be ruled out when looking at the powder spectrum in more detail. It is the lines at the shoulder in this spectrum that give rise to the asymmetry. A good spectral fit is possible with two doublets, each having equal line intensity. The main doublet (intensity = 84%) has a centroid shift $\delta = 0.490(2)$ mm/s and an electric quadrupole splitting $\Delta E_Q = 0.33(1)$ mm/s. The corresponding values for the small doublet are $\delta = -0.019(5)$ mm/s and $\Delta E_Q = 0.18(1)$ mm/s. The Lorentzian FWHM line width is 0.22 mm/s. The small asymmetry in the mosaic sample spectrum could be described using a centroid shift distribution linearly correlated with an electric quadrupole splitting distribution. The result from these fittings is $\delta = 0.493$ ($\sigma = 0.012$) mm/s and $\Delta E_Q = 0.33$ ($\sigma = 0.14$) mm/s in good accordance with the powder result. The impurity phase in the powder sample is not seen in the mosaic sample. Therefore we only deal further on with the result for the mosaic sample. The average thickness of the mosaic sample is 0.10 mm. The dimensionless Mössbauer resonance thickness T is then calculated to 3.2 assuming a recoil free factor of 0.7. Using the complete transition integral in the fitting of the 90° mosaic spectrum the line intensity ratio I_+/I_- was determined to 1.64(5). The theoretical ratio for $V_{zz} > 0$, according to [3],

$$\frac{I_+}{I_-} = \frac{4\sqrt{1 + \eta^2/3} + (3\cos^2\alpha - 1 + \eta\sin^2\alpha\cos 2\beta)}{4\sqrt{1 + \eta^2/3} - (3\cos^2\alpha - 1 + \eta\sin^2\alpha\cos 2\beta)}$$

is governed by the orientation of the principal axes system of the electric field gradient (EFG) tensor versus the γ-direction and also by the asymmetry parameter η of the tensor. Here α and β represent the polar and azimuthal angles, respectively. When $V_{zz} < 0$ the ratio has the inverse value. For the mosaic sample the γ-ray is parallel to the crystallographic c-axis. In the present crystal structure the principal axes would be parallel with the crystallographic axes and having an asymmetry factor $\eta = 0$. Furthermore the principal z-axis would be parallel with the crystallographic c-axis. Thus α and β would be 0° and undetermined, respectively, giving theoretical values 3 or 0.33. These values are not in agreement with the experimental findings of 1.64(5).

In the study of the isostructural compound TlFe$_2$Se$_2$ stoichometry was not achieved [4]. Instead iron vacancy ordering was found to take place with full ordering for the formula TlFe$_{1.6}$Se$_2$. The single Fe vacancy assumed to take place close to the Fe atom, changes the principal axes system for the EFG tensor so that

Table I. The I_+/I_- ratios for different cases

Case	I_+/I_- for $V_{zz} > 0$	I_+/I_- for $V_{zz} < 0$
1	0.6–0.4	1.67–2.52
2	0.6–1.0	1.67–1.0

the z-axis is in the ab-plane, which is the plane where the vacancies occur [4]. Assuming this also to be case for the present compound we can distinguish two cases. Case 1(2) with the $y(x)$-axis parallel to the c-axis and to the direction of the γ-radiation. The intensity ratio for these cases for $0 \leqslant \eta \leqslant 1$ are presented in Table I.

Our experimental results thus clearly reveal that V_{zz} is negative and gives a preference for case 2. It must, however, be stressed that this is an averaged result for the whole mosaic sample. It may be that not all of the doping atoms introduce Co vacancies. This will affect the asymmetry of the spectrum giving some extra uncertainties to the experimental ratio assumed to be valid for the vacancy modified situation.

3.2. THE LOW-TEMPERATURE SPECTRA

Below around 85 K the Mössbauer spectra change character from an asymmetric doublet to a magnetically split feature. This is in good agreement with the findings from magnetometry and neutron diffraction by Berger *et al.* [2]. The spectrum for the mosaic sample measured at an angle of incidence of 90° at 10 K is shown in Figure 2. The magnetic splitting is obviously very small meaning that the spectrum has to be analyzed with the full Hamiltonian formalism. In such an analysis the magnetic field directions (polar θ and azimuthal ϕ angles) versus the EFG principal z-axis are obvious parameters, besides the magnetic hyperfine field B_{hf} and the asymmetry parameter η of the EFG. Berger *et al.* found from their study [2], an incommensurate helical structure of the Co magnetic moments running along the c-axis with the moments (0.46 μ_B/Co atom) confined to the ab-plane. Adopting this structure to the Fe doped compound, the Fe hyperfine fields would be in the ab-plane. Such a hypothesis is supported by an analysis of neutron diffraction data on strongly doped material, $TlCo_{2-x}Fe_xSe_2$. Furthermore because of the helical turning the polar angle θ would vary between 0° and 90°. The azimuthal angle ϕ would in case 1(2) be fixed to 0° (90°). It was impossible to get good fits using one magnetic field value in the above models. Allowing the magnetic hyperfine field to vary freely as a function of the polar angle θ gave reasonable good fits for the two cases. The results from the fits are: $\delta = 0.61(1)$ mm/s, $eQV_{zz}/2 = -0.31$ mm/s, $\eta = 0.5(3)$ and B_{hf} varying from 12(1) T (for $\theta \approx 0°$ and 90°) to 1(1) T (for $\theta \approx 45°$) over 7(2)T (for $\theta \approx 25°$ and 65°). The average magnetic hyperfine field is 7.2(5) T which, using the conversion factor 15 T/μ_B (valid for pure α-Fe), gives an averaged Fe magnetic moment of 0.48(4)μ_B, a value very close to the Co moment. This finding of a magnetic moment modulated helical structure is analogous to what has been found earlier, e.g., FeP [5]. From the present study only it is not possible to give an answer about the exact magnetic structure in the parent compound since it seems, as discussed above, that the Fe doping introduces severe changes in the crystallographic structure close to the Fe atom. Assuming this to happen for all Fe introduced, the magnetic moment modulation found can partly or entirely be an effect of this Co vacancy. Unfortunately the neutron diffraction data cannot distinguish between a modulated or non-modulated helical structure.

Acknowledgement

Financial support from the Swedish Foundation for Strategic Research (SSF/FRAM) is gratefully acknowledged.

References

1. Ban, Z. and Sikirica, M., *Acta Cryst.* **18**(4) (1965), 594.
2. Berger, R., Fritzsche, M., Broddefalk, A., Nordblad, P. and Malaman, B., *J. Alloys Comp.* **343** (2002), 186.
3. Zory, P., *Phys. Rev. A* **140** (1965), 1410.
4. Häggström, L., Seidel, A. and Berger, R., *J. Mag. Mag. Mater* **98** (1991), 37.
5. Häggström, L., Sundqvist, T. and Fjellvåg, H., *Phys. Rev. B* **35** (1987), 6838.

Hyperfine Interactions **156/157**: 321–325, 2004.
© 2004 *Kluwer Academic Publishers. Printed in the Netherlands.*

Exploring the Verwey-Type Transition in GdBaFe$_2$O$_{5+w}$ Using ^{57}Fe Mössbauer Spectroscopy

J. LINDÉN[1,*], P. KAREN[2], H. YAMAUCHI[3] and M. KARPPINEN[3]

[1]*Åbo Akademi, Physics Department, FIN-20500 Turku, Finland; e-mail: jlinden@abo.fi*
[2]*Department of Chemistry, University of Oslo, N-0315 Oslo, Norway*
[3]*Materials and Structures Laboratory, Tokyo Institute of Technology, Yokohama 226-8503, Japan*

Abstract. ^{57}Fe Mössbauer spectroscopy was used to study the double perovskite GdBaFe$_2$O$_{5+w}$, which exhibits mixing of the integer valence states of iron. The valence mixing/separation process Fe^{2+} + Fe^{3+} ↔ 2Fe$^{2.5+}$ was investigated as a function of temperature. For nearly stoichiometric compositions of $w \approx 0$, a two-step Verwey-type transition is registered that separates Fe$^{2.5+}$ into intermediate valence- and spin states Fe$^{2.5-\epsilon}$ and Fe$^{2.5+\epsilon}$ and then into the integer valences Fe^{2+} and Fe^{3+}. Both steps are accompanied by a decrease in electrical conductivity, altogether by two orders of magnitude. Seebeck measurements identify holes as dominating charge carriers, with activation energy for hopping of ~0.10 eV in the valence-mixed state. It is inferred that the mixing electrons are not simply delocalized over the lattice, but rather form bridges connecting pairs of adjacent Fe atoms along the c axis.

Key words: ^{57}Fe Mössbauer spectroscopy, Verwey-type transition, charge ordering, valence mixing, double perovskite.

1. Introduction

GdBaFe$_2$O$_{5+w}$ belongs to the rare-earth (RE) based series REBaFe$_2$O$_5$, where ordering of the large Ba and small RE atoms stabilizes oxygen vacancies in the double-cell perovskite-type structure, in which Fe atoms reside in pyramidal oxygen coordinations similar to Cu(2) site in YBa$_2$Cu$_3$O$_7$ [1]. With the RE atom being trivalent [2] a mixed-valence situation is imposed upon iron, manifested by valence mixing at high temperatures and charge separation and -ordering at low temperatures. The phase transition between these states [3] is of the first order, accompanied by changes in entropy and electrical conductivity of the type described for the first time by Verwey [4] on magnetite.

At temperatures around 300 K and above, a single Fe$^{2.5+}$ state is observed in the ^{57}Fe Mössbauer spectrum [5]. The Verwey-type transition into two integer valence states proceeds in two steps. Upon cooling, Fe$^{2.5+}$ separates into two intermediate valence states Fe$^{2.5-\epsilon}$ and Fe$^{2.5+\epsilon}$, then the second step leads to a full charge separation and long-range ordering of the integer valences Fe^{2+} and Fe^{3+} [3, 6].

---

* Author for correspondence.

The phase is subject to a considerable oxygen nonstoichiometry given by the w parameter. The vacant oxygen site is located in the RE layer and may be partially filled. Increasing w rapidly decreases the concentrations of the valence-mixed/charge-ordered states. The intermediate states $Fe^{2.5-\epsilon}$ and $Fe^{2.5+\epsilon}$ are no longer observable beyond $w > 0.05$, but the remaining single step persists as a first-order Verwey transition up to $w \approx 0.25$ [6, 7].

2. Experiments

$GdBaFe_2O_{5+w}$ samples were synthesized from an amorphous citrate-based precursor following a procedure described elsewhere [6].

Mössbauer absorbers were made by mixing $\sim$70 mg of the powdered sample with epoxy resin on an Al foil, so that the Mössbauer sample density was $\sim$5 mg of pure Fe per cm^2. Spectra for all samples were recorded at 325 K in a transmission geometry using the maximum Doppler velocity of 11.15 mm/s in sinusoidal mode with a Cyclotron Co. $^{57}Co{:}Rh$ (25 mCi, Jan. 2002) source. For a sample close to the ideal mixed-valence ratio ($w = 0.019$), a series of isothermal scans was collected. In this series the fine scaling across both transitions took priority over high counting statistics for individual spectra while all hyperfine parameters could still be fitted well. Full Hamiltonian of combined electric and magnetic interactions was used to fit the spectra, with the internal magnetic field experienced by the Fe nucleus (B), the isomer shift relative to α-Fe at 300 K (δ), the quadrupole coupling constant (eQV_{zz}), the resonance line widths (Γ), and the relative intensities of the components (I) as parameters.

Sample resistivity was measured using a standard four-probe technique. Seebeck coefficient was obtained as a function of the overall temperature from thermovoltage data between thermojunctions at opposite ends of the sample that were kept at slightly unequal temperatures.

Differential scanning calorimetry (DSC) was used to register the endothermic absorption of heat upon warming the samples.

3. Results and discussion

Owing to antiferromagnetic order of iron moments at the investigated temperatures around the Verwey transition, Mössbauer spectra shown in Figure 1 have a sextet of peaks for each of the observed iron states; charge-separated, intermediate, and valence-mixed. Starting from the separation of the mixed $Fe^{2.5+}$ state the iron atoms that approach trivalence acquire a high internal field ($\sim$50 T). For the state that approaches divalence, unusually low internal field is eventually observed ($\sim$8 T), while the isomer shift of $\sim$1.0 mm/s confirms the 2+ valence assignment.

The DSC signal is drawn in Figure 2 as a background for plotting the Mössbauer parameters. The sharp peaks suggests that both transitions may be of the first order, a fact that is confirmed by the discontinuous change in both internal field and

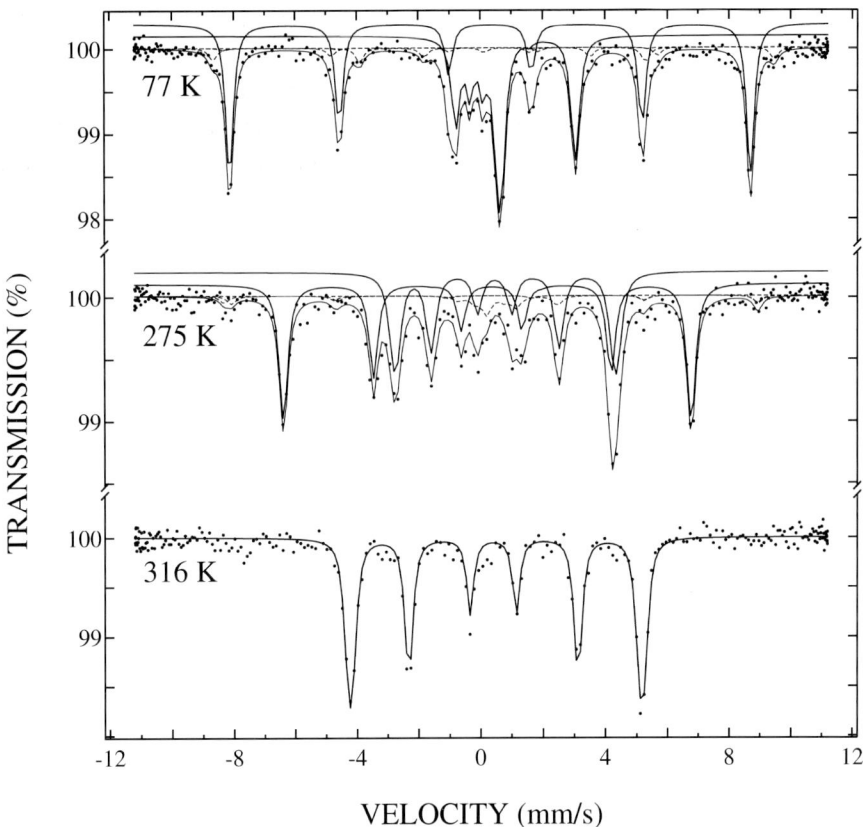

Figure 1. Examples of finely scaled isothermal ^{57}Fe Mössbauer scans across the two Verwey-type transitions in GdBaFe$_2$O$_{5.019}$: the charge-ordered state at 77 K is dominated by high-spin Fe^{3+} and a low-internal-field Fe^{2+}, the intermediate state at 275 K is dominated by Fe$^{2.5+\epsilon}$ (higher internal field) and Fe$^{2.5-\epsilon}$ (lower internal field), and the valence-mixed state at 316 K has only one component.

isomer shift. It should be noted that trace quantities from the major Mössbauer states spill over across the Verwey transition in both directions. The likely reason for that is the distribution of oxygen contents in the nonstoichiometric samples [6]. For clarity, these spill-over values have been omitted within the intermediate state in Figure 2. The local environments of iron could be determined up to the formation of the valence-mixed state, where eQV_{zz} became zero. In the intermediate temperature range between the two transitions, no long-range order of the intermediate states Fe$^{2.5-\epsilon}$ and Fe$^{2.5+\epsilon}$ could be detected [8] by diffraction methods. These methods, however, show that the long-range ordered Fe$^{2.5+}$ spins are aligned along the *b* axis, nearly perpendicular to the fourfold axis of the square-pyramidal coordination polyhedron [3]. Mössbauer refinement of the angles between the magnetic-field vector at the nucleus and the main component of the electric-field-gradient tensor shows that the same orientation is valid also for the intermediate range. In the

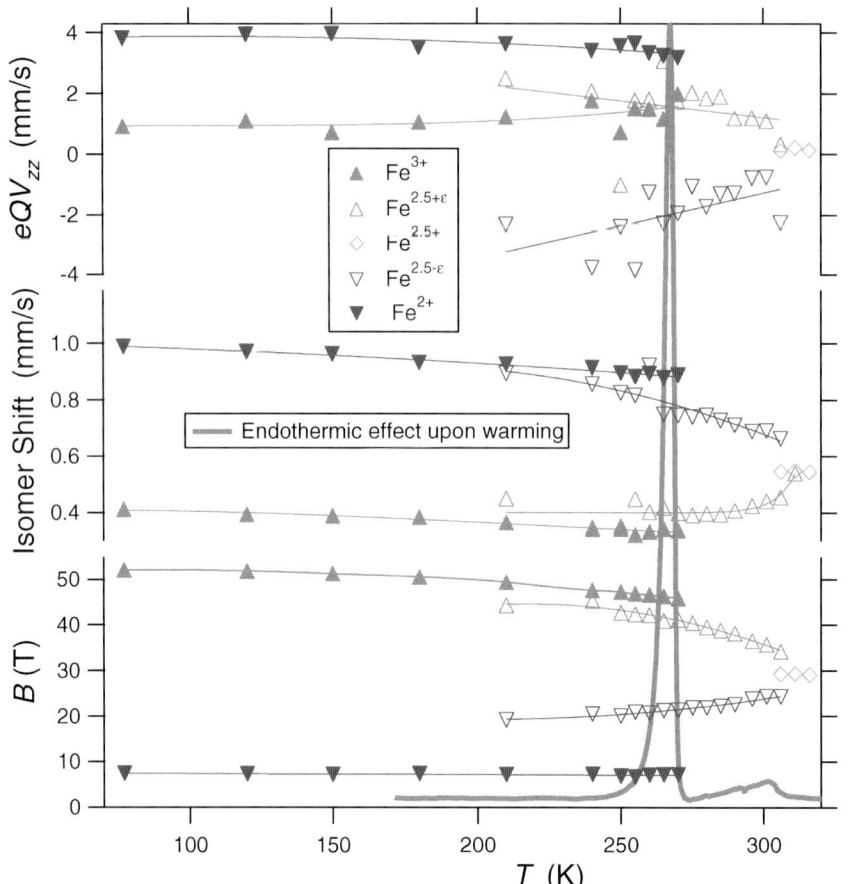

Figure 2. The internal field and isomer shift values obtained from the Mössbauer spectra of GdBaFe$_2$O$_{5.019}$ across the two Verwey-type transitions identified by the DSC signal in the background.

charge-ordered phase the quadrupole coupling constant is large due to the lattice distortion (~3.8 m/s for Fe^{2+} and ~1.0 m/s for Fe^{3+}). Due to the distortion, the V_{zz} axis of Fe^{2+} becomes parallel to b axis, whereas for the less-distorted environment around Fe^{3+} V_{zz} remains perpendicular to the local magnetic moment, i.e. parallel to the c axis.

Electrical-conductivity measurements reveal a non-metallic behavior. In the valence-mixed state, the fitted activation energy for hopping has a local maximum of 0.11 eV at $w \approx 0$ and decreases towards a minimum of 0.09 eV around $w = 0.11$. This indicates a complex contribution from both the integer and mixed valence states of iron identified by Mössbauer spectroscopy. It is also evident that the electrons involved in the valence mixing are not (fully) delocalized. Seebeck measurements identify holes as dominating charge carriers. However, the Seebeck coefficient in the valence-mixed state is relatively low (~70 μV/K), and this suggests a possible mixed hole–electron conductivity, realized, e.g., by a electron–hole excitation in an un-doped semiconductor. In the charge-ordered state, the electrical conductivity drops by two orders of magnitude as consequence of the localization and ordering of the mixing electrons. In contrast to the valence-mixed state, electrical conductivity in the charge-ordered state increases monotonically with increasing w.

4. Conclusions

A two-stage valence mixing-to-separation transition has been observed and studied using ^{57}Fe Mössbauer spectroscopy of GdBaFe$_2$O$_{5+w}$ double perovskite samples. A two-order of magnitude drop of electric conductivity is a consequence of the electron localization connected to the valence separation.

References

1. Karen, P. and Woodward, P. M., *Chem. Mater.* **9** (1999), 789.
2. Karen, P. and Lindén, J., unpublished.
3. Karen, P., Woodward, P. M., Lindén, J., Vogt, T., Studer, A. and Fischer, P., *Phys. Rev. B* **64** (2001), 214405.
4. Verwey, E. J. W. *et al.*, *Physica* **8** (1941), 979.
5. Lindén, J., Karen, P., Kjekshus, A., Miettinen, J., Pietari, T. and Karppinen, M., *Phys. Rev. B* **60** (1999), 15251.
6. Karen, P., *J. Solid State Chem.* **170** (2003), 9.
7. Karen, P. Woodward, P. M., Santhosh, P. N., Vogt, T., Stephens, P. W. and Pagola, S., *J. Solid State Chem.* **167** (2002), 480.
8. Woodward, P. M. and Karen, P., *Inorg. Chem.* **42** (2003), 1121.

Hyperfine Interactions **156/157**: 327–333, 2004.
© 2004 *Kluwer Academic Publishers. Printed in the Netherlands.*

First Principles Calculations of Mössbauer Spectra of Intermetallic Anodes for Lithium-Ion Batteries

P. E. LIPPENS*, J.-C. JUMAS and J. OLIVIER-FOURCADE
Laboratoire des Agrégats Moléculaires et Matériaux Inorganiques, UMR CNRS 5072,
Université Montpellier II, Place Eugène Bataillon, F-34095 Montpellier Cedex 05, France;
e-mail: lippens@univ-montp2.fr

Abstract. Changes in ^{119}Sn and ^{121}Sb Mössbauer spectra due to lithium insertion in tin and antimony based anode materials for lithium-ion batteries are analysed. Due to the complexity of the spectra linear augmented plane wave calculations of the electronic density were used to evaluate the electron density and the electric field gradients at the nucleus. The ^{119}Sn Mössbauer spectrum of SnO + 3.5 Li was evaluated from the theoretical spectra of the Li–Sn alloys. The observed good agreement between experimental and *ab initio* spectra is consistent with the reversible lithium insertion mechanism based on the formation of Li–Sn alloys. The analysis of the ^{121}Sb Mössbauer spectra for Li insertion into $CoSb_3$ is somewhat more complex but calculations of the Mössbauer parameters clearly indicate the existence of Li_3Sb at the end of the first discharge.

Key words: first-principles calculations, anodes, lithium.

1. Introduction

In order to improve the electrochemical performances of anode materials for lithium-ion batteries such as avoiding lithium plating or increasing specific capacity, new compounds have been proposed and investigated recently [1–7]. Such materials are interesting because of their high capacity, acceptable rate capability and low voltage. Unfortunately, they suffer a large volume expansion upon Li insertion which often leads to a mechanical destruction of the anode and limits the capacity. This effect can be strongly reduced by considering dispersed small particles that can be obtained by electrochemical reactions from pristine materials such as tin oxide based compounds or transition metal antimonides.

Different Li insertion mechanisms in tin and antimony based compounds have been proposed [1–7] and experimental evidences are still needed in order to clarify the observed discrepancies. Mössbauer spectroscopy is a powerful experimental tool to investigate both the electronic and structural local properties. Although ^{119}Sn and ^{121}Sb Mössbauer experiments are expected to be very efficient due to their sensitivity to changes in Sn and Sb local environments, respectively, there are at least two difficulties in the present case: the complexity of the spectra and the dif-

---

* Author for correspondence.

ficulty to obtain pure Li–Sn and Li–Sb crystalline phases as Mössbauer references. We propose to use theoretical Mössbauer spectra obtained from first principles calculations of the hyperfine parameters in order to interpret experimental data and obtain new insights in the understanding of Li insertion mechanisms in tin and antimony based anodes. Results are given here for SnO and $CoSb_3$.

2. Experimental and computational procedures

SnO and $CoSb_3$ were obtained by direct synthesis from the pure elements. Their purity and crystalline structures were checked by X-ray diffraction. Two electrode Swagelok cells with Li as anode and $LiPF_6$ as electrolyte were used for the electrochemical measurements. Discharges and charges were carried out under galvanostatic conditions using a Mac Pile II system at low rate. ^{119}Sn Mössbauer measurements were performed in transmission mode at room temperature with a $Ba^{119}SnO_3$ (10 mCi) source. ^{121}Sb Mössbauer measurements were performed with a $Ba^{121}SnO_3$ (0.5 mCi) source in transmission mode with both source and absorber at 4 K.

The theoretical values of the isomer shift δ and quadrupole splitting Δ were obtained from the calculated electronic density and electric field gradients (EFG) at the nucleus, respectively. The calibration procedures for ^{119}Sn and ^{121}Sb give the quadrupole moments $Q(^{119}\text{Sn}) = 10.5$ fm^2 [8] and $Q(^{121}\text{Sb}) = -66$ fm^2 [9]. The EFG were evaluated from the electronic density by considering the approach of Blaha et al. [10]. The electronic densities of the pristine materials (SnO, $CoSb_3$) and of the Li_xSn and Li_xSb crystalline phases were calculated with the linearized augmented plane wave (LAPW) method included in the WIEN97 package [11]. This method is based on the density functional theory (DFT) [12] and the generalized gradient approximation (GGA) with the parametrization of Perdew et al. [13] was used. The core states were treated fully relativistically whereas the scalar relativistic approximation was considered for the valence states. The following values of the muffin tin radius were used: $R_{mt}(\text{Li}) = 2.0$ bohr, $R_{mt}(\text{Sn}) = 2.6$ bohr, $R_{mt}(\text{O}) = 2.4$ bohr, $R_{mt}(\text{Sb}) = 2.6$ bohr, and $R_{mt}(\text{Co}) = 2.2$ bohr. The values of the cutoff parameter for the plane wave expansion ($R_{mt}K_{max} = 8–10$) and the number of k points in the irreducible Brillouin zone (100–500) were taken in order to obtain rather well converged values of the Mössbauer parameters.

3. Results and discussion

The electrochemical potential curve for SnO at 1 Li(per SnO)/15 h rate (Figure 1) shows that up to two inserted lithium there is an irreversible transformation which is usually described by the reaction [2]:

$$\text{SnO} + 2\,\text{Li} \rightarrow \beta\text{Sn} + \text{Li}_2\text{O}. \tag{1}$$

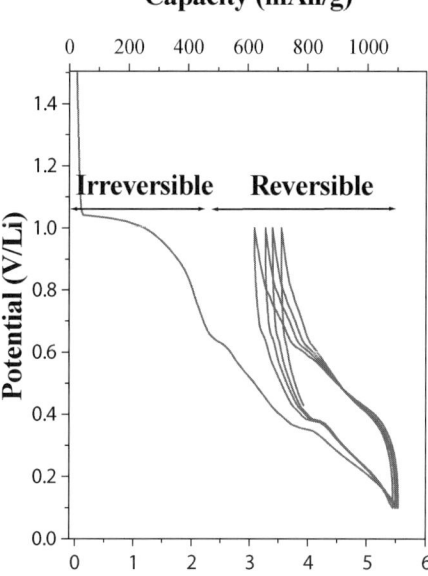

Capacity (mAh/g)

Figure 1. Voltage profile of SnO (vs Li) at 1 Li(per SnO)/15 h rate.

Then, insertion of additional lithium ions leads to the reversible formation of the Li_xSn crystalline phases with $x = 2/5, 1, 7/3, 5/2, 13/5, 7/2$. For example, the insertion of 3.5 Li into SnO is expected to form LiSn and $Li_{7/3}Sn$ as suggested by the reactions:

$$\beta Sn + 2/5\ Li \rightarrow Li_{2/5}Sn, \tag{2}$$

$$Li_{2/5}Sn + 3/5\ Li \rightarrow LiSn, \tag{3}$$

$$LiSn + 1/2\ Li \rightarrow 3/8\ Li_{7/3}Sn + 5/8\ LiSn. \tag{4}$$

For 3.5 Li, the X-ray diffraction characterization of the Li–Sn phases expected from the reaction (4) is not very accurate because of the small particle size and the poor crystallinity. In addition, ^{119}Sn Mössbauer spectroscopy provides rather complex spectra for the pure Li–Sn phases that are not enough reliable to be used as references. We have made use of the theoretical (*ab initio*) Mössbauer spectra of the Li–Sn crystalline phases obtained from the calculated values of δ and Δ in order to interepret the experimental ^{119}Sn Mössbauer data. The calculated values are $\delta_1 = 2.4$ mm/s, $\Delta_1 = 1$ mm/s (Sn_I), $\delta_2 = 2.4$ mm/s, $\Delta_2 = 0.5$ mm/s (Sn_{II}) for LiSn and $\delta_1 = 2.2$ mm/s, $\Delta_1 = 0.7$ mm/s (Sn_1), $\delta_2 = 2$ mm/s, $\Delta_2 = 1$ mm/s (Sn_2) $\delta_3 = 2.1$ mm/s, $\Delta_3 = 1.1$ mm/s (Sn_3) for $Li_{7/3}Sn$, where the atomic positions of the Sn sites are given in Refs [14] (LiSn) and [15] ($Li_{7/3}Sn$). The theoretical Mössbauer spectrum of SnO+3.5 Li obtained from the summation of the calculated spectra according to the reaction (4) is favourably compared to the experimental data (Figure 2). This clearly indicates the existence of the two phases LiSn and

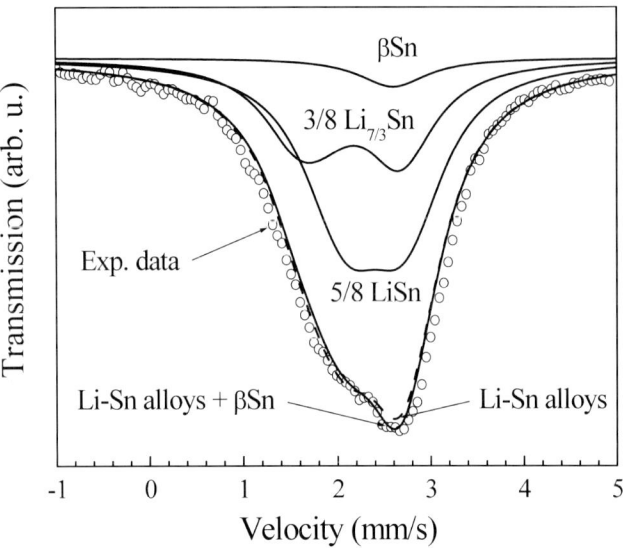

Figure 2. Experimental (circles) and calculated [119]Sn Mössbauer spectra of SnO + 3.5 Li. The calculated spectra were obtained from the calculated sub-spectra of LiSn and $Li_{7/3}$Sn according to the reaction (4) without βSn (dashed line) and with a small contribution of βSn (solid line).

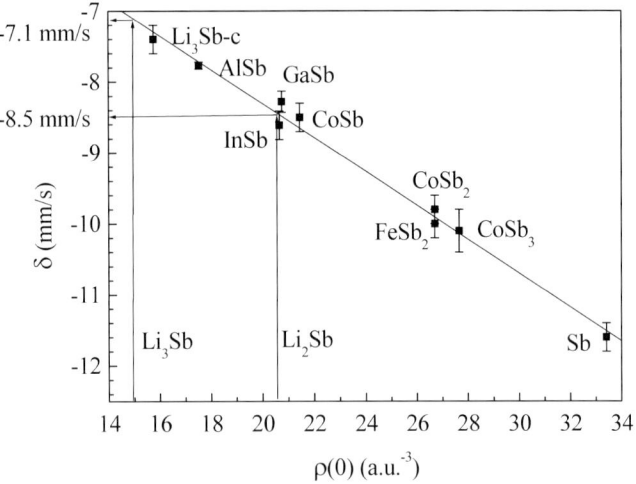

Figure 3. Linear correlation between the experimental values of the isomer shift (relative to $BaSnO_3$) and the calculated electron density at the nucleus shifted by 290700 a.u.$^{-3}$ for clarity.

$Li_{7/3}$Sn with the correct relative amounts and suggests that our analysis of the Mössbauer data is consistent with the proposed Li insertion mechanism. It can be noticed that agreement between experimental and theoretical data can be slightly improved by considering a component due to a small amount of unreacted βSn.

The electrochemical potential curve of $CoSb_3$ shows a first irreversible discharge at about 0.6 V followed by reversible reactions [4, 7]. Experimental works based on X-ray diffraction suggest that Li insertion in transition metal antimonide anodes generally leads to the formation of small Li_3Sb particles at the end of the first discharge. This is an irreversible reaction and Li_3Sb can be considered as the electrochemically active material involved in the alloying/de-alloying reversible reactions. However, such an interpretation is still controversial and requires additional evidences to be clearly established. The application of [121]Sb Mössbauer spectroscopy is rather difficult due to the manifold nuclear transitions $(5/2 \leftrightarrow 7/2)$, the existence of different active Mössbauer sites (different crystallographic sites and phases) and the difficulty to obtain pure Li_2Sb and Li_3Sb crystalline phases as Mössbauer reference materials. Thus, it is of importance to accurately evaluate both the isomer shift and the quadrupole coupling. The experimental values of the isomer shift are plotted as a function of the calculated electron density at the Sb nucleus in Figure 3 for different compounds in the range of values between metallic Sb and Li_3Sb. The observed linear correlation confirms the accuracy of the present LAPW calculations and provides a calibration scale. This allows to predict the isomer shift from the calculated electron density at the nucleus when no experimental data are available. For example, we can predict the following theoretical values: $\delta(Li_2Sb) = -8.5$ mm/s, $\delta(cubic\ Li_3Sb) = -7.4$ mm/s and $\delta(hexagonal\ Li_3Sb) = -7.1$ mm/s (relative to $BaSnO_3$). From the theoretical results it can be also shown that variations of the isomer shift of Li_xSb is linearly correlated to x $(0 < x < 3)$ although the structures of the compounds are different. Calculations of the atomic charges show that the observed decrease of the isomer shift from -7.4 mm/s (Li_3Sb) to -11.6 mm/s (Sb) is mainly due to the increase in the Sb 5s electron population and to a less extent to the decrease in the Sb 5p electron population. At the end of the first discharge the Mössbauer spectra obtained at two Li rates are found to be different (Figure 4). The spectrum obtained

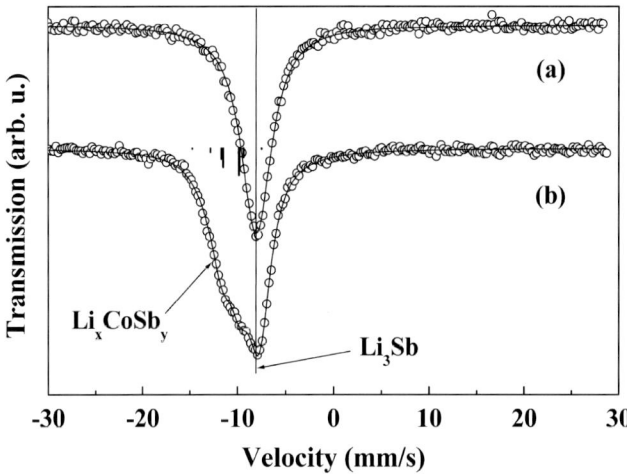

Figure 4. [121]Sb Mössbauer spectra of $CoSb_3$ at the end of the first discharge at 1 Li(per $CoSb_3$)/70 h (a) and 1 Li(per $CoSb_3$)/15 h (b) rates.

at the 1 Li(per $CoSb_3$)/70 h rate is formed by a rather sharp peak with isomer shift $\delta = -7.3$ mm/s in agreement with the calculated value for cubic Li_3Sb. This confirms the complete transformation of $CoSb_3$ into Li_3Sb and Co at very low lithium rate. The Mössbauer spectrum for 1 Li(per $CoSb_3$)/15 h rate reflects the existence of at least 2 different types of Sb sites among them one can be assigned to cubic Li_3Sb as expected from the basic mechanism. The other site does not correspond to $CoSb_3$ as it should be the case for partial lithiation and is tentatively attributed to a complex, maybe metastable, Li_xCoSb_y phase. Its exact determination requires further investigations.

4. Conclusion

First principles simulation of ^{119}Sn Mössbauer spectra provides a new approach to interpret the experimental data when commonly used fitting procedures based on the only experimental data fails. We have shown that Li insertion mechanisms based on the formation of Li–Sn alloys could explain Li insertion in tin oxides. For antimony anodes, Li insertion mechanisms depend on the experimental conditions and the ^{121}Sb Mössbauer spectroscopy provides very useful informations about changes in the Sb local environment due to Li insertion. The formation of Li_3Sb has been unambiguously shown from the comparison between theoretical and experimental values of the hyperfine parameters. However, additional phases can also occur depending on the Li rate.

References

1. Idota, Y., Kubota, T., Matsufuji, A., Maekawa, Y. and Miyasaka, T., *Science* **276** (1997), 1395.
2. Courtney, I. A. and Dahn, J. R., *J. Electrochem. Soc.* **144** (1997), 2045.
3. Chouvin, J., Branci, C., Sarradin, J., Olivier-Fourcade, J., Jumas, J. C., Simon, D. and Biensan, Ph., *J. Power Sources* **81–82** (1999), 277.
4. Alcantara, R., Fernandez-Madrigal, F. J., Lavela, P., Tirado, J. L., Jumas, J.C. and Olivier Fourcade, J., *J. Matter. Chem.* **9** (1999), 2517.
5. Fernandez-Madrigal, F. J., Lavela, P., Perrez-Vicente, C. and Tirado, J. L., *J. Electroanal. Chem.* **501** (2001), 205.
6. Vaughey, J. T., Johnson, C. S., Kropf, A. J., Benedek, R., Thackeray, M. M., Tostmann, H., Sarakonsri, T., Hackney, S., Fransson, L., Edström, K. and Thomas, J. O., *J. Power Sources* **97** (2001), 194.
7. Tarascon, J. M., Morcrette, M., Dupont, L., Chabre, Y., Payen, C., Larcher, D. and Pralong, V., *J. Electrochem. Soc.* **150** (2003), 732.
8. Lippens, P. E., Olivier-Fourcade, J. and Jumas, J. C., *Hyp. Interact.* **126** (2000), 137.
9. Lippens, P. E., Olivier-Fourcade, J. and Jumas, J. C., *Hyp. Interact.* **141/142** (2002), 303.
10. Blaha, P., Schwarz, K. and Herzig, P., *Phys. Rev. Lett.* **54** (1985), 1192.
11. Blaha, P., Schwarz, K. and Luitz, J., WIEN97, Vienna University of Technology, 1997 [improved and updated Unix version of the original copyright WIEN code, which was published by Blaha, P., Schwarz, K., Sorantin, P. and Trickey, S. B., *Comput. Phys. Commun.* **59** (1990), 399].
12. Hohenberg, P. and Kohn, W., *Phys. Rev.* **136** (1964), B864; Kohn, W. and Sham, L. J., *Phys. Rev.* **140** (1965), A1133.

13. Perdew, J. P., Burke, S. and Ernzerhof, M., *Phys. Rev. Lett.* **77** (1996), 3865.
14. Müller, W. and Schäfer, H., *Z. Naturforsch* **28b** (1973), 246.
15. Müller, W., *Z. Naturforsch* **29b** (1974), 304.

Hyperfine Interactions **156/157**: 335–340, 2004.
© 2004 *Kluwer Academic Publishers. Printed in the Netherlands.*

^{57}Fe Mössbauer Spectroscopy Study of LaFe$_{1-x}$Co$_x$O$_3$ ($x = 0$ and 0.5) Formed by Mechanical Milling

FRANK J. BERRY[1], XIAOLIN REN[1], J. RAMÓN GANCEDO[2] and
JOSÉ F. MARCO[2]

[1] *Department of Chemistry, The Open University, Walton Hall, Milton Keynes MK7 6AA,
United Kingdom*

[2] *Instituto de Química-Física 'Rocasolano', Consejo Superior de Investigaciones Científicas,
Serrano 119, 28006 Madrid, Spain*

Abstract. Perovskite-related phases of the type LaFe$_{1-x}$Co$_x$O$_3$ ($x = 0$ and 0.5) have been synthesised by milling techniques. The materials are of smaller particle size and more susceptible to reduction in a flowing hydrogen/nitrogen gas mixture than their counterparts made by conventional solid state calcination methods. The presence of cobalt enhances the reducibility of iron in the perovskite-related structure and, in contrast to LaFeO$_3$ where the oxide structure is at least partially retained even after treatment in the reducing atmosphere at 1150°C, treatment of LaFe$_{0.5}$Co$_{0.5}$O$_3$ in a reducing atmosphere leads to reduction of both Fe^{3+} and Co^{3+}, their segregation from the oxide matrix, and the formation of an iron–cobalt alloy.

Key words: ^{57}Fe Mössbuer spectroscopy, orthoferrites, mechanical milling.

1. Introduction

Perovskite-related rare earth orthoferrites have been investigated in the past because of their potential for use as logic- or memory-components or as laser and light modulators in optical materials [1, 2]. These solids are also catalytically active for hydrocarbon oxidation and combustion [3] and can be used in sensors [4–6] and solid electrolytes [7–9]. These materials, which contain only trivalent metal ions in the structure, are amenable systems for isovalent substitution and systems of the type LaFe$_{1-x}$Co$_x$O$_3$ have attracted interest in recent years [10–12] because of their potential for electroceramic and catalytic oxidation applications. More recently [13], these materials have been examined for use as automobile exhaust components. Applications such as these require operation under reducing conditions but the nature of the reduced phases in these systems is a matter of scarce investigation. We have previously reported [14] on the reduction properties of materials of the type LaFe$_{1-x}$Co$_x$O$_3$ ($x = 0$ and 0.5) formed by conventional heating processes. We report here on the reduction properties of similar materials made by mechanical milling.

2. Experimental

The materials were prepared by dry milling the reactant oxides in air using a Retsch PM 400 planetary ball mill with stainless steel balls and vials (250 ml) at a milling speed of 200 rpm.The powder to ball mass ratio was 1 : 20. The ^{57}Fe Mössbauer spectra were recorded with a constant acceleration spectrometer using a 50 mCi Co/Rh source. All the spectra were computer-fitted and the isomer shifts referred to the centroid of the α-Fe spectrum. Temperature programmed reduction (tpr) profiles were recorded from *ca.* 150 mg samples in flowing 10% hydrogen/90% nitrogen (15–20 ml/min) with the temperature being increased by 5°C/min.

3. Results and discussion

The X-ray powder diffraction patterns recorded from a mixture of La_2O_3 and α-Fe_2O_3 showed a gradual broadening and simultaneous decrease in intensity of the lines characteristic of the reactant oxides during the first 20 hours of milling and the formation of a virtually amorphous material after 40 hours. The occurrence of a phase corresponding to $LaFeO_3$ was observed after milling for 80 hours with a particle size, determined from the X-ray powder diffraction data, of *ca.* 10 nm which remained unchanged after milling for 110 h. The milling of a mixture of La_2O_3, CoO and α-Fe_2O_3 was also followed by X-ray powder diffraction. The results showed more facile formation of the amorphous materials after only 20 h of milling and the formation of $LaFe_{0.5}Co_{0.5}O_3$ after 40 h with a particle size of *ca.* 10 nm which remained unchanged after milling for 80 and 110 h.

The temperature programmed reduction profiles recorded from $LaFeO_3$ and $LaFe_{0.5}Co_{0.5}O_3$ prepared by ball milling techniques are collected in Figure 1. The profiles show the reduction phenomena to occur at lower temperatures as compared to their counterparts made by conventional methods [14]. The inclusion of cobalt into the $LaFeO_3$ structure enhanced the susceptibility of the material to reduction at lower temperatures.

The ^{57}Fe Mössbauer spectrum recorded from $LaFeO_3$ (Figure 2a) was composed of a broadened sextet pattern showing a magnetic hyperfine field distribution accounting for *ca.* 66% of the spectral area which we associate with a distribution of $LaFeO_3$ particles of different sizes, albeit sufficiently large enough to experience magnetic interactions at room temperature. This magnetic spectrum is broader than that shown by bulk $LaFeO_3$ made by conventional heating [14, 15] and its corresponding average hyperfine magnetic field is lower (48.9 T). In view of the X-ray powder diffraction data, we associate the broad singlet accounting for *ca.* 34% of the spectral area with small particle superparamagnetic $LaFeO_3$.

The ^{57}Fe Mössbauer spectrum recorded from the material following treatment in hydrogen at 600°C (Figure 2b) and after the first reduction peak in the tpr profile (Figure 1) continued to show the broadened sextet pattern which we have associated with $LaFeO_3$ ($H_{av} = 48.7$ T) and accounting for *ca.* 61% of the spectral area, together with a narrow line sextet ($H = 33.0$ T) characteristic of metallic iron

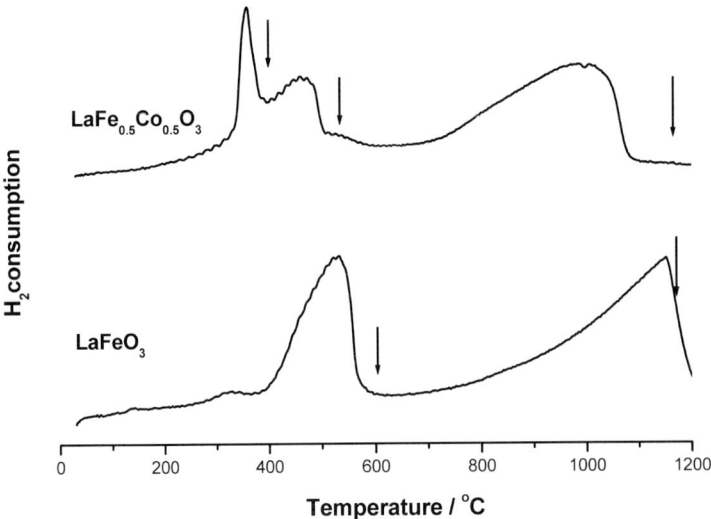

Figure 1. Temperature programmed reduction profiles recorded from LaFe$_{0.5}$Co$_{0.5}$O$_3$ (*top*) and LaFeO$_3$ (*bottom*) prepared by ball milling. The arrows indicate the temperatures at which the samples were taken for examination by [57]Fe Mössbauer spectroscopy.

(*ca.* 27% of spectral area) and a doublet ($\delta = 0.40$ mm s^{-1}, $\Delta = 1.44$ mm s^{-1}, *ca.* 12% of spectral area) characteristic of Fe^{3+} in octahedral oxygen coordination which we associate with a remaining fraction of small particle lanthanum ferrite. The results indicate that treatment at 600°C in a hydrogen/nitrogen gas mixture induces partial reduction of Fe^{3+} in the LaFeO$_3$ structure and its segregation to form metallic iron. The [57]Fe Mössbauer spectrum recorded from the material following treatment in the reducing atmosphere at 1150°C (Figure 2c) and after the final peak in the tpr profile (Figure 1) showed a more narrow line sextet with parameters ($\delta = 0.34$ mm s^{-1}, $2\varepsilon = -0.07$ mm s^{-1}, $H = 53.5$ T, *ca.* 55% of spectral area) that are more similar to those shown by bulk well-crystallized LaFeO$_3$ and which reflects the sintering of the smaller particles at the higher temperature, a sextet corresponding to metallic iron ($H = 33.0$ T, *ca.* 35% of spectral area) and a singlet ($\delta = \sim -0.13$ mm s^{-1}, *ca.* 10% of spectral area) which we associate with the presence of small superparamagnetic metallic alpha-iron particles [16]. The results demonstrate that complete reduction of Fe^{3+} to metallic iron is not achieved even at the elevated temperature of 1150°C, however, the extent of reduction is larger than that achieved when bulk LaFeO$_3$ made by conventional heating was treated under similar conditions [14] and presumably reflects the enhanced susceptibility to reduction of small particle LaFeO$_3$ made by milling.

The [57]Fe Mössbauer spectrum recorded from LaFe$_{0.5}$Co$_{0.5}$O$_3$ (Figure 3a) was dominated by a doublet ($\delta = 0.28$ mm s^{-1}, $\Delta = 0.59$ mm s^{-1}, *ca.* 71% of spectral area) and a narrow line sextet pattern ($\delta = 0.38$ mm s^{-1}, $2\varepsilon = -0.16$ mm s^{-1}, $H = 51.9$ T, *ca.* 29%). Given that X-ray powder diffraction showed the presence of a perovskite-related LaFeO$_3$-type structure similar to that previously identified

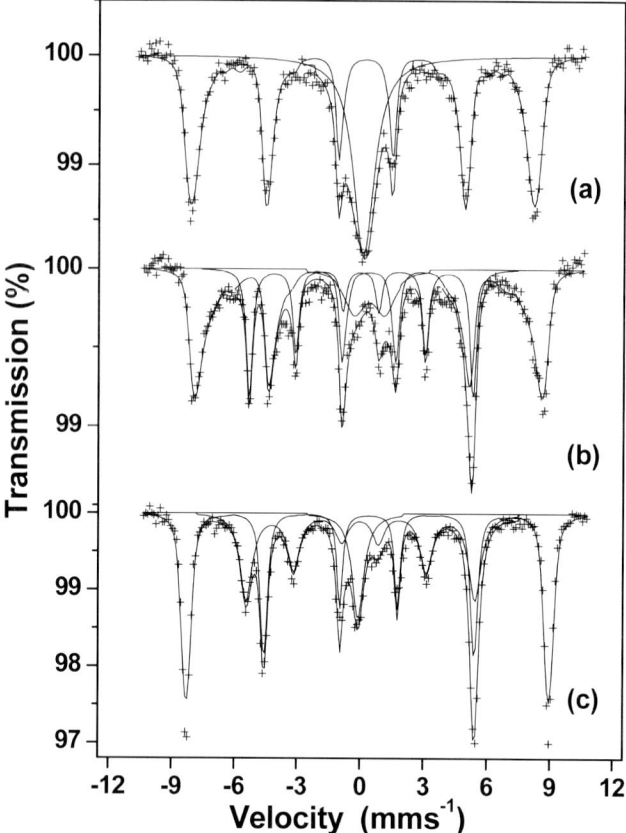

Figure 2. ^{57}Fe Mössbauer spectra recorded at room temperature from (a) freshly made LaFeO$_3$ and after tpr treatment at (b) 600°C and (c) 1150°C.

[14] as LaFe$_{0.5}$Co$_{0.5}$O$_3$, we associate the dominant doublet with superparamagnetic LaFe$_{0.5}$Co$_{0.5}$O$_3$. The sextet does not correspond to LaFe$_{0.5}$Co$_{0.5}$O$_3$ [14]. In view of the reduction properties of the material (see below and [14]) it cannot correspond to LaFeO$_3$. Therefore, we associate this sextet with an α-Fe$_2$O$_3$ impurity phase whose diffraction lines could not be seen in the corresponding X-ray powder diffraction pattern probably because they were masked by the intense background associated with the small particle LaFe$_{0.5}$Co$_{0.5}$O$_3$. The ^{57}Fe Mössbauer spectrum recorded from LaFe$_{0.5}$Co$_{0.5}$O$_3$ following treatment in hydrogen and nitrogen at 400°C (Figure 3b) and after the first peak in the tpr profile (Figure 1) showed a broadened sextet pattern ($\delta = 0.35$ mm s^{-1}, $2\varepsilon = -0.17$ mm s^{-1}, $H_{ave} = 25$ T, *ca.* 35% of spectral area) similar to that of bulk LaFe$_{0.5}$Co$_{0.5}$O$_3$ [14] together with the doublet observed in the original material and associated with small particle superparamagnetic LaFe$_{0.5}$Co$_{0.5}$O$_3$ (*ca.* 40% of spectral area) and two sextets characteristic of Fe$_3$O$_4$ (spectral area *ca.* 27%) resulting from reduction of the α-Fe$_2$O$_3$ impurity phase. Hence the features in the tpr profile recorded from

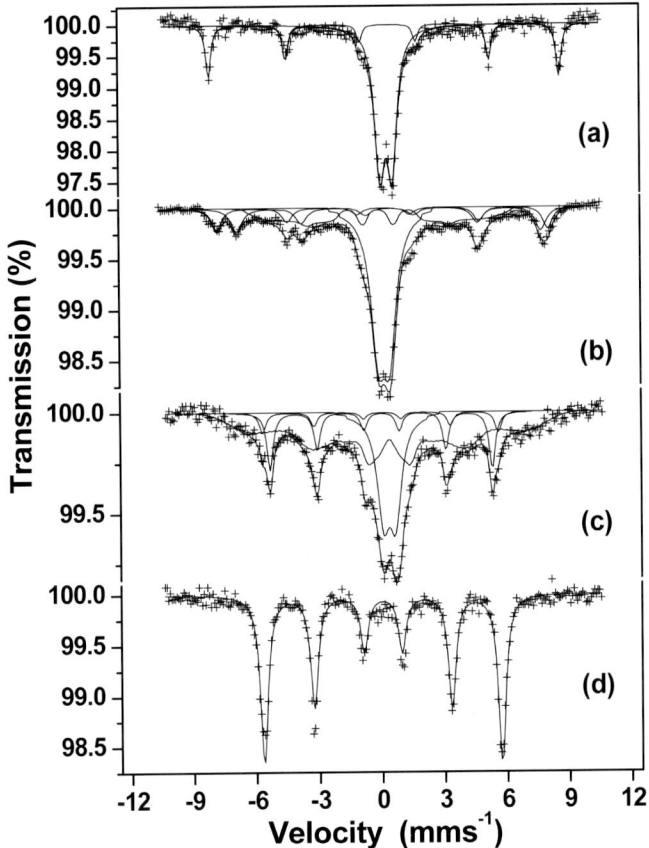

Figure 3. ^{57}Fe Mössbauer spectra recorded at room temperature from (a) freshly made LaFe$_{0.5}$Co$_{0.5}$O$_3$ and after tpr treatment at (b) 400°C, (c) 550°C and (d) 1150°C.

LaFe$_{0.5}$Co$_{0.5}$O$_3$ at *ca.* 300°C to 400°C appear to be associated with the reduction of the impurity α-Fe$_2$O$_3$ phase and the development of large particle LaFe$_{0.5}$Co$_{0.5}$O$_3$ presumably reflecting the sintering of some of the small particles of this phase at 400°C. The ^{57}Fe Mössbauer spectrum recorded from LaFe$_{0.5}$Co$_{0.5}$O$_3$ following treatment in the hydrogen and nitrogen gas mixture at 550°C (Figure 3c) and after the second peak in the tpr profile (Figure 1) was composed of a broad sextet pattern (*ca.* 55% of spectral area) characteristic of the large particle LaFe$_{0.5}$Co$_{0.5}$O$_3$ together with a doublet (*ca.* 24% of spectral area) associated with the superparamagnetic LaFe$_{0.5}$Co$_{0.5}$O$_3$, and two narrow sextets, one ($H = 33.0$ T, 15% of spectral area) corresponding to metallic iron and the other ($\delta = 0.01$ mm s^{-1}, $2\varepsilon = -0.06$ mm s^{-1}, $H_{ave} = 35.0$ T, 6% of spectral area) corresponding to an iron–cobalt alloy [17]. Following treatment in the hydrogen/nitrogen gas mixture at 1150°C and after the final peak in the tpr profile (Figure 1) the ^{57}Fe Mössbauer spectrum (Figure 3d) showed a narrow line sextet ($\delta = 0.07$ mm s^{-1}, $2\varepsilon = 0.00$ mm s^{-1}, $H = 35.4$ T) characteristic of an iron–cobalt alloy [17].

Taken together the results show that small particle $LaFe_{1-x}Co_xO_3$ phases made by milling techniques are more susceptible to hydrogen reduction than their counterparts made by conventional techniques. The presence of cobalt enhances the reducibility of iron in the perovskite-related structure and, in contrast to $LaFeO_3$ where the oxide structure at least was partially retained even after treatment in the reducing atmosphere at 1150°C, treatment of $LaFe_{0.5}Co_{0.5}O_3$ in a reducing atmosphere leads to reduction of both Fe^{3+} and Co^{3+}, their segregation from the oxide matrix and the formation of an iron–cobalt alloy phase.

References

1. Vasques, C., Kogerler, P. and Lopez-Quintela, M. A., *J. Mater. Res.* **13** (1998), 451.
2. Moskvin, A. S., Ovanesyan, N. S. and Trukhtarnov, V. A., *Hyp. Interact.* **1** (1975), 265.
3. McCarty, J. G. and Wise, H., *Catalysis Today* **8** (1990), 231.
4. Shimizu, Y., Shimabukuro, M., Arai, H. and Seiyama, T., *Chem. Lett.* (1985), 917.
5. Arakawa, T., Kurachi, H. and Shiokawa, J., *J. Mater. Sci.* **4** (1985), 1207.
6. Matsuura, Y., Matsushima, S., Sakamoto, M. and Sadoaka, Y., *J. Mater. Chem.* **3** (1993), 767.
7. Karlsson, F. G., *Electrochem. Acta* **30** (1985), 1555.
8. Nekamura, T., Petzow, G. and Gancker, L. J., *Mater. Res. Bull.* **14** (1979), 649.
9. Mizusaki, J., Sisamoto, T., Cannon, W. K. and Bowen, H. K., *J. Am. Ceram. Soc.* **65** (1982), 363.
10. Sadaoka, Y., Traversa, E., Nunziante, P. and Sakamoto, M., *J. Alloy Compounds* **261** (1997), 182.
11. Traversa, E., Nunziante, P., Sakamoto, M., Sodoaka, Y. and Montanar, R., *Mater. Res. Bull.* **33** (1998), 673.
12. Choudhary, V. R., Uphade, B. S. and Pataskar, S. G., *Fuel* **78** (1999), 919.
13. Nishihata, Y., Mizuki, J., Akao, T., Tanaka, H., Uenishi, M., Kimura, M., Okamoto, T. and Hamada, N., *Nature* **418** (2002), 164.
14. Berry, F. J., Gancedo, J. R., Marco, J. F. and Ren, X., *Hyp. Interact. (C)* **5** (2002), 273.
15. Eibschirz, M., Shtrikmand, S. and Treves, D., *Phys. Rev.* **156** (1967), 562.
16. Clausen, B. S., Morup, S., Nielsen, P., Traen, N. and Topsoe, H., *J. Phys. E* **19** (1979), 439.
17. Carles, V., Laurent, C., Brieu, M. and Rousset, A., *J. Mater. Chem.* **9** (1999), 1003.

Hyperfine Interactions **156/157**: 341–346, 2004.
© 2004 *Kluwer Academic Publishers. Printed in the Netherlands.*

A Crystallographic and Mössbauer Spectroscopic Study of $BaCo_{0.5x}Zn_{0.5x}Ti_x Fe_{12-2x}O_{19}$ (M-Type Hexagonal Ferrite)

T. M. MEAZ[1,*] and C. BENDER KOCH[2]
[1] *Physics Department, Faculty of Science, Tanta University, Tanta, Egypt;*
e-mail: tmeaz@yahoo.com
[2] *Chemistry Department, The Royal Veterinary and Agricultural University, Thorvaldsensvej 40,*
DK-1871 Frederiksberg C, Denmark

Abstract. A series of pure polycrystalline M-type hexagonal ferrite with the formula $BaCo_{0.5x}Zn_{0.5x}$ $Ti_x Fe_{12-2x}O_{19}$ ($x = 0.0, 0.4, 0.8, 1.2, 1.6, 2.0$) has been synthesized and studied by X-ray diffraction and Mössbauer spectroscopy. With increasing substitution the Mössbauer spectra change from magnetically ordered ($x = 0$) towards magnetically ordered with strong line broadening ($x = 0.4$ to 1.6) to nonmagnetic ($x = 2.0$), due to decreasing of Curie temperatures. Differential line broadening and relative area of components indicates that for small values of x, a substitution occurs preferentially in $4f_1$ and $4f_2$ sites indirectly affecting Fe in the 12k site.

Key words: hexagonal ferrite, Mössbauer spectroscopy, X-ray diffraction, barium ferrite.

1. Introduction

Barium ferrite (BaM) is well established for use as a permanent magnetic material [1]. BaM exhibits a fairly large magnetocrystalline anisotropy and high Curie temperature, and has in addition a relatively large saturation magnetization, chemical stability and corrosion resistivity [2]. It is also a suitable material for high-density magnetic recording applications [1–3]. BaM has a hexagonal ferrite structure ($P6_3/mmc$) which is symbolically described by RSR^*S^* where R is a three-layer block (containing two O_4 and one BaO_3) with composition $(BaFe(III)_6O_{11})^{2-}$ and S is a two O_4-layer block with composition $(Fe(III)_6O_8)^{2+}$, where the asterix imply that the corresponding block has been rotated 180° around the hexagonal c-axis. In this structure the metallic cations are distributed among five different sites (sublattice). These sites have different coordinations and are designated: tetrahedral ($4f_1$ and bipyramidal 2b) and three different octahedral (12k, $4f_2$, and 2a) sublattices. Opposite spin directions cause a ferrimagnetic structure.

Sr^{2+} or Pb^{2+} can replace the Ba^{2+} ion in the M-type hexagonal ferrites partly or completely, without changing the crystal structure. Substitutions of Fe^{3+} ion

---

* Author for correspondence.

and O^{2-} are also possible. For the various magnetic applications of BaM a high saturation, a suitable coercivity, and low temperature coefficients of coercivity and remanence are desired properties. Therefore, much work has been done to modify the magnetic properties through the substitutions of Fe^{3+} ions. Substitution of Fe^{3+} ions can be accomplished in two ways: Direct substitution by trivalent ions (Al^{3+}, Ga^{3+}, Mn^{3+}) or by combined substitution of divalent and tetravalent ions ($Co^{2+} + Ti^{4+}$, $Zn^{2+} + Ti^{4+}$, $Zn^{2+} + Sn^{4+}$). Various combinations of substituents have been reported (Co–Ti, Co–Sn, Zn–Ti, Co–Zn–Nd, Co–Ti–Sn, Co–Mo and Zn–Zr) Mössbauer studies [4–14] and others magnetic studies [15–30].

The magnetic properties of the substituted hexaferrites are strongly depend on electronic configuration of the substituted cations as well as on their site preference. Substitution by nonmagnetic ions are expected to lower the magnetic ordering temperatures and differences primarily in the charge of the substituents may cause different neighbor effects.

In this work coupled substitution of magnetic (Co^{2+}) and non-magnetic ions (Zn^{2+}–Ti^{4+}) for Fe^{3+} in BaM was investigated using X-ray diffraction and ^{57}Fe Mössbauer spectroscopy.

2. Experimental

The starting materials used for samples preparation were, high purity barium carbonate, cobalt oxide, zinc oxide, titanium oxide and ferric oxide. The compounds were mixed together in stoichiometric ratios to prepare a series of polycrystalline M-type hexagonal ferrites of composition $BaCo_{0.5x}Zn_{0.5x}Ti_xFe_{12-2x}O_{19}$ with ($x = 0.0$, 0.4, 0.8, 1.2, 1.6 and 2.0) using conventional ceramic technique. The powders were mixed for 5 hr using an agate-grinding machine. The final powders were pressed and pre-sintered for 6 hr at 950°C. After slowly cooling to room temperature, the reaction mixtures were again ground for 5 hr and then pressed into the shape of a disc. The discs were sintered at 1150°C for 6 hr, and the samples were slowly cooled to room temperature. The last heating step was repeated after reground and repressed the discs.

The X-ray diffractograms were recorded with a Siemens D-5000 X-ray diffractometer using Co K_α radiation.

The ^{57}Fe Mössbauer spectra were obtained at RT and 80 K using a constant acceleration spectrometers with a ^{57}Co source in Rh matrix. The spectrometers were calibrated using a thin foil of α-Fe at room temperature.

3. Results and discussion

The X-ray diffraction patterns of the $BaCo_{0.5x}Zn_{0.5x}Ti_xFe_{12-2x}O_{19}$ samples was used to confirm the existence of single-phase M-type hexagonal ferrites (not shown). They demonstrate that all the samples of this series have the magnetoplumbite structure, and no trace of other phases was detected. The line width of the

Table I. Lattice constants (*a* and *c*) of the BaCo$_{0.5x}$Zn$_{0.5x}$Ti$_x$Fe$_{12-2x}$O$_{19}$ samples as a function of the mole fraction (*x*)

x	*a* (Å)	*c* (Å)
0	5.8688 ± 0.008	23.19848 ± 0.043
0.4	5.87458 ± 0.019	23.23328 ± 0.082
0.8	5.83336 ± 0.073	23.25161 ± 0.091
1.2	5.87992 ± 0.0148	23.28674 ± 0.063
1.6	5.86254 ± 0.0168	23.31093 ± 0.072
2	5.87256 ± 0.0123	23.32603 ± 0.038

Table II. Mössbauer parameters of the spectra measured at room temperature for $x = 0.0$ and 0.4

Sample *x* value	Site	Isomer shift δ (mm s^{-1}) ±0.03	Quadrupole spliting δ (mm s^{-1}) ±0.03	Hyperfine field B_{hf} (Tesla) ±0.2	Line width Γ (mm s^{-1}) ±0.02	Relative area (%) ±2
0.0	12k	0.36	0.21	41.7	0.40	49
	4f$_1$	0.25	0.11	49.0	0.27	9
	4f$_2$	0.39	0.07	51.9	0.35	20
	2a	0.31	0.08	50.2	0.36	16
	2b	0.27	1.12	40.4	0.38	6
0.4	12k'	0.35	0.19	41.4	0.57	21
	12k''	0.35	0.15	37.3	1.12	35
	4f	0.39	0.12	49.8	0.40	13
	4f$_1$ + 2a	0.29	0.07	47.4	0.55	25
	2b	0.16	0.93	38.6	0.81	6

diffractions lines is insignificantly broadened (comparing with the standard), indicating that the samples consist of large crystallites. Although the lattice constant a does not change significantly with the mole ratio x, the constant c does increase monotonically as x increases, as seen from Table I. This may be attributed to the larger ionic radius of Ti^{4+} (0.68 Å), Zn^{2+} (0.74 Å) and Co^{2+} (0.72 Å) compared to Fe^{3+}(0.64 Å).

The Mössbauer spectrum of the BaFe$_{12}$O$_{19}$ end member consists at room temperature of five magnetically split and overlapping components. In Table II are given the parameters of a fit using five components. The parameters of the 4f$_2$ and the 2a sites are so similar that most frequently only four components are

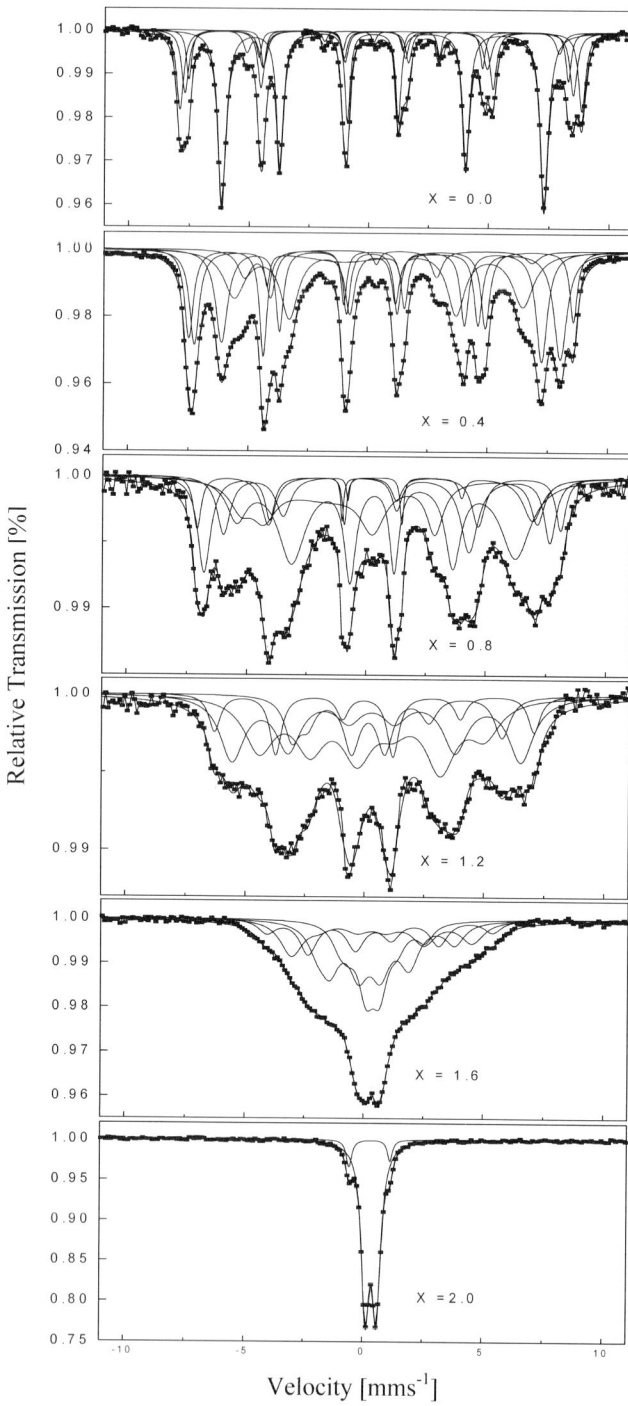

Figure 1. Mössbauer spectra of the $BaCo_{0.5x}Zn_{0.5x}Ti_xFe_{12-2x}O_{19}$ samples obtained at room temperature. Mole fractions are indicated in the figure.

used, taken these two sites as one component. The hyperfine parameters are in good agreement with previously published values [6, 8, 31–33]. The substitution of the iron ions by other cations leads to linc broadening due to perturbation of the magnetic interaction. In particular, the $x = 0.4$ sample exhibits a strong asymmetry in the component due to the 12k site. Introducing two components for the modified 12k site (into 12k$'$ and 12k$''$ sites) and treating 4f$_1$ and 2a as one component the derived hyperfine values are reported in Table II. These results demonstrate strong line broadening and some reduction in the magnetic hyperfine fields. It has previously been reported that Co substitution induces a splitting of the component due to the 12k site [13]. Increasing the substitution to 0.8 and 1.2 causes further line broadening and reduction of magnetic hyperfine field. At these substitution levels fitting becomes meaningless (in Figure 1) is indicated fits using five components for $x = 0.8$ to 1.6, but no assignment can be argued. The results indicate that at high substitution level all sites are affected. At highest levels of substitution ($x = 1.6$ and 2.0) magnetic order vanish indicating Curie temperatures around and below RT. Magnetic order in these samples is restored by lowering the temperature to 80 K (spectra not shown). The magnetic susceptibility at RT (not shown) is almost constant for $x = 0$ to 1.2, but decreases sharply for larger values of x, in agreement with a lowering of the Curie temperatures. Based on the intensity of the 12k sites for $x = 0.4$ is appears that the substitution for low x values occur not in the 12k site itself but rather in the neighboring 4f$_1$ and 4f$_2$ sites. It is plausible that 4f$_1$ host Ti^{4+} and Co^{2+} ions whereas Zn^{2+} may be preferred in 4f$_2$, but neutron diffraction and more detailed analysis of the X-ray data are needed.

4. Conclusion

The present investigation has extended the investigated ferrite composition to include BaCo$_{0.5x}$Zn$_{0.5x}$Ti$_x$Fe$_{12-2x}$O$_{19}$. High levels of substitutions expectedly change the magnetic properties of the ferrite. At low levels of substitution the line broadening is sufficiently small to allow a site assignment indicating that the substitution occurs in the 4f$_1$ and 4f$_2$ sites (Ti^{4+}, Co^{2+} and Zn^{2+}, respectively).

References

1. Kojima, H., In: E. P. Wohlfahrt (ed.), *Ferromagnetic Materials*, Vol. 3, North-Holland, Amsterdam, 1982, p. 305.
2. Liu, X., Wang, J., Gan, L.-M., Ng, S.-C. and Ding, J., *J. Magn. Magn. Mater.* **184** (1998), 344.
3. Wartewig, P., Krause, M. K., Esquinazi, P., Rösler, S. and Sonntag, R., *J. Magn. Magn. Mater.* **192** (1999), 93.
4. Zheng, Y., Yu, Z. and Shao, Y., *Hyp. Interact.* **94** (1994), 2035.
5. Melzer, K., Wartewig, P., Stumm, U., Gütlich, P. and Michalk, C., *Hyp. Interact.* **94** (1994), 2045.
6. Morrish, A. H., Zhou, X. Z., Yang, Z., Zeng, H.-X., *Hyp. Interact.* **90** (1994), 365.
7. Li, Z. W., Ong, C. K., Yang, Z., Wei, F. L., Zhou, X. Z., Zhao, J. H. and Morrish, A. H., *Phys. Rev. B* **62**(10) (2000), 6530.

8. Zhou, X. Z., Morrish, A. H., Yang, Z. and Zeng, H.-X., *J. Appl. Phys.* **75** (1994), 5556.
9. Batlle, X., Obradors, X., Rodriguez-Carvajal, J., Pernet, M., Cabañas, M. V. and Vallet, M., *J. Appl. Phys.* **70** (1991), 1614.
10. Hong, Y. K., Paig, Y. J., Agresti, D. G. and Shelfer, T. D., *J. Appl. Phys.* **61** (1987), 3872.
11. Rane, M. V., Bahadur, D., Mandal, S. K. and Patni, M. J., *J. Magn. Magn. Mater.* **153** (1996), L1.
12. Wartewig, P., Melzer, K., Krause, M. K. and Tellgren, R., *J. Magn. Magn. Mater.* **140–144** (1995), 2101.
13. Šimša, Z., Legu, S., Gerber, R. and Pollert, E., *J. Magn. Magn. Mater.* **140–144** (1995), 2103.
14. Rane, M. V., Bahadur, D., Nigam, A. K. and Srivastava, C. M., *J. Magn. Magn. Mater.* **192** (1999), 288.
15. Wang, C. S., Wei, F. L., Lu, M., Han, D. H. and Yang, Z., *J. Magn. Magn. Mater.* **183** (1998), 241.
16. Fang, H. C., Yang, Z., Ong, C. K., Li, Y. and Wang, C. S., *J. Magn. Magn. Mater.* **187** (1998), 129.
17. Fang, H. C., Ong, C. K., Zhang, X. Y., Li, Y., Wang, X. Z., Yang, Z., *J. Magn. Magn. Mater.* **191** (1999), 277.
18. Sugg, B. and Vincent, H., *J. Magn. Magn. Mater.* **139** (1995), 364.
19. Kreisel, J., Vincent, H., Tasset, F., Paté, M. and Ganne, J. P., *J. Magn. Magn. Mater.* **224** (2001), 17.
20. K. Kakizaki, K., Hiratsuka, N. and Namikawa, T., *J. Magn. Magn. Mater.* **176** (1997), 36.
21. Batlle, X., García del Muro, M., Tejada, J., Pfeiffer, H., Görnert, P. and Sinn, E., *J. Appl. Phys.* **74**(5) (1993), 3333.
22. Rane, M. V., Bahadur, D., Kulkarni, S. D. and Date, S. K., *J. Magn. Magn. Mater.* **195** (1999), L256.
23. Cabañas, M. V., González-Calbet, J. M. and Vallet-Raegí, M., *J. Sol. State Chem.* **115** (1995), 347.
24. Kubo, O., Ido, T., Yokoyama, H. and Koike, Y., *J. Appl. Phys.* **57**(1) (1985), 4280.
25. Chou, F., Feng, X., Li, J. and Liu, Y., *J. Appl. Phys.* **61**(8) (1987), 3881.
26. Clark, T. M., Evans, B. J. and Thompson, G. K., *J. Appl. Phys.* **85**(8) (1999), 5229.
27. Elkady, H. A., Abou-Sekkina, M. M. and Nagorny, K., *Hyp. Interact.* **116** (1998), 149.
28. Hernández-Gómez, P., de Francisco, C., Brabers, V. A. M. and Dalderop, J. H. J., *J. Appl. Phys.* **87**(7) (2000), 3576.
29. Díaz-Castañón, S., Leccabue, F., Watts, B. E. and Albanese, G., *J. Magn. Magn. Mater.* **196–197** (1999), 458.
30. Zhou, X. Z., Morrish, A. H., Li, Z. W. and Hong, Y. K., *IEE Trans. Magn.* **MAG-27** (1991), 4654.
31. Evans, B. J., Grandjan, F., Lilot, A. P., Vogel, R. H. and Gérard, A., *J. Magn. Magn. Mater.* **67** (1987), 123.
32. Campbell, S. J., Wu, E., Kaczmarek, W. A. and Jayasuriya, K. D., *Hyp. Interact.* **92** (1994), 933.
33. Sankaranarayanan, V. K. and Khan, D. C., *J. Magn. Magn. Mater.* **153** (1996), 337.

Hyperfine Interactions **156/157**: 347–352, 2004.
© 2004 *Kluwer Academic Publishers. Printed in the Netherlands.*

Laboratory Intercomparison on the Determination of the Fe(II)/Fe(III) Ratio in Glass Using Mössbauer Spectroscopy

H. MEHNER, M. MENZEL* and M. NOFZ
Federal Institute for Materials Research and Testing (BAM)
D-12489 Berlin, Richard-Willstätter-Str. 11, Germany; e-mail: Michael.Menzel@BAM.DE

Abstract. The Mössbauer community all over the world was given the opportunity to determine the Fe(II)/Fe(III) ratio in 4 glass samples of different total iron content, to classify themselves with respect to their measurement technique and especially to their kind of evaluation.

Key words: Mössbauer spectroscopy, iron, glass, intercomparison, Fe(II)/Fe(III) ratio.

1. Introduction

In the chemical analysis it is quite normal to use laboratory intercomparisons to compare the analytical results of different laboratories [1, 2]. The goal is, to determine the true content or concentration of an analyte in materials. These intercomparisons show often large differences in the analytical results of different groups and/or different analytical methods. It is necessary to overcome such deviations and to avoid them in the future.

2. Samples

The material glass plays an outstanding role for technical purposes. To reveal the correct Fe(II)/Fe(III) ratio without any destruction is a main goal in this respect. For Mössbauer spectroscopists the existence of superparamagnetic phases in the samples is a real challenge. To apply Mössbauer spectroscopy at a minimum expenditure it would be important to get "correct" results even at room temperature. Therefore, the comparison is of special interest.

The float glass samples were prepared from the substances shown in Table I.

The samples 1–3 are enriched in ^{57}Fe to improve the measurement effect of the Mössbauer measurements.

The element composition of the glass samples is shown in Table II.

---

* Author for correspondence.

Table I. Initial weight of the raw material for the glass samples

Sample 1	27.90 g Na_2CO_3	16.45 g $CaCO_3$	73.04 g SiO_2	0.3574 g $^{57}Fe_2O_3$
				1.0723 g Fe_2O_3
Sample 2	28.18 g Na_2CO_3	16.62 g $CaCO_3$	73.78 g SiO_2	0.2880 g $^{57}Fe_2O_3$
				0.1410 g Fe_2O_3
Sample 3	29.59 g Na_2CO_3	17.45 g $CaCO_3$	77.47 g SiO_2	0.1498 g $^{57}Fe_2O_3$
Sample 4	27.90 g Na_2CO_3	16.45 g $CaCO_3$	73.04 g SiO_2	1.4297 g Fe_2O_3

Table II. Compositions of the samples

Sample	Mol %					Ma %				
	Na	Ca	Si	Fe	O	Na	Ca	Si	**Fe**	O
1/4	10.9	3.4	25.3	0.4	60.0	12.1	6.6	34.1	**1.0**	46.2
2	11.0	3.4	25.5	0.1	60.0	12.2	6.7	34.5	**0.3**	46.3
3	11.0	3.4	25.5	(0.04)	60.0	12.3	6.7	34.6	**0.1**	46.3

3. Preparation and treatment of the raw material

The starting materials were mixed carefully and 80% of them heated up in a covered Pt/Rh crucible to 1300°C. When the temperature had increased up to 1500°C the remaining mixture was put in and kept 75 minutes at that temperature. Then 60 minutes later the wanted melting temperature of 1550°C was reached. After altogether 7 hours and 15 minutes the melt was quenched into water. The resulting coarse-grained material was powdered using a planet-ball-mill with highest speed during 1 minute in portions of 5 g because of the mill capacity. These single portions were collected and then mixed carefully.

4. Distribution of the samples and results of the intercomparison

To check the total iron mass fraction photometrical determinations of the Fe-contents were made. For each of the 4 glass samples 100 g material was produced. Finally 500 mg of each sample and an iron calibration foil were sent to 163 Mössbauer groups in 65 countries. 86 groups from 43 countries responded, 46 groups from 26 countries contributed results. Only 27 participants used a sufficient large Mössbauer velocity of $\geqslant 10$ mm/s to measure the samples, because they were aware of the presence of superparamagnetic species. Groups who registered the spectra at a velocity < 10 mm/s neglected this important contribution. All these results could not be taken into consideration. Figures 1 and 2 show Mössbauer spectra of sample 1 taken at the BAM at room temperature and at 78 K, respectively. The influence of the superparamagnetic species is obvious. The contributed results for the determination of the Fe(II)/Fe(III) ratio are shown in Figures 3–6 for room temperature and in Figures 7–10 for measurements at 78 K.

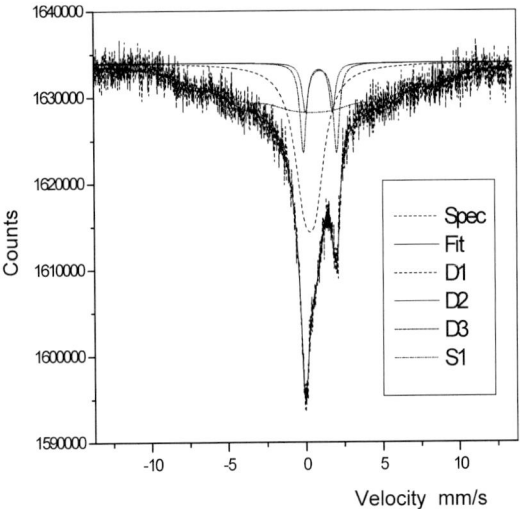

Figure 1. Mössbauer spectrum of sample 1 at room temperature.

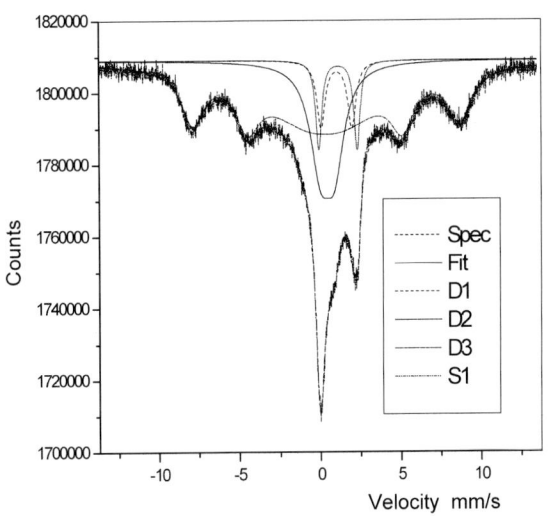

Figure 2. Mössbauer spectrum of sample 1 at 78 K.

In Table III the average values of the Fe(II)/Fe(III) ratio are summarised in order to compare them with the photometrical determinations according to regulations of the International Commission on Glass ICG/TC2/95-1039B (1995-01). For the interpretation of the spectra measured at the BAM a model with 4 multiplets was used. The doublet D1 and the sextet S1 represent Fe(III) and the doublets D2 and D3 represent Fe(II).

The intercomparison shows, that the determination of the Fe(II)/Fe(III) ratio is strongly influenced by choosing a Mössbauer velocity > 10 mm/s and by choosing a model which takes into consideration the superparamagnetic phase completely.

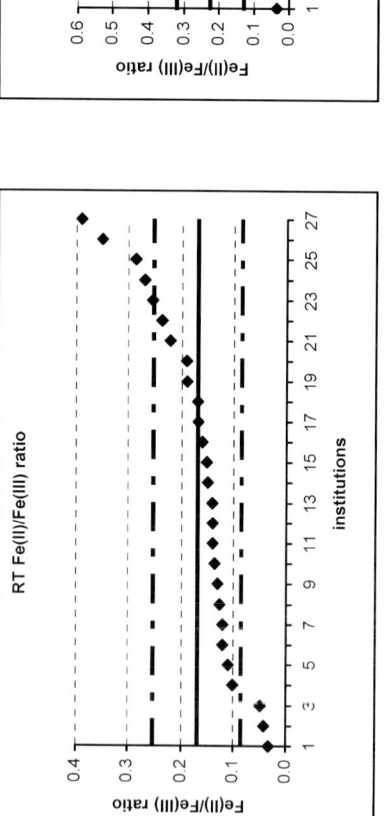

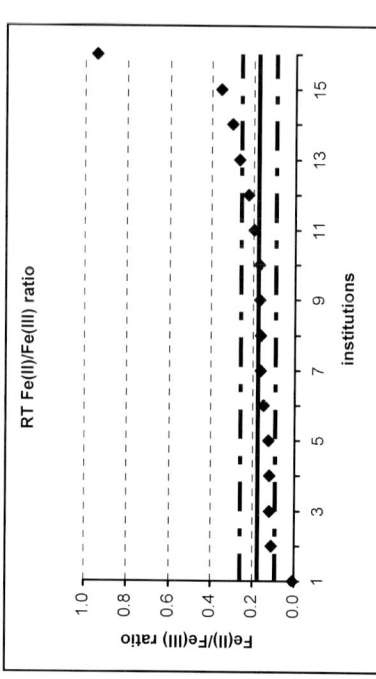

Figure 3. Fe(II)/Fe(III) ratio of sample 1 measured at room temperature.

Figure 4. Fe(II)/Fe(III) ratio of sample 2 measured at room temperature.

Figure 5. Fe(II)/Fe(III) ratio of sample 3 measured at room temperature.

Figure 6. Fe(II)/Fe(III) ratio of sample 4 measured at room temperature. The numbers on x-axis do not identify distinct institutions.

Please note: In Figures 3–10 the x-axis is always ordered after increasing values of the Fe(II)/(III) ratio.

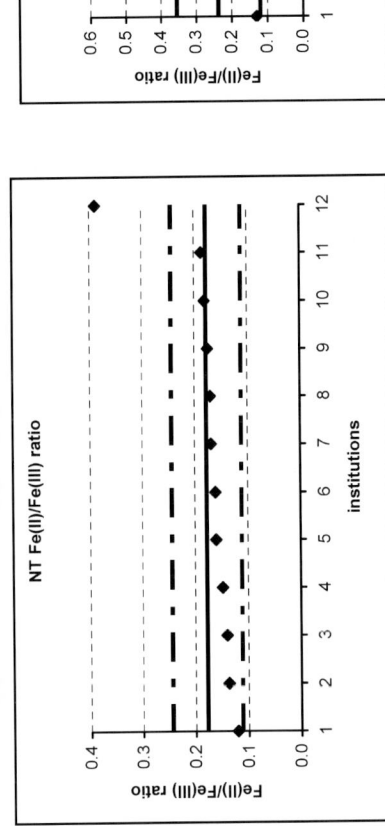

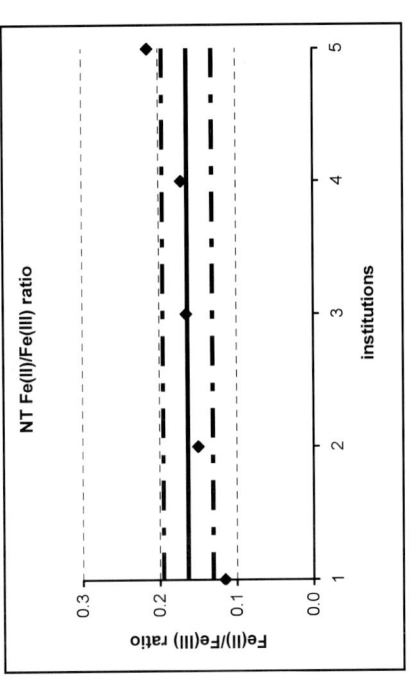

Figure 7. Fe(II)/Fe(III) ratio of sample 1 measured at liquid nitrogen temperature.

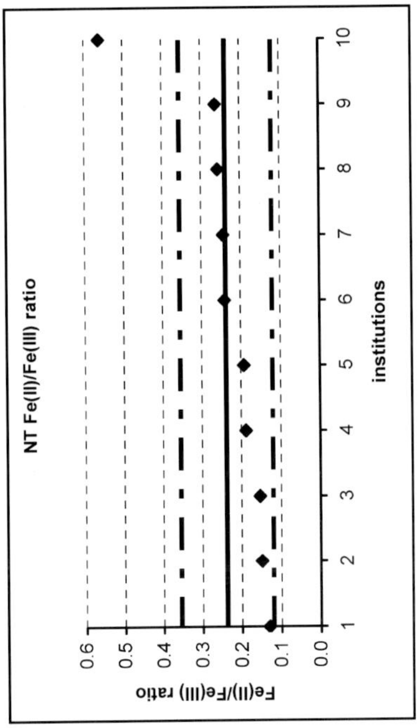

Figure 8. Fe(II)/Fe(III) ratio of sample 2 measured at liquid nitrogen temperature.

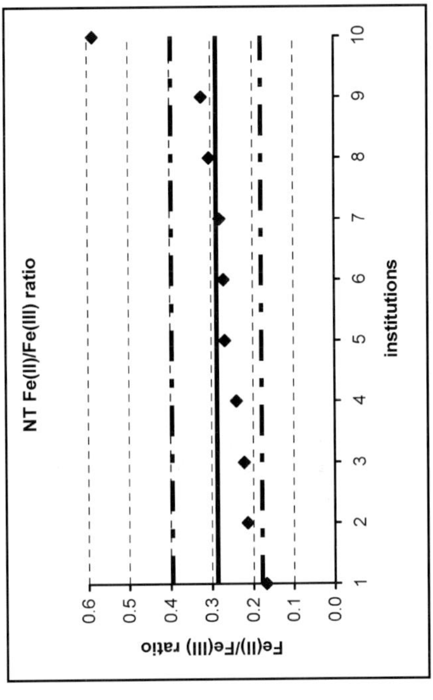

Figure 9. Fe(II)/Fe(III) ratio of sample 3 measured at liquid nitrogen temperature.

Figure 10. Fe(II)/Fe(III) ratio of sample 4 measured at liquid nitrogen temperature.

Table III. Overview of the determined Fe(II)/Fe(III) ratios

	Sample (absolute Fe content) relative Fe content			
	1 (1% Fe) 25% Fe-57 + 75% Fe-nat Fe(II)/Fe(III)	**2** (0.3% Fe) 67% Fe-57 + 33% Fe-nat Fe(II)/Fe(III)	**3** (0.1% Fe) 100% Fe-57 Fe(II)/Fe(III)	**4** (1% Fe) 100% Fe-nat. Fe(II)/Fe(III)
Room temperature	0.17 ± 0.08	0.22 ± 0.09	0.26 ± 0.13	0.18 ± 0.08
Temperature 78 K	0.18 ± 0.07	0.24 ± 0.12	0.29 ± 0.11	0.16 ± 0.03
Photometrical determination	0.16 ± 0.01	0.16 ± 0.01	0.16 ± 0.01	$0.17 + 0.01$

In the case of Mössbauer measurements the $\pm$ values are the standard deviations of all Mössbauer measurements.

A detailed statistical investigation of the data and the publication of these results is planned.

Acknowledgements

The authors are very grateful to all participants of this intercomparison for their support, for their engagement to measure the samples, for the efforts to evaluate the spectra and for the transfer of the results. To researchers, who are interested to compare the results of their analyses with the outcome of this intercomparison, the sample material can be shipped.

References

1. Malz, F., Ph.D. Thesis, Humboldt University, Berlin, 30 June 2003.
2. Dyar, M. D., *American Mineralogist* **69** (1984), 1127.

Hyperfine Interactions **156/157**: 353–358, 2004.
© 2004 Kluwer Academic Publishers. Printed in the Netherlands.

^{57}Fe Mössbauer Spectroscopic Study on the Assembled Iron Complexes

S. NAKASHIMA[1], Y. ASADA[2] and T. OKUDA[2]

[1] *Natural Science Center for Basic Research and Development (N-BARD), Hiroshima University, Kagamiyama, Higashi-Hiroshima 739-8526, Japan*
[2] *Department of Chemistry, Graduate School of Science, Hiroshima University, Kagamiyama, Higashi-Hiroshima 739-8526, Japan*

Abstract. ^{57}Fe Mössbauer spectroscopy was carried out to know the electronic states of the assembled iron complexes. The IS value revealed a high-spin Fe^{II} state in the bipyridine and pyrazine iron complexes with NCS or NCSe. The dissociation behavior of ligand was investigated by TG, and the resultant change in the coordination sphere around iron atom was reflected in the change in QS value.

Key words: Mössbauer spectroscopy, assembled iron complexes, thermal dissociation, spin state.

1. Introduction

Self-assembled coordination polymers containing transition metal ions and organic bridging ligands have attracted intensive interests because of their potential abilities for selective inclusion and transformation of ions and molecules. Thus, porous assembled complexes become candidates for artificial zeolites. The size of cavity can be controlled by ligands and also the thermal dissociation. $\{[M(NCS)_2(H_2O)_2(bpy)]$ $(bpy)\}_n$ is interesting because the coordinated water and one bpy (=bipyridine) are dissociated stepwise with heating, which results in the change of the size of cavity [1, 2]. However, the direct crystal structural analysis of the thermal dissociation products is not reported. The assembled iron complexes are also interesting from the point of spin crossover phenomena. For example, *trans*-1,2-bis (4-pyridyl)ethylene and *trans*-4,4′-azopyridine iron complexes with NCS show the spin crossover phenomena [3, 4].

In the present study, we carried out ^{57}Fe Mössbauer spectroscopy to know the spin state of the assembled bipyridine (bpy) and pyrazine (pyz) iron complexes with NCS or NCSe and to pursue the thermal dissociation of ligand.

2. Experimental

A ^{57}Co(Rh) source in a constant acceleration mode was used for ^{57}Fe Mössbauer spectroscopic measurements. ^{57}Fe Mössbauer spectra were obtained at 78 and 300 K by using a Wissel Mössbauer spectrometer and a proportional counter. The isomer shift was referred to a metallic iron foil. The Mössbauer parameters were obtained by a least-squares fitting to Lorentzian peaks.

Crystal structures were determined by the X-ray crystal structural analysis. All measurements were made on a Mac Science DIP 2030 imaging plate detector using graphite-monochromated Mo-Kα radiation. To compare the crystal structures, powder X-ray diffraction patterns were measured by using graphite-monochromated Cu-Kα radiation (RIGAKU) at room temperature.

3. Results and discussion

ORTEP drawings of $\{[Fe(NCS)_2(H_2O)_2(bpy)](bpy)\}_n$ **1a** and $\{[Fe(NCSe)_2(H_2O)_2 (bpy)](bpy)\}_n$ **1b** are shown in Figures 1 and 2, respectively. The X-ray structural

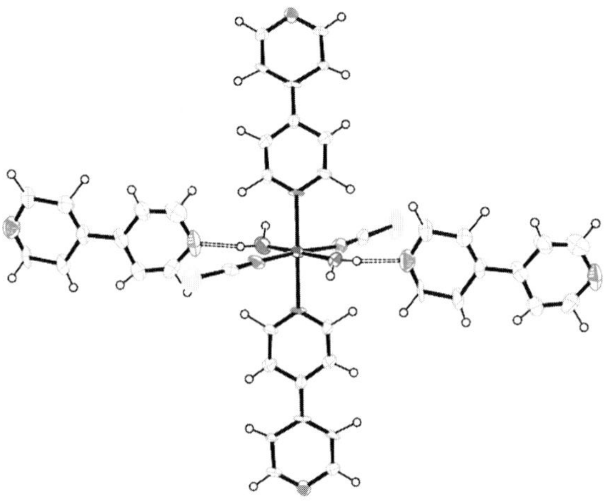

Figure 1. ORTEP drawing of **1a**.

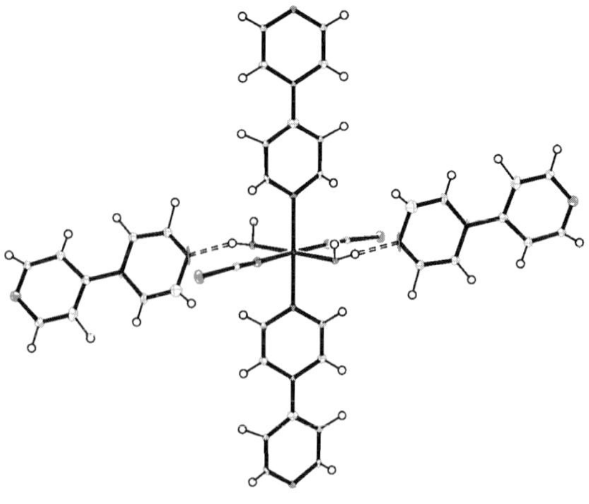

Figure 2. ORTEP drawing of **1b**.

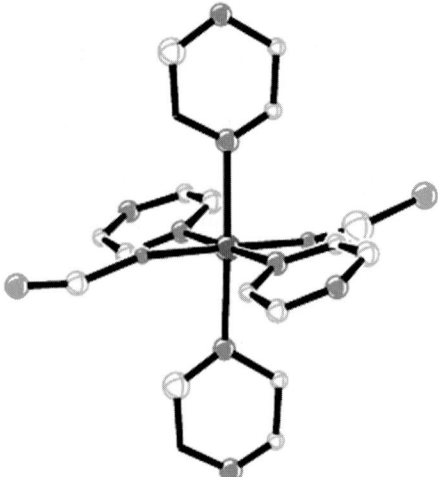

Figure 3. ORTEP drawing of **2a**.

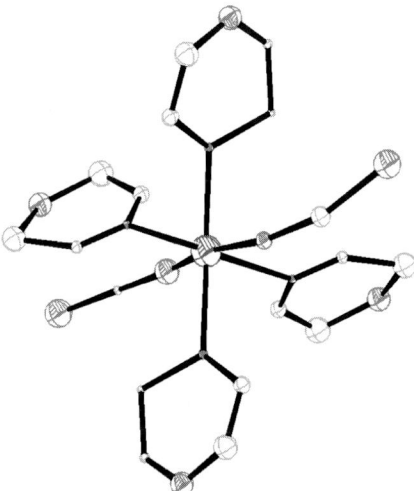

Figure 4. ORTEP drawing of **2b**.

analyses confirmed that **1a** and **1b** have an FeN_4O_2 arrangement; i.e., the iron atom has an octahedral environment with two thiocyanate (selenocyanate) nitrogen donors, two water ligands, and two pyridine nitrogen donors, which are the same with literatures [5, 6]. The crystal structures are made up of linear chains all running in a parallel direction. The chains consist of $[Fe(H_2O)_2(NCS)_2]$ or $[Fe(H_2O)_2(NCSe)_2]$ planar units bridged by bpy. These chains are bridged by hydrogen bonds between coordinated water and another bpy ligand.

ORTEP drawings of $\{Fe(NCS)_2(pyz)_2\}_n$ **2a** and $\{Fe(NCSe)_2(pyz)_2\}_n$ **2b** are shown in Figures 3 and 4, respectively. Although X-ray structural analyses are not perfect, the results revealed at least that **2a** and **2b** have an FeN_6 arrangement; i.e.,

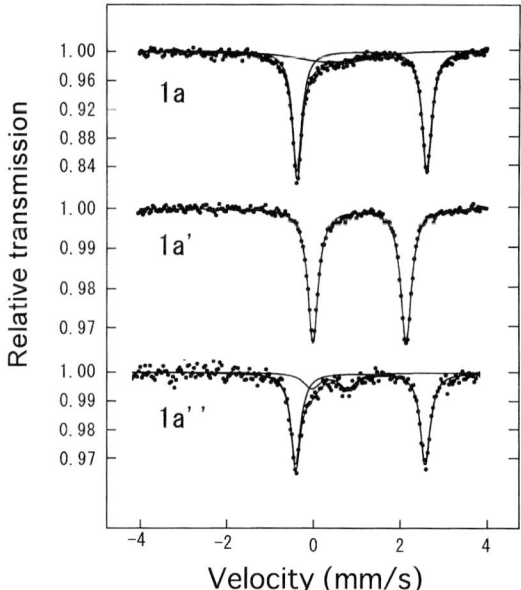

Figure 5. ^{57}Fe Mössbauer spectra at 78 K.

the iron atom has an octahedral environment with two thiocyanate nitrogen donors and four pyridine nitrogen donors. The crystal structures are made up of a grid structure, in which NCS and NCSe ligands are coordinated in an axial position.

TG of **1a** showed a stepwise weight loss, in which dissociation of water (3.3% around 100°C) and one bpy (33.2% around 225°C) took place successively. **2a** showed a weight loss of pyz. The dissociation phenomenon of NCSe complex was somewhat ambiguous.

Figure 5 shows the change in ^{57}Fe Mössbauer spectra by removing water and one bpy successively. The main doublet with a large quardrupole splitting (QS) value is observed with a small amount of impurity. It can easily be seen that the QS value changes depending on the thermal dissociation. Table I summarizes the Mössbauer parameters. All the spectra showed about 1.1 $mm\,s^{-1}$ of isomer shift (IS) value. The IS value around 1.1 $mm\,s^{-1}$ reveals a high-spin Fe^{II} state. **1a** has a large QS value because of the FeN_4O_2 arrangement. The QS value decreases by removing water molecule (**1a′**), suggesting an FeN_6 arrangement. The value increases again by removing one bpy (**1a″**), suggesting an FeN_4S_2 arrangement by bridging with NCS. The results support the proposed thermal dissociation phenomcna by TG.

We synthesized the Fe-bpy-NCS complex (**1a–2**) by using acetone as a solvent. The elemental analysis suggested the composition of $Fe(bpy)_2(NCS)_2(H_2O)_2$. But the sample showed different powder X-ray diffraction pattern from **1a**. Its QS value is larger than that of **1a**. The water of **1a–2** was removed with heating to obtain **1a–2′**. The dissociation temperature was slightly different from **1a**. **1a–2′** showed

Table I. ^{57}Fe Mössbauer parameters

Sample	T (K)	IS[a] (mm s^{-1})	QS (mm s^{-1})
1a	298	1.13	2.99
	78	1.18	3.33
1a′	298	1.08	2.15
	78	1.14	2.47
1a″	298	1.05	2.73
	78	1.09	2.98
1a–2	298	1.13	3.12
1b	298	1.12	3.22
	78	1.22	3.52
1b′	298	1.06	1.69
	78	1.18	2.37
2a	298	1.03	2.48
	78	1.12	2.65
2a′	298	1.03	2.51
	78	1.13	2.70
2b	298	1.03	2.36
	78	1.15	2.69

[a]Relative to iron foil.

the same powder X-ray diffraction pattern as **1a′** showed. These results suggest that one of possible structures of **1a–2** is a structure coordinated by water instead of NCS.

The corresponding NCSe complex (**1b**) showed a larger QS value compared with the NCS complex. The QS value decreased by removing water (**1b′**), suggesting an FeN$_6$ arrangement.

The QS value of Fe-pyz-NCS complex (**2a**) is relatively similar to that of **1a′**, because both complexes have FeN$_6$ arrangement. The QS value increased by removing one pyz ligand, suggesting an FeN$_4$S$_2$ arrangement by bridging with NCS.

The NCSe complex showed a larger QS value compared with the corresponding NCS complex in **1b**, while the NCSe complex showed a smaller QS value in **1b′** and **2b**. The difference is dependent on whether water molecules coordinate or not. That is, the complexes coordinated with water show a larger QS value in the NCSe complexes than in the NCS complexes, while the complexes without coordinated water show a smaller QS value in the NCSe complexes than in the NCS complexes. This trend is also observed in the *trans*-1,2-bis(4-pyridyl)ethylene complex [7]. And the trend in the FeN$_4$O$_2$ arrangement is also opposite to that in mononuclear FeN$_6$ complexes [8]. The difference might be explained by the position of NCS or NCSe; i.e., axial or equatorial position.

In the present study, all the present complexes were in a high-spin Fe^{II} state. The structural change with heating was pursued by the ^{57}Fe Mössbauer spectra.

The ^{57}Fe Mössbauer spectroscopy had a great role in determining the dissociation products.

Acknowledgement

This work was partially supported by The Iwatani Naoji Foundation's Research Grant.

References

1. Lu, J., Paliwala, T., Lim, S. C., Yu, C., Niu, T. and Jacobson, A. J., *Inorg. Chem.* **36** (1997), 923.
2. Czakis-Sulikowska, D. and Kaluzna-Czaplinska, J., *J. Therm. Anal. Cal.* **62** (2000), 821.
3. Real, J. A., Andrés, E., Muñoz, M. C., Julve, M., Granier, T., Bousseksou, A. and Varret, F., *Science* **268** (1995), 265.
4. Halder, G. J., Kepert, C. K., Moubaraki, B., Murray, K. S. and Cashion, J. D., *Science* **298** (2002), 1762.
5. Noro, S., Kondo, M., Ishii, T., Kitagawa, S. and Matsuzaka, H., *J. Chem. Soc., Dalton Trans.* (1999), 1569.
6. Moliner, N., Muñoz, M. C. and Real, J. A., *Inorg. Chem. Commun.* **2** (1999), 25.
7. Nakashima, S., Yamamoto, A., Asada, Y., Koga, N. and Okuda, T., unpublished results.
8. Little, B. F. and Long, G. J., *Inorg. Chem.* **17** (1978), 3401.

Hyperfine Interactions **156/157**: 359–364, 2004.
© 2004 *Kluwer Academic Publishers. Printed in the Netherlands.*

^{155}Gd Mössbauer Isomer Shifts and Quadrupole Coupling Constants of Gadolinium Complexes

MASUO TAKEDA, JUNHU WANG, TATSURU NISHIMURA,
KATSUYA SUZUKI, TAKAFUMI KITAZAWA and MASASHI TAKAHASHI
*Department of Chemistry, Faculty of Science, Toho University, Miyama, Funabashi,
Chiba 274-8510, Japan*

Abstract. We have examined the ^{155}Gd Mössbauer spectra at 12 K for 25 complexes having different coordination numbers and different ratios of coordinating oxygen to nitrogen atoms. The isomer shift decreases in the order: $GdO_{10} \approx GdO_9 > GdN_2O_7 \geqslant GdO_8 \approx GdN_6O_2 \geqslant GdO_7 \geqslant GdO_6 > GdN_8 \approx GdN_9$. The s electron density at the nucleus is larger as the coordination number decreases and as the coordinating atoms change from oxygen to nitrogen, suggesting the covalency in the bonding.

Key words: ^{155}Gd Mössbauer spectroscopy, Gd complexes, coordination configuration.

1. Introduction

Mössbauer spectroscopy is a powerful tool for studying the structure and bonding in compounds [1]. Many gadolinium compounds, including solid solutions, intermetallic compounds, alloys and oxides have been studied by ^{155}Gd Mössbauer spectroscopy using the 86.5 keV transition in ^{155}Gd [2, 3]. Among the ^{155}Gd Mössbauer parameters, the quadrupole coupling constants (e^2qQ) of several oxides and double oxides were successfully explained by a point charge model considering the lattice contribution (q_{latt}) only, since valence electron contribution (q_{val}) can be considered as negligible because of the $4f^7$ electron configuration of Gd(III) [4]. However, no Mössbauer studies on the bonding of Gd(III) in coordination compounds have been done, though information on the contributions of electrons with 6s and 5d (or 6p) character to the bonding may be obtained from the isomer shifts (δ), which give directly the electron densities at the Gd nuclei [1]. So we have started systematic studies on chelates, having Gd–O and Gd–N bonds.

We have already reported the Mössbauer spectra of eight coordinate $Gd[M(CN)_6] \cdot 4H_2O$ (M = Cr^{III}, Fe^{III} and Co^{III}), $KGd[M(CN)_6] \cdot 3H_2O$ (M = Fe^{II} and Ru^{II}) [5], and eight and seven coordinate Gd(III)-β-diketonato complexes [6a]. Here we extended the Mössbauer study to nine- and ten-coordinatated gadolinium compounds of tetraethylene glycol (EO4), pentaethylene glycol (EO5) and 4,4'-bipyridine N,N'-dioxide (dpdo). The gadolinium N,N-dimethylformamide (DMF) complex with eight-coordination was also studied. Furthermore ethylenediaminetetraacetic acid (H$_4$edta) and terpyridine (terpy) complexes with nine-coordination, and phthalocyanine (H$_2$Pc) complex with eight-coordination were investigated.

This paper shows that the isomer shift δ decreases as the coordination number (CN) decreases, showing an increase of 6s electron density with the decrease in CN. The increase of 6s electron density seems to parallel the decrease of Gd–O bond distance. Furthermore 6s electron density increases as nitrogen atoms of the ligands substitute for coordinating oxygen atom.

2. Experimental

^{155}Gd Mössbauer spectra of 86.5 keV transition were measured at 12 K using an ^{155}Eu/^{154}SmPd$_3$ source (about 231 MBq) prepared by us on a WissEl Mössbauer measuring system [5].

3. Results and discussion

Typical Mössbauer spectra at 12 K are shown in Figure 1. The spectra of **9** [6a] and **19** [5] were already reported. The Mössbauer parameters of various gadolinium

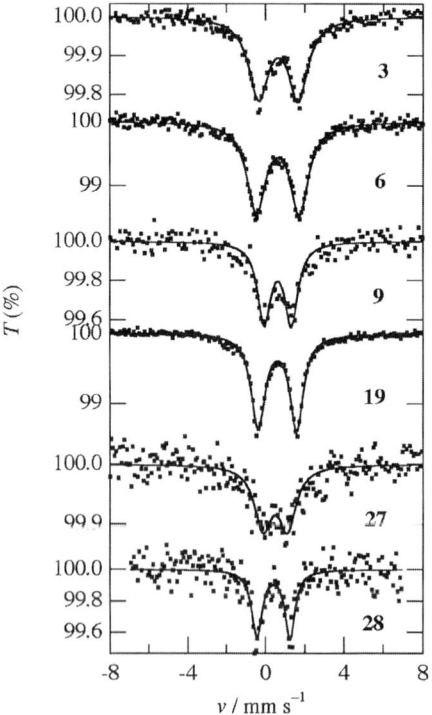

Figure 1. Mössbauer spectra at 12 K of [Gd(η^2-NO$_3$)$_2$(EO5)]NO$_3$ **3**, Na[Gd(edta)(H$_2$O)$_3$]·5H$_2$O **6**, [Gd(bfa)$_3$(H$_2$O)$_2$] **9**, Gd[Fe(CN)$_6$]·4H$_2$O **19**, [Gd(terpy)$_3$](ClO$_4$)$_3$ **27**, and [GdPc$_2$] **28**.

Table I. [155]Gd Mössbauer parameters at 12 K and coordination configurations around Gd(III) ions

Index	Compound	δ^* mm s^{-1}	e^2qQ mm s^{-1}	$2\Gamma^{**}$ mm s^{-1}	Coord. config.	Reference Möss-bauer	Reference Struc-ture
1	GdF$_3$	0.67	5.68	0.92	GdF$_9$	own	[7]
2	[Gd(η^2-NO$_3$)$_2$(η^1-NO$_3$)(EO4)]	0.68	3.53	1.19	GdO$_{10}$	"	[8]
3	[Gd(η^2-NO$_3$)$_2$(EO5)]NO$_3$	0.65	4.34	1.18	GdO$_{10}$	"	[8]
4	[Gd(EO5)(H$_2$O)$_3$](ClO$_4$)$_3$	0.67	3.37	1.13	GdO$_9$	"	[8]
5	Gd$_2$(NO$_3$)$_6$(μ-dpdo)$_3$]·2CH$_2$Cl$_2$	0.67	2.78	1.23	GdO$_9$	"	[9a]
6	Na[Gd(edta)(H$_2$O)$_3$]·5H$_2$O	0.62	4.72	1.10	GdN$_2$O$_7$	"	[10]
7	NH$_4$[Gd(edta)(H$_2$O)$_3$]·5H$_2$O	0.61	3.62	1.35	GdN$_2$O$_7$	"	[11]
8	Gd(bza)$_3$·2H$_2$O	0.64	4.43	1.11	GdO$_8$	[6a]	[6b]
9	[Gd(bfa)$_3$(H$_2$O)$_2$]	0.64	4.43	1.11	GdO$_8$	[6a]	[6b]
10	[Gd(pta)$_3$(H$_2$O)$_2$]	0.61	1.67	1.45	GdO$_8$	[6a]	[6b]
11	Gd(tta)$_3$·2H$_2$O	0.60	7.26	1.23	GdO$_8$	[6a]	[6b]
12	Gd(taa)$_3$·3H$_2$O	0.58	4.47	1.27	GdO$_8$	[6a]	[6b]
13	[Gd(acac)$_3$(H$_2$O)$_2$]·H$_2$O	0.57	5.64	1.38	GdO$_8$	[6a]	[6c]
14	Gd(fta)$_3$·3H$_2$O	0.55	7.56	1.21	GdO$_8$	[6a]	[6b]
15	Gd(fod)$_3$·H$_2$O	0.55	2.52	1.44	GdO$_8$	[6a]	[6b]
16	Zr$_2$Gd$_2$O$_7$(P type)	0.55	8.49	2.10	GdO$_8$	own	[15]
17	[Gd(dmf)$_4$(H$_2$O)$_3$(μ-CN)Fe(CN)$_5$]	0.66	3.40	1.02	GdNO$_7$	"	[9a, 9b]
18	Gd[Cr(CN)$_6$]·4H$_2$O	0.61	4.30	1.04	GdN$_6$O$_2$	[5]	
19	Gd[Fe(CN)$_6$]·4H$_2$O	0.61	4.07	0.90	GdN$_6$O$_2$	[5]	[12]
20	Gd[Co(CN)$_6$]·4H$_2$O	0.60	4.12	0.87	GdN$_6$O$_2$	[5]	
21	KGd[Fe(CN)$_6$]·3H$_2$O	0.59	4.68	1.01	GdN$_6$O$_2$	[5]	[13]
22	KGd[Ru(CN)$_6$]·3H$_2$O	0.60	4.81	0.93	GdN$_6$O$_2$	[5]	
23	[Gd$_2$(dpm)$_6$]	0.65	6.49	1.15	GdO$_7$	[6a]	[6b]
24	Gd(dbm)$_3$·H$_2$O	0.60	6.44	1.46	GdO$_7$	[6a]	[6b]
25	Gd$_2$O$_3$ (monoclinic)***	0.45	5.36		GdO$_7$	[3]	[4, 14]
		0.46	2.78		GdO$_7$	[3]	
		0.49	0.49		GdO$_7$	[3]	
26	Gd$_2$O$_3$ (cubic)						
	8b	0.51	10.84	0.80	GdO$_6$	own	[4]
	24d	0.51	5.54	0.80	GdO$_6$	"	
27	[Gd(terpy)$_3$](ClO$_4$)$_3$	0.40	1.55	1.55	GdN$_9$	"	[16]
28	[GdPc$_2$]	0.41	3.65	0.64	GdN$_8$	"	[17]

Experimental error: ±0.02 mm s^{-1} for δ, ±0.05 mm s^{-1} for e^2qQ and 2Γ.

*Relative to [155]Eu/[154]SmPd$_3$ source at 12 K.

**The full experimental linewidth at half maximum height.

***In **25**, there are three sites for Gd(III) ions.

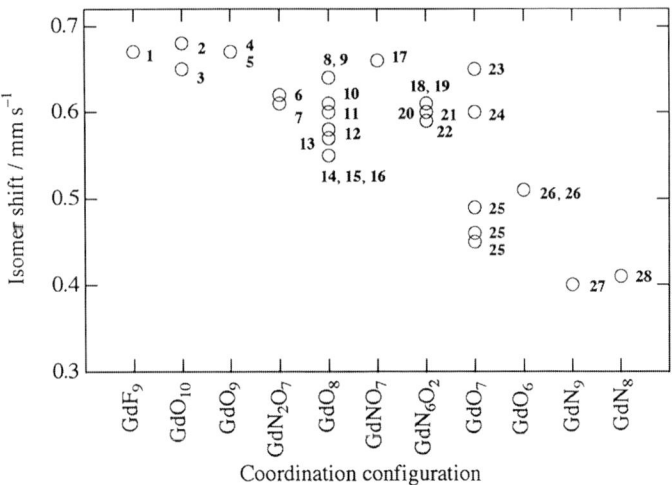

Figure 2. The plot of isomer shifts against coordination configuration.

compounds are presented in Table I which includes previously reported data and the configurations of the coordinated atoms around the Gd(III) ions.

Our data for GdF_3 (**1**) agree very well with the values of $\delta = 0.67$ mm s^{-1} and $e^2qQ = 5.64$ mm s^{-1} obtained by Cashion *et al.* [3]. The data of Gd_2O_3 (cubic) **26** are also in very good agreement with the data of Cashion *et al.*: $\delta = 0.53$ mm s^{-1}, $e^2qQ = 10.46$ mm s^{-1} for 8b site and $\delta = 0.50$ mm s^{-1}, $e^2qQ = 5.50$ mm s^{-1} for 24d site [3]. The δ values are plotted against coordination configuration in Figure 2. We can see there is a relationship between the δ values, and the number and type of coordinating atoms. The δ values decrease in the order $GdO_{10} \approx GdO_9 > GdN_2O_7 \geqslant GdO_8 \approx GdN_6O_2 \geqslant GdO_7 \geqslant GdO_6 > GdN_8 \approx GdN_9$. It is suggested that the actual arrangements of the gadolinium neighbours influence the δ values in these compounds. The smaller the CN, the lower the δ values. The substitution of nitrogen atoms for coordinating oxygen atoms lowers the δ values.

Since the nuclear parameter $\Delta R/R$ is negative for ^{155}Gd, the larger the s electron density at the nuclei, the smaller becomes the isomer shift of ^{155}Gd. The bonding nature of the ten- and nine-coordinated complexes of **2–5** is extremely ionic since their δ values are the same as that of GdF_3 which can be considered to be most highly ionic Gd compound. The δ values of other compounds except for **27** and **28**, bonded by only nitrogen atoms fall between those of GdF_3 and Gd_2O_3. These compounds seem to have a degree of covalency intermediate between GdF_3 and Gd_2O_3. The reduction of δ will arise from the addition of electron density to the 6s orbitals of the Gd(III) ion. The addition of electron density to the 4f, 5d and/or 6p orbitals of the Gd(III) ion would increase the shielding of 5s electrons in the ions and thus decrease the 5s electron density at the nucleus. Therefore this would lead to the increase in the δ values.

Glentworth *et al.* [18], and Mackey and Greenwood [19] measured the Möss-bauer spectra of 33 and 10 chelates of europium, respectively. They found that their δ values are greater than that of EuF_3 and smaller than that of Eu_2O_3. This result coincides with our results considering the positive $\Delta R/R$ for [151]Eu while the negative $\Delta R/R$ for [155]Gd.

Zaheer *et al.* [20] calculated a contribution to the isomer shift of 6s electron of 8.9 mm s^{-1} for the [151]Eu 21.5 keV transition. Based on this result, the [151]Eu isomer shift of the chelates, ranging from 0.15 to 0.51 mm s^{-1} relative to EuF_3 was found to correspond to the transfer of 0.017–0.057 of an electron to the 6s orbital of Eu(III) [20]. Therefore in our compounds the same order of electrons will be donated from oxygen atoms of the ligands to Gd(III) ions. The electron-donating power is generally larger for N atoms than for O atoms. This explains the smaller δ value of **27** and **28** compared to the oxygen-coordinated complexes.

We observed that the δ values of [155]Gd increase as the CN increases. For the nuclei with $\Delta R/R < 0$ like [155]Gd and [237]Np, δ values of [57]Fe [21] and [237]Np [22] increase with the increase of CN, whereas for the nuclei with $\Delta R/R > 0$ like [119]Sn and [151]Eu, the δ values decrease as the CN increases [23, 24]. These data show that the electron density at the nucleus decreases as the coordination number increases. The δ values of compounds with 4f and 5f electrons are also known to decrease with an increase of the Eu–O bond length [24–26] and to increase with an increase of the Np–O bond length [22, 27]. In future we will examine the relationship between the [155]Gd Mössbauer isomer shifts obtained and the Gd–O/Gd–N bond lengths.

The e^2qQ values of [155]Gd with 4f^7 configuration reflect the magnitudes of the electric field gradient ($eq = V_{zz}$) caused by the lattice terms, q_{latt}. Since poly-ethylene glycol is a neutral ligand, we assumed NO_3^- and ClO_4^- ions contribute to the q_{latt}. By using a point charge model we obtained V_{zz} values calculated (10^{20} V m^{-2}) of 1.64, 1.93, 0.327 and 1.36 for **2**, **3**, **4**, and **5**, respectively. The values for **2**, **3** and **5** explain the corresponding e^2qQ observed very well. The disagreement for **4** might suggest an asymmetrical distribution of 4f electrons.

Acknowledgements

The authors wish to thank Dr. F. Mohr for supplying us with sample **28** and Ms. Y. Nomoto-Kondo for preparing the samples **6** and **7**. This research was sup-ported in part by a Grant-in-Aid (No. 10440200) from the Ministry of Educa-tion, Culture, Sports, Science and Technology, Japan, and by the Inter-University Program for the Joint Use of JAERI Facilities.

References

1. Greenwood, N. N. and Gibb, T. C., *Mössbauer Spectroscopy*, Chapman and Hall, London, 1971.
2. Czjzek, G., In: Long, G. J. and Gradjean, F. (eds), *Mössbauer Spectroscopy Applied to Magnetism and Materials Science*, Vol. 1, Plenum Press, New York, 1993, p. 373.

3. Cashion, J. D., Prowse, D. B. and Vas, A., *J. Phys. C* **6** (1973), 2611.
4. Barton, W. A. and Cashion, J. D., *J. Phys. C* **12** (1979), 2897.
5. Wang, J., Abe, J., Kitazawa, T., Takahashi, M. and Takeda, M., *Z. Naturforsch.* **57a** (2002), 581.
6. (a) Wang, J., Takahashi, M., Kitazawa, T. and Takeda, M., *J. Radioanal. Nucl. Chem.* 255 (2003), 195; (b) Wang, J., PhD thesis, Toho University, 2002, (c) Shen, C., Fan, Y., Liu, G., Wang, Y. and Lu, P., *Chem. J. Chin Univ.* **4** (1983), 769.
7. Zalkin, A. and Templeton, D. H., *J. Am. Chem. Soc.* **75** (1953), 2453.
8. Nishimura, T., Takahashi, M. and Takeda, M., to be published.
9. (a) Unpublished data; (b) Figuerola, A., Diaz, C., Ribas, J., Tangoulis, V., Granell, J., Lloret, F., Mafia, J. and Maestro, M., *Inorg. Chem.* **42** (2003), 641.
10. Templeton, L. K., Templeton, D. H., Zalkin, A. and Ruben, H. W., *Acta Cryst. B* **38** (1982), 2155.
11. Takahashi, M., Nomoto, Y., Ikeda, T. and Takeda, M., to be published.
12. Mullica, D. M. and Sappenfield, E. L., *Acta Cryst. C* **47** (1991), 2433.
13. Petter, W., Gramlich, V., Hulliger, F. and Vetsch, H., *Eur. J. Solid State Inorg. Chem. T.* **29** (1992), 65.
14. Yakel, H. L., *Acta Cryst. B* **35** (1979), 564.
15. Moriga, T., Yoshisa, A., Kanamaru, F., Koto, K., Yoshimura, M. and Somiya, S., *Solid State Ionics* **31** (1989), 319.
16. Durham, D. A., Frost, G. H. and Hart, F. A., *J. Inorg. Nucl. Chem.* **31** (1969), 833.
17. Cian, A. D., Moussavi, M., Fischer, J. and Weiss, R., *Inorg. Chem.* **24** (1985), 3162.
18. Glentworth, P., Nichols, A. L. and Newton, D. A., *J. Chem. Soc., Dalton Trans.* (1973), 546.
19. Mackey, J. L. and Greenwood, N. N., *J. Inorg. Nucl. Chem.* **34** (1972), 1529.
20. Zaheer, A. H., Liss, I. B., Keck, N. B., Bos, W. G. and Ouseph, P. J., *J. Inorg. Nucl. Chem.* **36** (1974), 2515.
21. P. 150 and p. 154 in reference [1].
22. Saeki, M., Nakada, M., Nakamoto, T., Yamashita, T., Masaki, N. M. and Krot, N. N., *J. Radioanal. Nucl. Chem.* **239** (1999), 221.
23. P. 391 in reference [1].
24. Tanabe, S., Hirao, K. and Soga, N., *J. Non-Cryst. Solids* **113** (1989), 178.
25. Masaki, N. M., Otobe, H., Nakamura, A., Guillermo, N. R. D., Izumiyama, Y., Harada, D. and Hinatsu, Y., *Hyp. Interact. (C)* **5** (2002), 305.
26. Nakamura, A., Masaki, N. M., Nakada, M., Saeki, M., Tomomoto, K. and Akimitsu, J., *Ceramic Trans.* **71** (1996), 295.
27. Dunlap, B. D. and Kalvius, G. M., In: A. J. Freeman and G. H. Lander (eds), *Handbook on the Physics and Chemistry of the Actinides*, Vol. 2, North-Holland, 1985, p. 411.

Hyperfine Interactions **156/157**: 365–370, 2004.
© 2004 *Kluwer Academic Publishers. Printed in the Netherlands.*

Mössbauer Spectroscopy in the Characterisation of Polymetallic Cluster Compounds: a Triple Mössbauer Study of (PPh$_4$)[Fe$_2$Ir$_2$(CO)$_{12}$\{μ^3-Au(PPh$_3$)\}]

L. STIEVANO[1], R. DELLA PERGOLA[2] and F. E. WAGNER[3]

[1]*Laboratoire de Réactivité de Surface, Université Pierre et Marie Curie, Paris, France;*
e-mail: stievano@ccr.jussieu.fr
[2]*Dip. di Scienze dell'Ambiente e del Territorio, Università di Milano - Bicocca, Milano, Italy*
[3]*Physik-Department E15, Technische Universität München, Garching, Germany*

Abstract. The mixed-metal cluster compound (PPh$_4$)[Fe$_2$Ir$_2$(CO)$_{12}$\{μ^3-Au(PPh$_3$)\}], which has a trigonal bipyramidal core consisting of five atoms of three different Mössbauer isotopes, was studied by ^{193}Ir, ^{197}Au and ^{57}Fe Mössbauer spectroscopy. The nature and the chemical character of the atoms located at the different sites are discussed with respect to their Mössbauer spectra.

Key words: iron, gold, iridium, polymetallic cluster compounds, Mössbauer spectroscopy.

1. Introduction

Molecular mixed-metal clusters can be effectively used for the synthesis of supported metal catalysts [1, 2]. In fact, the polymetallic character of the original cluster can be retained, at least in principle, during the successive steps of immobilisation and activation on a support, resulting in the formation of highly dispersed particles with a homogeneous composition and a narrow particle size distribution [3]. Among other clusters, [Fe$_2$Ir$_2$(CO)$_{12}$]$^{2-}$ was recently used for the preparation of Fe–Ir/MgO catalysts that are remarkably active in the synthesis of methanol [4], and ^{193}Ir and ^{57}Fe Mössbauer spectroscopy together yielded valuable information on this system [5, 6].

Here results of a Mössbauer study of a trimetallic derivative of the [Fe$_2$Ir$_2$(CO)$_{12}$]$^{2-}$ cluster compound are presented: this cluster, [Fe$_2$Ir$_2$(CO)$_{12}$\{μ^3-Au(PPh$_3$)\}]$^-$, consists of the tetrahedral Fe$_2$Ir$_2$ bimetallic core of its [Fe$_2$Ir$_2$(CO)$_{12}$]$^{2-}$ precursor, to which a capping triphenylphosphinegold(I) moiety is added [7]. The trigonal-bipyramidal Fe$_2$Ir$_2$Au core thus formed is outstanding from the point of view of Mössbauer spectroscopy, since three different Mössbauer isotopes, namely ^{57}Fe, ^{193}Ir and ^{197}Au, can be studied in the same compound.

2. Experimental

Details on the preparation and characterisation of the cluster compound (PPh$_4$) [Fe$_2$Ir$_2$(CO)$_{12}${Au(PPh$_3$)}] were reported elsewhere [7]. In order to prevent decomposition, the sample material was kept immersed in liquid nitrogen during the preparation and handling of the Mössbauer absorbers. Due to the different requirements of the three Mössbauer transitions, absorbers of appropriate thickness were prepared for the individual Mössbauer experiments.

The ^{193}Os and the ^{197}Pt activities feeding the 73 keV Mössbauer transition of ^{193}Ir and the 77 keV Mössbauer transition of ^{197}Au were produced in the Munich Research Reactor by neutron irradiation of isotopically enriched ^{192}Os and ^{196}Pt metal, respectively. A commercial source of ^{57}Co:Rh was used for ^{57}Fe Mössbauer spectroscopy. All measurements were performed with both the source and the absorber at 4.2 K. An intrinsic Ge detector was used for the detection of the ^{193}Ir and ^{197}Au gamma rays, whereas a Kr/CO$_2$ proportional counter was used for those of ^{57}Fe. The spectrometer was operated with a sinusoidal velocity waveform and the spectra were fitted with appropriate combinations of Lorentzian lines. In the case of ^{193}Ir, due to the hexagonal lattice of osmium metal, the emission line of the source is a quadrupole doublet with a splitting of 0.48(2) mm s^{-1}, which was taken into account in the evaluation of the Mössbauer spectra. In this way, spectral parameters such as the isomer shift (δ), the electric quadrupole splitting (ΔE_Q), the linewidth (Γ) and the relative resonance areas (Area) of the different components of the absorption patterns were determined (Table I). The isomer shift scales are referred to iridium metal, natural α-iron and gold metal for ^{193}Ir, ^{57}Fe and ^{197}Au, respectively.

Table I. Hyperfine parameters at 4.2 K for (PPh$_4$)[Fe$_2$Ir$_2$(CO)$_{12}${Au(PPh$_3$)}] and for its precursor (PPh$_4$)$_2$[Fe$_2$Ir$_2$(CO)$_{12}$] (from [6])

Mixed-metal cluster	Mössbauer isotope	ΔE_Q [mm/s]	δ^a [mm/s]	Γ [mm/s]	Area [%]	Site
[Fe$_2$Ir$_2$(CO)$_{12}$]$^{2-}$	^{193}Ir	2.39(5)	0.13(2)	0.80(5)	61(3)	Ir(2)
(from [6])		0.66(9)	0.04(4)	0.80(5)	39(3)	Ir(1)
	^{57}Fe	1.01(1)	−0.15(1)	0.23(1)	87(2)	Fe(1,2)
		0.45(2)	−0.06(1)	0.35(3)	13(2)	Fe(*)
[Fe$_2$Ir$_2$(CO)$_{12}${μ_3-Au(PPh$_3$)$_3$}]$^-$	^{193}Ir	2.24(1)	−0.13(1)	0.83(2)	100	Ir(1,2)
	^{57}Fe	0.30(1)	−0.08(1)	0.24(1)	50(1)	Fe(1 or 2)
		0.21(1)	−0.07(1)	0.28(1)	50(1)	Fe(2 or 1)
	^{197}Au	6.89(2)	3.16(1)	2.11(5)	100	Au

^{a}The isomer shift (δ) is given relative to iridium metal, α-iron and gold metal for ^{193}Ir, ^{57}Fe and ^{197}Au, respectively.

3. Results and discussion

The mixed metal cluster $[Fe_2Ir_2(CO)_{12}\{Au(PPh_3)\}]^-$ is prepared by treating an acetone solution of $[Fe_2Ir_2(CO)_{12}]^{2-}$ with the stoichiometric amount of $Au(PPh_3)Cl$
at room temperature [7]. The structure of the $[Fe_2Ir_2(CO)_{12}\{Au(PPh_3)\}]^-$ anion, as
determined by X-ray diffraction on its tetraphenylphosphonium salt [7], is shown in
Figure 1. In contrast to $[Fe_2Ir_2(CO)_{12}]^{2-}$, both iridium atoms in $[Fe_2Ir_2(CO)_{12}\{Au$
$(PPh_3)\}]^-$ are equivalent and occupy an equatorial position together with the Fe(2)
atom, all three being bound to two bridging and two terminal carbonyls. Fe(1)
is on an apical position and has a quasi-octahedral coordination, being bound
to three terminal carbonyls and the three equatorial metal atoms. The $Au(PPh_3)$
group occupies the other apex, keeping a Au–P(1) distance equal to that observed
in $(PPh_3)Au(I)Me$, where the Au(I) is *sp* hybridised [7].

In agreement with this crystal structure, the ^{193}Ir Mössbauer spectrum (Figure 2) consists of a single quadrupole doublet with parameters quite similar to
those obtained for the equatorial iridium atom in $[Fe_2Ir_2(CO)_{12}]^{2-}$ (Table I), which
indicates that the addition of the $Au(PPh_3)$ moiety does not induce much change
in the electronic structure of the equatorial Ir atoms. The slight asymmetry in the
intensity of the two Mössbauer lines, which is also found in the ^{197}Au Mössbauer
spectrum (Figure 2), is most probably due to the Goldanskii–Karyagin effect, i.e.
to the influence of a directional anisotropy of the Lamb–Mössbauer f-factor on

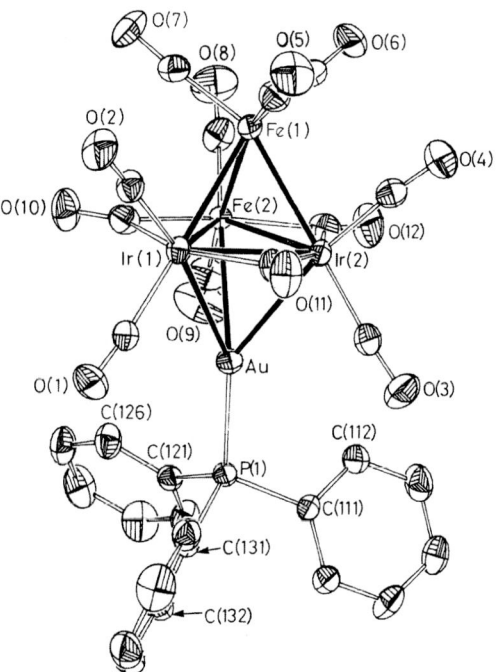

Figure 1. ORTEP drawing of the anion $[Fe_2Ir_2(CO)_{12}\{Au(PPh_3)\}]^-$ (from [7]).

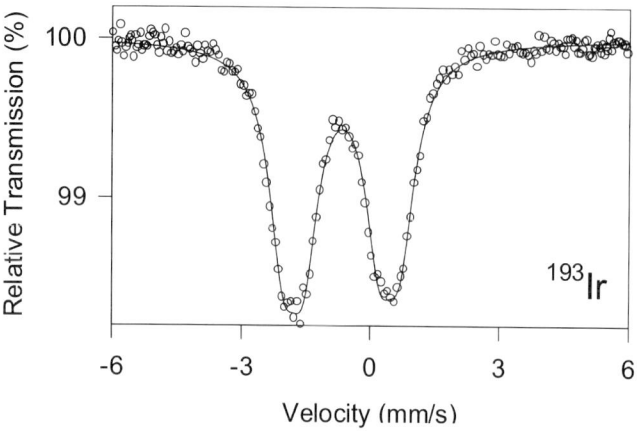

Figure 2. ^{193}Ir Mössbauer spectrum at 4.2 K of [PPh$_4$][Fe$_2$Ir$_2$(CO)$_{12}${Au(PPh$_3$)}].

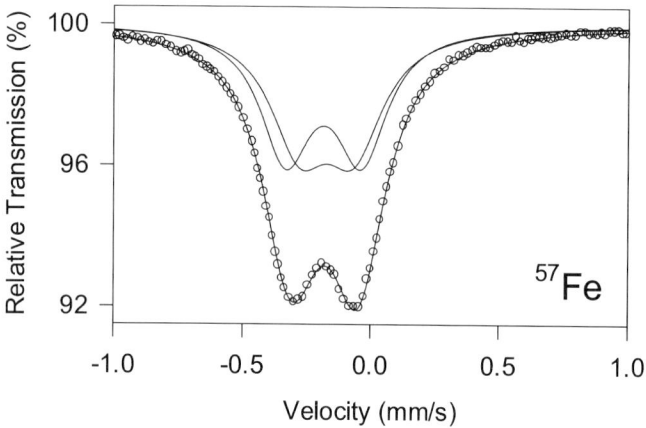

Figure 3. ^{57}Fe Mössbauer spectrum at 4.2 K of [PPh$_4$][Fe$_2$Ir$_2$(CO)$_{12}${Au(PPh$_3$)}].

the line intensities in the Mössbauer pattern, as already observed for other carbonyl cluster compounds [8]. In any case, a texture effect in the preparation of the sample, although quite improbable, cannot be completely excluded.

The ^{57}Fe Mössbauer spectrum (Figure 3) consists of two ill-resolved quadrupole doublets, both with a rather small quadrupole splitting (Table I). Two doublets arc, indeed, expected, since there are two iron sites with different coordination geometries, but it is impossible to say which doublet corresponds to Fe(1) and which to Fe(2). The small splitting of either doublet is in good agreement with the expectation for the apical pseudo-octahedral Fe(1) site, which should have parameters close to those observed for [Fe$_4$(CO)$_{13}$]$^{2-}$, [FeCo$_3$(CO)$_{12}$]$^-$ and [HFeCo$_3$(CO)$_{12}$], where the iron atom occupies a site with a similar highly symmetric C$_3$ co-ordination geometry [9, 10]. Moreover, this result confirms the earlier attribution of an additional weak iron component found in the Mössbauer study of [Fe$_2$Ir$_2$(CO)$_{12}$]$^{2-}$ to the partial substitution of the apical Ir by iron [6].

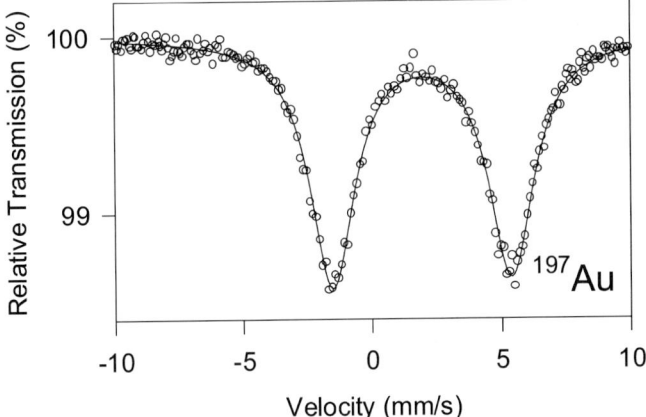

Figure 4. ^{197}Au Mössbauer spectrum at 4.2 K of [PPh$_4$][Fe$_2$Ir$_2$(CO)$_{12}${Au(PPh$_3$)}].

A small QS is, however, unexpected for the Fe(2) site, which has similar coordination geometry of the carbonyls as Fe(2) in the starting [Fe$_2$Ir$_2$(CO)$_{12}$]$^{2-}$ cluster, where its quadrupole splitting is quite large. Obviously the addition of Au(PPh$_3$) reduces the quadrupole splitting dramatically. This disagrees with the common idea that the ^{57}Fe quadrupole splitting in iron carbonyl clusters is mainly influenced by the geometry of the CO ligands [11], and could be a sign of a possible decomposition of the cluster compound, rather common for such carbonyl compounds. However, the good quality and the reproducibility of the accompanying ^{197}Au and ^{193}Ir Mössbauer spectra, which was tested after keeping the sample a few months in liquid nitrogen, agrees rather with a reasonably good stability of the measured sample.

Finally, the ^{197}Au Mössbauer spectrum (Figure 4) shows a quadrupole doublet with parameters (Table I) typical of linearly coordinated (PPh$_3$)Au(I)X (X = halogen) [12]. In fact, it is known that the main contribution to the quadrupole splitting in (PPh$_3$)Au(I)X arises from the bonding to the PPh$_3$ group. The Mössbauer parameters of gold in [Fe$_2$Ir$_2$(CO)$_{12}${Au(PPh$_3$)}]$^-$ are also very similar to those observed for bonding to PPh$_3$ in the [Au$_9$(PPh$_3$)$_8$](NO$_3$)$_3$ cluster [13], which generally agrees with the notion that the gold on the surface of clusters consisting of less than 11 atoms has an electronic configuration of the valence electrons typical of linearly coordinated monovalent gold [12]. This implies that in small clusters the bonding strength between the metal atoms and the ligands is much larger than that between the metal atoms.

This observation further confirms the notion that the *sp* hybridisation of the Au(I) atom remains largely unchanged by the bonding to the cluster. The value of the Au–P(1) distance [7] points in the same direction. With the first *sp* hybrid orbital directed towards the PPh$_3$ ligand, the second, opposite one will point towards the centre of the FeIr$_2$ triangle. This leaves the Au atom rather insensitive to the metal atoms directly coordinated to it [13]. For this reason, the Au(PPh$_3$) group should be considered as a μ_3-ligand rather than the gold atom as an additional member of the skeleton increasing the nuclearity of the cluster.

Acknowledgement

The present work was supported by CNR, MURST.

References

1. Gates, B. C., *Chem. Rev.* **95** (1995), 511.
2. Ichikawa, M., In: R. M. Lambert and G. Pacchioni (eds.), *Chemisorption and Reactivity on Supported Clusters and Thin Films*, Kluwer Academic Pub., Dordrecht, 1997, pp. 153.
3. Guczi, L. and Sarkany, A., In: *Catalysis*, Vol. 11, The Royal Society of Chemistry, Cambridge, 1994, p. 319.
4. Psaro, R., Dossi, C., Della Pergola, R., Garlaschelli, L., Calmotti, S., Marengo, S., Bellatreccia, M. and Zanoni, R., *Appl. Catal. A* **121** (1995), L19.
5. Stievano, L., *Hyp. Interact.* **126** (2000), 101.
6. Stievano, L., Calogero, S., Psaro, R., Guidotti, M., Della Pergola, R. and Wagner, F. E., *J. Cluster Sci.* **12** (2001), 123.
7. Della Pergola, R., Garlaschelli, L., Demartin, F., Manassero, M., Masciocchi, N. and Sansoni, M., *J. Chem. Soc., Dalton Trans.* (1990), 127.
8. Grandjean, F., Long, G. J., Benson, C. G. and Russo, U., *Inorg. Chem.* **27** (1988), 1524.
9. Cooke, G. C. and Mays, M. J., *J. Chem. Soc., Dalton Trans.* (1975), 455.
10. Benson, C. G., Long, G. J., Bradley, J. S., Kolis, J. W. and Shriver, D. F., *J. Am. Chem. Soc.* **108** (1986), 1898.
11. Farmery, K., Kilner, M., Greatrex, R. and Greenwood, N. N., *J. Chem. Soc. A* (1969), 2339.
12. Parish, R. V., In: G. J. Long (ed.), *Mössbauer Spectroscopy Applied to Inorganic Chemistry*, Vol. 1, Plenum Press, New York, 1984, p. 577.
13. Smit, H. H. A., Nugteren, P. R., Thiel, R. C. and De Jongh, L. J., *Physica B* **153** (1988), 33.

Hyperfine Interactions **156/157**: 371–377, 2004.
© 2004 *Kluwer Academic Publishers. Printed in the Netherlands.*

The Nonanuclear [Mo(IV){(CN)Fe(III)(3-ethoxy-saldptn)}8]Cl4 Complex Compound Exhibits Multiple Spin Transitions Observed by Mössbauer Spectroscopy [*]

F. RENZ[*] and P. KEREP
Institut für Anorganische Chemie und Analytische Chemie, Johannes Gutenberg-University,
Staudinger Weg 9, D-55099 Mainz, Germany; e-mail: Franz.Renz@uni-mainz.de

Abstract. The pentadentate ligand 3-EO-^5LH_2 = 3-ethoxy-saldptn = N,N′-bis(3-ethoxy-1-hydroxy-2-benzyliden)-1,7-diamino-4-azaheptane has been prepared by a Schiff base condensation between 1,7-diamino-4-azaheptane and the corresponding 3-ethoxy-salicyaldehyde. 3-EO-^5LH_2 is a sterical extention to $^5LH_2$. Its complexation with Fe(III) gave the high-spin ($S = 5/2$) complex of $[\text{Fe(III)}(3\text{-EO-}^5L)\text{Cl}]$. This precursor was combined with $[\text{Mo(CN)}_8]^{4-}$ and a blue nonanuclear cluster $[\text{Mo(IV)}\{(\text{CN})\text{Fe(III)}(3\text{-EO-}^5L)\}_8]\text{Cl}_4$ resulted. This starshaped nonanuclear compound is a high-spin system at room temperature. On cooling to 10 K some of the eight iron(III) centers switched to the low-spin state as proven by Mössbauer spectra, i.e. multiple electronic transitions.

Key words: nonanuclear complex, high-spin molecule, Fe(III)–Mo(IV), multiple spin transition.

1. Introduction

The electronic properties of high-spin molecules bear a potential for development in the field of information storage. The class of high-spin molecules containing transition metals has been enriched considerably over last ten years [1–5]. In parallel to the topology of iron(III) and manganese(II) wheels and other clusters like Mn_{12}, a considerable attention attracted the metal cyanides having at their octahedral vertices extra metal complexes – $[M(^5L)]$ surrounded by a suitable pentadentate blocking ligand [6–10]. The magnetic properties of these molecular analogues of the Prussian blue are intensively studied at present.

$^5LH_2$ = saldptn = N,N′-bis(1-hydroxy-2-benzyliden)-1,7-diamino-4-azaheptane represents one of the candidates to function as a pentadentate blocking ligand. The iron(III) complex $[\text{Fe}^{III}(^5L)\text{Cl}]$ is high-spin ($S = 5/2$). On the contrary, the pyridine complex $[\text{Fe}^{III}(^5L)\text{py}](\text{BPh}_4)$ exhibits the thermally induced spin crossover: the low-spin ($S = 1/2$) to high-spin ($S = 5/2$) transition [11–14].

---

[*] Dedicated to Professor Philipp Gütlich on the occasion of his 70th birthday.
[*] Author for correspondence.

In the present paper 3-EO-^{5}LH$_2$ = 3-ethoxy-saldptn = N,N'-bis(3-ethoxy-1-hydroxy-2-benzyliden)-1,7-diamino-4-azaheptane has been prepared, a sterically more hindered extention to ^{5}LH$_2$. Its Fe(III)Cl$_3$ complex and nonanuclear complex compounds have been prepared and investigated.

2. Experimental

2.1. LIGAND 3-EO-^{5}LH$_2$

The pentadentate ligand 3-EO-^{5}LH$_2$ = 3-ethoxy-saldptn = N,N'-bis(3-ethoxy-1-hydroxy-2-benzyliden)-1,7-diamino-4-azaheptane (see Figure 1) has been prepared by a Schiff base condensation between 1,7-diamino-4-azaheptane and the corresponding 3-ethoxy-salicylaldehyde at a ratio of 1 : 2. A mixture of the 3-ethoxy-salicylaldehyde (0.2 mol) and 1,7-diamino-4-azaheptane (0.1 mol) in 2-propanol (100 cm^3) was boiled for 10 min and the solution subjected to crowding. The yellow oily material, 3-ethoxy-saldptn, resulted and used without any purification. NMR spectra agree with the expected structure.

2.2. PRECURSOR [Fe(^{5}L)Cl]

A solution of anhydrous iron(III) chloride (10 mmol) in 2-propanol (50 cm^3) was added to a solution of 3-ethoxy-saldptn (10 mmol) in 2-propanol (40 cm^3). The mixture was stirred at 50°C for 10 min and then triethylamine (20 mmol) was added (see Figure 1). The resulting solution was stirred at 50°C for 1 h. After cooling, black-red-violet crystals precipitated. These were collected and recrystallized

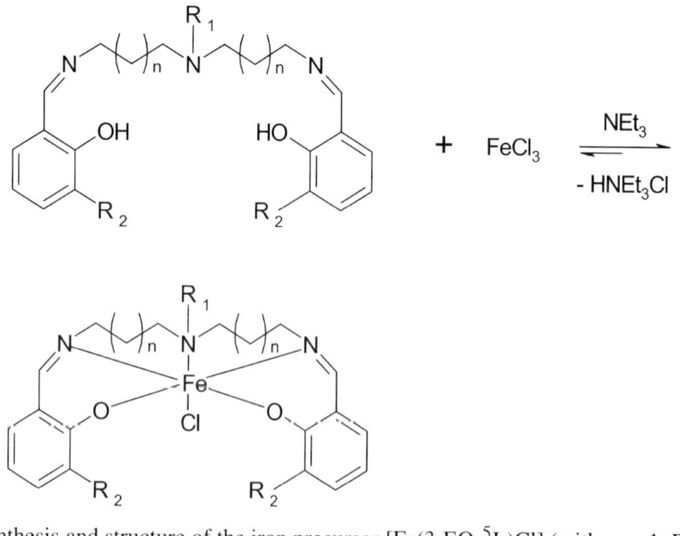

Figure 1. Synthesis and structure of the iron precursor [Fe(3-EO-^{5}L)Cl] (with $n = 1$; R$_2$ = ethoxy; R$_1$ = H).

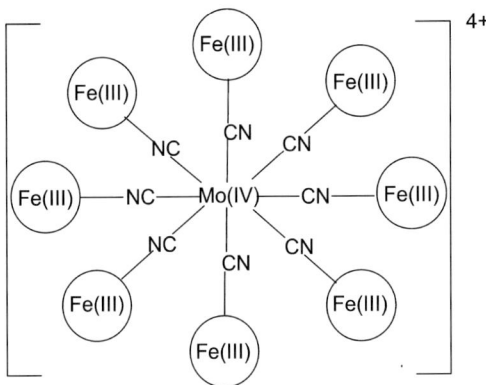

Figure 2. Schematic structure of the [Mo^{IV}{(CN)Fe^{III}(3-EO-^{5}L)}$_8$]$^{4+}$.

in 2-propanol, washed with 2-propanol and diethyl ether, and dried in vacuum. Yield: 76% rel. to Fe (4.9 g; 9.5 mmol; M = 516 g/mol). Analysis calculated: for C$_{24}$H$_{31}$ClFeN$_3$O$_4$ (%): C (55.78), N (8.13), H (6.05); found: C (54.39), N (7.93), H (6.29). IR (KBr): ν(cm^{-1}) = 3233 cm^{-1} (N–H, valence strech = val.), 2925 cm^{-1} (CH$_2$, val.), 2900–2700 cm^{-1} (O–CH$_3$, val.), 1623 cm^{-1} (C=N, val.), 1500–1600 cm^{-1} (aromat. C=C, val.), 1245 cm^{-1} (C–O, val.), 900–850 cm^{-1} (C–H, out of plane = o.o.p.). FD MS m/z = 428.0 (3.2%, [3-EtO-^{5}LFeCl]$^+$), 393.1 (100.0%, [3-EtO-^{5}LFe]$^+$).

2.3. PRECURSOR K$_4$[Mo(CN)$_8$]·2H$_2$O

K$_4$[Mo(CN)$_8$]·2H$_2$O was used as prepared by Renz *et al.* [16] in analogy to Furman and Miller [17].

2.4. NONANUCLEAR COMPLEXES [Mo^{IV}{(CN)Fe^{III}(3-EO-^{5}L)}$_8$]X$_4$

A aqueous 2-propanol (1 : 1) solution (10 cm^3) of K$_4$[Mo(CN)$_8$]·2H$_2$O (0.25 mmol) was added dropwise to a 2-propanol solution (100 cm^3) of [Fe(3-EO-^{5}L)Cl] (1 mmol). After 30 min of stirring at room temperature the dark-blue solution was filtered (remark: this fraction of the product was skipped because of its content of partial substituted complexes, i.e. lower then nonanuclear components) and reduced under pressure to 50 cm^3 (below 35°C). Addition of 150 ml distilled water precipitates the product, which was slowly filtered off on the frita funnel. The dark-bluish-black product was washed with distilled water and diethylether and dried under vacuum for 24 hours. Yield: 21 % rel. to Mo (90 mg; 0,021 mmol; M = 4290.5 g/mol). Analysis calculated for C$_{200}$H$_{248}$Cl$_4$Fe$_8$MoN$_{32}$O$_{32}$(%): C (55.91), N (10.43), H (5.82); found: C (51.29), N (11.91), H (5.89). Note that a too low carbon content is commonly observed and based on incomplete decomposition to CO$_2$ due to formation of stable metal carbides. IR (KBr): ν(cm^{-1}) = 3444 cm^{-1} (N–H, val.), 2928 cm^{-1} (CH$_2$, val.), 2115 cm^{-1} (C≡N, val., singlet), 1622 cm^{-1} (C=N, val.), 1600–1500 cm^{-1} (aromat. C=C, val.), 1246 cm^{-1} (C–O, val.), 900–850 cm^{-1} (C–H, o. o. p.).

2.5. MÖSSBAUER SPECTROSCOPY

A conventional spectrometer has been used in scanning the Mössbauer spectra at liquid helium and room temperature (^{57}Co/Rh source, calibration at α-Fe at room temperature; isomeric shifts are relative to the source).

3. Results and discussion

Figure 3 and Table I show the temperature-dependent Mössbauer spectra of the [Fe(3-EO-^{5}L)Cl] precursor compound. At 300 K the Mössbauer spectra exhibit a quadrupole split doublet which is slightly broadened due to an upcoming appearance of relaxation, common for Fe(III) in the HS state ($S = 5/2$) [15]. At 20 K the quadrupole split doublet is broadened due to relaxation and characteristic for Fe(III) in the HS state [15].
7

Figure 4 and Table II show the temperature-dependent Mössbauer spectra of the nonanuclear compound [Mo^{IV}{(CN)Fe^{III}(3-EO-^{5}L)}$_8$]Cl$_4$. At 300 K a quadrupole split doublet which is slightly broadened due to relaxation, common for Fe(III) in the HS state ($S = 5/2$) [15]. The quadrupole splitting is 0.07 mm/s smaller compared to its mononuclear precursor compound, similar to the analogous ^{5}L compound with a drop of 0.13 mm/s, caused by a change in the coordination sphere on going from Fe–Cl to Fe–NC–M. The difference is located in the 3rd

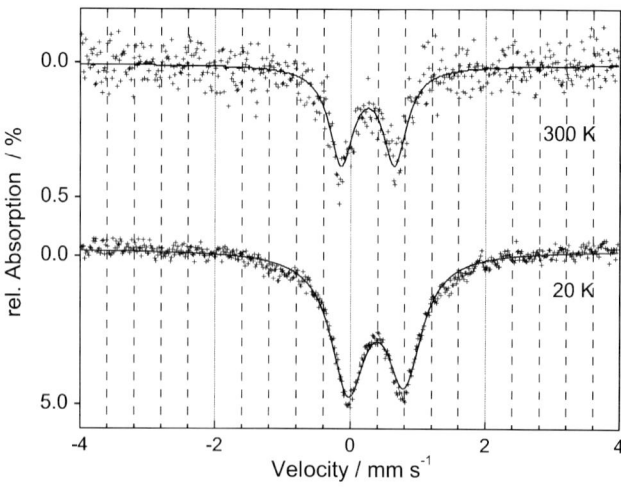

Figure 3. Mössbauer spectra of the precursor [Fe(3-EO-^{5}L)Cl] at 300 K and 20 K.

Table I. Mössbauer parameter of the precursor [Fe(3-EO-⁵L)Cl] at 300 K and 20 K

[3-EO-⁵LFeCl]	Fe(III) HS state		
T	δ (mm/s)	Δ (mm/s)	$\Gamma_{1/2}$ (mm/s)
300 K	0.251	0.790	0.216
20 K	0.381	0.830	0.313

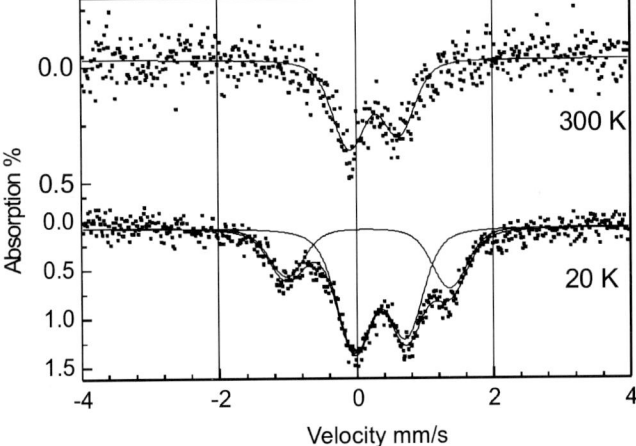

Figure 4. Mössbauer spectra of the nonanuclear complexe [Mo^{IV}{(CN) Fe^{III}(⁵L)}₈]X₄ at 300 K and 20 K.

Table II. Mössbauer parameter of the nonanuclear complexe [Mo^{IV}{(CN) Fe^{III}(3-EO-⁵L)}₈]Cl₄ and [Mo^{IV}{(CN) Fe^{III}(⁵L)}₈]Cl₄ [16] at 300 K and 20 K

	Fe(III) HS state			Fe(III) LS state		
	δ mm/s	Δ mm/s	Fraction %	δ mm/s	Δ mm/s	Fraction %
[(3-EO-⁵LFeNC)₈Mo]Cl₄						
300 K	0.258	0.723	100	–	–	–
20 K	0.337	0.752	67	0.171	2.384	33
[(⁵LFeNC)₈Mo]Cl₄						
300 K	0.208	0.58	100	–	–	–
20 K	0.337	0.695	68	0.161	2.283	32

position of the phenolic substructure, which is located close to coordination bridge as seen in Figure 1. Replacing the small H-atoms as in ^5L by bulky ethoxy-groups as in 3-EO-^5L introduced sterical hinderance. This repulsion term prevents optimal coordination and weakens the CN–Fe bonds in [Mo^{IV}{(CN)Fe^{III}(3-EO-^5L)}$_8$]Cl$_4$.

The prepared nonanuclear complex has been characterized by IR spectrum. No evidence for a splitting of the characteristic C–N stretching vibration band indicates a highly symmetric coordination sphere around the central Mo(IV) atom (Figure 2), i.e. either all eight or non-cyanides are bridged. The shift of the CN band from 2200 to 2115 cm^{-1} indicates a bridge.

At 20 K the Mössbauer spectrum shows in addition to the Fe(III) in the HS state a second doublet appearing which is characteristic for an Fe(III) in the LS state [15]. This indicates that in the nonanuclear complex.

[Mo^{IV}{(CN)Fe^{III}(3-EO-^5L)}$_8$]Cl$_4$ clearly a thermally induced spin transition between an HS to LS state exists. The area fraction of the respective doublets A(Fe^{III}-HS) : A(Fe^{III}-LS) is 67 : 33 and indicates that several Fe(III) centers switch, i.e. multiple electronic transition.

Acknowledgements

This work was partly funded by the "Fonds der chemischen Industrie", the University of Mainz "Forschungsfond" and "Material-Wissenschaftliches Forschungs-Zentrum" (MWFZ) and the "Deutsche Forschungsgemeinschaft" (DFG).

References

1. McCusker, J. K., Schmitt, E. A. and Hendrickson, D. N., In: D. Gatteschi, O. Kahn, J.S. Miller and F. Palacio (eds.), *Magnetic Molecular Materials, NATO ASI Series*, Vol. 198, Kluwer, Dordrecht, 1991, p. 297.
2. Kahn, O., *Molecular Magnetism*, VCH, New York, 1993.
3. Gatteschi, D., In: E. Coronado, P. Delhaes, D. Gatteschi and J. S. Miller (eds.), *Molecular Magnetism: From Molecular Assemblies to the Devices, NATO ASI Series*, Vol. 321, Kluwer, Dordrecht, 1996, pp. 141, 289.
4. Bencini, A., Benelli, C., Caneschi, A., Dei, A. and Gatteschi, D., *Inorg. Chem.* **25** (1986), 572.
5. Caneschi, A., Gatteschi, D. and Sessoli, R., *J. Am. Chem. Soc.* **113** (1991), 5873.
6. Gadet, V., Bujoli-Doeuff, M., Force, L., Verdaguer, M., ElMalkhi, K., Deroy, A., Besse, J. P., Chappert, C., Veillet, P., Renard, J. P. and Beauvillain, P., In: D. Gatteschi, O. Kahn, J. S. Miller and F. Palacio (eds.), *Magnetic Molecular Materials, NATO ASI Series*, Vol. 198, Kluwer, Dordrecht, 1991, p. 281.
7. Cano, J., Ruiz, E., Alvarez, S. and Verdaguer, M., *Comments Inorg. Chem.* **20** (1998), 27.
8. Verdaguer, M., Bleuzen, A., Marvaud, V., Vaissermann, J., Tournilhac, F., Train, C., Ouahes, R., Garde, R., Fabrizi de Biani, F., Desplanches, C. and Scuiller, A., In: G. Ondrejovič, A. Sirota (eds.), *Coordination Chemistry at the Turn of the Century*, Slovak Technical University Press, Bratislava, 1999, p. 67.
9. Scuiller, A., Ph.D. Thesis, University of P. and M. Curie, Paris, 1999.
10. Gembický, M., Boča, R. and Renz, F., *Inorg. Chem. Commun.* **3** (2000), 662.
11. Matsumoto, N., Ohta, S., Yoshimura, C., Ohyoshi, A., Kohata, S., Okawa, H. and Maeda, Y., *J. Chem. Soc., Dalton Trans.* (1985), 2575.

12. Ohyoshi, A., Honbo, J., Matsumoto, N., Ohta, S. and Sakamoto, S., *Bull. Chem. Soc. Jpn.* **59** (1986), 1611.
13. Ohta, S., Yoshimura, C., Matsumoto, N., Okawa, H. and Ohyoshi, A., *Bull. Chem. Soc. Jpn.* **59** (1986), 155.
14. Boča, R., Fukuda, Y., Gembický, M., Herchel, R., Jaroščiak, R., Linert, W., Renz, F. and Yuzurihara, J., *Chem. Phys. Lett.* **325** (2000), 411.
15. Gütlich, P. and Ensling, J., In: R. I. Solomon and A. B. P. Lever (eds.), *Inorganic Electronic Structure and Spectroscopy*, Vol. I, Wiley, New York, 1999, p. 161.
16. Renz, F. and Kerep, P., *Hyp. Interact.* (2003), submitted.
17. Furman, N. H. and Miller, C. O., *Inorganic Synthesis*, Vol. III, 1950, pp. 160.

Hyperfine Interactions **156/157**: 379–388, 2004.
© 2004 *Kluwer Academic Publishers. Printed in the Netherlands.*

Processes in Geophysics Studied by Mössbauer Spectroscopy

ÖRN HELGASON
Science Institute, University of Iceland, Dunhagi 3, IS-107 Reykjavik, Iceland

Abstract. Mössbauer spectroscopy has been appreciated in geoscience as a powerful tool to study magnetic and structural properties of a wide range of minerals and rocks. In this presentation the application of Mössbauer spectroscopy in different geophysical processes such as tracing the development of magma during volcanic eruptions and phase transitions of magnetic minerals due to thermal impact of dikes in earlier lava formation or hydrothermal alteration will be discussed.

Key words: Mössbauer spectroscopy, mantle-derived magma, oxidation state, rock magnetism, titanomagnetite, geothermal alteration.

1. Introduction

Since the first work on magnetite and hematite in the very early days of the Mössbauer effect, Mössbauer spectroscopy has been appreciated in geoscience as a powerful tool to study magnetic and structural properties of a wide range of minerals and rocks. More specific applications can be mentioned such as: characterization of minerals, site occupation in minerals, Fe^{2+}/Fe^{3+} relationship (oxidation state) in volcanic ashes, magnetization and magnetic minerals, tracing processes in clays and sediments, hydrothermal alteration, studies of meteorites, lunar rocks, tektites and high pressure inclusions in diamonds [1–4].

In some cases a series of Mössbauer spectra from sediments may depict a geophysical development over millions of years wheras in another case a spectral profile of volcanic ash may reflect the development in the magma chamber during a single eruption. The transition of magnetic minerals from titanomagnetite in primary basalt to ilmenite/magnetite and further oxidation to maghemite and hematite or iron (oxy)hydroxides in soil can be studied by Mössbauer spectroscopy and the results may shed light on interesting geophysical conditions during the time these transitions have taken place. The time scale of these processes varies enormously depending on different environmental conditions.

This presentation will mainly focus on the application of Mössbauer spectroscopy in geophysical processes where the time scale is relatively short on normal geological time scale. The examples are mainly related to processes in Iceland, which has only existed for 15 million years and is sited on an active plume at the Atlantic ridge on the boarder of the American/Euroasian plates. In this area there is a great variation in volcanic activities, which often gives excellent opportunities for studying geophysical processes and rock formations in "*status nascendi*". The presentation will mainly focus on, how Mössbauer spectroscopy can be used in the study of the partial pressure of oxygen (oxygen fugacity) in magma and phase transitions in magnetic minerals.

2. Iron oxidation state and oxygen fugacity

Iron is the only major constituent in basalt that occurs in two oxidation states, Fe^{2+} and Fe^{3+}. Owing to their different ionic radii and charge, the two species exhibit different geochemical behaviour, both in melt and in the minerals they enter [5–7]. Consequently the ferric/ferrous ratio in a melt affects both its physical properties (density, viscosity) and its evolution during crystallization.

Iron oxidation state and oxygen fugacity (partial pressure of oxygen, often assigned as f_{O_2}) are closely linked. Like pressure and temperature, oxygen fugacity is an intensive variable, i.e. it is independent of the amount of material present. The relation between iron valance state and oxygen fugacity is relatively simple in melt and amorphous systems where an increase in oxygen fugacity results in a larger Fe^{3+}/Fe_{total} ratio.

The Mössbauer technique is a powerful tool for studying the ferric/ferrous ratio. Amongst the advantages are (a) small sample required, (b) the fact that both Fe^{2+} and Fe^{3+} are determined simultaneously obviating a separate analysis of the total iron, and (c) that the Mössbauer technique enables the Fe^{3+}/Fe^{2+} ratio of the glass phase to be determined independent of crystals that may also be present in the sample.

The response of the ferric/ferrous ratio in silicate melts to variations in the temperature, oxygen fugacity, and composition has been the subject of several investigations [8, 9]. The rate of the redox reactions is also of interest, especially with reference to natural glasses because their oxidation state may reflect that of the magma before the eruption.

To address this question a special study was conducted to determine the minimum duration of runs to attain redox equilibrium for different types of basalts in a wide range of fugacity [10]. All experiments in that study were made at 1300°C, with the furnace atmosphere in either of two modes, oxidising (air) or reducing ($p_{O_2} = 10^{-4}$ Pa). In one series, the samples were oxidized for 4 hours to obtain equilibrium in the ferric/ferrous process. Then, the oxygen fugacity was abruptly changed to the reducing condition for different lengths of time before quenching the sample. Three spectra recorded from samples in equilibrium at different oxygen fugacity in this experiment are shown in Figure 1. The Fe^{3+}/Fe_{total} ratio as obtained from the Mössbauer spectra is shown in Figure 2 indicating that the process is fully reversible within the experimental error and the time constant for the assimilated "exponential decays" is less than 2 hours for samples of this size. But, the results also show that products of fast cooled magma, where the glass is formed in the

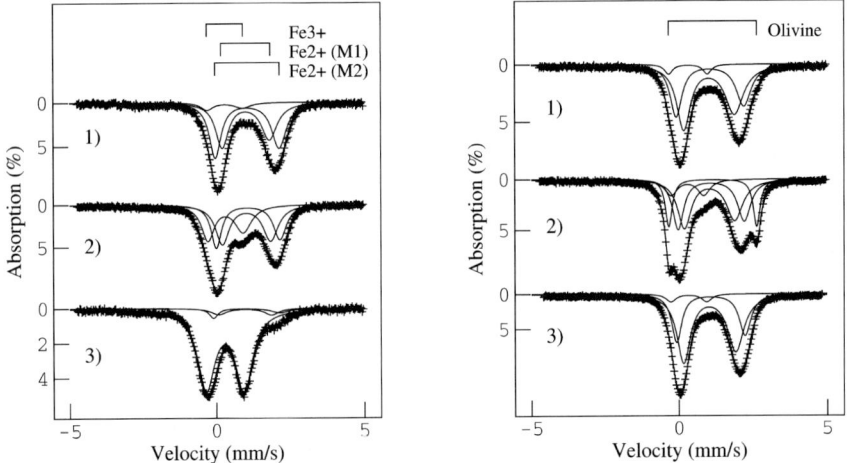

Figure 1. The three spectra to the left are recorded from samples in equilibrium at different oxygen fugacity (f_{O_2} 10^{-9}, 10^{-5} and 1 atmosphere, respectively). The spectra to the right are recorded from natural samples, (1) and (2) from the Mid-Atlantic ridge with different amount of olivine, but (3) from a subglacial eruption in 1996 in Southeast Iceland.

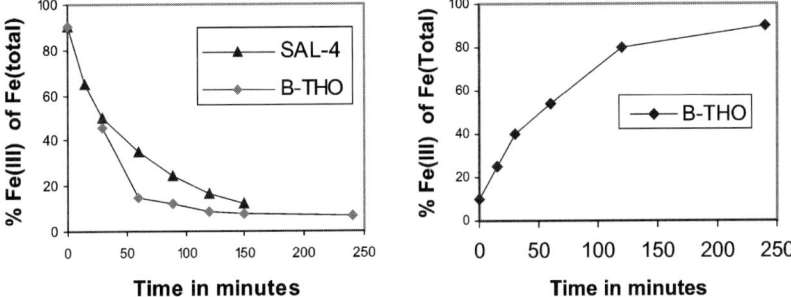

Figure 2. The graphs show the Fe(III)/Fe(total) ratio versus exposure time (minutes) for the samples kept in reducing atmosphere (to the left) and in oxidizing atmosphere (to the right) after equilibration in respectively, oxidizing and reducing atmospheres.

time scale of seconds, should reflect the ferric/ferrous ratio in the magma itself. Examples of this are volcanic ashes or glass formations on rock surface, when lava flow comes in contact with water as in subglacial or submarine eruptions.

In the following two different studies will be discussed in more details.

2.1. THE OXIDATION STATE OF MANTLE-DERIVED MAGMA

Olivine tholeiites are mantle-derived magmas that are formed by partial melting of their deep sources and which have equilibrated with mineral assemblages at slightly different subcrustal pressure-temperature-conditions prior to eruption. The minimum depth of the pre-eruptive reservoirs of these magmas is in the order of 10–15 km and their liquidus temperatures fall within the range 1180–1240°C. Three types of primitive olivine tholeiites are exposed along the rift zones in Iceland, which covers about a 500-km-long segment of the North-Atlantic rift system.

A study by Mössbauer spectroscopy has addressed whether the origin or the evolution of these magmas is associated with different oxygen fugacities and if there is a trend in the oxidation of iron that can be associated with their petrochemical evolution [11].

The oxidation state of iron in basaltic rocks, as inferred from bulk chemical analysis includes the sum of the iron species in all the phases of the rock. This fact makes natural glasses particularly important for estimating ferric/ferrous ratios. The Mössbauer technique also resolves microcrystallites of olivine and oxide minerals in minute amounts enabling highly reproducible ferric/ferrous analyses of the glass phase to be made (Figure 1*right*). The study dealt with chemical and Mössbauer data on 10 samples of basaltic glass from pillow rims, which represent the above-mentioned olivine tholeiites. Basaltic pillows form in subaquatic eruptions where rapid cooling chills the surfaces of bodies of silicate melt extruding into water. The present samples are all pillow rims formed during the latest epoch of the Quaternary Ice Age ending about 10,000 years ago.

For all the 10 samples the Fe(III) was determined to be within 10–15% of the total iron. At the liquidus temperature of these glasses this ferric/ferrous ratio corresponds to fugacity close to the fayalite-magnetite-quartz-oxygen (FMQ) buffer with an uncertainity of less than one log unit in f_{O_2}. The result confirms that there is no significant difference in the oxidation state of the three magma types.

2.2. STRATIFICATION OF THE OXYGEN FUGACITY IN A MAGMA CHAMBER

A Mössbauer study of volcanic glasses from the eruption in the Askja caldera 1875 is an example of studying a short time geophysical process [12].

Askja volcanic centre, situated in the eastern rift-zone in Iceland is the largest known active volcano along the global spreading ridges [13]. During the last 30,000 years the center has produced several tens of cubic kilometers of basaltic magma; mostly in subglacial eruptions while subaerial lava flows are being produced in the last 10,000 years. Few silicic eruptions have occurred within the Askja center indicating the temporary formation of shallow highly evolved magma reservoirs. The most recent silicic eruption occurred in March 1875 after a year of vigorous hydrothermal activity and basaltic eruptions. The large Plinian eruption gave rise to 1.85 cubic kilometers of dacitic pyroclastics which spread over NE-Iceland and the volcanic ash transported to the European continent is one of the volcanic marker horizons in the northern hemisphere. The eruption was composite; minor amounts of basaltic glass are found in the dacitic pumice. This is taken as an indication that the shallow dacitic magma chamber was intruded by hotter basaltic magma, which triggered the explosion [14].

To address this idea, samples from a 5.5 m thick profile through the pyroclastics was collected within the Askja caldera. The fallout was sampled in 15 units. Chemical analysis was performed to examine the degree of magma mixing between dacitic and basaltic end members. The material is a binary mixture of minor basaltic glass and highly evolved dacitic magma; the average silica content being close to 71 wt%. A conclusion drawn from the occurrence of uncontaminated basalt in the fallout is that the minor mixing took place shortly before the eruption.

The relation between %Fe(III) in the samples as determined from the Mössbauer spectra, and oxygen fugacity of magmas revealed that the dacite–magma chamber was stratified in oxygen fugacity. The eruption was triggered by basaltic magma, which was intruded into the base of the chamber. The observed oxidation state falls within the range of oxygen fugacities of H_2O/CO_2 mixtures at the liquidus temperature of the magma. It was concluded that oxygen fugacity of the chamber was effected by volcanic gases. The source of the gases was the degassing of superheated dacite at the dacite/basalt interface along the bottom of the chamber.

3. The role of titanomagnetite in basaltic rocks

Minerals known as titanomagnetite of composition $Fe_{3-x}Ti_xO_4$ form a solid solution series between Fe_3O_4 (magnetite) and Fe_2TiO_4 (ulvöspinel) and are common constituents of a wide variety of igneous rocks. Figure 3*left* shows a FeO–TiO_2–Fe_2O_3 ternary diagram, which relates the composition to magnetite, ulvöspinel, ilminite, pseudobrookite and hematite/maghemite. These oxides are of considerable interest in geophysics. Their magnetic properties are important to the field of paleomagnetism and transformations: such transformations inside the diagram may reveal interesting information about geophysical processes, which have occurred in the related area. Transformations, such as the subsolvus exsolution of magnetite/ilmenite and further oxidation to maghemite/hematite are time-temperature-dependent and may also depend on the amount of impurities in the oxides [15, 16].

[57]Fe Mössbauer spectroscopy is a unique technique to characterise iron oxides in basalt, giving information on the valence state of Fe and on magnetic interactions [17, 18]. The spectrum of pure magnetite consists of two characteristic sextets, usually labelled A and B (Figure 3*right*). The A sextet, assigned to Fe(III) on tetrahedral sites, has a magnetic hyperfine field of approximately 49 T. The B sextet, assigned to Fe(II) and Fe(III) on octahedral sites rapidly exchanging electrons, results in one unresolved sextet, which has a magnetic hyperfine field of approximately 46 T. The area ratio between the A sextet and the B sextet for pure magnetite is close to $1/2$. In titanomagnetite, there are characteristic changes in the hyperfine parameters that allow for rough estimates of the amount of titanium in the magnetite [19].

The oxidation of titanomagnetite to titanomaghemite γ-$Fe_{(2-x)}Ti_xO_3$, which like magnetite has a spinel related crystal structure, has received increased attention in recent years. Formation of (titano)maghemite as a surface weathering product of basalts has been intensively studied, especially in relation to the highly magnetised soil of Brazil [20–22]. The maghemite, like magnetite, can have a large magnetic moment (the spontaneous magnetisation for maghemite, σ_s is ca. 60 J/T/kg compared with ca. 100 J/T/kg for magnetite and 0.5 J/T/kg for hematite).

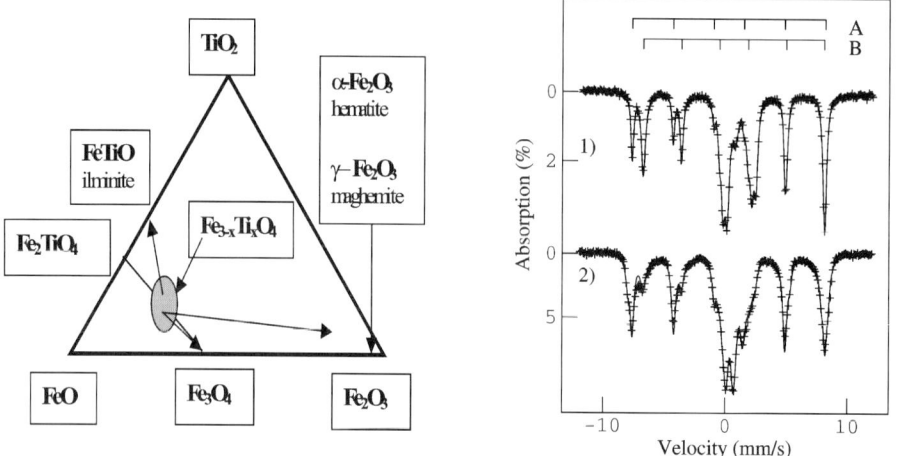

Figure 3. (*Left*) A FeO–TiO$_2$–Fe$_2$O$_3$ ternary diagram, which relates the composition to magnetite, ulvöspinel, ilminite, pseudobrookite and hematite/maghemite. (*Right*) Two spectra from samples, where the magnetic phase is low in titanium. The upper one is from Stardalur, SW-Iceland with a magnetic phase of nearly pure magnetite. The lower is also from strongly magnetic basaltic rocks, but the magnetic phase is far more transformed (oxidised) into maghemite [27].

The Mössbauer spectrum of maghemite can be represented by a single broad lined sextet, overlapping the A sextet of titanomagnetite (spectrum 2 in Figure 3*right*).

Magnetic survey across the mid-Atlantic ridge has shown a systematic pattern in the geomagnetic field, which has been correlated to changes in the direction and strength of the geomagnetic field at the time of the respective lava flows on the ocean floor. The drift of approximately 1 cm/year means that the distance across the rift is increasing by 10–20 km per million years. A more than less steady flow of lava through the rift has been going on for millions of years and, as it is solidifying, the rocks become magnetized and "record" the present geomagnetic field. Hence, the main magnetic pattern across the ridge will tell the story of the development of the geomagnetic field for some 10 to 50 million years. In 1970 Irving suggested that the ubiquitous decrease in palaeomagnetic intensities away from the mid-ocean ridges results from oxidation of titanomagnetite to titanomaghemite [23, 24]. This means that the transformation must also take place inside the solid rock formation.

The rift zone crosses Iceland as mentioned before and a similar paleomagnetic pattern can be seen in the lava formations, both horizontally as at the ridge, and also vertically. The combination of the movement at the rift and the extremely active hot plume beneath Iceland has produced large mountain formations from the late Tertiary stacked by series of lava layers (some cases 50–60 layers, 10–30 m thick each) spanning millions years and many paleomagnetic epochs. But, due to intensive volcanic activity, the pattern is often strongly modulated by intrusive dikes, hydrothermal alterations or subglacial eruption [25]. Two different processes will now be discussed in some details.

3.1. GEOTHERMAL ALTERATION

An example of the effect of geothermal alteration can be seen from a study from a profile of rock samples from an 1100 m deep productive well at Nesjavellir geothermal field, SW-Iceland [26]. Figure 4*left* shows a sequence of spectra recorded from bulk samples obtained at different depths. The well itself has a maximum output of about 53 MW thermal and the main active aquifer is at a depth of about 740 and another is at 1095 meters, both are utilised at temperatures of 300°C. The volcanic strata cut by the well, as inferred from the mineralogy of well cuttings, are dominantly basaltic lavas and hyaloclastites with iron bearing phases typical for basaltic rocks. The spectra clearly indicate the impact of the aquifers on the oxidation/redistribution of the magnetic iron oxides favouring the formation of maghemite/hematite, as seen in Figure 4*right*. The anomalous occurrence of maghemite in hydrothermal aquifers indicates that its presence may be used as a rate indicator for the oxidation of titanomagnetite. A well-known reaction path from titanomagnetite to hematite involves three epochs; (a) oxyexsolution of TiO_2 within titanomagnetite to produce successively more pure magnetite, (b) oxidation of iron within the cubic magnetite structure to produce maghemite, (c) crystallographic transition of maghemite to hematite. Knowing the reaction rate makes it

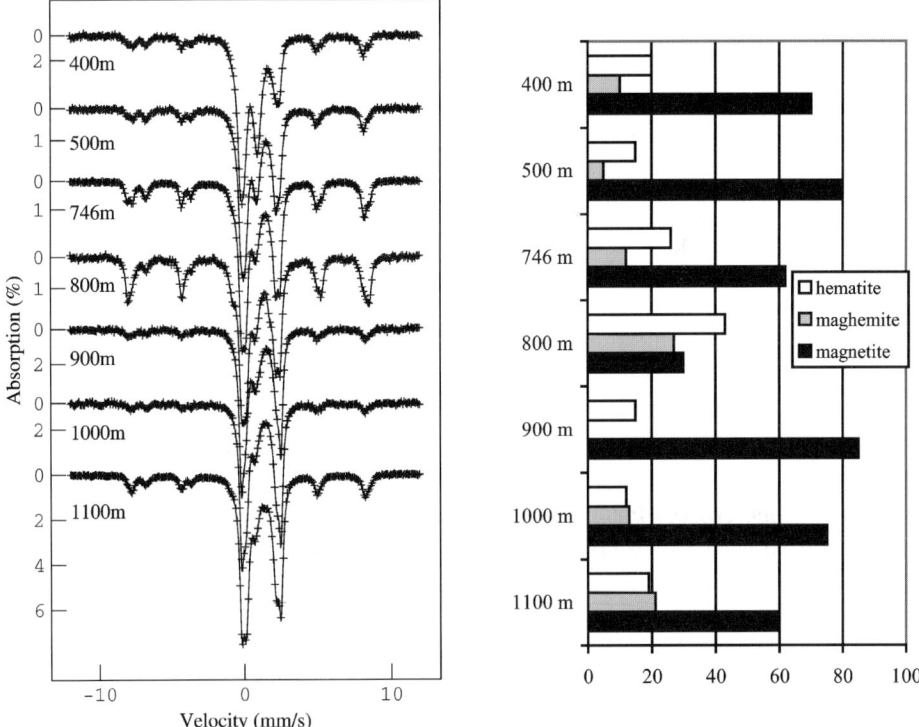

Figure 4. (*Left*) Mössbauer spectra of bulk samples from different depths of the well. (*Right*) The relative distribution of the magnetic minerals as determined from the bulk spectra.

possible to estimate the degree of alteration by examination of the relative abundance of the primary, intermediate and final phases within the hydrothermal strata. Since the reverse reaction from hematite to magnetite via maghemite seems not to occur, the presence of maghemite may also be a criterion for estimating the relative age of hydrothermal aquifers.

3.2. TITANOMAGNETITE–MAGHEMITE TRANSITION IN SOLID ROCKS

The question of the titanomagnetite–maghemite transition and its rate in igneous rocks is of great interest in geophysics. Not, only as mentioned before in relation to the time-dependent decrease of the magnetisation on the Atlantic ridge, but also to what extend it takes place before the erosive processes from the rock to soil formation become most important. This question was addressed in a multi-disciplinary reconnaissance study of the oxide minerals from eight samples of basaltic lava from Iceland with Js–T curves characteristic for nearly pure magnetite, ranging in geological age from 40,000 yrs to 15 m.y. The study grew out of our earlier investigations on highly magnetic basalts from the Stardalur caldera (spectrum 1 in Figure 3*right*), in which the samples showed characteristically high single Curie temperature and reversible magnetization–temperature curves [27]. In most of the samples titanium in the magnetite, as analyzed with the microprobe, ranged between 16 and 28 wt.%, indicating sub-microscopic solvus exsolution in the titanomagnetite beyond the power of resolution for the microprobe. More unexpectedly in view of the reversible Js–T curves, Mössbauer spectroscopy showed an appreciable proportion of maghemite in the magnetic fraction, leading to the conclusion that maghemite is much more common in Icelandic rocks than hitherto believed. Compared to the transition rate seen in the geothermal alteration, where the hot water/steam also plays an important role, the rate in the solid rock is far slower.

Studies of the temperature dependence of this transition in samples collected near by dikes are of importance. When hot lava intrusions forces their way into older rock formations forming dikes, the result is a strong thermal impact on the surrounding rock and in some cases the process seems to enhance the magnetisation of the rock by up to an order of a magnitude [28]. Samples obtained from different distances up to ca. 2–3 m from a dike have been studied with the Mössbauer effect and both transmission and CEMS spectra have been recorded together with other studies on the magnetic properties of the samples [29]. Sample of the same composition were also subjected to annealing at different temperatures and time periods for comparison. The main result was that samples taken far from the dike (>1 m) are unaltered by the intrusion and the magnetic phase is titanomagnetite with $x \sim 0.2$ according to Mössbauer spectroscopy. Close to the dike (<1 m), the thermal effect has led to exsolution of the titanomagnetite and the Mössbauer spectra show a mixture of relatively titanium-poor magnetite ($x \sim 0.05$) and titanomaghemite.

Other studies on both natural and synthetic samples have shown that impurities have a great effect on the transition rate and even block the process from the strong magnetic metal-doped maghemite to weak magnetic hematite [30, 31]. Formation of strongly magnetic phases of titanomaghemite in solid rock without interaction of water as described above could be possible clues for understanding the formation of the strongly magnetic phase in Marian soil and dust [29].

4. Conclusions

Mössbauer spectroscopy can be a powerful tool in multi-disciplinary studies in complicated processes of geophysics, especially where the iron plays an important part in the different rock and soil formations.

Acknowledgement

The author would like to thank K. Jonsson at the Science Institute of the University of Iceland for his extensive technical assistance in all phases of the work.

References

1. Mitra, S., *Applied Mössbauer Spectroscopy, Physics and Chemistry of the Earth*, Vol. 18, Part III–VI, Pergamon Press, Oxford, 1992.
2. Coey, C. M. D., In: G. J. Long and F. Grandjean (eds), *Mössbauer Spectroscopy Applied to Inorganic Chemistry*, Vol. I, Plenum, New York, 1984, p. 443.
3. Murad, E. and Johnston, J. H., In: G. J. Long and F. Grandjean (eds), *Mössbauer Spectroscopy Applied to Inorganic Chemistry*, Vol. II, Plenum, New York, 1987, p. 507.
4. McCammon, C. A., *Hyp. Interact.* **141/142** (2002), 73.
5. Sack, R., Carmichael, I., Rivers, M. and Ghiorso, M., *Contrib. Min. Petrol.* **75** (1980), 369.
6. Virgo, D. and Mysen, B. O., *Phys. Chem. Mineral* **12** (1985), 65.
7. Dyar, M. D., *American Mineralogist* **70** (1985), 304.
8. Helgason, Ö., Steinthorsson, S. and Mørup, S., *Hyp. Interact.* **45** (1989), 287.
9. Dyar, M. D., Naney, M. T. and Swanson, S. E., *American Mineralogist* **72** (1987), 972.
10. Helgason, Ö., Steinthórsson, S. and Mørup, S., *Hyp. Interact.* **70** (1992), 985.
11. Óskarsson, N., Helgason, Ö. and Steinthorsson, S., *Hyp. Interact.* **91** (1994), 733.
12. Helgason, Ö., Óskarsson, N. and Sigvaldason, G. E., *Hyp. Interact.* **70** (1992), 989.
13. Sigvaldason, G. E., *J. Petrol.* **15** (1974), 497.
14. Sigurdsson, H. and Sparks, R. S. J., *J. Petrol.* **22** (1981), 41.
15. Collyer, S., Grimes, N. W., Vaughan, D. J. and Longworth, G., *American Mineralogist* **73** (1988), 153.
16. Helgason, Ö., Berry, F. J., Jónsson, K. and Skinner, S. J., In: I. Ortalli (ed), *Proc. of the Internat. Conf. on Applications of the Mössbauer Effect*, S.I.F., Bologna, 1996, p. 59.
17. Vandenberghe, R. E., Barrero, C. A., da Costa, G. M., Van San, E. and De Grave, E., *Hyp. Interact.* **126** (2000), 247.
18. Vanderberghe, R. E. and De Grave, E., In: G. J. Long and F. Grandjean (eds), *Mössbauer Spectroscopy Applied to Inorganic Chemistry*, Vol. III, Plenum, New York, 1989, p. 59.
19. Tanaka, H. and Kono, M., *Geomag. Geoelectr.* **39** (1987), 463.
20. Allan, J. E. M., Coey, J. M. D., Sanders, I. S., Schwertmann, U., Friedrich, G. and Wiechowski, A., *Mineralogical Magazine* **53** (1989), 299.

21. Fabris, J. D., Coey, J. M. D., Qinian, Qi. and Mussel, W. N., *American Mineralogist* **80** (1995), 664.
22. Fabris, J. D., Coey, J. M. D. and Mussel, W. N., *Hyp. Interact.* **113** (1998), 249.
23. Irving, E., *Can. J. Earth Sci.* **7** (1970), 1528.
24. Smith, B. M., *Phys. Earth Planet. Int.* **46** (1987), 206.
25. Kristjansson, L. and McDougall, I., *Geophys. J. R. Astr. Soc.* **68** (1982), 273.
26. Helgason, Ö., Óskarsson, N., Jónsson, K. and Gunnlaugsson, E., In: I. Ortalli (ed.), *Proc. of the Internat. Conf on Applications of the Mössbauer Effect*, S.I.F., Bologna, 1996, 749.
27. Steinthorsson, S., Helgason, Ö., Madsen, M. B., Bender Koch, C., Bentzon, M. and Mørup, S., *Mineralogical Magazine* **56** (1992), 185.
28. Kristjansson, L. J., *Geophys. Res.* **90**(B12) (1985), 10129.
29. Gunnlaugsson, H. P., Weyer, G. and Helgason, Ö., *Planet. Space Sci.* **50** (2002), 157.
30. Helgason, Ö., Gunnlaugsson, H. P., Steinthorsson, S. and Mørup, S., *Hyp. Interact.* **70** (1992), 981.
31. Berry, F. J. and Helgason, Ö., *Hyp. Interact.* **126** (2000), 269.

Hyperfine Interactions **156/157**: 389–394, 2004.
© 2004 *Kluwer Academic Publishers. Printed in the Netherlands.*

High Pressure Mössbauer Studies on FCC $Fe_{53}Ni_{47}$ Alloy

Y. A. ABDU[1,*], H. ANNERSTEN[1], L. S. DUBROVINSKY[2] and
N. A. DUBROVINSKAIA[2]

[1]*Department of Earth Sciences, Uppsala University, SE-752 36, Uppsala, Sweden;*
e-mail: yassir.abdu@geo.uu.se
[2]*Bayerisches Geoinstitut, Universität Bayreuth, Germany*

Abstract. High pressure ^{57}Fe Mössbauer spectroscopy measurements (up to ∼41 GPa at room temperature) have been carried out for investigating the magnetic properties of γ (fcc) $^{57}Fe_{53}Ni_{47}$ alloy using diamond anvil cell (DAC) technique. The Mössbauer spectrum at 0 GPa shows a six line magnetic pattern with broad outer peaks and an average hyperfine field of ∼31 T characteristic of a disordered alloy. In the pressure range ($2 < P < 20$ GPa) we observe Mössbauer spectra with additional low hyperfine field component resembling spectra of γ (fcc) Fe–Ni Invar alloys (30–40 at % Ni). Our data indicate a pressure induced Invar effect for $^{57}Fe_{53}Ni_{47}$ alloy at ∼7–12 GPa. Above 20 GPa the hyperfine field breaks down and the alloy becomes non-magnetic showing only a single line Mössbauer spectrum.

Key words: Mössbauer spectroscopy, high pressure, Fe–Ni alloys, Invar.

1. Introduction

Apart from being technologically important, Fe–Ni alloys are of particular interest to material scientists as well as to Earth scientists. The Earth's core is mostly iron, but it may contains Fe–Ni alloys with fcc structure [1].

Connection between the average magnetic moment and molar volume is a common characteristic of γ (fcc) Fe–Ni alloys, and ∼35 at% Ni they show a near-zero thermal expansion coefficient in a wide temperature range (the Invar effect). This effect and other related abnormal physical properties have been a subject of intense theoretical and experimental research over a century. Theoretical calculations predicted that the magnetic moment of Ni in γ (fcc) Fe–Ni alloys appears to be nearly independent of the alloy composition or volume, whereas the moment of Fe changes with volume [2]. Furthermore, the magnetic structure is characterized (even at zero temperature) by a continuous transition from a high moment ferromagnetic state at high volumes to a low moment disordered non-collinear configuration at low volumes [2].

---

* Author for correspondence.

[57]Fe Mössbauer spectroscopy is a very valuable tool in high pressure research, and some times gives unique information regarding the magnetic properties of iron containing materials under extreme pressure. Abd-Elmeguid *et al.* [3] studied the weak ferromagnets γ (fcc) $Fe_{68.5}Ni_{31.5}$ and $Fe_{65}Ni_{35}$ Invar alloys at variable temperatures down to 2 K and high pressures up to ~8 GPa by [57]Fe Mössbauer spectroscopy using B_4C anvils. Their results showed that above a critical pressure, e.g., ~5.8 GPa, the ferromagnetic state in $Fe_{68.5}Ni_{31.5}$ alloy is destroyed and the system displayed antiferromagnetic ordering at low temperatures.

Here we study the strong ferromagnet γ (fcc) [57]$Fe_{53}Ni_{47}$ alloy by high-pressure [57]Fe Mössbauer spectroscopy (up to ~41 GPa at room temperature) using diamond anvil cell (DAC) in order to investigate its magnetic properties through the pressure dependence of the Mössbauer hyperfine parameters.

2. Experimental

The sample used in this investigation was prepared by arc melting technique as described elsewhere [4]. The fcc structure of the alloy was confirmed by X-ray diffraction (lattice parameter $a = 3.5802(3)$ Å). The composition of the sample was determined using microprobe analysis and its homogeneity was confirmed by SEM observations. High-pressure [57]Fe Mössbauer measurements were performed at room temperature in transmission geometry using DAC [5]. The spectrometer was calibrated using the spectrum of α-Fe at room temperature. NaCl was used as a pressure medium and several ruby chips were placed in the pressure chamber, sealed with a Re gasket, for pressure determination. Maximum pressure gradient across the pressure chamber at 20 GPa was less than 1 GPa, and at 40 GPa – less than 2 GPa.

3. Results and discussion

Figure 1 shows some selected Mössbauer spectra of [57]$Fe_{53}Ni_{47}$ alloy at different pressures collected on compression (a) and decompression (b). The spectrum at 0 GPa is fitted with a magnetic sextet having an average hyperfine field of 31.4 ± 0.5 T, characteristic of a high moment ferromagnetic alloy [6]. In the pressure range $2 < P < 20$ GPa, we have fitted the Mössbauer spectra using two sextet components with a high (B_H) and low (B_L) hyperfine fields (solid and dashed subspectra in Figure 1, respectively). For example, at 8.5 GPa, $B_H = 26.3$ T and $B_L = 18.7$ T (see Figure 1(a)). The linewidths and the centre shifts were constrained to be equal for the two sextets during the fitting routine. We have also used the parameter dB%, which is a measure of the relative hyperfine field distribution with respect to the average field B, and the texture parameter TP [7]. The spectra above 20 GPa are fitted with broad singlets.

In Figure 2(a) we show the pressure dependence of the normalized weighted average hyperfine field $[B(P)/B(0)]$ for the [57]$Fe_{53}Ni_{47}$ alloy. For comparison, we

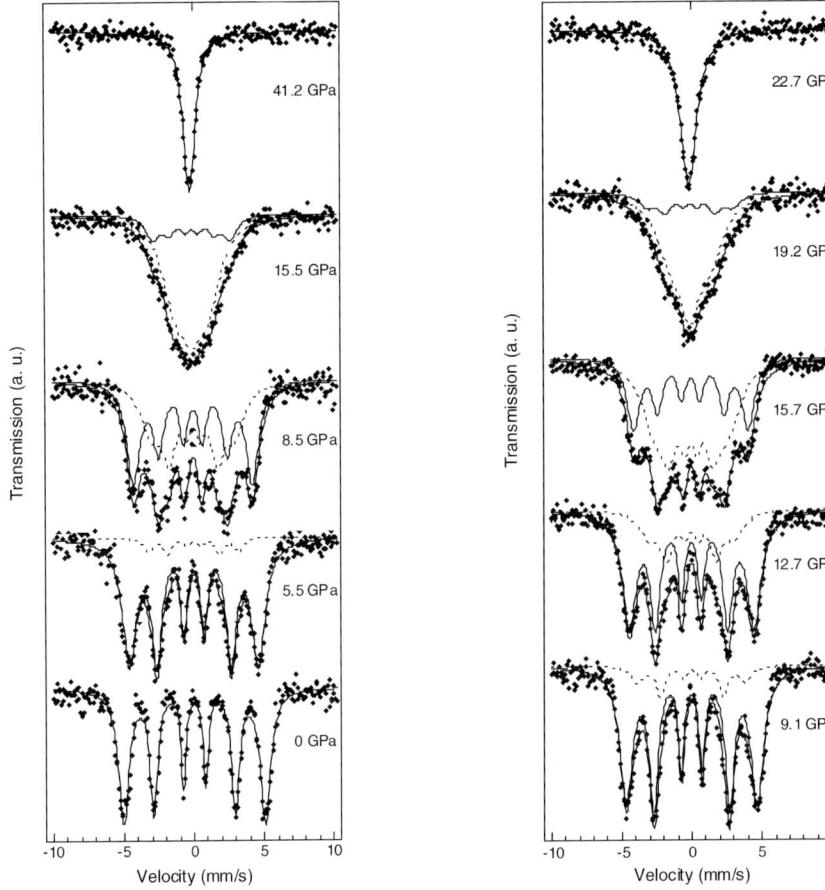

Figure 1. High-pressure Mössbauer spectra of $^{57}Fe_{53}Ni_{47}$ alloy collected at RT: (a) compression; (b) decompression.

show on the same figure the data for $Fe_{68.5}Ni_{31.5}$ and $Fe_{65}Ni_{35}$ Invar alloys (at 4.2 K) taken from Reference [3]. As seen from Figure 2(a) and following the compression curve, $B(P)/B(0)$ which is generally proportional to the average magnetic moment on Fe, does not change with pressure up to ~2 GPa. Above 2 GPa it shows a gradual decrease with increasing pressure followed by a large negative pressure dependence similar to that observed in weak itinerant ferromagnets [3]. This implies that the ferromagnetic state in $^{57}Fe_{53}Ni_{47}$ alloy has become unstable against pressure above ~2 GPa, as a result of the decrease in the Curie temperature with increasing pressure. Our high pressure Mössbauer spectra of $^{57}Fe_{53}Ni_{47}$ alloy in the range $2 < P < 20$ GPa, which show an additional hyperfine field component, see Figure 1, resemble the Mössbauer spectra of Fe–Ni alloys in the Invar region (30 < at% Ni < 40) at ambient conditions [6]. In Figure 2(b) we show for comparison the Mössbauer spectrum of the $^{57}Fe_{53}Ni_{47}$ alloy collected on compression at 7.6 GPa (lower) and that of a commercial Invar $Fe_{64}Ni_{36}$ alloy measured at ambient

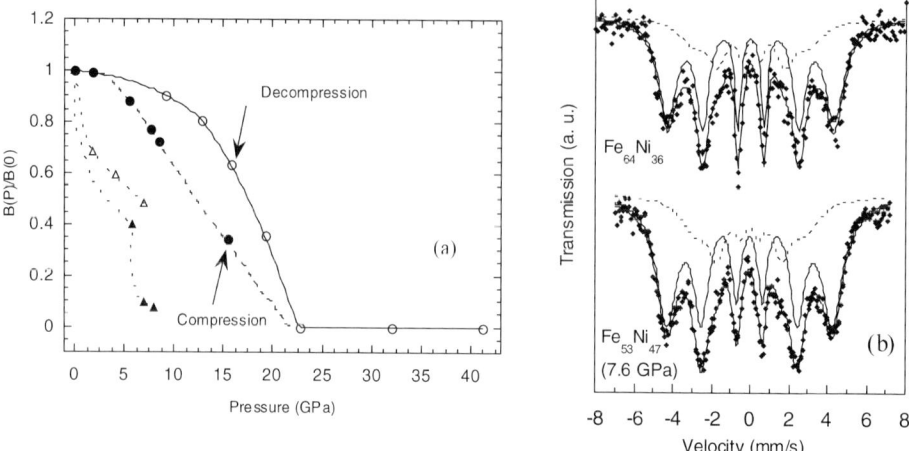

Figure 2. (a) Pressure dependence of the normalized weighted average hyperfine field for $^{57}Fe_{53}Ni_{47}$ alloy at RT: filled circles (compression), open circles (decompression). Open and filled triangles represent data from [3] (at 4.2 K) for Invar $Fe_{65}Ni_{35}$ and $Fe_{68.5}Ni_{31.5}$, respectively. The lines through the data points are only guides to the eye. (b) RT Mössbauer spectra for $^{57}Fe_{53}Ni_{47}$ alloy at 7.6 GPa (*lower*) and $Fe_{64}Ni_{36}$ Invar alloy at ambient pressure (*upper*).

pressure (upper). The two spectra are essentially identical and their hyperfine fields B_H and B_L are equivalent within errors. This is in excellent agreement with high-pressure X-ray diffraction studies on fcc Fe–Ni alloys, where a pressure induced Invar effect has been observed at 7.7 GPa (around RT) for $Fe_{55}Ni_{45}$ alloy [4].

On decompression, we observe a typical Invar spectrum at 12.6 GPa (see Figure 1(b)), which indicates a pressure hysteresis as evident from Figure 2(a). This hysteresis is also seen in the pressure dependence of the Mössbauer relative intensity of the low field component as shown in Figure 3(a), a feature that has been reported in $\alpha \to \varepsilon$ transition in pure Fe [8].

The pressure dependence of the centre shift (CS) of $^{57}Fe_{53}Ni_{47}$ alloy is shown in Figure 3(b). Note the decrease in the CS with increasing pressure as a result of volume decrease. The sudden decrease in the CS at $\sim$20 GPa may indicate a high moment to low moment transition. Above 20 GPa, the CS's are fitted smoothly to a straight line that has an intercept of -0.07 mm/s at 0 GPa. This value is very close to the CS's for mechanically alloyed γ (fcc) Fe-rich Fe–Ni alloys obtained at ambient conditions, where an antiferromagnetic ordering (with a low moment) indicated by the dramatic line broadening of the Mössbauer singlet was found to occur at $\sim$40 K for γ (fcc) $Fe_{76}Ni_{24}$ alloy (unpublished data). Abd-Elmeguid *et al.* [3], suggested a pressure induced antiferromagnetism in γ (fcc) $Fe_{68.5}Ni_{31.5}$ alloy at 7 GPa with T_N of $\sim$35 K, in agreement with the local magnetic moment model [9], where a pressure induced increase of "latent" antiferromagnetism leading to an antiferromagnetic ordering at high pressures is expected.

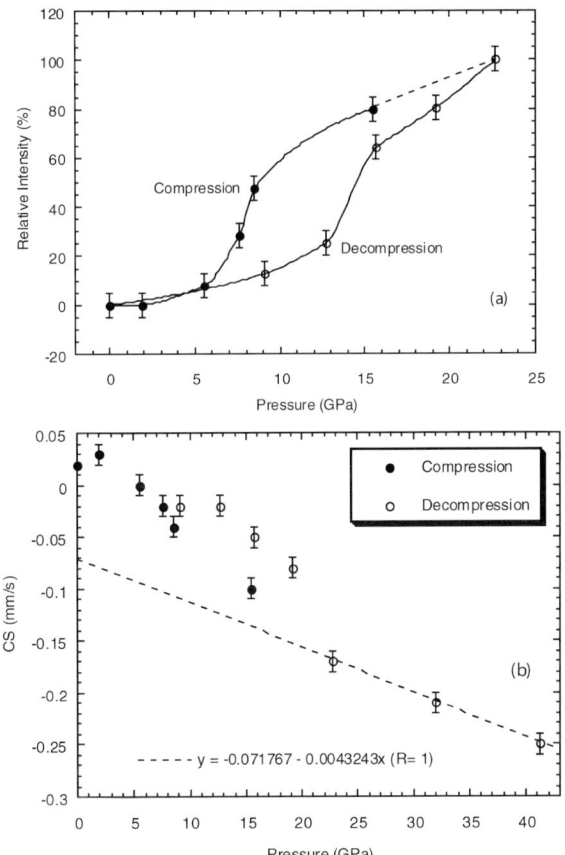

Figure 3. Pressure dependence of: (a) relative intensity of the low field component; (b) Mössbauer centre shift for ^{57}Fe$_{53}$Ni$_{47}$ alloy. The dashed line in Figure 3(b) is a linear fit to the CS's above 20 GPa.

It appears from the above discussion that as pressure increases and volume decreases we observe a continuous transition from a stable ferromagnetic state with a high moment to a less stable one and eventually to a low magnetic moment above 20 GPa. This is in agreement with theory [2] and it demonstrates the dependence of the magnetic moment on volume for γ (fcc) Fe–Ni alloys. It would be interesting to investigate the magnetic nature of the high pressure phase (above 20 GPa), and a suitable method is to conduct high pressure ^{57}Fe Mössbauer experiments at low temperatures and in external magnetic fields.

References

1. Huang, E., Bassett, W. A. and Weathers, M. S., *J. Geophys. Res.* **93** (1988), 7741.
2. van Schilfgaarde, M., Abrikosov, I. A. and Johansson, B., *Nature* **400** (1999), 46 and references there in.

3. Abd-Elmeguid, M. M., Schleede, B. and Micklitz, H., *J. Magn. Magn. Mater.* **72** (1988), 253.
4. Dubrovinsky, L. S., Dubrovinskaia, N. A., Abrikosov, I. A., Vennström, M., Westman, F., Carlson, S., van Schilfgaarde, M. and Johansson, B., *Phys. Rev. Lett.* **86**(21) (2001), 4851.
5. Dubrovinsky, L., Annersten, H., Dubrovinskaia, N., Westman, F., Harryson, H., Fabrichnaya, O. and Carlson, S., *Nature* **412** (2001), 527.
6. Ping, J. Y., Rancourt, D. G. and Dunlap, R. A., *J. Magn. Magn. Mater.* **103** (1992), 285.
7. Jernberg, P. and Sundqvist, T., Uppsala University report UUIP-1090, 1983.
8. Taylor, R. D. and Pasternak, M. P., *Hyp. Interact.* **53** (1990), 159.
9. Kondorsky, E. I. and Sedov, V. L., *J. Appl. Phys.* **31S**(5) (1960), 331.

Hyperfine Interactions **156/157**: 395–402, 2004.
© 2004 *Kluwer Academic Publishers. Printed in the Netherlands.*

Mössbauer and XRD Comparative Study of Host Rock and Iron Rich Mineral Samples from Paz del Rio Iron Ore Mineral Mine in Colombia

M. FAJARDO[1], G. A. PÉREZ ALCÁZAR[1], A. M. MOREIRA[2] and
N. L. SPEZIALI[2]
[1]*Departamento de Física, Universidad del Valle, A. A. 25360, Cali, Colombia*
[2]*Departamento de Física, Universidade Federal de Minas Gerais, C. P. 702, Belo Horizonte, Brazil*

Abstract. A comparative study between the host rock and the iron rich mineral samples from the Paz del Rio iron ore mineral mine in Colombia was performed using X-ray diffraction and Mössbauer spectroscopy. Diffraction results of the iron rich mineral sample show that goethite, hematite, quartz, kaolinite and siderite are the main phases, and that a small amount of illite is also present. By Mössbauer spectroscopy at room temperature (RT) the presence of all the above mentioned phases was detected except quartz as well as an additional presence of small amount of biotite. The goethite, which appears as four sextets with hyperfine fields of 33.5, 30.5, 27.5 and 18.5 T, respectively, is the majority phase. This result shows the different grades of formation of this oxyhydroxide. The Mössbauer spectrum of this sample at 80 K presents the same phases obtained at RT without any superparamagnetic effect. In this case the goethite appears as two sextets. Diffraction results of the host rock sample show a large amount of quartz and kaolinite and small amounts of illite and biotite, whereas by Mössbauer spectroscopy illite, kaolinite and biotite were detected.

Key words: iron mineral, Mössbauer spectroscopy, X-ray diffraction.

1. Introduction

Iron is one of the most abundant and ubiquitous elements of the earth's crust. Commonly the sedimentary rocks contain between 2 to 3% Fe, the basalts and graves contain nearly 8.5% Fe; the mean Fe content of the earth's crust is nearly 5%, and the economically exploited rocks have a range of Fe contents between 20 and 69%. The Fe ores economically exploited are formed by magmatic segregation, hydrothermal replacement, direct sedimentation and digenesis and metheorization in surfaces and subsoils [1].

Nearly 300 minerals contain iron as essential components, but only six of them can be considered as iron ore minerals and from these magnetite, hematite and goethite (limonite) are the most important ones [1].

The hypothesis about the genesis of the oolitic iron ore mineral is similar to that of the majority of the mines of the world. The oolites were formed by the precipitation of the soluble iron hydroxides, which were in an aqueous medium of low deep, in which angular quartz fragments and suspended cyan was initially deposited, then the oolites were deposited by saturation of the colloidal iron hydroxide in the presence of organic matter. In general the oolite core is formed by detritic quartz and fossil fragments. The oolitic iron is present in form of sedimentary seams interchanged with different lithologic types such as limestones, pizars, etc. The formation time of the oolitic iron seams is, in general, shorter than that of the sedimentary extracts. In Europe the seams were formed in the Jurassic, in United States in the Ordovician and Silurian, and in Colombia in the Tertiary.

The aim of the present work is the use of the Mössbauer spectroscopy and X-ray diffraction (XRD) to examine in detail the nature of the iron rich mineral and the corresponding host rock samples from the Paz del Rio mine in Colombia. This iron rich mineral is used as the principal prime matter for the steel production of the Acerías Paz del Río, the biggest steel factory in Colombia.

2. Experimental method

The iron mineral extracts used in the present study are from the Paz del Río zone in the Boyacá Department, Colombia. This deposit, as is shown in Figure 1, is extended by 20 km in the SW-NE direction (Sogamoso – La Uvita), with a width varying between 0.4 to 7.0 m [2, 3]. The samples of the iron rich mineral, with a brown red color, and of the host rock material, with a clear gray color, were manually collected in the open mining site named Uche, then they were milled to a grain size less than 50 μm. 100 g of sample was used to perform a chemical analysis

Figure 1. Localization of the Paz del Río iron ore formation in Colombia.

by atomic absorption spectroscopy, following the standard method described by Jeffery and Hutchison [4]. The total concentrations of Si^{+4}, Ca^{+2}, Al^{+3}, Mg^{+2}, Mn^{+2} and Fe were obtained. XRD patterns were obtained with a Rigaku 2200 diffractometer using the following parameters: Cu-Kα/40 kV/30 mA. A silicon standard was used for calibration. XRD patterns were refined by the Maud program, an improved version of Rietquan, procedure which is based on the Rietveld method combined with Fourier analysis [5]. Mössbauer spectra were recorded at room temperature (RT) in a conventional constant acceleration transmission spectrometer and with a ^{57}Co/Rh source. The spectra were fitted by using the Mosfit program and the reported isomer shifts are related to α-Fe [6].

3. Results and discussion

The obtained chemical analysis of the studied samples is shown in Table I, where LBC is the loss by calcinations. It can be noted that large quantities of Fe^{+3} and Fe^{+2} and of Si^{+3} and Al^{+3} are present in the iron ore mineral and host rock samples, respectively. This analysis confirms that the iron mineral samples of the Paz del Río zone have a medium content of iron (47.63%) and have an acceptable silicon level (4.0% Si). These contents are adequate for the steel production in the mentioned factory which uses materials of basic character for the high furnace and converters.

In Figure 2 the XRD pattern and the Rietveld refinement results obtained from the host rock sample are shown. The phases detected are illite, kaolinite, biotite, clinochlore and quartz [7]. Their corresponding lattice parameters and space groups are listed in Table II. According to the previous results the following can be concluded: the Fe detected by chemical analysis comes from the illite, biotite and kaolinite; Si comes principally from quartz and also in small quantities from illite, kaolinite and biotite; Al comes from kaolinite, illite and biotite and finally Mg and Mn come from biotite.

In Figure 3 the fitted Mössbauer spectrum of the host rock sample at 300 K is shown. The identified phases, fitted with only one doublet for each one, are illite, kaolinite and biotite [8]. These are the phases containing Fe which were also detected by XRD. It is possible to state from these results that the different phases in the host rock sample are either poor in iron or do not contain iron. The relative low width of the spectra shows the good crystallinity of the identified phases. This is a consequence of the sedimentary origin of the rocks which were formed under

Table I. Chemical analysis of the iron rich mineral and its corresponding host rock sample

	SiO$_2$	CaO	Al$_2$O$_3$	MgO	MnO	Fe$_2$O$_3$	FeO	PPC	P$_2$O$_5$
Mineral	8.56(1)	2.10(1)	5.91(1)	0.29(1)	0.19(1)	55.98(1)	10.90(1)	12.65(1)	1.17(1)
Host rock	58.32(1)	0.52(1)	21.26(1)	1.56(1)	0.05(1)	6.05(1)	5.80(1)	7.51(1)	0.21(1)

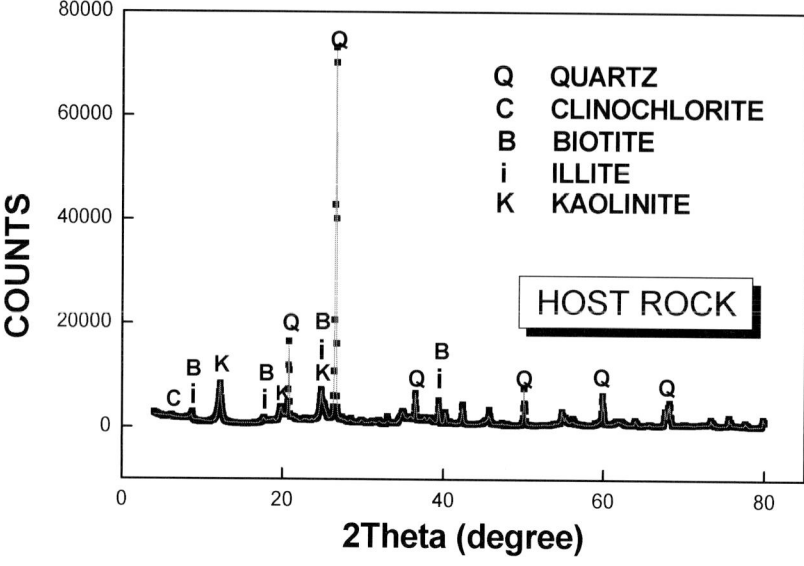

Figure 2. XRD pattern and its corresponding Rietveld refinement showing the different phases obtained from the host rock sample.

Table II. Lattice parameters obtained from the fit of the XRD pattern of the host rock sample

	a (Å)	b (Å)	c (Å)	α (deg)	β (deg)	γ (deg)	Symetry	S. group
Illite	5.16(3)	9.02(4)	9.99(4)	90.00(0)	90.00(0)	90.00(0)	Triclinic	P1
Kaolinite	5.14(1)	8.96(2)	7.39(1)	91.80(5)	104.50(3)	90.00(0)	Triclinic	P1
Biotite	5.22(2)	9.39(3)	10.19(2)	90.00(0)	99.74(3)	90.00(0)	Monoclinic	C2/m:b1
Clinochlore	6.23(2)	8.92(4)	14.22(2)	90.00(0)	93.07(3)	90.00(0)	Monoclinic	C2/m:b1
Quartz	4.91(2)	4.91(2)	5.41(3)	90.00(0)	90.00(0)	120.00(0)	Trigonal	P3221

Table III. Mössbauer parameters obtained from the fit of the host rock sample

	δ(Fe) (mm/s)	Δ(mm/s)	Area (%)	Γ(mm/s)
Illite	1.060(5)	2.500(2)	12.91	0.32(2)
Kaolinite	0.360(1)	0.613(2)	9.41	0.40(4)
Biotite	0.460(2)	1.400(5)	2.81	0.30(2)

high pressure and temperature conditions. The obtained Mössbauer parameters are reported in Table III. The spectrum of the host rock was also registered at 6 mm/s and their fitted parameters were identical to those of Table III.

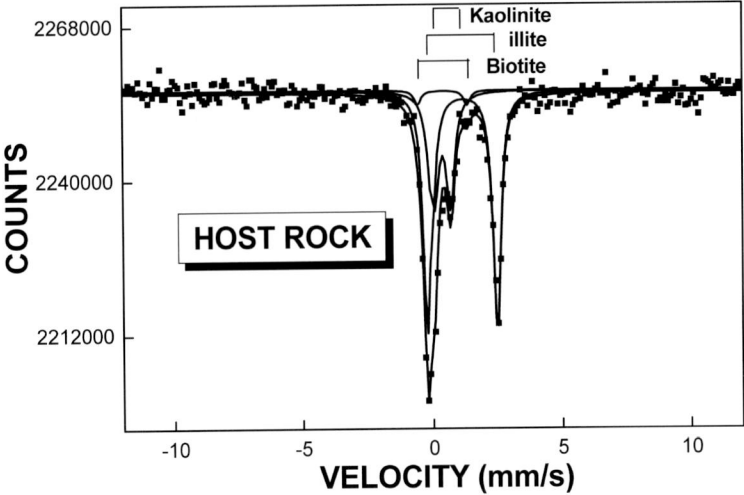

Figure 3. Mössbauer spectrum and the relevant fit of the host rock sample at 300 K.

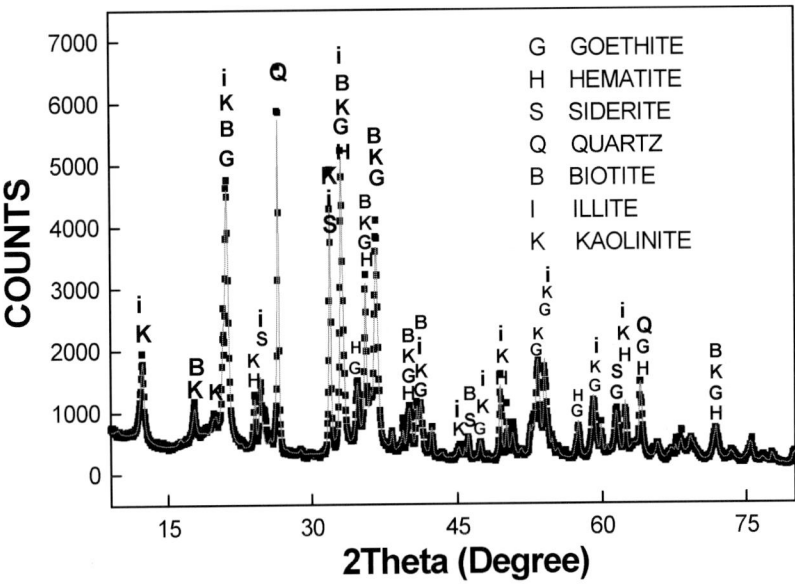

Figure 4. XRD pattern and its corresponding Rietveld refinement showing the different phases obtained from the iron rich mineral sample.

In Figure 4 the XRD pattern of the iron mineral sample is shown. The different identified phases are hematite, goethite, siderite, biotite, illite, kaolinite and quartz [7]. Their lattice parameters and space groups are listed in Table IV.

In Figure 5 the fitted Mössbauer spectrum of the iron mineral sample at 300 K is shown. The identified phases are hematite, goethite, siderite, biotite, illite and kaolinite [8]. Their Mössbauer parameters are collected in Table V. This result confirms the origin of the chemically detected Fe for this sample. It can be noted

Table IV. Lattice parameters obtained from the fit of the XRD pattern of the iron rich mineral sample

	a (Å)	b (Å)	c (Å)	α (deg)	β (deg)	γ (deg)	Symetry	S. group
Illite	5.16(2)	9.02(3)	9.99(3)	90.24(4)	90.43(3)	90.50(2)	Triclinic	P1
Kaolinite	5.21(3)	8.92(2)	7.38(3)	92.98(5)	104.88(3)	91.36(4)	Triclinic	P1
Biotite	5.35(4)	9.21(2)	10.08(5)	90.00(0)	100.16(3)	90.00(0)	Monoclinic	C2/m:b1
Quartz	4.91(2)	4.91(2)	5.41(3)	90.00(0)	90.00(0)	90.00(0)	Trigonal	P3221
Hematite	5.03(2)	5.03(2)	13.76(3)	90.00(0)	90.00(0)	90.00(0)	Trigonal	R-3c:H
Goethite	4.60(2)	9.94(3)	3.02(2)	90.00(0)	90.00(0)	90.00(0)	Orthorhombic	Pbnm:cab
Siderite	4.69(2)	4.69(2)	15.42(3)	90.00(0)	90.00(0)	90.00(0)	Trigonal	R-3c:H

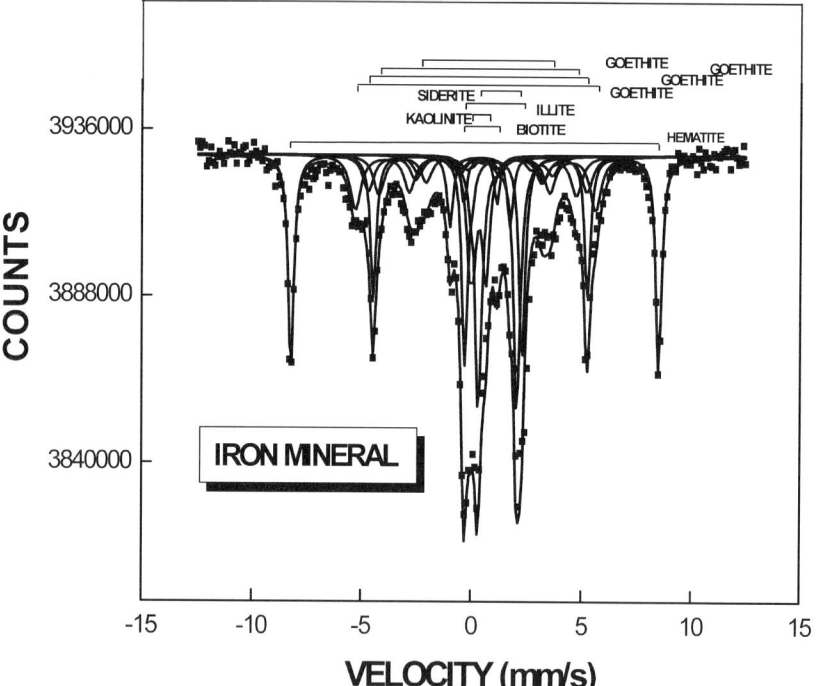

Figure 5. Mössbauer spectrum and the relevant fit of the iron rich mineral sample at 300 K.

that the goethite contribution appears as four broad sextets with hyperfine fields of 33.5, 30.5, 27.5 and 18.5 T, respectively. This shows the low crystallinity of the goethite and the possibility of particle sizes distribution of this mineral.

In Figure 6 the fitted Mössbauer spectrum of the iron mineral sample at 80 K is shown. The identified phases are hematite, goethite, siderite, biotite, illite and kaolinite [8]. In this case the fourth sextets of goethite appear with very high fields and the spectrum shows that superparamagnetic effects are not present. Their Mössbauer parameters are collected in Table VI.

Table V. Mössbauer parameters obtained at 300 K from the fit of the iron rich mineral sample

	δ(Fe) (mm/s)	Δ (mm/s)	Area (%)	Γ(mm/s)	H (kOe)
Illite	1.060(5)	2.500(2)	12.91	0.32(2)	–
Kaolinite	0.360(1)	0.613(2)	9.41	0.40(4)	–
Biotite	0.460(2)	1.400(5)	2.81	0.30(2)	–
Siderite	1.210(5)	1.750(8)	17.76	0.36(5)	–
Hematite	0.317(5)	−0.251(1)	26.73	0.32(1)	515.70(1)
Goethite	0.350(3)	−0.200(6)	10.45	0.60(3)	335.00(2)
Goethite	0.420(5)	−0.200(4)	8.30	0.60(3)	305.00(5)
Goethite	0.460(4)	−0.300(5)	9.10	0.60(3)	275.00(6)
Goethite	0.410(3)	−0.300(4)	3.02	0.60(3)	185.00(5)

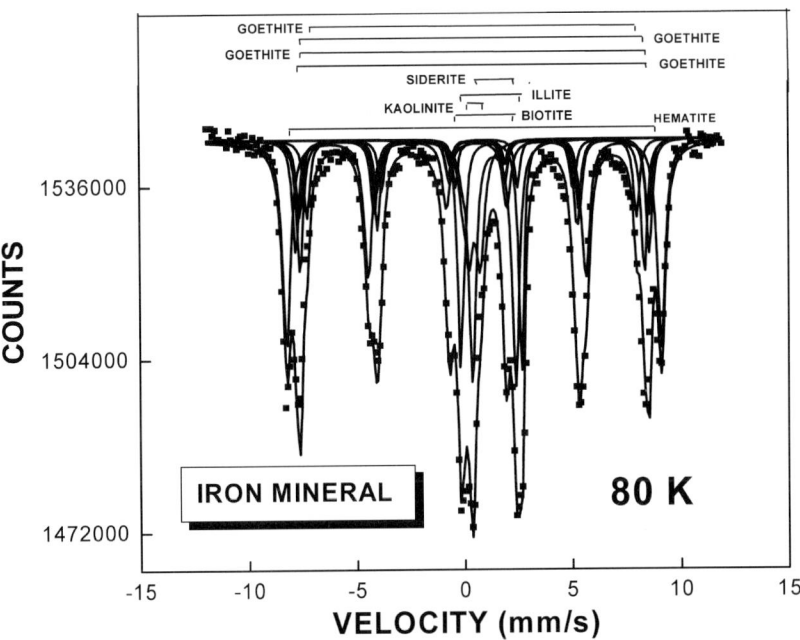

Figure 6. Mössbauer spectrum and the relevant fit of the iron rich mineral sample at 80 K.

Comparing with the phases obtained for the host rock we can note that the iron mineral sample has hematite (α-Fe$_2$O$_3$), goethite (α-FeOOH) and siderite (FeCO$_3$) as additional phases. These additional phases are rich in iron and contribute to the large increase of Fe content chemically detected and also, by reduction process, to the final steel produced in the high furnace. Improving of the Rietveld refinements is now in progress in order to complete the quantification of the different mineral

Table VI. Mössbauer parameters obtained at 80 K from the fit of the iron rich mineral sample

	δ(Fe) (mm/s)	Δ (mm/s)	Area (%)	Γ (mm/s)	H (kOe)
Illite	1.260(3)	2.760(2)	9.91	0.30(2)	
Kaolinite	0.450(2)	0.551(2)	8.12	0.40(4)	–
Biotite	1.070(2)	2.800(3)	2.07	0.30(2)	–
Siderite	1.361(4)	2.015(3)	13.99	0.40(5)	–
Hematite	0.531(5)	−0.200(2)	22.96	0.30(1)	535.42(2)
Goethite	0.501(3)	−0.208(4)	11.19	0.34(3)	504.89(3)
Goethite	0.502(2)	−0.252(3)	13.16	0.34(2)	491.89(2)
Goethite	0.501(3)	−0.282(2)	7.64	0.34(2)	470.89(3)
Goethite	0.571(2)	0.221(2)	7.64	0.34(2)	498.02(2)

phases in these two samples and principally in the iron mineral in the same way as was previously reported [9, 10].

Acknowledgements

The authors would like to thank the Universidad del Valle, Colciencias, Colombian Agency, and the Acerías Paz del Río steel factory for financial support and to the last one for the samples and the chemical analysis.

References

1. Klemic, H., James, H. L. and Eberlein, G. D., United States Mineral Resources. Iron. U.S. Geological Survey, Prof. Paper 820, 1973, p. 291.
2. Alvarado, B. and Sarmiento Soto, R., *Informe geológico general sobre yacimientos de hierro, carbón y caliza de la región de Paz de Río, Boyacá*, Ingeominas, Bogotá, Colombia, *Informe* **468** (1944) 68.
3. Mariño, J. A., *Ocurrencias minerales en el departamento de Boyacá*. Ingeominas, Bogotá, Colombia, *Informe* **1710** (1976).
4. Jeffery, P. G. and Hutchison, D., In: *Chemical Methods of Rock Analysis*, 3th edn, Pergamon Press Oxford, 1981, p. 379.
5. Lutterotti, L. and Scardi, P. J., *Appl. Crystallogr.* **23** (1990), 246.
6. Varret, F. and Teillet, J., Unpublished MOSFIT program.
7. Inorganic Crystal Structure Database, ICSD, Collection, Gmelin Institute, Fiz. Karlsruhe, Germany, 1997.
8. Stevens, J. G., Khasanov, A. M., Miller, J. W., Pollak, H. and Li, Z., *Mossbauer Mineral Handbook*, Mossbauer Effect Data Center, North Carolina, 1998.
9. König, U., Pöllmann, H. and Angelica, R. S., VII Simposio de Geologia da Amazonia, 2001, in press.
10. König, U., *Beith. Z. Eur. J. Mineral.* **11**(1) (1999), 130, Abstract.

Hyperfine Interactions **156/157**: 403–410, 2004.

© 2004 *Kluwer Academic Publishers. Printed in the Netherlands.*

Characterization and Thermal Behaviour of Garnets from Almandine–Pyrope Series at 1200°C

R. ZBORIL[1], M. MASHLAN[1], L. MACHALA[1], J. WALLA[1], K. BARCOVA[2]
and P. MARTINEC[3]
[1] *Departments of Physical Chemistry and Experimental Physics, Palacky University, Svobody 26, 771 46 Olomouc, Czech Republic*
[2] *Institute of Physics, VSB – Technical University of Ostrava, 17. listopadu 15, 708 33 Ostrava, Czech Republic*
[3] *Institute of Geonics, Academy of Sciences, Studentska 1768, 708 00 Ostrava, Czech Republic*

Abstract. The natural garnets from almandine ($Fe_3Al_2Si_3O_{12}$)–pyrope ($Mg_3Al_2Si_3O_{12}$) series with the iron to magnesium atomic ratio ranging from 0.2 to 1 were characterised and their thermal behaviour at 1200°C studied by ^{57}Fe Mössbauer spectroscopy, X-ray powder diffraction, X-ray fluorescence, DTA, TG and electron microprobe analysis. The pyrope-type samples with a dominant magnesium content at position 24*c* in the cubic garnet structure undergo oxidative decomposition at 1200°C resulting in the formation of the paramagnetic spinel $Mg(Al,Fe)_2O_4$ structure with a low iron content, enstatite $(Mg,Fe)SiO_3$ and anorthite $CaAl_2Si_2O_8$ as the host compound for calcium. Contrary to pyropes, the iron-rich garnets exhibit complete oxidation at 1200°C conforming to the formation of magnetically ordered nanocrystalline γ-Fe_2O_3 or $Mg(Fe,Al)_2O_4$ spinels depending on the initial chemical composition of the garnets. In the reaction products of iron-rich garnets, cordierite ($Mg_2Al_4Si_5O_{18}$) and anorthite were identified as non-ferrous phases.

Key words: maghemite, nanoparticles, thermal decomposition, superparamagnetic relaxation.

1. Introduction

Fe-bearing garnets in the pyrope–almandine solid solution series are often used in modern technological processes including plasma spraying and high-energy water jet cutting for abrasive materials. The knowledge of behaviour of garnet particles under high temperature and pressure conditions is very important for their effective applications [1]. Only a few authors have investigated the oxidative decompositions of garnets from almandine–pyrope series. Anovitz *et al.* [2] suggested two oxidation mechanisms for a synthetic almandine ($Fe_3Al_2Si_3O_{12}$) with the formation of magnetite, quartz and sillimanite or hercynite depending on the experimental conditions. However, Thiéblot *et al.* [3] identified hematite, sillimanite and cristoballite as decomposition products of almandine with a chemical composition close to the endmember. Zboril *et al.* [4] and Barcova *et al.* [5] specified the oxidation route of Fe^{2+} and proved the two-step formation mechanism of hematite via maghemite and ε-Fe_2O_3. Concerning the thermal behaviour of the pyrope garnets, the decomposition mechanism was studied only with pyrope close to the end-member ($Mg_3Al_2Si_3O_{12}$) at 1200°C, where corundum and enstatite were found as the reaction products [3].

The present study is aimed at the characterization of garnets from almandine–pyrope series showing the chemical compositions far from the endmembers and further, at the explanation of their oxidation-decomposition mechanisms at 1200°C.

2. Materials and methods

Five natural garnet samples with chemical composition varying between almandine ($Fe_3Al_2Si_3O_{12}$) and pyrope ($Mg_3Al_2Si_3O_{12}$) were investigated. The garnets originate from different localities including deposits in Same (A1), Songay (A2) and Tunduru (A3, P1) in Tanzania, and from the deposit in Podsedice (P2) in the Czech Republic. Red coloured, gem quality, garnet crystals were washed with dilute HCl and oxalate acid to remove secondary minerals such as carbonates and Fe-oxyhydroxides. The crystals were rinsed in distilled water and then dried at 100°C. In such treated garnet crystals, no chemical zoning or non-homogeneities were observed. The crystals were then crushed in an agate mill to a powder with particle size ranging from 100 to 300 μm. For thermal study, powdered garnets were isothermally heated at 1200°C for 3 hours.

The chemical composition and homogeneity of the samples were analysed using a 535 M type Phillips Scanning Electron Microscope, equipped with EDAX 9900 spectrometer. X-ray fluorescence analysis (XRF) was conducted using an Energy Disperse Spectrometer Spectro X-LAB. The transmission Mössbauer spectra were collected using the Mössbauer spectrometer in constant acceleration mode with a $^{57}Co(Rh)$ source. The measurements were carried out in a temperature range of 20–400 K using a cryostat with closed He-cycle (Janis Research Company) and a high-temperature furnace. The phase composition of samples was monitored by XRD using Seifert-FPM equipment with CuKα radiation and conventional θ–2θ geometry. Si was used as an external calibration standard. The individual phases were identified from XRD patterns by means of PDF2 database. Dynamic thermal analysis with simultaneous measurement of differential thermal analysis (DTA) and thermogravimetric (TG) curves was performed in air at a range of 25–1200°C with a temperature increase of 5°C/min using a SETARAM SctSys TG-DTA 16 instrument.

3. Characterization of garnets

The chemical compositions, structural and thermal characteristics of the starting garnets are covered in Table I. The chemical formulas of the starting garnets, as determined from the electron microprobe analysis, XRF and Mössbauer spectroscopy results, reveal three garnets (A1–A3) with high content of Fe at $24c$ position, the other two samples marked P1 and P2 correspond rather to pyrope

Table I. The chemical, structural and thermal characteristics of garnets

Chemical formulae	Fe/Mg	$\frac{Fe^{2+}}{\sum Fe}$	T_{min}	Δm
	[at.%]	[%]	[°C]	[wt%]
A1 $(Fe_{0.91}Mg_{0.95}Mn_{0.15}Ca_{0.34})(Al_{2.46}Fe_{0.04})Si_{2.95}O_{12}$	1	95.8	825	1.94
A2 $(Fe_{0.92}Mg_{0.98}Mn_{0.12}Ca_{0.20})(Al_{2.52}Fe_{0.03})Si_{2.98}O_{12}$	0.97	96.8	829	1.96
A3 $(Fe_{0.87}Mg_{1.04}Mn_{0.11}Ca_{0.33})(Al_{2.41}Fe_{0.05})Si_{2.98}O_{12}$	0.88	94.5	823	1.79
P1 $(Fe_{0.32}Mg_{1.68}Mn_{0.11}Ca_{0.32})(Al_{2.36}Fe_{0.03}Cr_{0.02}Ti_{0.02})Si_{2.92}O_{12}$	0.21	91.4	1109	0.30
P2 $(Fe_{0.33}Mg_{1.38}Mn_{0.10}Ca_{0.36})(Al_{2.30}Fe_{0.05}Cr_{0.09}Ti_{0.03})Si_{3.06}O_{12}$	0.28	86.8	1063	0.32

T_{min} – minimum oxidation temperature, Δm – total mass increase.

(major Mg content at $24c$ position). In all garnets, calcium and manganese appear as the main trace elements substituting at $24c$ position. In the pyrope samples, minor concentrations of titanium and chromium were found. The differences in the chemical compositions of garnets are reflected also in DTA/TG curves, where samples A1–A3 show significantly lower minimal oxidation temperature and higher total mass increase at 1200°C in contrast to pyrope garnets (see Table I). The XRD patterns show lines exclusively corresponding to the cubic garnet structure without any indications of the presence of other phases. Room temperature (RT) Mössbauer spectra of the starting garnets could be fitted using two doublets, an asymmetric outer doublet with isomer shift $\delta = 1.27$–1.28 mm/s and quadrupole splitting $\Delta E_Q = 3.53$–3.54 mm/s and a symmetric inner doublet with $\delta = 0.28$–0.33 mm/s and $\Delta E_Q = 0.30$–0.39 mm/s (Figure 1). The former doublet is assigned to Fe^{2+} at the $24c$ site and the later doublet to Fe^{3+} at the $16a$ site of the garnet structure [6–8]. The asymmetry of the Fe^{2+} doublet is commonly observed in the RT Mössbauer spectra of Fe-bearing garnets and can be explained as arising out of a paramagnetic relaxation of Fe^{2+} in the dodecahedral position [7–9] or by the Goldanskii–Karyagin effect [10].

In accordance with the Mössbauer data, pyropes exhibit higher oxidation stage due to a higher content of trivalent iron at the octahedral $16a$ position of the garnet structure.

4. Thermal behaviour of garnets at 1200°C

The influence of the chemical composition of garnets on the mechanism of their oxidative decomposition was studied at 1200°C. XRD patterns of the decomposed pyrope samples (P1/1200, P2/1200) are very similar and demonstrate the presence of the spinel $Mg(Al,Fe)_2O_4$ structure, enstatite $(Mg,Fe)SiO_3$ and anorthite $CaAl_2Si_2O_8$ as conversion products (Figure 2 *left*). The XRD identification of Fe-bearing phases is in full correspondence with the analyses of RT Mössbauer spectra showing one doublet of trivalent iron and two Fe^{2+} doublets (Figure 2 *right*). The

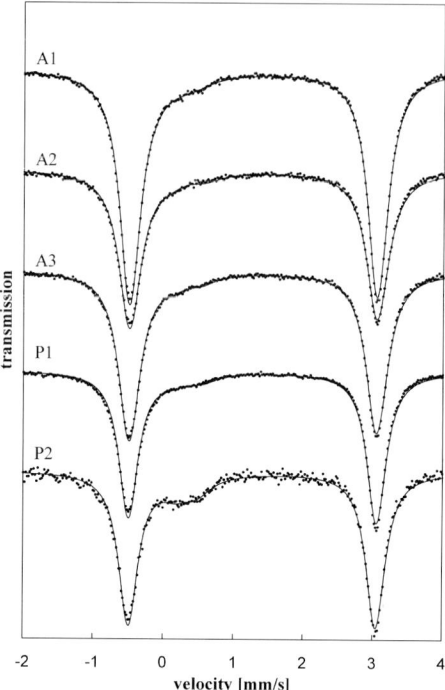

Figure 1. RT Mössbauer spectra of the untreated garnets.

hyperfine parameters of the Fe^{3+} doublet ($\delta = 0.31$ mm/s, $\Delta E_Q = 0.76$ mm/s) are characteristic of the paramagnetic aluminum-rich $MgAl_{2-x}Fe_xO_4$ spinels with the low iron content [11, 12].

On the basis of XRD data, the two Fe^{2+} doublets can easily be attributed to Fe^{2+} at the non-equivalent M1 and M2 sites of the enstatite structure. The hyperfine parameters that were obtained (M1: $\delta = 1.14$–1.16 mm/s, $\Delta E_Q = 2.34$–2.35 mm/s; M2: $\delta = 0.99$–1.0 mm/s, $\Delta E_Q = 2.02$–2.03 mm/s) agree well with those found by Srivastava [13] for orthoenstatite with low iron content. The relative spectrum areas corresponding to the Fe^{3+} ions in the spinel oxide and Fe^{2+} ions in the enstatite structure were found to be almost the same in the P1/1200 and P2/1200 samples: $Fe^{3+}/(Fe^{2+}_{M1} + Fe^{2+}_{M2}) = 92.3/(4.8 + 2.9) \cong 12$. Therefore, the content of ferrous ions stabilising the enstatite structure is the same in both samples and seems to be independent of the general content of iron and of the oxidation stage of the original pyrope structure (compare with Table I).

XRD patterns of the thermally treated iron rich A1–A3 samples (Figure 3 *top*) differ slightly from each other, however significantly from those corresponding to the decomposed P1 and P2 samples. As the host compounds for magnesium and calcium, cordierite $Mg_2Al_4Si_5O_{18}$ and anorthite $CaAl_2Si_2O_8$ were detected, respectively. Concerning the iron-bearing phase, the line positions indicate again the spinel oxide structure, however their shift to the lower 2Θ angles reflects the

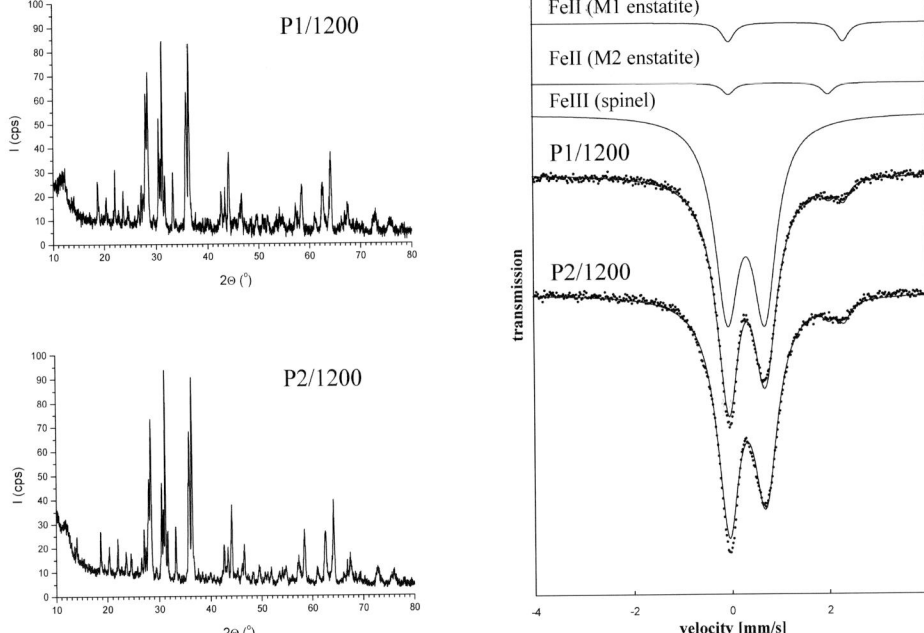

Figure 2. XRD patterns and RT Mössbauer spectra of the P1 and P2 samples heated at 1200°C.

higher content of the cubic unit cell in comparison with the spinels identified in heated P1 and P2 garnets (Figure 3 *bottom*). This comparison is interesting also among A1/1200–A3/1200 samples. The line positions of the spinels in the A1/1200 and A3/1200 samples correspond clearly to the maghemite or magnetite structure, while the evident shift of the spinel lines to the higher angles is seen in the A2/1200 sample (Figure 3 *bottom*). Moreover, the relative intensities of the spinel and cordierite lines are increased in advantage of the spinel in the A2/1200 sample in contrast to the A1/1200 and A3/1200 samples (Figure 3 *top*).

[t]

This fact demonstrates the higher weight content of the spinel phase in the A2/1200 sample evidently as a result of the incorporation of the part of magnesium and aluminium atoms to the spinel structure to the detriment of cordierite. Such conclusion is in clear relation with the initial chemical compositions of garnets (Table I) showing the near iron concentrations, however the A2 sample exhibits the highest content of aluminium and also the lowest content of calcium resulting in the reduced loss of aluminium bound in the anorthite structure.

RT Mössbauer spectra of the A1/1200 and A3/1200 samples (Figure 4 *left*) are quite similar and can be fitted using hyperfine magnetic field distributions with the low average fields (A1: 41.5; A3: 39.8 T) as typical for the ultrafine particles with broad particle size distribution. The isomer shift parameters (0.35–0.37 mm/s) exclude the presence of divalent iron (magnetite) and they are characteristic for iron(III) oxides. Moreover, at 25 K, the spectrum exhibits the slightly asymmetric

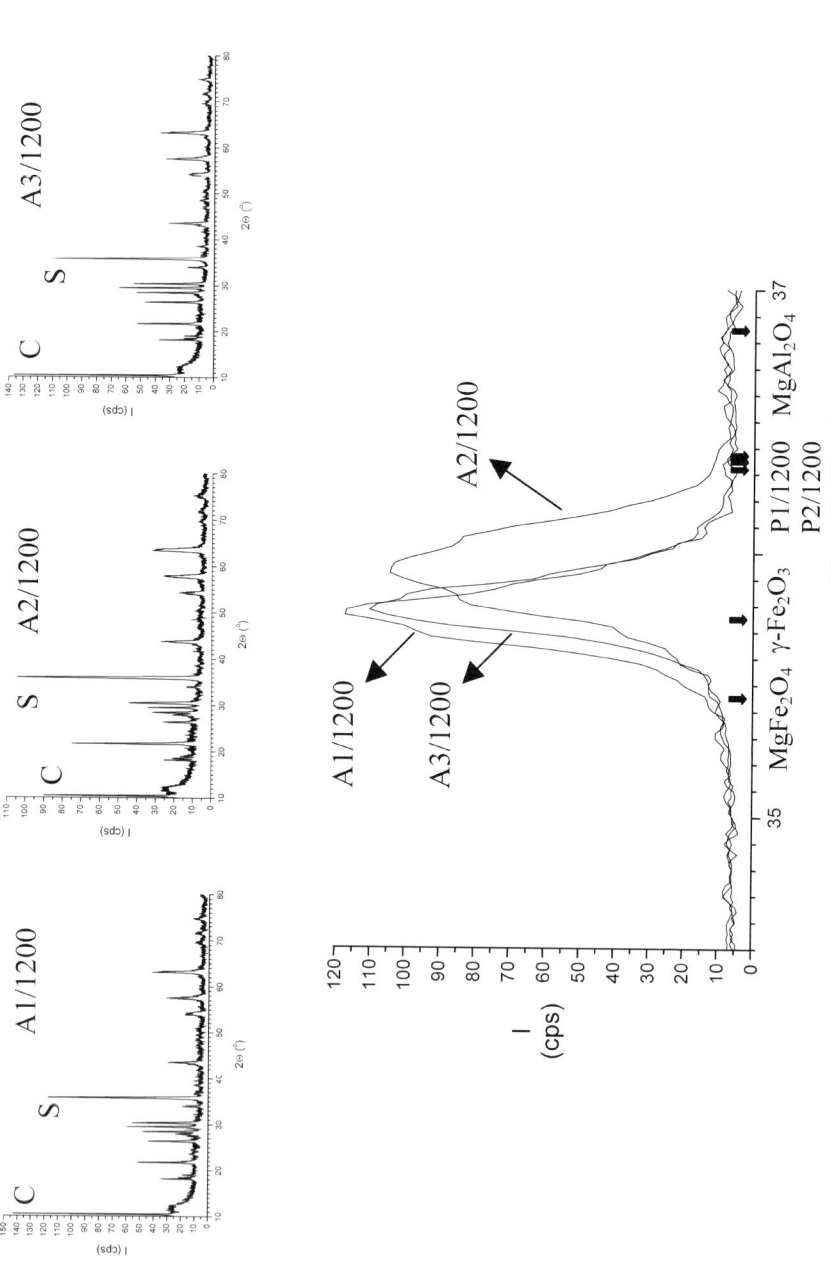

Figure 3. Top: XRD patterns of iron-rich garnets heated at 1200°C (*top*), C-cordierite, S-spinel. *Bottom*: the zoom of the most intensive lines and indications of the (100) line positions of other spinels with varying iron and aluminium contents.

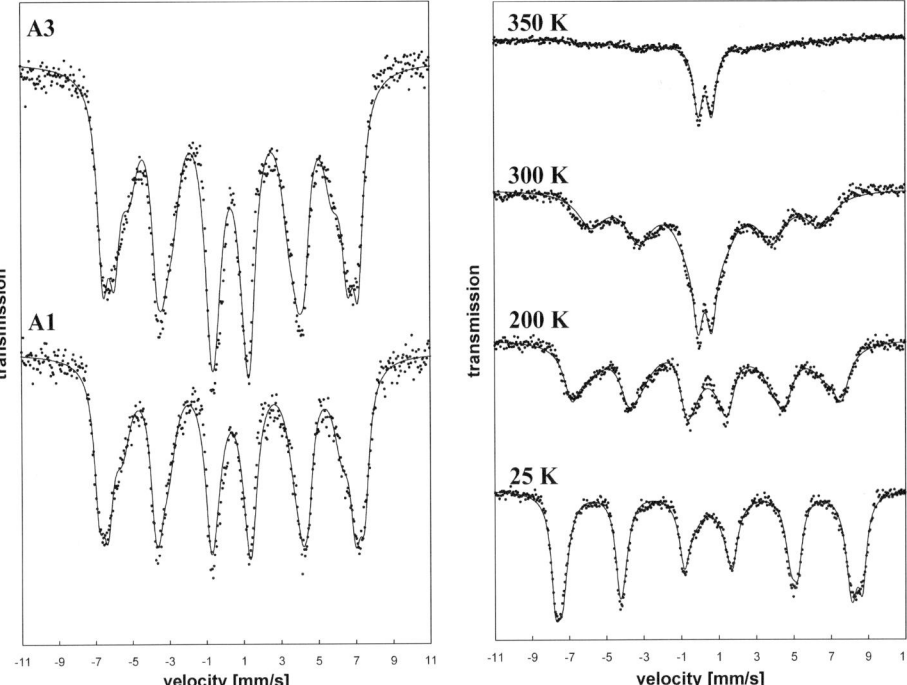

Figure 4. RT Mössbauer spectra of the A1/1200 and A3/1200 samples (*left*) and temperature dependent spectra of the A2/1200 sample illustrating the superparamagnetic relaxation (*right*).

sextet with unresolved tetrahedral and octahedral iron sites and with hyperfine parameters close to maghemite [14]. RT Mössbauer spectrum of the A2/1200 sample reveals the superposition of the doublet and sextet components illustrating that a part of the spinel particles occurs in superparamagnetic state at room temperature. The temperature dependent spectra of the A2/1200 sample confirm the superparamagnetic relaxation phenomenon with the blocking temperature lying between 300 and 350 K (Figure 4 *right*). As clearly shown, the superparamagnetic relaxation is suppressed at 25 K, and the Mössbauer spectrum of nanocrystalline $MgAl_{2-x}Fe_xO_4$ consists of broadened sextets with rather well resolved tetrahedral A ($\delta_A = 0.37$ mm/s, $\varepsilon_A = -0.01$ mm/s, $H_A = 48.2$ T, relative area: 62%) and octahedral B ($\delta_B = 0.46$ mm/s, $\varepsilon_B = -0.00$ mm/s, $H_B = 51.1$ T, relative area: 38%) subspectra. Such enhanced fraction of the Fe^{3+} cations in tetrahedral sites was recorded also for nanoparticles of magnesium and nickel ferrites prepared by high-energy milling [15, 16].

5. Conclusions

The spinel oxides with varying iron content, magnetic properties and crystallinity were identified as the products of thermally induced oxidation of garnets from almandine–pyrope series. The oxidation of pyrope-type samples is not complete at

$1200°C$ because a part of divalent iron remains stabilized in the M1 and M2 positions of the enstatite $(Mg,Fe)SiO_3$ structure. The ratio of ferrous ions incorporated into the enstatite structure to the overall number of iron ions in the system seems to be independent of the initial Mg/Fe or Fe^{2+}/Fe^{3+} ratios in the pyropes. The decomposition of iron-rich garnets at $1200°C$ leads to the complete oxidation of Fe^{2+} and to the formation of γ-Fe_2O_3 or $MgAl_{2-x}Fe_xO_4$ spinels showing the low crystallinity and the specific magnetic behaviour including the superparamagnetic relaxation.

Acknowledgements

Financial support from the Grant Agency of the Czech Republic (projects 202/03/P158, 202/03/P099) and from the project Kontakt ME 600 (MŠMT) is gratefully acknowledged.

References

1. Hlavac, L. and Martinec, P., In: M. Hashish (ed.), *Proceedings of 10th American Water Jet Conference*, Water Jet Technology Association, St. Louis Houston, 1999, p. 409.
2. Anovitz, L. M., Essene, E. J., Metz, G. W., Bohlen, S. R., Westrum, E. F. and Hemingway, B. S., *Geochim. Cosmochim. Acta* **57** (1993), 4191.
3. Thiéblot, L., Roux, J. and Richet, P., *Eur. J. Mineral.* **10** (1998), 7.
4. Zboril, R., Mashlan, M., Barcova, K. and Vujtek, M., *Hyp. Interact.* **139** (2002), 597.
5. Barcova, K., Mashlan, M., Zboril, R., Martinec, P. and Kula, P., *Czech. J. Phys.* **51** (2001), 749.
6. Zboril, R., Mashlan, M., Barcova, K., Walla, J., Ferrow, E. and Martinec, P., *Phys. Chem. Miner.* **30** (2003), 620.
7. Murad, E. and Wagner, F. E., *Phys. Chem. Miner.* **14** (1987), 264.
8. Mitra, S., *Applied Mössbauer Spectroscopy*, Pergamon Press, Oxford, 1992, p. 137.
9. Amthauer, G., Annersten, H. and Hafner, S. S., *Z. Kristallogr.* **143** (1976), 14.
10. Geiger, C. A., Armbruster, T., Lager, G. A., Jiang, K., Lottermoser, W. and Amthauer, G., *Phys. Chem. Miner.* **19** (1992), 121.
11. Marshall, I. and Dollase, W., *Am. Mineral.* **69** (1984), 928.
12. Wood, B. J. and Virgo, D., *Geochim. Cosmochim. Acta* **53** (1989), 1277.
13. Srivastava, K. K. P., *J. Phys. Solid State Phys. C* **20** (1987), 2161.
14. Zboril, R., Mashlan, M. and Petridis, D., *Chem. Mater.* **14** (2002), 969.
15. Šepelák, V., Schultze, D., Krumeich, F., Steinike, U. and Becker, K. D., *Solid State Ionics* **141–142** (2001), 677.
16. Šepelák, V., Baabe, D., Mienert, D., Schultze, D., Krumeich, F., Litterst, F. J. and Becker, K. D., *J. Magn. Magn. Mater.* **257** (2003), 377.

Hyperfine Interactions **156/157**: 411–415, 2004.
© 2004 *Kluwer Academic Publishers. Printed in the Netherlands.*

Mössbauer Study of Magnetite Formation by Iron- and Sulfate-Reducing Bacteria

N. I. CHISTYAKOVA[1], V. S. RUSAKOV[1], D. G. ZAVARZINA[1],
A. I. SLOBODKIN[2] and T. V. GOROHOVA[1]
[1]*M.V. Lomonosov Moscow State University, Leninskie gory, 119992 Moscow, Russia*
[2]*Institute of Microbiology, Russian Academy of Sciences, Prospect 60-letiya Oktyabrya 7/2,
117312 Moscow, Russia*

Abstract. Mössbauer investigations of the influence of physicochemical factors and the inert organic substances on the magnetite formation by thermophilic dissimilatory anaerobic Fe(III)-reducing bacteria (strain Z-0001) were carried out. The production of magnetite due to microbial sulfate reduction by hyperthermophilic dissimilatory sulfate-reducing microorganisms was investigated by Mössbauer spectroscopy methods.

Key words: Mössbauer spectroscopy, Fe(III)-reducing bacteria, magnetite Fe_3O_4, siderite $FeCO_3$, ferrihydrite $5Fe_2O_3 \cdot 9H_2O$, sulfate reduction.

1. Introduction

The interest to intensive investigations of anaerobic dissimilatory iron-reducing bacteria that can reduce the amorphous Fe(III)-oxides and hydroxides is caused by the suggestion on the possible participation of this group of microorganisms in constituting of Banded Iron Formations. To analyze this hypothesis, it is necessary to study mineral products of microbial activity, having separated biogenic stages from abiogenic ones. The aim of our work was to investigate by Mössbauer spectroscopy methods physicochemical factors affecting the process of mineral phase formation by iron-reducing bacterium as well as the inert organic substances (agar, microcrystalline cellulose and meat extract).

In contrast to direct enzymatic reduction of Fe(III), microbially mediated non-enzymatic processes that could lead to magnetite formation had received significantly less attention. Sulfide, the metabolic product of dissimilatory sulfate-reducing microorganisms, can react with Fe(III) compounds forming different iron sulfides, including magnetic greigite (Fe_3S_4) and pyrrhotite ($Fe_{1-x}S$). The production of magnetite due to microbial sulfate reduction has not been previously reported.

2. Experimental

Strain Z-0001 used amorphous Fe(III)-hydroxide as an electron acceptor and acetate (CH_3COO^-) as an electron donor for the anaerobic growth. An anaerobically prepared mineral medium dispensed into 60-ml flasks with a $N_2 + CO_2$ gas mixture was used for the experiment. Inoculated flasks were incubated at 60°C. The inert organic substances such as agar, microcrystalline cellulose and meat extract were added to a mineral medium. The content of amorphous Fe(III)-hydroxide was 90 mM, relative content of CO_2 was 20% and the ratio between liquid and solid phases was 20 ml/40 ml. The relative contents of organic substances were varied.

Archaeoglobus fulgidus DSM 4304[T] was able to grow in the medium supplemented with sulfate and amorphous Fe(III) oxide, with molecular hydrogen as the only electron donor. After 2–6 days of cultivation, amorphous Fe(III)-hydroxide was converted to black magnetic precipitate with high Fe(II) content. Precipitates was collected and dried under N_2 atmosphere.

The Mössbauer study was carried out at room temperature. To avoid oxidation investigated samples were prepared as paraffin tablets. The Mössbauer spectra were processed by use of the program DISTRI [1].

3. Results and discussion

Our investigations show that reduction of amorphous Fe(III)-hydroxide is the result of metabolic activity of anaerobic bacterium and do not occur without living cells. Magnetite (Fe_3O_4) is the only solid magnetically ordered phase that is formed during the growth of strain Z-0001 [2]. The increase of CO_2 pressure results in the predominant formation of siderite ($FeCO_3$). An excess of amorphous Fe(III)-hydroxide in the initial solution leads to a magnetite formation while its deficit to siderite formation [2].

Ferrihydrite ($5Fe_2O_3 \cdot 9H_2O$) is the most environmentally relevant electron acceptor for iron-reducing bacterium. Mössbauer investigations of the processes of natural ferrihydrite reduction by Fe(III)-reducing bacteria were carried out. When ferrihydrite was added into the mineral medium instead of synthetic amorphous Fe(III)-hydroxide, the formation of siderite (27%) was observed, and the formation of magnetite was not detected. Probably, the absence of magnetite as a product of Fe(III) reduction is caused by the presence of organic substances.

To determine the natural organic substance influence on mineral phase formation by iron-reducing bacterium, the organic substances that were not used as substrates by this microorganism were added to initial mineral medium. Mössbauer spectrum of precipitate obtained without organic substances is shown in Figure 1a. When microcrystalline cellulose (0.5–4.0 g/l) was added to the medium, the main mineral formed as a result of the Fe(III) reduction was magnetite (Figure 1b). The relative content of formed mineral phases did not depend on the amount of adding cellulose. The formation of magnetite was not also observed when agar (0.25–1.0%) was present in the cultivation medium (Figure 1c). The agar concentration

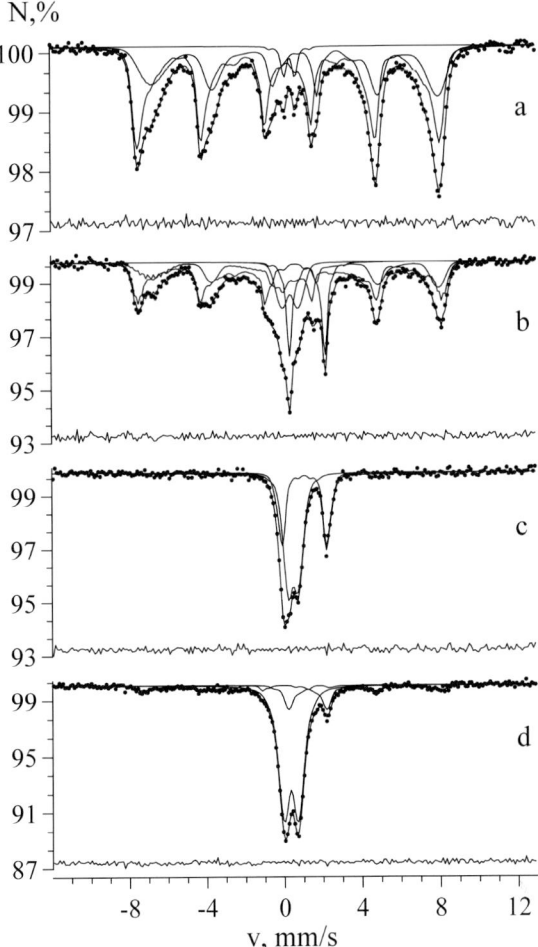

Figure 1. Mössbauer spectra at RT of 14-day incubation samples formed during bacterium (strain Z-0001) growing without organic substance (a) and with microcrystalline cellulose (b), agar (c) and meat extract (d) in the amorphous hydroxide medium.

did not significantly influenced the amount of siderite produced from amorphous Fe(III) hydroxide. When meat extract (0.5–3 g/l) was added, only partial reduction of amorphous Fe(III) hydroxide to siderite and magnetite was observed (Figure 1d). In this case, however, the increase of meat extract concentrations led to the decrease of magnetite content. Thereby, the presence of organic substances in a mineral medium essentially affects the process of mineral formation and the quantitative relations between obtained phases depend on these substances.

Mössbauer spectroscopy analysis of the magnetic precipitate formed during the growth of sulfate-reducing bacterium *Archaeoglobus fulgidus* shows that it contains magnetite. The subspectrum for magnetically ordered phase is the same as for magnetite formed during the growth of strain Z-0001 (Figure 2). For magnetically

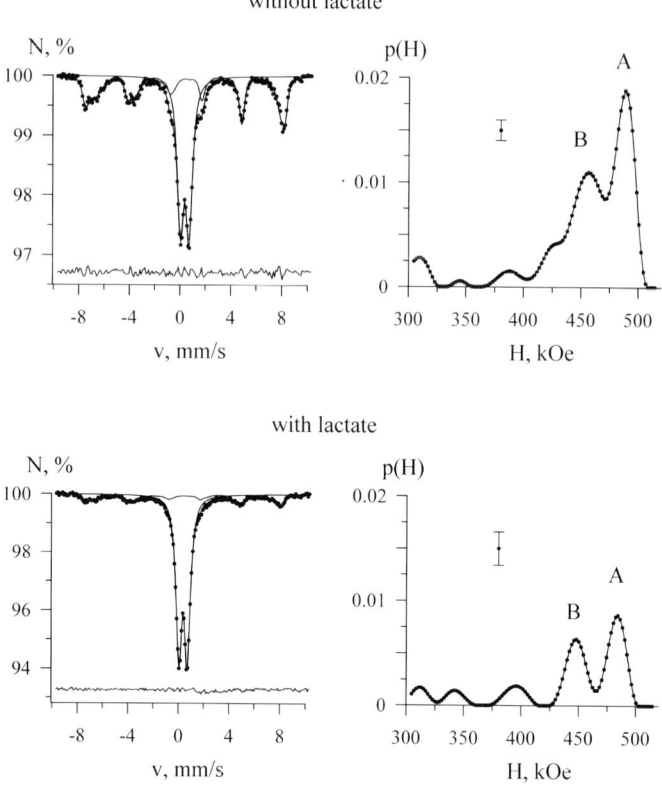

Figure 2. Mössbauer spectra and corresponding hyperfine field distribution functions $p(H)$ for the precipitates formed in the cultivation medium with 14 mM initial sulfate concentration in the absence and the presence of lactate.

ordered phase of these spectra the hyperfine magnetic field distribution function is calculated using the linear correlations between hyperfine field and Mössbauer line shift. The value of this correlation $\Delta H / \Delta \delta$ is negative that is typical for magnetite. The typical values of hyperfine magnetic field for A- and B-position of magnetite are shown in Figure 2.

The amount of magnetite in the solid phase increased with initial sulfate concentration increase in the cultivation medium (Table I). When lactate was added to the cultivation medium as an additional electron donor, Mössbauer investigations revealed that relative content of magnetite in the formed precipitate was lower than in precipitates obtained in the cultivation medium without lactate under the same concentrations of sulfate (Figure 2). The formation of magnetically ordered iron sulfides in the investigated system was not observed. *A. fulgidus* grows via sulfate reduction and not by reduction of Fe(III), because no growth was observed in the absence of sulfate.

To test the extent of Fe(III) reduction by sulfide produced in course of microbial dissimilation of sulfate, $Na_2S \cdot 9H_2O$ solution was injected to sterile non-inoculated

Table I. The relative contents of precipitates formed during the growth of sulfate-reducing bacterium for different synthesis conditions

Sulfate content, mM	Substratum	Fe(III)-hydroxide relative content in a precipitate, %	Fe_3O_4 relative content in a precipitate, %
0	H_2	100 ± 1.7	0 ± 1.7
0.5	H_2	80.6 ± 1.8	19.4 ± 1.8
2.0	H_2	70.9 ± 1.7	29.1 ± 1.7
14.0	H_2	54.3 ± 1.1	45.7 ± 1.1
2.0	H_2 + lactate	92.1 ± 1.3	7.9 ± 1.3
14.0	H_2 + lactate	77.9 ± 1.7	22.2 ± 1.7

medium to provide different sulfide concentrations. The formation of black magnetic precipitate was observed after 48–72 hours only with H_2 at 83°C, in variants without H_2 (N_2 in the gas phase) or at 20°C no magnetic precipitate was formed. The reduction of the significant amount of Fe(III) occurs even under sufficiently low concentrations of sulfide. Most probably sulfide or its oxidation products act as a catalyst for the reduction of Fe(III) by H_2. The results obtained in this study show that magnetite can be formed from amorphous Fe(III)-hydroxide in the presence of molecular hydrogen and enzymatically produced sulfide at increased temperatures.

4. Conclusions

(1) Complex organic substances influence the composition and mineral phase relative content during growth of dissimilatory Fe(III)-reducing bacterium strain Z-0001 with amorphous Fe(III) hydroxide as an electron acceptor.
(2) Use of ferrihydrite as an electron acceptor for Fe(III) reduction by strain Z-0001 led to formation of siderite instead of magnetite as a main product of ferric iron reduction.
(3) Magnetite can be formed from amorphous Fe(III) hydroxide due to activity of sulfate-reducing microorganisms.

References

1. Rusakov, V. S., *Izv. RAN Ser. Phys.* **63** (1999), 1380 (in Russian).
2. Chistyakova, N. I., Rusakov, V. S., Zavarzina, D. G. and Kozerenko, S. V., *The Physics of Metals and Metallography* **92** (2001), S138.
3. Chistyakova, N. I., Rusakov, V. S. and Zavarzina, D. G., *Hyp. Interact. (C)* **5** (2001), 397.

Hyperfine Interactions **156/157**: 417–422, 2004.
© 2004 *Kluwer Academic Publishers. Printed in the Netherlands.*

What Oxidation State of Iron Determines the Amethyst Colour?

S. K. DEDUSHENKO[1,*], I. B. MAKHINA[2], A. A. MAR'IN[2],
V. A. MUKHANOV[2] and YU. D. PERFILIEV[1]
[1]*Department of Chemistry, Moscow State University, Lenin Hills, Moscow 119992, Russia*
[2]*Russian Research Institute for Synthesis of Materials, Aleksandrov, Vladimir Region 601650, Russia*

Abstract. A colourless quartz crystal doped with $^{57}Fe^{3+}$ was obtained by hydrothermal synthesis in an NH_4F solution. The crystal was transformed into violet amethyst by gamma-irradiation. The change in colour was accompanied by changes in the Mössbauer spectrum that can be interpreted as the conversion of trivalent iron into the tetravalent state: $Fe^{3+} \rightarrow Fe^{4+}$.

Key words: quartz, amethyst, tetravalent iron, Mössbauer effect, hydrothermal synthesis.

1. Introduction

Artificial coloured quartz is of great importance for the present-day jewellery. In particular, iron-doped quartz allows one to obtain violet, yellow, green and brown gemstones. The location of the iron admixture in the crystal structure, the electronic structure of the metal ion as well as the general content of iron and the ratio of its different forms are the main factors determining the colour and thus influencing the value of the crystal.

Amethyst is the most popular variety of coloured quartz. This stone with the "magic" red-violet colour has been valued since antiquity but the nature of this coloration has not yet been established reliably. Amethyst can be synthesised artificially by growing quartz in the presence of Fe^{3+} ions followed by gamma-irradiation of the crystal obtained [1]. The appearance and disappearance of colour have been thoroughly investigated by numerous techniques including ESR and optical absorption [2–5]. However, the question "what oxidation state of iron determines the amethyst colour?" has not received a distinct answer.

Mössbauer spectroscopy is a very useful technique for the determination of the oxidation state of iron.

However, amethyst was not studied previously by this method because of the low content of iron in this mineral, which generally ranges from tenths to thousandths of mass percent.

---

* Author for correspondence.

This work presents results of a Mössbauer study on the transformation of quartz doped with trivalent iron into amethyst.

2. Experimental

Iron-doped quartz was grown from a hydrothermal ammonium fluoride solution. The synthesis was carried out in a 250 ml laboratory autoclave with contact fluoro-plastic lining. An inoculating plate (50 × 25 × 2 mm size) was cut from colourless synthetic quartz in parallel to the pinacoid plane. The autoclave was filled with 15% ammonium fluoride solution and a mixture of natural quartz (150 g) with iron(III) oxide powder (1 g) enriched with ^{57}Fe (95%). The temperature was monitored by thermocouples installed outside the autoclave. The synthesis was carried out at 300°C in the dissolution zone and at 260°C in the crystallisation zone. The calculated pressure in the autoclave was 10 MPa. Synthesis for 30 days gave a high-quality outgrowth of iron-containing quartz measuring 55 × 26 × 9 mm.

The iron content in the resulting quartz sample was analysed (after preliminary dissolution in a mixture of hydrofluoric and sulphuric acids) by emission spectrometry with inductively coupled plasma, using an IRIS Advantage instrument from Thermo Jarrell Ash. Analysis showed that the iron content in the quartz sample obtained was 1.6×10^{-2} mass%.

Gamma-irradiation was performed using ^{60}Co sources, which provided the absorption dose rate about 1 Gy·s^{-1}. The temperature in the irradiation zone was about 70°C.

Mössbauer absorption spectra were measured on a Perseus spectrometer working at constant velocities; the control and adjustment of the spectrometer-vibrator rate were performed by a laser interferometer. A standard γ-source of ^{57}Co in metallic rhodium matrix with the activity of 0.6 GBq (a product of Cyclotron, Co., Ltd., Obninsk, Russia) was employed. The spectrometer was equipped with a scintillation counter (0.1 mm plate of NaI doped by Tl).

In order to perform Mössbauer measurements, plates 1 mm thick were cut from the quartz outgrowth orthogonally to the growth direction.

In this work, chemical shifts are reported relative to α-Fe and correspond to room temperature.

3. Results and discussion

The Mössbauer spectrum of the quartz sample synthesised is shown in Figure 1. The spectrum is a superposition of numerous lines scattered over a broad range of velocities. The majority of these lines undoubtedly originate from magnetic interactions occurring in the substance. On the one hand, these interactions can be due to the presence of inclusions of magnetically ordered phases in the substance, e.g., admixtures of iron oxides and/or hydroxides. On the other hand, the appearance of

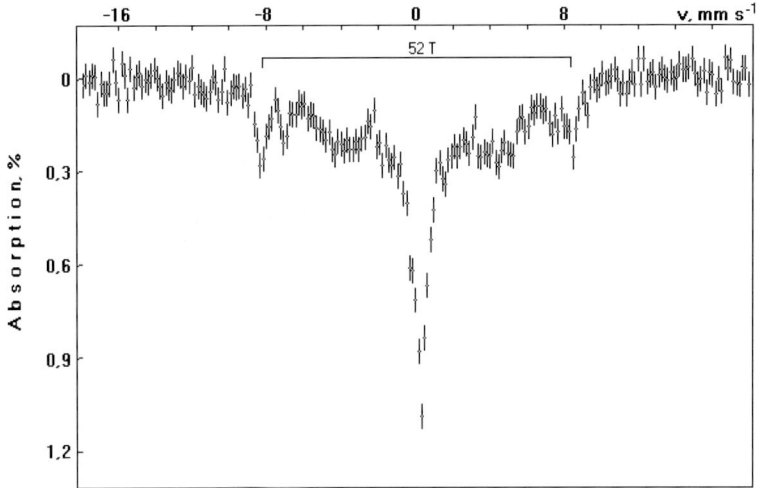

Figure 1. Mössbauer spectrum of quartz doped with Fe^{3+}.

hyperfine magnetic structure lines can be due to various phenomena of magnetic relaxation in the substance.

During gamma-irradiation, the colourless sample turned violet. The general view of the Mössbauer spectrum remained unchanged, except for its central part. These changes are well visible in Figure 2.

The central part of the Mössbauer spectrum of a non-irradiated sample (Figure 2a) contains an intense absorption line with maximum absorption near 0.4 mm·s^{-1}. As the γ-ray dose, and hence the colour intensity of the sample, increased, this line disappeared, but a new line with a maximum near 0.0 mm·s^{-1} appeared (Figure 2b).

Although the relationship between the amethyst colour and the presence of admixed iron ions in the quartz structure can be considered a proven fact [1, 2], presently there is no clear understanding on the exact location these ions occupy in the quartz structure. According to the most popular hypothesis, iron ions replacing silicon atoms in the quartz structure become the coloration centres [2]. The line with maximum absorption near 0.4 mm·s^{-1} (Figure 2a) can correspond to Fe^{3+} ions which have replaced Si^{4+} ions in the SiO_2 crystal structure. However, taking into consideration that the ionic radius of Fe^{3+} is somewhat larger than that of Si^{4+} [6], substitution may be accompanied by the formation of FeO_4 tetrahedra with Fe–O bonds shortened in comparison with ordinary bonds due to the effect of the matrix, i.e., the quartz crystal lattice. If an electron configuration of iron does not change, the shortening of bond distances should inevitably increase the electron density on the iron nucleus and hence increase the chemical shift. This can explain the overestimated chemical shift value ($\approx$0.4 mm·s^{-1}) in comparison with the expected value for Fe^{3+} in the tetrahedral oxygen environment (0.2–0.3 mm·s^{-1}) [7].

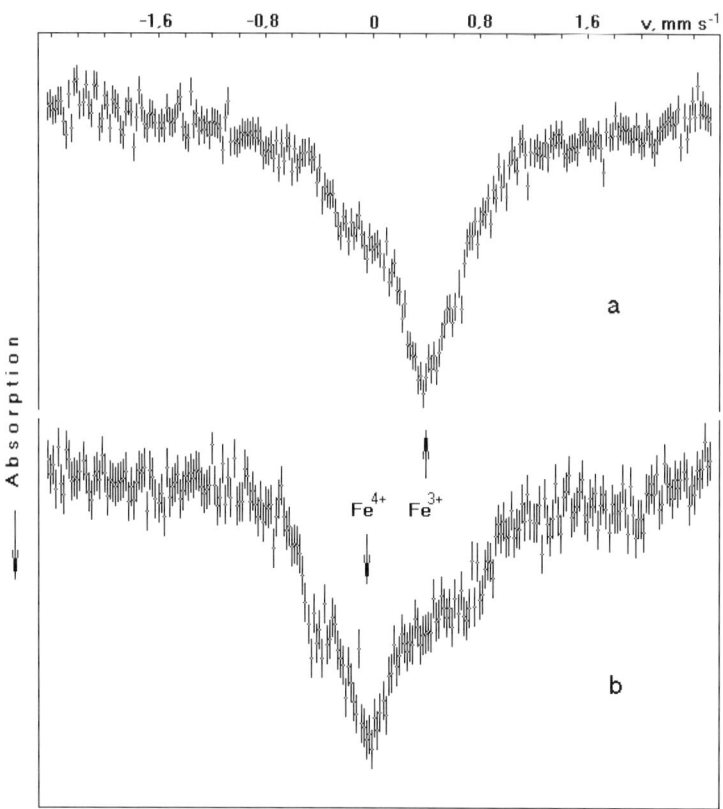

Figure 2. The central parts of Mössbauer spectra of quartz: (a) Fe^{3+}-doped quartz, (b) gamma-irradiated quartz.

The appearance of a line with a maximum near 0.0 mm·s^{-1} in the spectrum (Figure 2b) is well consistent with the hypothesis that it is the formation of Fe^{4+} ions that is responsible for the appearance of amethyst colour upon irradiation of iron-doped quartz. In fact, the value equal to 0.0 mm·s^{-1} belongs to the range of chemical shifts typical of tetravalent iron derivatives.

When considering the chemical shifts of iron(IV), one should take into account data for $CaFeO_3$ and related compounds (0.05–0.07 mm·s^{-1} [8]), in which the Fe^{4+} ion is located in a rigid oxygen cage, and the chemical shift of Na_4FeO_4 (-0.24 ± 0.02 mm·s^{-1} [9, 10]) with tetrahedral coordination of iron. Evidently, the value 0.0 mm·s^{-1} is more consistent with data for octahedral iron coordination. However, it should be taken into account that the quartz structure contains no alkaline or alkaline-earth metal ions that strongly polarise the Fe–O bond. Therefore, the chemical shift of iron in quartz should be more positive than the chemical shifts of ionic compounds with the same oxidation state and coordination number of iron. Moreover, the aforementioned effect of the matrix can also show itself.

Rotation of the sample relative to the direction of Mössbauer irradiation did not reveal any changes in the relative intensities of spectral lines characteristic of single-crystal samples. This is quite natural, assuming that iron isomorphically replaces silicon in the quartz structure. The three SiO_4 tetrahedra that a quartz unit cell comprises have different spatial orientations. Therefore, there is no reason to expect a pronounced anisotropy of the Mössbauer effect, irrespective of the extent of local distortion of the iron coordination polyhedron.

Unfortunately, we were unable to suggest a good model for description of the spectrum in total. However, the magnetic component of the spectrum does not undergo noticeable changes when the sample is being irradiated, unlike relaxation spectra that are sensitive to irradiation. Moreover, we could not detect any significant changes in this spectrum region under weak magnetic field imposed on the sample (for stabilising the relaxation pattern). This allows us to state that the intense peaks in the central part of the spectrum (Figure 2) and the lines of the magnetic structure correspond to different iron forms.

In order to propose an unambiguous model, it would be reasonable to try to simplify the spectrum by getting rid of iron admixtures that are not related to coloration. The content of these admixtures estimated by the relative areas of lines in the spectrum (Figure 1) can be significant. We believe that this can be achieved by decreasing the iron concentration in the hydrothermal solution and by purifying it thoroughly from colloid particles of iron oxides and hydroxides. A similar approach proved successful in a synthesis of V_2O_5 doped with Fe^{5+}, where a decrease in the iron content in the reaction mixture made it possible to decrease the amount of "interfering" iron in the interstices and vacancies in the vanadium oxide structure [11].

4. Conclusions

Thus, Mössbauer spectroscopy is suitable for studies on the traces of iron in quartz, including studies aimed at revealing the mechanism of the formation of amethyst coloration. In order to obtain reliable data, it is important to use stable and highly-efficient spectrometric equipment, as well as enriched iron for enhancing the sensitivity. However, it is no less important to optimise the synthesis conditions in order to obtain samples containing no interfering iron ions and colloid particles.

The Mössbauer spectroscopic measurements that we carried out show that the hypothesis about the presence of tetravalent iron in amethyst is well-founded.

Acknowledgement

This work was supported by the Russian Federal Program "Integracya".

References

1. Tsinober, L. I. and Chentsova, L. G., *Kristallografiya* **4** (1959), 633 [*Sov. Phys.-Crystallogr.* **4** (1959), 593].
2. Zaitov, M. M., Zaripov, M. M., Samoylovich, M. I., Khadgi, V. E. and Tsinober, L. I., *Kristallografiya* **19** (1974), 1090 [*Sov. Phys.-Crystallogr.* **19** (1974), 674].
3. Lehmann, G., *American Mineralogist* **60** (1975), 335.
4. Cohen, A. J., *American Mineralogist* **60** (1975), 338.
5. Cox, R. T., *J. Phys. C* **10** (1977), 4631.
6. Shennon, R. D., *Acta Cryst. A* **32** (1976), 751.
7. Menil, F., *Phys. Chem. Solids* **46** (1985), 763.
8. Takano, M., Nakanishi, N., Takeda, Y., Naka, S. and Takada, T., *Mat. Res. Bull.* **12** (1977), 923.
9. Dedushenko, S. K., Kholodkovskaya, L. N., Perfiliev, Yu. D., Kiselev, Yu. M., Saprykin, A. A., Kamozin, P. N. and Lemesheva, D. G., *J. Alloys Compd.* **262–263** (1997), 78.
10. Jeannot, C., Malaman, B., Gerardin, R. and Oulladiaf, B., *J. Solid State Chem.* **165** (2002), 266.
11. Dedushenko, S. K. and Perfiliev, Yu. D., *Bull. Russian Acad. Sci. Physics* **65** (2001), 1125.

Hyperfine Interactions **156/157**: 423–429, 2004.
© 2004 *Kluwer Academic Publishers. Printed in the Netherlands.*

Fe-Hydroxysulphates from Bacterial Fe^{2+} Oxidation

E. ENEROTH[1] and C. BENDER KOCH[2]

[1]*Department of Geology, GeoBiosphere Science Centre, Lund University, Sölvegatan 12, 223 62, Lund, Sweden; e-mail: erik.eneroth@geol.lu.se*
[2]*Chemistry Department, The Royal Veterinary and Agricultural University, Copenhagen, Thorvaldsensvej 40, DK-1871, Frederiksberg C, Denmark; e-mail: cbk@kvl.dk*

Abstract. Precipitates formed due to Fe(II)-oxidation catalysed by *Acidithiobacillus ferrooxidans* between pH 1.6 and 3.2 have been studied by powder X-ray diffraction, infrared spectroscopy and Mössbauer spectroscopy. The precipitates consist of well crystalline, ammonium-containing jarosite at pH 1.6, and mixtures of jarosite and poorly crystalline schwertmannite at pH 2.5 and 3.2. At low temperatures (10 and 20 K) the components of the two phases overlap strongly. Jarosites ordered (c. 45 K – defect antiferromagnetic) at temperatures well below the ordering temperature for schwertmannite (c. 80 K at pH 3.2 and c. 70 K at pH 2.5). Thus thermoscans measured between 100 and 10 K facilitates characterization of these two minerals when occurring in mixtures.

Key words: Mössbauer spectroscopy, jarosite, schwertmannite, acid mine drainage, bioleaching.

1. Introduction

Fe(III) hydroxysulphate-precipitates form mainly by bacterially mediated oxidation of ferrous iron under acidic conditions, e.g., during acid mine drainage or bioleaching of sulphide ore [1–4]. A number of different minerals can form depending on relative minor variations in solution compositions, and typical samples from such environments comprise mineral mixtures. A thorough understanding of the geochemistry of these processes depends on identification and characterization of all the components. This involves identification of both well and poorly crystalline minerals often having variable composition. In turn, this makes identification and characterisation very challenging, and combinations of many methods have to be used. Routine application of Mössbauer spectroscopy often relies on significant differences in the hyperfine parameters at low temperature (i.e. in the magnetically ordered state). However, many of the minerals encountered in these settings have similar magnetic hyperfine fields. In addition some minerals display inherently broad lines causing a decrease in spectral resolution and thus complicates distinction among different minerals. Solutions having pH between approximately 1 and 3 often contain mixtures of jarosite and schwertmannite. According to literature values these minerals differ by only c. 1.5 T in hyperfine field at 4.2 K [2]. However, they have different ordering temperatures (75 K for schwertmannite, 55–60 K for jarosite) and thus Mössbauer spectra measured in the vicinity of these temperatures would display mixtures of doublets and sextets and be much more useful in characterizing and discriminating among these minerals. Spectra of schwertmannite close to the ordering temperature have previously been published [3], but similar data for jarosite appear not to have been reported.

2. Materials and methods

A sterile ferrous sulphate medium was prepared according to [5]. The growth medium contained $FeSO_4 \cdot 7H_2O$ (33.3 $g\,dm^{-3}$) (energy source), $MgSO_4 \cdot 7H_2O$ (0.4 $g\,dm^{-3}$), $(NH_4)_2SO_4$ (0.4 $g\,dm^{-3}$) and K_2HPO_4 (0.1 $g\,dm^{-3}$). Three beakers with 1 dm^3 of medium were adjusted with 1 M H_2SO_4 to the desired pH. The beakers were inoculated with 1 cm^3 of a liquid culture of the chemolithoautotrophic acidophile *Acidithiobacillus ferrooxidans* (strain DSZM 1927) and incubated for 14 days at 28°C on a horizontally rotating shaker (100 rpm). During this period the pH of the medium was allowed to drift freely. Initial and final pH in the media was 3.2, 2.5, 1.6 and 1.7, 1.7, 1.9, respectively, and they are in the following designated by the initial pH. The measured changes in pH indicate that only acid consuming oxidation of Fe(II) occurred in the pH 1.6 experiment, whereas oxidation was followed by more thorough acid producing hydrolysis of Fe(III) in the experiments at pH 3.2 and 2.5. The precipitates were collected from glass plates placed approximately 3 cm above the bottom.

The samples were freeze-dried, and investigated using variable temperature MS at temperatures between room temperature and 10 K, infrared spectroscopy (IR) and powder X-ray diffraction (pXRD). Mössbauer spectra were acquired using a constant acceleration spectrometer calibrated by measuring a thin foil of α-Fe at room temperature and a source of ^{57}Co in Rh. Infrared (IR) spectra were measured with a Perkin Elmer 2000 FT-IR using the KBr pressed pellet technique. pXRD traces were measured using a Philips PW 1710 equipped with a diffracted beam graphite monochromator and Cu K_α radiation. The bacterial protein content of each precipitate was determined using the Peterson modification of the Lowry method for protein quantification, following dissolution in 10% (w/v) oxalic acid as suggested by [6].

3. Results and discussion

During the incubation period all media acquired a deep red colour and precipitates were observed. The colour change in the pH 1.6 media was delayed in comparison to the two other media, indicating a longer lag-period for the bacteria at low pH. At pH 3.2 a slow, but prompt reddening of the medium was observed, suggesting a slight amount of inorganic oxidation of Fe(II). More precipitate formed at pH 3.2 and 2.5 compared to pH 1.6, and there was a gradual change in the precipitate colour becoming more reddish with increasing pH. Bacterial protein constituted about 0.2% (w/w) for the three freeze dried samples, corresponding to c. 0.45% (w/w) of biomass (prior to freeze drying). This suggests that there is no influence by the presence of bacterial cell walls on the precipitation processes, which is also broadly in line with results from [6].

The pXRD patterns of the precipitates indicated marked differences in crystallinity (Figure 1a). At pH 1.6 the pattern could be identified as due to jarosite (general formula: $MFe_3(SO_4)_2(OH)_6$ M being a monovalent cation) of high purity (accounting for all peaks observed) and high crystallinity (sharp diffraction lines). The pattern of the precipitate formed at pH 3.2 is characterised by a series of

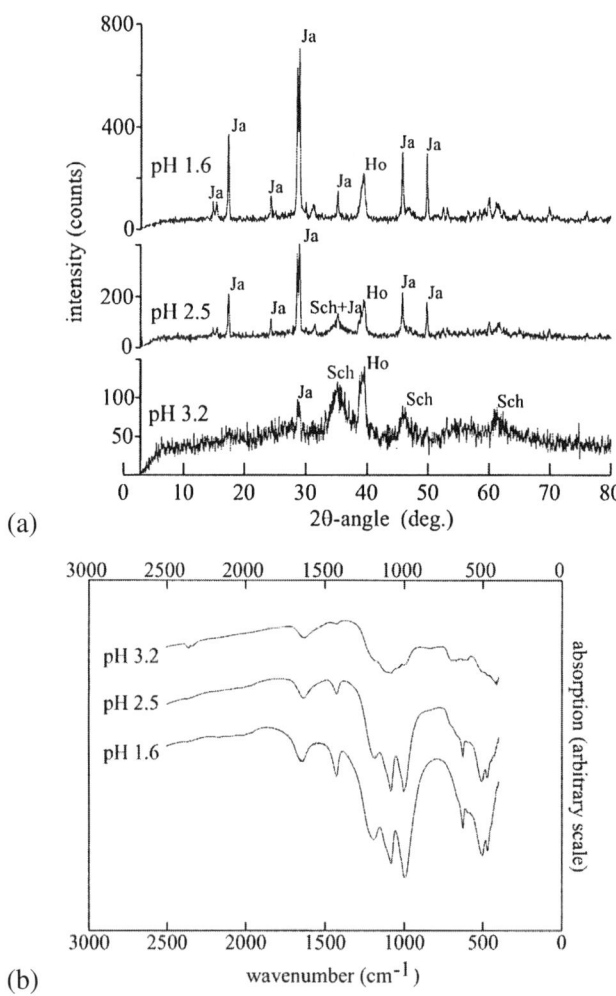

Figure 1. Results from pXRD (a) and IR (b). Designation of peaks in (a) is as follows Ja stands for jarosite, Sch for schwertmannite and Ho for sample holder. Note NH_4^+ deformation band at c. 1400 cm^{-1} in the pH 1.6 and 2.5 samples.

broad diffraction lines indicating poor crystallinity/small particle size. From their positions the presence of schwertmannite (ideal formula: $Fe_8O_8(OH)_6SO_4$) was indicated. In addition, this sample contains a small amount of jarosite as indicated by only one significant diffraction-line. The precipitate formed at pH 2.5 was dominated by jarosite with minor schwertmannite. Both jarosite and schwertmannite contain sulphate as confirmed by IR (Figure 1b). In jarosite a distinct splitting of v_3 was observed (positions: 1130 and 1020 cm^{-1}), whereas in schwertmannite these bands are broad and overlap. The jarosite in part contains ammonium as indicated by an NH_4^+ deformation band at c. 1400 cm^{-1} (for ref. see [7]).

The Mössbauer spectra measured at RT exhibit solely doublets due to Fe(III) (Figure 2). The spectrum of the sample prepared at pH 1.6 was somewhat asymmetric, whereas the spectra of the other two samples exhibit fine structure indicating that the Fe is present in two coordination environments. This interpretation is indeed supported by the pXRD and IR results. The spectra were fitted with one doublet (pH 1.6) and a combination of two doublets (pH 2.5 and 3.2) (one constrained from fitting of the pH 1.6 spectrum). In this way doublet parameters (given as isomer shift and quadrupole splitting) of $\delta = 0.38$ mm s^{-1} and $\Delta = 1.06$ mm s^{-1}, and $\delta = 0.38$ mm s^{-1} and $\Delta = 0.61$ mm s^{-1} were obtained. These values are in good agreement with previously published values for jarosite and schwertmannite, respectively [2, 8]. The relative areas of the jarosite component were 100, 65, and

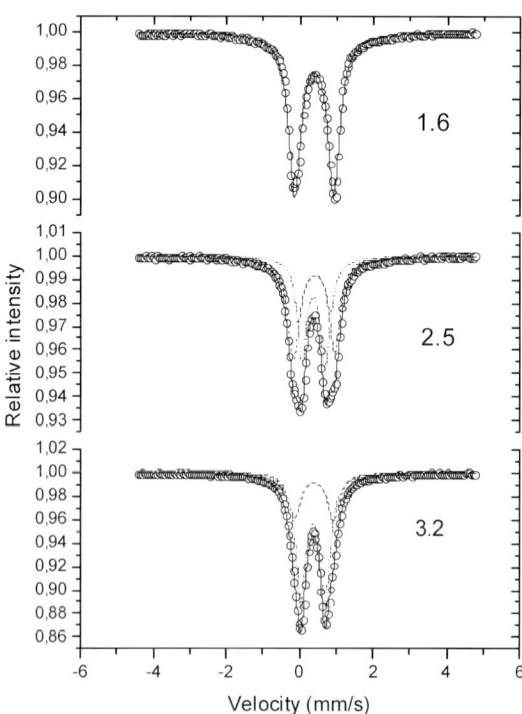

Figure 2. Low-amplitude Mössbauer spectra at room temperature.

25%, respectively for the precipitates formed at pH 1.6, 2.5 and 3.2. These results are qualitatively in good agreement with the pXRD results. Given the observed drift in pH it is possible that jarosite formed in the experiments at pH 2.5 and 3.2 in fact precipitated at a lower pH than the initial.

Spectra obtained at temperatures between 100 and 10 K are shown in Figure 3. The precipitate formed at pH 1.6 exhibit incipient magnetic order at 40 K and complete order occurs within 10 K. At low temperature a distinct new doublet could be discerned ($\delta = 0.54$ mm s^{-1} and $\Delta = 0.59$ mm s^{-1}). That doublet is not considered further here, however, due to dominance of jarosite this component was not resolved in the RT spectrum. Using a single sextet component to fit the 10 K spectrum, values in good agreement with the literature were obtained ($B_{hf} = 46$ T, $\varepsilon = -0.11$ mm s^{-1}, and $\delta = 0.48$ mm s^{-1}). However, the asymmetry of the sextet lines at 10 K and the extremely broad hyperfine field distribution just below the blocking temperature are indicative of non-ideal antiferromagnetic order. We suggest tentatively that these phenomena are associated with defects in the lattice.

The incipient magnetic ordering observed in the schwertmannite of the sample produced at pH 3.2 occurred at significantly higher temperature than for the jarosite-dominated sample at pH 1.6, and also at slightly higher temperatures compared to the pH 2.5 schwertmannite. At approximately 80 K magnetic ordering was onset (pH 3.2) and increased over a much larger temperature range than for the jarosite formed at pH 1.6. At low temperatures both schwertmannite and jarosite exhibit asymmetry in the individual lines as well as in the spectrum as a whole. Since samples with mixtures also have increased line widths, the two minerals are not clearly resolved in the spectrum even when they occur in approximately the same amounts (pH 2.5). However, the paramagnetic jarosite can be detected even in minor amounts at temperatures just above its magnetic transition (50 K for sample pH 3.2). Using the parameters of jarosite from the sample at pH 1.6 (10 K) we roughly estimated $B_{hf} = 43$ T, $\delta = 0.5$ and $\varepsilon = -0.1$ to -0.2 mm s^{-1} for schwertmannite in pH 2.5 sample (11 K).

The lower ordering temperature of the pH 2.5 sample compared to pH 3.2, probably reflects a more thorough incorporation of SO_4^{2-} into the schwertmannite structure (Figure 3). This is consistent with increased protonation at pH 2.5, which is needed for the necessary attachment of sulphate to hydroxide during precipitation in solutions with low SO_4^{2-}/Fe^{2+} ratio [3, 9]. Thus, we interpret the lower ordering temperature of this sample as beeing due to the "super-exchange screening effect" of a more ideal SO_4^{2-} stoichiometry. It seems conceivable however, that a more frustrated arrangement of the spins could occur at higher temperatures if some of the structural SO_4^{2-} is lacking, since superexchange forces would act more freely in sulphate free "iron hydroxide islands" of arbitrarily varying sizes in a SO_4^{2-} deficient schwertmannite. We suggest that this phenomenon is reflected by broadening of the H_{fs} distribution upon magnetic ordering at a higher temperature in the pH 3.2 sample compared to pH 2.5 (Figure 2). Indeed, that view is consistent with the presence of atomic scale relaxation close to the ordering temperature in a

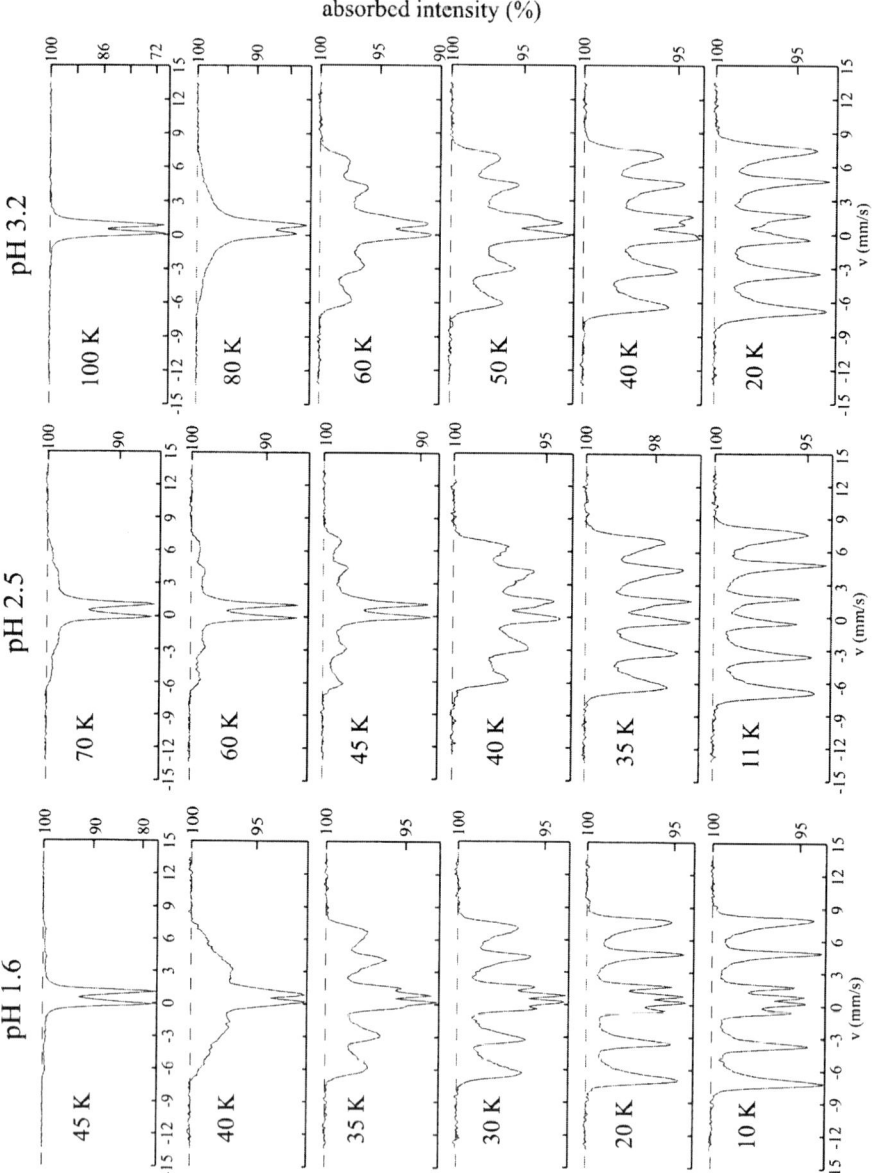

Figure 3. Mössbauer thermoscans of the three samples.

likely SO_4^{2-} deficient schwertmannite, as suggested by spectra measured in applied magnetic fields [2].

4. Conclusion

Our observations suggest that proper characterisation of jarosite–schwertmannite mixtures can be achieved by low temperature thermoscans in combination with small-amplitude spectra at room temperature. Subtle details on the crystal chemistry of both jarosite (well crystalline structure yet defective magnetic ordering) and schwertmannite (poorly crystalline) may be achieved provided the causes of spectral asymmetry can be explained. Being characterised by low crystallinity, variable composition, and SO_4^{2-} both structurally bound and in the surface adsorbed state, the structure and crystal chemistry of schwertmannite is still enigmatic. Yet, the MS data presented here suggest that studies on the magnetic transition in schwertmannites formed under slightly different conditions (varying pH) may contribute to a more thorough understanding of its modifications, presumably seen in relation to different amounts of structural SO_4^{2-}. We also conclude that the physical presence of bacterial cells did not affect the precipitation process (it is rather an indirect result of the bacterial oxidation), and based on our IR results we argue that the incorporation of salts (such as NH_4^+) in jarosite should be taken into account when growth media for *A. ferroxidans* are discussed.

Acknowledgement

We thank the two anonymous referees for their comments on the manuscript.

References

1. Kirby, C. S., Thomas, H. M., Southam, G. and Donald, R., *Appl. Geochem.* **14** (1999), 511.
2. Bigham, J. M. and Nordstrom, D. K., In: C. N. Alpers, J. L. Jambor and D. K. Nordstrom (eds.), *Sulfate Minerals, Crystallography, Geochemistry and Environmental Significance. Reviews in Mineralogy & Geochemistry*, Vol. 40, Mineralogical Society of America, Washington, DC, 2000, p. 351.
3. Bigham, J. M., Schwertmann, U., Carlson, L. and Murad, E., *Geochim. Cosmochim. Acta* **54** (1990), 2743.
4. Pogliani, C. and Donati, E., *Proc. Biochem.* **35** (2000), 997.
5. Fortin, D., Davis, B. and Beveridge, T. J., *FEMS Microbiol. Ecol.* **21** (1996), 11.
6. Ramsay, B., Ramsay, J., de Tremblay, M. and Chavarie, C., *Geomicrobiol. J.* **6** (1988), 171.
7. Lazaroff, N., Sigal, W. and Wasserman, A., *Appl. Environ. Microbiol.* **43** (1982), 924.
8. Takano, M., Shinjo, T., Kiyama, M. and Takada, T., *J. Phys. Soc. Japan* **25** (1968), 902.
9. Parfitt, R. L. and Smart, R. St. C., *Soil Sci. Soc. Am. J.* **42** (1978), 48.
10. Murad, E., *J. Magn. Magn. Mat.* **74** (1988), 153.

Hyperfine Interactions **156/157**: 431–437, 2004.
© 2004 *Kluwer Academic Publishers. Printed in the Netherlands.*   431

Mössbauer Spectroscopic Study of a Mural Painting from Morgadal Grande, Mexico

A. KUNO[1], M. MATSUO[1], A. PASCUAL SOTO[2] and K. TSUKAMOTO[3]

[1] *Graduate School of Arts and Sciences, The University of Tokyo, 3-8-1 Komaba, Meguro, Tokyo 153-8902, Japan*
[2] *Instituto de Investigaciones Estéticas, Universidad Nacional Autónoma de México, Circuito Mario de la Cueva, Ciudad Universitaria, 04510 México D.F., Mexico*
[3] *Escuela Nacional de Antropología e Historia, México D.F., Mexico*

Abstract. In this study, ^{57}Fe Mössbauer spectroscopy has been applied to fragments of a mural painting excavated at Morgadal Grande, Mexico, to characterize the pigments used. A sextet attributable to hematite (α-Fe$_2$O$_3$) was clearly detected in the red fragments. The spectra of orange fragments showed a doublet attributable to paramagnetic high-spin Fe^{3+}, which presumably originates from goethite (α-FeOOH) exhibiting superparamagnetic relaxation due to its small particle size. The blue fragments contained little iron. The scattered X-ray Mössbauer spectra revealed that the thickness of the pigments was larger than 20 μm.

Key words: Mössbauer spectroscopy, Morgadal Grande, mural painting, pigment.

1. Introduction

Morgadal Grande in north-central Veracruz on the Gulf coast of Mexico is located 3 km from El Tajín, a great pre-Columbian center, and considered as its satellite center. Both centers flourished from the early Classic period (300–600 A.D.) to the local Epiclassic period (900–1100 A.D.) [1, 2]. Morgadal Grande, as well as El Tajín and Cerro Grande, is located in the flood plains of the Cazones and Tecolutla Rivers [3]. Buildings were decorated by relieves and mural paintings, and the characterization of the pigments used for the mural paintings would give an insight into the technique of making murals at that time. Consequently, its historical change, social background and relationship with other regions can be investigated. From such an archaeological point of view, ^{57}Fe Mössbauer spectroscopy has been applied to the fragments of the mural painting excavated at Morgadal Grande in this study. In order to clarify the depth profile of iron species on the surface of the mural nondestructively, the measurement was performed in conversion electron and scattered X-ray modes in addition to the conventional transmission mode.

Mössbauer spectroscopy in combination with other analytical techniques has proved to be a powerful tool to study archaeological artifacts. Some results have demonstrated the uniqueness of Mössbauer spectroscopy in obtaining information not accessible by other techniques [4, 5]. Although it might be ironic to investigate the trace iron compounds excavated at the region that was finally conquered by the steel civilization [6], the aim of this study is to characterize the pigments used for the mural painting from Morgadal Grande. The samples are also analyzed by X-ray absorption fine structure (XAFS).

2. Experimental

Fifteen fragments excavated at Morgadal Grande were subjected to the analysis. The fragments consisted of colored plaster and were thinned to 1 mm in thickness. The mural was so damaged that it cannot be seen what was painted. Still, the color of pigments, red, orange, blue, etc., remained vivid. The themes of mural painting excavated at El Tajín can be found elsewhere [7].

Mössbauer spectra were measured with an Austin Science S-600 Mössbauer spectrometer using a 1.11 GBq ^{57}Co/Rh source at room temperature. The spectrum of the plaster without pigments was also measured in transmission mode to check its influence. Conversion electrons and scattered X-rays were detected by using our specially designed proportional counter [8]. Q-gas (90% He–10% CH_4) and PR-gas (90% Ar–10% CH_4) were used as flow gas at a rate of 40 cm^3/min for conversion electron and scattered X-ray modes, respectively [9]. The curve fitting of the obtained spectra was performed, assuming that the spectra were composed of peaks with Lorentzian line shapes. Half-widths and peak areas within each quadrupole doublet were constrained to be equal. Isomer shifts were expressed with respect to the centroid of the spectrum of a metallic iron foil. The XAFS measurements were made using synchrotron radiation with a Si(111) double crystal monochromator at beam line 9A [10] of the Photon Factory, Tsukuba, Japan.

3. Results and discussion

The transmission Mössbauer spectra (TMS) of fragments of the mural painting (Figure 1) mainly consisted of two doublets ascribable to paramagnetic high-spin Fe^{3+} and Fe^{2+}. In addition, a sextet corresponding to hematite (α-Fe_2O_3) was detected in the red fragments. This compound was responsible for the red color. The fit curves are shown as solid lines, and the chi-squared values were sufficiently small. However, the spectrum of the plaster without pigments also showed two doublets ascribable to paramagnetic high-spin Fe^{3+} and Fe^{2+} with relative peak areas of 73% and 27%, respectively (Table I). Therefore, the transmission spectra were overlapped by the spectrum of the plaster under the pigments. In such cases, conversion electron and scattered X-ray Mössbauer spectroscopies are quite useful means to analyze only the pigments on the surface of the plaster in a nondestructive manner [11]. According to the penetrating length of electrons and X-rays, Möss-

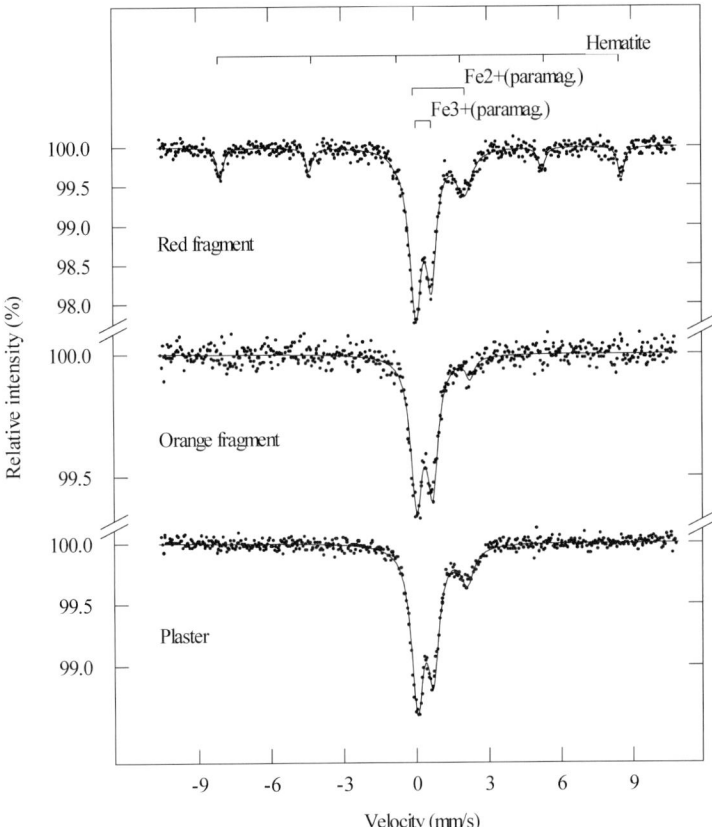

Figure 1. TMS of one of the red fragments (*top*), one of the orange fragments (*middle*), and the plaster (*bottom*) measured at room temperature.

bauer spectra for surface regions with depths of $\lesssim 0.4$ μm and 20 μm are obtained by detecting conversion electrons and X-rays, respectively [12].

Figure 2 shows the conversion electron Mössbauer spectrum (CEMS) and the X-ray Mössbauer spectrum (XMS) of one of the red fragments measured at room temperature. As shown at the top of the spectra, a sextet ascribable to hematite and a doublet ascribable to paramagnetic high-spin Fe^{3+} were detected in both spectra, but the relative peak area of the paramagnetic Fe^{3+} in the XMS was smaller than that in the CEMS. A possible component corresponding to the paramagnetic Fe^{3+} is iron in the plaster, although its Mössbauer parameters contained much statistical error. Even if so, the spectra clearly indicate that the thickness of the pigment is larger than 20 μm, because the ratio of paramagnetic Fe^{3+} in the surface <20 μm detected by XMS was very small compared to that detected in transmission mode and the iron in the plaster was mainly composed of paramagnetic Fe^{3+}. For the analysis of this kind of mural painting, the scattered X-ray mode is the most effective, since the signals from a larger amount of iron in the thicker surface region can be collected compared to the conversion electron mode even though

Table I. Mössbauer parameters of the fragments of the mural painting measured at room temperature in transmission (TMS), conversion electron (CEMS), and scattered X-ray (XMS) modes

Sample	Mode	Species	Area (%)	δ (mm/s)	Δ (mm/s)	H_i (T)
Red fragment	TMS	Fe^{3+} (paramag.)	56(1)	0.37(1)	0.64(1)	
		Fe^{2+} (paramag.)	27(1)	1.00(2)	2.13(4)	
		Hematite	17(1)	0.38(1)	−0.20(2)	51.3(1)
Orange fragment		Fe^{3+} (paramag.)	89(1)	0.37(1)	0.64(1)	
		Fe^{2+} (paramag.)	11(1)	1.07(4)	2.36(7)	
Plaster		Fe^{3+} (paramag.)	73(1)	0.36(1)	0.65(3)	
		Fe^{2+} (paramag.)	27(1)	1.04(5)	2.10(9)	
Red fragment	CEMS	Fe^{3+} (paramag.)	36(2)	0.36(3)	0.75(5)	
		Hematite	64(2)	0.37(1)	−0.20(2)	51.7(1)
Orange fragment		Goethite	100	0.28(3)	0.44(6)	
Red fragment	XMS	Fe^{3+} (paramag.)	11(1)	0.35(3)	0.99(7)	
		Hematite	89(1)	0.38(1)	−0.20(1)	51.3(1)
Orange fragment		Goethite	100	0.23(3)	0.38(7)	

The errors in the least significant figure are given in parentheses.

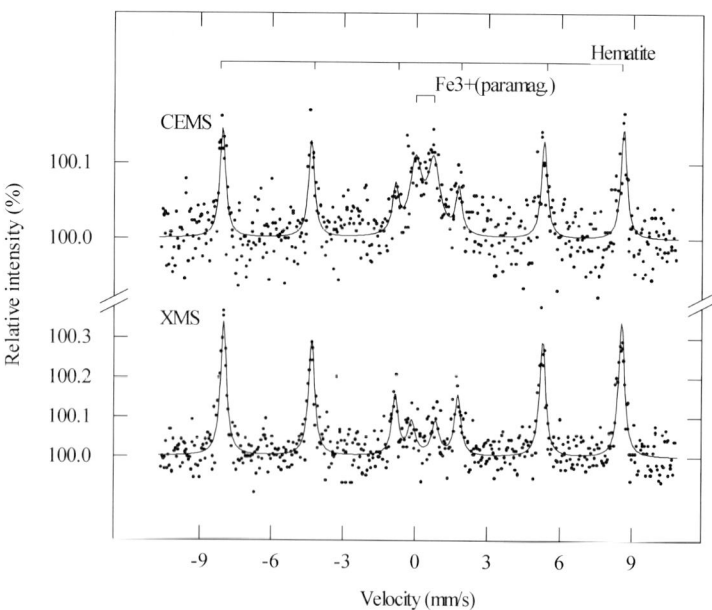

Figure 2. CEMS (*top*) and XMS (*bottom*) of one of the red fragments measured at room temperature.

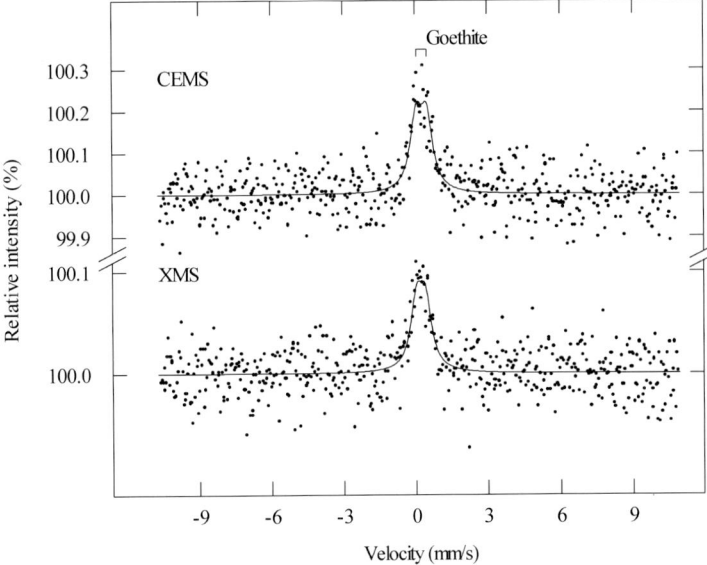

Figure 3. CEMS (*top*) and XMS (*bottom*) of one of the orange fragments measured at room temperature.

the possibility of emission of X-rays is lower than that of electrons. Therefore, better statistics were obtained in the scattered X-ray mode than the conversion electron mode, taking the measurement time into consideration. Another possible component contributing to the paramagnetic Fe^{3+} is an iron species originating from altered hematite in the surface region. The decrease of the paramagnetic Fe^{3+} in the XMS compared to the CEMS indicates that it does not originate from the plaster.

Figure 3 shows the CEMS and XMS of one of the orange fragments measured at room temperature. A paramagnetic Fe^{3+} doublet with large half-widths was detected in both spectra and no significant difference between the CEMS and XMS was observed. The spectral features were very similar to those of cappagh brown (an umber-like material) that were reported by Keisch [13]. Naturally occurring yellow iron-bearing earths such as cappagh brown, ochre, sienna, and umber contain varying amounts of silica and alumina that owe their color principally to goethite (α-FeOOH). Basically, goethite shows a sextet at room temperature. However, natural goethite-bearing pigments show a doublet because of superparamagnetism due to the small particle size [14], while synthetic goethite-bearing yellow pigments show a sextet [15]. Some of this phenomenon may also be due to the presence of clay in the natural material [16]. The Fe K-edge X-ray absorption near-edge structure (XANES) of the orange fragments was similar to that of goethite.

For the blue fragments, peaks were hardly detected owing to its low iron content. It could have been colored by pigments similar to Maya blue, a blue pigment composed of palygorskite ($Mg_5Si_8O_{20}(OH)_2 \cdot 8H_2O$) clay and indigo. Polette *et al.* [17] detected trace iron compounds in Maya blue mural chip samples from Chichén Itzá and Tulum in the Yucatán peninsula, Mayan regions. Such iron compounds could not be identified in our blue fragments from Morgadal Grande by Mössbauer spectroscopy, possibly due to its low iron content.

As a whole, Mössbauer spectroscopy was successfully applied to the characterization of the pigments used for the mural painting excavated at Morgadal Grande. For the estimation of the relationship with other regions, further studies on the pigments excavated at other regions are necessary.

4. Conclusions

The iron speciation in the fragments of the mural painting excavated at Morgadal Grande, Mexico, was carried out using ^{57}Fe Mössbauer spectroscopy. Different spectra were observed for the fragments with different colors. A sextet attributable to hematite was clearly detected in the red fragments. The spectra of orange fragments showed a doublet attributable to paramagnetic high-spin Fe^{3+}, which presumably originates from goethite exhibiting superparamagnetic relaxation due to its small particle size. The blue fragments contained little iron. The scattered X-ray Mössbauer spectra revealed that the thickness of the pigments was larger than 20 μm. It was demonstrated that the scattered X-ray mode was the most effective for this kind of mural painting.

Acknowledgement

This study was partly supported by the Saneyoshi Scholarship Foundation.

References

1. Wilkerson, S. J. K., *Natl. Geogr.* **158** (1980), 202.
2. Pascual Soto, A., *Revista de la Universidad Nacional Autónoma de México* **590** (2000), 30.
3. Pascual Soto, A., *Voices of Mexico* **60** (2002), 79.
4. da Costa, G. M., Cruz Souza, L. A. and de Jesus Filho, M. F., *Hyp. Interact.* **67** (1991), 459.
5. da Costa, G. M. and Viana, R. R., *Am. Miner.* **86** (2001), 1053.
6. Diamond, J., *Guns, Germs, and Steel: The Fates of Human Societies*, W. W. Norton & Co., New York, 1997.
7. Uriarte, M. T. and Falcón, T., In: B. de la Fuente (ed.), *The Pre-Columbian Painting: Murals of the Mesoamerica*, Editoriale Jaca Book Spa, Milan, 1999, p. 177.
8. Matsuo, M., Sato, H. and Tominaga, T., *Radiochem. Radioanal. Lett.* **48** (1981), 253.
9. Swanson, K. R. and Spijkerman, J. J., *J. Appl. Phys.* **41** (1970), 3155.
10. Nomura, M. and Koyama, A., In: S. S. Hasnain (ed.), *X-ray Absorption Fine Structure*, Ellis Horwood, Chichester, 1991, p. 667.
11. Yagnik, C. M., Mazak, R. A. and Collins, R. L., *Nucl. Instrum. Methods* **114** (1974), 1.
12. Wagner, F. E., *J. Phys.* **37** (1976), C6-673.
13. Keisch, B., *J. Phys.* **35** (1974), C6-151.
14. Kündig, W., Bömmel, H., Constabaris, G. and Lindquist, R. H., *Phys. Rev.* **142** (1966), 327.

15. Keisch, B., *Archaeometry* **15** (1973), 79.
16. Yassoglou, N. J. and Peterson, J. B., *Proc. Soil Sci. Soc. Am.* **33** (1969), 967.
17. Polette, L. A., Meitzner, G., Yacaman, M. J. and Chianelli, R. R., *Microchem. J.* **71** (2002), 167.

Hyperfine Interactions **156/157**: 439–443, 2004.
© 2004 *Kluwer Academic Publishers. Printed in the Netherlands.*

Quantification of Secondary Fe-Phases Formed During Sorption Experiments on Chlorites

H. REUTHER[1], T. ARNOLD[2] and E. KRAWCZYK-BÄRSCH[2]

[1] *Forschungszentrum Rossendorf e.V., Institut für Ionenstrahlphysik und Materialforschung, Postfach 510119, D-01314 Dresden, Germany*
[2] *Forschungszentrum Rossendorf e.V., Institut für Radiochemie, Postfach 510119, D-01314 Dresden, Germany*

Abstract. During batch sorption experiments of heavy metals on chlorite not only sorption reactions take place, but also reactions of chemical weathering leading to mineral dissolution and the formation of secondary phases, in particular the Fe-oxy-hydroxide ferrihydrite. Despite of its minor mass but because of its large specific surface area, ferrihydrite plays a major role in removing aqueous uranium(VI) from solution. To accurately model the sorption and transport on or through geological materials it is necessary to precisely determine the mass of the newly-formed Fe-phase. As the relative mass of the ferrihydrite within the geological matrix was too small, it was impossible to use powder X-ray diffraction or some other spectroscopic techniques, e.g., Raman spectroscopy, for its identification and quantification.

Because of the ability to discriminate different sites and oxidation states of iron, Mössbauer spectroscopy at room temperature was used. At first spectra of pure chlorite and pure ferrihydrite were measured. In a second step, simulated spectra were compared with spectra from powders with known chlorite/ferrihydrite ratios. There was a good agreement between the predicted values and those obtained by the spectrum fit. Finally, the calibration spectra were used to investigate real geological material and to estimate the fraction of the secondary Fe-phase, which has been formed during the sorption experiments. Changes of less than 2% (absolute) could be detected. It should be noted that there is a strong overlap between the different subspectra and that a good counting statistics is required.

Key words: ferrihydrite, chlorite, Mössbauer spectroscopy.

1. Introduction

Chemical weathering of the sheet silicate chlorite leads to an evolution of altered and newly-formed secondary iron phases as coatings and Fe-colloids [1]. Such newly-formed phases often represent the most reactive fractions of soils due to their small particle sizes and associated large specific surfaces [2]. In respect to sorption reactions such secondary phases, in particular iron phases, often dominate the overall sorption behaviour of rocks [3]. However, their identification and quantification is almost impossible and if attempted afflicted with a large analytical error. However in respect to sorption reactions, it is essential to quantify even minor amounts of secondary iron phases, since they may provide a major component to adequately model sorption reactions. To overcome this problem a new method was developed in which Mössbauer spectra of pure chlorite and pure ferrihydrite were recorded at room temperature and used to simulate Mössbauer spectra of defined chlorite and ferrihydrite ratios. Then the simulated spectra were taken as calibration standards for chlorite samples, used in mixed flow reactor experiments, to identify and to quantify minor amounts of newly-formed ferrihydrite within narrow analytical errors.

2. Experiments and discussion

The chlorite used in our investigations was an unaltered ripidolithe chlorite from Flagstaff Hill, which was obtained from the Source Clays Repository of the Clay Mineral Society. The chemical formula of this chlorite was determined on the basis of electron microprobe analysis results as $(Mg_{5.5}Al_{2.48}Fe_{3.96})[(Si_{5.33}Al_{2.66})O_{20}]$-$(OH)_{16}$ [4]. Titration of completely dissolved chlorite samples may be used to determine the concentration of Fe^{2+} and Fe^{3+} [5]. However oxidation of some Fe^{3+} during the H_2SO_4/HF digestion step cannot be completely excluded in this method.

Ferrihydrite was precipitated in air from 1 mM iron(III) nitrate solution by slowly raising the pH to 7. The suspension was aged for about 60 minutes before the pH was lowered to 5 and the ionic strength adjusted to 0.1 $NaClO_4$. Afterwards, the aging of the ferrihydrite proceeded by continuously stirring at room temperature for 65 h.

The various mixtures of chlorite and ferrihydrite, listed in Table I, were obtained by adding a certain amount of a ferrihydrite suspension (1 mM Fe) to 500 mg of chlorite powder and drying the samples at 40°C. The resulting powders were used for the Mössbauer spectroscopic investigations.

A specimen obtained from a mixed flow reactor experiment with chlorite was used as real geological sample in which some ferrihydrite as a secondary phase

Table I. Mixtures of ferrihydrite and chlorite

Sample	Fe contribution from chlorite (%)	Fe contribution from ferrihydrite (%)
Freeze dried ferrihydrite powder	0	100
500 mg chlorite + 400.63 ml 1 mM Fe suspension (20F/80C)	80	20
500 mg chlorite + 178.06 ml 1 mM Fe suspension (10F/90C)	90	10
500 mg chlorite + 84.34 ml 1 mM Fe suspension (5F/95C)	95	5
500 mg chlorite	100	0

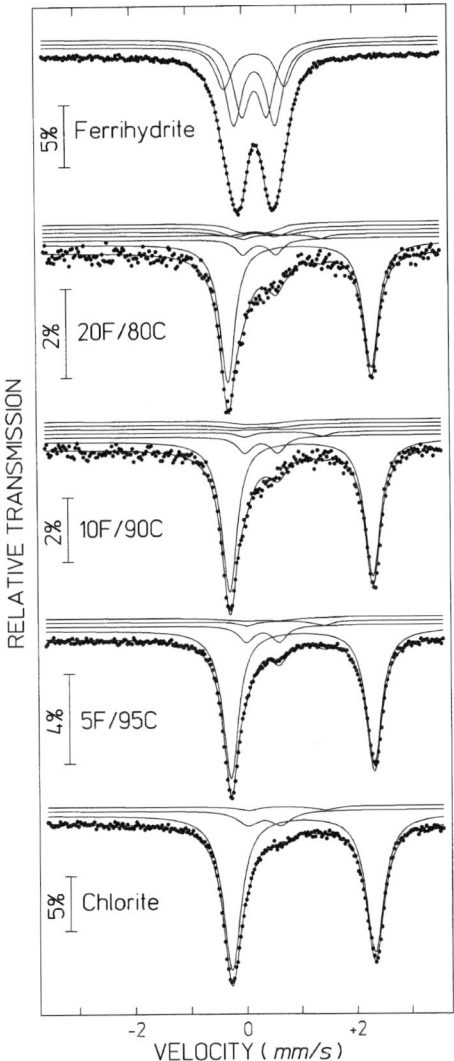

Figure 1. Mössbauer spectra of ferrihydite, chlorite and the mixed powders according Table I.

has been formed during the experiment. The mixed flow reactor experiment was performed with a mass of 500 mg through which for three weeks a 0.1 N NaClO$_4$ solution was pumped at a flow rate of 19 ml/h. The pH value of the solution was monitored and pH values of around 7 were recorded.

Mössbauer spectra were measured at room temperature with a conventional constant acceleration spectrometer in transmission geometry with a ^{57}Co(Rh) source with an activity of nominally 3.7 GBq and a krypton filled proportional counter. The evaluation of the spectra was performed by using the NORMOS least square fitting program of R. A. Brand (University Duisburg) [6].

Table II. Hyperfine parameters and spectrum areas of ferrihydrite and chlorite

	Doublet	Isomer shift δ in ref. to α-Fe (mm/s)	Quadrupole splitting Δ (mm/s)	Relative spectrum area (%)
Ferrihydrite	1 (Fe^{3+})	0.346	1.101	21.5
	2 (Fe^{3+})	0.350	0.744	41.7
	3 (Fe^{3+})	0.346	0.456	36.8
Chlorite	1 (Fe^{2+})	1.133	2.600	86.8
	2 (Fe^{3+})	0.425	0.603	10.1
	3 (Fe^{2+})	0.819	1.398	3.1

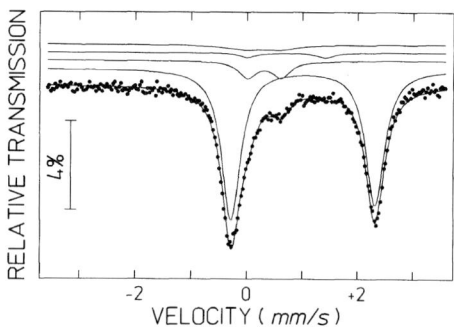

Figure 2. Mössbauer spectrum of the chlorite sample obtained from the mixed flow reactor experiment.

The measured spectra are shown in Figure 1. The spectra of the pure ferrihydrite and the pure chlorite were used for calibration. According to refs. [7–10] the spectra were decomposed by several doublets. Best fits were obtained by assuming three doublets both for the ferrihydrite (likely due to its poor crystallinity) and for the chlorite (one Fe^{2+} place with relative low isomer shift of 0.819 mm/s). The hyperfine parameters are given in Table II.

The Mössbauer spectra of the mixed powders were fit with these hyperfine parameters by varying only the fractions of the two compounds and keeping fixed all other parameters. For the mixture with a small amount of ferrihydrite it made sense to combine its three doublets to one with $\delta = 0.237$ mm/s and $\Delta = 0.700$ mm/s because subspectra with relative areas in the range of 1% cannot be surely resolved. The agreement between the origin fractions and those obtained by the fit was good:

mixture 20% ferrihydrite/80% chlorite,   from spectrum fit 19.8%/80.2%,
mixture 10% ferrihydrite/90% chlorite,   from spectrum fit 10.0%/90.0%,
mixture 5% ferrihydrite/95% chlorite,   from spectrum fit 4.5%/95.5%.

The standard deviation ranges from 3% for the high to 10% for the low ferrihydrite content. Because of this good agreement between the powder synthesis and the calculated fractions the method can be applied to chlorites with unknown content of secondary phases now. As an example, Figure 2 shows the Mössbauer spectrum of a chlorite sample treated in a mixed flow reactor experiment. In comparison to the spectrum of the untreated chlorite the formation of the secondary phase is obvious due to the pronounced growth of the shoulder at about 1.0 mm/s. Applying the above described algorithm we obtained a fraction of this phase of $5.0 \pm 0.9\%$.

Such experiments require a good counting statistics. In this case, even changes of 2% can be detected. Smaller ferrihydrite contents can be determined by Mössbauer spectroscopy at temperatures below 20 K where the ferrihydrite is magnetically ordered and produces magnetically split spectra [7]. Of course, measurements at low temperatures could be performed successfully at the higher ferrihydrite contents too, but the necessary equipment is not available in every laboratory.

3. Conclusion

In conclusion, we could demonstrate how to calculate the fraction of the newly formed secondary iron phase ferrihydrite evolved during sorption experiments with chlorites by using Mössbauer spectroscopy at room temperature. This method enables the quantitative determination of the Fe(III) increase as the iron mineral ferrihydrite during aging and sorption studies.

References

1. Krawczyk-Bärsch, E., Arnold, T., Reuther, H., Brandt, F., Bosbach, D. and Bernhard, G., *Applied Geochemistry* **19** (2004), 1403.
2. Scott, A. D. and Amonette, J., *Iron in Soils and Clay Minerals. NATO ASI Series* **217** (1988), 537.
3. Arnold, T., Zorn, T., Bernhard, G. and Nitsche, H., *Chemical Geology* **151** (1998), 129.
4. Brandt, F., Bosbach, D., Krawczyk-Bärsch, E., Arnold, T. and Bernhard, G., *Geochimica et Cosmochimica Acta* **67** (2003), 1451.
5. Herrmann, G. A., *Praktikum der Gesteinsanalyse*, Springer, Berlin, 1975.
6. Brand, R. A., *Nucl. Instrum. Methods B* **28** (1987), 398.
7. Murad, E. and Schwertmann, U., *American Mineralogist* **65** (1980), 1044.
8. Arshed, M., Butt, N. M., Siddique, M. and Anwar-Ul Islam, M., *Phys. Stat. Sol. A* **137** (1993), K33.
9. Kodama, H., Longworth, G. and Townsend, M. G., *Canadian Mineralogist* **20** (1982), 585.
10. Borggaard, O. K., Lindgren, H. B. and Morup, S., *Clays and Clay Minerals* **30** (1982), 353.

Hyperfine Interactions **156/157**: 445–451, 2004.
© 2004 *Kluwer Academic Publishers. Printed in the Netherlands.*

Synthesis by Coprecipitation of Al-Substituted Hydroxysulphate Green Rust $Fe_4^{II}Fe_{(2-y)}^{III}Al_y^{III}(OH)_{12}SO_4, nH_2O$

R. AISSA, C. RUBY*, A. GEHIN, M. ABDELMOULA and J.-M. R. GÉNIN
Laboratoire de Chimie Physique et Microbiologie pour l'Environnement (LCPME), UMR 7564 CNRS-Université Henri Poincaré, Nancy 1, Equipe Microbiologie et Physique, 405 rue de Vandoeuvre, F-54600 Villers-lès-Nancy, France; e-mail: ruby@lcpe.cnrs-nancy.fr

Abstract. Al-substituted hydroxysulphate green rust (Al-GR{SO$_4$}) were synthesised by the co-precipitation of Fe^{II}, Fe^{III} and Al^{III} cations. The Al-GR{SO$_4$} crystals ($\sim$50 nm) are significantly smaller than the hydroxysulphate green rust GR{SO$_4$} crystals ($\sim$500 nm). The Mössbauer spectrum of Al-GR{SO$_4$} was adjusted with two ferrous doublets D$_1$ and D$_3$ and one ferric doublet D$_2$. Doublet D$_3$ is attributed to Fe^{II} ions that have Al^{III} ions as a first neighbour.

Key words: green rust, coprecipitation, aluminium substitution, Mössbauer spectroscopy, TEM.

1. Introduction

Aluminium-substituted iron oxides were often identified in soils [1–4]. It was shown that the degree of Al-substitution of goethites may reflect the environment in which the minerals have formed and serves as an indicator of soil forming processes [1]. Synthetic Al-substituted iron oxides were extensively studied with Mössbauer spectroscopy by De Grave *et al.* [5] and the use of external fields was determined to be an important tool for a precise determination of the hyperfine parameters of such samples. Synthetic Al^{III}–Fe^{II} and Fe^{II}–Fe^{III} layered double hydroxides (LDHs) were prepared by Taylor *et al.* [6]; it was proposed that "a more careful examination of the green horizons of certain gleys should lead to the identification of one of these phases as a metastable pedogenic mineral". This was performed later and a mineral located in a hydromorphic gley soil of the forest of Fougères (Brittany) that has structural properties very close to synthetic green rusts (GRs) was identified [7, 8]. LDHs are very flexible compounds concerning the nature of the cations present in the structure. Therefore, the existence of cations other than Fe^{II} and Fe^{III} in natural GRs may occur easily. Nevertheless, the dilution of the mineral found in the soil of Fougères makes very difficult their identification. In this work, it will be shown that Al-substituted hydroxysulphate green rust (Al-GR{SO$_4$}) can be synthesised by coprecipitation and the properties of such samples

---

* Author for correspondence.

will be determined by X-ray diffraction (XRD), transmission electron microscopy (TEM) and transmission Mössbauer spectroscopy (TMS).

2. Experimental

Four samples M_0, M_1, M_2 and M_3 that contain increasing amount of Al^{III} were prepared by using the coprecipitation method described in previous studies [9, 10]. $FeSO_4 \cdot 7H_2O$, $Fe_2(SO_4)_3 \cdot 5H_2O$ and $Al_2(SO_4)_3 \cdot 5H_2O$ salts were dissolved in 100 ml of distilled water ($C(Fe^{II}) + C(Fe^{III}) + C(Al^{III}) = 0.4$ M). The initial mixture was precipitated at room temperature by adding a NaOH solution ($C = 0.8$ M). A magnetic stirring ensured a complete dissolution at a rotation velocity of 500 rpm. After a few seconds, the mixture was introduced in a glass flask and sheltered from the air to avoid any oxidation. Then, the precipitate settled down. The formula of the expected product is $Fe_4^{II}Fe_{(2-y)}^{III}Al_y^{III}(OH)_{12}SO_4, nH_2O$ so that molar ratios $n(Fe^{II})/[n(Fe^{III})+n(Al^{III})] = 2$ and $n(OH^-)/[n(Fe^{II})+n(Fe^{III})+n(Al^{III})] = 2$ were maintained constant. The values of the molar ratio $n(Al^{III})/n(Fe^{III})$ of the samples are listed in Table I. The samples were aged 24 hours before analysis.

TMS spectra were measured by means of a constant-acceleration spectrometer with a 50 mCi source of ^{57}Co in Rh. The spectrometer was calibrated with a 25 μm foil of α-Fe at room temperature. The samples were filtered on a paper under inert atmosphere, set in the sample holder and introduced in the cryostat for Mössbauer measurements. Computer fittings were done using Lorentzian-shape lines. Products were analysed by XRD using a monochromatized Co-Kα_1 wavelength ($\lambda = 0.1789$ nm). The final products to be analysed were filtered and rapidly coated with glycerol to avoid any oxidation. TEM (CM20/STEM Philips) coupled with an energy dispersive X-ray system (EDX) was performed using a voltage of 200 kV. One drop of the suspension was deposited on a copper grid that was introduced in the microscope under a 10^{-8} Torr vacuum.

3. Results

3.1. X-RAY DIFFRACTION

Typical GR{SO_4} lines [11] are observed for the non-substituted M_0 sample (Figure 1a). For Al-GR{SO_4} samples, only the (000l) lines are clearly observed. Figure 1 shows that there is a gradual decrease of the intensity of these lines with increasing mount of Al^{III}. Concomitantly, an increase of the full width at half maximum (FWHM) of the (000l) lines is measured, e.g., FWHMs of ~0.4° and ~0.7° of the (0003) lines are measured for M_0 and M_2 samples, respectively. No significant shift of the (000l) lines is observed in Figure 1. Therefore the presence of Al^{III} cations does not induce a significant change of the interlayer distance along the [001] direction ($d = 1.1$ nm), which is mainly governed by the nature of the intercalated anion, here SO_4^{2-}.

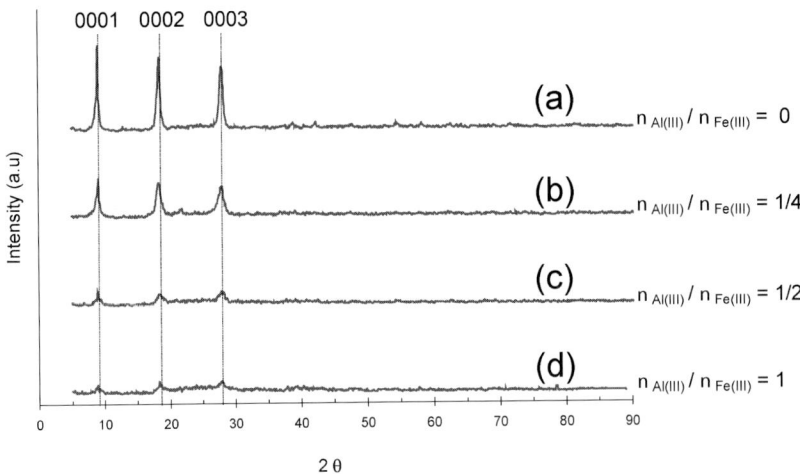

Figure 1. XRD patterns of sample M_0 (a), M_1 (b), M_2 (c) and M_3 (d).

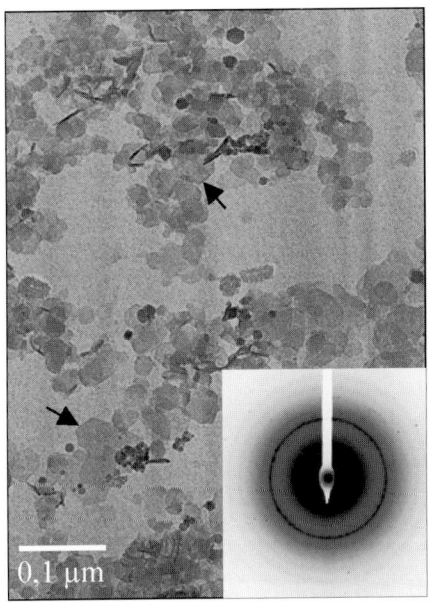

Figure 2. TEM image and diffraction pattern of sample M_1.

3.2. TRANSMISSION ELECTRON MICROSCOPY

The TEM image of sample M_1 shows small crystals of size ~50 nm (Figure 2). The hexagonal shape of some crystals is pointed by arrows. The crystals are much smaller than the $GR(SO_4)$ crystals observed in a previous study [9], i.e. size ~500 nm. The d_{hkl} distances that correspond to the rings of the diffraction pattern were measured. From these data, a cell parameter value $a = 0.315 \pm 0.004$ nm was deduced. It can be compared to the value $a = 0.314 \pm 0.004$ nm of $GR(SO_4)$

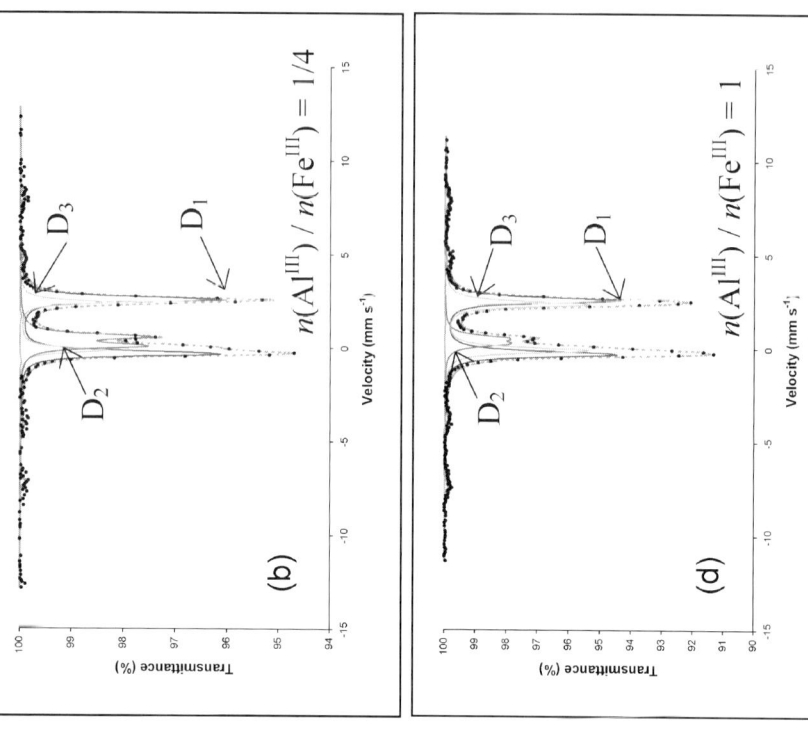

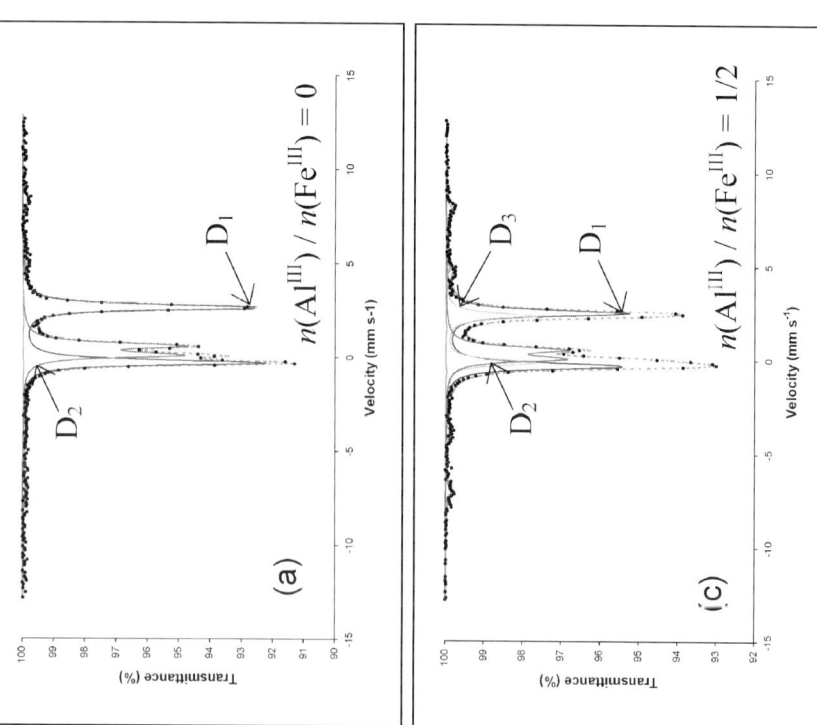

Figure 3. TMS spectra measured at 150 K: (a) sample M_0, (b) sample M_1, (c) sample M_2, (d) sample M_3.

measured with TEM [9]. Fe, Al, S, and O were detected by energy dispersive X-ray (EDX) analyses performed on various groups of crystals.

3.3. TRANSMISSION MÖSSBAUER SPECTROSCOPY

The TMS spectra measured at 150 K display the two Fe^{II} and Fe^{III} paramagnetic components that characterise GRs (Figure 3). A low intensity magnetic component of relative area $\sim$5% is also observed. The paramagnetic components of the spectrum of sample M_0 were adjusted with two doublets D_1 and D_2. The hyperfine parameters and the values of the molar ratio $n(Fe^{II})/n(Fe^{III})$ measured with TMS are given in Table I. The molar ratio $n(Fe^{II})/n(Fe^{III})$ increases gradually from 1.7 to 4.5 for increasing values of the molar ratio $n(Al^{III})/n(Fe^{III})$. It was necessary to add a second ferrous doublet D_3 in order to achieve a satisfactory fit for the TMS spectra of the Al-GR{SO_4} samples. The relative abundance of doublet D_3 increases gradually with increasing quantity of Al^{III} (Table I). Doublet D_3 is therefore assigned to Fe^{II} cations that has Al^{III} cations in their local environment.

Table I. Designation of the samples, $n(Al^{III})/n(Fe^{III})$ molar ratio and hyperfine parameters of the components presented in Figure 3

Sample		δ	Δ or 2ε	H	Γ	RA	Fe^{II}/Fe^{III} (TMS)
M_0	D_1	1.3	2.93	–	0.34	60	
$n(Al^{III})/n(Fe^{III}) = 0$	D_2	0.49	0.5	–	0.34	35	1.7
	S	0.69	0	47	0.7	5	
M_1	D_1	1.29	2.91	–	0.33	44	
$n(Al^{III})/n(Fe^{III}) = 1/4$	D_2	0.49	0.48	–	0.33	28	2.5
	D_3	1.29	2.57	–	0.33	25	
	S	0.48	0	47	0.42	3	
M_2	D_1	1.27	2.86	–	0.37	39	
$n(Al^{III})/n(Fe^{III}) = 1/2$	D_2	0.49	0.49	–	0.37	28	2.6
	D_3	1.27	2.49	–	0.37	26	
	S	0.51	0.02	47	0.75	7	
M_3	D_1	1.30	2.86	–	0.35	43	
$n(Al^{III})/n(Fe^{III}) = 1$	D_2	0.51	0.47	–	0.35	17	4.5
	D_3	1.29	2.47	–	0.35	34	
	S	0.53	0	47	0.82	6	

δ (mm s^{-1}), isomer shift with respect to metallic α-iron at room temperature; Δ (mm s^{-1}), quadrupole splitting; 2ε (mm s^{-1}), quadrupole shift; H (Tesla), hyperfine field; Γ (mm s^{-1}), full width at half maximum; RA (%), relative abundance; Fe^{II}/Fe^{III}, molar ratio corresponding to the ratio $[RA(D_1) + RA(D_3)]/RA(D_2)$.

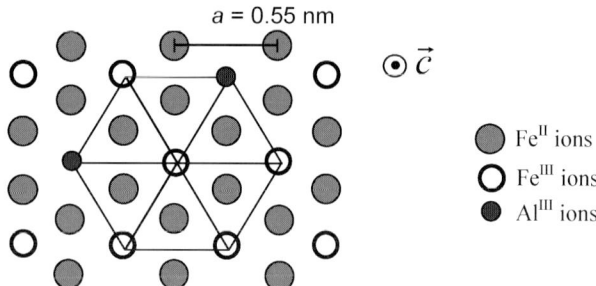

Figure 4. Structure of the (0001) cation layer of Al-GR{SO$_4$} where Al^{III} cations replace Fe^{III} cations.

4. Discussion

Both the TMS and EDX results support the fact that Al^{III} cations are present inside the GR structure. According to the ordered structure of the GR{SO$_4$} (0001) cation layer proposed by Simon *et al.* [11], Al^{III} is a first neighbour of solely Fe^{II} cations if some Fe^{III} are replaced by Al^{III} cations (Figure 4). This is in agreement with the observed broadening of the ferrous doublet and thus the presence of doublet D$_3$.

As already observed for Al-substituted goethite [12], TEM show a significant decrease of the Al-GR{SO$_4$} crystals size. The crystals are however much more well defined and larger than the nanoparticles of Mg^{II}–Al^{III} and Ni^{II}–Al^{III} LDHs synthesised by coprecipitation [13, 14]. Therefore, in natural environment where the substitution of Fe^{II} and Fe^{III} cations may occur, small crystal size of GR(s) is expected. These crystals may have high adsorption capacity towards pollutants.

Acknowledgement

We are grateful to J. Ghanbaja who performed the TEM experiments at the Faculty of Science of the University Henri Poincaré (Nancy I).

References

1. Fitzpatrick, R. W. and Schwertmann, U., *Geoderma* **27** (1982), 335.
2. Mirabella, A. and Carnicelli, S., *Geoderma* **55** (1992), 95.
3. Ramanaidou, E., Nahon, D., Decarreau, A. and Melfi, A. J., *Clays Clay Miner.* **44** (1996), 22.
4. Mongelli, G., *Catena* **50** (2002), 43.
5. De Grave, E., Barrero, C. A., Da Costa, G. M., Vandenberghe, R. E. and Van San, E., *Clay Miner.* **37** (2002), 591.
6. Taylor, R. M. and McKenzie, R. M., *Clays Clay Miner.* **28** (1980), 179.
7. Trolard, F., Génin, J.-M. R., Abdelmoula, M., Bourrié, G., Humbert, B. and Herbillon, A. J., *Geochim. Cosmochim. Acta* **61** (1997), 1107
8. Refait, P., Abdelmoula, M., Trolard, F., Génin, J.-M. R., Ehrhardt, J. J. and Bourrié, G., *Am. Miner.* **86** (2001), 731.
9. Géhin, A., Ruby, C., Abdelmoula, M., Benali, O., Ghanbaja, J., Refait, Ph. and Génin, J.-M. R., *Sol. State Sci.* **4** (2002), 61.

10. Ruby, C., Géhin, A., Abdelmoula, M., Génin, J.-M. R. and Jolivet, J. P., *Sol. State Sci.* **5** (2003), 1055.
11. Simon, L., François, M., Refait, P., Renaudin, G., Lelaurain, M. and Génin, J.-M. R., *Sol. State Sci.* **5** (2003), 327.
12. Schulze, D. G. and Schwertmann, U., *Clay Miner.* **19** (1984), 521.
13. Carja, G., Nakaruma, R., Aida, T. and Niiyama, H., *Micropor. Mesopor. Mater.* **47** (2001), 275.
14. Prinetto, F., Ghiotti, G., Graffin, P. and Tichit, D., *Micropor. Mesopor. Mater.* **39** (2000), 229.

Hyperfine Interactions **156/157**: 453–458, 2004.
© 2004 *Kluwer Academic Publishers. Printed in the Netherlands.*

Mössbauer Study of Ancient Albanian Ceramics

R. RÜFFLER[1], E. GJYLAÇI[2] and K. NAGORNY[1]

[1] *Job Foundation & University of Hamburg, Institute of Physical Chemistry, Grindelallee 117, D-20146 Hamburg, Germany*
[2] *Institute of Nuclear Physics, P.O.Box 85, Tirana, Albania*

Abstract. Ceramic finds from three different excavation sites in Albania (Belsh, Apollonia, and Durrës) were studied by Mössbauer spectroscopy and X-ray diffraction. After identification and discussion of the mineral phases found in the sherds the results were used for a first attempt to reconstruct the ancient firing technique.

Key words: Mössbauer spectroscopy, ceramics, archaeometry.

1. Introduction

Pottery is one of the most common and therefore important remains of ancient civilisations all over the world. Because iron is generally present in unpurified clays as raw material and therefore also in the archaeological ceramics ^{57}Fe Mössbauer spectroscopy is a very effective tool for studying the firing process. During firing the iron-bearing minerals undergo characteristic changes determined by process parameters like the kiln atmosphere, the firing temperature and the duration of firing. Aim of the investigations is the reconstruction of the original production process by combining the results of an extensive phase analysis of the ancient pottery by Mössbauer spectroscopy and additional techniques with those of laboratory and field firing experiments and the archaeological evaluation of the finds. Laboratory firing experiments have been done on clay samples from different countries like Greece [1], Egypt [2], and Lebanon [3] but the most extensive and systematic investigation was performed with Peruvian clays for reconstruction of the Precolumbian firing technology [4–6]. Wagner *et al.* [5–7] also had the unique opportunity to use well-preserved original pottery kilns for field experiments. It could be demonstrated that ceramic specimens from laboratory firing and from field firing essentially yield the same Mössbauer patterns.

In the present study the first results of an investigation of sherds from three different archaeological sites in Albania (Belsh, Apollonia, and Durrës; 4th to 2nd century B.C.) are presented.

2. Experimental

Twelve powdered sherd samples were examined by Mössbauer spectroscopy in transmission geometry at room temperature; spectra of selected samples also were measured at liquid nitrogen temperature or after chemical extraction of iron oxides by dithionite treatment (citrate-bicarbonate-dithionite (CBD) extraction) [8]. Additionally, X-ray powder diffractometry was used as complementary method for identifying the other mineral components.

The spectra were fitted with the minimum number of Lorentzian doublets and sextets necessary; but in case of the low temperature spectra the use of hyperfine field distributions often was unavoidable.

3. Results

In a first step the mineral phases found in the archaeological ceramics should be identified. The room temperature Mössbauer spectra of all samples (e.g., Figures 1a and 2a) were dominated by a Fe^{3+} quadrupole pattern which had to be fitted by at least two doublets; also an asymmetrical and strongly broadened magnetic hyperfine pattern was observed whose contribution to the total spectral area range from that of a minor component up to 43%. Most of the spectra additionally show a Fe^{2+} quadrupole doublet in the range from approximately 5 to 25%.

The Fe^{2+} quadrupole doublet originates from paramagnetic Fe^{2+} in octahedral lattice sites of the phyllosilicates, the predominant mineral constituents of clays. The Fe^{3+} quadrupole pattern may be caused by structural Fe^{3+} in the siliceous clay matrix as well as by small particles of iron oxides or oxyhydroxides which exhibit superparamagnetism. Therefore, measurements at low temperatures were necessary to distinguish both species because cooling to adequate temperatures will block the oxidic particles. For example the fractional intensity of the magnetic components in a sherd from Belsh increases substantially from 19.0% at RT to 28.6% at 78K (Figure 1b and Table I) while the Fe^{3+}/Fe^{2+} ratio of the nonmagnetic components decreases from 2.71 to 1.98; therefore part of the Fe^{3+} pattern at RT indeed arises from superparamagnetic oxidic particles (with sizes large enough to be blocked at liquid nitrogen temperature). For investigating the completeness of the blocking process measurements at liquid helium temperature are necessary which are in progress. Another way of distinguishing between structural and oxidic iron is to extract the oxidic phases by dithionite treatment. As an example the RT spectrum of the CBD-extracted sample from Belsh is shown in Figure 1c. The hyperfine pattern is strongly diminished after chemical extraction but also the Fe^{3+}/Fe^{2+} ratio decreases from 2.71 to 2.39 which confirms the oxidic character of the magnetically split components as well as part of the nonmagnetic Fe^{3+} species. The value of the area-averaged Fe^{3+} quadrupole splitting, however, shows no remarkable change.

Part of the magnetically split pattern at room-temperature arises from relatively well-crystallized hematite with a magnetic field of ~ 50 T. The reduction of the

Relative transmission

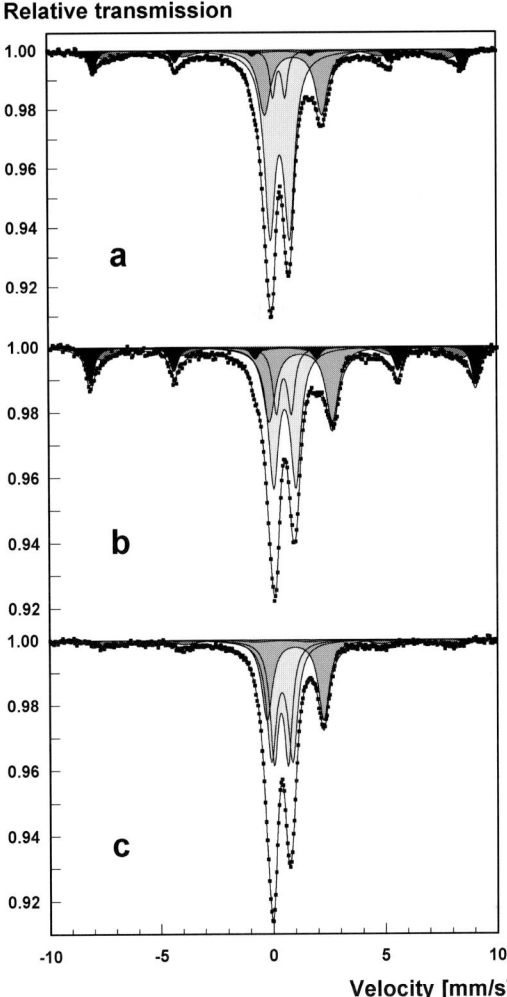

Figure 1. Transmission Mössbauer spectra of a sherd from Belsh measured at RT (a), 78 K (b) and at RT after CBD-extraction (c).

hyperfine field compared to bulk hematite ($H_{hf} = 51.8$ T [9]) could be caused by a reduced particle size leading to collective magnetic excitations and/or substitution of aluminium for iron. A second component with a smaller hyperfine field (which is often better distinguishable at low temperature (Figure 2b)) presumably could be attributed to iron oxyhydroxides.

The low-temperature Mössbauer spectra show an increase of both magnetic components at the expense of the nonmagnetic Fe^{3+} components, i.e. a considerable amount of hematite *and* iron oxyhydroxides has particle sizes small enough to show superparamagnetism. The increase of the magnetic pattern with the smaller hyperfine field is especially strong in the case of two sherds excavated in a lake

Table I. Corresponding hyperfine parameters of the sample from Belsh

Sample	Fe^{3+} (doublets)		Fe^{2+} (doublet)		Fe^{3+} (sextet 1)		Fe^{3+} (sextet 2)		A_{mag} [%]	Fe^{3+}/Fe^{2+}
	$\langle\delta\rangle$ [mm/s]	$\langle\Delta\rangle$ [mm/s]	δ [mm/s]	Δ [mm/s]	δ [mm/s]	H_{hf} [T]	δ [mm/s]	H_{hf} [T]		
RT	0.38	0.82	0.99	2.53	0.36	50.7	0.36	46.8	19.1	2.71
78 K	0.50	0.91	1.22	2.80	0.49	53.3	0.44	49.7	28.6	1.98
CBD extracted	0.39	0.81	1.00	2.52	0.36 f.	50.7 f.	0.36 f.	46.8 f	9.8	2.39

δ – isomer shift (relative to iron metal), Δ – quadrupole splitting, H_{hf} – magnetic hyperfine field, $\langle\rangle$ – area-averaged value, A_{mag} – area of the magnetic fraction, Fe^{3+}/Fe^{2+} – Fe^{3+}/Fe^{2+} ratio of the nonmagnetic components, f. – fixed.

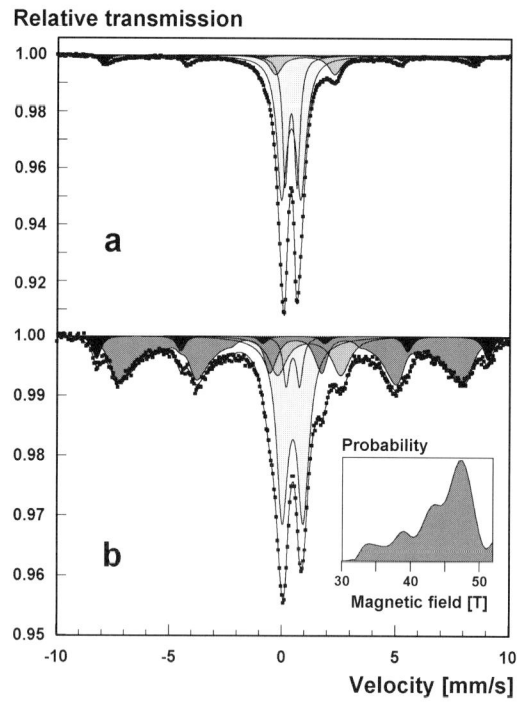

Figure 2. Mössbauer spectra of a sherd found in a lake near Belsh measured at RT (a) and 78 K (b).

(region of Belsh) (Figures 2a and 2b). This increasing occurrence of iron oxyhydroxides is an indication of a strong weathering process during the burial in a moist environment.

The predominant phase found by powder diffractometry was α-quartz; minor components were feldspars, micas, etc. The iron-containing phases have too small concentrations and/or too small particle sizes to be seen. Therefore, the both methods complement each other.

4. Discussion

The mineralogical composition of the sherds is determined by the raw materials used for the production of the ancient ceramics, the firing procedure, and the weathering conditions. The firing technique involves many parameters like the original firing temperature, the duration of firing, and the kiln atmosphere. Three main types can be distinguished: (i) firing under oxidizing conditions in an open furnace, (ii) firing under reducing conditions in a closed furnace, (iii) firing under reducing conditions, but with an intentional or accidental oxidising step at the end of the firing cycle.

The most relevant Mössbauer parameters in this context are the quadrupole splitting of the Fe^{3+} pattern ($\langle\Delta\rangle$), the total fractional area of the nonmagnetic Fe^{3+} and Fe^{2+} components (A_{nm}) and the fraction of Fe^{2+} ($A_{Fe^{2+}}$) [4]. For a first attempt to reconstruct the ancient Albanian firing technique the results of the extensive studies of Wagner et al. [4–7] are used: When clay is directly fired in air (process (i)) a typical "camel's back" variation of the Fe^{3+} quadrupole splitting relative to the firing temperature was observed with a maximum value up to 1.4 mm/s around 700°C. The strong increase of the quadrupole splitting compared to fresh clay is due to the distortion of the symmetry of the iron sites caused by dehydroxilation of the octahedral layers in the clay minerals. Firing in air after a preceding reduction (process (iii)), however, is reflected by nearly constant Fe^{3+} quadrupole splittings around 0.9 mm/s. The relatively small $\langle\Delta\rangle$ values around 0.9 mm/s for the Fe^{3+} species measured in the Albanian sherds together with the occurrence of a relatively small but observable Fe^{2+} fraction (Table I) are a strong indication of a more or less moderate oxidation process after a preceding reduction. For a more detailed reconstruction of Albanian firing techniques refiring experiments of appropriate ancient sherd samples at different temperatures with subsequent characterization by Mössbauer spectroscopy are in progress.

5. Conclusion

In the present study, the following iron-bearing phases could be identified by Mössbauer spectroscopy in sherds from three different Albanian excavation sites: structural Fe^{3+} and Fe^{2+} in the siliceous clay matrix as well as hematite and iron oxyhydroxide particles which partly are small enough to exhibit superparamagnetism. The Mössbauer parameters especially the quadrupole splitting of the Fe^{3+} species at RT indicate for the examined samples a firing process under oxygen-deficient conditions with a more or less moderate oxidation step at the end of the cycle.

Acknowledgements

It is a pleasure to thank J. Ludwig for the XRD measurements and Dr. O. Krentz, D. Schnabel, and U. Tromsdorf for assistance in the sample preparation. We also are very thankful for the financial support of the International Atomic Energy Agency.

References

1. Simopoulos, A., Kostikas, A., Sigalas, I., Gangas, N. H. and Moukarika, A., *Clay Clay Miner.* **23** (1975), 393.
2. Eissa, N. A., Gomaa, S. Sh., Hassan, M. Y. and Sallam, H. H., *Hyp. Interact.* **41** (1988), 775.
3. El-Hage, Y., Wagner, U., Wagner, F. E., Riederer, J., Echt, R. and Hachmann, R., *Hyp. Interact.* **57** (1990), 2173.
4. Wagner, U., Wagner, F. E., Stockklauser, A., Salazar, R., Riederer, J. and Kaufmann-Doig, F., *Hyp. Interact.* **29** (1986), 1113.
5. Wagner, U., Gebhard, R., Murad, E., Grosse, G., Riederer, J., Shimada, I. and Wagner, F. E., *Hyp. Interact.* **110** (1997), 166.
6. Wagner, U., Gebhard, R., Grosse, G., Hutzelmann, T., Murad, E., Riederer, J., Shimada, I. and Wagner, F. E., *Hyp. Interact* **117** (1998), 323.
7. Wagner, U., Gebhard, R., Häusler, W., Hutzelmann, T., Riederer, J., Shimada, I., Sosa, J. and Wagner, F. E., *Hyp. Interact.* **122** (1999), 163.
8. Mehra, O. P. and Jackson, M. L., *Clays Clay Miner.* **7** (1960), 317.
9. Murad, E., *Hyp. Interact.* **111** (1998), 251.

Hyperfine Interactions **156/157**: 459–464, 2004.
© 2004 *Kluwer Academic Publishers. Printed in the Netherlands.*

Variation of Some Physical Properties of Brownmillerite Doped with a Transition Metal Oxide

M. Y. HASSAAN*, F. M. EBRAHIM and S. H. SALAH
*Physics Department, Faculty of Science, Al-Azhar University 11884-Cairo, Egypt;
e-mail: Yousry@tedata.net.eg*

Abstract. Cement clinker is the main component of Portland cement. It is composed of four main phases. One of them is the brownmillerite or the ferrite phase of cement clinker. It is prepared according to the formula $(4CaO)(Al_2O_3)(Fe_2O_3)_{1-x}(M)_x$, where M represents transition metal oxides (TMO): TiO_2, Cr_2O_3, Mn_2O_3 and WO_3, where $x = 1, 2, 3, 4$ and 5 mol%. Each mixture was fired at 1300°C for 30 minutes in a platinum crucible. The samples were pulverized for Mössbauer spectroscopy, X-ray diffraction and a.c. conductivity measurements. A shift in the position of the characterized peaks of pure brownmillerite appears in the X-ray diffraction patterns of brownmillerite doped with a transition metal oxide. The a.c. conductivity showed a maximum value for the samples containing 3 mol% TiO_2, Cr_2O_3 and Mn_2O_3, and 2 mol% WO_3. The Mössbauer parameters for the sample containing 5 mol% M showed a gradual increase in the isomer shift values. The number of electrons in d-orbital for the doped transition atoms, as the nearest neighbor atoms increased from 2 to 5 electrons. The hyperfine magnetic field at Fe^{3+} (Oh) iron nucleus decreases with increasing M content. This may be due to the decrease of the particle size of brownmillerite.

Key words: brownmillerite, physical properties, cement, Mössbauer.

1. Introduction

It is well known that cement clinker is composed of four main phases which are C_2S, C_3S, C_3A and C_4AF**. These phases produced from oxides of calcium, silicon, aluminum and iron which are found naturally in limestone, shale, clay and basalt rock. Due to the presence of other compounds in these raw materials, the produced cement clinker may contain some impurities which affect its forming temperature and its characteristics. These impurities may be alkali oxides, fluorides, sulfides or transition metal oxides [1–7]. Alkalis in clinker phases modify their crystal structure, affect the clinkering process and deteriorate the cement characteristics [5]. This work is concerned with the effect of some transition metal ions as impurities in brownmillerite (C_4AF).

---

* Author for correspondence.
** Conventional cement chemistry notations are used: C = CaO, S = SiO_2, A = Al_2O_3 and F = Fe_2O_3.

2. Experimental

The samples under investigation are pure brownmillerite (C$_4$AF) and twenty samples of brownmillerite with different additions of different TMO. Pure C$_4$AF sample was prepared according to the formula (4CaO)(Al$_2$O$_3$)(Fe$_2$O$_3$). The resulting mixture was ground to a very fine powder then fired in a platinum crucible at 1300°C for 30 minutes.

The crucible was withdrawn from the furnace and the sample was ground, then returned again for another 30 minutes. Each sample was quenched in air and ground again. The four groups of brownmillerite samples doped with different TMO according to the formula C$_4$AF$_{(100-x)}$M$_x$, where M represents TiO$_2$, Cr$_2$O$_3$, Mn$_2$O$_3$ and WO$_3$, and $x = 1, 2, 3, 4$, and 5 mol%. These samples were prepared by the same previous method. The samples were measured using Philips analytical X-ray diffraction system, using CuK$_\alpha$ radiations. The a.c. conductivity $\sigma(\omega)$, was obtained by varying the temperature between 340 K and 790 K using a computer controlled Stanford LCR Model SR 720, at four different frequencies namely 0.12, 1, 10 and 100 kHz. The measurements were done with the four points prop method by using a silver paste on two flat surfaces of the sample.

The Mössbauer measurements were performed at room temperature using a 512 channels analyzer. Co-57 source in Rh matrix was used. Isomer shift are given relative to Fe foil.

3. Results and discussion

3.1. X-RAY DIFFRACTION DATA

The pure ferrite phase formation is confirmed by the X-ray diffraction where its characterizing peaks of diffraction appeared in the diffraction pattern. Doping the TMO with C$_4$AF shifted these peaks to lower 2θ, indicating a slight distortion in the structure of C$_4$AF due to doping the TMO.

3.2. A.C. CONDUCTIVITY

The relation between $\ln(\sigma)$ and the reciprocal of the absolute temperature $(1/T)$ for all the investigated samples showed a linear behavior. The activation energy was calculated from the slopes of the linear relations. Figure 1 exhibit the dependence of conductivity on temperature for the ferrite phase samples doped with different TMO at different frequencies. The electrical conductivity of the present samples increased with increasing their temperature like the conductivity of semiconductors. Each relation presents two different slopes, with transition temperature T_c within a narrow range for most of the samples. At high temperature region there is no change with frequency, so we can calculate the activation energy from the relation:

$$\sigma = \sigma_o \exp(-W/KT),$$

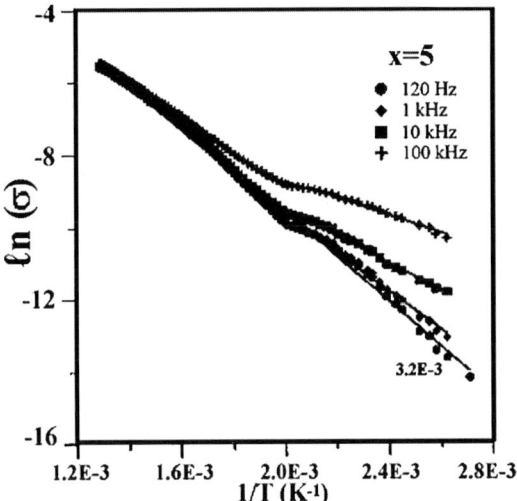

Figure 1. Representative relation between $\ln(\sigma)$ and $1/T$ of C_4AF doped with TMO at different frequencies.

Table I. Activation energies for C_4AF doped with different TMO at frequency 1 kHz

TMO	1%	2%	3%	4%	5%
$C_4AF_{100-x}(Cr_2O_3)_x$	0.461	0.482	0.531	0.579	0.485
$C_4AF_{100-x}(TiO_2)_x$	0.531	0.419	0.432	0.55	0.512
$C_4AF_{100-x}(WO_3)_x$	0.407	0.388	0.410	0.343	0.38
$C_4AF_{100-x}(Mn_2O_3)_x$	0.464	0.398	0.391	0.57	0.544

where W is the thermal activation energy. The temperature dependence of conductivity implies one activation energy in the high temperature region. The activation energy values of the different samples are listed in Table I.

A general feature for the semi-conducting behavior is the coexistence of transition metal ions in the different samples in more than one valence state. Then conduction can take place by hopping electrons from low to high valence state [9]. Electrical conductivity of pure C_4AF results from both electronic and ionic conduction.

3.3. MÖSSBAUER DATA

Representative Mössbauer spectra of $C_4AF_{100-x}M_x$ samples, shown in Figure 2, exhibit four sub spectra. The inner doublet represents Fe^{3+} in tetrahedral (Td) site and the outer doublet Fe^{3+} in octahedral (Oh) site, whereas the inner and outer sextets can also be assigned to Fe^{3+} (Td) and Fe^{3+} (Oh), respectively. The Mössbauer parameters are listed in Tables II, III.

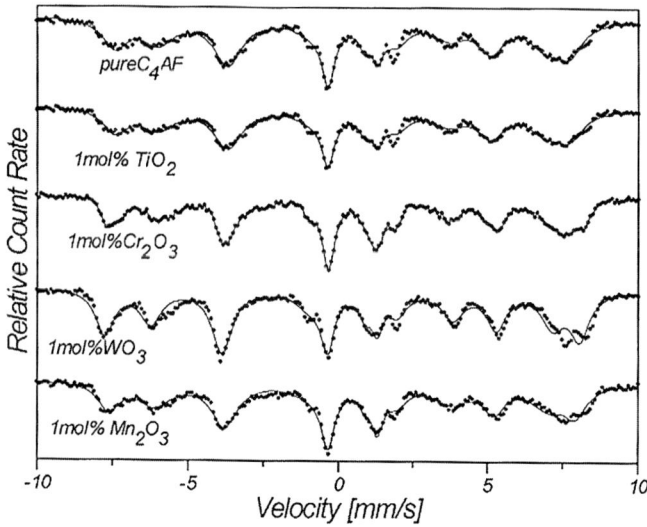

Figure 2. Representative Mössbauer spectra of C_4AF doped with 5 mol% TMO.

Table II. Isomer shift values δ for C_4AF doped with different TMO

C_4AF doped with TMO	δ (mm s^{-1}) for paramagnetic sites		δ (mm s^{-1}) for ferrimagnetic sites	
	Fe^{3+} (Td)	Fe^{3+} (Oh)	Fe^{3+} (Td)	Fe^{3+} (Oh)
Pure C_4AF	0.103	0.369	0.104	0.318
1 mol% TiO_2	0.167	0.33	0.104	0.318
1 mol% Cr_2O_3	0.103	0.369	0.104	0.318
1 mol% Mn_2O_3	0.103	0.369	0.104	0.318
1 mol% WO_3	0.132	0.377	0.176	0.342
5 mol% TiO_2	0.058	0.267	0.14	0.31
5 mol% Cr_2O_3	0.0717	0.303	0.1	0.269
5 mol% Mn_2O_3	0.143	0.338	0.125	0.324
5 mol% WO_3	0.093	0.284	0.111	0.315

The inner doublet represents the iron in tetrahedral site Fe^{3+} (Td) and the outer doublet represents the iron in octahedral sites Fe^{3+} (Oh). While the inner and outer sextets represent the ferri-magnetic Fe^{3+} (Td) and Fe^{3+} (Oh), respectively.

There are no remarkable differences in the isomer shift values for C_4AF samples that doped with 1 mol% TMO and the pure C_4AF. Most likely 1 mol% doping is not enough to produce an observable changes as the atoms of the doped transition metal oxides lay in the interstitial positions and substitution process does not take place yet. While substitution of TMO with different charges in C_4AF still has no differences due to its very small contribution. On the other hand, the isomer shift values for the samples doped with 5 mol% are lower than those of the 1 mol%

Table III. Quadrupole splitting Δ, quadrupole shift ε and magnetic field H for C_4AF doped with different TMO

C_4AF doped with TMO	Δ (mm s^{-1}) for paramagnetic		ε (mm s^{-1}) for ferrimagnetic		H (Oe) for	
	Fe^{3+} (Td)	Fe^{3+} (Oh)	Fe^{3+} (Td)	Fe^{3+} (Oh)	Fe^{3+} (Td)	Fe^{3+} (Oh)
Pure C_4AF	1.264	1.59	0.486	−0.58	404	478
1 mol% TiO_2	1.139	1.63	0.48	−0.58	406	480
1 mol% Cr_2O_3	1.264	1.59	0.486	−0.58	404	478
1 mol% Mn_2O_3	1.264	1.59	0.486	−0.58	408	482
1 mol% WO_3	1.29	1.58	0.49	−0.545	420	495
5 mol% TiO_2	1.167	1.57	0.65	−0.56	404	477
5 mol% Cr_2O_3	1.32	1.6	0.556	−0.6	395	469
5 mol% Mn_2O_3	1.31	1.59	0.491	−0.61	398	473
5 mol% WO_3	1.39	1.61	0.499	−0.55	400	477

doped ones which can be explained by the increase in the intensity. The samples which are include 5 mol% (Cr_2O_3, TiO_2, Mn_2O_3 and WO_3) gave a gradual increase in the isomer shift values, as the number of electrons in (d) orbitals, for the transition metal atoms, as nearest neighbor atom, increases from 2 electrons, in case of Ti atom, to 4 electrons in case of Cr and W atoms, and to 5 electrons in case of Mn atoms. The quadrupole splitting values of the two sextets for all the samples have positive value except that of Fe^{3+} (Oh).

The negative sign of quadrupole shift comes from the angle θ between the direction of the hyperfine field and the principle axis of the axial symmetric electric field gradient tensor [10].

It has been observed that no effect on values of quadrupole splitting of Fe^{3+} (Oh) while there is a slight effect on Fe^{3+} (Td), this is may be due to the tendency of Al^{3+} (Td) to substitute Fe^{3+} (Td), because the ionic radius of the first one is slightly smaller than that of the second one.

From Table III it was observed that for all the investigated samples, the nuclear magnetic field is considerably smaller for the Fe^{3+} (Td) ions than Fe^{3+} (Oh) ions. This difference cannot be attributed only to the slight difference in the covalency of the tetrahedral ions, but it is also due to the nature of the intersuplattice magnetic bands.

The hyperfine magnetic field at Fe^{3+} (Oh) nucleus decreases with increasing the transition metal percentage indicating a decrease of the particle size of C_4AF phase [7].

The hyperfine magnetic field intensity H for the samples doped with 5 mol% (Cr_2O_3, Mn_2O_3 and WO_3) were slightly less than that doped with 1 mol%, whereas the two samples doped with 1 and 5 mol% TiO_2 have no different hyper fine fields. This may be due to a little distortion in orientation happened to oriented electron of iron ions in ferromagnetic crystals. The added metal ions tend to substitute Fe^{3+} (Td), the hyperfine field at the Fe^{3+} (Oh) ions nucleus decreases with an increase of the number of added transition metal in the tetrahedral site which is the nearest neighbor to the iron nucleus [11].

4. Conclusion

The characteristic peaks of C_1AF shifted to lower values of 2θ indicating that a slight distortion in the structure takes place. The electrical conductivity showed maximum value at 3 mol% TiO_2, Cr_2O_3 and Mn_2O_3, while it was at 2 mol% only in case of WO_3. The samples which were doped with 5 mol% TMO gave a gradual increase in the isomer shift as the number of d-electrons, of the doped transition metal atoms as nearest neighbor, increased. The quadrupole shift values of ferrimagnetic Fe^{3+} (Oh) have a negative value. The TMO have no effect on Fe^{3+} (Oh) ions, while there is a slight effect on Fe^{3+} (Td) ions indicating the tendency of the TM ions to substitute Fe^{3+} (Td). The hyperfine magnetic field at Fe^{3+} (Oh) nucleus decreases with increasing the transition metal percentage indicating a decrease of the particle size of C_4AF phase. A.C. conductivity and Mössbauer effect have high sensitivity to estimate the quality of cement clinker.

References

1. Jawed, I. and Skainy, J., *Cem. Conc. Res.* **7** (1977), 719.
2. Odler, I. and Wonnemann, R., ibid **13** (1983), 477.
3. Hassaan, M. Y., Moutawie, M. A. and Eissa, N. A., *Hyp. Interact.* **46** (1989), 739.
4. Hassaan, M. Y., Eissa, N. A. and Salah, S. H., *TIZ International Powder Magazine* **114**(5) (1990), 346.
5. Odler, I. and Abdul-Maula, S., *J. Amer. Ceram. Society* **63**(11–12) (1980), 654.
6. Odler, I. and Schmidt, O., ibid **63**(1–2) (1988), 13.
7. Hassaan, M. Y., Panteleev, D. and Salah, S. H., *Hyp. Interact.* **71** (1992), 1389.
8. Tareev, B., *Physics of Dielectric Materials*, Mir Publishers, Moscow, 1979.
9. Sayer, M. and Mansingh, A., *Phys. Rev. Rev. B* **6** (1972), 4629.
10. Bhide, V. J., *Mössbauer Effect and Its Applications*, McGraw-Hill, New Delhi, 1973, p. 182.
11. Salah, S. H., PhD Thesis, Al-Azhar University, Cairo, Egypt, 1979.

Hyperfine Interactions **156/157**: 465–469, 2004.

© 2004 *Kluwer Academic Publishers. Printed in the Netherlands.*

Iron-Containing Adsorbents in Great Nile Sediments

T. M. MEAZ[1], M. A. AMER[1] and C. BENDER KOCH[2]

[1] *Physics Department, Faculty of Science, Tanta University, Tanta, Egypt*
[2] *Chemistry Department, The Royal Veterinary and Agricultural University, Frederiksberg, Denmark*

Abstract. A sample from the Great Nile (near Khartoum) has been investigated by Mössbauer spectroscopy, infrared spectroscopy, analytical electron transmission microscopy, and X-ray diffraction aiming to evaluate potential iron-containing sinks for heavy metal adsorption. The phyllosilicate fraction is dominated by highly-oxidized illite/smectite and the oxide fraction is dominated by defective goethite and hematite (as evidenced from low ordering temperatures and line broadening effects).

Key words: iron oxides, phyllosilicates, Great Nile River, Mössbauer spectroscopy.

1. Introduction

The Great Nile River has two major water and sediment sources: the Blue Nile and the White Nile. Water and sediment transport is mainly facilitated though the Blue Nile originating in Ethiopia. Broadly speaking the Blue Nile transported sediment reflects the weatherings of the basalt dominated bedrocks and are thus expected to contain secondary iron-containing minerals. Among these, particular interest is devoted to the ultra fine iron oxides known to function as adsorbents in various geochemical settings.

The Great Nile is essentially the only waterway though the intensely populated Nile Valley and functions as source of water and recipient of pollutants. Thus the adsorption properties and the dynamics of the adsorbents becomes a crucial point in understanding and evaluating the Great Nile geochemical system. Mössbauer spectroscopy has previously been applied to biogeochemical problems involving Fe [1–11].

We here report a study of the sediments entering the Great Nile River system by investigating a sample from the River bed sampled at Khartoum, Sudan, i.e. after the confluence of the Blue and White Rivers.

2. Material and experimental

A sediment sample was obtained from the Great Nile River bed downstream of Khartoum after the confluence of the Blue and White Nile in Sudan. The sample

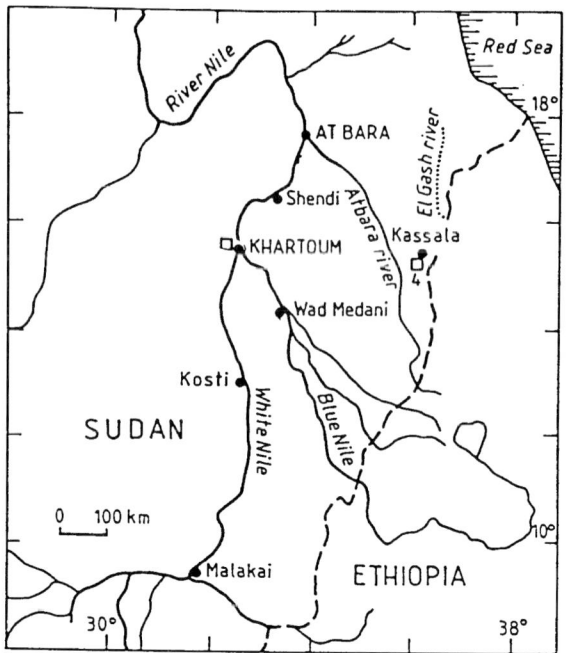

Figure 1. Sampling site from the Great Nile River near Khartoum, Sudan.

was dried at room temperature prior to analysis by X-ray powder diffraction, analytical transmission electron microscopy, infrared spectroscopy and Mössbauer spectroscopy. The sampling location is shown in Figure 1. Mössbauer spectra were obtained using a constant acceleration spectrometer with 50 mCi ^{57}Co in Rh. The spectrometer was calibrated using a thin foil of natural Fe at room temperature. Spectra were measured between RT and 5 K.

The transmission electron microscope was equipped with an X-ray energy dispersive detector allowing quantifying elements with Z larger than or equaling to Na.

3. Results and discussion

The X-ray diffraction pattern of the random powder sample is shown in Figure 2. It shows a number of sharp and broad diffraction lines caused by differences in crystal size and perfection. Among the sharp lines peaks due to the primary minerals quartz (Q) and feldspars (F) are detected, and among the broad lines the secondary minerals smectite (Sm), kaolinite (K) and illite (I) can be identified (common lines designated to phyllosilicates (P)). Infrared spectroscopy (not shown) verified kaolinite, illite, and smectite as the dominant phyllosilicates. The iron oxides are present only below the detection limits of these methods.

Analytical transmission electron microscopy revealed highly fluctuating analysis as expect from such a multi component system. The element analysis by EDX revalued the presence of Na, Mg, Al, Si, K, Ca, Ti and Fe and the results from dif-

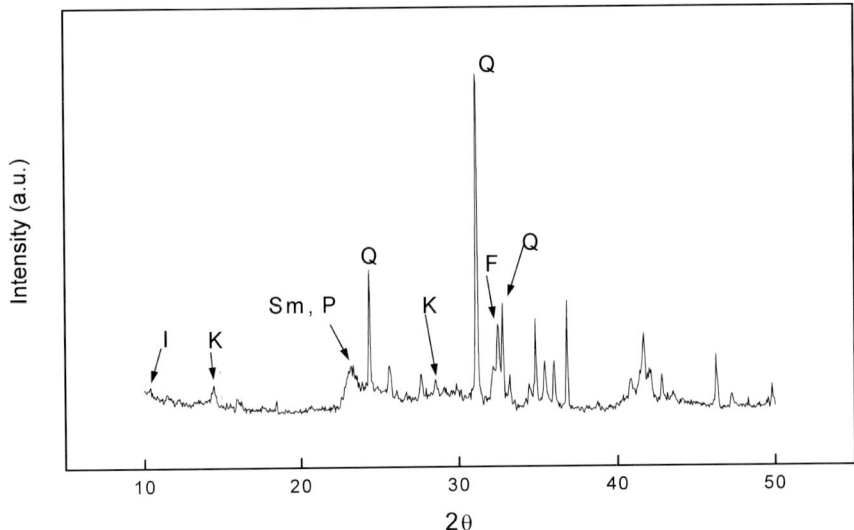

Figure 2. X-ray powder diffraction pattern of the Great Nile River sample (Co K$_\alpha$).

Table I. Major elements from analysis of four composite particles

Element	L1 (wt%)	L2 (wt%)	L3 (wt%)	L4 (wt%)	Av. (wt%)	St. dev.
Na	3.11	4.69	4.58	–	4.13	0.72
Mg	1.85	3.68	1.06	5.11	2.92	1.57
Al	7.69	14.74	25.55	20.33	17.0	6.63
Si	80.06	45.43	48.47	48.31	55.57	14.19
K	0.26	0.88	1.08	2.02	1.06	0.63
Ca	3.82	17.74	15.69	8.68	11.48	5.55
Ti	–	3.19	0.31	0.94	1.48	1.23
Fe	3.18	9.60	3.21	14.58	7.64	4.78

ferent particles are given in Table I. It is interesting to note the ubiquitous presence of Fe in point analysis of quartz (approx. 3 wt%) and fairly high contents of Fe in the phyllosilicates (15 wt%). No iron oxides were detected as separate particles. These observations indicate an intimate association of the iron oxides with both the primary and secondary silicates, probably dominantly via surface associations.

The Mössbauer spectra of the sample are shown in Figure 3. At all temperatures magnetically ordered sextets and ferric and ferrous doublets coexists. The very small amount of ferrous components detected points to weathering in a highly oxidized environment in the basalt. The relative intensity of the magnetically ordered components increases with lowering of the temperature, indicating influence by substitution and size effects on the spectra. The parameters of the ferric and ferrous doublet at 5 K (isomer shifts of 0.47 mm s^{-1} and quadrupole splitting

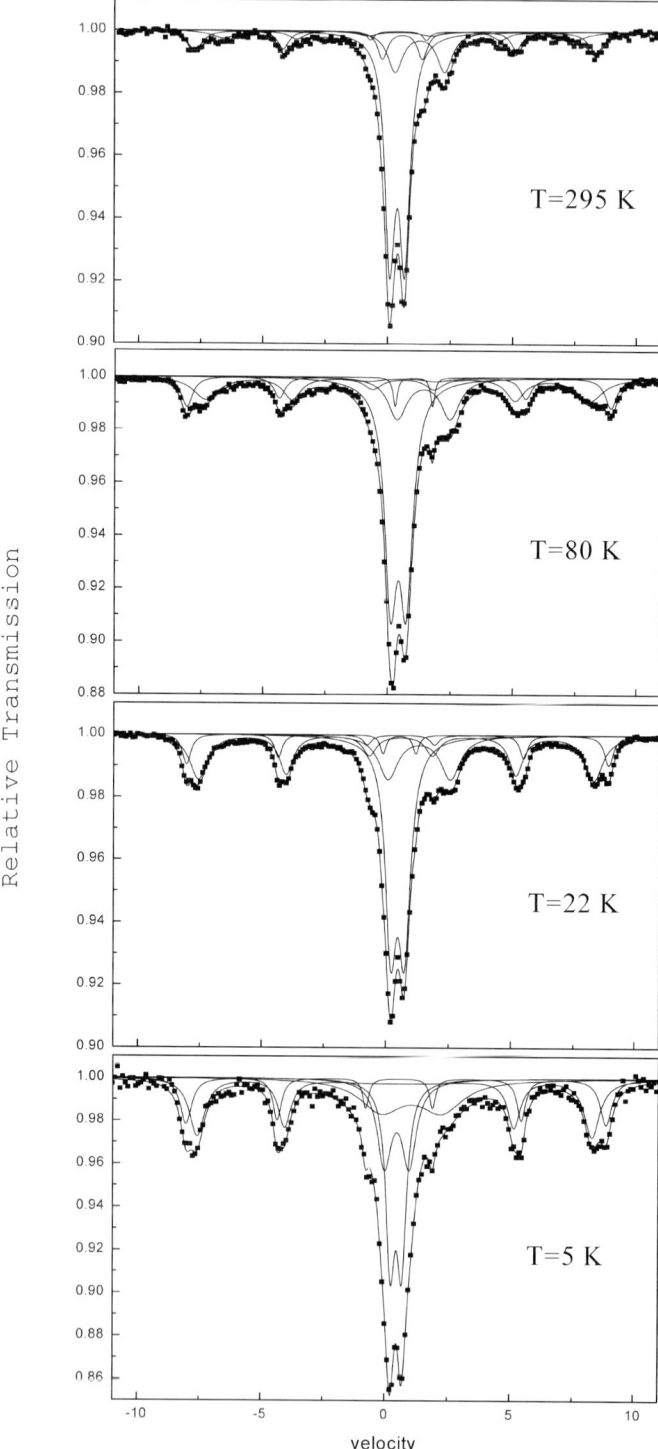

Figure 3. Mössbauer spectra of the Great Nile River sample obtained at the indicated temperatures.

of 0.57 mm s^{-1} for the ferric component and isomer shifts of 1.32 mm s^{-1} and quadrupole splitting of 2.57 mm s^{-1} for the ferrous component) are within the parameters previously reported for phyllosilicates [9]. At 5 and 20 K the magnetically ordered part seems to consist of two overlapping sextets with fairly sharp lines. Magnetic hyperfine fields of 48.8 and 51.7 T, respectively, indicate the presence of substituted hematite (α-Fe$_2$O$_3$) and goethite (α-FeOOH) in almost equal amounts as indicated from the intensities. The total amount of Fe approx. 50% is in the oxide fraction. At higher temperatures strong asymmetries develop in the lines of both components. At 80 K in particular the low-field component (goethite) shows extensive line broadening. The finding of magnetic hyperfine field reduction together with lowering of ordering temperature and absence of a Morin transition indicates that the broadening is related to the substitution in the lattice. Moreover, it is known that Al substituted iron affect the Morin transition. At room temperature the hematite component consists of two components: a rather sharp lined one causing the appearance of a marked maximum and another one causing asymmetric lines. This suggests the possible presence of hematite with two different levels of substitution/crystal sizes. We suggest tentatively that the well crystalline hematite is inherited directly from the basaltic rock, whereas the poorly crystalline hematite formed during weathering.

4. Conclusion

The fine-grained iron oxide in the sediments from the Great Nile has been identified in this pilot study by Mössbauer spectroscopy. We find that goethite and hematite are present in almost equal amounts. Both minerals are expected to be highly surface reactive due to their dominantly small crystal size.

References

1. Mijovilovich, A., Morras, H., Causevic, H. and Saragovi, C., *Hyp. Interact.* **122** (1999), 83.
2. Muggler, C. C., Van Loef, J. J., Buurman, P. and Van Doesburg, J. D. J., *Geoderma* **100** (2001), 147.
3. Rancourt, D. G., *Hyp. Interact.* **117** (1998), 3.
4. Murad, E. *Hyp. Interact.* **117** (1998), 39.
5. Bender Koch, C., *Hyp. Interact.* **117** (1998), 131.
6. Sahi Ram, Patel, K. R., Sharma, S. K. and Tripathi, R. P., *Fuel* **76**(14/15) (1997), 1369.
7. Murad, E. *Hyp. Interact.* **111** (1998), 251.
8. Bancroft, G. M., *Mössbauer Spectroscopy – an Introduction for Inorganic Chemists and Geochemists*, McGraw-Hill, London, 1973, p. 232.
9. Coey, J. M. D., In: G. J. Long (ed.), *Mössbauer Spectroscopy Applied to Inorganic Chemistry*, Vol. 1, Plenum Press, New York, 1984, p. 443.
10. Russo, U., Carbonin, S. and Della Giusta, A., In: G. J. Long and F. Grandjean (eds), *Mössbauer Spectroscopy Applied to Magnetism and Material Science*, Vol. 2, Plenum Press, New York, 1996, p. 207.
11. Eissa, N. A., Gomaa, Sh. S., Hassan, M. Y. and Sallam, H. A., *Hyp. Interact.* **41** (1988), 775.

Hyperfine Interactions **156/157**: 471–485, 2004.
© 2004 *Kluwer Academic Publishers. Printed in the Netherlands.*

Fe(II–III) Hydroxysalt Green Rusts;
from Corrosion to Mineralogy and Abiotic to Biotic
Reactions by Mössbauer Spectroscopy

J.-M. R. GÉNIN
*Laboratoire de Chimie Physique et Microbiologie pour l'Environnement, UMR 7564
CNRS-Université Henri Poincaré-Nancy 1, Equipe Microbiologie et Physique and Département
Matériaux et Structures, ESSTIN, 405 rue de Vandoeuvre, F-54600 Villers-lès-Nancy, France*

Abstract. Fe(II)–Fe(III) hydroxysalts commonly called green rusts are layered double hydroxides of formula $[Fe^{II}_{(1-x)}Fe^{III}_x(OH)_2]^{x+} \cdot [(x/n)A^{n-} \cdot (m/n)H_2O]^{x-}$ constituted of brucite-like layers containing Fe cations in the centres of OH^- octahedrons and interlayers, which anions and water molecules belong to. They play a key role in corrosion and environmental sciences as well as mineralogy since they are, on the one hand, intermediate products between Fe(II) and Fe(III) states and, on the other hand, can be the major iron-bearing mineral in hydromorphic gley soils. Their crystal structure, Mössbauer spectra, methods of synthesis, abiotic as well as biotic, and some applications are presented here.

Key words: Fe(II–III) hydroxysalt green rust, LDH, corrosion, mineralogy.

1. Introduction

Layered double hydroxides (LDH) are usually composed of di- and trivalent metallic cations of different elements such as Mg, Co, Ni or Fe. Of special interest are those that comprise only Fe(II) and Fe(III) ions and that are called green rusts (GRs) because of their celadon-like blue to green colour. These mixed-valence Fe hydroxides appeared on oxidation of $Fe(OH)_2$ or by mixing Fe(II) and Fe(III) hydroxides. They were known long ago [1–3] but it is Feitknecht [4] that understood first the nature of layered hydroxides followed by Bernal *et al.* [5] who studied their crystal structure and coined the term "green rust". They have been found as intermediate transient compounds during the corrosion of iron-based materials but also during iron oxide transformations in hydromorphic gley soils when appropriate redox conditions are met such that both di- and trivalent oxidation states of Fe can exist. Coupling to respiration and therefore to biological processes causes the reduction of Fe(III) to Fe(II) in natural anoxic environments. In geological conditions GRs are able to reduce pollutants such as nitrates, chlorinated solvents, nitroaromatic compounds, selenate, chromate and may cause sorption of metal cations or anions essentially oxyanions, through precipitation and the role of anion exchange is open to discussion. Anyhow, GRs intervene in transitions between anoxic and oxic conditions determining the nature of Fe oxides that form. Finally, Mössbauer spectroscopy (MS) is the ideal tool for characterising the iron (oxyhydr)oxides especially since they belong often to a mixture of Fe(II) and Fe(III) states.

2. Structure, composition and characterisation

In early days, Fe(II–III) hydroxides were assumed to be some kind of hydrated magnetite, $Fe^{II}Fe^{III}_2(OH)_8$, often called basic ferroferrites or ferroso-ferric hydroxides [2, 3, 6–9]. Working in a chlorinated medium, Feitknecht and Keller [10] made the first systematic study showing that the green hydroxides, which are formed on oxidation of $Fe(OH)_2$ were layered compounds with variable compositions; the structure was obtained by shifting layers of $Fe(OH)_2$ and FeOCl, xH_2O; this model persisted untill the 90s with some slight modifications [11, 12]. In the meantime, Bernal *et al.* [5] distinguished two types of GRs from XRD patterns; green rust one (GR1) with a rhombohedral unit cell, and green rust two (GR2) with a hexagonal cell. If indexed in the hexagonal system, GR1 has three brucite-like hydroxide layers in the repeat c unit, whereas GR2 has only one, which correspond to polytypes $3R$ and $1H$, respectively, according to the scheme of Bookin and Drits [13]. Stampfl showed that the powder pattern of hydroxycarbonate $GR1(CO_3^{2-})$ was the same as that of pyroaurite [14]. Here, the structure of two typical GR1, i.e. the hydroxychloride and -carbonate, $GR1(Cl^-)$ and $GR1(CO_3^{2-})$, respectively, and the prototype $GR2(SO_4^{2-})$, the hydroxysulphate, is proposed.

The stacking sequences in GR1 and GR2 can be understood by starting with the structure of $Fe(OH)_2$. The sequence is $AcBVA\ldots$ where A and B are the hexagonal layers of OH^- ions and c that of the Fe(II) ions lying in the centres of the octahedral sites, whereas V represents vacant sites. The stacking of OH^- ions is thus hexagonal close-packed. In GR1, vacant sites are filled with anions such as Cl^- or CO_3^{2-} ions and water molecules that must be counterbalanced by Fe(III) ions in the brucite-like hydroxide layers; the stacking sequence becomes $AcBiBaCjCbAkA\ldots$ where A, B and C represent the OH^- ion layers, a, b, and c the Fe ion layers, and i, j and k the intercalated anion layers. Space group of GR1 is thus $R\bar{3}m$.

In contrast, the structure of $GR2(SO_4^{2-})$ was proposed recently from Rietveld analysis and space group set as $P\bar{3}m1$ [15]. The OH^- ion layer stacking remains hcp as in $Fe(OH)_2$, whereas there exist two anion interlayers. Thus, the overall sequence can be written $AcRijA\ldots$ Figure 1 displays the stacking sequences of the hydroxy-chloride, -carbonate and -sulphate. The c parameters along the hexad axis are 2.385(6), 2.23(6) and 1.1011(3) nm for $GR1(Cl^-)$ [16], $GR1(CO_3^{2-})$ [17] and $GR2(SO_4^{2-})$ [15], respectively. Parameter a of the hexagonal packing is 0.3190(1) and 0.317(1) nm for $GR1(Cl^-)$ [16] and $GR1(CO_3^{2-})$ [17], respectively, which must be compared to the a_0 parameter of 0.326 nm in $Fe(OH)_2$, whereas a long range order exists among Fe(III) cations and SO_4^{2-} anions in $GR2(SO_4^{2-})$ [15] leading to $a = 0.5524(1)$ nm i.e. $\sim a_0 \times \sqrt{3}$. In contrast, no long-range order was observed among Fe(III) cations or intercalated anions in GR1s. However, more detailed mod-

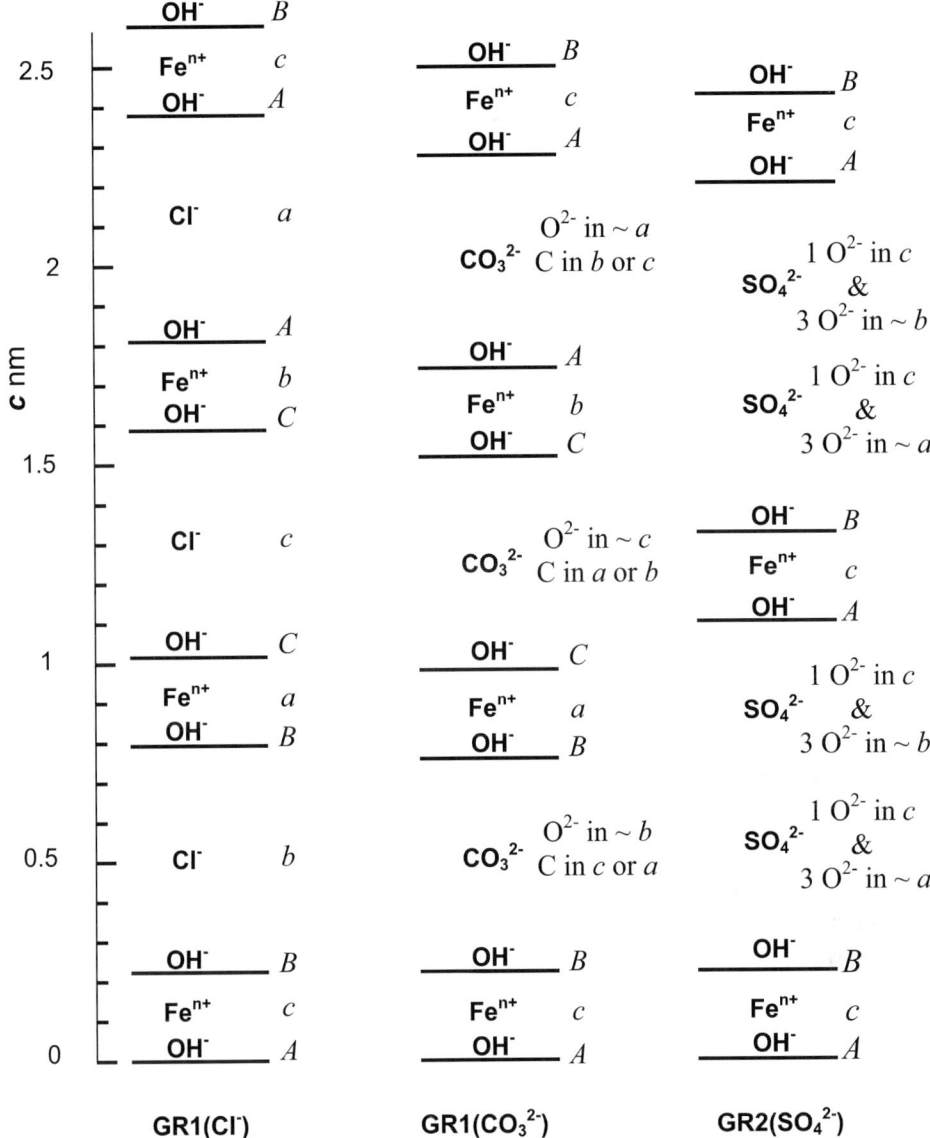

Figure 1. Stacking sequences of GR1(Cl$^-$), GR1(CO$_3^{2-}$) and GR2(SO$_4^{2-}$) along the *c* axis.

els can be developed for GR1 as illustrated in Figure 2 where are represented (001) projections of the successive layers and interlayers. Rules are simple: (i) halogen anions, here Cl$^-$, are surrounded by empty sites in the hexagonal interlayer for steric reasons whereas the negative charge is shared by 2 Fe(III) cations, one above and one below, (ii) CO$_3^{2-}$ occupy 3 sites but only 2 Fe(III) cations out of 3 ions counterbalance the anion charge inducing that the Fe ion in register above or below the carbon is left as a Fe(II) ion.

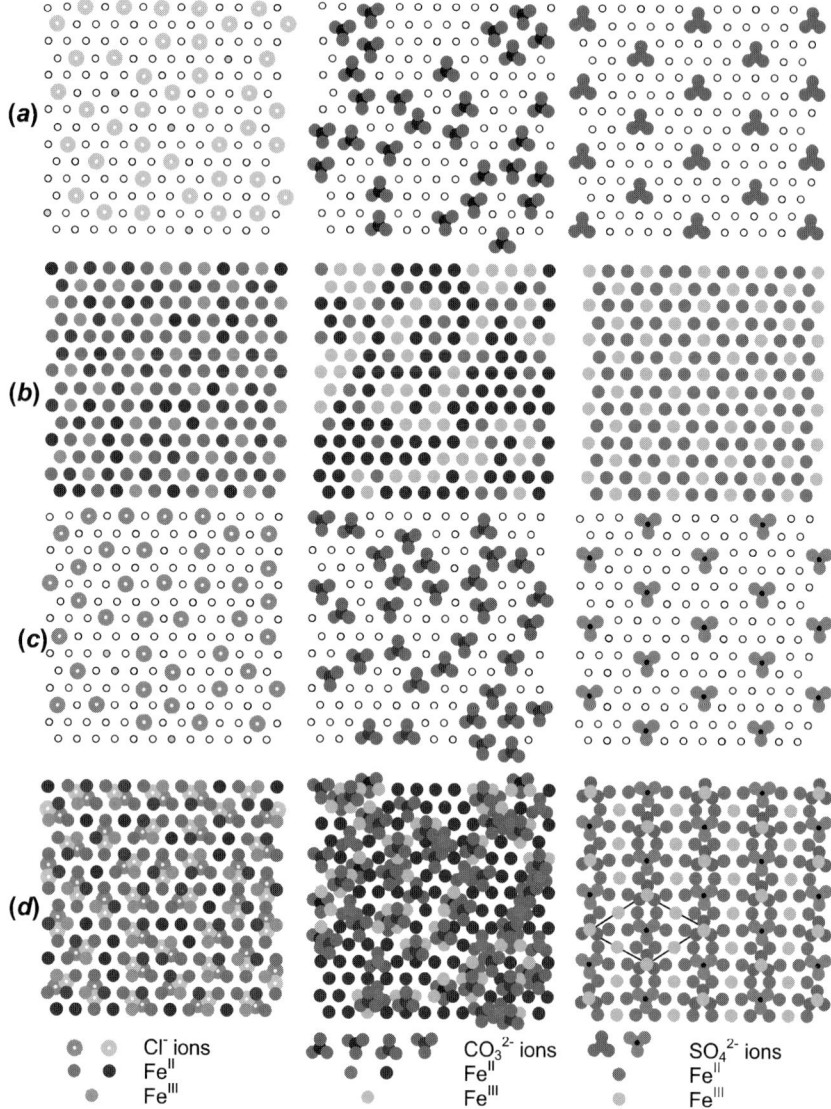

Figure 2. (001) projections perpendicular to the c axis of the GR structure of, from left to right, GR1(Cl$^-$), GR1(CO$_3^{2-}$), and GR2(SO$_4^{2-}$). OH$^-$ ion layers are not taken into account and only 3 adjacent layers made of one Fe layer in the way between two anion interlayers are shown for spotting the Fe sites seen in Mössbauer spectra. (a) The first anion interlayer. (b) One Fe layer with Fe(II) and Fe(III) cations. (c) The next anion interlayer. (d) Superimposition of (a), (b) and (c). Water molecules in interlayers are not represented.

From these models, $x = $ [Fe(III)]/[Fe$_{\text{total}}$] must be limited by the upper value 1/3 for GR1(Cl$^-$) [18] whereas $1/3 \leqslant x \leqslant 2/3$ for GR1(CO$_3^{2-}$). In contrast, ordering cations in GR2(SO$_4^{2-}$) yields a definite formula set at $x = 1/3$ [15, 19]. Therefore, the general formula of a GR is [Fe$^{\text{II}}_{(1-x)}$Fe$^{\text{III}}_x$(OH)$_2$]$^{x+}$ · [(x/n)A^{n-} ·

(m/n) H$_2$O]$^{x-}$ where A^{n-} is the inserted anion; this becomes for GR1(Cl$^-$) [Fe$^{II}_{\sim 3}$ Fe^{III}(OH)$_8$]$^+$ · [Cl$^-$ · y H$_2$O]$^-$ [18], whereas it is for GR1(CO$_3^{2-}$) [Fe$^{II}_{(1-x)}$Fe$^{III}_x$ (OH)$_2$]$^{x+}$ · [$(x/2)$CO$_3^{2-}$ · z H$_2$O]$^{x-}$ where $1/3 \leqslant x \leqslant 2/3$ with $z \leqslant [1 - (3x/2)]$ [24], and for GR2(SO$_4^{2-}$) [Fe$^{II}_4$Fe$^{III}_2$(OH)$_{12}$]$^{2+}$·[SO$_4$·$\sim$8H$_2$O]$^{2-}$ [15, 19].

Gancedo *et al.* published the first Mössbauer spectrum of GR1(CO$_3^{2-}$) [20]. Since then, MS was used to characterise GRs. Figure 3 displays spectra measured at 77 or 20 K of GR1(Cl$^-$) with $x = 0.25$ [11] and 0.33 [21], GR1(CO$_3^{2-}$) with $x = 0.33$ [17] and GR2(SO$_4^{2-}$) [19]. All spectra look alike and are constituted of one or two quadrupole doublets D_1 and D_2, with large splitting attributed to Fe(II) ions, and one quadrupole doublet D_3 with small splitting attributed to Fe(III) ions. Peak intensity is assumed proportional to the abundance of sites if Lamb–Mössbauer f factor correction is ignored. Hyperfine parameters (Table I) can distinguish between ferrous doublets from models of Figure 2. In GR1s some Fe(II) ions are close to anions, Cl$^-$ or CO$_3^{2-}$ (doublet D_2) whereas others are close to water molecules (doublet D_1). Figure 4 represents the relative abundance of

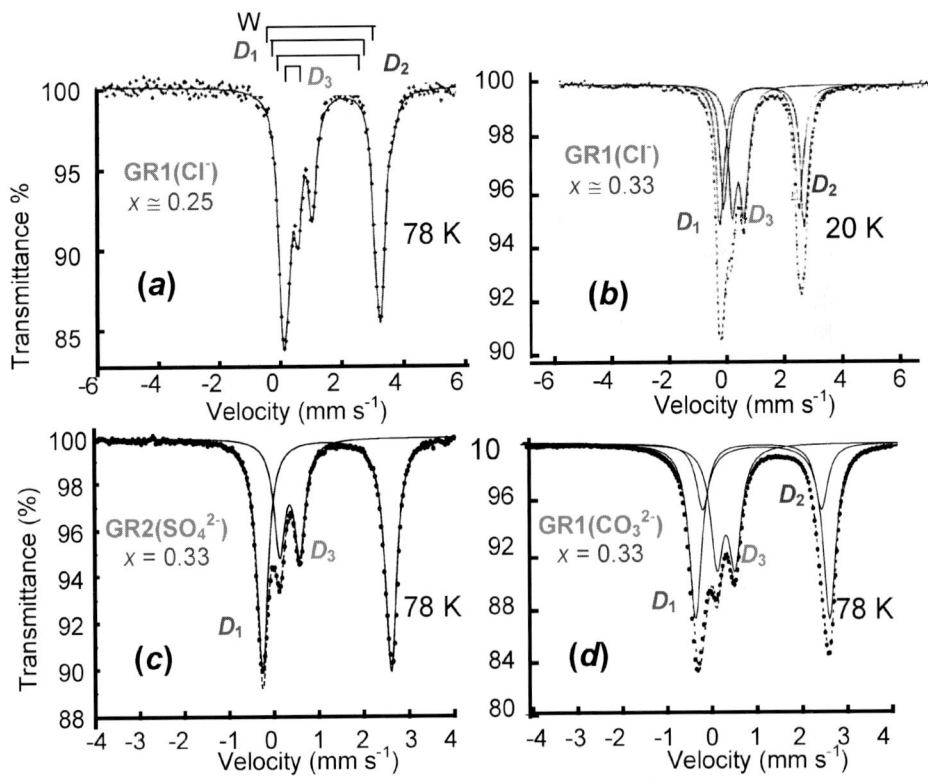

Figure 3. Mössbauer spectra recorded at 77 or 20 K of GR1(Cl$^-$) with $x = 0.25$ and 0.33 [21], GR1(CO$_3^{2-}$) with $x = 0.33$ [17] and GR2(SO$_4^{2-}$) where $x = 0.33$ [19].

Table I. Hyperfine parameters of Mössbauer spectra measured at various temperatures from Figure 3. δ: isomer shift (α-iron as reference at ambient); Δ: quadrupole splitting; *RA*: relative abundance. The temperature of measurement is noted with the corresponding reference

	GR1(Cl$^-$) $x = 0.25$			GR1(Cl$^-$) $x \cong 0.33$			GR2(SO$_4^{2-}$) $x = 0.33$			GR1(CO$_3^{2-}$) $x = 0.33$		
Ref.	[11]			[21]			[17]			[19]		
T	77 K			20 K			77 K			77 K		
	δ (mm s^{-1})	Δ (mm s^{-1})	*RA* (%)	δ (mm s^{-1})	Δ (mm s^{-1})	*RA* (%)	δ (mm s^{-1})	Δ (mm s^{-1})	*RA* (%)	δ (mm s^{-1})	Δ (mm s^{-1})	*RA* (%)
D_1	1.26	2.88	48	1.31	2.89	37	1.27	2.88	66	1.27	2.93	51
D_2	1.25	2.60	24	1.29	2.57	32				1.28	2.64	15
D_3	0.47	0.41	24	0.40	0.40	31	0.47	0.44	34	0.47	0.42	34
W	1.37	3.36	4									

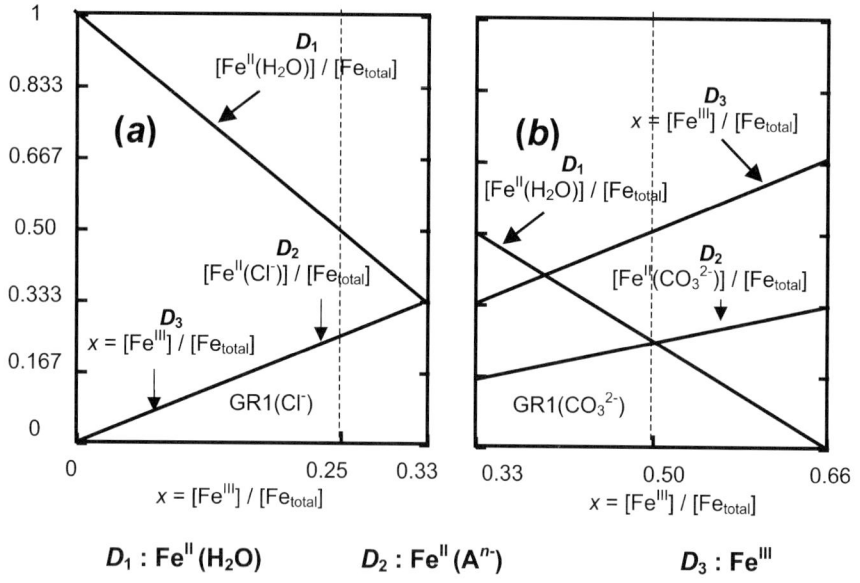

D_1 : Fe^{II} (H$_2$O) D_2 : Fe^{II} (A^{n-}) D_3 : Fe^{III}

Figure 4. Relative abundance of the 3 Fe sites found in GR1(Cl$^-$) and GR1(CO$_3^{2-}$), which correspond to the 3 doublets D_1, D_2 and D_3 of the Mössbauer spectra in Figure 3. The doublets D_1, D_2 and D_3 correspond to Fe^{II}(H$_2$O), Fe^{II}(A^{n-}) and Fe^{III}, respectively.

the 3 Fe sites, D_1, D_2 and D_3, vs. x. They match perfectly for GR1(Cl$^-$) with $x = 0.25$ [11] and 0.33 [21], and for GR1(CO$_3^{2-}$) with $x = 0.33$ [17]. For GR2(SO$_4^{2-}$) [19], D_2 disappears due to anion ordering [15].

3. Abiotic synthesis of green rusts

Three major modes for preparing Fe(II)–Fe(III) hydroxysalt GRs in the laboratory have been elaborated: (i) controlled oxidation of a precipitate of $Fe(OH)_2$ in the presence of an anion, (ii) coprecipitation of Fe(II) and Fe(III) ions in an anion-containing solution, and (iii) bacterial reduction of Fe(III) oxyhydroxide in anoxic conditions.

The classical method to synthesise GRs is done by oxidising a precipitate of $Fe(OH)_2$ ferrous hydroxide with oxygen [5, 10]. We developed extensively this method by assessing the pH and electrode potential E_h of the solution, which gets oxidised by stirring in a beaker with a magnetic rod. Typical curves are illustrated in Figure 5 for the case of $GR2(SO_4^{2-})$, where 3 domains can be distinguished [19]: (i) a first plateau A ending at inflection point t_g correspond to the transformation of $Fe(OH)_2$ into $GR2(SO_4^{2-})$ according to the reaction: $6Fe^{II}(OH)_2 + SO_4^{2-} + (1/2)O_2 + H_2O \Rightarrow Fe^{II}_4Fe^{III}_2(OH)_{12}SO_4 + 2OH^-$ where intercalated water molecules are not taken into account; then a second plateau B that ends at inflection point t_f corresponds to the oxidation of $GR2(SO_4^{2-})$ into γ-FeOOH lepidocrocite, according to: $Fe^{II}_4Fe^{III}_2(OH)_{12}SO_4 + (3/4)O_2 \Rightarrow 5Fe^{III}OOH + SO_4^{2-} + Fe^{2+} + (7/2)H_2O$. Finally, the last plateau C corresponds to the oxidation of remaining Fe^{2+} in solution that both precipitates into FeOOH and oxidizes into Fe^{3+} according to: $3Fe^{2+} + (1/4)O_2 + (3/2)H_2O \Rightarrow Fe^{III}OOH + 2Fe^{3+} + H_2$. The overall reaction is thus: $6Fe^{II}(OH)_2 + (4/3)O_2 + SO_4^{2-} \Rightarrow (16/3)Fe^{III}OOH + (2/3)Fe^{3+} + SO_4^{2-} + (1/3)H_2 + 2OH^-$. The intermediate compound $GR2(SO_4^{2-})$ incorporates temporarily the SO_4^{2-} anions that are at the end released back into solution. Mössbauer spectroscopy was used for characterising compounds at each step during the course of the oxidation process [22].

The coprecipitation of di- and trivalent cations for preparing LDH was often used, but successful only recently for GRs since the oxidation of Fe(II) ions must be avoided [23, 24]. Figure 6a displays the hexagonal shape of $GR2(SO_4^{2-})$ crystals observed by TEM, which were obtained by coprecipitating solutions of

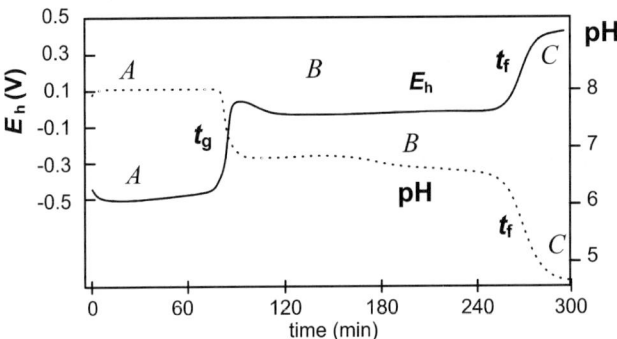

Figure 5. E_h and pH variations with time during the formation and oxidation of $GR2(SO_4^{2-})$ by aerial oxidation of $Fe(OH)_2$.

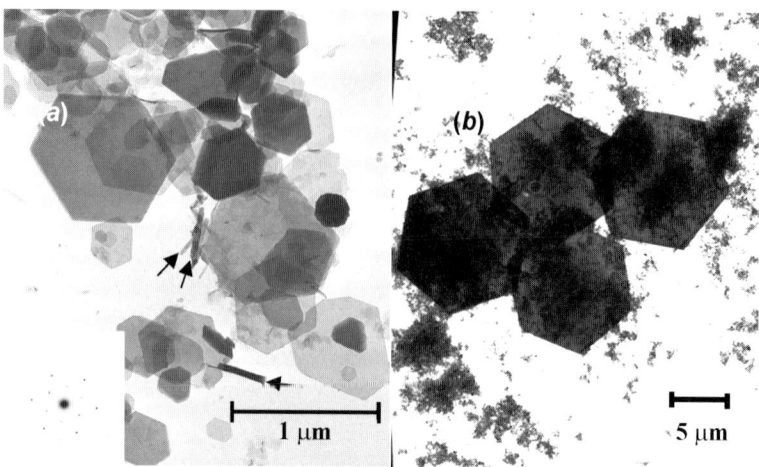

Figure 6. TEM micrographs of (a) GR2(SO_4^{2-}) obtained by coprecipitation of a mixture of ferrous and ferric solutions of $FeSO_4^{2-} \cdot 5H_2O$ and $Fe_2(SO_4^{2-})_3 \cdot 7H_2O$ in the stoichiometry of GR2 [23]. (b) GR2(CO_3^{2-}) obtained by reducing γ-FeOOH lepidocrocite with *Shewanella putrefaciens* bacteria.

$FeSO_4^{2-} \cdot 5H_2O$ ferrous sulphate in the presence of a $Fe_2(SO_4^{2-})_3 \cdot 7H_2O$ ferric sulphate when the proportions were kept those in GR2(SO_4^{2-}) [23].

Then, the precipitates were prepared by a controlled slow addition of NaOH with a peristaltic pump to an acidic solution of Fe(II)$_{aq}$ and Fe(III)$_{aq}$ cations in a sulphated aqueous solution [24]. The pH and electrode potential E (V_{SHE}) of the solution were monitored with respect to the quantity of base added (Figure 7). Titration curves comprised 3 plateaux separated by 2 equivalent points E_1 and E_2. In the first plateau (pH $\approx$ 2.5), Fe(III) cations precipitated as basic ferric salt, schwertmannite, $Fe_{16}O_{16}(OH)_y(SO_4)_z$ [$16 - y = 2z$ where $2 \leqslant z \leqslant 3.5$] from Mössbauer characterisation. In the second plateau (pH $\approx$ 7), the ferric compound reacted with Fe^{2+} in solution forming GR2(SO_4^{2-}). Fe(OH)$_2$ and/or magnetite Fe_3O_4 precipitated later depending on the initial {[Fe(II)]/[Fe(III)]} ratio. In the third plateau (pH $\approx$ 11), GR2(SO_4^{2-}) is no longer stable and transformed into a mixture of Fe_3O_4 and Fe(OH)$_2$. All titration curves are plotted and a change of slope is detected at E^* in the second plateau, which is attributed to the complete dissolution of FeOOH transforming into GR2(SO_4^{2-}) as checked by Mössbauer spectra (Figure 8). A mass-balance diagram was plotted, i.e. $R = \{[OH^-]/[Fe_{total}]\}$ vs. $x_{Fe(III)}$. Every species is represented by $\{x_{Fe(III)}, R\}$ where Fe(II), Fe(III), Fe(OH)$_2$, GR2(SO_4^{2-}), Fe_3O_4 and FeOOH are found at positions {0; 0}, {1; 0}, {0; 2}, {0.33; 2}, {0.66; 2.66} and {1; 3}, respectively (Figure 9). In the diagram, a mixture of GR2(SO_4^{2-}) with Fe(II) ions followed an $R = 7 \times x$ straight line as precisely determined from Mössbauer analysis and not $R = 6 \times x$ as one could expect it. The process of formation of GR2(SO_4^{2-}) is made by adsorption of Fe(II) ions upon the FeOOH surface and the excess of consumed OH^- ions that is used to re-

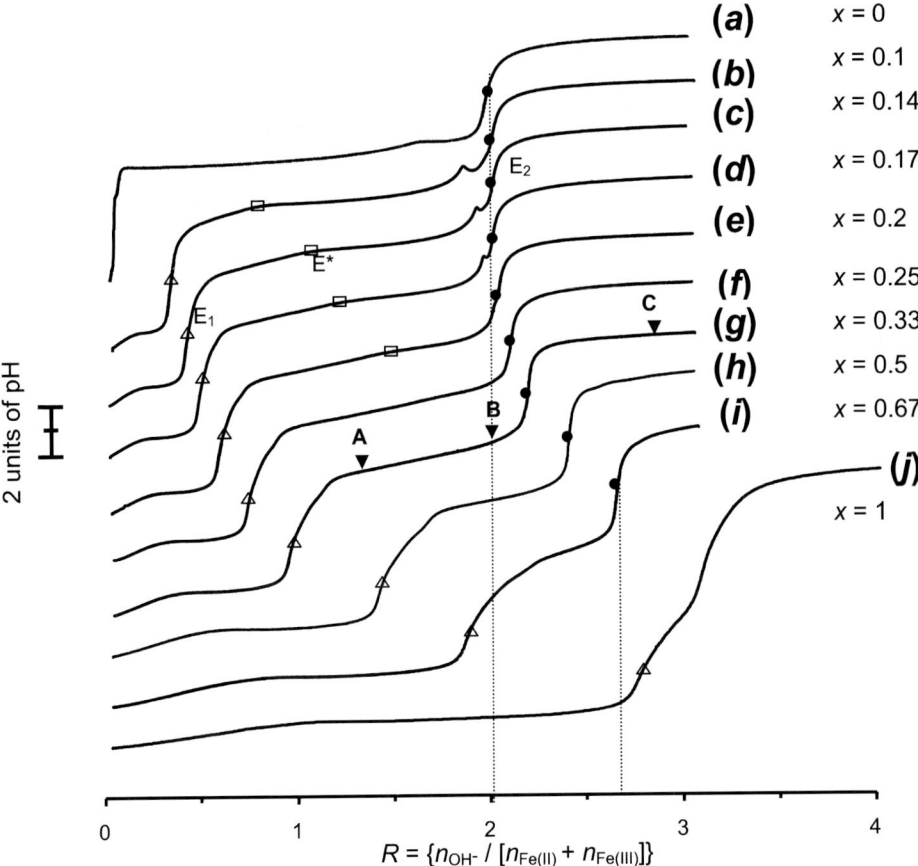

Figure 7. Titration curves pH vs $R = \{[OH^-]/[Fe_{total}]\}$ with $x_{Fe(III)} = \{[Fe(III)]/[Fe_{total}]\}$. E* corresponds to GR2($SO_4^{2-}$) formation and γ-FeOOH complete dissolution [22].

built the surface of the dissolving FeOOH according to: $\{[Fe^{II}_4(OH)_8], SO_4^{2-}\}_{ads} +$ $2FeOOH_{(s)} + 2H_2O_{(1)} \Rightarrow Fe^{II}_4Fe^{III}_2(OH)_{12}SO_{4(s)} + 2[OH^-]_{new\ surface}$.

Thus, the green complex, $[Fe^{II}_xFe^{III}_yO_z(OH)_t]^{(2x+3y-2z-t)+}$, prior GR2($SO_4^{2-}$) precipitation as proposed by Misawa [25], is irrelevant; Fe(II) ions are in solution whereas Fe(III) ions that belong to the solid phase within FeOOH are never transferred into solution.

4. Biogenesis of green rusts

GR1(CO_3^{2-}) is formed during bioreduction by *Shewanella putrefaciens*, a dissimilatory iron reducing bacterium (DIRB), of well-crystallised ferric oxyhydroxide, γ-FeOOH lepidocrocite. XRD, TEM (Figure 6b) and transmission MS (Figure 10, Table II) characterised the GR [26]. The bacterial activity stabilised it in a steady state, but the GR-lepidocrocite mixture is transformed into a mixture of mag-

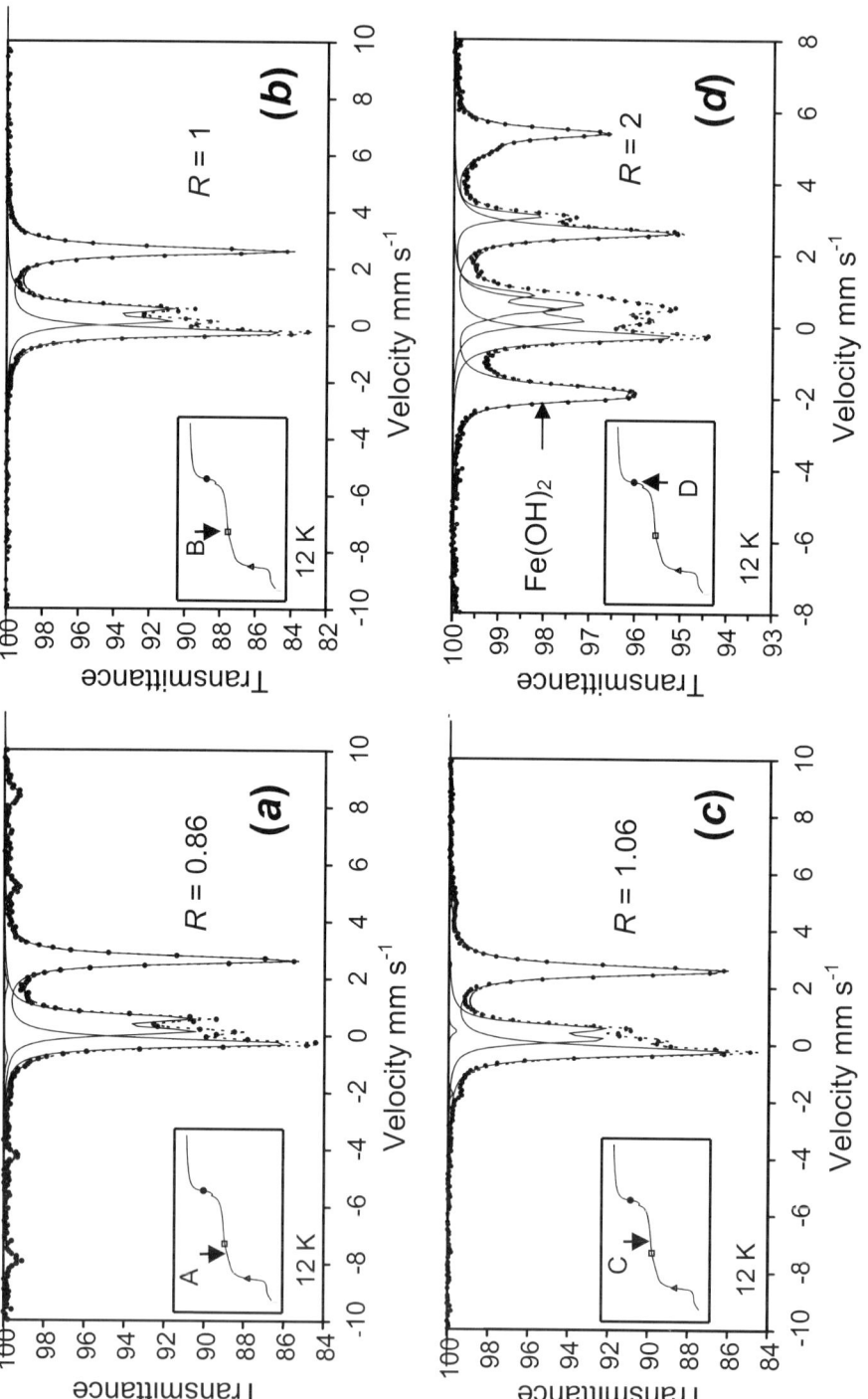

Figure 8. Mössbauer spectra for $x_{Fe(III)} = 0.14$, showing precisely that $R = 7 \times x$ is followed at E^* [22].

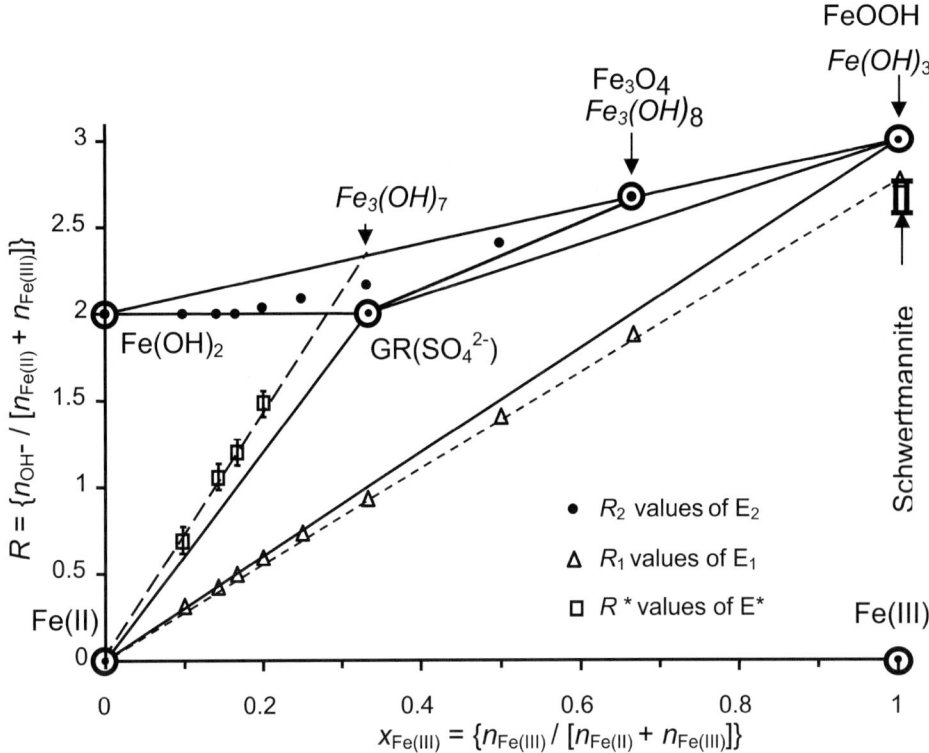

Figure 9. Mass-balance diagram presenting Fe species by plotting $R = \{[OH^-]/[Fe_{total}]\}$ vs. $x = \{[Fe(III)]/[Fe_{total}]\}$. Experimental values of E_1, E_2 and E^* are reported according to [22].

netite, siderite and lepidocrocite when bacterial activity has stopped. Thus, the bluish green mineral turning ochrous on exposure to the air, evidenced recently in hydromorphic soils and in equilibrium with the dissolved iron in soil solutions and aquifers [8, 9], would be the hydroxycarbonate $GR1(CO_3^{2-})$ generated by the dissimilatory microbial reduction of ferric (oxyhydr)oxides. Note that the crystals obtained by biogenesis are much larger than those obtained from abiotic synthesis (Figures 6a & 6b).

5. Corrosion

Understanding the corrosion of iron is one of the direct applications of the studies concerning GRs. Abdelmoula *et al.* [27] showed in an *in situ* transmission Mössbauer study that an iron coupon set in an electrochemical cell at a potential value consistent with the E–pH diagram [17] was progressively covered by the green rust $GR1(CO_3^{2-})$ at the surface before transforming into ferrihydrite and goethite. Benali *et al.* [28] demonstrated also by MS the efficiency of corrosion inhibitors such as phosphate by limiting the oxidation of $GR1(CO_3^{2-})$ into ferrihydrite that never transformed into goethite. Finally, if the formation of $GR2(SO_4^{2-})$

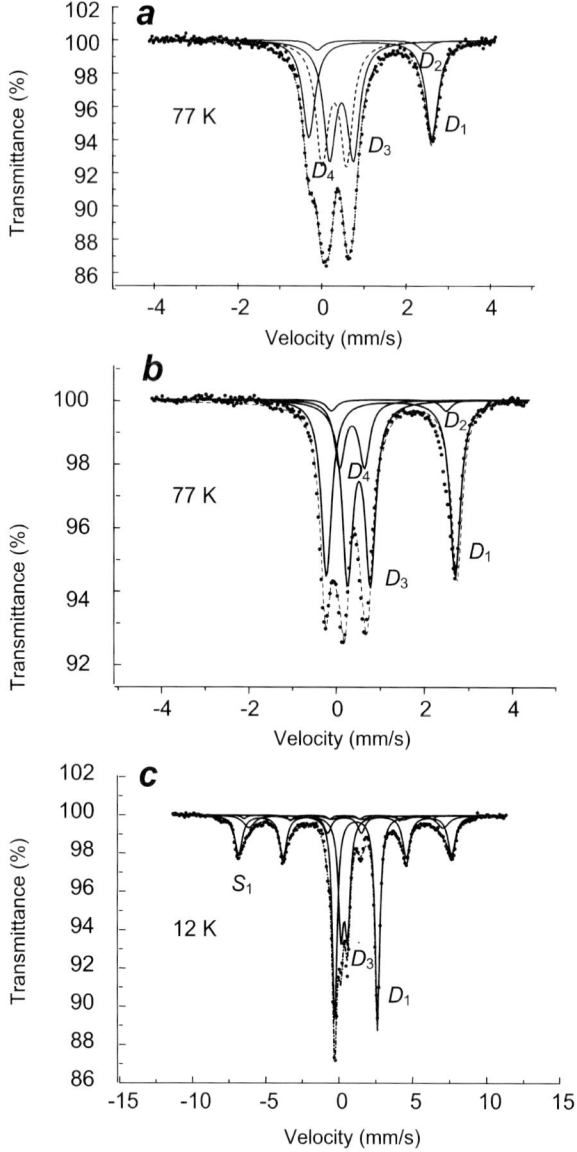

Figure 10. Mössbauer spectra of precipitates forming during the reduction of γ-FeOOH by *Shewanella putrefaciens*. Measurements at 77 K after (a) 20 hours and (b) 117 hours of culture and (c) measurement at 12 K after 117 hours of culture.

was established some time ago in the microbially influenced corrosion of steels in sea-water [29, 30] as reported in [31], a recent paper by Refait *et al.* [32] carefully described the sequence of rust stratification on steel sheet piles left 25 years at the anoxic level in mud suggesting that it was due to a bacterial reduction of previously formed lepidocrocite (Figure 11). It is proposed that $GR2(SO_4^{2-})$ forms firstly by

Table II. Mössbauer hyperfine parameters of spectra measured at 77 K of products obtained during reduction of γ-FeOOH lepidocrocite in bacterial cultures. After (a) 20 hours and (b) 117 hours of culture (Figures 2a & 2b) [24]. δ: isomer shift taking α-iron as reference at ambient, Δ quadrupole splitting and *RA*: relative abundance

	(a)			(b)		
	δ (mm s^{-1})	Δ (mm s^{-1})	*RA* (%)	δ (mm s^{-1})	Δ (mm s^{-1})	*RA* (%)
D_1	1.27	2.89	29	1.27	2.90	41
D_2	1.27	2.51	3	1.18	2.58	3
D_3	0.59	0.56	33	0.51	0.51	41
D_4	0.42	0.57	35	0.35	0.56	15

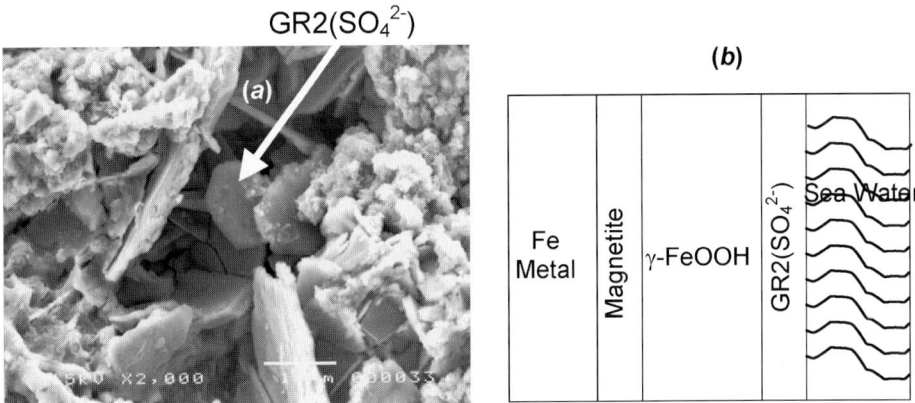

Figure 11. (a) SEM micrograph showing hexagonal shaped crystals of GR2(SO$_4^{2-}$) upon corroded steel sheet left 25 years in sea water and (b) sequence of the rust layers: metal–magnetite–lepidocrocite-GR2(SO$_4^{2-}$) [32].

DIRB reduction of the protective γ-FeOOH and then the produced sulphate reservoir is more fully reduced by sulphate reducing bacteria (SRB) reaching the acidic conditions of ferrous sulphides.

6. The green rust mineral

The environmental implication of GRs in the cycle of iron in Nature took all its importance when a GR mineral was definitively identified in 1996 [8, 9] by MS. As stressed already [9], only MS was able to detect in the forest of Fougères (Brittany) the exact compound under which iron was involved in hydromorphic soils because of the incapacity of XRD to be sensitive to a mineral that was too dilute, less than 5% Fe. This result was comforted by an EXAFS study comparing synthetic GRs with a mineral sample [33]. However these methods did not reveal the exact nature of the incorporated anion and, at that time, we assumed that it could be a GR1(OH$^-$) with variable composition [8, 9, 34].

Many facts tend now to discard this assumption since we know that GR1(CO$_3^{2-}$) may also have a variable composition written as [Fe$^{II}_{(1-x)}$Fe$^{III}_x$(OH)$_2$]$^{x+}$ · [(x/2) CO$_3^{2-}$ · zH$_2$O]$^{x-}$ where $1/3 \leqslant x \leqslant 2/3$ with $z \leqslant [1 - (3x/2)]$ [9]. All occurrences of the mineral studied by MS demonstrated that $1/3 \leqslant x \leqslant 2/3$ [8, 9]. Moreover the EXAFS study [33] was done with synthetic GR1(CO$_3^{2-}$). But the major argument is that the bioreduction by dissimilatory iron reducing bacteria (DIRB) [26] gave rise to GR1(CO$_3^{2-}$) of variable composition.

A most recent study of the GR mineral at the place where it was first found in the forest of Fougères was done by Feder [35] using the miniaturised Mössbauer spectrometer MIMOS developed by G. Klingelhöfer to be launched to Mars but adapted for terrestrial applications. The *in situ* Mössbauer spectra measured at room temperature in the soil demonstrated clearly that Fe was engaged in a GR mineral that could not be confused with others, e.g., clays: isomer shift and quadrupole splitting values for Fe(II) as well as Fe(III) ion sites lie in different domains. The fluctuations with time and depth of the {[Fe(II)]/[Fe(III)]} ratio confirmed that x remained in the [1/3, 2/3] interval and in particular the trend was that {[Fe(II)]/[Fe(III)]} increased with depth till some lepidocrocite began to appear. This is another argument for favouring the biogenesis formation of the GR mineral in gley soils and *in situ* Mössbauer spectrometer is indeed a marvellous tool for environmental science.

7. Conclusion

The interest of Mössbauer spectroscopy has been enhanced since most of the structural features of GRs were obtained by coupling information from MS and XRD. The importance of GRs in the process of corroding iron-based materials was clearly demonstrated during the oxidation as an intermediate Fe(II–III) compound but also as the result of microbially influenced corrosion using bacterial reduction for respiration. Finally, the massive existence in Nature of GRs, most likely GR1(CO$_3^{2-}$), in waterlogged anoxic grounds plays a major role in the cycle of iron by redox processes and can be thought for soil remedy or water treatment. In that respect, many efforts have been done recently indicating that synthetic GRs, as a matter of fact GR2(SO$_4^{2-}$) for experimental facility, can be utilised to reduce anionic pollutants such as nitrate, selenate and dichromate [36–39].

References

1. Ostwald, W., *Grundlinien der Anorganischen Chemie*, Verlag W. Engelmann, Leipzig, 1900, p. 577.
2. Deiss, E. and Schikkor, G., *Z. Anorg. Allg. Chem.* **172** (1928), 32.
3. Girard, A. and Chaudron, G., *Compt. Rend. Acad. Sci.* **200** (1935), 127.
4. Feitknecht, W., *Fortschr. Chem. Forsch.* **2** (1953), 670.

5. Bernal, J. D., Dasgupta, D. R. and Mackay, A. L., *Clay Miner. Bull.* **4** (1959), 15.
6. Arden, T. V., *J. Chem. Soc.* (1950), 882.
7. Ponnamperuma, F. N., Estrella, M. T. and Loy, T., *Soil Sci.* **103** (1967), 374.
8. Bourrié, G., Trolard, F., Génin, J.-M. R., Jaffrezic, A., Maître, V. and Abdelmoula, M., *Geochim. Cosmochim. Acta* **63** (1999), 3417.
9. Génin, J.-M. R., Bourrié, G., Trolard, F., Abdelmoula, M., Jaffrezic, A., Refait, Ph., Maître, V., Humbert, B. and Herbillon, A., *Environ. Sci. Technol.* **32** (1998), 1058.
10. Feitknecht, W. and Keller, G., *Z. Anorg. Allg. Chem.* **262** (1950), 61.
11. Génin, J.-M. R., Rézel, D., Bauer, P., Olowe, A. and Béral, A., In: Duprat (ed.), *Electrochemical Methods in Corrosion Research*, Vol. 8: *Materials Science Forum*, Trans. Tech. Publ, Switzerland, 1986, p. 477.
12. Olowe, A. A., Génin, J.-M. R. and Bauer, P., *Hyp. Interact.* **46** (1989), 437.
13. Bookin, A. S. and Drits, V. A., *Clays Clay Miner.* **41** (1993), 551.
14. Stampfl, P. P., *Corr. Sci.* **9** (1969), 185.
15. Simon, L., François, M., Refait, Ph., Renaudin, G., Lelaurain, M. and Génin, J.-M. R., *Solid State Sci.* **5** (2003), 327.
16. Refait, Ph., Abdelmoula, M. and Génin, J.-M. R., *Corros. Sci.* **40** (1998), 1547.
17. Drissi, S. H., Refait, Ph., Abdelmoula, M. and Génin, J.-M. R., *Corros. Sci.* **37** (1995), 2025.
18. Refait, Ph. and Génin, J.-M. R., *Corros. Sci.* **39** (1997), 539.
19. Génin, J.-M. R., Olowe, A. A., Refait, Ph. and Simon, L., *Corros. Sci.* **38** (1996), 1751.
20. Gancedo, R., Martinez, L. and Otton. J. M., *J. Physique (Colloq. C6)* **37** (1976), 297.
21. Refait, Ph. and Génin, J.-M. R., *Clay Miner.* **32** (1997), 32.
22. Refait, Ph., Bon, C., Simon, L., Bourrié, G., Trolard, F., Bessière, J. and Génin, J.-M. R., *Clay Miner.* **215** (1999), 313.
23. Géhin, A., Ruby, C., Abdelmoula, M., Benali, O., Ghanbaja, J., Refait, Ph. and Génin, J.-M. R., *Solid State Sci.* **4** (2002), 61.
24. Ruby, C., Géhin, A., Abdelmoula, M., Génin, J.-M. R. and Jolivet, J. P., *Solid State Sci.* **5** (2003), 1055.
25. Misawa, T., Hashimoto, K. and Shimodaira, S., *J. Inorg. Nucl. Chem.* **35** (1973), 4167.
26. Ona-Nguema, G., Abdelmoula, M., Jorand, F., Benali, O., Géhin, A., Block, J.-C. and Génin, J.-M. R., *Environ. Sci. Technol.* **36** (2002), 16.
27. Abdelmoula, M., Refait, Ph., Drissi, S. H., Mihé, J.-P. and Génin, J.-M. R, *Corros. Sci.* **38** (1996), 623.
28. Benali, O., Abdelmoula, M., Refait, Ph. and Génin, J.-M. R., *Geochim. Cosmoschim. Acta* **65** (2001), 1715.
29. Olowe, A. A., Bauer, P., Génin, J.-M. R. and Guézennec, J., *Corrosion* **45** (1989), 229.
30. Génin, J.-M. R., Olowe, A. A., Benbouzid-Rollet, N. D., Prieur, D., Confente, M. and Resiak, B., *Hyp. Interact.* **69** (1991), 875.
31. Génin, J.-M. R., Refait, Ph. and Abdelmoula, M., *Hyp. Interact.* **139–140** (2002), 119.
32. Refait, Ph., Memet, J.-B., Bon, C., Sabot, R. and Génin, J.-M. R., *Corros. Sci.* **45** (2003), 833.
33. Refait, Ph., Abdelmoula, M., Trolard, F., Génin, J.-M. R., Ehrhardt, J.-J. and Bourrié, G., *Amer. Mineralogist* **86** (2001), 714.
34. Génin, J.-M. R., Refait, Ph., Bourrié, G., Abdelmoula, M. and Trolard, F., *Applied Geochem.* **16** (2001), 559.
35. Féder, F., Ph.D. thesis, University of Aix-Marseille, September 2001.
36. Hansen, H. C. B., Bender-Koch, C., Nancke-Krogh, H., Borgaard, O. K. and Sørensen, J., *Environ. Sci. Technol.* **30** (1996), 2053.
37. Myneni, S. C. B., Tokunaga, T. K. and Brown, Jr., G. E., *Science* **278** (1997), 1106.
38. Refait, Ph., Simon, L. and Génin, J.-M. R., *Environ. Sci. Technol.* **34** (2000), 819.
39. Loyaux-Lawniczak, S., Refait, Ph., Ehrhardt, J.-J., Lecomte, P. and Génin, J.-M. R., *Environ. Sci. Technol.* **34** (2000), 438.

Hyperfine Interactions **156/157**: 487–492, 2004.
© 2004 *Kluwer Academic Publishers. Printed in the Netherlands.*

Identification of Corrosion Products Due to Seawater and Fresh Water

A. GISMELSEED, M. ELZAIN, A. YOUSIF, A. AL RAWAS, I. A. AL-OMARI,
H. WIDATALLAH and A. RAIS
Department of Physics, College of Science, Box 36, Al-Khoud, 123, Oman

Abstract. Mössbauer and X-ray diffraction (XRD) measurements were performed on corrosion products extracted from the inner surface of two different metal tubes used in a desalination plant in Oman. One of the tubes corroded due to the seawater while the second was corroded due to fresh water. The corrosion products thus resulted due to seawater were scrapped off in to two layers, the easily removable rust from the top is termed outer surface corrosion product and the strongly adhered rust as internal corrosion product. The Mössbauer spectra together with the XRD pattern of the outer surface showed the presence of magnetite (Fe_3O_4), akaganeite (β-FeOOH), lepidocrocite (γ-FeOOH), goethite (α-FeOOH) and hematite (Fe_2O_3). The inner surface however showed the presence of akaganite, goethite, and magnetite. On the other hand, the corrosion products due to the fresh water showed only the presence of goethite and magnetite. The mechanism of the corrosion process will be discussed based on the significant differences between the formation of the iron components of the corrosion products due to seawater and the fresh water.

Key words: corrosion, seawater, fresh water, Mössbauer, oxyhydroxides.

1. Introduction

Corrosion is a complex electrochemical process involving the interaction between metallic materials and their environments producing rust and leading to degradation of materials. Various factors affect the corrosion process. For example, the presence of water vapor and anions such as SO_4^-, Cl^-, CO_3^-, etc. in the environmental media effectively accelerate the processes that lead to the atmospheric corrosion of steel [1]. To understand such process we should systematically study the formation and transformation of the corrosion products and see how to relate them to the environmental parameters like temperature, humidity, chloride and sulfur dioxide concentrations and the exposure periods.

Various complementary spectroscopic techniques have been used in identification of most of the atmospheric corrosion products. For instance, X-ray diffraction has been used thoroughly in the identification of many iron oxides in corrosion products; however, in some cases X-rays might be unable to easily distinguish between some of these products [2]. On the other hand, the Mössbauer measurements at different temperature ranges (room temperature and below) are able to determine both the oxide phases and their fractions. Additional support in distinguishing minerals in corrosion products is obtained through Raman spectroscopy [3].

The composition of the corrosion products in steels changes with time. In general, it was observed that initially iron oxidizes to Fe^{2+}, that the intermediate corrosion products contain both Fe^{2+} and Fe^{3+} and the final products include Fe^{3+} [4]. On time scales of days green rust is formed. The green rust transforms initially to lepidocrocite (γ-FeOOH), followed by goethite (α-FeOOH) and magnetite (Fe_3O_4). Akaganite (β-FeOOH) are formed in Cl^--rich environments [5].

2. Samples preparation and experimental methods

Samples were taken from water pipes used in a desalination plant. One set of samples (SW1, SW2) was extracted from pipes used in loading seawater into the plant, while the other set (FW1) was extracted from pipes used for discharging fresh water out of the plant. The pipes were laid out abandoned and left to rust at the site of the plant for a couple of years. Parts of the abandoned pipes were also kept in open atmosphere in our laboratory for about five years. The pipes were made from cast-iron. The composition of the pipes was tested using scanning electron microscopy. Apart from pockets of graphite, the composition was close to that of carbon steel. In particular, neither chromium nor copper was detected in the material of the pipes.

The fresh water pipe had a single layer of dark rust, whereas the seawater pipe showed two distinct layers. The sample extracted from the outer layer of the seawater tube SW1, has a soft texture and is a mixture of yellow and red rust. The outer sample was very easy to powder in contrast to the sample from the inner layer SW2, which was strongly adherent to the surface of the tube. The inner sample obtained from the seawater tube is dark-red and required care to scrap it off from the steel substrate.

The three samples were powdered for the XRD and Mössbauer measurements. The powder X-ray diffraction (XRD) patterns of the three samples were recorded on a Philips diffractometer (model pw 1820) with a Co Kα source. Mössbauer measurements were performed on the samples in a continuous liquid nitrogen flow cryostat using a 50 mCi ^{57}Co(Rh) source with a spectrometer in the transmission mode. The spectrometer was calibrated with α-Fe foil spectrum at room temperature.

3. Results and discussion

X-ray powder diffraction pattern for all samples were recorded in the angular range 10–70 degrees to obtain a basic identification of the oxides present due to corrosion. The spectrum of the seawater inner layer identified the presence of varied amount of akaganeite, goethite, and magnetite, while the spectrum collected from outer layer sample showed the presence of lepidocrocite and hematite in addition to the oxides existing in the inner layer. On the other hand, the fresh water sample showed sharp diffraction peak matching only goethite and magnetite with the latter in a larger proportion. The presence of the other oxides in the seawater samples and especially the akaganeite is ascribed to the high level concentration of chloride in the seawater.

To complete the identification of the detected oxides and to measure their partial fractions in each of the three samples, transmission Mössbauer measurement performed at room temperature and liquid nitrogen are presented in Figure 1 and their least square fitting parameters are collected in Table I.

The spectrum of the outer surface sample SW1 (Figure 1*left*, a) consists of a broadened magnetic component and a paramagnetic doublet. The Mössbauer parameters of the doublet shown in Table I are in good agreement with those published for superparamagnetic component of goethite, akaganite and lepidocrocite at RT [5]. It was not easy to fit the magnetic component of the spectrum since its field distribution is attributed to more than one phase and different particle sizes. In order to identify the superparamagnetic components the Mössbauer spectrum of the sample was recorded at 78 K (Figure 1*right*, a). The resolution of the magnetic component is much better compared to that of RT. The spectrum was fitted with one doublet and six magnetic sub-spectra assigned as follows: magnetic goethite, magnetically ordered goethite particles, magnetite (two non-equivalent sites), akaganite and hematite. The percentage of the doublet reduces by almost one-third at 78 K due to the magnetic ordering of both superparamagnetic goethite and akaganite at this temperature.

A dominant doublet with weak intensity broadened magnetic component characterizes the Mössbauer spectrum of the inner layer sample SW2 produced at RT. The hyperfine interaction parameters obtained for the doublet are attributed to the superparamagnetic goethite, and akaganite since the XRD data excluded the presence of lepidocrocite. The shape of the spectrum at 78 K shows a tremendous change compared to the RT spectrum with a remarkable increase in the relative intensity of the magnetic component. The 5% contribution of the paramagnetic component as obtained from the fitting results is attributed to the superparamagnetic component of goethite not yet resolved even at 78 K. According to Janot *et al.* [6] the persistence of this component at 78 K indicates that the size of its particle is below 8 nm. The fitting of the magnetic component at 78 K was achieved by assuming five magnetic sextet, two of them attributed to the magnetite phase in its tetrahedral (A) and octahedral [B] sites, one ascribed to magnetic goethite, one to superparamagnetic goethite and the fifth to akaganite phase. The identification of the iron oxide phases in the sample as determined by the analysis of 78 K Mössbauer spectra are in good agreement with XRD data and their area percentages are collected in Table I.

The Mössbauer spectrum of the third specimen (fresh water corroded layer) showed completely different pattern compared to the other two showing a typical spectrum of magnetite at RT in its two non-equivalent sites (A) : [B] = 1 : 2,

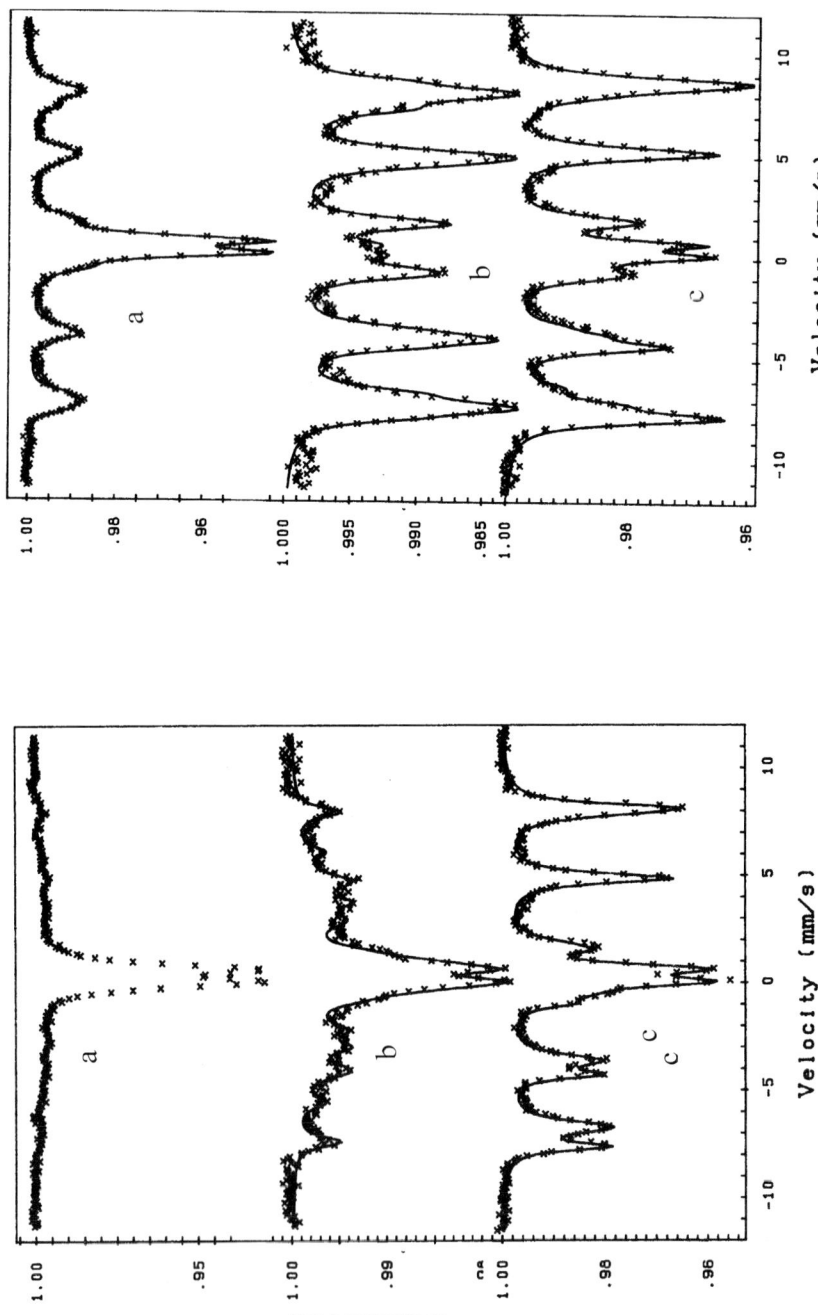

Figure 1. Mössbauer spectra of (a) outer layer, (b) inner layer and (c) fresh water layer measured at 300 K (*left*) and 78 K (*right*).

Table I. Mössbauer hyperfine interaction parameters of the three corroded samples SW1, SW2, and FW1. Pm for paramagnetic component and m for magnetic component

	295 K				78 K			
	δ (±0.02) (mm/s)	Δ (±0.01) (mm/s)	B (±0.1) (T)	A (±0.1) (%)	δ (±0.02) (mm/s)	Δ (±0.01) (mm/s)	B (±0.1) (T)	A (±0.1) (%)
SW1								
α-FeOOH (pm)	0.35	0.63		80.4	0.48	0.61		52.2
α-FeOOH (m)	0.38	−0.24	35.6	3.9	0.54	−0.23	47.22	16.5
β-FeOOH					0.47	−0.32	43.44	11.8
α-Fe$_2$O$_3$	0.30	−0.13	50.0	5.4	0.6	−0.3	50.25	5.2
Fe$_3$O$_4$ (A)	0.30	−0.01	48.5	3.6	0.31	0.16	45.72	5
Fe$_3$O$_4$ [B]	0.56	−0.01	45.6	6.7	0.42	−0.12	48.83	9.4
SW2								
α-FeOOH (pm)	0.37	0.57		21.1	0.38	0.52		5
α-FeOOH (m)	0.37	−0.16	34.0	29.1	0.44	−0.1	48.3	47.4
β-FeOOH	0.40	1.05		20.3	0.44	−0.19	43.3	19.7
α-Fe$_2$O$_3$								
Fe$_3$O$_4$ (A)	0.30	0.01	45.9	9.7	0.4	0.04	46.8	9.3
Fe$_3$O$_4$ [B]	0.42	−0.06	48.8	19.8	0.44	−0.07	51.5	18.6
FW1								
α-FeOOH (pm)	0.35	0.6		26.5	0.45	0.57		15.5
α-FeOOH (m)					0.83	−0.03	46.8	12.7
Fe$_3$O$_4$ (A)	0.30	−0.01	49.1	25.4	0.35	−0.05	50.9	24.5
Fe$_3$O$_4$ [B]	0.64	−0.01	45.6	48.1	0.63	−0.3	50.4	47.2

in addition to a paramagnetic doublet. Based on the parameters of the other two specimens the doublet is fitted with parameters corresponding to superparamagnetic goethite and the magnetic components of the magnetite. On lowering the temperature to 78 K, about 10% of the superparamagnetic goethite has been magnetically ordered while the rest remains in the paramagnetic phase. The other three components are assigned to magnetite and goethite. The results of the analysis are presented in Figure 1*left, c* & 1*right, c*.

It is worth to mention that for the three samples the subspectra of magnetite at 78 K, which is below the Verwey transition temperature ($\sim$120 K) have been fitted assuming average magnetic field distribution.

4. Conclusion

From the above qualitative and quantitative phases analysis of the corrosion product formed on the two pipes we can draw the following remarks. The presence of some oxides/oxyhydroxides on the sea water pipe, which are not formed on the fresh water pipe can be attributed to the difference in the alloying elements in the structure of the two pipes though these differences are not strongly considered in the formation of the of the corrosion product. Within the sea water pipe the outer layer product showed no akaganite β-FeOOH phase, which is significantly observed in the inner layer product. This can be understood in term of its transformation to goethite α-FeOOH as the relative percentage of the later is much higher than in the inner layer.

References

1. Graedel, T. E. and Frankenthal, R. P., *J. Electrochem. Soc.* **137** (1990), 2385.
2. Oh, S. J., Cook, D. C. and Townsend, H. E., *Corrosion Science* **41** (1999), 1687.
3. Cook, D. C., Oh, S. J., Balasubramanian, R. and Yamashita, M., *Hyp. Interact.* **122** (1999), 57.
4. Cook, D. C., Van Orden, A. C., Carpio, J. J. and Oh, S. J., *Hyp. Interact.* **113** (1998), 319.
5. Marco, J. F., Gracia, M., Gancedo, J. R., Martin Luengo, M. A. and Joseph, G., *Corrosion Science* **42** (2000), 753.
6. Janot, C., Gibert, H. and Tobias, C., *Bull. Soc. Fr. Mineral. Cristtallogr.* **96** (1973), 281.

Hyperfine Interactions **156/157**: 493–496, 2004.
© 2004 *Kluwer Academic Publishers. Printed in the Netherlands.*

Atmospheric Corrosion on Steel Studied by Conversion Electron Mössbauer Spectroscopy

AKIO NAKANISHI and TAKAYUKI KOBAYASHI
Department of Physics, Shiga University of Medical Science, Otsu, Shiga 520-2192, Japan

Abstract. In order to investigate initial products on steel by atmospheric corrosion, conversion electron Mössbauer measurements were carried out at temperatures between 15 K and room temperature. From the results obtained at low temperatures, it was found that the corrosion products on steel consisted of ferrihydrite.

Key words: atmospheric corrosion, ferrihydrite, CEMS.

1. Introduction

Corrosion on steel has been widely studied by Mössbauer spectroscopy [1, 2]. Both the transmission Mössbauer spectroscopy and the conversion electron Mössbauer spectroscopy (CEMS) have been applied to the corrosion science. For the transmission Mössbauer measurement it is necessary to separate the corrosion products from the substrate material, which implies that the corrosion products have grown up enough to be separated. It is not the case if the amount of corrosion products is small. On the other hand, CEMS is useful for such a sample because the spectrum can be obtained without separating the products from the substrate material. Conversion electron Mössbauer (CEM) measurements are usually carried out at room temperature (RT). Since the corrosion products consist of a lot of small particles, ferric doublets are mostly observed in a CEM spectrum at RT. In order to investigate the properties of corrosion products in detail, it is, therefore, effective to measure a CEM spectrum at lower temperatures.

For the studying initial products on steel by the atmospheric corrosion, both a natural iron foil and a weathering steel were exposed for two months in atmosphere. NO_x and SO_x in the air accelerate a corrosion rate of the steel. Rain also contains a small amount of SO_4^{2-}. When the steel is affected by these molecules, α-FeOOH and γ-FeOOH are expected to be the resultant species of corrosion product [3, 4]. The purpose of this study is to characterize initial states of atmospheric corrosion on the steel without the effect of these molecules, NO_x and SO_4^{2-}. For reducing the effect of these molecules samples were prevented from rain and were exposed in a rural area. CEM measurements of corrosion products were carried out at temperatures between 15 K and RT.

2. Experimental

Natural iron foil and weathering steel were used as substrate materials. The surface of the steel plates were polished by a sand paper and washed by acetone. They were hanged by a string and covered by a plastic vessel in order to avoid the influence of rain and dust in the air, and they were exposed in a rural environment for two months. Corrosion products were grown up on the steel.

CEM spectra were taken at temperatures between 15 K and RT, with the gas filled proportional counter described elsewhere [5, 6]. Hydrogen gas and the 98% He–2% CH_4 gas mixture were used as the counter gas at low temperatures and RT, respectively. The velocity calibration was carried out with a natural iron foil. X-ray diffraction analyses were also performed using a Philips diffractometer (PW1710) with Co Kα radiation.

3. Results and discussion

The surface of steel plate was not fully covered with corrosion products and the metallic part was still remained. Mössbauer spectra of the natural iron foil are shown in Figure 1. Similar spectra are also observed in the weathering steel. In the spectrum at RT, a ferric doublet in addition to the sextet of the substrate was observed. The isomer shift (IS) and the quadrupole splitting (QS) were 0.37 mm/s and 0.67 mm/s, respectively. At 70 K, the ferric doublet was also observed but the line width became broader. At 15 K, the substrate sextet and a sextet with broad lines were noticed. The IS, QS and the hyperfine field (HF) of the sextet were 0.47 mm/s, −0.06 mm/s and 46.6 T, respectively, for one sextet fit. Its line width was larger than 1.0 mm/s. Comparing these values with those measured at 4.2 K reported in the literatures [7], the HF of the sextet was larger than that of γ-FeOOH and slightly smaller than that of α-FeOOH. The HF of ferrihydrite was similar to that of observed sextet. In the XRD pattern of the sample, α-FeOOH and γ-FeOOH peaks were not observed. A weak peak with broad line width was noticed at $d = 0.25$ nm, which is similar to the reported value of ferrihydrite [7, 8]. Since ferrihydrite is not sensitive to XRD analysis and yields broad lines [8], it is possible to conclude that the corrosion products consist of ferrihydrite.

The broad line widths of the sextet at low temperature suggest a distribution of the hyperfine field. The hyperfine field distributions of the samples were calculated with the method of Wivel and Mørup [9] at temperatures between 15 K and 40 K. The mean hyperfine field was estimated from the distribution, which is smaller than the HF fitted by supposing a single sextet. A plot of the mean hyperfine fields as a function of temperature is shown in Figure 2. From Figure 2, it is found the temperature dependence of the mean hyperfine field for these two samples are similar each other. Since the temperature dependence of the mean hyperfine field depends on the average particle size of the corrosion products, the average particle size of the corrosion products on the weathering steel is similar to that on the natural iron

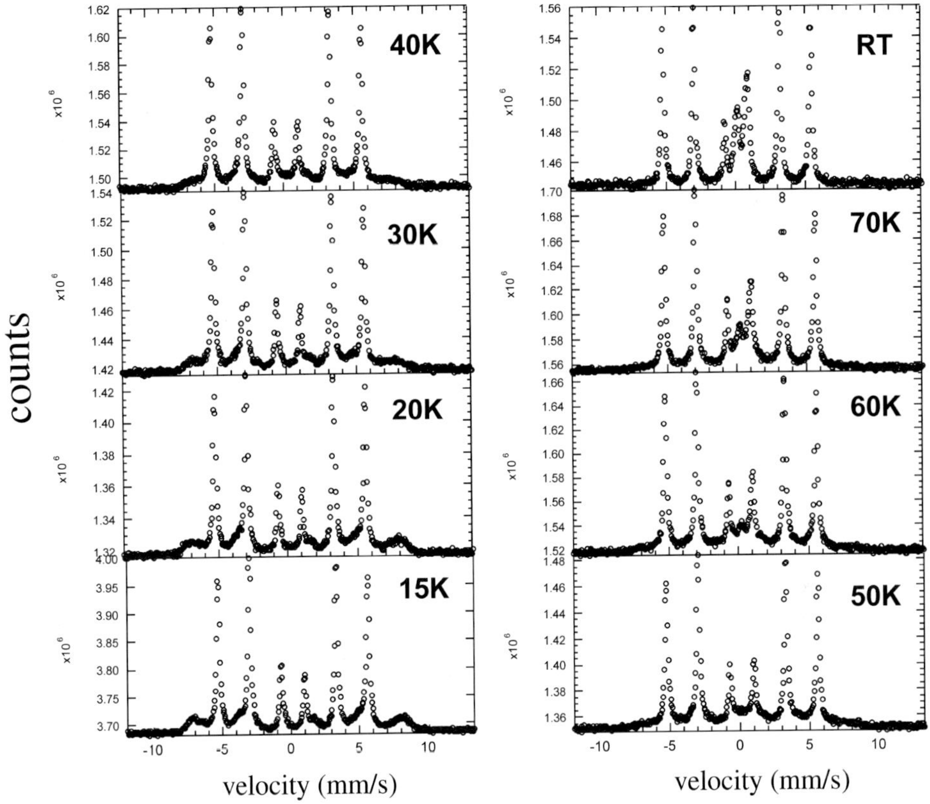

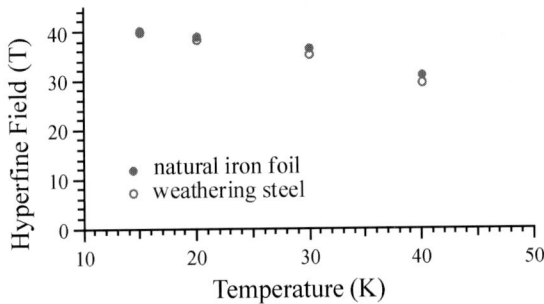

Figure 1. CEM spectra of the natural iron foil.

Figure 2. Temperature dependence of the mean hyperfine field of the samples.

foil. Therefore, it is found that there is little difference between the initial products of corrosion on the weathering steel and the natural iron foil.

Ferrihydrite is an intermediate corrosion product formed in the humid environment [10, 11]. When the steel was exposed in the air, it was left under day/night cycle. In night, the surface was expected to be wet with dew condensation. On the other hand, the surface was dried in the daytime. Thus, the water content on the steel surface in night has the influence on the amount of produced ferrihydrite.

Leidheiser and Czkó-Nagy [10] were suggested that the ferrihydrite in time was converted to other minerals, such as α-FeOOH and γ-FeOOH. In this study, almost all of the corrosion products were ferrihydrite, which indicates the conversion of ferrihydrite into other mineral did not occur due to the short exposure time.

4. Conclusion

The initial products on steel by atmospheric corrosion were investigated. From CEM spectra measured at low temperatures it is found that corrosion products consist of ferrihydrite. It is also found that there is little difference between the initial products of corrosion on the weathering steel and the natural iron foil, if the molecules, which accelerate the corrosion rate, do not affect the corrosion process. The differences between corrosion products on the weathering steel and the natural iron foil will be appeared after the surface of steel is fully covered with corrosion products.

Acknowledgement

The authors are indebted to Prof. K. Nomura for providing the weathering steel.

References

1. Kamimura, T. and Nasu, S., *Materials Transactions, JIM* **41** (2000), 1208.
2. Okada, T., Ishii, Y., Mizoguchi, T., Tamura, I., Kobayashi, Y., Takagi, Y., Suzuki, S., Kihara, H., Itou, M., Usami, A., Tanabe, K. and Masuda, K., *Jpn. J. Appl. Phys.*, **39** (2000), 3382.
3. Nakanishi, A., Fukumura, K. and Kobayashi, T., In: I. Ortalli (ed.), *Conf. Proc.*, Vol. 50, SIF, Bologna, 1996, p. 669.
4. Nakanishi, A., Kobayashi, T. and Fukumura, K., *Hyp. Interact.* **112** (1998), 43.
5. Fukumura, K., Katano, R., Kobayashi, T., Nakanishi, A. and Isozumi, Y., *Nucl. Instr. Meth. A* **301** (1991), 871.
6. Fukumura, K., Nakanishi, A. and Kobayashi, T., *Nucl. Instr. Meth. B* **86** (1994), 387.
7. Murad, E. and Johnston, J.H., In: G. J. Long (ed.), *Mössbauer Spectroscopy Applied to Inorganic Chemistry*, Vol. 2, Plenum Press, New York, 1987, p. 507.
8. Murad, E. and Schwertmann, U., *Am. Mineral.* **65** (1980), 1044.
9. Wivel, C. and Mørup, S., *J. Phys. E* **14** (1981), 605.
10. Leidheiser, H., Jr., and Czkó-Nagy, I., *Corr. Sci.* **24** (1984), 569.
11. Arshed, M., Butt, N.M., Siddique, M. and Anwar-Ul Islam, M., *Phys. Stat. Sol. (a)* **137** (1993), K33.

Hyperfine Interactions **156/157**: 497–503, 2004.
© 2004 *Kluwer Academic Publishers. Printed in the Netherlands.*

Mössbauer Spectrometry as a Powerful Tool to Study Lithium Reactivity Mechanisms for Battery Electrode Materials

L. ALDON, P. KUBIAK, A. PICARD, P. E. LIPPENS,
J. OLIVIER-FOURCADE* and J.-C. JUMAS
*Laboratoire des Agrégats Moléculaires et Matériaux Inorganiques (UMR 5072 CNRS),
Université Montpellier II, CC 15, Place E. Bataillon, 34095 Montpellier Cedex 5, France;
e-mail: jolivier@univ-montp2.fr*

Abstract. The use of ^{57}Fe as a local Mössbauer probe is of high interest for studying mechanisms induced by lithium insertion. In this way the substitutions Ti/Fe and Li/Fe have been carried out for $Li_4Ti_5O_{12}$ to obtain Fe substituted spinel and $Li_2Ti_3O_7$ ramsdellite. In the case of $Li_4Ti_5O_{12}$ iron ions are reduced ($Fe^{III} \rightarrow Fe^{II}$), then migrate from tetrahedral to octahedral sites allowing us to establish the spinel $\leftrightarrow$ rocksalt phase transition. Such phase transition definitively explains the well-defined plateau observed in the electrochemical potential curves. In the case of $Li_2Ti_3O_7$ ramsdellite, all the iron ions are located on octahedral sites and the quadrupole splittings are related to the number of lithium in the neighbourhood of probed atoms.

Key words: electrode materials, lithium insertion mechanism, titanium-based oxides, ^{57}Fe Mössbauer data.

1. Introduction

At the present time the lithium-ion technology is the preferable portable power source but the development of high energy and power batteries for other applications, such as portable power tools or hybrid vehicles, leads to intensive world-wide research about new electrode materials and electrolytes [1]. Among the new anode materials, the spinel titanate $(Li)_3[LiTi_5]O_{12}$ allows to insert three lithium atoms per formula unit at a potential of 1.55 V leading to a theoretical capacity of 175 A h/kg [2, 3]. Recently a new type of lithium-cell based on the combination of this high voltage spinel anode with a high voltage spinel cathode has been evidenced [4]. An alternative for anodic material is the $Li_2Ti_3O_7$ ramsdellite.

---

* Author for correspondence.

2. Experimental

The spinel $Li_4Ti_5O_{12}$ and ramsdellite $Li_2Ti_3O_7$ phases was obtained by sol-gel as well as ceramic synthesis routes, using various precursors (Li_2CO_3, LiOH, TiO_2, $Ti(OiPr)_4$, $Li(CH_3CO_2)\cdot 2H_2O$). Substituted compounds have been precipitated from solutions of titanium iso-propoxide and various lithium and dopant precursors ($FeCl_3\cdot 6H_2O$). Iron doping was alternatively carried out from metal powder enriched in the isotope ^{57}Fe in order to introduce a probe for ^{57}Fe Mössbauer spectroscopy. Electrochemical lithiation was carried out with (Li/LiPF6 1M (EC:DMC)/ lithium titanate) Swagelock^{TM} test cells as described in [5]. Galvanostatic discharge/charge curves were obtained with cycling rate of 1 Li/5 h. Electrochemically inserted samples for Mössbauer characterisation were prepared using slow rate of 1 Li/20 h.

Purity of pristine and lithiated samples were characterised by X-ray powder diffraction (XPD).

Iron doped materials and some samples at several depths of discharge have been characterised by ^{57}Fe Mössbauer spectroscopy at room temperature with classical EG&G constant accelerator spectrometer in transmission mode using a ^{57}Co in a Rh matrix as γ-ray source. The velocity scale was calibrated using the magnetic sextuplet spectrum of a high purity iron foil absorber and the origin of isomer shift scale was determined from the centre of the α-Fe spectrum. For the electrochemically inserted samples the measurements were performed *ex situ* from Swagelock electrodes. Experimental data were analysed by evaluation of hyperfine parameters with the procedure described in [6].

3. $Li_4Ti_5O_{12}$

Electrochemical tests of $Li_4Ti_5O_{12}$ have been done in Swagelock cell. Typical discharge/charge curves are shown in Figure 1. Mössbauer spectroscopy gives valuable informations concerning the local environment of iron atoms (^{57}Fe probe) in iron-doped spinel. From the pristine material, the fitting of the experimental spectrum evidenced the occurrence of two doublets at isomer shift values which agree well with the presence of Fe^{III} in both tetrahedral ($\delta = 0.22$ mm/s, $\Delta = 0.29$ mm/s) and octahedral ($\delta = 0.65$ mm/s, $\Delta = 0.65$ mm/s) (Table I and Figure 2). Fraction of iron atoms located in tetrahedral sites depends on the Li/Ti ratio.

During the first discharge one can observe the $Fe^{III} \rightarrow Fe^{II}$ reduction and the migration of Fe atoms from tetrahedral to octahedral sites. At the end of the first discharge (1 V) all the Fe atoms are located in octahedral sites in agreement with a spinel $\rightarrow$ rocksalt phase transition according to the reaction:

$$(Li_3)_{8a}[LiTi_5]_{16d}O_{12} + 3Li \rightarrow [Li_6]_{16c}[LiTi_5]_{16d}O_{12}.$$

Defects present on tetrahedral site prevents the insertion mechanism [2].

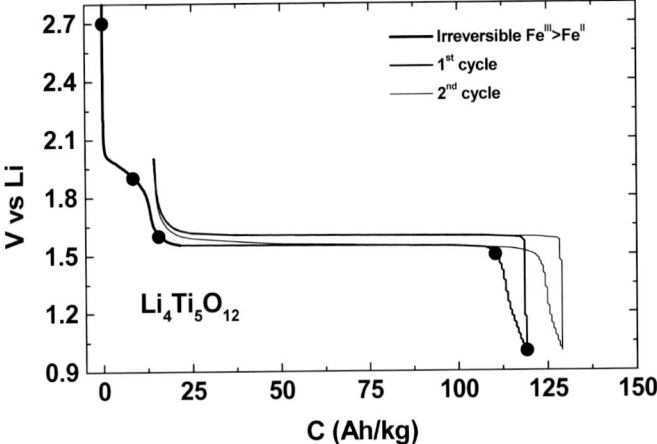

Figure 1. Discharge/charge curve of pristine $Li_4Ti_5O_{12}$ upon Li-insertion. Mössbauer experiments were performed at different discharge depths labelled a, b, c, d.

Table I. ^{57}Fe hyperfine parameters of $Li_{4.25+x}Ti_{4.75}Fe_{0.25}O_{12}$, with x, the number of lithium electrochemically inserted. IS = Isomer shift relative to α-Fe; QS = weighted average values of the quadrupole splitting; Γ = line width at half maximum; A = contribution of sub-spectra to the spectrum

x	Sample	IS (mm/s)	QS (mm/s)	Γ (mm/s)	A (%)	Attribution
0	–	0.22(1)	0.29(1)	0.35(1)	33(1)	Fe^{III} tetra
		0.65(1)	0.65(1)	0.66(1)	67(1)	Fe^{III} octa
0.15	a	0.29(1)	0.26(1)	0.40(3)	26(1)	Fe^{III} tetra
		0.35(1)	0.67(1)	0.29(2)	13(1)	Fe^{III} octa
		0.92(1)	1.16(1)	0.85(2)	61(1)	Fe^{II} octa
0.25	b	0.20(1)	0.19(1)	0.29(2)	18(1)	Fe^{III} tetra
		1.01(1)	0.96(2)	0.37(2)	82(1)	Fe^{II} octa
2.03	c	1.01(1)	1.21(1)	0.41(3)	100	Fe^{II} octa
2.20	d	1.01(1)	1.19(1)	0.43(2)	100	Fe^{II} octa

Specific heat treatment of the pristine $Li_4Ti_5O_{12}$ (spinel) gives at high temperature (above 1000°C) the pure ramsdellite structure $Li_2Ti_3O_7$ [7] (space group Pbnm) with vacancies on the metallic atom sites according to the $[Li_{2.29}\square_{1.71}]_c$ $[Ti_{3.43}\square_{0.57}]_fO_8$ structural formulae ($\square$ = vacancies, c = channel, f = framework). By doping with iron atoms, ramsdellite is obtained at lower temperature (800°C) [8].

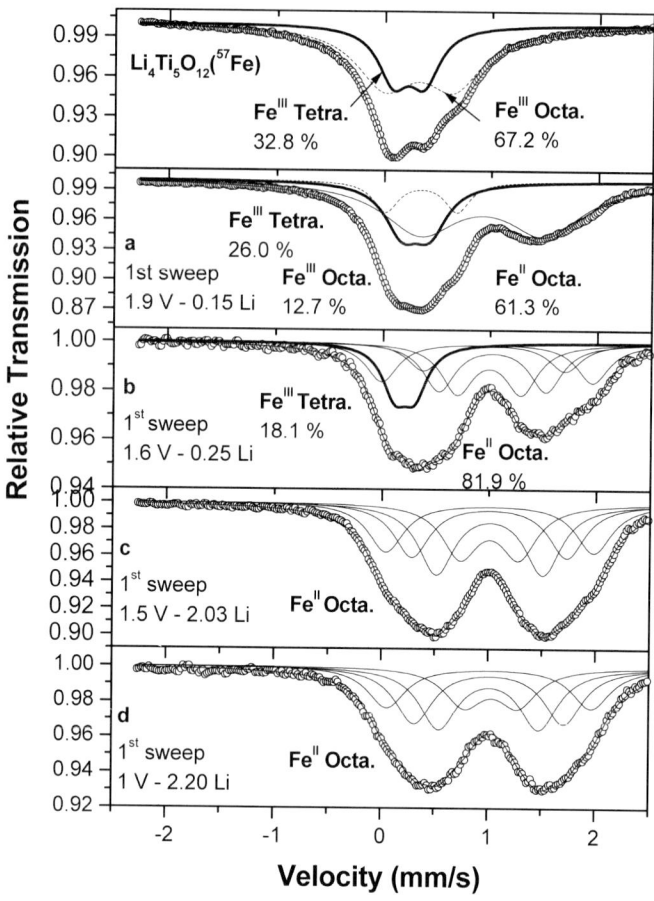

Figure 2. ^{57}Fe Mössbauer spectra of pristine iron doped $Li_4Ti_5O_{12}$ before Li-insertion and after various discharge depths.

4. $Li_2Ti_3O_7$

As shown on Figure 3, no phase transition occurs during insertion resulting in a charge/discharge 'S'-shaped curve characteristic of a one-phase system. Li atoms are located in the partially filled channels of the ramsdellite structure. Charge and discharge in this material correspond to extraction and insertion into these channels in agreement to a one-phase mechanism:

$$[Li_{2.29}\square_{1.71}]_c[Ti_{3.43}\square_{0.57}]_f O_8 + xLi \rightarrow Li_{2.29+x}Ti_{3.43}O_8.$$

In Figure 4, we compare at the same velocity scale, the spectra of the pristine $Li_2Ti_3O_7$, in which sub-spectra correspond to iron atoms located into octahedral sites with an oxidation state Fe^{III} and of the lithiated compounds. The different observed quadrupole splittings are related to the probability to find Li atoms within the channels in the neighbourhood of the probed iron atoms (Table II).

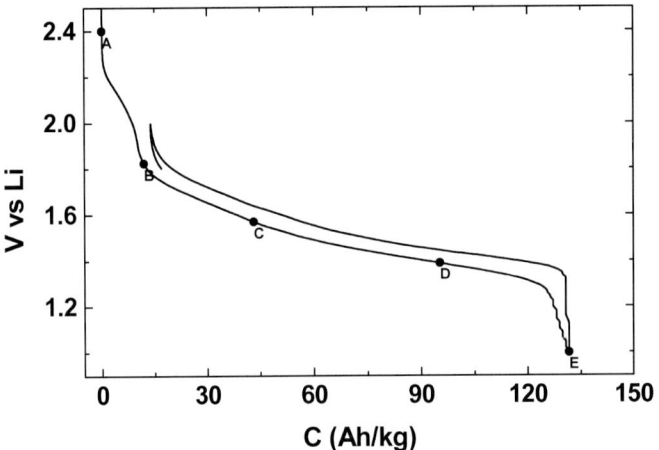

Figure 3. Discharge/charge curve of iron doped $Li_2Ti_3O_7$ showing different samples A, B, C, D, E investigated by ^{57}Fe Mössbauer spectroscopy.

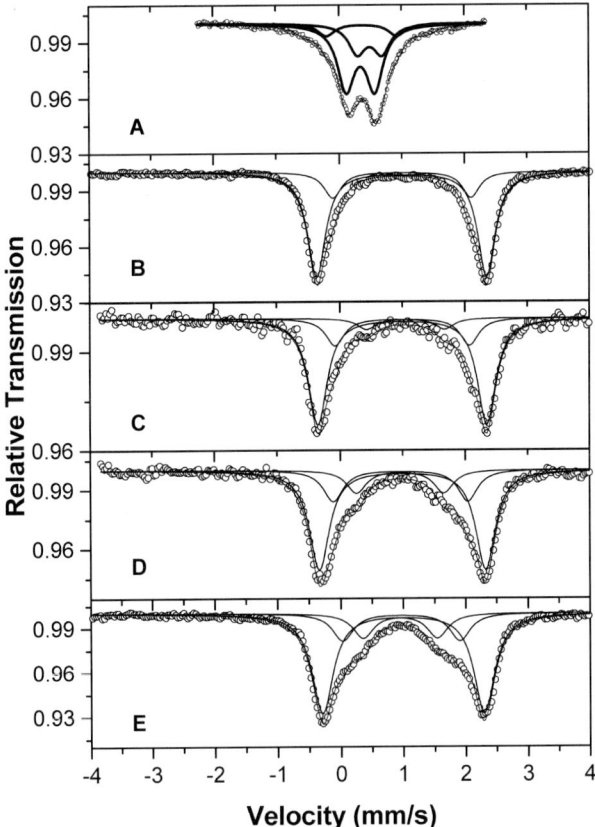

Figure 4. ^{57}Fe Mössbauer spectra of pristine iron containing $Li_2Ti_3O_7$ before (*top*) and after Li-insertion/extraction.

Table II. [57]Fe hyperfine parameters of $Li_{2.29+x}Ti_{2.83}Fe_{0.13}O_7$, with x, the number of lithium electrochemically inserted. IS = Isomer shift relative to α-Fe; QS = quadrupole splitting; Γ = line width at half maximum; A = contribution of sub-spectra to the spectrum

x	Sample	IS (mm/s)	QS (mm/s)	Γ (mm/s)	A (%)	Attribution
0	A	0.49(1)	0.40(1)	0.31(1)	32(1)	Fe^{III} octa
		0.33(3)	0.45(1)	0.31(1)	55(1)	Fe^{III} octa
		0.37(1)	1.12(1)	0.31(1)	13(1)	Fe^{III} octa
0.15	B	0.99(1)	2.71 (1)	0.34(3)	80(2)	Fe^{II} octa
		1.00(1)	2.21(2)	0.34(3)	20(2)	Fe^{II} octa
0.70	C	1.00(1)	2.70(1)	0.36(4)	74(3)	Fe^{II} octa
		0.99(2)	2.15(3)	0.36(4)	19(3)	Fe^{II} octa
		1.05(4)	1.24(5)	0.36(4)	7(3)	Fe^{II} octa
1.30	D	0.99(1)	2.66(1)	0.38(2)	65(5)	Fe^{II} octa
		0.96(2)	2.15(3)	0.38(2)	20(5)	Fe^{II} octa
		0.95(1)	1.39(2)	0.38(2)	15(5)	Fe^{II} octa
1.45	E	1.00(1)	2.59(1)	0.40(3)	66(5)	Fe^{II} octa
		0.97(2)	1.87(3)	0.40(3)	18(5)	Fe^{II} octa
		0.95(3)	1.18(4)	0.40(3)	16(5)	Fe^{II} octa

Then, upon discharge, Fe^{III} is reduced into Fe^{II}, corresponding to the shoulder observed in the discharge curve in Figure 3 between 2.2 and 1.8 V. As can be seen in Mössbauer spectra, Fe^{II} remains present upon Li-insertion. The increasing small contributions are due to increasing number of Li atoms in the neighbourhood of iron atoms.

5. Conclusion

Electrochemical behaviour of doped material has been investigated by Mössbauer spectroscopy. Information on local environments in terms of vacancies or lithium occupation has been achieved. Mössbauer spectroscopy has been used to give a more precise description of the Li-insertion mechanisms involved in two different cases chosen among the lithium titanate family.

Acknowledgements

The authors would like to thank EC for funding (NEGELiA, contract ENK6-CT-2000-00082). The authors are grateful to Ph. Biensan and C. Tessier from SAFT and J. Scoyer, A. Audemer and S. Levasseur from UMICORE for their fruitful discussions.

References

1. *11th International Meeting on Lithium Batteries, IMLB-11*, Monterey, California, June 23–28, 2002.
2. Kubiak, P., Garcia, A., Womes, M., Aldon, L., Olivier-Fourcade, J., Lippens, P. E. and Jumas, J. C., *J. Power Sources* **119–121** (2003), 626.
3. Ozhuku, T., Ueda, A. and Yamamoto, N., *J. Electrochem. Soc.* **142** (1995), 1431.
4. Panero, S., Satolli, D., Salomon, M. and Scrosati, B., *Electrochem. Comm.* **2** (2000), 810.
5. Denis, S., Baudrin, E., Touboul, M. and Tarascon, J. M., *J. Electrochem. Soc.* **144** (1997), 4099.
6. Ruebenbauer, K. and Birshall, T., *Hyp. Interact.* **7** (1979), 125.
7. Morosin, B. and Mikkesen, Jr., J. C., *Acta Cryst. B* **35** (1979), 798.
8. Ma, S. and Noguchi, H., *Electrochem.* **69**(7) (2001), 526.

Hyperfine Interactions **156/157**: 505–521, 2004.
© 2004 *Kluwer Academic Publishers. Printed in the Netherlands.*

Arrangements of Interstitial Atoms in *fcc* Fe–C and Fe–N Solid Solutions

J. DESIMONI
Departamento de Física, Facultad de Ciencias Exactas, UNLP, IFLP-CONICET, C.C. 67,
1900 La Plata, Argentine; e-mail: desimoni@fisica.unlp.edu.ar

Abstract. The distribution of C and N atoms in the octahedral interstitial sites of the face-centred-cubic austenite phase of the Fe–C and the Fe–N alloys is controversial. In this work, Mössbauer experiments, the quasichemical approximation, the hard-blocking excluded-sites model, the chemical activity data, electron charge calculations and Monte Carlo simulations have been combined to advance in its understanding. A database is developed, with analyses of Mössbauer spectra using models assuming either ordered or random distributions of the interstitial atoms in the interstices around an Fe atom. The data are compared as a function the fraction of occupied sites, and various striking differences between Fe–N and Fe–C alloys are discussed. The experimental trends are confronted with predictions of combined theoretical approaches.

Key words: solute distribution, iron austenite, Monte Carlo simulations.

1. Introduction

Interstitial solutes like N and C affect the mechanism and kinetics of phase transformations in iron-based alloys. The determination, assessment and explanation of
the properties of the Fe–X (X = C, N) alloys have been an issue of long-standing
theoretical and practical interest. The austenite solid solution phase has attracted
considerable attention, in connection with the modelling of thermodynamic properties, the prediction of phase diagrams, the diffusion controlled reactions and the
martensitic phase transitions [1–8].

In austenite, the iron atoms are arranged in a close-packed face-centred cubic
(*fcc*) lattice, and the smaller X atoms occupy the crystallographically equivalent
octahedral interstices located at the centres and at the midpoints of the edges of the
unit cell [9]. Two sublattices are distinguished, one for the Fe atoms and the second
one for the mixture of interstitial X atoms and vacant (V) octahedral interstices. In
spite of the several attempts performed to obtain the interaction between solute
atoms from the Mössbauer [10–24] and activity [25, 26] data and the numerous
articles dealing with the distribution of solute atoms in the interstitial sites, up to
now, there has been no full understanding on the way the interstitial atoms are
distributed; different authors drew different conclusions about the atomic structure
of Fe based C and N alloys and a wide range of X–X energies have been reported.

In this article, the analysis of the different approaches in the interpretation and deconvolution of the Mössbauer spectra of the *fcc* Fe–C and Fe–N phases will be presented. The hard-blocking excluded-sites model [8, 27] and the quasichemical approximation [28, 29] were employed to analyse the thermodynamic data. Monte Carlo simulations performed to obtain the relative fractions of the different iron environments in the austenite lattice [17, 30, 31] or the energy interactions between the pairs of interstitial atoms [16, 22–24] will be discussed. Electron density charge calculations [32, 33] will be summarized.

2. Thermodynamic models and database

The activity (a_X) of the interstitial element as a function of composition is the key property considered in thermodynamic studies. The ideal solution model for thermodynamic properties of austenite assumes that X and V distribute themselves at random in the octahedral interstitial sites, the amount of which is equal to the number of Fe atoms. On these basis, the thermodynamic activity of X in an ideal mixture of N_X solute atoms with N_{Fe} iron atoms is shown to be proportional to the ratio $y/(1 - y)$, where $y = N_X/N_{Fe}$ and $1 - y$ represent the fraction of occupied and of empty interstitial sites, respectively [34–37]. Since the experimental a_C and a_N in austenite deviates positively from the ideal solution model, many approaches have been proposed to account for the non-ideal behaviour [8, 25–27, 38–44].

The quasichemical approximation [28, 29, 37, 45–47] assumes that all interstitial sites are available for occupation, but the X atoms are regarded as exerting a repulsive force on each other, entering in adjacent interstitial positions less frequently as would be the case if their distribution were at random. The energy of formation of a X–X pair, $\Delta\varepsilon_X = 2\varepsilon_{X-V} - \varepsilon_{X-X}$, where ε_{X-X} and ε_{X-V} are the interaction energies of the X–X and X–V pairs, can be determined by analysing experimental a_X data in terms of the approximation developed by Bhadeshia [28, 29]. A linearized form,

$$RT \ln\left[a_X \frac{(1 - y)}{y}\right] = yZ\Delta\varepsilon_{X-X} + \Delta G_X$$

appropriate for describing the dilute solution range was fitted to C activity data [25] measured at 1423 K and to N activity [26] determined in the range 933–1083 K (Figure 1). These analyses yielded values for $\Delta\varepsilon_X$ and for the Gibbs energy difference ΔG_X between C and N in austenite and their stable form. The parameter values for the Fe–N system are $\Delta\varepsilon_N = 770_{26}$ cal/mol and $\Delta G_N = 13611_{27}$ cal/mol [30]. In the Fe–C system Laneri *et al.* [17] obtained a similar value, $\Delta\varepsilon_C = 1492_{39}$ cal/mol. The ΔG parameter, on the other hand, is significantly different, viz., $\Delta G_C = 4451_{25}$ cal/mol. The $\Delta\varepsilon_N$ value is slightly lower that those reported in [7] ($\Delta\varepsilon_N = 950$ cal/mol).

The approach known as the hard-blocking excluded-sites model assumes that the presence of a solute atom blocks the occupancy of a certain number (**b**) of the

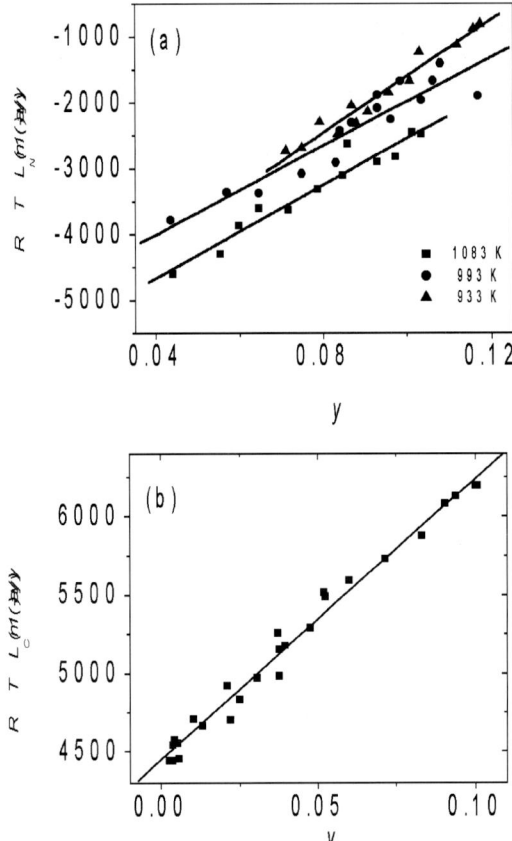

Figure 1. First order quasichemial approximation analysis of the activity data of: (a) Fe–C [25] and (b) Fe–N [26] systems, respectively.

nearest neighbour interstitial sites, so that a site is either blocked or is available for the mixing of X atoms and V empty sites [8, 9, 27, 39]. Further, if the mixing in the non-blocked sites occurs at random, the a_X in austenite becomes proportional to $y/[1-(1+b)y]$ [8]. Frequently, b has been treated as a fitting parameter, identified with the value to be inserted in the expression for a_X in order to reproduce the experimental data [25, 26]. Alternatively, some theoretical studies have suggested that b is composition dependent [42]. The reduction of activity data to:

$$\frac{y}{a_X} = 1 - (b+1)y$$

allows to directly extract the average of b. An analysis of the activity data [25, 26] plotted in Figure 2 for both systems indicates that the blocking effects are less important in Fe–C austenite ($b = 4.3 \pm 0.2$) than in Fe–N ($b = 6.0 \pm 0.4$).

Both models arrive to contradictory results, i.e. the quasichemical approximation should indicate a slightly N–N pair formation energy and the hard-blocking excluded-site model a higher blocking effect for the Fe–N system.

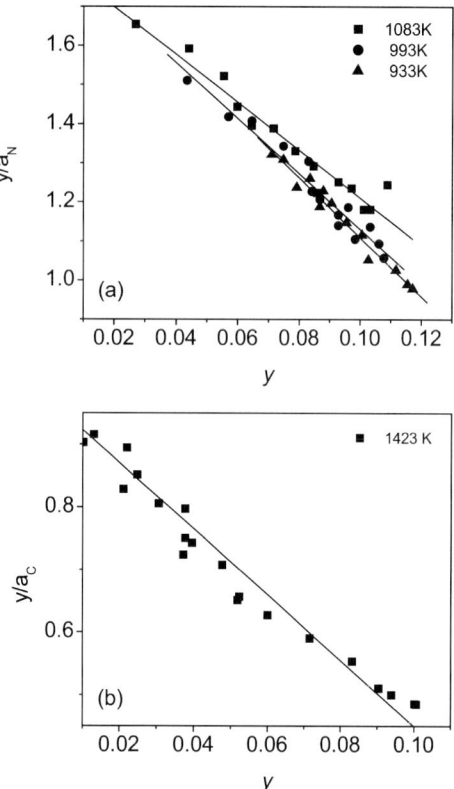

Figure 2. Hard-blocking excluded-site model analysis of the activity data of: (a) Fe–C [25] and (b) Fe–N [26] systems, respectively.

3. The Mössbauer pattern and database

The Mössbauer spectrum of austenite can be decomposed in two kinds of components. The single lines are ascribed to Fe atoms without solute atoms in the first interstitial shell and the quadrupole interactions arise from the change of the charge density of the cubic symmetry, due to the presence of interstitial atoms in the neighbourhood. The magnitude of the hyperfine parameters depends on the type of interstitial atom and on their number, i.e. in the differences of the electronic structure. In general, the quadrupole splitting values are smaller for Fe–N than in the Fe–C case, which suggests that the perturbation of the electronic structure of an iron atom by the presence of a neighbouring interstitial is greater for C than for N [32].

Two alternative models have been proposed to describe the distribution of X atoms in the Fe–X austenite solutions. The first one assumes an ordered structure (OSM) [10–12] picture while the second proposes a random structure (RSM) [13–16].

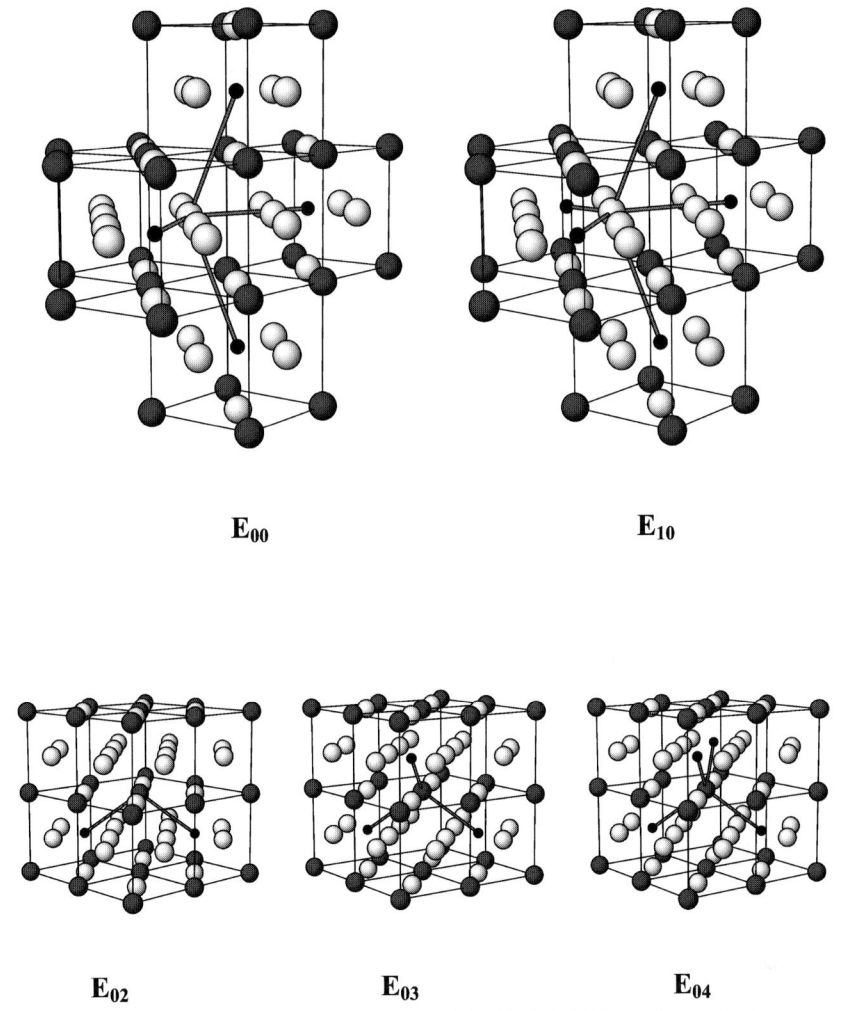

E_{00} E_{10}

E_{02} E_{03} E_{04}

Figure 3. Fe environments in the order structure model [10–12]. White and gray circles correspond to Fe atoms. Small black circles represented X atoms.

In the OSM [10–12], the solute atoms are placed at the center of the cube in the *fcc* Fe lattice, only and the occupation by the solute atoms of the first (p) and second (q) interstitial shells are considered. Figure 3 sketched three different iron environments:

E_{00}: Fe atoms without nearest neighbour and next nearest neighbour interstitial atoms, associated to a singlet.

E_{0n}: Fe atoms without nearest neighbour but with n next nearest neighbour X atoms ($n = 1$–4), ascribed to a singlet.

E_{10}: Fe atoms with one solute atom nearest neighbour but without next nearest interstitial neighbors, related to a doublet.

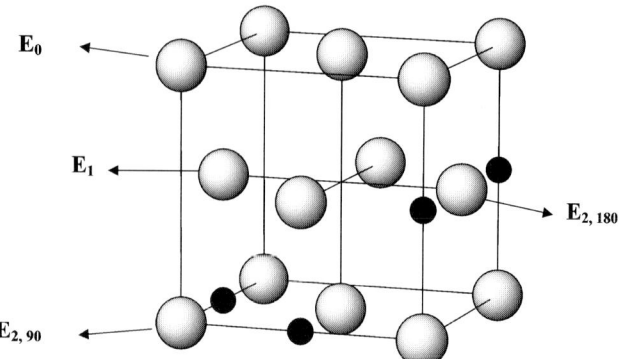

Figure 4. Fe environments in the random structure model [13–16]. Gray circles correspond to Fe atoms. Small black circles represented X atoms.

In the RSM [13–16] only the occupation of the first interstitial shell is treated, and all the octahedral sites are available for occupation by interstitial atoms. The four Fe environments proposed are displayed in Figure 4:

E_0: Fe atoms without nearest neighbours solute atoms, associated to a singlet.

E_1, $E_{2,90}$: Fe atoms with one nearest neighbour solute atom, or Fe atoms with two interstitial atoms nearest neighbours at 90° from each other, related to a doublet.

$E_{2,180}$: Fe atoms with two solute atoms placed at opposite nearest sites (180°), ascribed to a doublet.

The Mössbauer spectra of Fe–C alloys obtained for various C contents are shown in Figure 5 [10, 11, 17]. The samples corresponding to the data in Figure 5(a) also contain the ferromagnetic martensite and ferrite phases [48], hence the spectra show the paramagnetic signal of austenite placed between the internal lines of the sextet associated to martensite and ferrite [49, 50]. Similar spectra for austenite were obtained by Oda *et al.* [13–16] in transmission geometry (Figure 5(b)), and by Uwakweh *et al.* [11] using conversion electron Mössbauer spectroscopy (Figure 5(c)).

The alloys studied by Gènin *et al.* [10, 11] were prepared by carburizing in CH_4/H_2 mixtures $1333 < T < 1510$ K (Figure 5(c)). Their results were interpreted in terms of the OSM, with the ordered Fe_8C structure [10, 11] using three interactions E_{00}, E_{0n} and E_{10} to reproduce the MS. Oda *et al.* [13–16] studied electropolished ribbons obtained by melt-spun on a rotating wheel in He atmosphere. The data were analysed using the RSM. Finally, Laneri *et al.* [17] recently analysed the Fe–C spectra showed in Figure 5(a) using both the OSM [10–12] and the RSM [13–16]. The hyperfine parameters and the relative fractions obtained from the different experiments are listed in Table I.

Figure 6 presents the Mössbauer spectra corresponding to Fe–N alloys with $y = 0.10$ to $y = 0.11$, prepared by nitriding Fe with a NH_3/H_2 gas mixture at $866 < T < 973$ K [12–16, 18–22]. Fall and Génin [12], adopted the OSM to

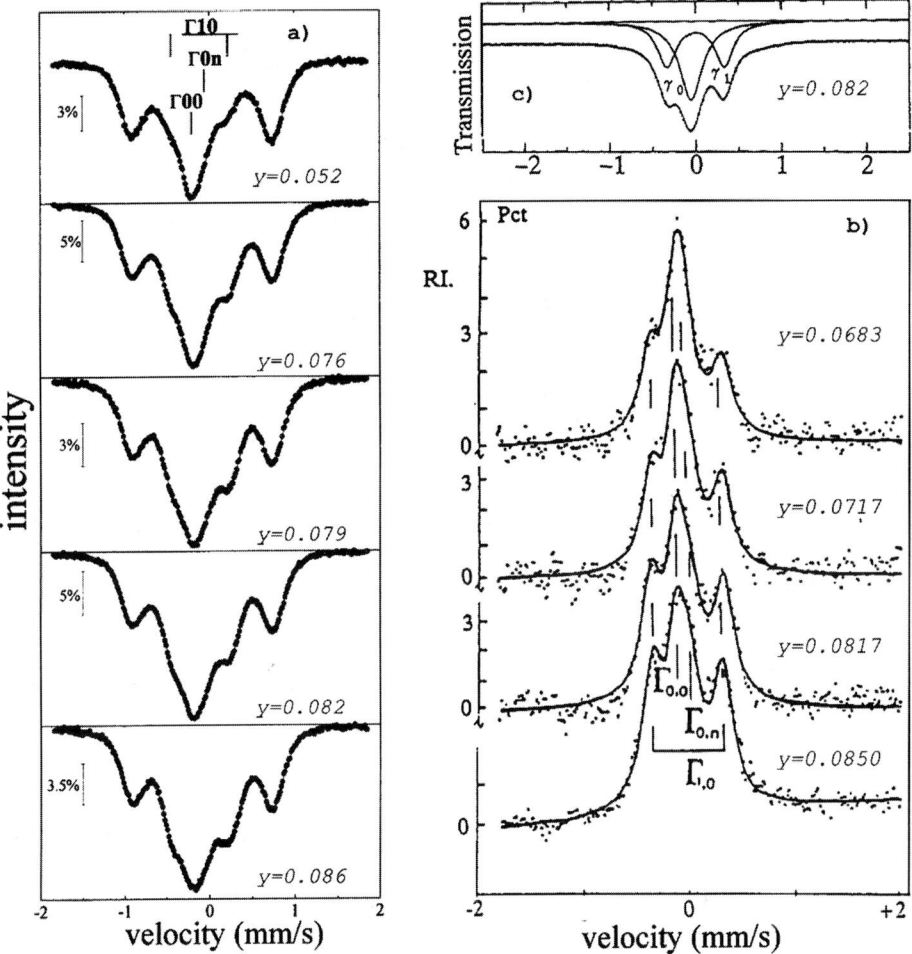

Figure 5. Mössbauer spectra recorded on Fe–C austenite: (a) [17], (b) [10, 11] and (c) [16].

analyze the spectra in terms of five Fe environments, adding the E_{1m} and E_{2m} environments to those used in the analysis Fe–C data [10, 11]. Giellen and Kaplow [18] and De Cristofaro and Kaplow [19] used only two hyperfine interactions in the framework of the RSM. Foct *et al.* [20, 21] improved this model by adding a third interaction to account for Fe atoms with two next nearest neighbor N atoms. However, Oda *et al.* [13–16] attributed such interaction to atoms located at 180°. The results of these various treatments are compared in Table I.

The main differences in the analysis of the Mössbauer pattern of Fe–C and Fe–N alloys are reflected in inclusion of additional hyperfine interactions to reproduce the Fe–N spectra (E_{1m} and E_{2m} for the analysis with the OSM [10–12] and E_2 for the case of the RSM [13–16]). A noticeable increase in the isomer shift values of the interaction associated to E_{10} and the decrease of the quadrupole splitting of E_1 is observed. Slightly higher the isomer shift values of the Fe sites with no

Table I. The isomer shift (δ), quadrupole splitting (Δ) and relative fractions of the different Fe environments after applying OMS [10–12] and RSM [13–16] on the set of data of [10–17] and [19–22]. The isomer shifts are referred to α-Fe at room temperature

OMS

y	E_{00} δ (mm/s)	E_{00} f_{00} (%)	E_{0n} δ (mm/s)	E_{0n} f_{0n} (%)	E_{10} δ (mm/s)	E_{10} Δ (mm/s)	E_{10} f_{10} (%)	E_{1m} δ (mm/s)	E_{1m} Δ (mm/s)	E_{1m} f_{1m} (%)	E_{2m} δ (mm/s)	E_{2m} Δ (mm/s)	E_{2m} f_{2m} (%)	Reference
					Fe–C system									
0.052	−0.10	43	0.05	16	0.01	0.66	41							[17]
0.0683	−0.10	39	−0.018	20	−0.007	0.655	41							[10, 11]
0.0717	−0.10	36	0.007	21	−0.003	0.655	43							[10, 11]
0.076	−0.10	33	0.05	23	0.01	0.67	44							[17]
0.079	−0.10	29	0.06	23	0.01	0.67	48							[17]
0.0817	−0.10	33	0.027	18	−0.006	0.67	49							[10, 11]
0.082	−0.10	27	0.05	21	0.01	0.66	52							[17]
0.0850	−0.10	30	0.029	19	−0.004	0.67	51							[10, 11]
					Fe–N system									
0.102	−0.11	4.4	0.02	31.2	0.20	0.67	7.4	0.09	0.37	54.8	0.29	0.95	2.2	[12]
0.105	−0.11	1.1	0.05	28.4	0.20	0.67	9.9	0.10	0.37	57.4	0.26	0.95	6.8	[12]
0.111	−0.11	0.3	0.06	21.8	0.20	0.67	8.5	0.11	0.37	66.4	0.20	0.95	3.0	[12]

RMS

y	E_0 δ (mm/s)	E_0 f_0 (%)	E_1 δ (mm/s)	E_1 Δ (mm/s)	E_1 f_1 (%)	E_2 δ (mm/s)	E_2 Δ (mm/s)	E_2 f_2 (%)	Reference
					Fe–C system				
0.052	−0.07	57	−0.01	0.61	43				[17]
0.076	−0.05	52	0.01	0.62	48				[17]
0.079	−0.04	49	0.02	0.63	41				[17]
0.082	−0.04	50	0.02	0.63	50				[17]
0.082	−0.05	54.5	0.00	0.67	45.5			<0.6	[16]
					Fe–N system				
0.096	−0.16	45	0.05	0.29	55	0.30	0.40		[19]
0.098	−0.10	23	0.15	0.25	75	0.20	0.72	2	[20, 21]
0.099	0.01	47	0.08	0.39	50	0.09	0.66	3	[13–16]
0.110	−0.09	37	−0.12	0.38	59			4	[22]

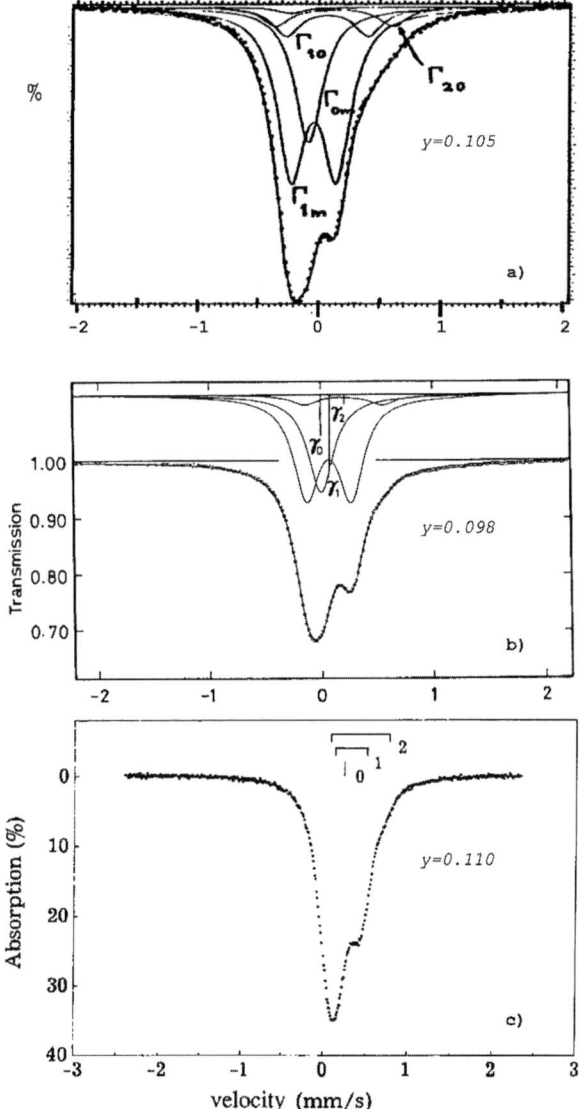

Figure 6. Mössbauer spectra recorded on Fe–N austenite: (a) [12], (b) [13–16] and (c) [20, 21].

interstitial atoms in the first coordination shell (E_0 and E_{00}) for Fc–N alloy arc detected indicating a decrease in the density of charge of s-electrons.

The composition dependence of the f_{pq} and f_q fractions obtained by analysing Fe–C and Fe–N spectra using the OSM [10–12] and the RSM [13–16] is displayed in Figures 7 and 8, respectively. In the Fe–C system f_{10} is the largest, but becomes lower that 10% in Fe–N austenite. On the other hand, the f_{1m} contribution, which is absent in the Fe–C results, is the dominant one in Fe–N. These results support a picture of the Fe–N austenite where each Fe atom has one N atom in the first

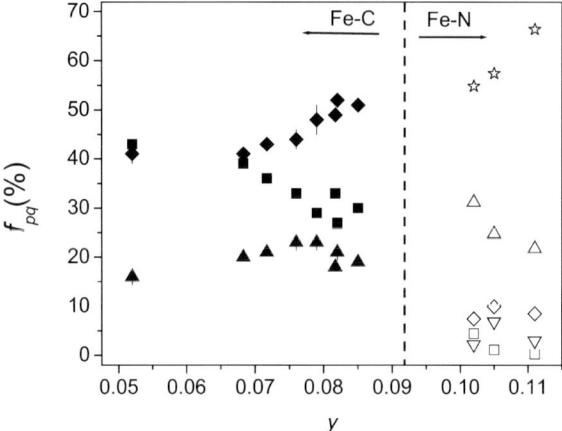

Figure 7. f_{pq} relative fractions obtained from the analysis of the spectra considering the OSM [10–12] of the Fe–N and Fe–C systems. (a) Squares: f_{00}, (b) rhombs: f_{10}, (c) up triangles: f_{0n}, (d) stars: f_{1m} and (e) down triangles: f_{2m}.

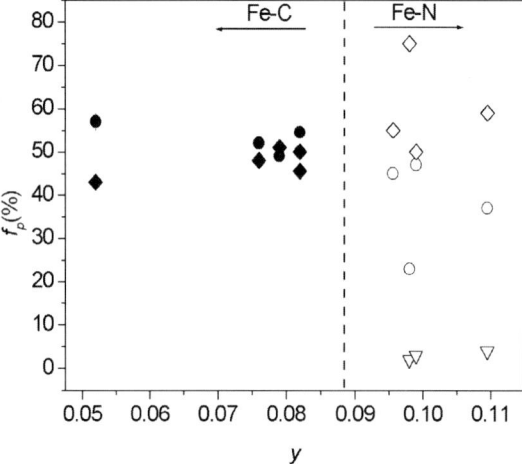

Figure 8. f_q relative fractions obtained from the analysis of the spectra considering the RSM [13–16] of the Fe–N and Fe–C systems. (a) Circles: f_0, (b) rhombs: f_1, and (c) down triangles: f_2.

shell, and one or more in the second. This suggests a weaker blocking effect of the atoms in the first shell upon those in the second, which in general should favor a more random distribution. This possibility has already been suggested by Fall and Genin [12], on the basis of a large f_{1m} together with f_{2m}. However, the contrary is derived from the analysis of activity data using the hard-blocking excluded sites model [8, 9, 27, 39].

After the analysis of the spectra with the RSM [13–16], the most important contributions in both the Fe–C and Fe–N are the same, namely, f_0 and f_1, with very similar values (see Figure 8). This fact supports a picture where the Fe–C and the Fe–N austenite do not differ markedly in their degree of randomness as suggested by the thermodynamic data. Indeed, the $E_{2,180}$ environments have been considered in the Fe–N treatments, and Oda *et al.* [13–16] interpreted it as indications of a more ordered structure of the γ'-Fe$_4$N type [51]. However, the magnitude of such contribution, less than 5%, is hardly significant, and should not be taken as a strong evidence of ordering.

4. Monte Carlo simulations

Two main lines have been followed to perform the Monte Carlo simulations. The first set of simulations was performed refining the X–X pair energies up to the relative fractions obtained from the simulations agree with those determined from the analysis of the Mössbauer spectra [16, 22–24]. The resulting interactions energies were used to calculate the activity and compare it with the experimental data. The other simulations were done obtaining the X–X pair formation energy from the thermodynamic data analysed using the first order quasichemical approximation [17–30]. The simulated and experimental relative fractions were compared to study the validity of the interstitial atom distributions proposed in the OSM [10–12] and the RSM [13–16].

The first attempt to estimate the C–C and N–N interaction energies using Monte Carlo simulations from the Mössbauer data, obtained from Fe 8.0 at%C and Fe 9.0 at%N samples, was performed by Oda and co-workers [16]. The random model was proposed with interactions up to the second interstitial shell. The values obtained from the simulations resulted more repulsive than those obtained from the activity data by assuming the dilute approximation. The interaction energies in the nearest and second nearest C–C pairs were strongly repulsive contrary to the weak interaction in the second nearest N–N pair.

Sozinov *et al.* [23] used the lattice gas model with C–C interactions up to the third coordination shell and 773 K as freezing temperature to perform Monte Carlo simulations of C distribution in the Fe–C solid solution with $0.683 < y < 0.0825$. The interactions energies were varied up to the calculated relative fractions agree with the experimental Mössbauer ones reported in [11], subsequently, the activity was calculated. Based on the results of activity calculations, the ordered structured model [10–12] was disregarded. The thermodynamic data can be successfully described considering a weak C–C repulsion in the first coordination shells and a stronger one in the second shell. Later, a estimation of the interactions energies between interstitial atoms was obtained for Fe–7.66 at%C and Fe–9.87 at%N systems considering the strain-induced contribution (X–X interactions up to the 6th coordination shell) added to the two near coordination shell interactions [22]. The calculations based on the statistical-thermodynamic approach [52, 53] considering only interaction in the first shell did not reproduce the Mössbauer pattern, then interactions up to the second shell was accounted. The strain-induced contribution resulted more important in Fe–C than in Fe–N alloy. In addition, the interaction between N atoms in the first shell was stronger than between C atoms as was claimed by Oda *et al.* [16]. The N–N interaction compatible with the Mössbauer data was also estimated using Monte Carlo simulations and tested against the thermodynamic data set [23]. Based in the random model, a weak repulsion in the second coordination shell and a stronger in the first one were found. This interaction energy values are consistent with the absence of N pairs at $90°$ and with the presence of N pairs at $180°$, approaching the N distribution to that of an ordered structure of the γ'-Fe$_4$N phase [51]. However, the activity data [25, 26] cannot be reproduced with the interaction energies obtained from the Mössbauer analysis.

The strain-induced interaction of $X–X$ pairs was numerically calculated for 7.7 at% and 8 at% Fe–C austenites [31]. The evaluation of the chemical activity and the relative fractions of the different iron sites showed that the model is applicable but a strong C–C repulsion should be included to fully describe the system.

Finally, a comparison of the relative fractions obtained from Monte Carlo simulations [17, 30] with the experimental were performed covering the whole concentration set of Fe–C and Fe–N alloys studied using Mössbauer spectroscopy [10–17]. The pair energy formation entering in the simulations was obtained from the analysis of the activity [25, 26] data with the first order approximation of the quasichemical model [28, 29]. The validity of the simulations was tested against the results obtained from the quasiqhemical calculations of the number of $X–X$, $X–V$ and $V–V$ pairs. Figure 9(a) presents the f_{pq} values from MC simulations obtained in the range $0 < y < 0.25$, whereas Figure 9(b) focuses on the composition range where Mössbauer experiments have been performed. In such region the main contributions come from the E_{0n} and the E_{1m} environments. The f_{00} and f_{10} relative fractions decrease with increasing X content, whereas f_{2m} increases from 2 to 10%. The resulting composition dependence of the theoretical f_0 and f_{00} fractions is compared in Figure 10 with Mössbauer results for Fe–C [10, 11, 15–17] and Fe–N [12–16, 19, 20, 22] austenite. The Monte Carlo calculated f_0 follows the experimental data as much as one could expect, in view of the discrepancy already [17] found in the Fe–C system. However, the simulations overestimate the f_{00} by approximately 20%, which is natural, because the theoretical f_{00} versus y relation for Fe–N in the right-hand side of Figure 10 almost coincides with a smooth extrapolation of the calculated line in the Fe–C side [17]. Consequently, it was claim that the C distribution in the *fcc* phase is better reproduced with the OSM [10–12] while in the case of Fe–N alloys, none of the models resulted satisfactory, in spite that from the thermodynamic point of view the systems are quite similar.

5. Electronic structure calculations

Timoshevki *et al.* [32] modelled the austenite with Fe$_8X$ ordered structures using full-potential linearized augmented-plane-wave. Only the occupation of the first interstitial shell and two types of Fe environments (Fe atoms with no near neighbours and one interstitial near neighbour) were considered as well as the local magnetic moments. These results proved that only Fe–C alloy could represent by the or-

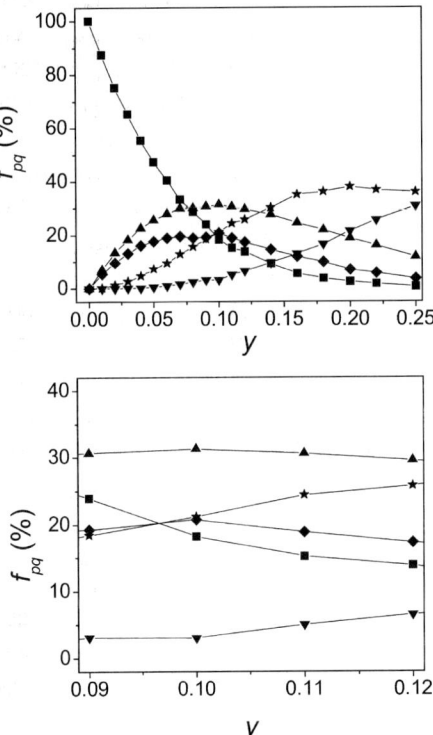

Figure 9. Dependence of f_{pq} relative fractions obtained from the Monte Carlo simulations [17, 30]. (a) Squares: f_{00}, (b) rhombs: f_{10}, (c) up triangles: f_{0n}, (d) stars: f_{1m} and (e) down triangles: f_{2m}.

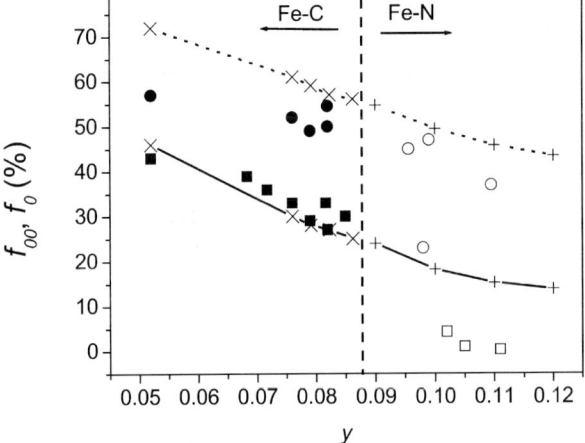

Figure 10. The relative fractions f_{00} (squares) and f_0 (circles) obtained from the analysis of the Mössbauer spectra of Fe–C [10, 11, 13–17] and Fe–N [12–16, 19–21] austenite compared with the Monte Carlo results (f_{00}: solid line and f_0: dashed line).

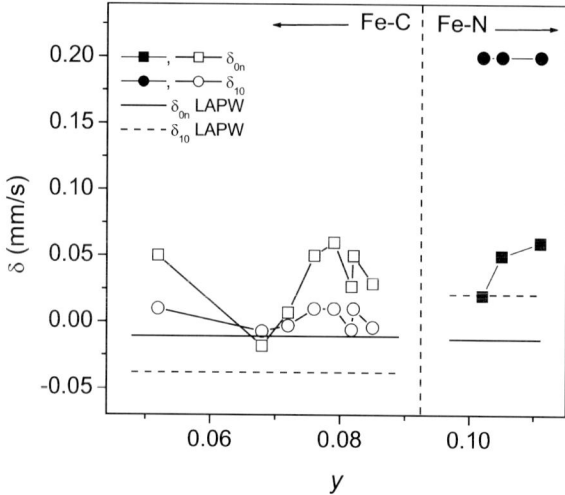

Figure 11. Comparison of experimental (see Table I) and LAPW calculated [54] isomer shifts associated to E_{0n} and E_{10} iron environments in Fe–C and Fe–N alloys.

dered structure. A reasonable agreement between the calculated and experimental quadrupole splitting was observed in Fe–C system and no information about the isomer shift values were presented.

The electron exchange in Fe–N and Fe–C alloys was also calculated using *ab initio* norm-conserving pseudopotential method [33]. It was found that the N atoms increase the concentration of free electrons and the electrons of C atoms contributed to the *d*-band. This fact lend to a short range atomic ordering in Fe–N austenite while Fe–C alloys is revealed as a random solid solution. In agreement with the positive values of the experimental isomer shifts, a decrease of the density of *s*-electrons at the iron site due to the presence of N and C atoms was observed. The calculations are also consistent with the more positive values for the Fe–N solid solution isomer shift.

The LAPW method has been also applied for both systems to calculate the isomer shift values of the different iron environments [54]. A superstructure of the Fe_8X type has been employed. For this high solute concentration, only two iron sites are distinguished: E_{10} and E_{0n}. The resulting isomer shift values are: $\delta_{10} = -0.038$ mm/s and $\delta_{0n} = -0.011$ mm/s and $\delta_{10} = 0.021$ mm/s and $\delta_{0n} = -0.013$ mm/s, for Fe–C and Fe–N systems, respectively. For Fe–C, $\delta_{10} < \delta_{0n}$ and the opposite is observed for Fe–N. The same trend was observed in the experimentally determined values. A comparison of the experimental and calculated isomer shift parameters is presented in Figure 11. The agreement between experimental and calculated parameters is better for the Fe–C than for the Fe–N solid solution. The LAPW calculations reproduce quite well the Fe–C experimental data [10, 11, 17]. However, the calculated δ_{10} parameter for Fe–N are far away from those obtained by the analysis of the Mössbauer spectra [12]. This fact would support that the N distribution among the interstitial sites is random.

6. Summary and conclusions

Mössbauer spectroscopy, thermodynamic data, hard-blocking excluded-sites models, quasichemical calculations, Monte Carlo simulations and electronic structure calculations have been discussed to gain insight on the understanding of the distribution of the interstitial atoms in Fe–N and Fe–C.

The analysis of the activity data with both approximations gave different trends concerning the interaction between the solute atoms. According to the quasichemical approximation the energy of formation of a N–N pair favouring the formation of such a pair. On the contrary, the blocking parameters indicate a harder blocking effect in Fe–N system.

Significant qualitative differences are detected already in the assessment of the Mössbauer spectroscopy database. In particular, the relative importance of the f_{10} and the f_{1m} contributions entering into the OSM change strikingly when passing from the Fe–C to the Fe–N system. The former is the largest in the Fe–C but becomes unimportant in Fe–N austenite. The f_{1m} contribution, absent in the Fe–C results, is the dominant one in Fe–N.

The interactions energies obtained from the Monte Carlo simulations performed adopting the random structure model do not allow reproducing the chemical activity data for Fe–N. In the case of Fe–C austenite, these simulations disregard the ordered structure model. According to the results obtained using the formation energy obtained from the activity data, an Fe_8C structure reproduces better the Mössbauer relative fractions. In the case of Fe–N, no conclusions can be extracted. These results are also supported by the full-potential linearized augmented-plane-wave electronic structure calculations. However, *ab initio* norm-conserving pseudopotential method calculations suggested short range ordering in Fe–N austenite and a more random solid solution for Fe–C. The isomer shift values calculated using LAPW method in the frame of an Fe_8X superstructure reproduce quite well the experimental values for Fe–C alloy.

The present comparison between experimental and theoretical results indicates that the distribution of the solute atoms in the interstitial sites of the *fcc* phase as well as the pair energy interactions nature and magnitude are still open questions. Probably, additional insight would be gained from the interaction between experimental, phenomenological and theoretical approaches.

Acknowledgements

This work was partially supported by Consejo Nacional de Investigaciones Científicas y Técnicas (CONICET), PICT 2853, Fundación Antorchas and Fundación Rocca. Profs. A. G. Fernández Guillermet, R. C. Mercader and G. J. Zarragoicoechea are gratefully thanked for the valuable help and illuminating discussions. K. Laneri and L. Vergara are acknowledged.

References

1. Rees, G. I. and Bhadeshia, H. K. D., *Mat. Sci. Tech.* **8** (1992), 985.
2. Babu, S. S. and Bhadeshia, H. K. D. H., *J. Mater. Sci. Lett.* **14** (1995), 314.
3. Van der Ven, A. and Delaey, L., *Progress Mat. Sci.* **40** (1996), 181.
4. Fernández Guillermet, A. and Du, H., *Z. Metallkd.* **85** (1994), 3.
5. Malinov, S., Böttger, A. J., Mittemeijer, E. J., Pekelharing, M. I. and Somers, M. A. J., *Mat. Trans. A* **32** (2001), 59.
6. Ågren, J., *Met. Trans. A* **10** (1979), 1847.
7. McLellan, R. B. and Alex, K., *Scrip. Metall.* **4** (1970), 967.
8. Hillert, M., *Z. Metallkd.* **90** (1999), 1.
9. Kaufman, L., Radcliffe, S. V. and Cohen, M., In: V. F. Zackay and H. I. Aronson (eds), *Decomposition of Austenite by Diffusional Processes*, Interscience Publishers, New York, 1962, p. 313.
10. Bauer, Ph., Uwakweh, O. N. C. and Genin, J. M. R., *Hyp. Interact.* **41** (1988), 555.
11. Uwakweh, O. N. C., Bauer, J. P. and Génin, J. M., *Metall. Trans. A* **21** (1990), 589.
12. Fall, I. and Genin, J. M. R., *Hyp. Interact.* **69** (1991), 513.
13. Oda, K., Kojima, N., Ito, K. and Ino, H., *Hyp. Interact.* **54** (1990), 853.
14. Oda, K., Umezu, K. and Ino, H., *J. Phys.: Condens Matter* **2** (1990), 10147.
15. Oda, K., Fujimura, H., Mae, K. and Ino, H., *Hyp. Interact.* **69** (1991), 533.
16. Oda, K., Fujimura, H. and Ino, H., *J. Phys.: Condens Matter* **6** (1994), 679.
17. Laneri, K., Desimoni, J., Zarragoicoechea, G. J. and Fernández Guillermet, A., *Phys. Rev. B* **66** (2002), 134201.
18. Giellen, P. M. and Kaplow, R., *Acta Metall.* **15** (1967), 49.
19. DeCristofaro, N. and Kaplow, R., *Mettal. Trans. A* **8** (1977), 35.
20. Foct, J., *J. Physique C* **6** (1974), C6–487.
21. Foct, J., Rochegude, P. and Hendry, A., *Acta Metall.* **36** (1988), 501.
22. Balanyuk, A. G., Bugaev, V. N., Nadutov, V. M. and Sozinov, A. L., *Phys. Stat. Sol. B* **207** (1998), 3.
23. Sozinov, A. L., Balanyuk, A. G. and Gavriljuk, V. G., *Acta Mater.* **47** (1999), 927.
24. Sozinov, A. L., Balanyuk, A. G. and Gavriljuk, V. G., *Acta Mater.* **45** (1997), 225.
25. Ban-ya, S., Elliott, J. F. and Chipman, J., *Trans. AIME* **245** (1969), 1199.
26. Atkinson, D. and Bodsworth, C., *J. Iron Steel Inst.* **208** (1970), 587.
27. Oates, W. A., Lambert, J. A. and Gallagher, P. T., *Trans. TMS-AIME* **245** (1969), 47.
28. Bhadeshia, H. K. D. H., *Met. Sci.* **16** (1982), 167.
29. Bhadeshia, H. K. D. H., *Mater. Sci. Technology* **14** (1998), 273.
30. Vergara, L., Desimoni, J., Fernádcz Guillermet, A. and Zarragoicoechea, G. J., *ICAME* (2003).
31. Blanter, M. S., *J. of Alloys and Compounds* **291** (1999), 167.
32. Timoshevskii, A. N., Timoshevskii, V. A. and Yanchistsky, B. Z., *J. Phys.: Condens. Matter.* **13** (2001), 1051.
33. Gavriljuk, V. G., Shanina, B. D. and Berns, H., *Acta Mater.* **48** (2000), 3879.
34. Lacher, J. R., *Proc. R. Soc. (London) A* **161** (1937), 525.
35. Lacher, J. R., *Proc. Camb. Phil. Soc.* **33** (1937), 518.
36. Gurney, R. W., *Introduction to Statistical Thermodynamics*, McGraw Hill, New York, 1949.
37. Fowler, R. and Guggenheim, E. A., *Statistical Thermodynamics*, Cambridge Univ. Press, Cambridge, 1956.
38. Speiser, R. and Spretnak, J. W., *Trans. ASM* **47** (1955), 493.

39. Moon, K. A., *Trans. AIME* **227** (1963), 1116.
40. McLellan, R. B., Garrard, T. L., Horowitz, S. J. and Sprague, J. A., *Trans. AIME* **239** (1967), 528.
41. McLellan, R. B., In: P. S. Rundman, J. Stringer and R. I. Jaffee (eds), *Phase Stability in Metals and Alloys*, McGraw Hill Co, New York, 1968.
42. Gallagher, P. T., Lambert, J. A. and Oates, W. A., *Trans. AIME* **245** (1969), 887.
43. Hillert, M. and Staffansson, L.-I., *Acta Chem. Scand.* **24** (1970), 3618.
44. Lee, H. M., *Metall. Trans.* **5** (1974), 787.
45. Guggenheim, E. A., *Mixtures*, Oxford Univeristy Press, 1952.
46. Darken, L. S. and Smith, R. P., *J. Am. Chem. Soc.* **68** (1946), 1172.
47. Mc Lellan, R. B. and Dunn, W. W., *J. Phys. Chem. Solids* **30** (1969), 2631.
48. Ron, M., In: R. L. Cohen (ed.), *Applications of Mössbauer Spectroscopy*, Vol. II, Academic Press, New York, 1976, p. 329.
49. Desimoni, J., Gregorutti, R., Laneri, K., Sarutti, J. L. and Mercader, R. C., *Met. Mat. Trans. A* **30** (1999), 2745.
50. Laneri, K. F., Desimoni, J., Mercader, R. C., Gregorutti, R. W. and Sarutti, J. L., *Met. Mat. Trans. A* **32** (2001), 51.
51. Wriedt, H. A., Gokcen, N. A. and Nafziger, R. H., *Bull. of Alloys and Phase Diagrams* **8** (1987), 355.
52. Mac Lellan, R. B. and Co, C., *Acta Metall* **35** (1987), 2151.
53. Shiflet, G. J., Bradley, J. and Aronson, H. I., *Metals Trans.* **15A** (1984), 1287.
54. Peltzer y Blancá, E. L., Desimoni, J. and Laneri, K., to be published.

Hyperfine Interactions **156/157**: 523–529, 2004.
© 2004 *Kluwer Academic Publishers. Printed in the Netherlands.*

A Dilute-Limit Heat of Solution of 3d Transition Metals in Iron Studied with ^{57}Fe Mössbauer Spectroscopy

JAN CHOJCAN

Institute of Experimental Physics, University of Wrocław, Pl. M. Borna 9, 50-204 Wrocław, Poland;
e-mail: chojcan@ifd.uni.wroc.pl

Abstract. The room-temperature ^{57}Fe Mössbauer spectra for binary iron-based solid solutions $Fe_{1-x}D_x$ with D = V, Cr, Mn and Co, were analysed in terms of binding energy E_b between two D atoms in the Fe–D system. The extrapolated values of E_b for $x = 0$ were used for computation of the dilute-limit heat of solution of D metals in iron. The results were compared with those derived from calorimetric data concerning the heat of formation of the systems mentioned as well as with those resulting from the Miedema's model of alloys. The comparison shows that our Mössbauer spectroscopy findings are in a qualitative agreement with the available calorimetric data and they are at variance with corresponding Miedema's values for Fe–Mn and Fe–Co systems.

Key words: heat of solution, iron alloys, Mössbauer spectroscopy.

1. Introduction

It has been proved that the ^{57}Fe Mössbauer spectroscopy is a useful tool for the study of interactions of impurity atoms dissolved in iron [1–3]. The technique is especially powerful when the impurity neighbours of the Mössbauer probe have a sufficiently large effect on the hyperfine field generated at the probe, to yield distinguishable components in the Mössbauer spectrum attributed to different configurations of the probe neighbours. Among the 3d transition metal impurities for example, the favourable ones are Ti, V, Cr and Mn atoms whereas Co and Ni atoms seem to be less suitable for the mentioned studies [4]. At the same time from the findings concerning the impurity interactions one can easily derived the dilute-limit heat of solution of the impurity elements in iron [5]. The latter considerably increases importance of the studies as the experimental values of the heat play an essential role in developing and testing different models of binary alloys as well as methods for calculating the alloy parameters [6–11]. The main source of such experimental data are calorimetric studies of the heat of formation of binary systems [12]. Unfortunately sometimes there are significant discrepancies in the data obtained by different authors. This is observed, e.g., for the Fe–V alloys [13, 14]. Moreover, the calorimetric investigations are performed in relatively high temperatures at which some of iron systems are in their high-temperature γ (fcc) phases. Such situation exists for instance in the case of the Fe–Mn system. Consequently, there is no calorimetric data concerning the heat of solution of Mn in the low-temperature α (bcc) phase of Fe. All the above encouraged us to use the ^{57}Fe Mössbauer spectroscopy for supplying the experimental heats of solution of different elements in α-Fe. In our previous studies we collected the proper spectra for the Fe–Cr [2] and Fe–V [3] alloys. In this work we have extended the investigations to the Fe–Co and Fe–Mn systems. The present systems are essentially different from the previously studied ones. In the case of Fe–Co samples the Mössbauer spectrum components are practically unresolved whereas for the Fe–Mn specimens the shape of a spectrum is directly dependent on atoms located in the first coordination shell of the probing nuclei [15]. For clarity one should add that in the case of iron-based Fe–Cr and Fe–V solid solutions the spectrum components are quite well distinguishable and moreover the shape of the spectrum measured is directly effected by atoms located in the first two coordination shells of the nuclear probes. The latter concerns the Fe–Co system as well.

2. Experimental and results

The samples of iron–manganese and iron–cobalt alloys were prepared by melting the Aldrich 99.999% pure iron, 99.98% pure manganese and 99.995% pure cobalt in an arc furnace filled with argon. Resulting ingots were cold-rolled to the final thickness of about 0.05–0.07 mm and then the foils were gradually annealed in vacuum in two steps: (1) at 1170 K for 2 h and (2) at 1270 K for 2 h. After each step specimens were slowly cooled to room temperature during 6 h. Under these conditions, diffusion effectively stops at about 700 K [16], so the observed distributions of atoms in the annealed samples should be the frozen-in state corresponding to 700 K. The atomic compositions of all the specimens were checked by an analysis of the energy distribution of X-rays (EDX) induced by the 20 keV electron beam. The EDX data showed that the x values for the $Fe_{1-x}Mn_x$ samples amounted to 0.022, 0.031, 0.040, 0.050, 0.058, 0.071, 0.075, 0.085 and 0.098 whereas in the case of the $Fe_{1-x}Co_x$ specimens the x values were 0.012, 0.023, 0.035, 0.044, 0.055, 0.060, 0.075, 0.087 and 0.098.

The room temperature measurements of the ^{57}Fe Mössbauer spectra were performed in transmission geometry by means of a constant-acceleration POLON spectrometer of standard design. Some spectra are presented in Figures 1 and 2. All the spectra were analysed, as in our previous papers [1–4], in terms of three or four six-line patterns corresponding to different hyperfine fields B at ^{57}Fe nuclei generated by different numbers of Fe and Mn atoms located in the first coordination shell of the probing nuclei or generated by dissimilar numbers of Fe and Co atoms existing in the first and second coordination shells of the nuclear probes. It was done under assumption that the effect of non-iron (D) atoms on B is additive and independent of their positions in the surroundings of the probe. Moreover, it

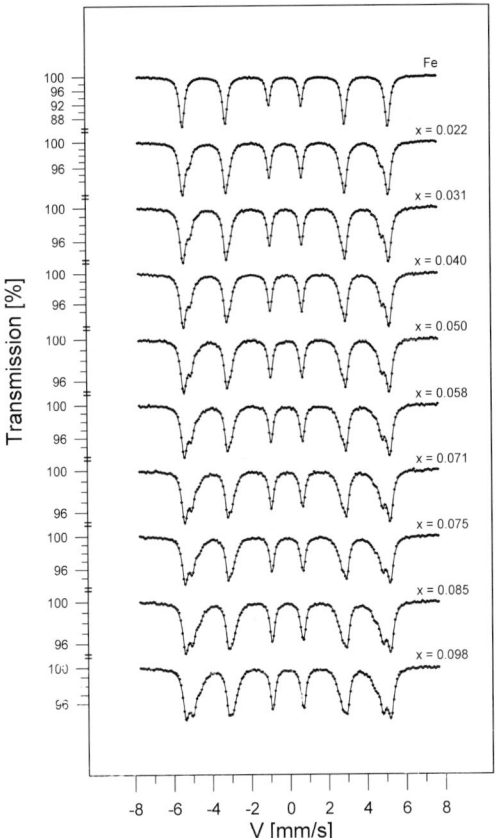

Figure 1. ^{57}Fe Mössbauer spectra for the $Fe_{1-x}Mn_x$ alloys measured at room temperature after the annealing process at 1270 K.

was accepted that the hyperfine field and the corresponding centre shift IS of the subspectrum are linear functions of the number n of D neighbours of ^{57}Fe with the forms: $B = B_0 + n\Delta B$ and $IS = IS_0 + n\Delta IS$, where B_0, ΔB, IS_0 and ΔIS are fitted parameters; $\Delta B(\Delta IS)$ stands for the change of $B(IS)$ with one D atom in the neighbourhood of the Mössbauer probe. Finally, it was assumed that the three linewidths Γ_{16}, Γ_{25} and Γ_{34} as well as the two line area ratios I_{16}/I_{34} and I_{25}/I_{34} are the same for all six-line components of the given spectrum. It is worth noticing that the fits obtained under these assumptions are very good – see Figures 1 and 2. Besides that, the found values of the best-fit parameters ΔB and ΔIS (displayed in Tables I and II) are in a good agreement with corresponding data given in the literature; e.g., in Ref. [15] one can find that ΔB amounts to -2.30 T for D = Mn and for D = Co it is 0.96 T. As the main result of the analysis the relative areas c_1 and c_2 of the second and third components of each spectrum were determined. The components are related to the existence of one D atom and two D atoms in the neighbourhood of ^{57}Fe, respectively. Assuming that the Lamb–Mössbauer factor is

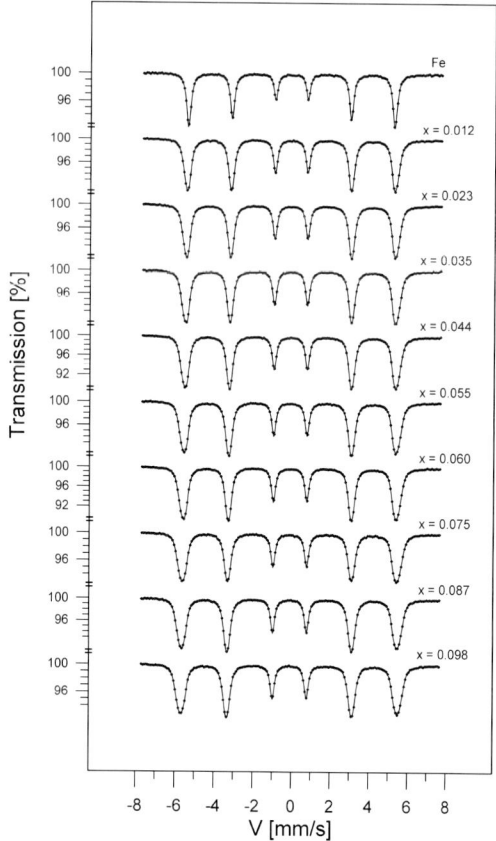

Figure 2. ^{57}Fe Mössbauer spectra for the $Fe_{1-x}Co_x$ alloys measured at room temperature after the annealing process at 1270 K.

independent of the configuration of atoms in the surroundings of the ^{57}Fe nucleus, the c_1 and c_2 values describe intensities of the components mentioned above. The results are listed in Tables I and II.

The c_1 and c_2 values were used to calculate the binding energy E_b for pairs of Mn and Co atoms in the studied materials. The computations were performed, as in [1–4], on the basis of the modified Hrynkiewicz–Królas formula [4, 17]:

$$E_b = -kT_d \cdot \ln\left[(1 + 2 \cdot c_2/c_1) \cdot (c_2/c_1) \cdot \left(1 + 2 \cdot p(2)/p(1)\right)^{-1}\right.$$
$$\left. \times \left(p(2)/p(1)\right)^{-1}\right], \tag{1}$$

where k is the Boltzmann constant, T_d denotes the "freezing" temperature for the atomic distribution in the sample ($T_d = 700$ K), $p(n) = [N!/((N-n)!n!)]x^n(1-x)^{N-n}$ is the probability for the existence of n D atoms among all N atoms located in the first (D = Mn) or the first and second (D = Co) coordination shells of the Fe atom in a random Fe–D alloy, finally x stands for the concentration of D atoms and N, in the case of annealed samples being at the room temperature, is practically 8

Table I. The best-fit parameters of the assumed model of the ^{57}Fe Mössbauer spectrum measured for $Fe_{1-x}Mn_x$ alloys

x	B_0 [T]	ΔB [T]	IS_0 [mm/s]	ΔIS [mm/s]	c_1	c_2
0	33.062(2)		0			
0.022	33.062(4)	−2.33(2)	0.0015(4)	−0.0086(12)	0.228(2)	0.001(2)
0.031	33.066(5)	−2.37(2)	0.0026(5)	−0.0102(13)	0.271(3)	0.014(3)
0.040	33.087(5)	−2.35(1)	0.0027(5)	−0.0085(10)	0.299(3)	0.032(2)
0.050	33.066(7)	−2.38(2)	0.0031(7)	−0.0096(13)	0.324(4)	0.048(3)
0.058	33.051(6)	−2.40(1)	0.0051(6)	−0.0106(10)	0.331(3)	0.075(3)
0.071	33.049(7)	−2.39(2)	0.0050(7)	−0.0111(12)	0.343(4)	0.099(3)
0.075	33.005(6)	−2.36(1)	0.0046(6)	−0.0079(9)	0.347(3)	0.115(2)
0.085	32.972(7)	−2.36(1)	0.0050(7)	−0.0083(10)	0.352(3)	0.140(3)
0.098	32.935(7)	−2.36(1)	0.0070(7)	−0.0096(7)	0.370(3)	0.153(2)

Table II. The best-fit parameters of the assumed model of the ^{57}Fe Mössbauer spectrum measured for $Fe_{1-x}Co_x$ alloys

x	B_0 [T]	ΔB [T]	IS_0 [mm/s]	ΔIS [mm/s]	c_1	c_2
0.012	32.92(1)	0.78(2)	0.0013(4)	0.0016(6)	0.366(10)	0.161(11)
0.023	33.01(1)	0.89(1)	0.0024(3)	0.0019(4)	0.386(5)	0.167(5)
0.035	33.06(1)	0.98(2)	0.0027(5)	0.0023(6)	0.394(6)	0.194(6)
0.044	32.97(2)	0.82(1)	0.0046(4)	0.0024(3)	0.347(5)	0.250(4)
0.055	33.01(2)	0.86(1)	0.0062(5)	0.0022(3)	0.332(3)	0.263(4)
0.060	33.13(2)	0.88(1)	0.0070(7)	0.0024(3)	0.340(4)	0.271(4)
0.075	33.22(2)	0.93(2)	0.0089(6)	0.0027(4)	0.317(5)	0.284(5)
0.087	33.34(2)	0.95(1)	0.0095(5)	0.0030(3)	0.331(4)	0.292(3)
0.098	33.45(2)	0.96(1)	0.0125(6)	0.0024(4)	0.317(5)	0.298(4)

(D = Mn) or $8 + 6 = 14$ (D = Co) – the total number of the lattice sites in the first or the first and second coordination shells of an atom in the bcc lattice. The E_b values are presented in Figure 3.

Finally, the extrapolated values of E_b for $x = 0$ were used for computation of the dilute-limit heat H_{FeD} of solution of D metals in iron. The calculations were performed on the basis of the Królas model [5] for the binding energy according to which $H_{FeD} = -z \cdot E_b/2$, where z is the coordination number of the crystalline lattice ($z = 8$ for α-Fe). The results are displayed in Table III together with both the findings based on calorimetric measurements and the values resulting from Miedema's model of alloys.

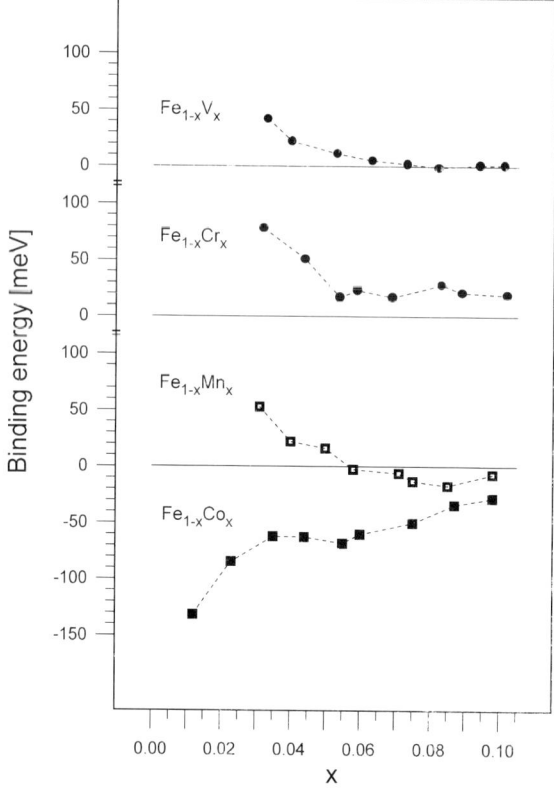

Figure 3. The binding energy E_b between a pair of 3d transition metal atoms in iron alloys. The data for Fe–Cr and Fe–V systems was taken from [2] and [3], respectively.

Table III. The dilute-limit heat H_{FeD} [eV/atom] of solution of 3d metals in iron

Solute D	Calorimetric data	Miedema's model [6]	This work (α-Fe)
V	−0.31 (α-Fe) [14]	−0.31	−0.55(8)
Cr	−0.22 (α-Fe) [12]	−0.06	−0.60(12)
Mn	+0.14 (γ-Fe) [12]	+0.01	−0.64(27)
Co	+0.30 (α-Fe) [12]	−0.02	+0.73(14)

3. Conclusions

The values of the heat H_{FeD} of solution of vanadium, chromium and cobalt in α-iron, determined from the ^{57}Fe Mössbauer spectra are in a good agreement with corresponding values obtained by using calorimetric methods. Such conclusion can be drawn if it is taken into account that the calorimetric absolute values of H_{FeD} are underestimated. The latter results from the way of computing the H_{FeD} values on the basis of the available heats of formation for the proper alloys. The above can suggest that the Mössbauer value for the heat of solution of manganese in α-iron is correct as well.

As far as the predictions of the Miedema's model are concerned it is clear that they are reasonable in the case of solution of V in Fe only. The model absolute value of H_{FeCr} seems to be to small whereas the model H_{FeMn} and H_{FeCr} values are in qualitative disagreement with our findings.

Acknowledgements

The author would like to thank Marcin Mróz and Joanna Ochla for their assistance in the measurements. This work was supported by the State Committee for Scientific Research in Poland (KBN) under the Grant No 5 P03B 129 20.

References

1. Chojcan, J. and Roztocka, G., *Phys. Stat. Sol. B* **204** (1997), 829.
2. Chojcan, J., *Phys. Stat. Sol. B* **219** (2000), 375.
3. Chojcan, J., *J. Alloys Compd.* **350** (2003), 62.
4. Chojcan, J., *J. Alloys Compd.* **264** (1998), 50.
5. Królas, K., *Phys. Lett.* **85A** (1981), 107.
6. Miedema, A. R., *Physica B* **182** (1992), 1.
7. Bozzolo, G., Ferrante, J. and Smith, J. R., *Phys. Rev. B* **45** (1992), 493.
8. Bozzolo, G. and Ferrante, J., *Phys. Rev. B* **45** (1992), 12191.
9. Daw, M. S. and Baskes, M. I., *Phys. Rev. B* **29** (1984), 6443.
10. Johnson, R. A., *Phys. Rev. B* **39** (1989), 12554.
11. Bangwei, Z. and Yifang, O., *Phys. Rev. B* **48** (1993), 3022.
12. Hultgren, R., Desai, P. D., Hawkins, D. T., Gleiser, M. and Kelley, K. K., *Selected Values of Thermodynamic Properties of Binary Alloys*, American Society for Metals, Metals Park, OH, 1973.
13. Myles, K. M. and Aldred, A. T., *J. Physical Chem.* **68** (1964), 64.
14. Kubaschewski, O., Probst, H. and Geiger, K. H., *Z. Phys. Chem. Neue Folge* **104** (1977), 23.
15. Vincze, I. and Campbell, I. A., *J. Phys. F* **3** (1973), 647.
16. Cranshaw, T. E., *J. Phys.: Condensed Matter* **1** (1989), 829.
17. Hrynkiewicz, A. Z. and Królas, K., *Phys. Rev. B* **28** (1983), 1864.

Hyperfine Interactions **156/157**: 531–539, 2004.
© 2004 *Kluwer Academic Publishers. Printed in the Netherlands.*

Distribution of N Atoms in the *fcc* Fe–N Interstitial Solid Solution

L. VERGARA[1], J. DESIMONI[1], A. FERNÁNDEZ GUILLERMET[2] and
G. J. ZARRAGOICOECHEA[3]
[1]*Departamento de Física, Facultad de Ciencias. Exactas, UNLP. IFLP-CONICET C.C. N° 67,
1900 La Plata, Argentina; e-mail: desimoni@fisica.unlp.edu.ar*
[2]*Consejo Nacional de Investigaciones Científicas y Técnicas, Centro Atómico Bariloche,
8400 Bariloche, Argentina*
[3]*Instituto de Física de Líquidos y Sistemas Biológicos, Comisión de Investigaciones Científicas de
la Provincia de Buenos Aires, UNLP C.C. N° 565, 1900 La Plata, Argentina*

Abstract. Monte Carlo simulations, analytical calculations and thermodynamic activity data have been combined in a study of the distribution of interstitial nitrogen atoms in the octahedral sites of the Fe–N solid solution. Two models are compared in the analysis of the Mössbauer spectra. Thermodynamic activity data are analyzed in terms of the quasichemical approximation to the statistical mechanics of interstitial solutions. The formation energies of the N–N pairs thus assessed are used as input in both Monte Carlo simulations and the quasichemical calculations. A significant discrepancy between Monte Carlo predictions and the Mössbauer results is detected. This striking effect becomes particularly important when the ordered structure model is applied in the analysis of Mössbauer data.

Key words: Monte Carlo simulations, solid solutions, Mössbauer spectroscopy.

1. Introduction

In the austenite phase of the Fe–N system, the iron atoms are arranged in a close-packed face-centred cubic (*fcc*) lattice, and the N atoms occupy a limited number of the octahedral interstices which are located at the centres and at the mid-points of the edges of the unit cubes, these two positions being crystallographically equivalent [1]. Various models have been proposed to account for the N distribution in the interstitial sites, which consider two sublattices: one for the Fe atoms, and the other for the mixture of N atoms and vacant octahedral interstices (V). In an alloy with N_N nitrogen atoms and N_{Fe} iron atoms $y = N_N/N_{Fe}$ and $1 - y$ represent the fraction of occupied and of empty interstitial sites, respectively [2–5]. Even though it has long been accepted that the N atoms occupy the octahedral interstitial sites in the *fcc* lattice of the Fe–N austenite phase, the details of the interstitial distribution among the available sites as well as the origin and the magnitude of the interactions in the mixture remain as open questions of theoretical and practical interest.

The purpose of the present paper is to analyse and compare various current models for N distribution used to analyse the Mössbauer spectra of the Fe–N austenite phase. With this aim, the relative fractions of the Fe environments obtained from the Mössbauer spectra (MS) are compared with those obtained by Monte Carlo (MC) simulations based on energies extracted from the N activity of N in austenite.

2. The Mössbauer pattern and the models

Fe–N alloys with high content of N, viz., from $y = 0.095$ to $y = 0.111$, prepared by nitriding pure Fe in a mixture of NH_3 and H_2 gas at temperatures between 866 K and 973 K were studied using Mössbauer effect by different research groups [6–15]. In all cases, the N concentration, very close to the solubility limit [16], was determined by X-ray diffraction using the known composition dependence of the austenite lattice parameter [17].

The spectra, shown in Figure 1, are somewhat different from that of the Fe–C alloys. In general, the quadrupole splitting values are smaller than in the Fe–C case, which indicates that the perturbation of the electronic shell of an Fe atom by the neighbouring interstitials is greater in the case of C. Despite the similarity of the spectra, two quite different models for the distribution in the interstitial sites have proposed to reproduce the austenite MS [7–10, 15, 18, 19], which are briefly reviewed in the following.

The ordered structure model (OSM) assumes a superstructure of the Fe_8C_{1-x}-type, with $x = 1 - 8y$, in which the solute atoms are allowed to occupy only the interstices located at the centres of the cubes in the *fcc* Fe lattice [6]. In this model, the occupation of the first and second interstitial shells around the Fe atoms was considered, and Fe environments E_{pq} (p, q = number of N atoms in the first and second interstitial shells, respectively) were taken into account.

According to the second model [7–10], the random structure model (RSM), all the interstitial sites of the *fcc* lattice can be occupied by the solute atoms. However, only the occupation of the first interstitial shell was considered, and different Fe environments E_i ($i = 0, 1, 2$) were proposed. In addition, point charge calculations suggested that the interactions arisen from E_1 and $E_{2,90}$ (two N atoms in the first interstitial shell at 90°) are similar.

Fall and Genin [6] analysed the spectra obtained in transmission geometry in terms of five different Fe environments arising from the OSM. The hyperfine parameters and the relative fractions are reported in Table I.

In the framework of the RSM, Giellen and Kaplow [11] and afterwards De-Cristofaro and Kaplow [12] used only two hyperfine interactions, one (E_1) associated to Fe atoms with a nearest neighbor N atom, and the other ascribed to the rest of the Fe atoms (E_0). By analyzing the relative fractions (see Table I), it was concluded that the N atoms are randomly distributed in the octahedral interstitial sites. Foct *et al.* [13, 14] improved this model by adding a third interaction, in order to account for the Fe atoms with two next near N atoms, which can be placed at 180° or at 90°. Finally, Oda *et al.* [7–10] attributed such interaction to atoms located at 180°. The results of these various treatments are compared in Table I.

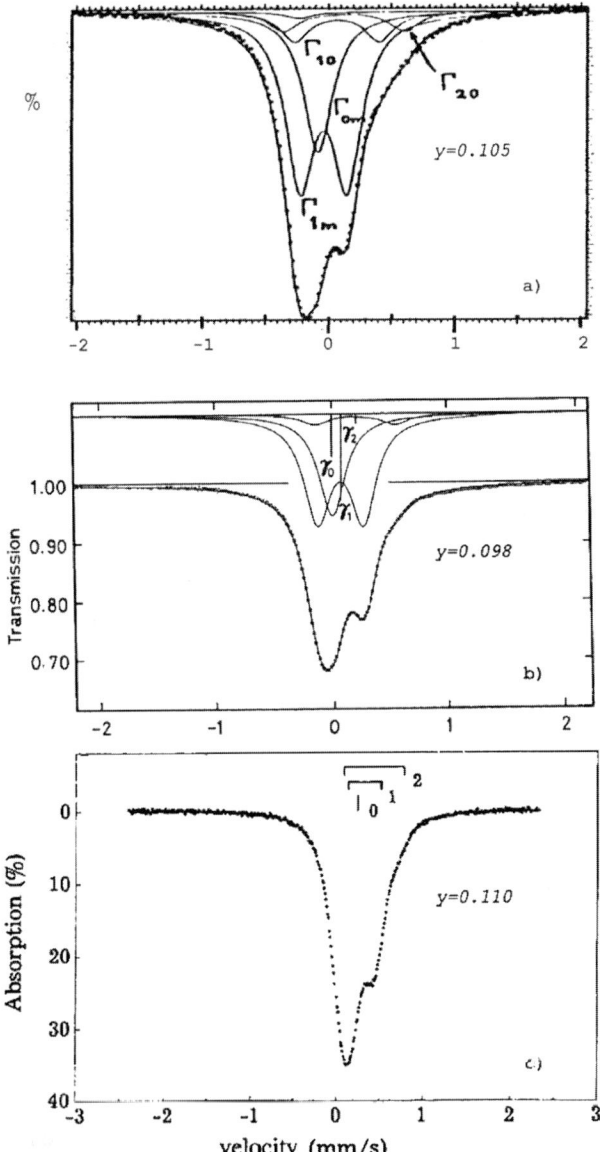

Figure 1. Mössbauer spectra recorded on Fe–N austenite: (a) [6], (b) [7–11] and (c) [13, 14].

Concerning the hyperfine parameters obtained in the framework of the OSM, it can be noticed that the isomer shift associated to the E_{10} environment for Fe–N austenite is more positive than that reported for the Fe–C solid solution [20]. In the framework of the RSM the hyperfine parameters obtained for both Fe–C and Fe–N austenite are, however, quite similar. The reduction in the quadrupole splitting value of the interactions E_1 may be understood by accounting for the effective charge density.

Table I. Isomer shift (δ), quadrupole splitting (Δ) and relative fractions of the different Fe environments after (a) applying OMS on the set of data of [6], and (b) the RSM on the set of data of [7–14]. The isomer shifts are referred to α-Fe at room temperature

	OMS						
y	E_{00}		E_{0n}		E_{10}		
	δ (mm/s)	f_{00} (%)	δ (mm/s)	f_{0n} (%)	δ (mm/s)	Δ (mm/s)	f_{10} (%)
0.102	−0.11	4.4	0.02	31.2	0.20	0.67	7.4
0.105	−0.11	1.1	0.05	28.4	0.20	0.67	9.9
0.111	−0.11	0.3	0.06	21.8	0.20	0.67	8.5

y	E_{1m}			E_{2m}			Reference
	δ (mm/s)	Δ (mm/s)	f_{1m} (%)	δ (mm/s)	Δ (mm/s)	f_{2m} (%)	
0.102	0.09	0.37	54.8	0.29	0.95	2.2	[6]
0.105	0.10	0.37	57.4	0.26	0.95	6.8	[6]
0.111	0.11	0.37	66.4	0.20	0.95	3.0	[6]

	RMS						
y	E_0		E_1			E_2		Reference
	δ (mm/s)	f_0 (%)	δ (mm/s)	Δ (mm/s)	f_1 (%)	δ (mm/s)	Δ (mm/s)	f_2 (%)	
0.096	−0.16	45	0.05	0.29	55				[12]
0.098	−0.10	23	0.15	0.25	75	0.30	0.40	2	[13, 14]
0.099	0.01	47	0.08	0.39	50	0.20	0.72	3	[7–10]
0.110	−0.09	37	−0.12	0.38	59	0.09	0.66	4	[15]

The evolution of the relative fraction of each Fe environment with the occupation of the interstitial sites, obtained by using both models of the MS is shown in Figure 2. These results lend support to a picture of the Fe–N austenite where each Fe atom has one N neighbor in the first shell, and one or more in the second. This suggests a weaker blocking effect of the atoms in the first shell upon those in the second, which in general should favor a more random distribution. This possibility has already been suggested by Fall and Genin [6], on the basis of a large f_{1m} together with f_{2m}.

The results in Figure 2(b) indicates that the most important contributions to the Mössbauer spectra of Fe–N alloy are f_0 and f_1, with similar values to those of Fe–C system [7–10]. This fact suggests a picture where the Fe–C and the Fe–N austenite do not differ markedly in their degree of randomness. Indeed, the $E_{2,180}$ environments have been considered in the Fe–N treatments, and Oda *et al.* [7–10] interpreted it as indications of an ordered structure of the γ'-Fe$_4$N type [21].

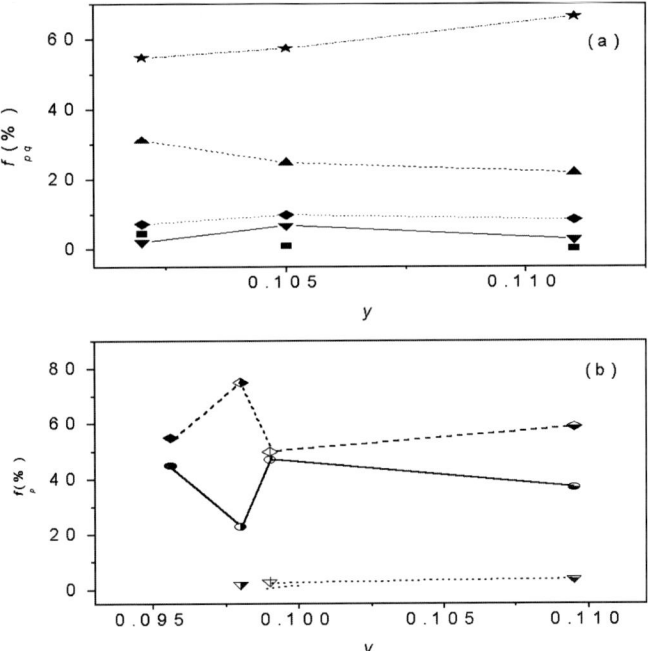

Figure 2. (a) f_{pq} relative fractions obtained from the analysis of the spectra considering the OSM [6, 18, 19] of the Fe–N system. (i) Squares: f_{00}, (ii) rhombs: f_{10}, (iii) up triangles: f_{0n}, (iv) stars: f_{1m} and (v) down triangles: f_{2m}. (b) f_q relative fractions obtained from the analysis of the spectra considering the RSM [7–10] of the Fe–N system. (i) Circles: f_0, (ii) rhombs: f_1, and (iii) down triangles: f_2. Filled symbols belong to results of [12], vertical half filled symbols of [13] and [14], crossed symbols of [7–10] and horizontal half filled symbols to [15].

3. The Monte Carlo simulations

A description of the MC method used in the present study has recently been reported [20]. Here we only summarize the main points. The austenite interstitial solid solution is described as a lattice gas of N_N interstitial nitrogen atoms and N_V vacant interstitial sites, distributed in the $N_N + N_V = N_i = N_{Fe}$ octahedral interstitial sites of an *fcc* structure comprising N_{Fe} iron atoms. The occupancy of the first and second interstitial shells was accounted for in three-dimensional MC simulations performed at 933 K. The calculated quantities include the number n_{ij} (i, j = N or V) of N–N, N–V and V–V pairs and the relative fractions f_{pq} (p, q = number of N atoms in the first and second interstitial shell, respectively) associated to the different Fe environments. A Fortran 77 routine using the MC method, an Ising-type Hamiltonian and periodic boundary conditions was developed. The Metropolis method was used to define the jump probability of the N atoms, which was calculated as explained in [20].

The pair energy formation entering in the MC simulations was determined by analysing N activity data [22] in the range 933–1083 K in terms of a linearized form of the quasichemical approximation (QCA) [23, 24] to interstitial solutions. The parameters thus obtained are $\Delta\varepsilon_N = 770_{26}$ cal/mol and $\Delta G_N = 136111_{27}$ cal/mol. The $\Delta\varepsilon_N$ value is slightly higher than that reported in [25], $\Delta\varepsilon_N = 950$ cal/mol.

4. Results and discussion

The numbers of pairs n_{ij} in Fe–N austenite with $0 < y < 1$ calculated using the QCA [23, 24] are compared in Figure 3 with the values given by MC in which the same $\Delta\varepsilon_N$ value ($= 770$ cal/mol) were adopted. The inset amplifies the composition range corresponding to the experimental solubility of N in austenite, $y < 0.12$. A very good agreement between the QCA and the MC predictions for n_{ij} is found, as in the case of Fe–C alloys [20].

Figure 4(a) presents the f_{pq} values from MC simulations for the range $0 < y < 0.25$, and Figure 4(b) focuses on the composition range where Mössbauer experiments have been performed. In such region the main contributions come from the E_{0n} and the E_{1m} environments. The f_{00} and f_{10} relative fractions decrease with increasing N content, whereas f_{2m} increases from 2% to 10%.

The main qualitative difference between the results in Figures 3 and 5(b) is that the majority of the contributions in the latter fall in a relatively small range of values, $10\% < f_{pq} < 30\%$. In addition, the relations between the theoretical f_{pq}, viz., $f_{0n} > f_{1m} > f_{10} > f_{00}$, differ from the experimental ones, viz., $f_{1m} > f_{0n} > f_{10} > f_{00}$, in the relative importance of f_{0n} and f_{1m}. The simulations also predict a contribution of the E_{2m} environments which is comparable with that from MS. It

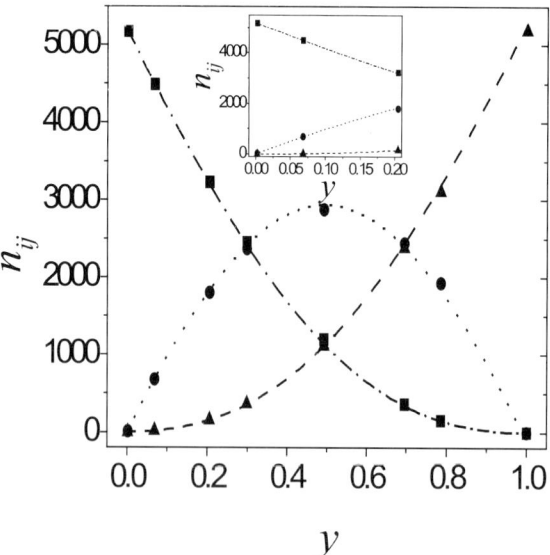

Figure 3. The n_{ij} ($i, j = N, V$) number of pairs obtained using $\Delta\varepsilon_N = 770_{26}$ cal/mol from: (a) Monte Carlo simulations (squares, triangles and circles are the of $V-V$, $N-N$ and $N-V$, respectively) and (b) using the quasichemical approximation [23, 24] (dash-dotted line, dashed line and dotted line corresponds to $V-V$, $N-N$ and $N-V$ pairs, respectively).

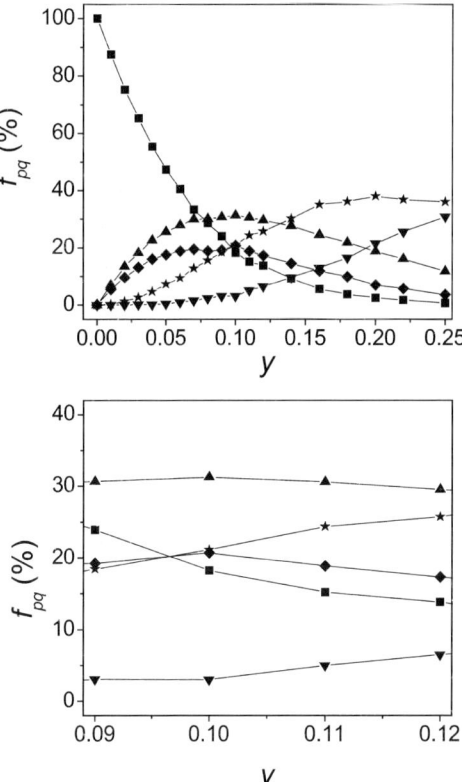

Figure 4. N content dependence of f_{pq} relative fractions obtained from the Monte Carlo simulations. (a) Squares: f_{00}, (b) rhombs: f_{10}, (c) up triangles: f_{0n}, (d) stars: f_{1m} and (e) down triangles: f_{2m}.

is interesting to note that such contribution is also similar to the one coming from the $E_{2,180}$, Figure 3. The f_{1m} contribution, which is absent in the Fe–C results, is the dominant one in Fe–N. The Fe–N data also includes a small contribution of the E_{2m} environment, which is again not included in the Fe–C analysis [18, 19].

The quantitative comparison between the MC and MS results was performed using the fractions f_{00} and f_0, which are considered as the key variables in the OSM [6, 18, 19] and the RSM [7–10], respectively. The f_{00} fraction is presented in Figure 4, whereas f_0 was calculated as $f_0 = f_{00} + \Sigma f_{0n}$ as in the previous Fe–C work [20]. The resulting composition dependence of the theoretical f_0 and f_{00} fractions are compared in Figure 5 with the MS results for Fe–N austenite [6–15]. The MC calculated f_0 approaches the experiments as much as one could expect, in view of the discrepancy already found in the Fe–C system [20]. However, the MC simulations overestimate the f_{00} by approximately 20%.

In previous MC simulations, Oda *et al.* [7–10] treated the N–N pair energies interactions as free parameters, which were varied until the f_0, f_1 and f_2 fractions were reproduced satisfactorily. However, the resulting energy values failed to reproduce the activity data [22]. MC simulations considering a long-range interaction

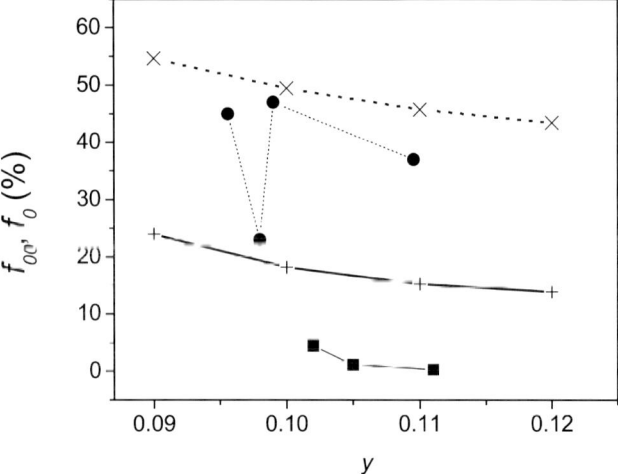

Figure 5. The relative fractions f_{00} (squares) and f_0 (circles) obtained from the analysis of the Mössbauer spectra of Fe–N [6–15] austenite compared with the Monte Carlo results (f_{00}: solid line and f_0: dashed line).

between the solute atoms up to six interstitial shells were performed by Balanyuk [15] in the framewok of the RSM. A stronger repulsion between N atoms in the first coordination shell were included. In the second interstitial shell N–N interactions also turned out to be repulsive. However, those results are not fully supported by the MC simulations reported by Sozinov [26, 27], who accepted a strong repulsion in the first interstitial shell, but a weak interaction in the second one. Furthermore, they claimed that the RSM [7–10] accounts better for the Mössbauer data, and the possibility of an ordered structure of the γ'-Fe$_4$N type [21] was speculated about, but this treatment was not able to reproduce the available Fe–N activity data [22]. Gavriljuk *et al.* [28] defended a picture of the Fe–N austenite involving short range ordering.

Finally, *ab initio* norm-conserving pseudopotential method calculations suggested short range ordering in Fe–N austenite and a more random solid solution for Fe–C [29].

5. Conclusions

The MC predictions for the number of n_{ij} pairs agree satisfactory with the QCA calculations. Important discrepancies between the MC simulations and the results of adopting either the RSM or OSM in analysing the MS has been found. The MC predictions deviate from the Mössbauer spectroscopy configurations for Fe–N alloys significantly more than those on the Fe–C system, more than the MC results of using the OSM. First and second interstitial shells in Fe–N system are more populated than in the Fe–C system, which suggests a weaker blocking effect and thus a more random distribution.

Additional insight upon the current problems would be gained from the interaction between experimental, phenomenological and theoretical approaches. In particular, systematic *ab initio* calculations might shed some light on the microscopic origin of the interaction energies in the Fe–N and Fe–C solid solutions.

Acknowledgements

This work was partially supported by Consejo Nacional de Investigaciones Científicas y Técnicas (CONICET), PICT 2853, and PIP No. 02612, by Fundación Antorchas and by Fundación Rocca.

References

1. Kaufman, L., Radcliffe, S. V. and Cohen, M., In: V. F. Zackay and H. I. Aronson (eds), *Decomposition of Austenite by Diffusional Processes*, Interscience Publishers, New York, 1962, p. 313.
2. Lacher, J. R., *Proc. Royal Soc. (London) A* **161** (1937), 525.
3. Lacher, J. R., *Proc. Camb. Phil. Soc.* **33** (1937), 518.
4. Gurney, R. W., *Introduction to Statistical Thermodynamics*, McGraw Hill, New York, 1949.
5. Fowler, R. and Guggenheim, E. A., *Statistical Thermodynamics*, Cambridge Univ. Press, Cambridge, 1956.
6. Fall, I. and Genin, J. M. R., *Hyp. Interact.* **69** (1991), 513.
7. Oda, K., Kojima, N., Ito, K. and Ino, H., *Hyp. Interact.* **54** (1990), 853.
8. Oda, K., Umezu, K. and Ino, H., *J. Phys.: Condens Matter* **2** (1990), 10147.
9. Oda, K., Fujimura, H., Mae, K. and Ino, H., *Hyp. Interact.* **69** (1991), 533.
10. Oda, K., Fujimura, H. and Ino, H., *J. Phys.: Condens Matter* **6** (1994), 679.
11. Giellen, P. M. and Kaplow, R., *Acta Metall.* **15** (1967), 49.
12. DeCristofaro, N. and Kaplow, R., *Mettal. Trans. A* **8** (1977), 35.
13. Foct, J., *J. Physique C* **6** (1974), C6-487.
14. Foct, J., Rochegude, P. and Hendry, A., *Acta Metall.* **36** (1988), 501.
15. Balanyuk, A. G., Bugaev, V. N., Nadutov, V. M. and Sozinov, A. L., *Phys. Stat. Sol. B* **207** (1998), 3.
16. Fernández Guillermet, A. and Du, H., *Z. Metallkd.* **85** (1994), 3.
17. Wriedt, H. A., Gokcen, N. A. and Nafziger, R. H., *Bull. of Alloys and Phase Diagrams* **8** (1987), 355.
18. Bauer, Ph., Uwakweh, O. N. C. and Genin, J. M. R., *Hyp. Interact.* **41** (1988), 555.
19. Uwakweh, O. N. C., Bauer, J. P. and Génin, J. M., *Metall. Trans. A* **21** (1990), 589.
20. Laneri, K., Desimoni, J., Zarragoicoechea, G. J. and Fernández Guillermet, A., *Phys. Rev. B* **66** (2002), 134201.
21. Wriedt, H. A., Gokcen, N. A. and Nafziger, R. H., *Bull. of Alloys and Phase Diagrams* **8** (1987), 355.
22. Atkinson, D. and Bodsworth, C., *J. Iron Steel Inst.* **208** (1970), 587.
23. Bhadeshia, H. K. D. H. *Met. Sci.* **16** (1982), 167.
24. Bhadeshia, H. K. D. H., *Mater. Sci. Technol.* **14** (1998), 273.
25. McLellan, R. B. and Alex, K., *Scrip. Metall.* **4** (1970), 967.
26. Sozinov, L., Balanyuk, A. G. and Gavriljuk, V. G., *Acta Mater.* **47** (1999), 927.
27. Sozinov, A. L., Balanyuk, A. G. and Gavriljuk, V. G., *Acta Mater.* **45** (1997), 225.
28. Gavriljuk, V. G., Shanina, B. D. and Berns, H., *Acta Mater.* **48** (2000), 3879.
29. Timoshevskii, A. N., Timoshevskii, V. A. and Yanchistsky, B. Z., *J. Phys.: Condens. Matter.* **13** (2001), 1051.

Hyperfine Interactions **156/157**: 541–545, 2004.
© 2004 *Kluwer Academic Publishers. Printed in the Netherlands.*

Distribution of Mn Atoms in a Substitutional bcc-FeMn Solid Solution

M. MIZRAHI*, A. F. CABRERA, S. M. COTES, S. J. STEWART,
R. C. MERCADER and J. DESIMONI
*Departamento de Física, Facultad de Ciencias Exactas, UNLP, IFLP-CONICET C.C. 67,
1900 La Plata, Argentine; e-mail: mizrahi@fisica.unlp.edu.ar*

Abstract. We have taken conversion electron Mössbauer spectra to Fe–Mn alloys with 7.5, 10.5 and 13.7 wt.% Mn, that allowed us to obtain the relative fractions of the different iron environments in the bcc structure. We compare these populations with those obtained from Monte Carlo simulations. The magnetic interactions between Fe–Fe, Fe–Mn and Mn–Mn atoms have been modeled using an Ising-type Hamiltonian in a 3D bcc lattice with periodic boundary conditions. The simulations were performed using a Metropolis algorithm to determine the jump probability of Mn atoms.

Key words: Monte Carlo simulations, Fe–Mn alloys.

1. Introduction

Some features of the Fe–Mn metastable phase diagram are still a research field. The system displays various metastable phases – like the ε- and α-martensite phases of hcp and bcc crystalline structure respectively, and γ-austenite having an fcc structure – that can transform into each other under suitable thermal treatments [1] or plastic deformations [2]. The role played by the intricate related physical parameters, not clarified yet, hinder its study by experimental, theoretical and numeric methods. Then, we have undertaken a research project that applies the Monte Carlo (MC) method [3] to simulate the conditions that determine the relative existence of its different phases.

It is known that the bcc-FeMn phase is a ferromagnetic solid solution at room temperature. It has been observed in alloys up to 16 wt% Mn [2], where it coexists with the paramagnetic fcc or hcp phases, depending on the preparation method.

As a first step to meet our goal, in this work – to assess the applicability of the MC method to the simulation of the atomic arrangements in the bcc structure – we compare our MC simulated results with the relative fractions of the different Fe sites in Mössbauer spectra prepared with different Mn concentrations.

---

* Author for correspondence.

2. The Ising Hamiltonian

The Hamiltonian of the system could be written as the sum of two contributions:

$$H = H_{conf} + H_{mag} \tag{1}$$

one being configurational (H_{conf}) and the other magnetic (H_{mag}).

The configurational energy is related to the mixing enthalpy (ΔH_{mix}) [4]. Different values of ΔH_{mix} are found in literature for bcc Fe–Mn. They spread from 1 to -9 kJ/mol at high temperatures (>1800 K) [5].

As a preliminary analysis and due to the lack of accurate data of the configurational energy, we only consider the magnetic contribution to the Hamiltonian [6]. Consequently, to perform the MC simulations, interactions between Fe–Fe, Fe–Mn and Mn–Mn atoms, assuming localized magnetic moments on Fe and Mn atoms [6], have been modeled using an Ising-type Hamiltonian defined as:

$$H_{mag} = - \sum_{i,j} J_{ij} \sigma_i \sigma_j.$$

The sum runs over the nearest-neighbor sites $\langle i, j \rangle$ and $\sigma_i = \pm 1$ depending on the spin state. Since the bcc-FeMn is ferromagnetic at room temperature, then the magnetic energy of the system is:

$$E = - \sum_{i,j} J_{ij}. \tag{2}$$

The exchange interaction J_{ij} was considered dependent on site i and j occupation state,

$$J_{ij} = \varepsilon_i \varepsilon_j J_{FeFe} + \varepsilon_i (1 - \varepsilon_j) J_{FeMn} + (1 - \varepsilon_i) \varepsilon_j J_{MnFe}$$
$$+ (1 - \varepsilon_i)(1 - \varepsilon_j) J_{MnMn}, \tag{3}$$

where $\varepsilon_i = 1$ ($\varepsilon_i = 0$) if the site is occupied by Fe (Mn) and $J_{FeMn} = J_{MnFe}$. The used values for J_{ij} were $J_{FeFe} = 12.98$ meV, $J_{FeMn} = 3.245$ meV, $J_{MnMn} = -0.649$ meV reported in [6].

3. Monte Carlo simulations

The FeMn alloy was described as a lattice of N_{Mn} manganese atoms and N_{Fe} iron atoms distributed in the $N_{Mn} + N_{Fe} = N$ sites of the *bcc* structure. The occupancy by Mn atoms of the nearest neighbor shell around the Fe atoms was accounted for in 3D MC simulations to calculate the relative fractions f_p (p = number of Mn atoms near neighbors) associated to the different Fe environments.

A Fortran 77 routine using the Monte Carlo method and a 3D bcc lattice with periodic boundary conditions was developed in order to simulate an infinite solid. The Metropolis method [3] was used to define the probability of the Mn jumps. A randomly chosen Mn atom has the probability P to exchange positions with a lattice-neighbor Fe atom, also randomly chosen, viz.,

$$P = \begin{cases} e^{-(E_A - E_B)/kT}, & \text{if } E_A > E_B, \\ 1, & \text{if } E_A \leqslant E_B, \end{cases} \tag{4}$$

where E_B and E_A, are the total energies before and after the atomic exchange, respectively, calculated using the Equation (2), k is the Boltzmann constant and T is the temperature in Kelvin (300 K). If the atomic exchange decreases the total energy, the jump is allowed ($P = 1$), while if the total energy increases, the jump is allowed with a probability $P = \exp[-(E_A - E_B)/kT]$.

In order to study the convergence of the results, arrangements of dimension $L \times L \times L$ body centered cubic cells, with $L = 4, 6, 8$ and 10 primitive cells were used, the number of atomic sites being $2 \times L^3$. The f_p fractions did not vary for arrangements of $L = 8$ cells and higher. Hence arrangements of $N = 1024$ ($L = 8$) atomic sites were employed to optimize the calculation time. The number of Fe and Mn atoms, respectively, depended on the selected composition. For all Mn concentrations, the equilibrium of the system was attained approximately at 500 MC steps, where a MC step is defined as the exchange between an Fe and Mn atom. The fractions were calculated between 500 and 3000 MC steps. The final average fractions were obtained averaging the results of 20 program runs.

4. Experimental procedure

Alloys of different compositions were prepared by arc-melting from pure elements. After solidification, the ingots were thermal treated at 1273 K and then quenched. The final compositions, established by wavelength dispersion spectra microprobe analysis, were 7.5, 10.2 and 13.7 wt% of Mn. The samples were characterized by X-ray diffraction (XRD) and Mössbauer Spectroscopy (MS) using a 5 mCi ^{57}CoRh source and recorded in a standard constant acceleration spectrometer. The Mössbauer spectra were taken in backscattering geometry using a flow detector (He–5% methane) to detect conversion electrons (CEMS). The quoted isomer shifts are relative to α-Fe.

5. Results and discussion

According to X-ray diffractograms (not shown here), the bcc crystalline structure is observed in the three alloys. The Mössbauer spectra obtained on these alloys are shown in Figure 1. The spectra are characteristic for bcc-disordered alloys with magnetic arrangement [2]. They were reproduced by three interactions (see Figure 1), which can be associated to different iron environments. Since iron phase is very sensitive to the surrounding of the atom and it has been reported that each Mn near neighbour (nn) produce a change in the hf value of $\Delta B \cong 2.5$ T [7], each magnetic interaction was assigned to Fe atoms with 0 nn, 1 nn, 2 or more Mn atoms in the first coordination shell, respectively. The results of the fitting procedure are shown in Table I as a function of alloy composition.

The resulting f_0, f_1 and f_{2+} relative fractions obtained by MC simulations are compared with experimental results in Figure 2. A good agreement is found between calculated and experimental relative fractions.

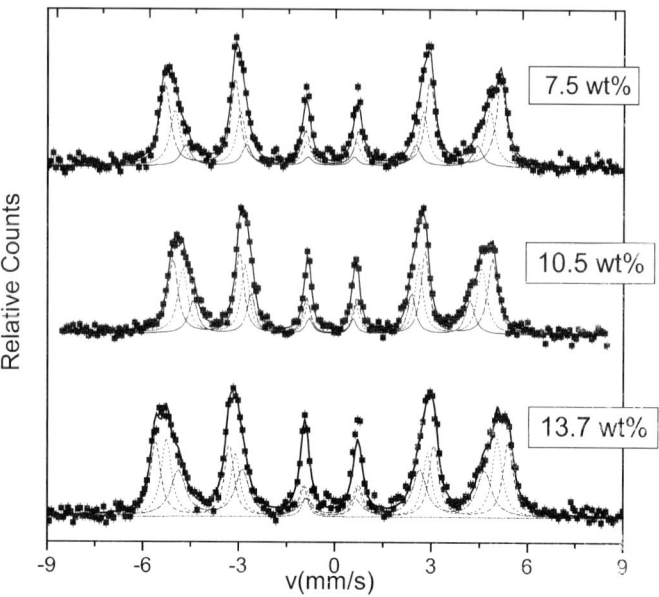

Figure 1. Mössbauer spectra taken at room temperature for different Mn concentration.

Table I. Hyperfine parameters of bcc phase obtained from the ME spectra corresponding to the three alloys

Wt% Mn	B_0 (T)	δ_0 (mm/s)	f_0	B_1 (T)	δ_1 (mm/s)	f_1	B_2 (T)	δ_2 (mm/s)	f_{2+}
7.5	33.5_1	0.005_4	0.53_3	31.4_1	-0.015_7	0.34_3	29.1_1	-0.037_{17}	0.13_2
10.5	33.5_1	0.002_4	0.44_2	31.3_1	-0.007_4	0.37_2	29.0_1	-0.028_8	0.19_2
13.7	32.9_1	0.009_4	0.34_5	30.7_1	-0.002_4	0.33_7	28.1_1	-0.014_9	0.33_6

Despite of the fact that the configuration energy term was not considered in the present approximation, our experimental results are well reproduced by MC simulation only considering the magnetic contribution to the energy. This may be related to the fact that Fe and Mn atoms are chemically similar. However, the relative importance of both contributions to the Hamiltonian deserves a further study.

6. Conclusions

Alloys with different Mn concentration have been studied by MS and by MC simulations. Within the errors, the MC calculated iron-site relative fractions are found in good agreement with the experimental ones. In spite of considering only the magnetic contribution to the Hamiltonian of the system, as a first approach

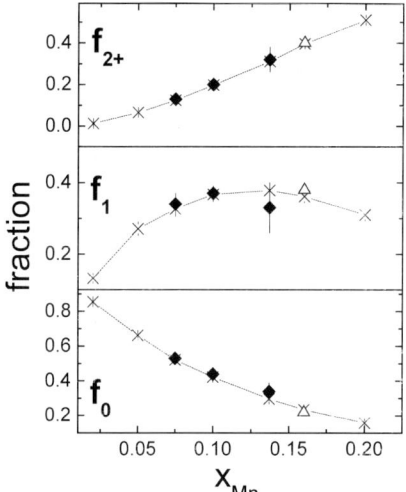

Figure 2. f_0, f_1, f_{2+} relative fractions extracted from Mössbauer spectra (filled rhombs) and that obtained from MC. Simulations (crosses). Experimental results (up empty triangles) taken from [2] are also included.

to the problem, the agreement between experimental and MC simulated relative fractions results surprising. This adds confidence to the MC method as a powerfull tool to analyse substitutional solid solution systems, in particular, on compositional regions where the phases exhibit very similar properties and hence are almost impossible to distinguish experimentally.

Hence, MC simulations accounting for the complete description of the problem are needed.

References

1. Cotes, S., Baruj, A., Sade, M. and Guillermet, A. F., *J. Phys. IV* **5** (1995), 373.
2. Gauzzi, F., Verdini, B., Principi, G. and Badan, B., *J. Mater. Sci.* **18** (1983), 3661.
3. Fishman, G. S., In: *Monte Carlo*, Springer, New York, 1995.
4. Pierron-Bohnes, V., Cadeville, M. C., Finel, A., Caudron, R. and Solal, F., *Phys. B* **180–181** (1992), 811.
5. de Boer, F. R., Boom, R., Mattens, W. C. M., Miedema, A. R. and Niessen, A. K., In: F. R. de Boer and D. G. Pettifor (eds), *Cohesión and Structure*, Vol. 1, North-Holland, Amsterdam, 1989, p. 219.
6. Paduani, C., Plascak, J. A. and Galvão da Silva, E., *Phys. Rev. B* **44**(17) (1991), 9715.
7. Stearns, M. B., *Phys. Rev.* **147** (1966), 439.

Hyperfine Interactions **156/157**: 547–553, 2004.
© 2004 *Kluwer Academic Publishers. Printed in the Netherlands.*

Effect of Sulfur Addition on the Redox State of Iron in Iron Phosphate Glasses

M. M. EL-DESOKY[1,*], A. AL-HAJRY[1], M. TOKUNAGA[2], T. NISHIDA[2] and M. Y. HASSAAN[3]
[1]*Physics Department, Faculty of Science, King Khalid University, PO Box, 9004 Abha, Saudi Arabia; e-mail: mmdesoky@yahoo.com*
[2]*Department of Biological and Environmental Chemistry, Kyushu School of Engineering, Kinki University, Kayanomori 11-6, Iizuka 820 8555, Japan*
[3]*Physics Department, Faculty of Science, Al-Azhar University, Nasr City, Cairo, Egypt*

Abstract. Mössbauer, IR, and electrical conductivity were investigated in iron phosphate glasses, $40Fe_2O_3$–$60P_2O_5$ (in mol%), containing sulfur ranging from 0 to 2, 4, 6, and 8 mass%. Sulfur proved to act as a reducing agent for the redox reaction during the glass preparation. Mössbauer spectroscopy was used in order to determine the relative fraction of Fe^{2+}, i.e. Fe^{2+}/Fe(total), isomer shift (δ), and quadrupole splitting (Δ). Mössbauer results revealed that the relative fraction of Fe^{2+} increases with an increasing sulfur content. Electrical conductivity showed a similar composition dependency as the fraction of Fe^{2+}. These results indicate that higher electrical conductivity of sulfur-containing iron phosphate glasses is due to small polaron hopping (SPH) between iron atoms of different valance states, i.e., a step-by-step electron hopping from Fe^{2+} to Fe^{3+}. IR spectra of these glasses are very similar to those of sulfur-free iron phosphate glasses, proving that the structure of sulfur-containing iron phosphate glasses is essentially the same as that of sulfur-free phosphate glasses.

Key words: iron phosphate glasses, sulfur, Mössbauer, IR, dc conductivity.

1. Introduction

Iron phosphate glasses typically have a relatively poor chemical durability [1], which often limits their industrial applications. However several studies have shown that chemical durability of phosphate glasses can be improved by the addition of various oxides such as Al_2O_3 and Fe_2O_3 [2]. As result, iron phosphate glasses are now of interest for several technological and biological applications. Structural and physical properties of iron phosphate glasses have been investigated so far [1–3], but the effect of sulfur on the physical properties, structure, and redox state of iron phosphate glasses have rarely been reported to our knowledge.

Mössbauer spectroscopy is a useful technique for characterizing the coordination symmetry and redox states of iron in glasses. Mössbauer spectra of glasses generally have broader linewidth than those of crystalline materials, because of dis-

---

* Permanent address: Physics Department, Faculty of Education, Suez Canal University, El-Arish, Egypt.

ordered nature of the environment at around the absorbing atoms, such as different chemical bond lengths and bond angles. Reliable average hyperfine parameters, isomer shift (δ), and quadrupole splitting (Δ), can be obtained by fitting the Mössbauer spectra with a minimum number of broadened Lorentzians required to match the absorption envelope [4].

Electrical conductivity of iron phosphate glasses is originated from the electron hopping from an atom of lower valency state to that of higher valency state, e.g., from Fe^{2+} to Fe^{3+}. This is usually termed "small polaron hopping" (SPH) [5–10] and the electrical conductivity strongly depends upon the distance between two iron atoms [13]. In iron phosphate glasses, the electrical conductivity is related to the total concentration of iron atoms and to the ratio of Fe^{2+} to the total number of iron atoms: $Fe^{2+}/(Fe^{2+} + Fe^{3+})$ [3]. In the present paper, Mössbauer, IR, and dc conductivity were investigated for iron phosphate glasses, $40Fe_2O_3–60P_2O_5$ (in mol%) containing additional amounts (x) of sulfur (S), where x is 0, 2, 4, 6, and 8 in mass%.

2. Experimental

Glass samples were prepared from analytical reagent grade chemicals according to the formula $xS–(40Fe_2O_3–60P_2O_5)$ (mol%), where $x = 0, 2, 4, 6$ and 8 (mass%). Batches were prepared by mixing reagent grade Fe_2O_3, P_2O_5 and S crystalline powders, which were melted in platinum crucibles in air at 1200°C for 1 hour with occasional stirring. Each melt was poured onto a polished copper block kept at room temperature and immediately pressed with another copper block. Amorphous nature of the glasses was ascertained from X-ray diffraction analysis. Density of the glasses was measured by the Archimedes method using toluene as the immersion liquid. Mössbauer measurements were performed at room temperature with a $^{57}Co(Rh)$ source of 925 MBq using Wissel spectrometer of a constant acceleration type. IR spectra of the glass samples were measured from 450 to 2000 cm^{-1} by a conventional KBr pellet method on a Fourier transform infrared (FT-IR) spectrometer (Perkin-Elmer 1760 X). Each pellet was prepared by mixing about 4 mg of glass powder with 150 mg of anhydrous KBr. A background of the IR spectra was corrected with a spectrum of KBr. The dc conductivity (σ) of the as-quenched glass samples was measured at temperatures between 303 and 473 K. Silver paste electrodes were deposited on both faces of the polished samples. A satisfactory I–V characteristic was observed in these samples.

3. Results and discussion

3.1. MÖSSBAUER SPECTRA

Room-temperature (RT) Mössbauer spectra of the glass samples are similar to each other, as illustrated in Figure 1; a quadrupole doublet due to octahedral Fe^{3+} is observed with a weak peak due to octahedral Fe^{2+} in every case. Mössbauer

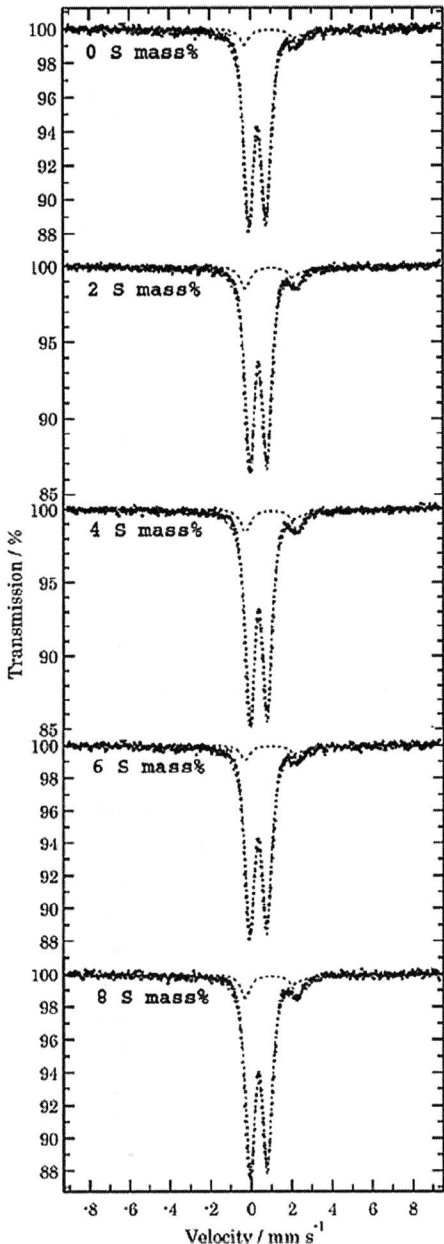

Figure 1. Mössbauer spectra of glass samples with different sulfur content.

hyperfine parameters of each site show that Fe^{2+} has an isomer shift (δ) of 1.21 to 1.23 $\pm$ 0.005 mm s^{-1}, while a quadrupole splitting (Δ) decreases from 2.32 to 2.24 $\pm$ 0.01 mm s^{-1} along with a sulfur content. This means that the iron environment at this site becomes more symmetrical with an increasing sulfur content ranging from 0 to 8 mass%. At the same time, the absorption area (A) increases from 10.2 to 13.4 $\pm$ 0.5% with an increasing sulfur content. The Fe^{3+} site has nearly constant δ values of 0.34 to 0.37 $\pm$ 0.005 mm s^{-1}, and an identical Δ value of 0.87 $\pm$ 0.01 mm s^{-1}. This means that the iron environment at a given site is independent of the Fe^{2+} content and the sulfur content. It should be noted that the reduction effect of sulfur is very small compared with our previous results for borate glass [11], basalt glass [12] and sodium phosphate glass containing iron [13].

3.2. IR SPECTRA

Typical IR spectra of sulfur iron phosphate glasses are shown in Figure 2. The IR spectra of these glasses show the most prominent bands at 515, 770, 918 and 1070 cm^{-1}.

The band between 1060 and 1080 cm^{-1} is attributed to a symmetric and an asymmetric vibration of PO_2^- and PO_3^{2-} terminal groups [8, 9]. A band between 915 and 930 cm^{-1} is due to an asymmetric vibration of P–O–P bridge [14, 15]. A band around 770 cm^{-1} is due to a symmetric stretching mode of the P–O–P bridge [18, 19] and a band at around 515 cm^{-1} may be due to an overlapping vibrations of iron–oxygen polyhedra and P_2O_7 groups [16]. The IR spectra of all the present samples have a shoulder peak at about 630 cm^{-1}; this may be due to vibration of some iron–oxygen bands [17]. IR spectra of these glasses are very similar to those of sulfur-free iron phosphate glasses, proving that the structure of sulfur-containing iron phosphate glasses is essentially the same as that of sulfur-free phosphate glasses.

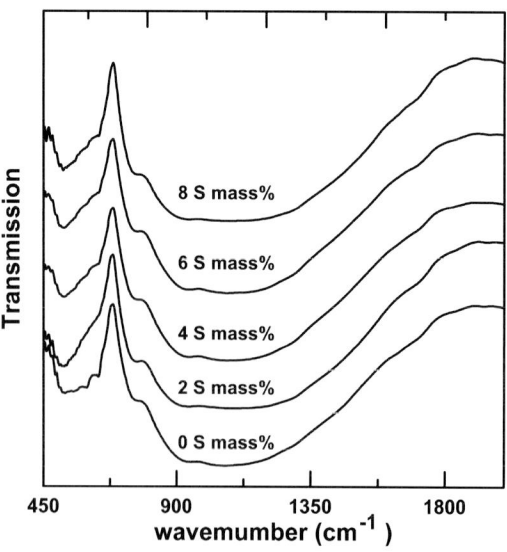

Figure 2. IR spectra for glass samples with different sulfur content.

3.3. DC CONDUCTIVITY AND ACTIVATION ENERGY

Figure 3 shows a relationship between dc conductivity, $\log \sigma$, for the glasses and reciprocal temperature. These glasses have σ values from 1.75×10^{-10} to 1×10^{-5} S m^{-1} at temperature from 303 to 473 K. Figure 3 shows a linear relationship between $\log(\sigma)$ and $1/T$. The slope of the curves gives an activation energy for the conduction, which increases at higher temperatures. The value of σ is apparently expressed as:

$$\sigma = \sigma_0 \exp(-W/kT), \tag{1}$$

where σ_0 is the pre-exponential factor, W the activation energy and k the Boltzmann constant. The activation energy (W) and the pre-exponential factor (σ_0) were obtained from a least-squares method for the experimental data above 335 K. The compositional dependence of the conductivity (σ) at 405 K and the activation energy are shown in Figure 4. A general trend observed in Figure 4 is that the magnitude of the conductivity at fixed (405 K) tends to be highest in those compositions having smallest activation energy. This result is consistent with the small polaron hopping (SPH) mechanism [8]. In Figure 3, dc conductivity varies linearly with $1/T$ in two temperature regions. Such phenomenon was found in mixed calcium and barium iron phosphate glasses [3] and sodium iron phosphate glasses [18]. This phenomenon is attributed to the change of the conduction mode from the SPH to an intermediate variable-range hopping (VRH) [8, 19]. We speculate that the change of the slope is caused by a transition from SPH to VRH, where the polaron binding energy is small and the static energy of the system plays a dominant role

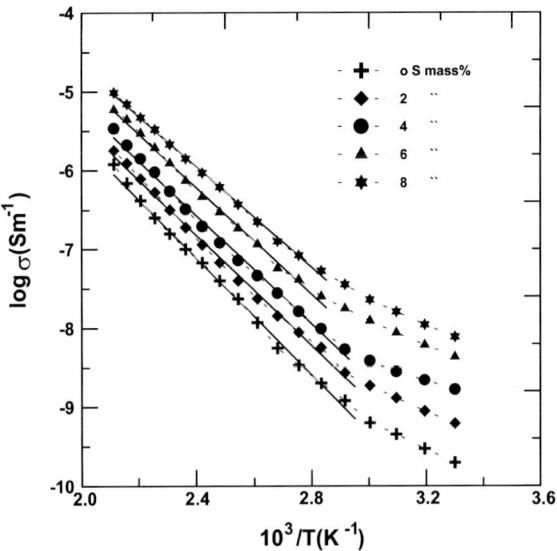

Figure 3. Temperature dependence of the dc conductivity (σ) for glass samples with different sulfur content. Solid lines are calculated by using a least-squares method.

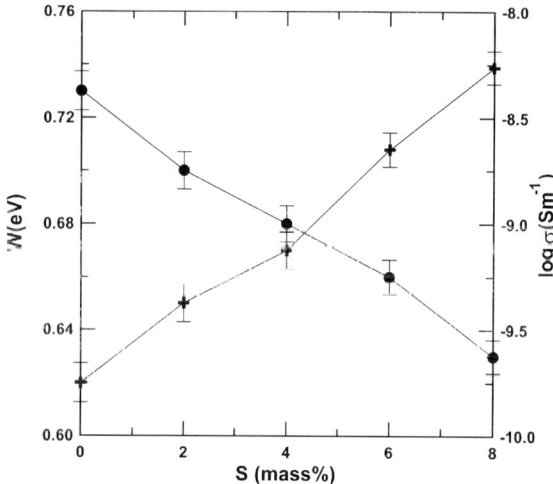

Figure 4. Effect of sulfur (S) content on the dc conductivity (+) at $T = 405$ K on the activation energy (●).

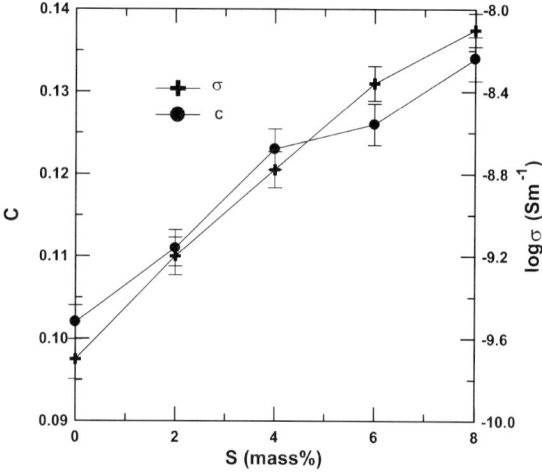

Figure 5. The dc conductivity (σ) at 303 K and the TM ion ratio (C) for different sulfur content. Lines are drawn as a guide for the eyes.

in the conduction mechanism [3, 7]. Figure 5 shows the electrical conductivity (σ) and the fraction of reduced transition metal (TM) ion ($C = Fe^{2+}/Fe_{total}$) as a function of sulfur (S) content (in mass%) in the present glasses. Figure 5 shows the electrical conductivity (σ) of several glasses with different compositions. The σ values are within an order of magnitude of that of iron phosphate glasses [3, 18]. Since the electrical conduction in iron phosphate glasses [3] was assumed to be due to an electron hopping from Fe^{2+} to Fe^{3+} sites, it is reasonable to assume that the electron hopping also occurs in the sulfur-containing iron phosphate glasses. Also, it is clear that the C increases with an increasing S content (in mass%), as the electrical conductivity increases in the same manner (see Figures 3 and 4). These results indicate that the electrical conductivity of sulfur-containing iron phosphate glasses can be understood by the SPH between iron ions of different valency states [8–10]. The small polaron and variable-range hopping parameters will be discussed in a forthcoming paper.

4. Conclusion

Semiconducting oxide glasses, $S-Fe_2O_3-P_2O_5$ system, are fabricated with S, Fe_2O_3 and P_2O_5 raw materials by a melt-quenching method. The IR spectra of the S-containing glass samples are similar to that of S-free iron phosphate glasses, indicating that the local structure of these glass samples is not affected by S-content. Sulfur seems to act as a reducing agent during glass synthesis and affects the electrical conductivity. Mössbauer spectra revealed that the Fe^{2+} ratio increases with an increasing sulfur content. The dc conductivity measurements leads to a conclusion that the electrical conductivity is dominated by the electron hopping between Fe^{2+} to Fe^{3+} sites which increases with an increasing sulfur content.

References

1. Day, D. E., Wu, Z., Ray, C. S. and Hrma, P., *J. Non-Cryst. Solids* **241** (1998), 1.
2. Moguš-Milanković, A., Šantić, B., Ray, S. C. and Day, D. E., *J. Non-Cryst. Solids* **263–264** (2000), 229.
3. El-Desoky, M. M. and Kashif, I., *Phys. Stat. Sol. (a)* **194**(1) (2002), 89.
4. Hassaan, M. Y., El-Desoky, M. M., Salem, S. M. and Salah, s. H., *J. Radio. Nucl. Chemistry* **249** (2001), 595.
5. Moguš-Milanković, A. and Day, D. E., *J. Non-Cryst. Solids* **162** (1993), 275.
6. Moguš-Milanković, A., Šantić, B., Ray, C. S. and Day, D. E., *J. Non-Cryst. Solids* **119** (2001), 283.
7. Mott, N. F., *J. Non-Cryst. Solids* **1** (1968), 1.
8. Austin, I. G. and Mott, N. F., *Adv. Phys.* **18** (1969), 41.
9. Murawski, L., *J. Mater. Sci.* **17** (1982), 2155.
10. Moguš-Milanković, A., Gajovic, A., Šantić, A. and Day, D. E., *J. Non-Cryst Solids* **289** (2001), 204.
11. Hassaan, M. Y., El-Desoky, M. M., Salem, S. M. and Salah, S. H., *J. Radio. Nucl. Chem.* **249** (2001), 595.
12. Hassaan, M. Y., El-Desoky, M. M., Gomaa, N. S. and Eissa, N. A., *Hyp. Interact. C* **5** (2001), 61.
13. Karabulut, M., Marasinghe, G. K., Ray, C. S., Day, D. E., Waddill, G. D., Booth, C. H., Allen, P. G., Bucher, J. J., Caulder, D. L. and Shuh, D. K., *J. Non-Cryst. Solids* **306** (2002), 182.
14. Samuneva, B., Tzvetkova, P., Gugov, I. and Dimitrov, V., *J. Mater. Sci. Lett.* **15** (1996), 2180.
15. Efimov, A. M., *J. Non-Cryst. Solids* **209** (1997), 209.
16. Wang, G., Wang, Y. and Jin, B., *SPIE* **2287** (1994), 214.
17. Fang, X., Ray, C. S., Marasinghe, G. K. and Day, D. E., *J. Non-Cryst. Solids* **263&264** (2000), 293.
18. El-Desoky, M. M., Tahoon, K. and Hassaan, M. Y., *Mater. Chem. Phys.* **69** (2001), 180.
19. Greaves, G. N., *J. Non-Cryst. Solids* **11** (1973), 427.

Hyperfine Interactions **156/157**: 555–561, 2004.
© 2004 *Kluwer Academic Publishers. Printed in the Netherlands.*

Structural and Electronic Features of Sb-Based Electrode Materials: ^{121}Sb Mössbauer Spectrometry

C. M. IONICA, L. ALDON, P. E. LIPPENS, F. MORATO,
J. OLIVIER-FOURCADE and J.-C. JUMAS*
Laboratoire des Agrégats Moléculaires et Matériaux Inorganiques (UMR 5072 CNRS)
Université Montpellier II, CC 15, Place E. Bataillon, 34095 Montpellier Cedex 5, France;
e-mail: Jumas@univ-montp2.fr

Abstract. Lithium insertion mechanisms in two antimony based compounds: $CoSb_3$ and $CoSb$ have been studied by means of ^{121}Sb Mössbauer spectrometry. Structural and electronic modifications induced by insertion of lithium have been characterised for different depths of discharge. In all cases the insertion mechanisms can be described from several steps. In the first step antimony is partially dispersed in the metallic matrix with amorphisation of the electrode material and in a second step we can observe the alloy forming (Li_3Sb). However this amorphous alloy remains in interaction with the matrix allowing then a good reversibility.

Key words: electrode materials, lithium insertion mechanism, antimony-based compounds, ^{121}Sb Mössbauer data.

1. Introduction

In the recent years, lithium-ion batteries have become the major power sources for portable equipment because of their specific energy and high working voltage. The conventional anode material of lithium-ions batteries is graphite but Li storage capacity is low. Early efforts were placed on binary lithium–metal systems Li_xM (M = Al, Sb, Si, Sn) [1] and tin composite oxides [2, 3]. The disadvantage with the alloys studied so far is their drastic volume expansion during Li insertion/extraction cycles. The cycling performance of lithium alloys can be significantly improved if intermetallic hosts are used instead of pure metals [4–10]. The lithium insertion in these alloys takes softened place through multi-phase mechanisms and the X-ray diffraction patterns of the discharged electrodes are characteristic of amorphous compounds. The use of alternative technique such as Mössbauer spectrometry is indispensable to complete the characterisation. Mössbauer spectrometry is a hyperfine technique that can be used as a probe to investigate changes in the electronic state of the Sb contained within these compounds as a function of the degree of lithiation.

---

* Author for correspondence.

Here we present the study of lithium insertion mechanism in two antimony-based compounds (CoSb and $CoSb_3$) from ^{121}Sb Mössbauer measurements. The hyperfine Mössbauer parameters isomer shift δ (IS) and quadrupole splitting Δ (QS), which are respectively linearly dependent on the s electron density and on the electric field gradient (EFG) at the nucleus, are related to the oxidation states and environments of the probed element (here ^{121}Sb). For ^{121}Sb these two hyperfine parameters (IS and QS) are derived from the analysis of experimental data using transmission-integral procedures [11].

2. Experimental

$CoSb_3$ and CoSb samples were prepared by direct synthesis from pure elements in silica tubes sealed under vacuum. After annealing one week at 750°C for $CoSb_3$ and 8 h at 1200°C for CoSb the tubes samples were slowly cooled in the furnace to room temperature. To identify phase purity of the samples, X-ray powder diffraction was carried out.

Electrodes containing 82 wt% sample, PTFE binder 12 wt% and carbon black were pressed in 7 mm diameter pastilles. The electrochemical properties of the materials were measured with Li||1M $LiPF_6$ (PC : EC : DMC = 1 : 1 : 3, v/v)|$CoSb_3$ (or CoSb) cells using a Mac Pile II system. Two constant charge-discharge current densities were used for the $CoSb_3$ tests: 9 mA cm^{-2} and 15 mA cm^{-2} between 0.05 and 1.1 V versus Li$^+$/Li. The CoSb cells were cycled at 3 and 12 mA cm^{-2} between 0.05 and 1.2 V.

^{121}Sb Mössbauer measurements were performed using a ^{121m}Sn in $BaSnO_3$ source of nominal activity 0.5 mCi on an EG & G constant acceleration spectrometer in transmission mode. During the measurements, both source and absorbers were simultaneously cooled down to 4 K in order to increase the fraction of recoil-free absorption and emission processes. The velocity scale was calibrated using the magnetic sextuplet spectrum of a high purity Fe absorber and a $^{57}Co(Rh)$ as source. The origin of isomer shift scale was determined from the centre of the InSb spectrum at 4 K (-8.70 mm s^{-1} relative to the Ba$^{121m}SnO_3$ source). Measurements were performed ex situ at several depths of discharge and charge. The Swagelok cells were opened inside the glove box and the electrode materials containing the active material were placed on specific sample holder transparent for the γ rays.

3. Results and discussions

The pristine powders, containing 1 to 25 μm diameter particles were identified like single phase compounds. The refinement of the unit cell parameters leads to cubic $CoSb_3$ with $a = 9.032$ Å which is in good agreement with the reported literature value of $a = 9.035$ Å [12] and hexagonal CoSb with $a = 3.893$ Å and $c = 5.189$ Å as compared with literature data [13].

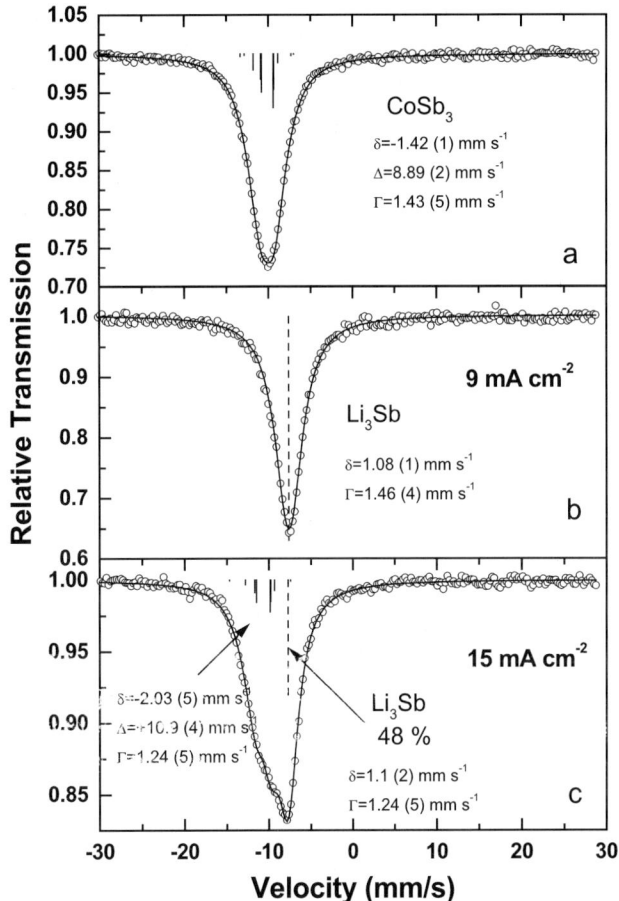

Figure 1. ^{121}Sb Mössbauer spectra of CoSb$_3$ (a) and at the end of the first discharge at 0.05 V with different current densities of 9 mA cm^{-2} (b) and 15 mA cm^{-2} (c). The velocity scale has not been corrected for InSb reference ($\delta = -8.70$ mm/s relative to the source).

The study of the ^{121}Sb Mössbauer effect in alloys of antimony with transition metals led to interesting conclusions concerning the local electron density of the antimony site. The ^{121}Sb Mössbauer spectra of the pristine materials CoSb$_3$ (Figure 1a) and CoSb (Figure 2a) show one hyperfine site for Sb whose IS values (-1.42 mm s^{-1} and -0.24 mm s^{-1} respectively) are characteristic of intermetallic compounds [14]. The large QS values ($+8.89$ mm s^{-1} and -6.24 mm s^{-1}) are in agreement with the distortion of the Sb environment.

Figure 3 gives the voltage profiles of the first discharge (insertion of Li$^+$) and charge (extraction of Li$^+$) profiles of the CoSb$_3$ and CoSb samples. It can be seen that Li-inserting/deinserting capacity in the first cycle are 674/371 mA h g^{-1} and 550/398 mA h g^{-1} for the CoSb$_3$ and CoSb electrodes, respectively. The discharge potential for the CoSb$_3$ electrode drops rapidly to 0.8 V vs. Li process associated

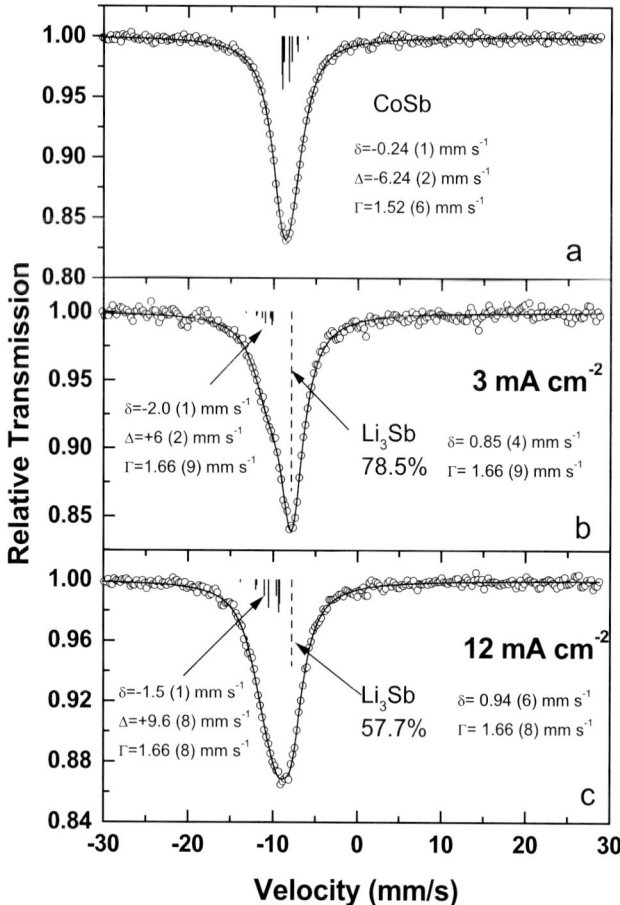

Figure 2. [121]Sb Mössbauer spectra of CoSb (a) and at the end of the first discharge at 0.05 V with different current densities of 3 mA cm^{-2} (b) and 12 mA cm^{-2} (c). The velocity scale has not been corrected for InSb reference ($\delta = -8.70$ mm/s relative to the source).

with the formation of a SEI (solid electrolyte interface) passivating film as a result of the reduction reactions of Li with electrolyte onto the carbon particles [15]. The passivating film is mainly composed of Li_2CO_3 and $ROCO_2Li$ [16]. The first discharge proceeds through a voltage plateau corresponding to a two phase reaction, whose potential 0.57 V is lower than for metallic Sb (0.9 V) [17]. At the end of the plateau 9 Li$^+$ were inserted and about 1.5 Li$^+$ are inserted while discharge at 0.05 V.

Similar to CoSb$_3$ alloy, together with other antimony based compounds and even pure Sb, the CoSb compound shows similarities in the charge/discharge curves. The potential drops to 0.8 V, then a plateau at 0.32 V which corresponds to a biphasic system, can be observed and 3.7 Li$^+$ were inserted while complete discharge at 0.05 V. This amount depends on the electrochemical discharge regime.

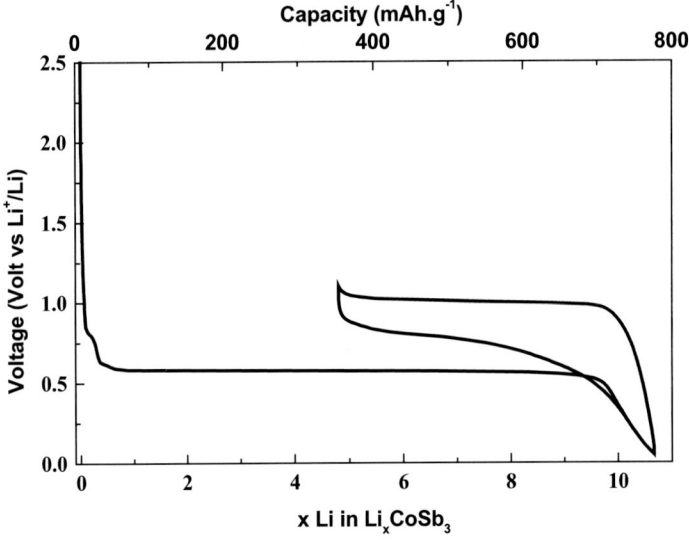

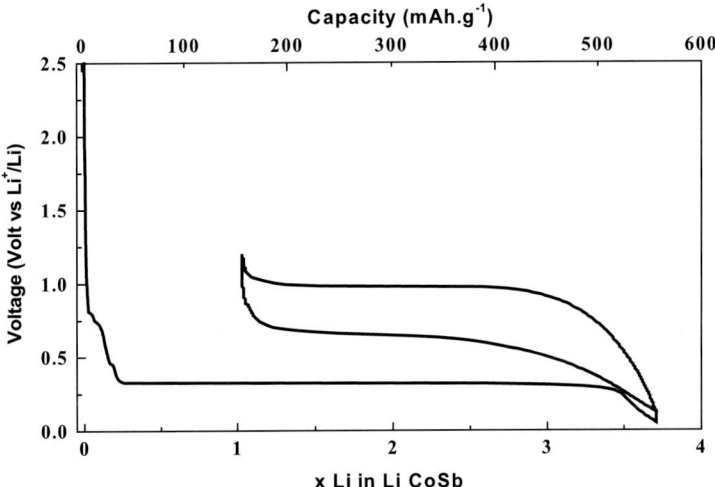

Figure 3. Charges/discharge curves of $CoSb_3$ and $CoSb$ electrode materials.

The *ex situ* [121]Sb Mössbauer spectra of CoSb samples which were discharged at 3 and 12 mA cm^{-2} (Figures 2b and 2c, respectively), confirm the presence of a component whose parameters ($\delta_1 = 0.85$ mm s^{-1}, $\Delta_1 = 0$, $\Gamma_1 = 1.66$ mm s^{-1} and 78.5% relative contribution for the cell discharge with 3 mA cm^{-2} and $\delta_1 = 0.94$ mm s^{-1}, $\Delta_1 = 0$, $\Gamma_1 = 1.66$ mm s^{-1} and 57.7% relative contribution for the cell discharge with 12 mA cm^{-2}) are close to those which have been observed for Li_3Sb ($\delta = 1.3$ mm s^{-1} and $\Delta = 0$) [14]. Previous studies which have been performed in our group on the insertion mechanism of Li in $CoSb_3$ by means of Mössbauer spectroscopy [14] showed that during the first discharge with 9 mA cm^{-2} regime, the signal of the pristine CoSb$_3$ compound disappears, and a new signal appears, centered at ca. 1–1.3 mm s^{-1} (relative to InSb). The comparison with the IS of pure Li–Sb compounds allows us to assign this signal to Li$_3$Sb formation unequivocally. Two components were identified in the spectra of the sample discharged at 0.6 V: one of them, which appear at ca. 1 mm/s, can be assigned to Li$_3$Sb and the other on, at ca. -2.4 mm/s, is assigned to a modified environment of Sb in Li$_x$CoSb$_3$ as compared with the pristine compound, and also ascribable to a possible insertion phenomenon. At the end of the first discharge the Mössbauer spectra shows only one signal for Sb which corresponds to the Li$_3$Sb alloy (Figure 1b). ^{121}Sb Mössbauer spectrum which has been obtained for the completely discharged CoSb$_3$ electrode with 15 mA cm^{-2} is shown on Figure 1c. In the rapid regime the Mössbauer spectrum of the fully discharged sample remains a combination of two components: Li$_3$Sb ($\delta_1 = 1.1$ (2) mm s^{-1}, 48% relative contribution) and a new phase Li$_x$CoSb$_y$ attributed to a highly dispersed electrochemical active matrix ($\delta_2 = -2.03$ (5) mm s^{-1}, $\Delta_2 = +10.9$ (4) mm s^{-1} and 52% relative contribution).

Two reactions can be proposed from these results:

(1) lithium dispersion in the metallic matrix and progressive amorphisation of CoSb$_3$ electrode:

$$x\mathrm{Li} + \mathrm{CoSb}_3 \rightarrow \mathrm{Li}_x\mathrm{CoSb}_y + (3 - y)\mathrm{Sb}.$$

(2) Li$_3$Sb formation and phase separation:

$$\mathrm{Li}_x\mathrm{CoSb}_y + z\mathrm{Li} \rightarrow \mathrm{Li}_{x+z}\mathrm{Co}_{1-m}\mathrm{Sb}_y + m\mathrm{Co} \leftrightarrow \mathrm{Co} + 3\mathrm{Li}_3\mathrm{Sb}.$$

The reversibility of the reaction is provided by the Li$_3$Sb alloy and the active matrix (Li$_{x+z}$Co$_{1-m}$Sb$_y$ + mCo) which has been finely dispersed.

4. Conclusions

The electrochemical properties of CoSb$_3$ and CoSb have been investigated by using lithium anode model cells. The insertion mechanism of Li in CoSb$_3$ and CoSb has been analysed by ^{121}Sb Mössbauer spectroscopy. The lithium insertion induces a displacement of Sb atoms which leads to the formation of an intermediate material. Further insertion of lithium into the active matrix leads to the formation of Li$_3$Sb which remains in interaction with the matrix for good reversibility. The mechanism of lithium insertion depends on the electrochemical discharge regime. An increase in the discharge current density determines a decrease in the Li$_3$Sb amount which has been formed. At present, additional studies are in progress in order to analyse the influence of the particle size on insertion mechanism and electrochemical performances.

Acknowledgements

The authors acknowledge financial support from SAFT, Bordeaux, France (Contract No. 752295/00 MP). One of the authors (C. M. Ionica) thanks ADEME (France) for the award of a PhD scolarship.

References

1. Huggins, R. A. In: J. O. Besenhard (ed.), *Handbook of Battery Materials*, Part III.1, Wiley/VCH, Wienheim, Germany, 1999, p. 359.
2. Idota, Y., Kubota, T., Matsufuji, A., Maekawa, Y. and Miyasaka, T., *Science* **276** (1997), 1395.
3. Chouvin, J., Olivier-Fourcade, J., Jumas, J. C., Simon, B., Biensan, Ph., Fernandez-Madrigal, F. J., Tirado, J. L. and Pérez Vicente, C., *J. Electroanal. Chem.* **494** (2000), 136.
4. Mao, O. and Dahn, J. R., *J. Electrochem. Soc.* **146** (1999), 414.
5. Kepler, K. D., Vaughey, J. T. and Thackeray, M. M., *J. Power Sources* **8182** (1999), 383.
6. Sakaguchi, H., Honda, H. and Esaka, T., *J. Power Sources* **81–82** (1999), 229.
7. Alcántara, R., Fernandez-Madrigal, F. J., Lavela, P., Tirado, J. L., Jumas, J. C. and Olivier-Fourcade, J., *J. Mater. Chem.* **9** (1999), 2517.
8. Cao, G. S., Zhao, X. B., Li, T. and Lu, C. P., *J. Power Sources* **94** (2001), 102.
9. Larcher, D., Beaulieu, L. Y., Mao, O., George, A. E. and Dahn, J. R., *J. Electrochem. Soc.* **147** (2000), 1703.
10. Hewitt, K. C., Beaulieu, L. Y. and Dahn, J. R., *J. Electrochem. Soc.* **148** (2000), A402.
11. Rubenbauer, K. and Birchall, T., *Hyp. Interact.* **7** (1979), 125.
12. Mandrus, D., Mighiori, A., Darling, T. W., Hundley, M. F., Peterson. E. J. and Thompson, J. D., *Phys. Rev. B* **52** (1995), 4926.
13. Siegrist, T. and Hulliger, F., *J. Solid State Chem.* **63** (1986), 23.
14. Aldon, L., Garcia, A., Olivier-Fourcade, J., Jumas, J. C., Fernandez-Madrigal, F. J., Lavela, P., Pérez Vicente, C. and Tirado, J. L., *J. Power Sources* **119–121** (2003), 585.
15. Tarascon, J.-M., Morcrette, M., Dupont, L., Chabre, Y., Payen, C., Larcher, D. and Pralong, V., *J. Electrochem. Soc.* **150**(6) (2003), A732.
16. Li, J. Z., Li, H., Wang, Z. X., Chen, L. and Huang, X. J., *J. Power Sources* **107** (2002), 1.
17. Grugeon, S., Laruelle, S., Dupont, L. and Tarascon, J.-M., *Solid State Science* **5** (2003), 895.

Hyperfine Interactions **156/157**: 563–567, 2004.
© 2004 *Kluwer Academic Publishers. Printed in the Netherlands.*

Study of Disordered $Fe_2Cr_{(1-x)}Mn_xAl$ Alloys

N. LAKSHMI, K. VENUGOPALAN and V. K. AGARWAL
Department of Physics, M.L. Sukhadia University, Udaipur 313 001, India

Abstract. A series of ferromagnetic alloys $Fe_2Cr_{(1-x)}Mn_xAl$ were prepared by arc melting for $x = 0$ to 0.05. The alloys are single phased, but disordered with B2 type structure. Variation of the lattice parameter with concentration of Mn is smooth with a maximum occurring at $x = 0.02$. Highest Curie temperature (T_C) corresponds to sample with $x = 0.02$. Mössbauer spectra for these alloys show the presence of a paramagnetic peak along with the magnetic hyperfine field. This co-existence can be explained by clustering of Mn atoms. The hyperfine field distribution shows the presence of a single, broad component of field. The variation of this component as well as saturation magnetization follows the variation in the lattice parameter with concentration of Mn. Mössbauer and X-ray studies point to the disorder present in these alloys.

Key words: [57]Fe Mössbauer, disordered alloys, hyperfine fields.

1. Introduction

Heusler alloys are ternary alloys of stoichiometric composition bearing the general formula X_2YZ. In this class of alloys, X and Y are generally transition elements and Z an sp element. These alloys offer excellent systems for studying magnetic interactions. Large volumes of studies have been devoted to Heusler alloys bearing the general formula Pd_2YZ and Co_2YZ, with Y mostly being Mn [1, 2]. Not many studies for Fe-based Heusler alloys have been reported.

In general, it can be seen that on preparing Heusler alloys X_2YZ with Pd as X, ordering is easily possible. It can even tolerate some amount of off-stoichiometric composition [3]. It was observed that there was more of a tendency for disorder in Co based alloys as compared to Pd based ones. When X is Fe, few of these alloys show good ordering. In particular, in most Heusler-like alloys, irrespective of what element X is, when Z is a III B element, the possibility for forming disordered systems is greater [4].

Iron-based transition metal alloys show a great sensitivity to environmental effects [5], for instance, the definite site preference of Cr, Co or V in Fe–Al or Fe–Si systems even in the presence of disorder [6–9]. Our studies on the Heusler like alloy Fe_2CrAl showed this alloy to be difficult to order chemically [10]. In addition, the study of hyperfine fields in this alloy showed a co-existence of a paramagnetic and magnetic portion. In order to study the effect of introduction of manganese into Fe_2CrAl, a series of alloys $Fe_2Cr_{(1-x)}Mn_xAl$ was prepared for $x = 0.01, 0.02$,

0.03, 0.04 and 0.05. This study reports the results of Mössbauer and magnetization studies made on this series of alloys.

2. Experimental

The alloys were made by arc melting together Fe, Cr, Mn and Al in stoichiometric quantities in argon atmosphere for x corresponding to 0.01, 0.02, 0.03, 0.04 and 0.05. The materials used were of at least 99.99% purity (obtained from M/s Spex Inc., USA). The iron used was in the form of a foil whereas other materials were in the form of ingots. These were then remelted several times to ensure homogeneity. The hard but brittle buttons were then crushed and sealed into quartz ampoules evacuated to 10^{-5} torr and kept for annealing. Annealing was done at 800°C for 72 hours and then quenched to room temperatures. Weight loss for all samples was negligible (less than 1%).

The alloys were powdered to have grain size corresponding to No. 400 mesh and used for X-ray analysis. All spectra were recorded at room temperature using Cu $K\alpha_1$ radiation. Parameters obtained for the $Fe_2Cr_{(1-x)}Mn_{(x)}Al$ diffraction patterns for $x = 0$ to 0.05 are given in Table I.

Magnetization curves and Curie temperatures (T_C) for the alloys were made on an Ortec Model 155 vibrating sample magnetometer (VSM). Mössbauer spectra of these alloys showed their centroids were very close to the zero velocity channel. Therefore measurement of Curie temperatures using the zero velocity thermal scan method [6] was also made. Results of these measurements are given in Table I.

Samples for Mössbauer studies were made to have a thickness of 10 mg/cm^2 of natural iron. A standard Austin drive and controller assembly was used in the fly-back mode. The source used was a 25-mCi Co^{57} in rhodium matrix. Velocity calibration was done using laser velocity calibrator. The line width of inner peaks of natural iron was found to be 0.26 mm/sec.

Table I. Lattice constants and Curie temperatures for $Fe_2Cr_{(1-x)}Mn_{(x)}Al$

Mn Concentration	a_0 (Å)	T_C (K) (VSM) (± 1 K)	T_C (K) (zero velocity thermal scan) (± 1 K)
0	5.807 ± 0.001	376	348
0.01	5.850 ± 0.001	412	355
0.02	5.861 ± 0.001	416	378
0.03	5.841 ± 0.003	385	358
0.04	5.843 ± 0.003	385	368
0.05	5.675 ± 0.002	385	373

3. Results and discussion

X-ray diffractograms show that these alloys have formed in a single phase with a B2 type structure for x from 0 to 0.04. In the well ordered state, Fe$_2$CrAl and the Mn substituted samples should have had the L2$_1$ Heusler-like structure. This therefore showed the presence of considerable chemical disorder in these samples. From Table I we see that the lattice parameter a_0 increases for Mn concentrations corresponding to $x = 0$ to 0.02 then decreases. The $x = 0.05$ sample had a simple cubic structure with Fe, Mn, Cr and Al randomly distributed among the available sites. Magnetization measurements showed all these alloys to be ferromagnetic in nature.

Mössbauer spectra at room temperature for Fe$_2$Cr$_{(1-x)}$Mn$_{(x)}$Al samples for $x = 0.01$ to 0.05 are shown in Figure 1(a). Their field distributions are shown in Figure 1(b). The spectra of all the Mn substituted samples retained the paramagnetic portion initially seen in Fe$_2$CrAl.

At room temperature, the Mössbauer spectrum for Fe$_2$CrAl showed largely the presence of a paramagnetic part co-existing with a magnetic hyperfine portion [10]. Attempts were made to fit the spectra with a sextet and a paramagnetic peak. However, the line widths were seen to be more than twice that of iron. Therefore the

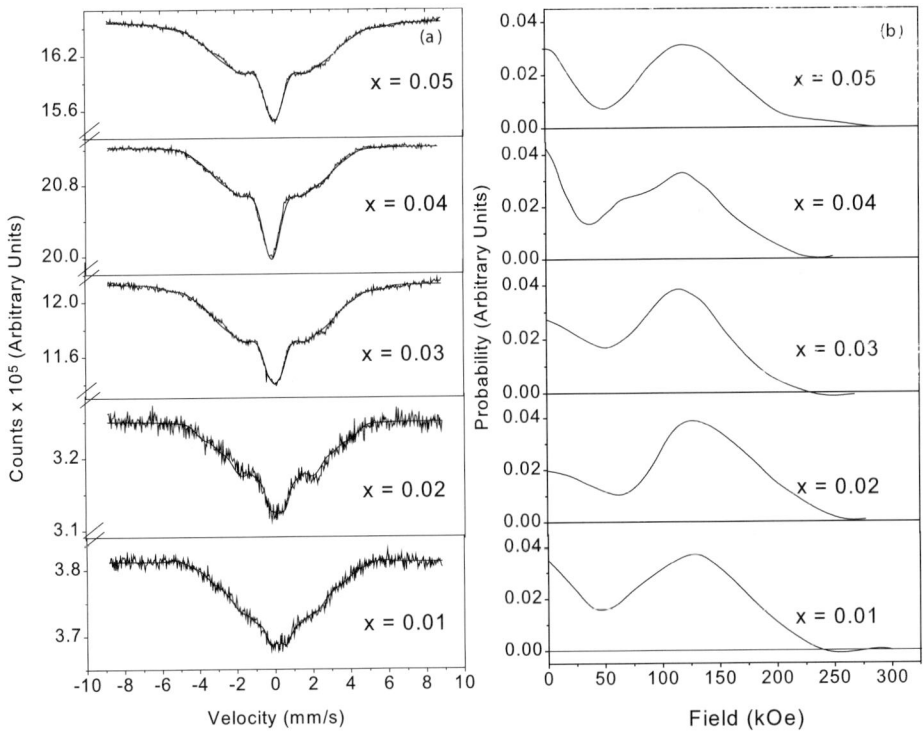

Figure 1. (a) Mössbauer spectra of Fe$_2$Cr$_{(1-x)}$Mn$_x$Al series for different concentration of Mn. (b) Hyperfine field distributions of Fe$_2$Cr$_{(1-x)}$Mn$_x$Al series for different concentration of Mn.

spectra were fitted again for a distribution of magnetic hyperfine fields using the Window's [11] method. The sextet line width was constrained to be that of Fe. The probability distribution for the hyperfine field showed a zero field component (paramagnetic portion) as well as a high field component at one distinct field position. It was seen that the fractional width of the peak (i.e. $\Delta H/H$) corresponding to this position is considerable, coming to above 30%. This pointed to the considerable disorder present in theses samples. The variation of the Curie temperature with the concentration of Mn measured using the Mössbauer zero velocity thermal scan method showed the same trend as that of Curie temperatures measured using the VSM. Both showed the alloy with $x = 0.02$ to have the maximum T_C. However, the values of T_C were seen to be higher for the VSM measurements, ranging from 376 K to 416 K, whereas it ranged from 348 K to 378 K for measurements made by the zero velocity thermal scan method. This difference arises because when the zero velocity thermal scan method is used, only short range order comes to play and so the probe Fe sees a transition at an earlier temperature. However, in the case of bulk measurements, long range order comes into play. The consideration of long range order results in an averaged out value of magnetization so that as compared to the zero velocity scan method, the transition occurs at a higher temperature.

It was observed that the addition of Mn had increased the overall hyperfine field with this value showing a maximum at $x = 0.02$. Correspondingly from Figure 1(b) it can be seen that both the field value as well as the probability of the single peak in the hyperfine field distribution show a maximum at $x = 0.02$.

The energy of magnetization depends on the value of R/r, where R is the radius of the atom or half the inter-nuclear distance and r the radius of the 3-d shell responsible for the magnetization [12]. Therefore an increase in the lattice constant would lead to an increase in the interaction energy followed by an increase in the Curie temperature [13]. In the system studied, we can see that replacement with Mn has increased the value of the lattice parameter till $x = 0.04$. At $x = 0.05$, as remarked before, a structural change has occurred which accounts for the decrease in the lattice parameter. But X-ray measurements are not sensitive enough to show the changes in a_0 within the range for $x = 0.01$ to $x = 0.04$ whereas the Mössbauer measurements are far more sensitive. The trend in the magnetic properties, viz., values of hyperfine fields, saturation magnetization as well as Curie temperatures showed a steady increase with increase in Mn concentration with a maximum at $x = 0.02$, following the trend in the variation of a_0 for these alloys.

Thus it was seen that the substitution of Mn into Fe_2CrAl gave rise to a series of ferromagnetic alloys. This substitution is only possible upto about $x = 0.04$ while retaining the initial structure. Introduction of Mn has resulted in an increase in the volume of the unit cell. This increase is reflected in an overall increase of magnetic hyperfine field and Curie temperatures of these alloys.

Acknowledgement

This work has been supported by the UGC-DRS and COSIST schemes of the Department of Physics, M.L. Sukhadia University, Udaipur, India.

References

1. Webster, P. J., *J. Phys. Chem. Solids* **32** (1971), 1221.
2. Le Dang Khoi, Veillet, P. and Campbell, I. A., *Hyp. Interact.* **4** (1978), 379.
3. Dunlap, R. A., March, R. H. and Stroink, G., *Canad. J. Phys.* **50** (1981), 1577.
4. Suits, J. C., *Phys. Rev. B* **14** (1976), 4131.
5. Cadeville, M. C. and Moran-Lopez, J. L., *Phys. Reports* **153**(6) (1987), 1153.
6. Lakshmi, N., Venugopalan, K. and Varma, J., *Phys. Rev. B* **47** (1993), 14054.
7. Waliszewski, J., Dobrzynski, L., Malinowski, A., Satula, D., Szymanski, K., Prandl, W. Brückel, Th. and Shärpf, O., *J. Magn. Magn. Mater.* **132** (1994), 349.
8. Tuszynski, M., Zarek, W. and Popiel, E. S., *Hyp. Interact.* **59** (1990), 369.
9. Szymanski, K., Biernacka, M., Dobrzynski, L., Perzynska, K., Recko, K., Satura, D., Waliszewski, J. and Zaleski, P., *J. Magn. Magn. Mater.* **210** (2000), 150.
10. Lakshmi, N., Venugopalan, K. and Varma, J., *Pramana J. Phys.* **59** (2002), 531.
11. Window, B., *J. Phys. E* **4** (1984), 401.
12. Bethe, H., *Handbuch der Physik*, Vol. 24 pt 2, (1933), p. 595.
13. Bozoroth, R. M., *Ferromagnetism*, D. Von Nostrand Co. Inc., Princeton, New Jersey, 1961.

Hyperfine Interactions **156/157**: 569–574, 2004.
© 2004 *Kluwer Academic Publishers. Printed in the Netherlands.*

Mössbauer and X-ray Study of $Fe_{1-x}Al_x$, $0.2 \leqslant x \leqslant 0.5$, Samples Produced by Mechanical Alloying

D. OYOLA LOZANO[1,*], Y. ROJAS MARTÍNEZ[1], H. BUSTOS[1] and
G. A. PÉREZ ALCÁZAR[2]
[1]*Departamento de Física, Universidad del Tolima, A. A. 546, Ibagué, Colombia;*
e-mail: doyola@ut.edu.co
[2]*Departamento de Física, Universidad del Valle, A. A. 25360, Cali, Colombia*

Abstract. In this work we report the magnetic and structural properties obtained by Mössbauer spectroscopy and X-ray diffraction, of the $Fe_{1-x}Al_x$, $0.2 \leqslant x \leqslant 0.5$, alloys produced by mechanical alloying. Alloys with $x = 0.2, 0.3, 0.4$ and 0.5, were for milled 12, 24, 36, and 48 hours. All the obtained alloys are in the bcc phase. The obtained Mössbauer spectra are characteristic of disordered ferromagnetic system. The lattice parameter remains nearly constant ($\sim$2.91 Å) for all the milling times and compositions. The mean grain sizes in the (110) and (211) direction are nearly constants with the milling time but vary from 15.5 to 11 nm and from 10.5 to 8.5 nm when Al content grow between $x = 0.2$ to $x = 0.4$, respectively. The difference between the mean grain sizes in these two directions shows that the grains are of prolate spheroid form.

Key words: mechanical alloying, X-ray diffraction, FeAl, Mössbauer spectroscopy.

1. Introduction

Mechanical alloying (MA) is one of the most efficient methods to produce materials with nanocrystalline size. An advantage of mechanical alloying over many other techniques is that it is a solid state technique and consequently problems associated with melting and solidification are bypassed. The Fe–Al system is of interest due to the potential commercial application of these alloys as structural or magnetic materials. In the recent years a number of studies were carried on mechanically alloyed FeAl alloys [1–3] milling during 36 hours or more, the majority devoted to the magnetic behaviour around $x = 0.5$, composition in which the system changes from ferro- to paramagnetic. In this work we study samples of the $Fe_{1-x}Al_x$ system with $x = 0.2, 0.3, 0.4$ and 0.5. The samples were prepared by using a high energy planetary ball milling. We studied the magnetic and structural properties dependency on the concentration and the milling time.

---

* Author for correspondence.

2. Experimental description

Samples of the $Fe_{1-x}Al_x$ system with $x = 0.2, 0.3, 0.4$ and 0.5 were prepared by mechanical alloying using high purity iron and aluminium powders in a Frish planetary ball mill of high energy at 280 rpm in an argon environment. The milling was made in hardened stainless steel vials provided with four balls of the same material with 2 cm diameter. A total of 8 grams of mixture of powders was prepared. The ball to powder mass ratio used was 16 : 1. The milling times were 12, 24, 36 and 48 hours. The powders of the milled samples were measured in a transmission Mössbauer spectrometer using a radioactive Co-57/Rh source. The spectra were fitted with hyperfine field distribution (HFD) and doublets using the NORMOS program. The X-ray analysis to establish the structure of the lattice were performed at room temperature for all samples using a RINT2000l diffractometer with the Cu Kα radiation. From the fit of pattern lines it were obtained the lattice parameters and with the width of each line and the Scherrer equation, the mean grain sizes were obtained.

3. Results and discussion

Room temperature (RT) Mössbauer spectra for Fe–Al samples with $x = 0.2, 0.3, 0.4$ and 0.5, respectively, and milling times of 12, 24, 36 and 48 hours, are shown in Figure 1. It can be observed that the samples with $x = 0.2$ and 0.3, for all the milling times, are disordered and ferromagnetic with high hyperfine fields and that the samples with $x = 0.4$ and 0.5 additionally present a wide paramagnetic site for all the milling times, as a result of the greater Al content, which behaves as magnetic hole. The magnetic behaviour obtained here for the $x = 0.5$ samples differ from those reported for ordered [5] and disordered [4] samples produced by melting which are totally paramagnetic.

Table I report the Mössbauer parameters of the HFD obtained from the fits. It can be noted that the mean hyperfine field (MHF), the isomer shift (δ) and the quadrupolar splitting (ΔQ) are nearly constant with the milling time, but the MHF decreases, the δ is nearly constant and the ΔQ decreases with the Al increases. Figure 2 shows the X-rays diffraction pattern of $x = 0.4$ sample milled during 48 hours. All the patterns are similar and their X-rays analysis indicates that in all the samples the phase is BCC. The lattice parameter remains nearly constant (~ 2.91 Å) for all the milling times and compositions.

Figures 3 and 4 show the dependence of the mean grain size with the milling time and Al concentration, respectively, along the (110) direction. The mean grain size along the (110) direction for all the samples are nearly constant with the milling time. However it decreases with the Al concentration from ~ 15.5 nm for $x = 0.2$ to ~ 11.0 nm for $x = 0.4$ and then remains nearly constant with a value ~ 11.0 nm. The decrease of the mean grain size show the fragile character of the alloys and this is induced by the Al atoms. Figures 5 and 6 show the dependence of

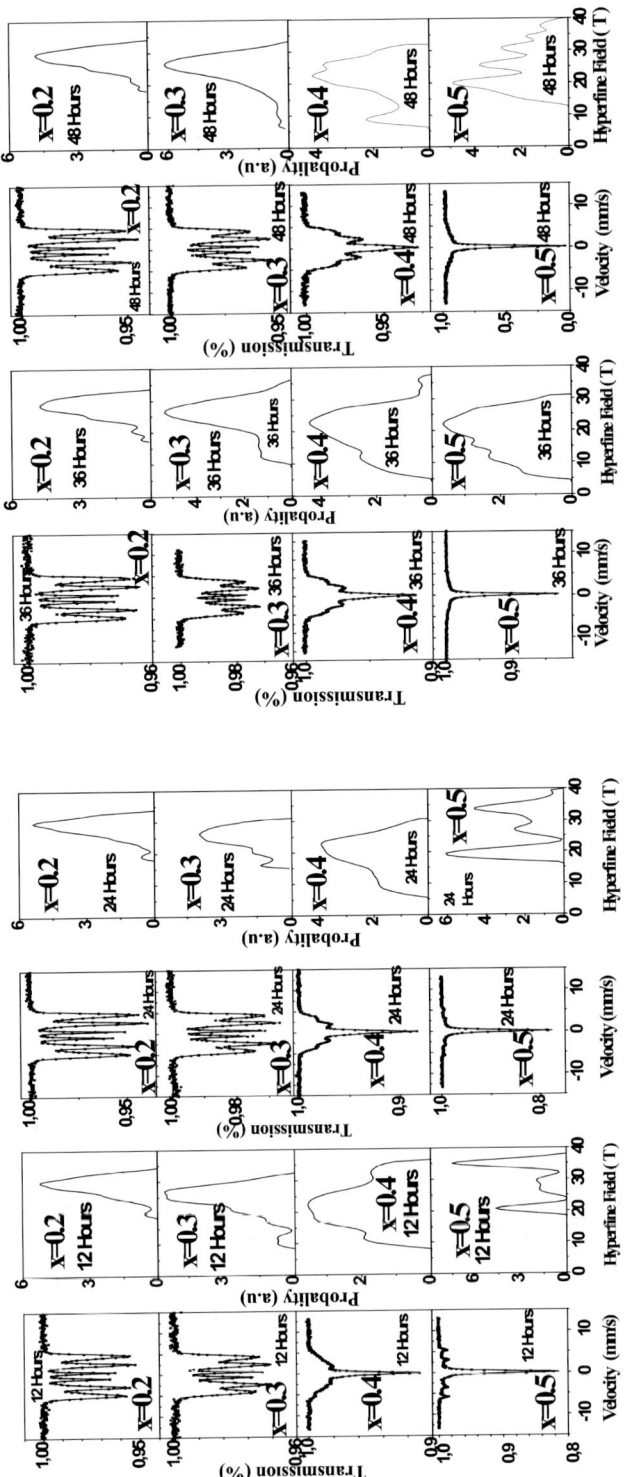

Figure 1. Mössbauer spectra and hyperfine field distributions of $Fe_{1-x} Al_x$ alloys ($0.2 \leqslant x \leqslant 0.5$) for 12, 24, 36 and 48 hours of milling.

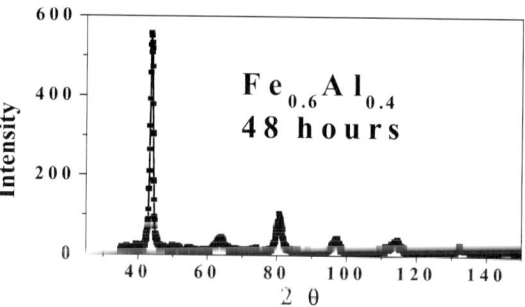

Figure 2. X ray diffraction of $Fe_{1-x}Al_x$ with $x = 0.4$ for 48 hours of milling.

Table 1. Mössbauer parameters of the samples of the $Fe_{1-x}Al_x$ system with $x = 0.2, 0.3, 0.4$ and 0.5. Hyperfine field (HF) values are in Tesla, the isomer shift (δ) and the quadrupolar splitting (ΔQ) are in mm/s

Hours	HF			
	$x = 0.2$	$x = 0.3$	$x = 0.4$	$x = 0.5$
12	28.64 ± 0.04	23.97 ± 0.04	21.32 ± 0.03	29.23 ± 0.08
24	28.49 ± 0.02	24.45 ± 0.10	18.23 ± 0.01	26.23 ± 0.01
36	27.76 ± 0.04	24.32 ± 0.10	18.71 ± 0.08	17.93 ± 0.02
48	28.06 ± 0.09	24.05 ± 0.02	20.12 ± 0.08	23.74 ± 0.02

Hours	δ			
	$x = 0.2$	$x = 0.3$	$x = 0.4$	$x = 0.5$
12	0.02 ± 0.01	0.09 ± 0.01	0.06 ± 0.02	-0.06 ± 0.00
24	0.08 ± 0.01	0.15 ± 0.01	0.08 ± 0.01	-0.03 ± 0.02
36	0.03 ± 0.01	-0.03 ± 0.01	0.07 ± 0.02	0.08 ± 0.02
48	0.03 ± 0.01	0.09 ± 0.01	0.09 ± 0.02	0.08 ± 0.02

Hours	ΔQ			
	$x = 0.2$	$x = 0.3$	$x = 0.4$	$x = 0.5$
12	0.01 ± 0.00	0.01 ± 0.00	0.04 ± 0.02	0.22 ± 0.01
24	0.08 ± 0.00	0.03 ± 0.01	0.02 ± 0.00	0.25 ± 0.01
36	0.02 ± 0.00	0.01 ± 0.01	0.01 ± 0.01	0.01 ± 0.01
48	0.01 ± 0.00	0.00 ± 0.00	0.00 ± 0.01	0.24 ± 0.01

the mean grain size with the milling time and Al concentration, respectively, along the (211) direction.

The mean grain size in the (211) direction for the sample with $x = 0.2$ decreases from 10.5 up to 8.5 nm when the milling time changes from 12 up to 24 hours, then it remains nearly constant. For other Al contents it remains nearly constant. This

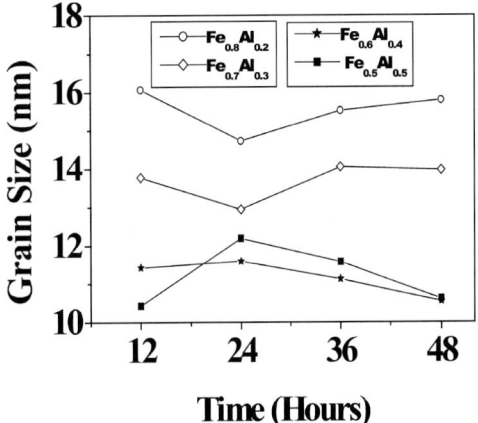

Figure 3. Dependence of the mean grain size with the milling time along the (110) direction.

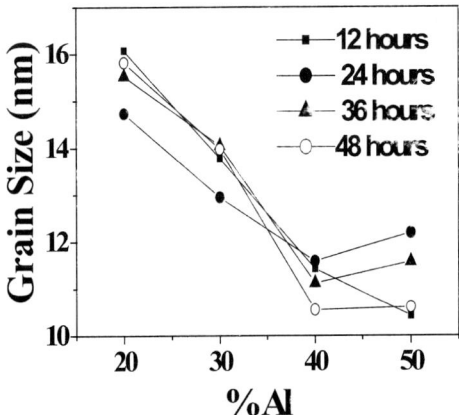

Figure 4. Dependence mean grain size with the Al concentration along the (110) direction.

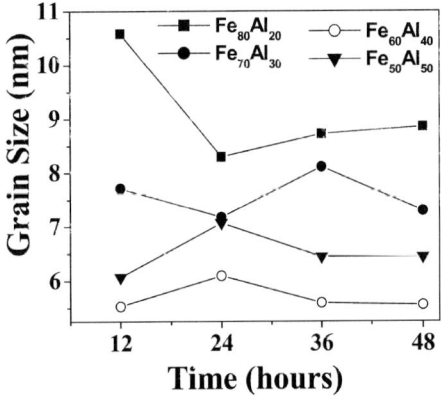

Figure 5. Dependence of the mean grain size with the milling time along the (211) direction.

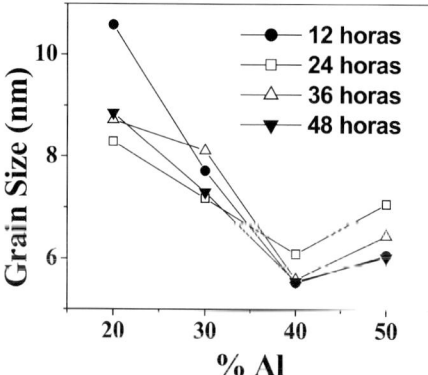

Figure 6. Dependence of the mean grain size with the Al concentration along the (211) direction.

mean grain size decreases from 10.5 nm at 12 hours, for the sample with $x = 0.2$, and from 8.5 nm, for the other milling times, up to ~6 nm for the sample with $x = 0.4$, and then remains nearly constant. The difference between the mean grain sizes in the two reported directions show that the grains are of prolate spheroid form.

4. Conclusion

Mechanical alloying has been used to produce $Fe_{1-x}Al_x$, with $x = 0.2, 0.3, 0.4$ and 0.5, alloys. All the obtained alloys are in the bcc phase and behave as a ferromagnetic disordered system with a tendency to a paramagnetic behaviour for $x \geqslant 0.3$. This is a consequence of the magnetic diluting character of the Al atoms. The difference between the mean grain sizes in the (110) and (211) directions shows that the grains are of prolate spheroid form.

Acknowledgements

We are grateful to the Central Committee of Investigations of the University of Tolima for its financial support and to the Laboratory Mössbauer of the University of Valle for the achievement of this work.

References

1. Pekala, M. and Oleszak., D., *J. Magn. Magn. Mater* **157** (1996), 231.
2. Eelman, D. A., Dahn, J. R., Mackay, G. R. and Dunlap, R. A., *J. Alloys and Compounds* **266** (1998), 234.
3. Pochet, P., Tominez, E., Chaffron, L. and Martin, G., *Phys. Rev. B* **52** (1995), 4006.
4. Perez Alcazar, G. A. and Galvão da Silva, E., *J. Phys. F.* **17** (1987), 2323.
5. Vincze, J., *Phys. Status Solidi A* **7** (1971), K43.

Hyperfine Interactions **156/157**: 575–579, 2004.
© 2004 *Kluwer Academic Publishers. Printed in the Netherlands.*

Investigation of Steel Surfaces Treated by a Hybrid Ion Implantation Technique

H. REUTHER[1], E. RICHTER[1], F. PROKERT[1], M. UEDA[2], A. F. BELOTO[2] and
G. F. GOMES[2]

[1] *Forschungszentrum Rossendorf e.V., Institut für Ionenstrahlphysik und Materialforschung,
Postfach 510119, D-01314 Dresden, Germany*
[2] *Instituto Nacional de Pesquisas Espaciais, Av. dos Astronautas, 1758 Caixa Postal 515,
Sao Jose dos Campos, Brasil*

Abstract. Implantation of nitrogen ions into stainless steel in combination with oxidation often results in a decrease or even complete removal of the chromium in the nitrogen containing outermost surface layer. While iron nitrides can be formed easily by this method, due to the absence of chromium, the formation of chromium nitrides is impossible and the beneficial influence of chromium in the steel for corrosion resistance cannot be used. To overcome this problem we use the following hybrid technique. A thin chromium layer is deposited on steel and subsequently implanted with nitrogen ions. Chromium can be implanted by recoil into the steel surface and thus the formation of iron/chromium nitrides should be possible. Both beam line ion implantation and plasma immersion ion implantation are used. Due to the variation of the process parameters, different implantation profiles and different compounds are produced. The produced layers are characterized by Auger electron spectroscopy, conversion electron Mössbauer spectroscopy and X-ray diffraction. The obtained results show that due to the variation of the implantation parameters, the formation of iron/chromium nitrides can be achieved and that plasma immersion ion implantation is the most suitable technique for the enrichment of chromium in the outermost surface layer of the steel when compared to the beam line implantation.

Key words: steel, ion implantation, nitrides, Mössbauer spectroscopy, Auger electron spectroscopy.

1. Introduction

Ion treatment of surfaces has been established as very suitable method to produce thin alloyed layers which can modify the wear, friction or corrosion behaviour of the whole workpiece in a very beneficial manner [1–3]. For instance, this technique is used for the improvement of tools for cutting, drilling, turning or punching of metals and plastics. Especially, implantation with nitrogen ions can improve the properties of steel tools [4]. However, when nitrogen is implanted into nickel or chromium containing steels and if oxygen is present, nickel or chromium will go aside into the bulk and the formation of nickel/chromium nitride is impossible [5]. To overcome this problem we use a hybrid technique. A thin chromium layer is deposited on mild steel and subsequently implanted with nitrogen ions. Chromium can be implanted by recoil into the steel surface. Thus the formation of iron/chromium nitrides should be possible. Both beam line ion implantation and plasma immersion ion implantation are used as ion treatment techniques and compared with each other.

2. Experimental

SAE 1020, a mild construction steel, is frequently used as mortar reinforcement in buildings and small machine parts, as bolts, screws, gears and so on. But aside from good mechanical and tribological properties as ductility, hardness and wear resistance, its surface is prone to severe corrosion. As is known, the presence of chromium in excess of 12% in the Fe alloys increases the resistance to several corrosive attacks. So we proposed to introduce a feasible amount of Cr into the surface of AISI 1020 steel.

Before the ion treatment, samples of this steel were covered by electron beam deposition with 50 and 100 nm chromium films, respectively. In the case of the beam line implantation, samples were implanted with 2×10^{17} or 5×10^{17} N^+/cm^2. Energies of 50 and 100 keV were chosen such that the maximum nitrogen concentration was expected either near the Cr/steel boundary or just below the steel surface. In the case of the plasma immersion ion implantation, a frequency of 400 Hz, pulse lengths of 5 μs, a power of 30 W and a nitrogen atmosphere were used. Durations between 30 to 120 min lead to ion doses of more than 10^{17} cm^{-2}. Ions accelerated by the maximum voltage of 40 kV should come to rest either in the chromium layer (100 nm thickness) or reach just the Cr/steel boundary (50 nm thickness).

The produced layers were characterized by different methods: depth profiles were determined by Auger electron spectroscopy (AES) with a scanning Auger electron spectrometer Microlab 310F (Fisons Instruments). Phase analysis was performed both by conversion electron Mössbauer spectroscopy (CEMS) with a ^{57}Co(Rh) source and a conventional constant acceleration spectrometer with a proportional gas flow counter (He/CH$_4$ counting gas) and by X-ray diffraction (XRD) with a D5000 diffractometer (Siemens).

3. Results

The results of AES are shown in Figure 1. The depth profiles of the beam line implanted samples show that no mixing or migration of the chromium occurred due to the nitrogen implantation. The Cr/steel layer structure has been kept and there is a normal nitrogen depth profile as predicted by the TRIM code for the calculation of ion implantation profiles [6]. The peaks of the nitrogen profiles were expected at about 60 nm and 110 nm for 50 keV and 100 keV ions, respectively, which is roughly the case. For the plasma immersion ion implantation, the depth profiles are quite different. A nitrogen containing chromium/iron alloy surface layer is produced. It is obvious that mixing and migration processes took place during the ion treatment.

Beam line implanted   Plasma immersion implanted

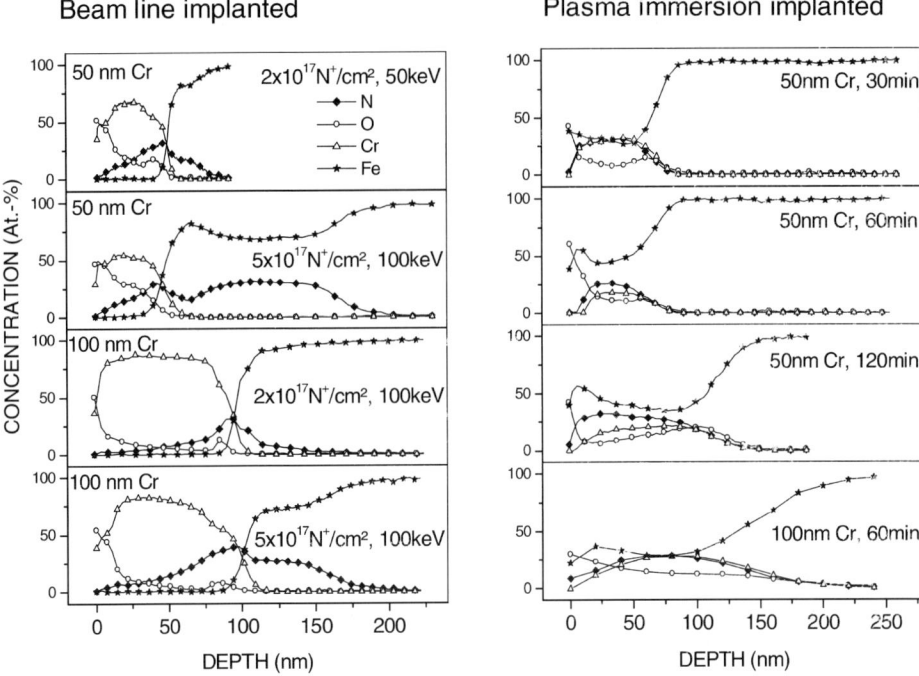

Figure 1. AES depth profiles of the chromium covered and nitrogen beam line implanted (*left*) and nitrogen plasma immersion implanted (*right*) steel samples.

The obtained CEM spectra are given in Figure 2. They are magnetically split six-line pattern. All but that of the sample covered with 100 nm Cr and 60 min plasma treatment could be evaluated best by assuming a crystal subspectrum of α-Fe, a magnetic hyperfine field distribution and in some cases a quadrupole doublet. Phase analysis was performed by the aid of the hyperfine parameters of the pure iron nitrides known from literature [7, 8]. While the main components of the distributions could be assigned to γ-Fe_4N the doublet could be assigned to ε-Fe_2N. The spectrum of the sample covered with 100 nm Cr and 60 min plasma treatment consisted of four sextets, the strongest from α-Fe (70%), the other from γ-Fe_4N (30%).

CEMS is not sufficient for a complete phase analysis because of the chromium content of the sample. The additionally performed XRD analysis confirmed the CEMS results and proved the formation of chromium nitride in all ion treated samples. But in the beam line implanted samples considerable amounts of chromium oxide were detected. Iron oxide was not found.

4. Discussion and conclusions

Comparing the two applied ion beam treatments, only plasma immersion ion implantation enables the formation of iron/chromium nitrides in the same depth of the steel sample. Mixing and migration processes seem to take place. Beam line

Beam line implanted   Plasma immersion implanted

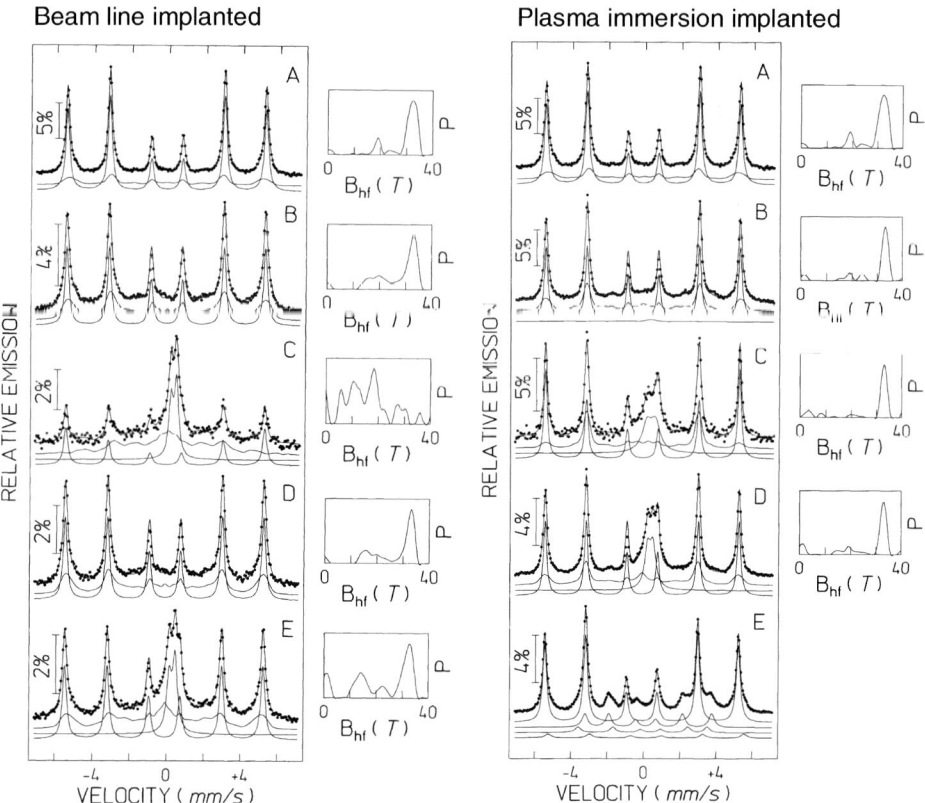

Figure 2. CEM spectra and calculated hyperfine field distributions B_{hf} (T) of the chromium covered and nitrogen beam line implanted samples (*left*, spectrum A: uncovered and untreated reference, B: 50 nm Cr, 2×10^{17} N$^+$/cm^2, 50 keV, C: 50 nm Cr, 5×10^{17} N$^+$/cm^2, 100 keV, D: 100 nm Cr, 2×10^{17} N$^+$/cm^2, 100 keV, E: 100 nm Cr, 5×10^{17} N$^+$/cm^2, 100 keV) and nitrogen plasma immersion ion implanted steel samples (*right*, spectrum A: uncovered and untreated reference, B: 50 nm Cr, 30 min plasma treatment, C: 50 nm Cr, 60 min, D: 50 nm Cr, 120 min, E: 100 nm Cr, 60 min). Both uncovered and chromium covered steel samples were used as reference however there was no difference in the CEM spectra but the counting rate.

implantation with the used parameters is suitable for doping without destroying the layered sample structure.

For sample characterization the utilization of the different analysis methods was necessary and sufficient. CEMS and XRD could detect the formed phases but AES delivered the information about the depth distribution of the elements and thus the depth where the phases are formed. In the case of the beam line implantation, chromium and iron nitrides were formed in single layers one upon the other but with a pronounced chromium oxide at the top. In the case of the plasma immersion implantation, all phases were formed at the same depth and the formation of ternary phases of chromium/iron nitride can be assumed. It is obvious that due to the variation of the process parameters different nitrides, especially the ε- and the γ-phases can be formed which may influence the mechanical or chemical properties in different manner. It is possible to produce both pure phases and phase mixtures.

For practical applications, the mechanical or corrosion properties are of interest. Micro hardness measurements showed that only for the beam line implanted samples an improvement could be reached. Such results are already known from normal steel samples without special chromium deposit on the surface. For the plasma treated samples no changes or even a change for the worse were obtained. On the other hand, the corrosion tests (potentiometric measurements) showed strong improvements for the latter technique while the beam line implanted samples did not show any difference when compared with the untreated ones.

In conclusion, this new hybrid ion implantation technique brings advantages for special applications and expands the possibilities of materials treatment with ion beams.

References

1. Smidt, F. A. and Hübler, G. H., *Nucl. Instrum. Methods Phys. Res. B* **80/81** (1993), 207.
2. Richter, E., Günzel, R., Parascandola, S., Telbizova, T., Kruse, O. and Möller, W., *Surf. Coat. Techn.* **128/129** (2000), 21.
3. Ueda, M., Berni, L. A., Castro, R. M., Beloto, A. F., Abramof, E., Rossi, J. O., Barroso, J. J. and Lepienski, C. M., *Surf. Coat. Technol.* **156** (2002), 71.
4. Ueda, M., Berni, L. A., Gomes, G. F., Beloto, A. F., Abramof, E. and Reuther, H., *J. Appl. Phys.* **86** (1999), 4821.
5. Ueda, M., Gomes, G. F., Abramof, E. and Reuther, H., *Nucl. Instrum. Methods Phys. Res. B* **206** (2003), 749.
6. Ziegler, J. F., *The Stopping and Range of Ions in Matter*, Vols. 2–6, Pergamon Press, 1977–1985.
7. Longworth, G. and Hartley, N. E. W., *Thin Solid Films* **48** (1978), 93.
8. Schaaf, P., *Hyp. Interact.* **111** (1998), 113.

Hyperfine Interactions **156/157**: 581–593, 2004.
© 2004 *Kluwer Academic Publishers. Printed in the Netherlands.*

Ferromagnetic Planar Nanocomposites

M. CARBUCICCHIO and M. RATEO
Department of Physics, University of Parma, and INFM, Parma, Italy

Abstract. Modern permanent magnets require a high coercive field on account of a strong magnetocrystalline anisotropy, as well as a high saturation magnetization and high Curie temperature. The achievement of so different characteristics in a unique phase is the present main difficulty. In principle, this problem can be solved combining the high saturation magnetization of a soft phase with the high magnetic anisotropy of a hard phase, via the exchange coupling on a nanometric scale. The first attempts showed the feasibility of planar magnetic nanocomposites, where soft and hard magnetic layers are intercalated, but on the other hand they also stressed the difficulties still existing. The present paper reviews some theoretical aspects and experimental results, pointing out the potentiality of Mössbauer spectroscopy in determining the spin configuration, as well as the nature and thickness of interfaces, which strongly influence the exchange interaction in these systems.

Key words: magnetic multilayers, hard/soft nanocomposites, Mössbauer spectroscopy.

1. Introduction

The reasons of the increasing scientific interest in nanostructured systems, surfaces, thin films and multilayers are certainly related to the discovery of different magnetic properties with respect to the bulk materials which open in principle new possibilities for technological applications. In particular, the nanocomposites represent completely new physical systems where basic phenomena like the domain wall structure and nucleation, the coercivity mechanisms and the magnetic viscosity have to be reconsidered.

Nowadays, the permanent magnets based on magnetic oxides and rare-earth intermetallic compounds, are characterized by high coercive field on account of the strong magneto-crystalline anisotropy. Besides this, permanent magnetic materials require further intrinsic characteristics, namely high saturation magnetization and high Curie temperature. The achievement of so different characteristics in a unique phase is the present principal difficulty and the existing permanent magnets are indeed far below the physical limit of energy density estimated to be about 1 MJ/m^3 [1].

Recently, after the successes obtained in the field of nanocrystalline magnetic materials both soft (*very low effective anisotropy*) and hard (*enhanced remanence*), a new concept came out [2], which puts together the advantages of both nanostructures and composites. The basic idea is to combine soft and hard ferromagnetic phases via the reduction of the dimensions of the grains. In principle, a high exchange coupling energy between the different phases should be established and the whole material should display a single-phase magnetic behaviour.

This phenomenon can be considered an extension to a mesoscopic scale of what happens on an atomic scale in rare-earth intermetallic compounds. The exchange coupling between the two magnetic sublattices of the transition metal (TM) and of the rare-earth (RE) gives rise to their peculiar characteristics of combined high magnetocrystalline anisotropy and high saturation magnetization. In magnetic nanocomposites the interaction is between neighboring mono-phases instead of between single atoms. The dimensional scale parameter is the ferromagnetic exchange length roughly proportional to the average thickness of the domain wall [3–5], which is for example of a few nanometers for hard RE-TM intermetallic compounds, and of some tens of nanometers for soft magnetic materials such as α-Fe.

Magnetic nanocomposites in the form of multilayer, where soft and hard magnetic layers are intercalated, can represent a good choice to realize high performance permanent magnets [1, 6]. A multilayer is indeed an intrinsically simple system whose structure can be controllable and reproducible. Moreover, it exhibits interesting aspects concerning basic physics related to interfacial phenomena, which can give rise to peculiar properties not present in massive systems.

Possible applications of magnetic nanocomposites may regard high density magnetic recording, as well as high quality and high performance devices in the field of magneto-electronics and Micro-Electro-Mechanical Systems (MEMS) technology [7–10]. In particular, milli-size engines and micro-actuators [11] have great potentialities in many fields such as bioengineering medical applications.

The present paper briefly reviews the theoretical basis of exchange coupled hard/soft systems, and reports experimental results concerning some planar ferromagnetic nanocomposites. The role of Conversion Electron Mössbauer Spectroscopy (CEMS) in providing microstructural and magnetic information on this kind of systems is emphasized.

2. Theoretical aspects

The initial studies of the spin reversal processes in layered systems where soft material film is coupled to a hard material layer, were carried out assuming a completely rigid behaviour for the hard phase and a total magnetic isotropy for the soft one [12]. Further theoretical approaches based on micromagnetic and first-principles calculations allowed to explain the magnetization reversal in real systems [13, 14].

Assuming that (i) an exchange coupling between the soft and hard layers is established, and (ii) the thickness of hard and soft phases are lower than their relative domain wall widths, w_h and w_s, the following cases can be considered.

For a thickness of the soft layer, t_s, smaller than w_h, the magnetic moments in the soft layer remain hooked to the spins of the hard material, thanks to the

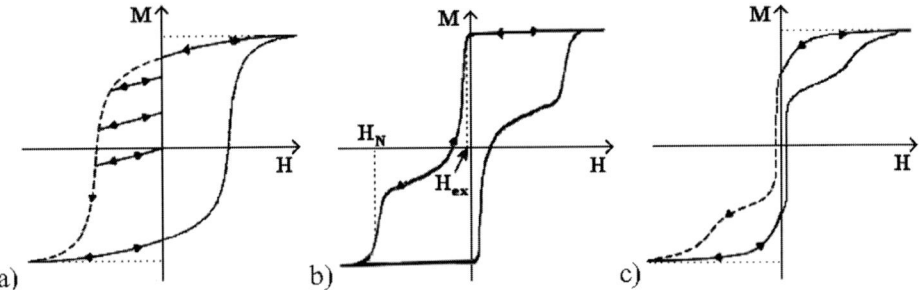

Figure 1. Typical hysteresis loops for the three types of magnets: (a) rigid nanocomposite, (b) exchange-spring nanocomposite, (c) two uncoupled phases.

exchange field, H_{ex}, which depends on the exchange interaction, the thickness and saturation magnetization, M_s, of the soft layer,

$$H_{ex} = \frac{\pi^2 A_s}{2 M_s t_s^2},$$

where A_s is the soft exchange constant. The reversal of the whole system occurs applying a magnetic field, H_{ap}, greater than the nucleation field, H_N, which is depending on geometrical and magnetic properties of both phases

$$H_N = \frac{2(t_h K_h + t_s K_s)}{t_h M_h + t_s M_s},$$

where t, K and M are respectively the thickness, the anisotropy constant and the magnetization of the hard (h) and soft (s) layers. The magnetic behaviour is similar to that of classical permanent magnets, and the systems are called "*rigid magnetic nanocomposites*" (Figure 1(a)).

For $t_s > w_h$, three cases can be considered.

(1) $H_{ap} < H_{ex}$. The spins in the soft layer remain aligned with those of the hard layer.

(2) $H_{ex} < H_{ap} < H_N$. A twisting of the spins occurs in the soft layer, because they are strongly pinned only at the interfaces, and the centre of the layer is free to follow the applied field (Figure 2). In this range of applied field the demagnetization curve displays a reversible behaviour, and the systems are called "*exchange-spring nanocomposites*" [15] (Figure 1(b)).

This picture was experimentally verified by nuclear resonant scattering of synchrotron radiation on a Fe/FePt exchange-spring magnet bilayer, where a tilted ultra-thin film of ^{57}Fe has been deposited within the Fe soft layer [16] (Figure 3).

(3) $H_{ap} > H_N$. The reversal of the spins of hard layer also occurs in an irreversible way.

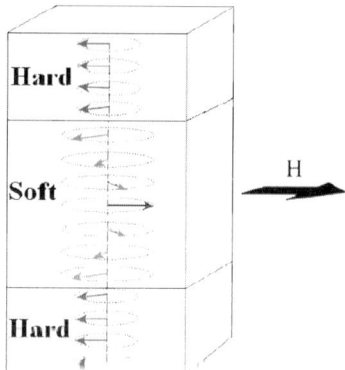

Figure 2. Sketch of the exchange-spring behavior in a hard/soft trilayer.

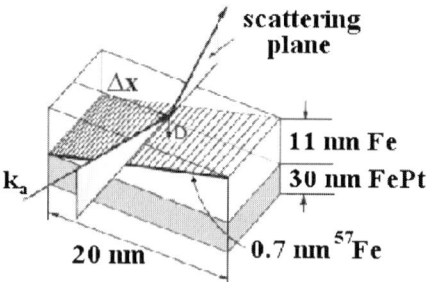

Figure 3. Fe/FePt bilayer for nuclear resonant scattering experiments [16].

3. Experimental issues

In the last few year a considerable effort has been devoted to realize planar magnetic nanocomposites where soft and hard magnetic materials are exchange coupled. The first attempts regarded the realization of SmCo/FeCo bilayers [17] and NdFeB/Fe/NdFeB trilayers [18], both obtained by sputtering. In the case of bilayers, the authors report the dependence of coercivity and saturation magnetization on the thickness of the soft or the hard layer. For the trilayers the authors obtain an enhanced remanence and a magnetic single-phase behavior for a thickness soft/hard ratio around unity (Figure 4).

One of the first attempts to obtain a nanocomposite in the form of multilayer was performed intercalating α-Fe and NdFeB [19]. However, the demagnetization curves, particularly for a high soft/hard ratio, showed an incomplete reversible behaviour, thus indicating a weak exchange coupling between the magnetic phases.

In order to obtain single-phase magnetic behaviour, as well as to stabilize the RE-containing hard phase, some authors subjected their multilayers to thermal treatments. This gave rise to the destruction of the system periodicity with a mixing of the components, the random structural orientation of the hard phase, and the precipitation of spurious phases (Figures 5(a),(b)) [20, 21]. Therefore, the repro-

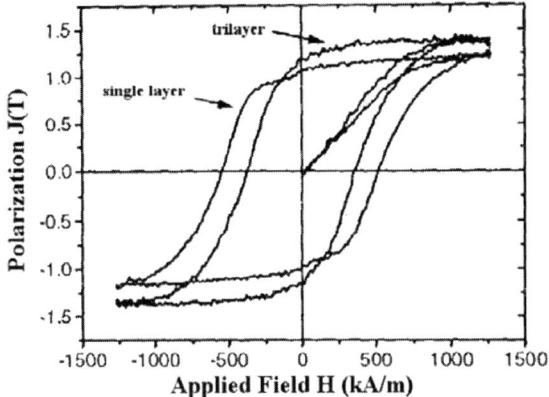

Figure 4. Hysteresis loop of a $NdFeB_{27\,nm}/Fe_{24\,nm}/NdFeB_{27\,nm}$ trilayer compared to a $NdFeB_{80\,nm}$ single layer [18].

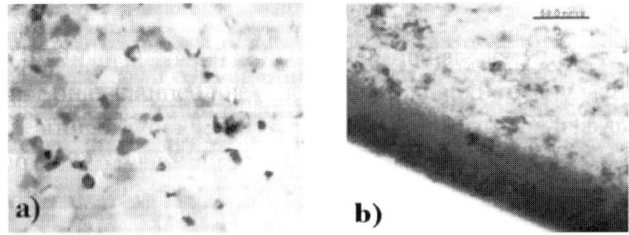

Figure 5. (a) TEM bright-field image of the $Si/Ti_{20\,nm}/NdDyFeCoNbB_{15\,nm}FeCo_{6\,nm}/Ti_{20\,nm}$ multilayer thin film annealed at 625°C for 1 min [20]. (b) Cross-sectional image of TEM for the film $SmCo_{21\,nm}/FeCo_{5\,nm}$ annealed at 630°C [21].

ducibility as well as the correlation between structural and magnetic properties result to be in this case very difficult.

A different approach to obtain planar magnetic nanocomposites is represented by epitaxial growth techniques. It allows the composition of the phases to be well controlled and, moreover, the easy axis alignment of the hard component to be obtained. This way, a largely reversible demagnetization curve was obtained for SmCo/Co and SmCo/Fe bilayers [22–24]. However when the number of repetition was increased, the demagnetization curves became more and more irreversible. This suggests that the structural coherence of the layers (i) is loosing on getting out of the substrate, and (ii) is strictly connected with the exchange coupling strength.

It is to be noted that in the above reported cases, the soft/hard ratio is generally well below (or at most equal to) the unity, meaning that the soft phase is diluted into the hard one. However, in order to realize a multilayer with the magnetic energy density superior to that of the hard component, it is convenient that the soft phase is the dominant one [25].

In order to clarify the mechanisms at the basis of the magnetic nanocomposites behaviour and to gain insight into the involved phenomena, a model system has been realized by alternating layers of simple materials, namely Fe as the soft com-

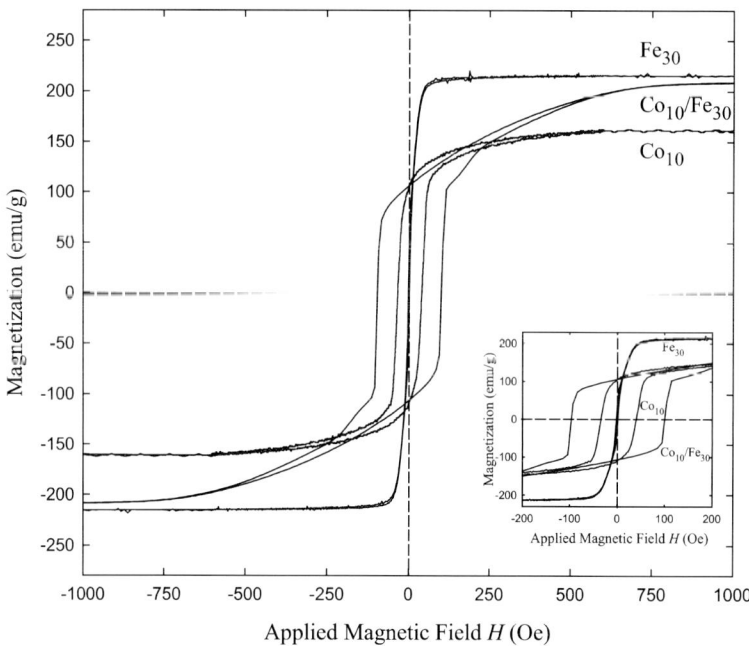

Figure 6. Alternating gradient force magnetometry hysteresis loops for $Co_{10\,nm}$ and $Fe_{30\,nm}$ films, and $Co_{10\,nm}/Fe_{30\,nm}$ multilayer. The inset shows a particular of the low magnetic field region [26].

ponent and Co as the intrinsically hard one [15]. The multilayers were constituted by a constant number of Co/Fe bilayers, covered by a Co capping layer. High purity materials were electron beam evaporated in ultra-high vacuum at very low deposition rates, and multilayers with very low contamination and clean interfaces were obtained. The soft/hard ratio equal to three was chosen taking into account that the ratio of the domain wall thickness for Co and Fe can be considered as the characteristic scale length. The thickness of both layers was well below their relative bulk domain width.

In spite of the high soft/hard ratio, the system shows a single-phase magnetic behavior, due to a strong exchange coupling established between the layers [27, 28]. The hysteresis loops show a high squareness, with a saturation magnetization close to that of the soft phase and a coercive field unexpectedly greater than that of the hard one (Figure 6). The last result was explained taking into account that the interfaces can be considered as large defects reducing the domain wall mobility. This behaviour is similar to that of a classical permanent magnet, and the system can be considered a "*rigid magnetic nanocomposite*" [27, 29].

The strength of the exchange coupling between the layers was investigated by means of remanence switching field distributions, SFD, and interaction-based deviation parameters plotted against the applied magnetic field, ΔM-plot (Figure 7) [30].

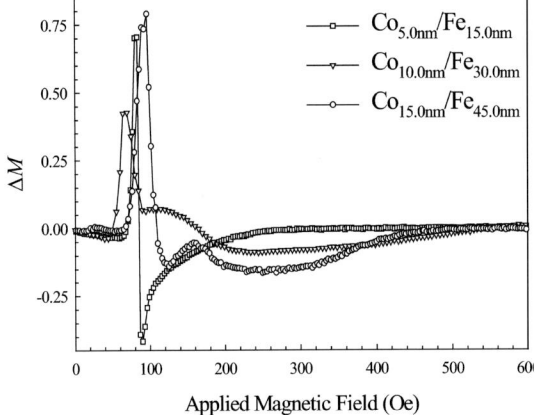

Figure 7. ΔM-plot for Co/Fe multilayers [30].

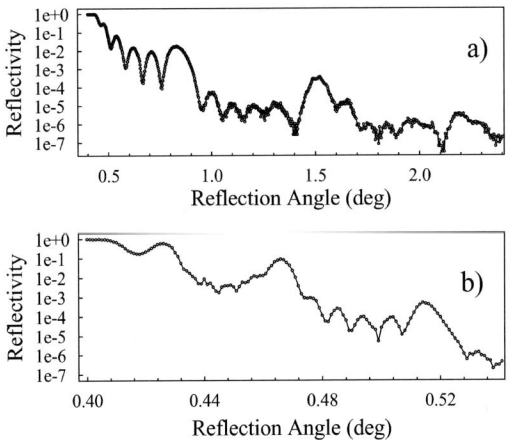

Figure 8. Grazing incidence X-ray reflectivity patterns for (a) $Co_{5\,nm}/Fe_{0.5\,nm}$ and (b) $Co_{10\,nm}/Fe_{16\,nm}$ multilayers [31].

The crystal structure and orientation in respect to the film plane, the dimension and shape of the grains, as well as the nature and morphology of the interfaces are of fundamental importance in determining the magnetic properties of planar nanocomposites. Interfaces, in particular, play an important role in establishing the exchange coupling between the layers.

Grazing incidence X-ray diffraction measurements showed that the multilayers are polycrystalline with hcp Co and bcc Fe, and that the average grain sizes are comparable to those of the respective thicknesses. Grazing incidence X-ray reflectivity patterns are typical of a superlattice structure (Figure 8) [31]. TEM observations (Figure 9) in turn showed that, even for a very low thickness of iron layer, the periodicity is maintained and the interfaces are very flat.

The study of the interfaces was carried out by means of Conversion Electron Mössbauer Spectroscopy, CEMS. The measurements were performed on Co/Fe

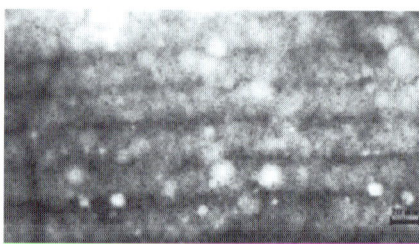

Figure 9. Plane view TEM image for $Co_{5\,nm}/Fe_{0.5\,nm}$ multilayer [26].

multilayers where the cobalt layer was 10 nm thick, and that of iron was very thin and enriched in ^{57}Fe isotope [26].

For ^{57}Fe layer 2 nm thick, the CEMS spectrum (Figure 10(a)) was fitted superimposing (i) a sextet due to pure iron and (ii) a series of sextets due to iron atoms at the interfaces, whose hyperfine magnetic field distribution shows a main contribution peaked at 35 T, and a very small and broadened one at lower fields. Taking into account the narrow line width of the main peak ($\sim$1 T), it was attributed to iron atoms in very similar lattice sites. This suggests that a sharp Fe concentration gradient has formed at the interfaces. In agreement grazing incidence X-ray reflectivity measurements showed that the interface thickness is within the limit of the technique, i.e., $\sim$1 nm. The low field contribution to the CEMS spectrum, in turn, was attributed to a few iron atoms jumping into the Co layer [32].

For a nominal Fe layer thickness of 0.5 nm, the Mössbauer spectrum (Figure 10(b)) becomes very broadened and was interpreted as due only to a series of sextets with a large distribution of hyperfine magnetic fields, peaked at $\sim$36 T. The spectrum was attributed to an iron rich Fe–Co solid solution. Its formation was justified considering that the very thin Fe layer is no longer continuous (Figure 9), highly defective and stressed, and this favoured, during evaporation, the Co atom diffusion.

From the line intensity ratios, 3 : 4 : 1 : 1 : 4 : 3, of both spectra (Figure 10) it follows that the magnetization is lying in the film plane. The corresponding hysteresis loops show a highly squared shape when applying the field parallel to the easy axis. The loops become almost a straight line with little hysteresis applying the field perpendicular to the easy axis [31, 33] (Figure 11). This in-plane uniaxial magnetic anisotropy can be also investigated by CEMS measurements in non-perpendicular geometry, i.e., with an oblique incidence of the γ-ray, and rotating the sample around its perpendicular axis [24, 34]. The causes that determine this in-plane uniaxial anisotropy are not well understood, several hypotheses have been advanced and specific experimental approaches have been performed in order to clarify its origin [31].

By increasing the Fe layer thickness, the magnetization starts to rise up, reaching for a 24 nm thickness an out-of-plane angle of $\sim$10°, as deduced by the line intensity ratios of the Mössbauer spectra (Figure 12). For thicker layers, the magnetization jumps abruptly out of the plane forming an angle of $\sim$40°. This perpendic-

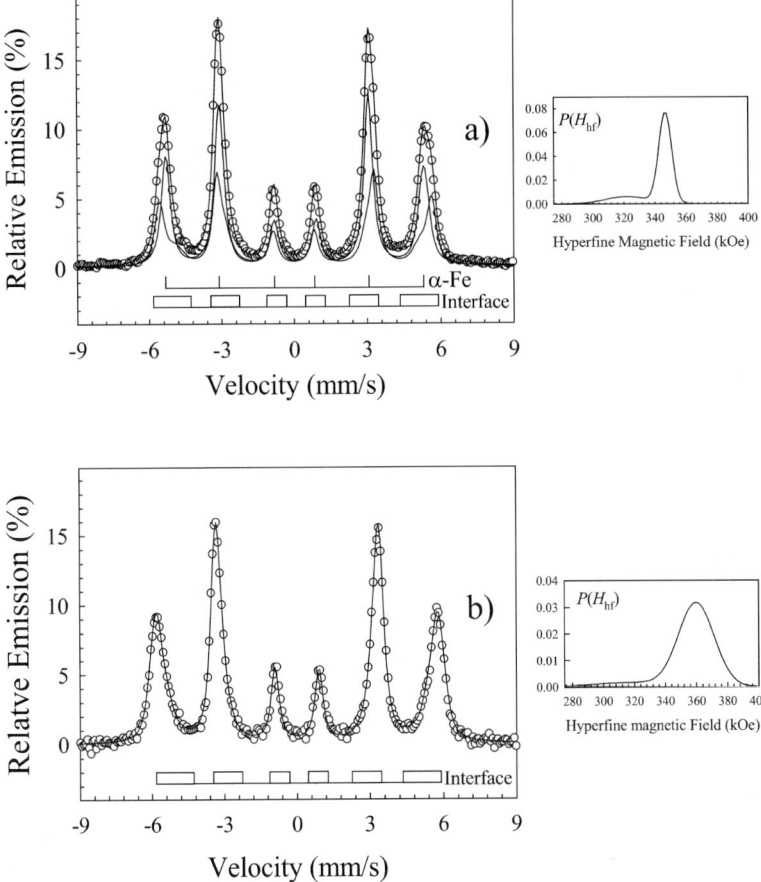

Figure 10. CEMS spectra for (a) $Co_{5\,nm}/Fe_{2\,nm}$ and (b) $Co_{5\,nm}/Fe_{0.5\,nm}$ multilayers. The hyperfine magnetic field distributions relative to the interface are also reported [26].

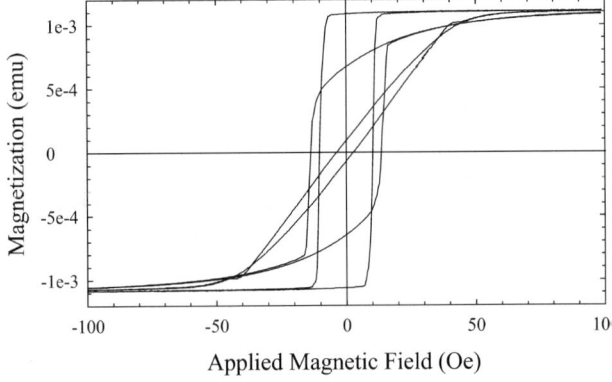

Figure 11. AGFM hysteresis loops at different applied magnetic field angles in the film plane for $Co_{5\,nm}/Fe_{2\,nm}$ multilayer.

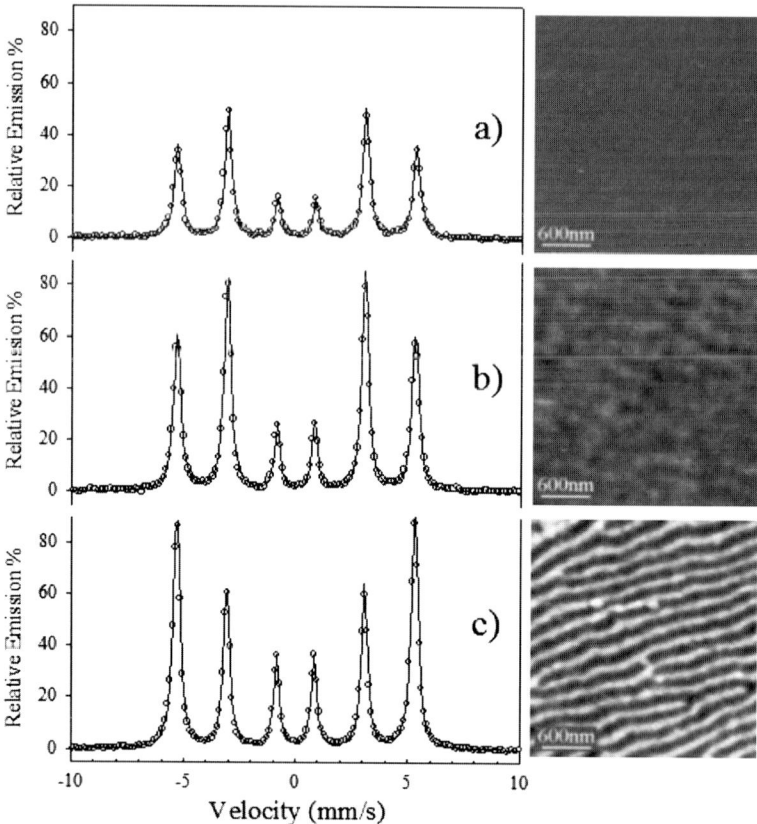

Figure 12. CEMS spectra (*left*) and magnetic force microscopy images (*right*) for (a) $Co_{10\,nm}/Fe_{10\,nm}$, (b) $Co_{10\,nm}/Fe_{24\,nm}$, and (c) $Co_{10\,nm}/Fe_{30\,nm}$ multilayers.

ular magnetic anisotropy results in the formation of magnetic domains detectable by magnetic force microscopy. They are absent for very thin multilayers, and evolve from domains stretched along a preferred direction to very long and parallel magnetic domains, "stripe domains" (Figure 12). In the last case the magnetization is pointing alternately up and down, and the in-plane uniaxial magnetic anisotropy is strongly reduced. Figure 13 shows the remanence ratios *vs.* the in-plane angle of the applied magnetic field for thin and thick Co/Fe multilayers.

This perpendicular anisotropy was explained taking into account the planar stress states at the interfaces between the layers [33, 35–38]. In fact, considering that the Co magnetostrictive constant is negative and the Co–Fe lattice mismatch is also negative on account of the difference between bcc Fe and hcp Co atomic volumes, the resulting negative magnetoelastic energy density leads to a perpendicular magnetic anisotropy. This anisotropy is competitive with the shape anisotropy predominant in very thin films, which constrains the magnetization in the film plane.

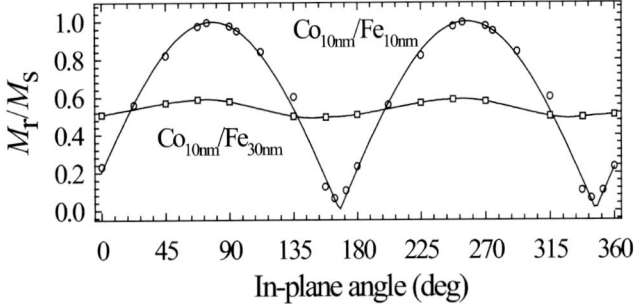

Figure 13. Remanence ratio M_r/M_s vs the in-plane applied magnetic field angle for Co/Fe multi-layers.

These first endeavors have shown the feasibility of a multilayer nanocomposite magnet, but on the other hand they have also stressed the difficulties still existing. First of all, bulk hard materials also with a very high coercive field, can present a soft behavior when deposited in the form of thin film. A possible solution is to use materials with intrinsic anisotropy, i.e., with an anisotropy independent from morphology and/or microstructure. Other problems are concerning (i) the obtainment of a single phase magnetic behavior at the required high soft/hard ratio, (ii) the alignment of the easy axis for the hard component, and (iii) the control of interface morphology and composition, which can influence the exchange coupling between the layers.

4. Conclusions

Planar magnetic nanocomposites, constituted by magnetically hard and soft materials, show a single-phase magnetic behaviour thanks to the exchange coupling established between the components. The soft phase provides the high saturation magnetization, and the hard one provides the high coercive field, this way allowing a high performance permanent magnet to be obtained.

The first attempts reported in literature are concerning the growth by sputtering of bi- and trilayers showing an enhanced remanence and a single-phase magnetic behaviour. In the case of multilayers an incomplete reversibility of the demagnetization curves was measured, thus indicating a weak exchange coupling between the layers. Recently, planar nanocomposites have been obtained by thermal treatments of hard/soft multilayers. The treatments gave rise to the destruction of periodicity, the random orientation of the hard component, and the precipitation of spurious phases.

Layered systems where the hard phase is characterized by a high coercive field and a structural orientation have been obtained by epitaxial growth techniques. However, the exchange coupling strength between components have been found loosing as the number of layers increases.

The feasibility of nanocomposites in the form of well-defined multilayers with the soft component as the predominant one has been demonstrated growing, by electron beam in ultra-high vacuum, a model system where elemental iron and cobalt are intercalated. The system shows a strong exchange coupling between soft and hard layers, and, moreover, it displays a magnetic anisotropy that changes by increasing the soft layer thickness, from an in-plane uniaxial to a perpendicular geometry.

Considering that iron and its alloys are the best soft constituents of planar magnetic nanocomposites, Conversion Electron Mössbauer Spectroscopy represents a powerful tool for analyzing the spin orientation as well as the composition and thickness of the interfaces that strongly influence the exchange coupling between the layers.

References

1. Coey, J. M. D. and Skomski, R., *Physica Scripta* **T49** (1993), 315.
2. Kneller, E. F. and Hawig, R., *IEEE Trans. Magn.* **MAG-27** (1991), 3588.
3. Herzer, G., *IEEE Trans. Magn.* **MAG-26** (1990), 1397.
4. Hernando, A., *J. Magn. Magn. Mater.* **117** (1992), 154.
5. Fullerton, E. E., Jiang, J. S. and Bader, S. D., *J. Magn. Magn. Mater.* **200** (1999), 392.
6. Skomski, R., *J. Appl. Phys.* **76** (1994), 7059.
7. Yamashita, S., Yamasaki, J., Ikeda, M. and Iwabuchi, N., *J. Appl. Phys.* **70** (1991), 6627.
8. Awschalom, D. D. and DiVincenzo, D. P., *Phys. Today* **48** (1995), 43.
9. Le Gall, H., Ben Youssef, J., Vokadinivic, N. and Ostorero, J., *IEEE Trans. Magn.* **28** (2002), 2526.
10. Grünenberg, P., *Phys. Today* **54** (2001), 31.
11. Asti, G., Ghidini, M. and Solzi, M., *J. Magn. Magn. Mater.* **242–245** (2002), 973.
12. Goto, E., Hayashi, N., Miyashita, T. and Nakagawa, K., *J. Appl. Phys.* **36** (1965), 2951.
13. Schrefl, T. and Fidler, J., *J. Magn. Magn. Mater.* **177–181** (1998), 970.
14. Mibu, K., Nagahama, T., Shinjo, T. and Ono, T., *Phys. Rev. B* **58** (1998), 6442.
15. Asti, G., Carbucicchio, M., Rateo, M. and Solzi, M., *J. Magn. Magn. Mater.* **196–197** (1999), 59.
16. Röhlsberger, R., Thomas, H., Schlage, K., Burkel, E., Leupold, O. and Rüffer, R., *Phys. Rev. Lett.* **89** (2002), 237201.
17. Al-Omari, I. A. and Sellmyer, D. J., *Phys. Rev. B* **52** (1995), 3441.
18. Parhofer, S., Gieres, G., Wecker, J. and Schultz, L., *IEEE Trans. Magn.* **32** (1996), 4437.
19. Shindo, M., Ishizone, M., Sakuma, A., Kato, H. and Miyazaki, T., *J. Appl. Phys.* **81** (1997), 4444.
20. Liu, W., Zhang, Z. D., Liu, J. P., Li, X. Z., Sun, X. K. and Sellmyer, D. J., *J. Appl. Phys.* **91** (2002), 7890.
21. You, C., Yang, C. J., Zhang, Z. D., Han, J. S., Liu, W. and Sun, X. K., *J. Phys. D* **36** (2003), 423.
22. Fullerton, E. E., Jiang, J. S., Sowers, C. H., Pearson, J. E. and Bader, S. D., *Appl. Phys. Lett.* **72** (1998), 380.
23. Jiang, J. S. and Bader, S. D., *Scripta Mater.* **47** (2002), 563.
24. Kuncser, V. E., Doi, M., Keune, W., Askin, M. and Spies, H., *Phys. Rev. B* **68** (2003), 064416.
25. Amato, M., Rettori, A. and Pini, M. G., *Physica B* **275** (2000), 120.

26. Carbucicchio, M., Grazzi, C., Lanotte, L., Rateo, M., Ruggiero, G. and Turilli, G., *Hyp. Interact.* **139/140** (2002), 553.
27. Asti, G., Carbucicchio, M., Ghidini, M., Rateo, M., Ruggiero, G., Solzi, M., D'Orazio, F. and Lucari, F., *J. Appl. Phys.* **87** (2000), 6689.
28. Carbucicchio, M., Grazzi, C., Lanotte, L., Rateo, M., Ruggiero, G. and Turilli, G., *Hyp. Interact.* **139/140** (2002), 553.
29. Asti, G., Solzi, M. and Ghidini, M., *J. Magn. Magn. Mater.* **226–230** (2001), 1464.
30. Agazzi, L., Bennett, S., Berry, F. J., Carbucicchio, M., Rateo, M., Ruggiero, G. and Turilli, G., *J. Appl. Phys.* **92** (2002), 3231.
31. Carbucicchio, M., Bennett, S., Berry, F. J., Prezioso, M., Rateo, M. and Turilli, G., *J. Appl. Phys.* **93** (2003), 7631.
32. Häggström, L., Kalska, B., Nordström, E., Blomqvist, P. and Wäppling, R., *J. Alloys Comp.* **347** (2002), 252.
33. Carbucicchio, M., Rateo, M., Ruggiero, G. and Turilli, G., *J. Magn. Magn. Mater.* **242–245** (2002), 601.
34. Keune, W., Kuncser, V. E., Doi, M., Askin, M., Spies, H., Sahoo, B., Duman, E., Acet, M., Jiang, J. S., Inomata, A. and Bader, S. D., *J. Phys. D* **35** (2002), 2352.
35. Ausanio, G., Iannotti, V., Lanotte, L., Carbucicchio, M. and Rateo, M., *J. Magn. Magn. Mater.* **226–230** (2001), 1740.
36. Carbucicchio, M. and Rateo, M., *Hyp. Interact.* **141/142** (2002), 441.
37. D'Orazio, F., Lucari, F., Carlotti, G., Gubbiotti, G., Carbucicchio, M. and Ruggiero, G., *J. Magn. Magn. Mater.* **226–230** (2001), 1767.
38. Labrune, M. and Carbucicchio, M., *J. Magn. Magn. Mater.* **269** (2004), 203.

Hyperfine Interactions **156/157**: 595–606, 2004.
© 2004 *Kluwer Academic Publishers. Printed in the Netherlands.*

Surface and Interface Investigations by Nuclear Resonant Scattering with Standing Waves

M. A. ANDREEVA
Department of Physics, M.V. Lomonosov Moscow State University, Moscow 119992, Russia

Abstract. The main features of the reflectivity spectroscopy ("*coherent spectroscopy*") in energy, time and angular scales are considered. They are the enhancement or suppression of different hyperfine contributions to the spectra, strong dependence of these effects from the order of the Bragg reflectivity and from the slight angular shifts in the vicinity of the Bragg maximum, phase shifts of the quantum beats in the time spectra, shift of the angular position of the delayed reflectivity maximum relative prompt Bragg peak. These effects are caused by the depth distribution of the resonant nuclei, standing wave creation in the reflecting multilayers or thin films and also by the unhomogeneity along the surface of the resonant nuclei with different hyperfine parameters. The simple explanation is presented on the bases of the kinematical approach of the reflectivity theory. Experimental results are discussed on the bases of the kinematical approach of the reflectivity theory.

Key words: nuclear resonant scattering, Mössbauer reflectometry, hyperfine interactions.

1. Introduction

When nuclear resonant scattering with synchrotrons appeared as a new method for hyperfine interaction investigations [1–4] it was essential to assure people that in such a way we can get exactly the same information as in conventional Mössbauer spectroscopy (hyperfine field parameters) – through a quantum beat pattern. The advantages of the high brilliance of the synchrotron source provide possibilities to work with very small samples and in extreme conditions (low temperatures, high pressures, high external fields, biological objects, monolayers and so on). Anyhow it is coherent scattering but not absorption spectroscopy and even in the forward scattering geometry many new phenomena should be taken into account: polarization of the hyperfine transitions, dynamical beats, speed-up effect, transverse coherence length and so on* (see, e.g., reviews [5, 6]).

Ultrahigh collimation of synchrotron radiation (SR) makes it very suitable for diffraction and reflection experiments, which were very difficult with radioactive

---

* Here we discuss only the hyperfine spectroscopy with synchrotron radiation but do not touch the other interesting experiments with nuclear resonant scattering, such as quasi-elastic and inelastic scattering, small angle scattering, pure nuclear Bragg diffraction and GIAR-film monochromatization, TDPAC on Mössbauer levels (single-nucleus quantum beats), light-house effects, US and radio-frequency modulations of the radioactive decay, heterodyne detection of synchrotron radiation and a lot of others.

sources. Total external reflection of Mössbauer radiation was known since 1963 [7–9], and with simultaneous measurement of the conversion electrons it has proven a very effective method for the surface and thin film depth selective analysis [10, 11]. However, only with synchrotron radiation, Mössbauer reflectometry has become a widely used method.

Up to now two phenomena in synchrotron Mossbauer reflectometry attracted attention. The first one is the appearance of the maximum of nuclear resonant reflectivity near the critical angle of the total reflection in the integral delayed reflectivity curve [12, 13] – because it supplies the mostly effective registration of the time spectra at grazing angles and allows to detect nuclear resonant scattering even from single monolayer [14–16]. And the second one is the existence of the magnetic superstructure maxima on the delayed reflectivity curve, which gives the possibility to investigate the antiferromagnetic interlayer exchange coupling [13, 17] in the same way as in, e.g., the neutron scattering.

Investigations of the time spectra of reflectivity measured at different grazing angles confirmed that they are very sensitive to the depth distribution of the hyperfine fields [18–24]. Simultaneously it has become clear that these spectra are quite complicated for the "spectroscopic" analysis and the observed distortion of the quantum beats is not the only dynamic and speed-up effects. It seems that nowadays we can introduce the new term *"coherent spectroscopy"* which simply means that contrary to the absorption spectra, the different spectrum contributions in the reflectivity spectra can appear or disappear depending on the interference conditions, structure of the radiation field inside the film and uniformity of layers and surfaces. The other reason for such definition is the desirable result of the spectrum analysis: it is not only the spectrum contributions but mainly the depth position for each contribution.

There are several reasons for the specific character of the reflectivity spectra. In the present paper we would like to select and partially explain some peculiarities of the reflectivity spectra in the energy and time representation.

2. Kinematical limit in the reflectivity theory

Dynamical essence (multibeam interference) of the total external reflection makes it very difficult for qualitative analysis. The theory is based on the Parratt recurrent procedure [25] and for the anysotropic Mössbauer multilayer medium it becomes essentially more complicated [26, 27]. In the angular region near the critical angle one can only use the penetration depth value (typically several nanometers at 2–4 mrad for 14.4 keV radiation) for interpretation of experimental data. It is usually evaluated for an ideal semi-infinite space, besides for the resonant interaction it is rather different for different energies and after Fourier transform (for the time representation) it has become very uncertain.

For slightly larger glancing angles and in particular for the Bragg peaks of the periodical multilayers the general theory of reflectivity can be simplified (at least for the qualitative analysis) [28]. In the kinematical approximation, where we consider the simple sum of the waves reflected by all sublayers with proper phases, a lot of peculiarities in the energy or time spectra of reflectivity can be easily understood. So we start from the main expression [29, 30] for the reflectivity amplitude from a periodical multilayer

$$R = L_N F(\vartheta, d_j, \chi_j, \omega), \tag{1}$$

where $L_N(\vartheta, D)$ and $F(\vartheta, d_j, \chi_j, \omega)$ is the famous Laue function and the structure factor. The famous Laue function has a form:

$$L_N = \left(1 + e^{i\varphi} + e^{2i\varphi} + e^{3i\varphi} + \cdots + e^{(N-1)i\varphi}\right) = \frac{1 - e^{iN\varphi}}{1 - e^{i\varphi}}, \tag{2}$$

where φ is the phase difference of the reflected wave which it acquires during propagation through one repetition period – D:

$$\varphi = QD + \frac{\kappa}{\sin \vartheta} \sum_{k=1}^{K} \chi_k(\omega) d_k, \tag{3}$$

$Q = (4\pi/\lambda) \sin \vartheta$, $\kappa = 2\pi/\lambda$, the sum in (3) over $k = 1, \ldots, K$ refers to selected sublayers in one repetition period with thickness d_k and susceptibility $\chi_k(\omega)$ (which is resonant in general case), ϑ is the glancing angle, λ is the wavelength. If $\chi/\sin \vartheta \ll 1$ and we can neglect the second term in (3), then the Lauer function will describe only the position and shape of the Bragg peak $L_N = L_N(\vartheta, D)$. The structure factor

$$F(\vartheta, d_j, \chi_j, \omega) = \frac{i\lambda}{\sin \vartheta} \sum_{j=1}^{K} \rho_j f_j(\omega) e^{i(Q\xi_j + (\kappa/\sin \vartheta) \sum_{k=1}^{j-1} \chi_k(\omega) d_k)} \tag{4}$$

describes the interference of the waves reflected by sublayers (having the coordinate ξ_j, the surface density ρ_j and the scattering amplitude $f_j(\omega)$) in one repetition period. This structure factor essentially determines the "weight" which each resonant contribution will have in the reflectivity spectrum.

Let us condider the simplest examples and at first neglect the second term in the exponent in (4). Suppose we have two identical sublayers in one repetition period at a distance $D/2$. Than they be added with the phase difference π for the Bragg maximum of the first order (see Figure 1) and their contributions to the reflectivity amplitude will cancel each other. For the Bragg maximum of the second order their phase difference becomes 2π and they will enhance each other – Figure 2.

The enhancement of the hyperfine component beats connected with the middle part of the resonant layer was for the first time observed in [18, 20] but it was mistakenly explained as due to the effect of the standing waves only (later we shall explain the difference).

The simple interference effect appears even for very weak Bragg reflections (it takes place even for one repetition period – if we would be able to observe

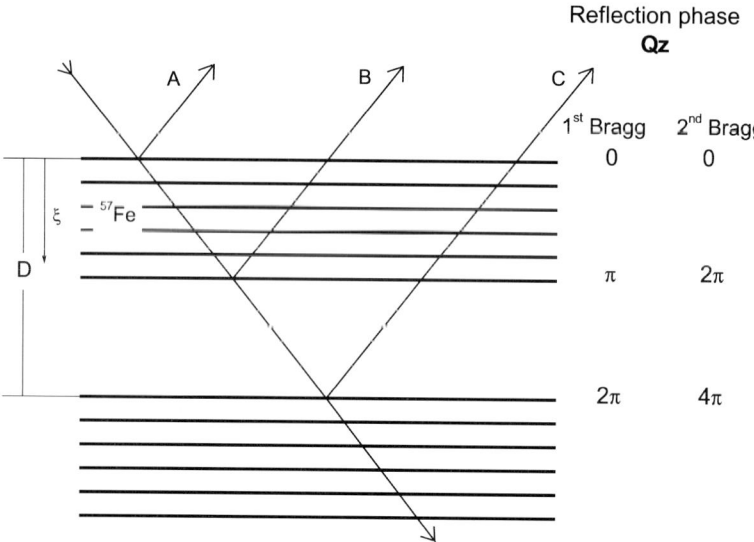

Figure 1. The relative phases for the wave addition at the exact Bragg conditions. For a half of the period distance between layers (A and B) the interference is destructive for the first order Bragg peak.

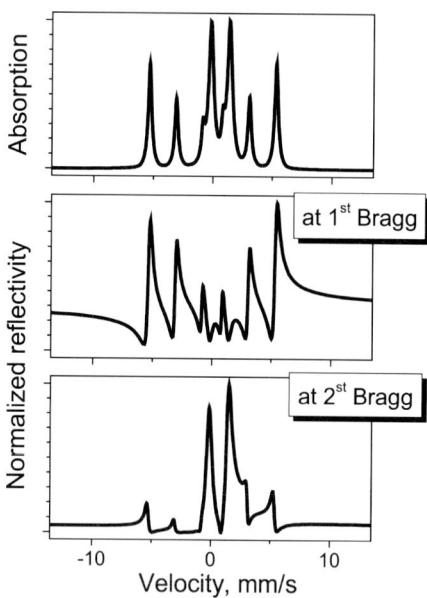

Figure 2. Calculated spectra of reflectivity for periodical multilayer $[^{57}Fe/V]*30$ where just for illustration we suppose that ^{57}Fe nuclei have doublet resonance spectrum in the interfaces but magnetic sextet spectrum in the middle part of the ^{57}Fe layers. In spite of the distortion of the line shapes in reflectivity signal the suppression and enhancement of the doublet contribution in the 1st and 2nd order Bragg peak is clearly seen.

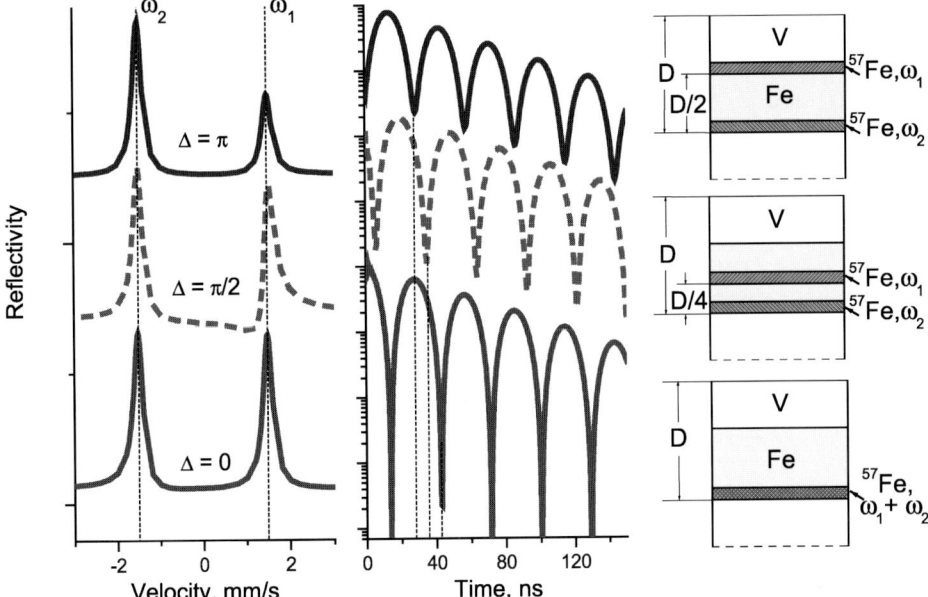

Figure 3. The energy and the time spectra of reflectivity at the first Bragg peak ($\vartheta = 14.75$ mrad for $D = 3$ nm, [Fe/V] multilayer, 20 repetitions) calculated for three different relative positions of the resonant sublayers in the repetition period, giving the phase shift Δ between the scattered waves π, $\pi/2$ and 0, respectively (from top to the bottom). Resonant frequency (single) of the first layer is $\omega_1 = 1.5$ mm/s, of the second layer is $\omega_2 = -1.5$ mm/s (in units of the velocity scale of the Mössbauer spectrum). The spectra are in relative units and shifted relative each other.

it). For example, such idea of suppression of the interface contributions was used for the interpretation of the Bragg reflectivity time spectrum from ^{57}Fe/Co multi-layer (the Bragg reflectivity was low), and it was possible to prove that the lower hyperfine magnetic field is associated with the interface layer (contrary to the ex-pectations) [22]. Notice that such depth selectivity of the Bragg reflections has been effectively used in X-ray magnetic resonance scattering [31]. The reflectivity spec-tra measured near $L_{2,3}$ edges for nine Bragg peaks in a Ce/Fe and La/Fe multilayers provide information for the depth profile of 5d magnetic moments.

If hyperfine interactions in two interfaces are not identical we shall have an another situation. Let's suppose that we have two thin resonant sublayers in the repetition period and each of them has only one resonant frequency denoted by ω_1 and ω_2 – Figure 3. The corresponding structure factor is then given as

$$F(\omega) = F^{\text{el}} - \frac{id}{2\sin\vartheta}\mu^{\text{nuc}}\left(\frac{\Gamma/2\hbar}{\omega - \omega_1 + i\Gamma/2\hbar}e^{iQ\xi_1} + \frac{\Gamma/2\hbar}{\omega - \omega_2 + i\Gamma/2\hbar}e^{iQ\xi_2}\right),$$

(5)

where F^{el} is the structure factor for the scattering by the electronic density, μ^{nuc} is the known factor of the nuclear scattering amplitude [5], Γ is the line width, ξ_1 and ξ_2 are the positions of the resonant layers in the repetition period. The time dependence of the structure factor is calculated by applying the Fourier transform of (5) and it gives:

$$F(t) \propto e^{-(\Gamma/2\hbar)t}\left(e^{iQ_1\xi_1 - i\omega_1 t} + e^{iQ_2\xi_2 - i\omega_2 t}\right). \qquad (6)$$

The intensity of the reflected wave will be

$$\left|F(t)\right|^2 \propto 2e^{-t/\tau_0}\left(1 + \cos\left((\xi_1 - \xi_2)Q + (\omega_2 - \omega_1)t\right)\right), \qquad (7)$$

where $\tau_0 = \hbar/\Gamma$ is a decay time. So we have the obvious quantum beat pattern with $(\omega_2 - \omega_1)$ frequency. Contrary to the forward scattering the beats in reflectivity spectrum have a shifted phase, which is determined by the $(\xi_1 - \xi_2)$ distance between the two resonant layers (Figure 3). This shift becomes the source of the structure information [32].

3. Standing wave effects

Up to now we have neglected the energy dependence in (3). This simplification means that the energy and time spectra of reflectivity do not depend on the Laue function and consequently do not depend on the number of repetitions. But we can easily check by performing exact calculations that even for relatively low reflectivity (when kinematical approach is still valid) this supposition is not correct – Figure 4. The slight variation of the angle in the vicinity of Bragg peak causes the

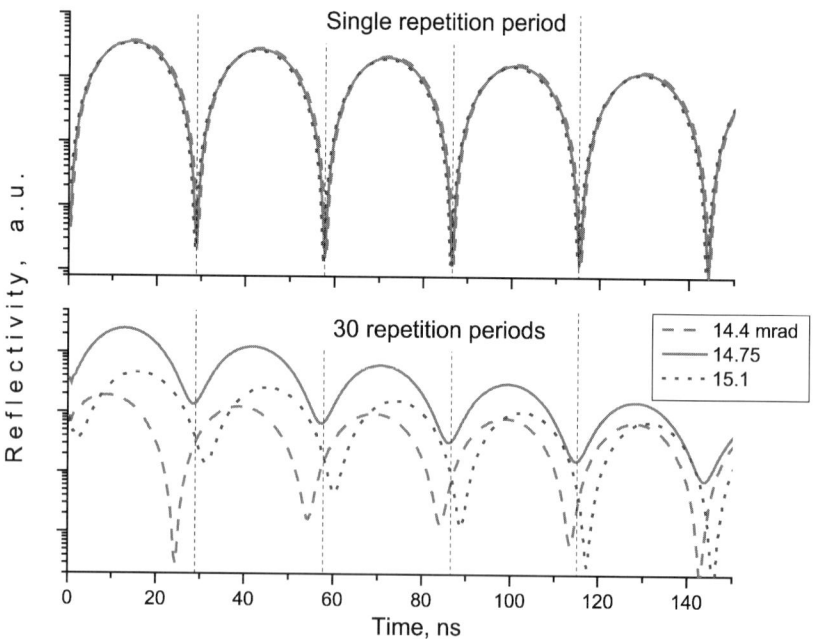

Figure 4. Time spectra of reflectivity for different angular shifts in vicinity of the Bragg peak, calculated for the same model as in Figure 3 ($\Delta = \pi$) but with different number of repetitions.

change of the beat phase but only if the number of repetitions is larger than 1. This means that the contributions to the reflectivity spectrum from different "unit cells" in periodical structure are not identical. The effect arises through the ω-dependence of (3), which leads to the resonant refraction and reflection corrections of the transmitted wave: it acquires the resonant dependence inside the medium. In other words, the resonant standing waves appear. Their influence on the reflectivity spectrum can be noticed only for relatively high Bragg reflectivity and reveals itself as the spectrum change during the slight angle variation. In refs. [21, 23, 24] it was demonstrated how such spectrum variation can be used for determination of the hyperfine field depth distribution across single repetition period.

Recently we have studied the standing wave influence on the reflectivity [33] on the bases of the exact theory. At a first sight in the kinematical approximation, when we use the simple sum of the reflected waves, the contribution from each layer to the reflectivity is independent from the others. But let us consider the contribution to the total reflectivity amplitude R^{tot} from one additional layer (with a thickness d) on the top of multilayer or other substrate having the reflectivity R. The result will be:

$$R^{tot} = Re^{iQd+i(2\lambda\rho/\sin\vartheta)f} + r \approx R' + r(1 + 2R').$$ (8)

In (8) we took into account the phase and refraction corrections of the total reflectivity caused by the added layer and the exact expression for the reflectivity by thin layer $r = i(\lambda/\sin\vartheta)\rho f$. The above consideration can easily be repeated for the thin layer placed inside the multilayer. R' in (8) is the reflectivity R at the distance d from the top of the multilayer (relative the incident wave at the same height) $R' = Re^{iQd}$. For very thin layers we can neglect this phase factor. However, it can be generalized for the case when the added layer is placed at some distance H from the surface. It is this phase factor, which determines the angular or depth variations of the standing waves.

For illustration we consider a resonant ^{57}Fe layer on the top of nonresonant superstructure [Si/W]*20. The calculated spectrum of Mössbauer reflectivity $R^{tot}(\omega)$ is determined by $r(\omega)$ and by the position H of the resonant layer relative the surface of the superstructure (through the phase shift between the amplitudes of scattering R' and $r(\omega)$). The standing wave influence appears through the complex factor $(1 + 2R')$ modulating $r(\omega)$ – Figure 5. This factor enhances the contrast of the spectrum when ^{57}Fe layer is placed in the antinode of standing wave (Figure 5a) and suppresses it in the node (Figure 5b). It is clear that the effect becomes more complicated for a resonant substrate.

For angles close to the critical angle the kinematical approximation of the reflectivity theory is not valid. We should use here the exact Parratt recursive formula for reflectivity amplitude:

$$R_j = \frac{r_j + R_{j+1}e^{i\varphi_j}}{1 + r_j R_{j+1}e^{i\varphi_j}},$$ (9)

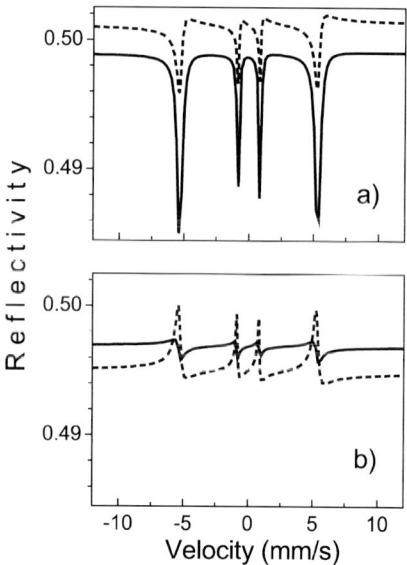

Figure 5. Calculated Mössbauer spectra of reflectivity at the exact Bragg angle 14.75 mrad from a thin ^{57}Fe layer placed at the distance $H = D/2$ (a) and $H = 0$ (b) from the surface of a multilayer with period $D = 3.0$ nm (solid lines). Dash lines correspond to the pure interference of $r(\omega)$ and R without the standing wave factor $(1 + R)^2$.

where r_j is the Fresnel reflectivity amplitude, R_{j+1} the multiple reflectivity amplitude at the previous boundary, the phase shift φ_j should now be calculated exactly:

$$\varphi = \frac{4\pi}{\lambda}\sqrt{\sin^2\vartheta + \frac{\lambda^2}{\pi}\rho_v f}\, d. \tag{10}$$

For calculation of the reflectivity from a thin layer by (9) we suppose that $i\varphi$ is small. In order to simplify the calculations we insert an artificial vacuum buffer layer, with zero thickness, between the tested layer and substrate, although the algebra can be carried out also without this layer. Neglecting $(i\varphi)^2$ term we obtain

$$R^{\text{tot}} = \frac{r_{01} + \tilde{R}}{1 + r_{01}\tilde{R}} + \frac{\tilde{R}(1 - r_{01}^2)}{(1 + r_{01}\tilde{R})^2}2i\varphi_1 \tag{11}$$

and

$$\tilde{R} = \frac{r_{10} + R}{1 + r_{10}R}, \tag{12}$$

where r_{01} is the Fresnel reflectivity amplitude from the tested layer and R is the reflectivity amplitude from the substrate. Finally we get:

$$R^{\text{tot}} = R(1 + iQd) + (1 + R)^2 r. \tag{13}$$

The obtained result is surprisingly simple and clear. The layer reflectivity amplitude r in the presence of the other reflecting layers is modulated by the "squared standing wave" $(1 + R)^2$, (for the intensity the influence of the radiation field describes by 4th power of its amplitude) because the standing wave structure of the radiation field reveals itself for the incident (absorption process) as well as for reflected wave (emission process). That is the essential difference of the standing wave appearance for the reflectivity and secondary radiations. The kinematical approach (8) is true because when R is small then $(1 + R)^2 \approx 1 + 2R$.

The strongest influence of the standing waves on the single-layer resonant reflectivity was demonstrated in ref [34] where the large enhancement of the radiation field was obtained in so-called wave-guide mode.

4. The shift of the delayed reflectivity Bragg peak relative the prompt Bragg peak position

The derived analytical expression describing the standing wave influence on the reflectivity can be effectively used for the physical interpretation of some experimental phenomena. Firstly it easily explains the appearance of the maximum in the angular dependence of the integral delayed nuclear resonant reflectivity at the critical angle of the total reflection [12, 13] because at the critical angle the radiation field has a maximum near the surface (incident and reflected waves are added in phase) [33]. Next, it can predict that this maximum will be shifted if the resonant layer is placed not at the surface but at some depth.

The standing wave can be created by the electronic scattering and partially by the nuclear resonant scattering as well. When ^{57}Fe/Cr periodical multilayer having a rather high Bragg reflectivity was investigated, it was noticed that the Bragg maximum for the delayed reflectivity is not coincide with the prompt Bragg maximum, but at that time the interpretation was not clear [24]. Obviously, the maximum in delayed reflectivity appears at the angle when the antinode of the standing wave crosses the resonant layer. The situation is quite analogous to the familiar standing wave method with registration of the secondary radiation. Nuclear resonant reflectivity plays this role of the "secondary radiation" but because it is also a coherent scattering the influence of the standing waves becomes more intense (4th power of the radiation field).

The very bright example of the standing wave influence on the nuclear reflectivity was obtained for the Nb(70 nm)/^{57}Fe/[Mo/Si]*45/Si sample because the Bragg reflectivity from nonresonant periodical multilayer [Mo/Si]*45 was very high and the nuclear resonant scattering arose only from one thin ^{57}Fe layer [35]. The interpretation of the obtained shift and shape of the integral delayed nuclear reflectivity maximum near the Bragg angle allows us to determine exactly the position and thickness of the ^{57}Fe layer (Figure 6).

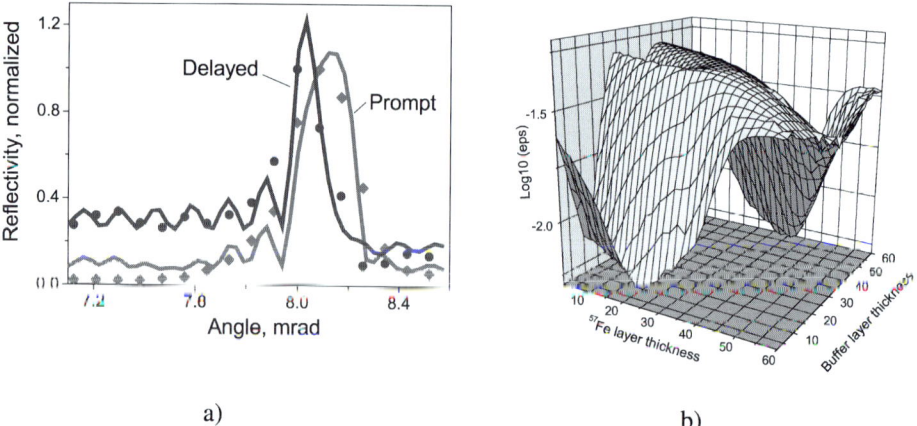

a)   b)

Figure 6. (a) Experimental prompt and delayed first Bragg peak. Dots are the experimental points, solid lines are the theoretical fit. (b) The mean-squared deviation between the theoretical and experimental delayed nuclear reflectivity curve for different resonant layer thicknesses and positions relative the top heavy layer in (Si/Mo) periodical structure.

5. Interface and layer unhomogeneities and the reflectivity spectrum

We have above discussed the influence of the hyperfine field depth distribution on the reflectivity spectrum. Real samples, however, can be not uniform along the surface. Initial stages of the film growing are often characterized by the island formation, monocrystalline films have steps, terraces and so on. Such unhomogeneities are always characterized by specific hyperfine fields (that's why the Mössbauer spectra of thin films and even monolayers have so complicated content). It is well known that the surface and interface roughness as well as the volume unhomogeneities (such as magnetic domains) create the diffuse scattering and decrease the specular reflectivity. Such influence on reflectivity can be described by Debay or Nevot–Croce damping factors or by introducing the interface layer with gradient of density between sublayers. Such description is good in the common X-ray reflectivity, but it is not sufficient for the coherent reflection spectroscopy. If we are interesting in the different resonant contributions to the reflectivity spectrum we should realize that one and the same amount of the given hyperfine contribution will be present in the reflectivity spectrum with different weights regarding on the degree of the distribution smoothness along the surface of the corresponding nuclei.

Figure 7 illustrates this idea. If resonant nuclei are collected in small irregular grains (with zero correlation length), then their contribution to the coherent reflectivity signal will be negligible (Figure 7b). The contributions from the surface islands will be more essential (Figure 7c) but smaller than from the absolutely uniformly distributed along the surface nuclei (Figure 7a).

Additionally we should remember that the size of the registration slits could essentially change the relative amount of different resonant contribution by collecting some part of the diffuse scattering. The exact theory of this reason of suppression

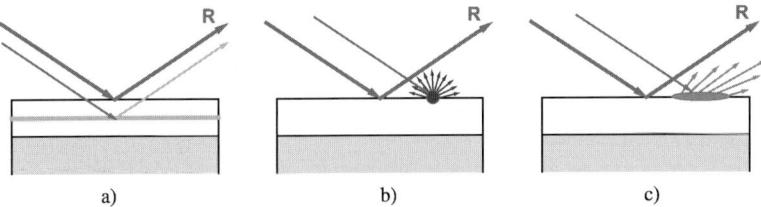

a)   b)   c)

Figure 7. Illustration of the difference in the hyperfine spectrum contributions originated from the non uniform lateral distribution of the scattering nuclei.

or enhancement of the different hyperfine contributions has not been developed yet, but qualitatively we can interpret the decreasing of some contribution to the reflectivity spectrum relative corresponding CEMS spectrum as due to the non-uniform surface distribution of this component.

6. Conclusions

Coherent spectroscopy differs (as we have demonstrated here) from the normal Mössbauer spectroscopy in many respects and the interpretation of the experimental nuclear resonant reflectivity spectra is a rather complicated task. The advantage of this spectroscopy is the sensitivity of the reflectivity to the space distribution of the different hyperfine fields and it can be the alternative method for the depth selective investigations to the ^{57}Fe probe monolayer method [36].

Acknowledgements

The work was supported by RFBR (grant No. 01-02-17541 and No. 03-02-17168), INTAS (grant No. 01-0822), the Royal Swedish Academy of Sciences and by the Department of Physics, Uppsala University.

References

1. Rüffer, R., Gerdau, E., Grote, M., Hollatz, R., Röhlsberger, R., Rüter, H. D. and Sturhahn, W., *Nucl. Instrum. Meth. Phys. Res.* **303** (1991), 495.
2. Sturhahn, W. and Gerdau, E., *Phys. Rev. B* **49** (1994), 9285.
3. Rüffer, R. and Chumakov, A. I., *Hyp. Interact.* **97/98** (1996), 589.
4. Sturhahn, W., Alp, E. E., Toellner, T. S., Hession, P., Hu, M. and Sutter, J., *Hyp. Interact.* **113** (1998), 47.
5. Smirnov, G. V., *Hyp. Interact.* **123/124** (1999), 31.
6. Siddons, D. P., Bergman, U. and Hastings, J. B., *Hyp. Interact.* **123/124** (1999), 681.
7. Bernstein, S. and Campbell, E. C., *Phys. Rev.* **132** (1963), 1625.
8. Wagner, F. E., *Z. Physik* **210** (1968), 361.
9. Frost, J. C., Cowie, B. C. C., Chapman, S. N. and Marshall, J. P., *Appl. Phys. Lett.* **47** (1985), 581.
10. Irkaev, S. M., Andreeva, M. A., Semenov, V. G., Belozerskii, G. N. and Grishin, O. V., *Nucl. Instrum. Meth. Phys. Res. B* **74** (1993), 545, 554; **103** (1995), 351.

11. Andreeva, M. A., Irkaev, S. M. and Semenov, V. G., *Hyp. Interact.* **97/98** (1994), 605.
12. Baron, A. Q. R., Arthur, J., Ruby, S. L., Chumakov, A. I., Smirnov, G. V. and Brown, G. S., *Phys. Rev. B* **50** (1994), 10354.
13. Toellner, T. S., Sturhahn, W., Röhlsberger, R., Alp, E. E., Sowers, C. H. and Fullerton, E. E., *Phys. Rev. Lett.* **74** (1995), 3475.
14. Niesen, L., Mugarza, A., Rosu, M. F., Coehoorn, R., Jungblut, R. M., Roozeboom, F., Baron, A. Q. R., Chumakov, A. I. and Rüffer, R., *Phys. Rev. B* **58** (1998), 8590.
15. Carbone, C., Dallmeyer, A., Malagoli, M. C., Maiti, K., Wingbermühle, J., Eberhardt, W., Nagy, D. L., Bottyán, I., Deák, L., Szilágy, E. und Rüffer, R., ESRF Highlights, 1999, p. 60.
16. Röhlsberger, R., *Hyp. Interact.* **123/124** (1999), 455.
17. Bottyán, L., Dekoster, J., Deák, L., Baron, A. Q. R., Degroote, S., Moons, R., Nagy, D. L. and Langouche, G., *Hyp. Interact.* **113** (1998), 295.
18. Andreeva, M. A., Irkaev, S. M., Semenov, V. G., Prokhorov, K. A., Salashchenko, N. N., Chumakov, A. I. and Rüffer, R., *J. Alloys & Compounds* **286** (1999), 322.
19. Nagy, D. L., Bottyán, L., Deák, L., Dekoster, J., Langouche, G., Semenov, V. G., Spiering, H. and Szilágy, E., In: M. Miglierini and D. Petridis (eds), *Proc. Mössbauer Spectroscopy in Materials Science*, Kluwer Academic Publishers, 1999, p. 323.
20. Andreeva, M. A., Irkaev, S. M., Semenov, V. G., Prokhorov, K. A., Salashchenko, N. N., Chumakov, A. I. and Rüffer, R., *Hyp. Interact.* **126** (2000), 343.
21. Kalska, B., Häggström, L., Lindgren, B., Blomquist, P., Wäppling, R., Andreeva, M. A., Nikitenko, Yu. V., Proglyado, V. V., Aksenov, V. L., Semenov, V. G., Chumakov, A. I., Leupold, O. and Rüffer, R., *Hyp. Interact.* **136/137** (2001), 295.
22. Lindgren, B., Andreeva, M. A., Häggström, L., Kalska, B., Semenov, V. G., Chumakov, A. I., Leupold, O. and Rüffer, R., *Hyp. Interact.* **136/137** (2001), 439.
23. Andreeva, M. A., Semenov, V. G., Häggström, L., Lindgren, B., Kalska, B., Chumakov, A. I., Leupold, O., Rüffer, R., Prokhorov, K. A. and Salashchenko, N. N., *The Physics of Metals and Metallography* **91**(Suppl. 1) (2001), 22.
24. Andreeva, M. A., Semenov, V. G., Lindgren, B., Häggström, L., Kalska, B., Chumakov, A. I., Leupold, O., Rüffer, R., Prokhorov, R. R. and Salashchenko, N. N., *Hyp. Interact.* **141/142** (2002), 119.
25. Parratt, L. G., *Phys. Rev.* **95** (1954), 359.
26. Andreeva, M. A. and Rosete, C., *Vestnik Moscovskogo Universiteta, Fizika* **41** (1986), 65 (English transl. by Allerton Press, Inc.).
27. Röhlsberger, R. *Hyp. Interact.* **123/124** (1999), 301.
28. Hamley, I. W. and Pedersen, J. S., *J. Appl. Cryst.* **27** (1994), 29.
29. Andreeva, M. A. and Lindgren, B., In: M. Mashlan, M. Miglierini and P. Schaaf (eds), *Material Research in Atomic Scale by Mössbauer Spectroscopy*, NATO Science Series, II. Mathematics, Physics and Chemistry, Vol. 94, 2002, p. 217.
30. Andreeva, M. A. and Lindgren, B., *Surface Investigations* **1** (2003), 12 (in Russian).
31. Séve, L., Jaouen, N., Tonnerre, J. M., Raoux, D., Bartolome, F., Arend, M., Felsch, W., Rogalev, A., Goulon, J., Gautier, C. and Bérar, J. F., *Phys. Rev. B* **60** (1999), 9662.
32. Andreeva, M. A. and Lindgren, B., *Nucl. Instrum. Meth. Phys. Res.*, in press.
33. Andreeva, M. A. and Lindgren, B., *JETP Lett.* **76** (2002), 704.
34. Röhlsberger, R., Thomas, H., Schlage, K., Burkel, E., Leupold, O. and Rüffer, R., *Phys. Rev. Lett.* **89** (2002), 237201.
35. Andreeva, M. A., Vdovichev, S. N., Nozdrin, Yu. N., Pestov, E. E., Salashchenko, N. N., Semenov, V. G., Lindgren, B., Häggström, L., Nordblad, P., Kalska, B., Leupold, O. and Rüffer, R., *Izv. Ros. Acad. Nauk, Ser. Fiz.*, in press.
36. Shinjo, T. and Keune, W., *J. Magn. Magn. Mater.* **200** (1999), 598.

Hyperfine Interactions **156/157**: 607–613, 2004.
© 2004 *Kluwer Academic Publishers. Printed in the Netherlands.*

Nuclear Resonant Reflectivity Investigations of a Thin Magnetic ^{57}Fe Layer Adjacent to a Superconducting V Layer

M. A. ANDREEVA[1], L. HÄGGSTRÖM[2], B. LINDGREN[2], B. KALSKA[2,3],
A.-M. BLIXT[2], S. KAMALI[2], O. LEUPOLD[4] and R. RÜFFER[4]
[1]*Department of Physics, M.V. Lomonosov Moscow State University, Moscow 119992, Russia*
[2]*Department of Physics, Uppsala University, Uppsala, Box 530, 751 21, Sweden*
[3]*University of Bialystok, Institute of Chemistry, 15-399 Bialystok, Poland*
[4]*European Synchrotron Radiation Facility, Grenoble, BP 220, F-38043, France*

Abstract. With nuclear resonant scattering of synchrotron radiation we have investigated the magnetic hyperfine fields of a thin ($\sim$2 nm) ^{57}Fe layer below and above the superconducting temperature of an adjacent thick ($\sim$50 nm) V cap layer. A standing wave enhancement of the nuclear resonant reflectivity from the buried ^{57}Fe layer was achieved by a periodic superstructure [^{56}Fe/V]$_{25}$ created below the resonant layer. No visible variation of the magnetization in the ^{57}Fe layer below and above the superconducting transition temperature was observed (neither at the critical angle, nor at the Bragg peak). The strong exchange interaction in the investigated ^{57}Fe layer can explain the result.

Key words: thin film magnetism, superconductivity, nuclear resonant scattering, reflectometry, hyperfine interactions.

1. Introduction

Superconductivity and magnetism are usually incompatible because the exchange interaction E_{ex} destroys the coupling Δ_S of the Cooper pairs although the recent discovery of magnetic materials that are also superconducting has reconciled these phenomena [1]. In ordinary (BCS-type) bulk superconductors the normal state is recovered when $E_{ex} > (1/\sqrt{2})\Delta_S$ but when Cooper pairs are injected from a superconductor (S) into a ferromagnet (F) through an S/F interface, the superconducting correlations persist in F even for exchange energies much higher than Δ_S (proximity effect). They survive for a time corresponding to a traveled length on the order of the coherence length scale in F, which is independent of the energy gap [2, 3]. S/F and S/F/S junctions have been studied extensively, mostly focusing on the spin selectivity introduced in the conductance, the influence of the ferromagnet on the superconductivity and the strong implication for devices. The reconstruction of the magnetic order below the superconducting transition temperature T_C was only considered in a few papers [4–6].

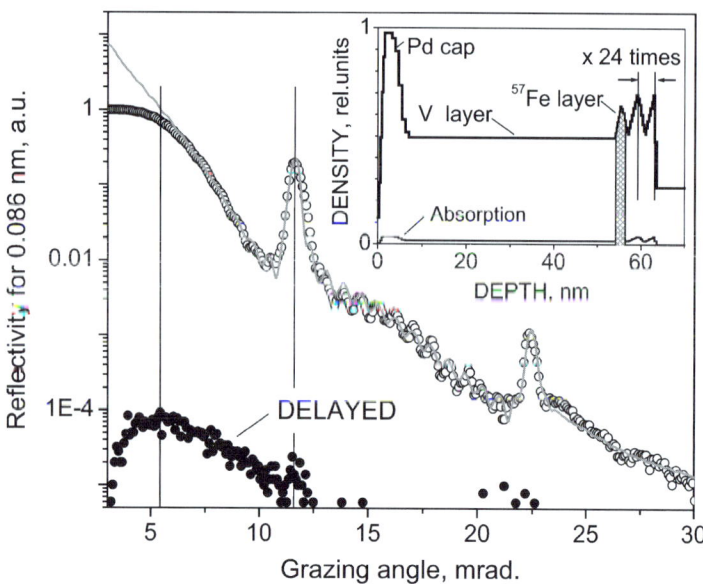

Figure 1. Reflectivity curves for prompt and delayed response. Vertical lines denote the angles where the time spectra of the nuclear resonant reflectivity were measured. The solid line represents the fit with obtained model shown in the insert. The repeating part of the multilayer (×24) is marked by vertical lines.

Mühge *et al.* [4] studied Nb/Fe bilayers using different experimental techniques and observed a reduction of the average Fe magnetic moment below T_C. It was interpreted as an evidence of a domainlike magnetic structure called cryptoferromagnetism or due to a spatial modulation of the spin direction in the Fe-layer caused by the demagnetizing field in the geometrically rough interface. The former interpretation was opposed by Bergeret *et al.* [5] who theoretically investigated the cryptoferromagnetic state in S/F bilayers and found that multilayers with much weaker ferromagnets must be used, e.g., $V_{1-x}Fe_x$/V multilayers where the exchange energy is varied by alloying.

Hyperfine interaction methods, such as Mössbauer spectroscopy and nuclear resonant reflectivity with synchrotron radiation (synchrotron Mössbauer reflectometry) do not measure only the average of the magnetic moments but the magnitude and direction of Fe moments from various sites in the superstructure. Moreover, the measurements do not need any applied field, hence there is no influence from demagnetizing effects. Here we report on such experiments on a similar structure as used by Mühge *et al.* [4] but with superconducting V instead of Nb.

2. Sample characterization and nuclear resonant reflectivity analyses

A multilayer $MgO(001)/[^{56}Fe_{13ML}/V_{13ML}]_{24}/^{57}Fe_{13ML}/V_{50\,nm}/Pd_{5\,nm}$ was epitaxially grown by magnetron sputtering. The 50 nm thick V layer was confirmed by

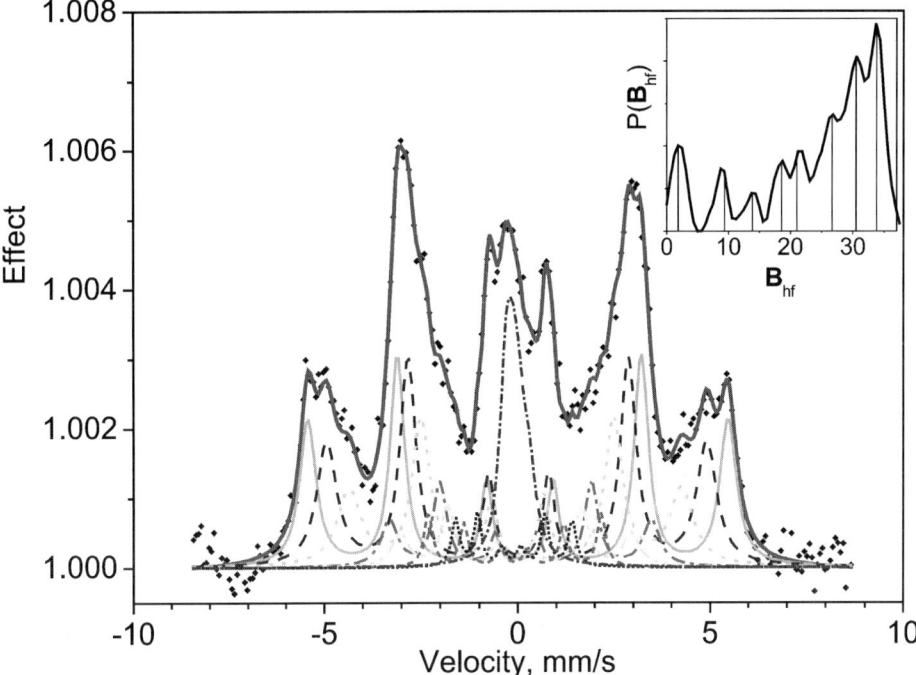

Figure 2. CEMS data of our sample. The field distribution, shown in the insert, was initially fitted with Rusakov's program DISTRI (available from rusakov@moss.phys.msu.ru) but then refitted with 8 sextets corresponding to the marked peaks in the field distribution. These sextets are shown together with the CEMS data.

resistivity measurements to be superconducting below 3.8(1) K. The purpose of the non-resonant $[^{56}Fe/V]_{24}$ multilayer below the resonant ^{57}Fe layer is to create an X-ray standing wave at Bragg conditions which enhance the signal from the interface region close to the superconducting V-layer. By shifting the grazing angle in the vicinity of the exact Bragg condition it is also possible to tune the depth selectivity [7, 8]. Figure 1 shows the X-ray reflectivity as measured for the resonant wavelength 0.086 nm at ESRF. Roughness was accounted for by introducing inter-face layers with different densities instead of using the Nevot–Croce correction [9] which are difficult to implement in the anisotropic case of nuclear resonant reflectivity [10]. From Figure 1 it is clear that the intensity near the critical angle of total reflection is much larger than at the superstructure Bragg angle, despite the enhancement from the standing wave, so most of our synchrotron Mössbauer reflectometry measurements were actually performed near this angle.

CEMS measurement at room temperature reveals a magnetic hyperfine splitting corresponding to $\mathbf{B}_{hf} \sim 33$ T for a large fraction of the ^{57}Fe nuclei, see Figure 2. In order to make the CEMS analysis compatible with our program for analyzing the time spectra from the nuclear resonance reflectivity we fitted the CEMS data with 8 sextets with the $\mathbf{B}_{hf}$ given in Table I. These parameters describe quite well the

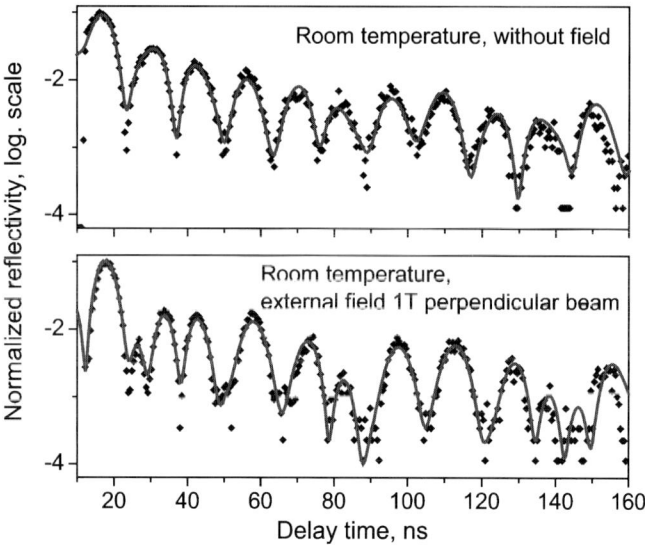

Figure 3. Time spectra of the nuclear resonant reflectivity measured near the critical angle ($\theta = 5.41$ mrad.) at room temperature without and with the external magnetic field. The fit (solid lines) was performed with the multilayer structure and hyperfine field parameters obtained from the reflectivity curve (Figure 1) and CEMS spectrum (Figure 2). For the lower spectrum with an applied field the only change in parameters were the direction and the 1 T decrease of $|\mathbf{B}_{\mathrm{hf}}|$.

Table I.

$\mathbf{B}_{\mathrm{hf}}$ at room temperature (T)	33.8	30.5	26.7	21.0	18.7	13.6	7.8	1.2
$\mathbf{B}_{\mathrm{hf}}$ at 2.3 K (T)	34.9	31.8	28.3	23.7	20.5	17.6	9.3	1.3

synchrotron Mössbauer reflectometry time spectra measured at room temperature (Figure 3). Only the fractional depth distribution of the different field components and their orientation were adjusted in this fit. Despite the sample was magnetized in a field of 1T along the beam direction (parallel to the bcc $\langle 110 \rangle$ direction) before the measurement, the magnetization had relaxed to one of the easy axis of magnetization $\langle 100 \rangle$, 45° relative to the beam direction (except for a small fraction of random orientation of the low field components). This was confirmed by a 90° rotation of the sample around the surface normal which gave the same spectrum. If the fields were in an antiparallel domain configuration ($\langle 100 \rangle$, $\langle \bar{1}00 \rangle$) or along both easy axis of magnetization ($\langle 100 \rangle$, $\langle 010 \rangle$) the time spectrum should show more beat frequencies as in Figure 3 lower panel, where an external field of 1 T was applied perpendicular to the beam during our final measurement. For the time spectrum measurements at low temperatures the hyperfine fields increased by ~6% and their fitted values are also given in Table I.

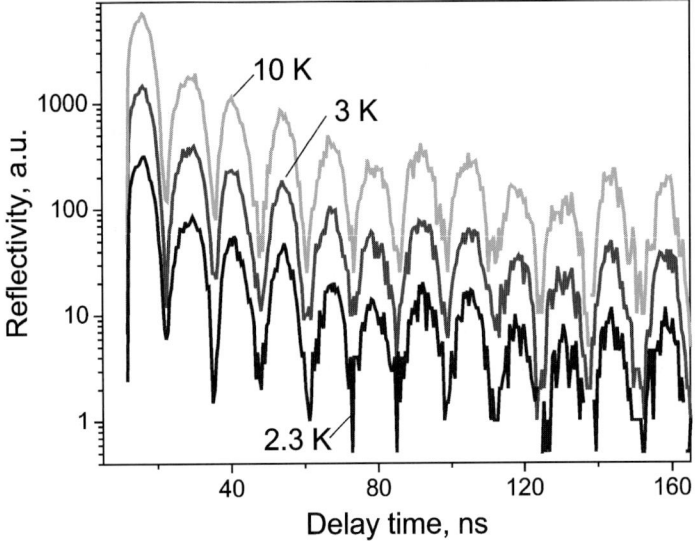

Figure 4. Comparison of the experimental time spectra measured near the critical angle at three different temperatures close to the superconducting transition temperature of the adjacent V layer.

Time spectra measured at different temperatures above and below the superconducting transition temperature (3.8 K) are shown in Figure 4. However, we could not reveal any visible change in either the magnitude or the direction of the fields. A reduction or a 90–180° rotation of the fields in a 2 Å interface region should easily be detected due to new frequency components, similar to the difference between the time spectra in Figure 3. Neither did spectra measured at the 1st order Bragg peak show any change, although here statistics are less good.

The shape of the time spectra measured at the critical angle and at the Bragg angle are quite different due to different depth selectivity (see Figure 5). A simultaneous fit shows that the largest hyperfine fields are predominantly located in the upper part of the resonant layer.

3. Conclusions

No spatial modulation of the magnetic structure (cryptoferromagnetism) was observed in the ^{57}Fe layer adjacent to a superconducting V layer, at least not within the coherent scattering length along the surface (1–10 μm). This is in agreement with the theoretical estimations of Bergeret *et al.* [5] who predicted that a cryptoferromagnetic state is only expected for the exchange fields an order of magnitude smaller than found in α-Fe. Hence, the by Mühge *et al.* [4] observed reduction of the magnetization is probably an artifact of the applied field and an inhomogeneous demagnetization.

However, by passing T_C we have observed a very small variation of the time spectra for another sample, a ^{57}Fe/Nb structure [11] were the magnetic ordering

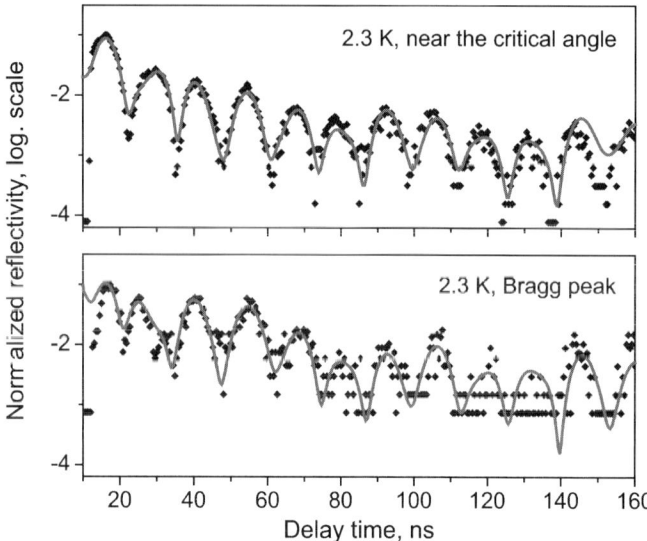

Figure 5. Comparison of the time spectra measured at 2.3 K, near the critical angle and at the Bragg peak. The full lines show the result from a simultaneous fit.

turned out to be quite weak (magnetic only at low temperatures) due to a partial oxidation of the ^{57}Fe layer. Unfortunately the interpretation of this oxide layer signal is difficult and no definite conclusions can be made at the moment. A better way of creating a smaller exchange interaction is by alloying. The V/Fe interface is indeed not sharp. Some intermixing could also reduce the exchange interaction but the presence of the exchange from the Fe further away from the interface is probably too strong. Replacing the resonant Fe-layer by a resonant FeV alloy is maybe better.

Acknowledgements

The work was supported by RFBR (No. 01-02-17541) and by the Royal Swedish Academy of Sciences.

References

1. Flouquet, J. and Buzdin, A., *Physics World* **15** (2002), 41.
2. Kontos, T., Aprili, M., Lesueur, J. and Grison, X., *Phys. Rev. Lett.* **86** (2001), 304.
3. Izyumov, Yu. A., Proshin, Yu. N. and Khusainov, M. G., *Physics – Uspekhi* **45** (2002), 109.
4. Mühge, Th., Garif'yanov, N. N., Goryunov, Yu. V., Theis-Bröhl, K., Westerbolt, K., Garifullin, I. A. and Zabel, H., *Physica C* **296** (1988), 325.
5. Bergeret, F. S., Efetov, K. B. and Larkin, A. I., *Phys. Rev. B* **62** (2000), 11872.
6. Garifullin, I. A., Tikhonov, D. A., Garif'yanov, N. N., Fattakhov, M. Z., Theis-Bröhl, K., Westerbolt, K. and Zabel, H., *Appl. Magn. Reson.* **22** (2002), 439.
7. Andreeva, M. A., Irkaev, S. M., Semenov, S. M., Prokhorov, K. A., Salashchenko, N. N., Chumakov, A. I. and Rüffer, R., *Hyp. Interact.* **126** (2000), 343.

8. Andreeva, M. A. and Lindgren, B., *JETP Lett.* **76** (2002), 704.
9. Nevot, L. and Croce, P., *Rev. Phys. Appl.* **15** (1980), 761.
10. Röhlsberger, R., *Hyp. Interact.* **123/124** (1999), 301.
11. Andreeva, M. A., Vdovichev, S. N. Nozdrin, Yu. N., Pestov, E. E., Salashchenko, N. N., Semenov, V. G., Lindgren, B., Häggström, L., Nordblad, P., Kalska, B., Leupold, O. and Rüffer, R., *Izv. Ros. Acad. Nauk*, Ser. Fiz. **68**(4) (2004), 487.

Hyperfine Interactions **156/157**: 615–621, 2004.
© 2004 *Kluwer Academic Publishers. Printed in the Netherlands.*

CEMS Investigations of Swift Heavy Ion Irradiation Effects in Tb/Fe Multilayers

J. JURASZEK[1,*], J. TEILLET[1], A. FNIDIKI[1] and M. TOULEMONDE[2]
[1]*Groupe de Physique des Matériaux, Université de Rouen, Avenue de l'Université, BP 12,*
F-76801 Saint Etienne du Rouvray Cedex, France; e-mail: jean.juraszek@univ-rouen.fr
[2]*Centre Interdisciplinaire de Recherches Ions Lasers, BP 5133, F-14070 Caen Cedex 05, France*

Abstract. In nanometric Tb/Fe multilayers, the effect of the electronic energy loss of fast heavy ions on the interfaces has been selectively investigated by [57]Fe conversion electron Mössbauer spectrometry, thanks to deposition of [57]Fe probe layers at the bottom or top of the natural iron layers. The as-deposited Tb-on-Fe interfaces are sharp while the Fe-on-Tb ones are amorphous and diffuse. Modifications of the environments of the [57]Fe probe atoms were investigated as a function of the electronic stopping power (Sn and U ions) and the ion fluence. An effect of recrystallization of iron is evidenced at the initially diffuse Fe-on-Tb interface while only mixing between Fe and Tb atoms is observed at the sharp Tb-on-Fe interface. For very high rates of electronic energy deposition (stopping power ~60 keV/nm), severe mixing leads to the interdiffusion of the layers.

1. Introduction

The energy loss of fast heavy ions (above 1 MeV/nucleon) in a solid occurs predominantly through inelastic interactions with the electrons of the target. The main feature of this process is that high local doses of energy (few tens of keV/nm) are deposited close to the ion path within a very short time of 10^{-17} to 10^{-14} s. The relaxation of the lattice (in the 10^{-13}–10^{-12} s range) may induce on a nanometer scale structural modifications such as point defect creation or phase transformation in insulators but also in metallic materials above a threshold of Se stopping power [1]. In pure iron, annealing of defects induced by electronic excitation has been evidenced for $10 \leqslant S_e \leqslant 40$ keV/nm, whereas damage creation occurs for $S_e \geqslant 40$ keV/nm [2]. Up to now, only a few studies have been devoted to the effect of electronic energy loss on the interfaces of multilayers. In the bcc-Fe/Tb multilayers, amorphization by interdiffusion occurs for $S_e \geqslant 30$ keV/nm [3]. Below this threshold, segregation of Fe and Tb at the interfaces has been found to increase the crystalline iron layer thickness [3, 4].

[57]Fe conversion electron Mössbauer spectrometry technique (CEMS) is a very suitable technique to study quantitatively ion-beam-mixing effects in multilayers. In all previous studies, informations on the Fe environments were obtained from the whole volume of iron layers. The aim of this study is to explore at a mono-

---

* Author for correspondence.

layer scale the effects of swift-heavy-ion irradiation on the interfaces of Tb/Fe multilayers in a very large electronic slowing-down range. The Tb-on-Fe and Fe-on-Tb interfaces will be selectively analyzed by CEMS thanks to the use of two monolayers of ^{57}Fe deposited first either at the top or at the bottom of natural iron layers.

2. Experimental

(Tb 1.9 nm/Fe 3.8 nm) multilayers have been grown at 150 K by thermal evapo-ration in an ultra-high vacuum system (see reference [5] for preparation details). Twenty bilayers were deposited onto Si (111) wafers with a 5 nm thick buffer SiO layer. Two monolayers of ^{57}Fe were deposited either at the top (sample A) or at the bottom (sample B) of the natural iron layers (Figure 1). The layered structure and the bilayer period were checked by grazing X-ray reflectometry (GXR). To avoid contamination and oxidation of the multilayer, all the Fe/Tb stack was covered by a 5 nm-thick SiO film. However, to neglect the detection by Mössbauer spectrometry of the Fe/SiO interfaces as compared to the ^{57}Fe probe layers signal, the first and last Fe layers were only composed of natural iron.

The multilayers have been irradiated at room temperature under normal inci-dence at the IRASME facilities of the GANIL accelerator (located in Caen, France) with Sn and U ions of respective energy of 4.635 and 3.407 MeV/nucleon and respective electronic stopping power $S_e = 36$ and 63.5 keV/nm [6]. The beam flux Φ was kept below 3×10^8 ion cm^2 s^{-1}, so that beam heating does not exceed a few tens of degrees. It is worth noticing that the thickness of the Tb/Fe stacks is much smaller than the projected range (~few tens micrometers), so that stopping regions

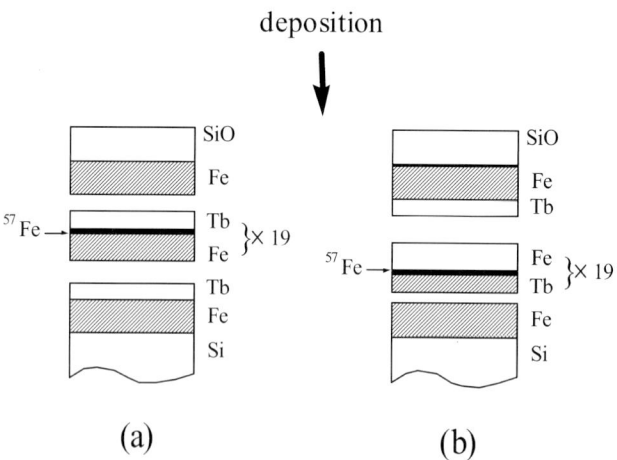

Figure 1. Description of the prepared samples with two monolayers of ^{57}Fe deposited at the top (a) or at the bottom (b) of natural iron layers.

are avoided and that the electronic stopping power is almost constant throughout the multilayer thickness.

[57]Fe conversion électron Mössbauer spectra were recorded at room temperature using a home-made helium/methane proportional counter with a source of [57]Co in a rhodium matrix. The samples were set perpendicular to the γ-beam. The fitting procedure used a least-square technique based on the histogram method [7]. Isomer shifts are given with respect to standard α-iron at 300 K.

3. Ion irradiation results

The spectra of the as-deposited samples (Figures 2(a) and 3(a)) consist of the superposition of a sharp sextet and a magnetic component with broad lines. They have been fitted by a distribution of hyperfine fields correlated with a distribution of isomer shifts to take into account the slight assymetry of the lines. The peak on the distributions corresponds to the sharp sextet with hyperfine field and isomer shift values close to those of bulk iron ($B_{hf} = 33$ T, $\delta = 0$ mm s^{-1} and comes from the sites of [57]Fe in the bcc-Fe phase. The distribution of weaker hyperfine fields is associated with the broad component on the spectra. It corresponds to the [57]Fe atoms having terbium neighbors. Comparison between A and B spectra shows that (i) The relative area of the α-Fe sextet (F_α) is much higher for sample A than for sample B. For the latter, the F_α value is close to that of the natural iron layer signal, estimated to 14% of the total spectrum area. Therefore, both [57]Fe

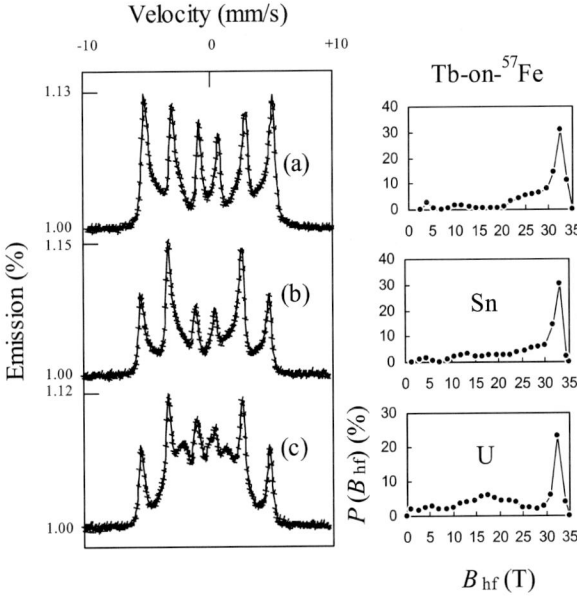

Figure 2. Mössbauer spectra and hyperfine field distributions at 300 K for sample A (Tb-on-[57]Fe) before and after irradiation with Sn (a) and U (b) ions at the fluence 5×10^{12} ions cm^{-2}.

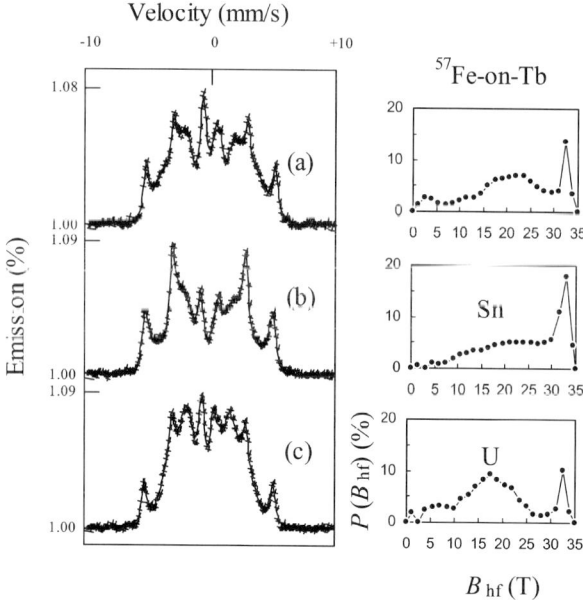

Figure 3. Mössbauer spectra and hyperfine field distributions at 300 K for sample B (^{57}Fe-on-Tb) before and after irradiation with Sn (a) and U (b) ions at the fluence 5×10^{12} ions cm^{-2}.

monolayers deposited on the Tb layers are entirely mixed with terbium atoms. For sample A, the corrected α-Fe fraction reaches the value 0.5, indicating that only one ^{57}Fe monolayer on the both deposited is implied with the Tb layers at the Tb-on-Fe interface. (ii) The shape of the hyperfine field distributions of the non-crystalline components (i.e. below 30 T) is different for the two probe layers. For the sample A, it presents few components at low fields and a maximum close to the α-Fe peak centered at 32.5 T, indicating a sharp interface. For the sample B, the distribution is spread out towards the low fields due to a strong composition gradient at the Fe-on-Tb interface. Thus, the Fe-on-Tb interface is amorphous and diffuse while the Tb-on-Fe interface is sharp.

For the multilayers irradiated by Sn ions (S_e = 36 keV/nm) at the fluence 5×10^{12} ions cm^{-2}, the increase of the relative intensity of the intermediate lines on the spectra indicates a rotation of the magnetic moments towards the plane of the layers (Figures 2(b) and 3(b)). For sample A, a slight reduction of the area under the peak centered at 32.5 T to the expense of weaker field components shows that, at the sharp Tb-on-Fe interface, a small fraction of the probe atoms initially in the α-Fe phase has been mixed with Tb atoms. This mixing effect is confirmed by the decrease of the mean hyperfine field and isomer shift values ($\langle B_{hf} \rangle$ = 26.6 T, $\langle \delta \rangle$ = -0.04 mm/s), evolving towards those of amorphous FeTb alloys. This mixing process is in agreement with an electronic stopping power value above the threshold for Fe–Tb mixing [3]. For sample B, a striking result is obtained: the Mössbauer spectrum exhibits an increase of the α-Fe component from 23 to 39 %

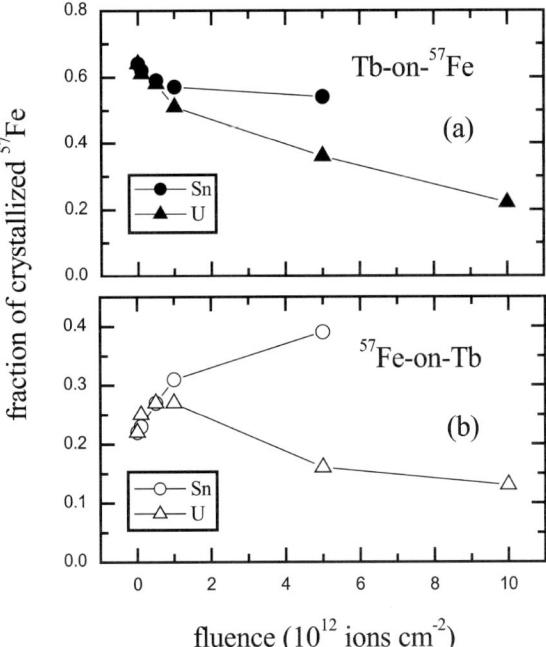

Figure 4. Evolution versus the ion fluence of the fraction of ^{57}Fe atoms in the cristalline α-phase as compared to all Fe atoms, for the ^{57}Fe-on-Tb (a) and Tb-on-^{57}Fe (b) interfaces irradiated with Sn ions and U ions.

of the total spectrum area and the average hyperfine parameters ($\langle B_{hf} \rangle = 22.9$ T, $\langle \delta \rangle = -0.06$ mm/s) increase towards the α-Fe values. Thus, a recrystallization of a fraction of the probe atoms has occured at the initially diffuse Fe-on-Tb interface, as the result of the Sn ion irradiation.

After U ion irradiation ($S_e = 63.5$ keV/nm) at the same fluence, drastic changes are observed on the Mössbauer spectra of the probe layers (Figures 2(c) and 3(c)). For sample A, irradiation results in the growth of a broad component (concerning 64% of the probe atoms) typical of an amorphous Fe/Tb alloy [8] and the cristalline Fe sextet is strongly reduced in intensity. Thus a very efficient atomic mixing between ^{57}Fe and Tb atoms takes place at the Tb-on-Fe interface during irradiation with very heavy ions. For the sample B, the reduction of the α-Fe peak on $P(B_{hf})$ for this fluence indicates a global loss of crystallinity of the natural iron layer. The present results are in agreement with those evidenced by high-angle X-ray diffraction (XRD) and CEMS in similar samples without probe layers [9].

The modification of the environments of the probe atoms induced by the electronic energy loss was investigated by following the fluence dependence of the fraction of ^{57}Fe atoms in the crystalline phase as compared to all ^{57}Fe atoms, noted F_α. This is illustrated in Figure 4. For the sample A, a continuous decrease of F_α with the fluence is observed, evidencing for a mixing effect between ^{57}Fe and Tb atoms at the initially sharp Tb-on-Fe interface. For a given fluence, the higher the S_e value, the higher the mixing effect. However, similar irradiation made on a (Tb/natural Fe) multilayer has led to complete interdiffusion of the layers. Therefore the remaining α-Fe fraction indicates that, in addition to the Fe–Tb intermixing, a self-diffusion of a fraction of the probes into the remaining natural iron layer is induced by ion irradiation. Obviously, at high fluences, this ion-irradiation-induced diffusion would have the result to homogenize the distribution of the ^{57}Fe atoms inside the remaining iron layers. For the diffuse Fe-on-Tb interface, different behaviors are observed according to the electronic stopping power value. For $S_e = 36$ keV/nm, the continuous increase of F_α with the fluence indicates recrystallization of a fraction of the probes. In the range of applied fluences, only one stage of recrystallization is evidenced, without later stage of mixing. For strong value of the electronic stopping power ($S_e = 63.5$ keV/nm), the recrystallization effect is confirmed by an increase of F_α at low fluence. However, for fluences above 1×10^{12} ions cm^{-2}, the decrease of F_α while the fluence is rising indicates that mixing between iron and terbium layers is the dominant process.

4. Conclusion

In this study, we showed that in nanometric multilayers, the asymmetry of the interfaces resulting from growth conditions has to be taken into account for studying ion-beam-mixing mechanisms. Ion irradiation effects can induce mixing between the probe and Tb atoms, but also a diffusion of the ^{57}Fe atoms inside the cristalline Fe layer at the sharp Tb-on-Fe interface. Finally, our experiments show that the recrystallization effect induced by heavy-ion irradiation previously evidenced in Tb/Fe multilayers is due to a segregation of Fe and Tb which occurs only at the diffuse Fe-on-Tb interface.

Acknowledgements

The authors are grateful to F. Richomme, U. von Oersten, and W. Keune (Laboratorium fur Angewandte Physik, Duisburg, Germany) for providing the samples. F. Levesque and the technical staff of CIRIL laboratory (Caen, France) are greatly acknowledged for their valuable assistance during irradiation experiments at GANIL.

References

1. Balanzat, E. and Bouffard, S., In: *Solid State Phenomena*, Vols 30–31, 1993, p. 7.
2. Dunlop, A., Lesueur, D., Legrand, P. and Dammak, H., *Nucl. Instrum. Methods Phys. Res. Sect. B* **90** (1994), 330.
3. Teillet, J., Richomme, F., Fnidiki, A. and Toulemonde, M., *Phys. Rev. B* **55** (1997), 11560.
4. Gupta, A., Paul, A., Gupta, R., Avasthi, D. K. and Principi, G., *J. Phys.: Condens. Matter* **10** (1998), 9669.

5. Richomme, F., Scholz, B., Brand, R. A., Keune, W. and Teillet, J., *J. Magn. Magn. Mater.* **156** (1996), 195.
6. Ziegler, J. F., Biersack, J. P. and Littmark, U., *The Stopping and Range of Ions in Solids*, Pergamon Press, New York, 1985.
7. Teillet, J. and Varret, F., Mosfit program, unpublished.
8. Rusakov, V., Vedensky, B., Gadetsky, S., Voropaeva, E., Kochetov, V., Stupnov, A. and Nikolaev, E., *J. Magn. Soc. Jpn.* **17-S1** (1993), 35.
9. Juraszek, J., Fnidiki, A., Teillet, J., Toulemonde, M. and Keune, W., *Appl. Phys. Lett.* **74** (1998), 2378.

Hyperfine Interactions **156/157**: 623–628, 2004.
© 2004 *Kluwer Academic Publishers. Printed in the Netherlands.*

Thermally Induced Processes of Intermetalloid Phase Formation in Laminar Systems Fe–Sn

K. K. KADYRZHANOV[1], V. S. RUSAKOV[2], B. O. KORSHIYEV[1],
T. E. TURKEBAEV[1] and M. F. VERESCHAK[1]
[1]*Institute of Nuclear Physics NNC RK, Almaty 480082, Kazakstan*
[2]*Moscow State University, Moscow 119992, Russia*

Abstract. Phase transformations at series of thermal annealings in a laminar system of Sn–Fe–Sn were investigated by means of Mössbauer spectroscopy. Sequence and characteristic temperatures of phase transformations have been revealed. The values of hyperfine parameters of Mössbauer spectra for the forming phases as a function of annealing temperature have been obtained.

Key words: magnetron sputtering, Mössbauer spectroscopy, conversion electron, absorption geometry, hyperfine parameters, restoration of distribution function, subsequent isochronous annealing, phase transformation, laminar system.

1. Introduction

Development of methods to obtain protective high temperature resistive coatings [1] is of importance for practical application of such coatings at heightened temperatures. In this connection it is necessary to have faithful understanding of physical processes at interface protective layer – substrate.

This work reports on Mössbauer investigations on nuclei ^{119}Sn and ^{57}Fe of thermally induced processes of phase formation in a laminar system Sn–Fe–Sn obtained by magnetron sputtering.

2. Experimental

Thin film samples were prepared of α-Fe with purity of 99.98% rolled to thickness of $8\pm1\mu$m and annealed at 850°C during 2 h to eliminate the crystalline lattice defects induced at rolling. The prepared foils were covered with tin layer of 0.5 ± 0.1 μm by magnetron sputtering. The samples were sequentially annealed at temperature range of 200–700°C with a step 50°C; duration of annealing at each step was 10 h.

In our research thicknesses of tin layers (0.5 μm) and iron layer (8 μm) were selected to assure mean concentration of tin in the sample bulk ($C_{Sn} \cong 6$ at.%) to be in diphase areas of the phase diagram [2], which consist of pure α-Fe or solution of Sn in α-Fe (α-Fe(Sn)) and corresponding intermetallic compounds: at temperatures up to $607°C$ it is an intermetallic compound of FeSn, at $T = 607–765°C – Fe_3Sn_2$, at $T = 765–838°C – Fe_3Sn$, and at $T = 838–910°C – Fe_5Sn_3$.

After each annealing, Mössbauer spectra at room temperature were obtained on nuclei ^{119}Sn and ^{57}Fe in absorption geometry. Calibration of a spectrometer has been performed on standard specimens of $BaSnO_3$ and $α$-Fe. The experimental spectra were processed by methods of restoring hyperfine parameters distribution functions and of model-fitting, which were realized in a software complex MSTools [3].

3. Results

Figure 1 presents characteristic Mössbauer spectra of nuclei ^{119}Sn obtained after series of isochronous thermal annealings of the laminar system $Sn(0.5 \mu m)$–$Fe(8 \mu m)$–$Sn(0.5 \mu m)$. As a result of model interpretation of ^{119}Sn spectra, we identified partial Mössbauer spectra and determined the trend of phase transformations in a sample. So, spectrum of initial sample represents a singlet corresponding to $β$-Sn [4], and in annealing range $200–500°C$ changes of the spectrum are conditioned mainly by formation of $FeSn_2$ and FeSn phases. Spectrum obtained after annealing at $500°C$ corresponds to the compound FeSn. Parameters of spectra for annealing at $20–500°C$ are in agreement with [4–7]. One can see from the figure that at annealing within the temperature range $600–700°C$ new lines of the sextet, with an effective magnetic field of about 80 kOe, appear. This sextet corresponds to solution of $α$-Fe(Sn) studied in [9]. Simultaneously there appears a singlet with value of shift of ~1.5 mm/s. This singlet, the intensity of which has increased considerably after annealing at $650°C$, has not been possible to interpret as one of probable intermetallic compounds of the system Fe–Sn. However, presented below (Figure 2) dependence of relative intensities for partial spectra (made available from investigation of ^{57}Fe), and also data on the phase diagram of the system Fe–Sn [2] made it possible to suspect the contribution of intermetallic compound Fe_3Sn_2, which structure is not fully formed so, consequently, the hyperfine field at ^{119}Sn nuclei is not observed.

As a result of interpretation of all ^{119}Sn experimental Mössbauer spectra the dependencies of relative intensities of partial spectra for all forming phases on temperature of annealing, shown in Figure 2 were obtained. Since spectral intensity for sufficiently thin films is proportional to number of Mössbauer nuclei, in the assumption about equality of probabilities of Mössbauer effect for different nonequivalent positions, the relative intensity of partial spectrum is proportional to the number of tin atoms in each crystallographic position. Therefore, obtained relations can be interpreted as dependencies of abundance of phases in nuclear units. From the figure it is clear, that even after the first annealing at $200°C$ the content of $β$-Sn in a sample decreased from 100 to 42 at.% Sn. At further increase of annealing temperature T_{ann} this contribution continue to decrease and at $350°C$ completely disappears. At annealing within the temperature range $200–300°C$ abundance of

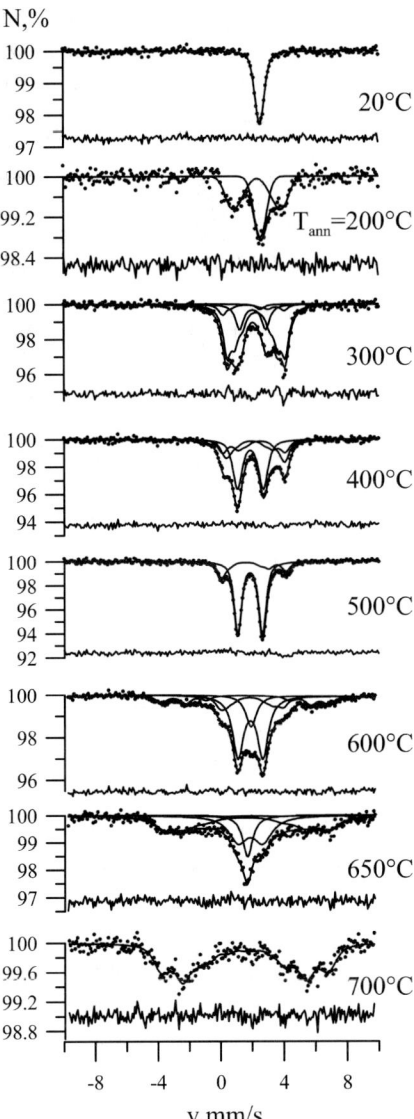

N,%

20°C

T_{ann}=200°C

300°C

400°C

500°C

600°C

650°C

700°C

-8   -4   0   4   8

v,mm/s

Figure 1. Characteristic Mössbauer spectra of ^{119}Sn nuclei obtained after consequent isochronous annealings.

phase FeSn$_2$ increases. At further increase of annealing temperature one can observe its disintegration and this phase completely disappears at 450°C. Phase FeSn forms at annealing temperature 250°C and it reaches its maximum abundance in a sample after annealing at 450°C. Feature of crystal lattice structure of FeSn is that tin in this compound takes two crystallographicly nonequivalent positions 1(a) and 2(d). Accordingly, the spectrum consists of two partial components: sextet – 1(a) and doublet – 2(d) [7]. Within the temperature range 250–400°C there is an

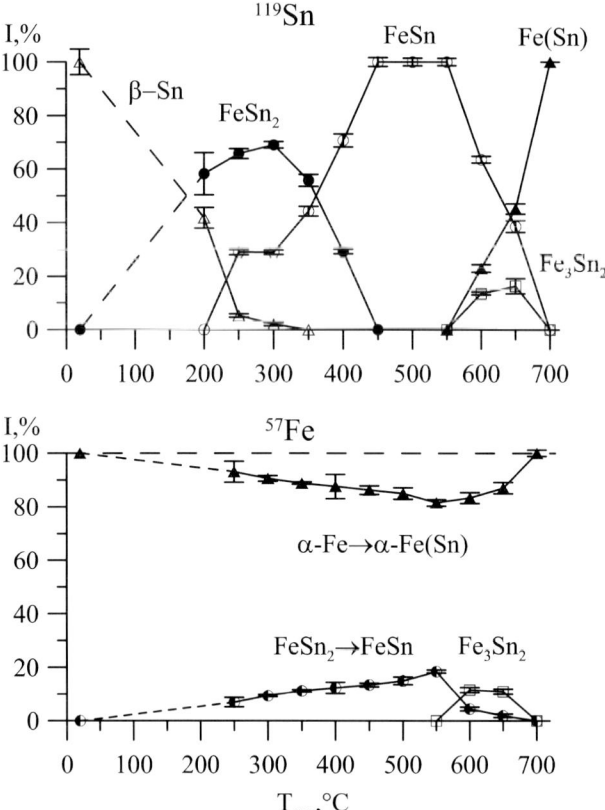

Figure 2. Dependence of the partial spectra intensity of the formed phases as a function of the annealing temperature.

intensive diffusion of tin atoms contributing to decrease of concentration gradient, reduction of local concentration of tin in intermetallic compounds and phase transformation of $FeSn_2$ to FeSn. At annealing temperatures from 450 to 550°C there are only two phases in a sample – FeSn and α-Fe. After annealing at 600°C there appears the phase Fe_3Sn_2 and solution of tin in α-Fe(Sn). Thus, the FeSn compound starts to disintegrate. At annealing within the temperature range 600–650°C disintegration of FeSn occurs due to destruction of structure of tin in a position 1(a). Quadrupole doublet that corresponds to position of tin in 2(d) can be observed till annealing temperature of 650°C. Pure solution α-Fe(Sn) with no intermetallic impurities forms after annealing at 700°C. Thus, the sequence of phase transformations in the temperature range 200–700°C is determined by temperature of consequent thermal annealings and, according to local concentration of Sn atoms, has the following sequence: β-Sn $\rightarrow$ $FeSn_2$ $\rightarrow$ FeSn $\rightarrow$ Fe_3Sn_2 $\rightarrow$ α-Fe(Sn).

Upon analysis of the distribution functions for hyperfine parameters of obtained experimental spectra for ^{57}Fe, there was plotted similar dependence of intensities for phases formed at annealing temperature (Figure 2). From the figure it is clear that the intensity of phases $FeSn_2$ and $FeSn$ increases up to the annealing temperature 550°C and decreases to zero after consequent annealing at 600°C. Starting from the annealing temperature 600°C the phase of Fe_3Sn_2 is formed, which exists up to annealing at temperature 650°C. The hyperfine parameters of $FeSn_2$, $FeSn$ and Fe_3Sn_2 compounds are in good agreement with previous works [6–8].

Analysis of relative intensities dependence on annealing temperature for partial spectra shows agreement with the phase diagram data. Indeed, according to the phase diagram, at annealing temperatures up to 513°C and large local concentration of tin atoms in iron only $FeSn_2$ can be obtained. We studied this process after annealing at 200°C. At subsequent annealings, due to diffusion processes, local concentration of tin in the iron layer should decrease. With the decrease of tin concentration, the system passes in the region of existence of $FeSn$ and solid solution in α-Fe. For this moment $FeSn_2$ is disintegrated and $FeSn$ appears.

Such nature of phase transformations in this system presents after annealing at 250°C and further, at subsequent annealings, when disintegration of $FeSn_2$ is observed. The phase of $FeSn$ is stable up to 607°C, above which there is an area of intermetallic compound Fe_3Sn_2 existence. In our case disintegration of $FeSn$ is induced by annealing at 600°C, after which the phase of Fe_3Sn_2 appears. Since at full dissolution of initial quantity of tin in iron substrate, its concentration comprises ~6 at.% and, according to data from the phase diagram, at annealing at 700°C tin dissolves completely; this is what we have observed in our investigations.

4. Conclusions

As a result of investigations of the laminar systems $Sn(0.5\ \mu m)$–$Fe(8\ \mu m)$–$Sn(0.5\ \mu m)$ subjected to sequential isochronous annealing:

- the sequence of phase transformations and relative amount of formed intermetalloid phases at each annealing step were determined;
- it was shown that the direction of phase transformations is determined by the change in local concentration of tin in the sample during the diffusion;
- a physical explanation of the phase-formation processes in the laminar systems Sn–Fe–Sn taken into account the peculiarities of the phase diagram Fe–Sn was provided;
- the hyperfine parameters of [119]Sn and [57]Fe Mössbauer spectra relating to intermetalloids phases of the system Fe–Sn formed at each annealing step were determined.

References

1. Kadyrzhanov, K. K., Turkebaev, T. E. and Udovsky, A. L., *Physical Foundations of Ionic Technologies of Stable Multilayer Metallic Materials Creation*, Almaty, 2001, 316 p.
2. Lakisheva, N. P. (ed.), *Phase Diagrams of Binary Metallic Systems: The Reference Book*, Vol. 2, Engineering, M., 1997, 1024 p.

3. Rusakov, V. S., *Mössbauer Spectroscopy of Locally Inhomogeneous Systems*, Almaty, 2000, 431p.
4. Suzdalev, I. P., Gene, M. Y., Goldansky, V. I. and Makarov, E. F., *JETP edition.1 (7)* **15** (1966), 118.
5. Le Caer, G., Malaman, B., Venturini, G., Fruchart, D. and Roques, B., *J. Phys. F* **15** (1985), 1813.
6. Nikolaev, V. I., Sherbina, Y. I. and Krachevsky, A. I., *JETP* **44** (1963), 775.
7. Kulshreshtha, S. K. and Raj, P., *J. Phys. F* **11** (1981), 281.
8. Ichiba, S., Sakai, H. and Negita, H., *Bulletin of the Chemical Society of Japan* **41**(11) (1968), 2191.
9. Nikolaev, I. N., Potapov, V. P. and Deliagin, N. N., *JETP edition.1* **70** (1976), 241.

Hyperfine Interactions **156/157**: 629–636, 2004.
© 2004 *Kluwer Academic Publishers. Printed in the Netherlands.*

DCEMS Study of Thin Oxide Layers and Interface of Stainless Steel Films Deposited by Sputtering Austenitic AISI304

K. NOMURA[1], K. TAKAHASHI[2], M. TAKEDA[2], K. SHIMIZU[3],
H. HABASAKI[4] and E. KUZMANN[5]
[1]*School of Engineering, The University of Tokyo, 7-3-1 Hongo, Bunkyo-ku, Tokyo 113-8656, Japan*
[2]*Department of Chemistry, Toho University, 2-2-1 Miyama, Funabashi, Chiba 274-8510, Japan*
[3]*University Chemical Laboratory, Keio University, 4-1-1 Hiyoshi, Yokohama 223-8521, Japan*
[4]*School of Engineering, Hokkaido University, N13-W8, Sapporo 060-8628, Japan*
[5]*Research Group for Nuclear Methods in Structural Chemistry, Hungarian Academy of Sciences,
Eötvös Loránd University, Budapest, Hungary*

Abstract. Thin stainless steel films were deposited on surface oxidized Si plate using austenitic AISI304 stainless steel as target with a RF magnetron Ar sputtering method. The deposited films and the oxidized films with about 15 nm in thickness were characterized by depth selective conversion electron Mössbauer spectroscopy (DCEMS) using a 2π gas proportional counter. The as-deposited film consisted of ferromagnetic phase. The average hyperfine magnetic fields increased from 25 to 28 T by heating. A relative large amount of iron oxide (Fe_2O_3) was produced on the top surface layer upon heating at 400°C. After heating at 500°C the relative amount of iron oxide decreased and chromium oxide layers grew in the interface between the iron oxide and substrate layers. The ferromagnetic phase in the deposited stainless steel film was partially converted into austenitic phase at 500°C and largely at 600°C. DCEMS is effective for non-destructive characterization of both surface and interface layers of thin stainless steel films with several 10 nm thickness.

Key words: stainless steel films, Ar sputtering Austenitic steel, AISI304, thin oxide layers, CEMS, GDOES.

1. Introduction

Stainless steel is widely used in many fields. We have shown that the oxide film (about 100 nm thick) of stainless steel becomes a practical pH sensor with a quick response for pH 1 to 13 solutions [1], where a magnetite coated pH sensor shows Nernst response for pH 4 to 12 solutions [2]. The pH responses of oxide films on stainless steel treated at lower temperatures were a little disturbed in low pH solutions, whereas the pH responses of oxide films on stainless steel treated at high temperatures were not affected and tolerable against coexisting other ions. Thin oxide films on previously heat-treated stainless steel are useful for vacuum vessels because of little gas absorption [3]. Ferritic stainless steel is one of the candidate materials for high-temperature container and heat exchange vessel of nuclear reactors [4].

On the other hand, conversion electron Mössbauer spectroscopy (CEMS) is useful nuclear absorption by X-rays (100%) for characterization of surface states of steels [5, 6]. After Mössbauer effect of ^{57}Fe, 7.3 keV K-electrons (80%), 13.6 keV L-electrons (8%) and 5.5 keV Auger electrons (63%) are emitted together with the secondary electrons of reemitted 6.3 keV $K\alpha$ X-rays (23%) and 14.4 keV γ-rays (10%). Integral CEMS, which detects all conversion electrons emitted from solid surface, is most prevailing. Kuzmann et al. have shown that the main phase of thick electrodeposited Fe–Ni–Cr samples is ferromagnetic [7], and have studied the metastable phase formation of Fe–Cr–Ni multilayer prepared by high vacuum deposition and subsequent irradiation by 246 MeV Xe ions [8]. Boubeker et al. [9] have investigated the crystallographic structure and morphology of as-sputtered Fe–Cr–Ni films and phase transformation by integral CEMS. We have character-ized the surface oxide layers of austenitic AISI316 and AISI304 plates by integral CEMS [10]. CEMS spectra of oxide layers formed by heating at low temperatures gave the sextets of spinel compounds and hematite, whereas CEMS spectra of oxide layers produced at high temperatures showed paramagnetic doublets.

Recently, we could observe the interface change between thin oxide and sub-strate layers of ferritic stainless steel (SUS430: 18Cr + <0.8C + Fe balance) by depth selective CEMS (DCEMS) using a He gas proportional counter [4]. DCEMS using a He counter is useful for thin surface and interface characterization. Thin stainless steel films deposited on Si plate were prepared using austenitic AISI304 stainless steel (18Cr + 8Ni + 0.8C + Fe balance) as sputtering target. The phase transformation and surface oxidation of the stainless steel deposited films heated at various temperatures were investigated by DCEMS.

2. Experimental

Stainless steel films (about 200 nm thick) were prepared on oxidized Si wafer by a RF magnetron sputtering method with Ar gas, using austenitic AISI304 steel as a target material. The as sputtered films were oxidized in air at 400, 500, and 600°C.

Mössbauer spectra were recorded at room temperature, using radiation source of 1.85 GBq $^{57}Co(Cr)$. The incident γ-rays were perpendicular to sample plane in a home-made gas counter with 2π back scattering geometry, flowing $He + 5\%$ CH_4 gas at 15 cm^3/min and applying high voltage at 950 V. Three Mössbauer spectra were simultaneously obtained by discriminating the electron energy into three re-gions such as the high (>6.5 keV), middle (3–6.5 keV) and low (1–3 keV) energy. The hyperfine field distributions of the ferritic phase were obtained by an analytical method developed by Hesse and Rubartsh [11], using the Mosswinn program [12]. Glow discharge optical emission spectroscopy (GDOES) (Jobin Yvon RF5000, Horiba Co.) was used for the depth profiles of elements, under the conditions of 13.56 MHz frequency, 40 W power, and 6.8×10^1 Pa Ar pressure. Transmission electron microscope (TEM) and atomic force microscopy (AFM) were used for estimation of thickness of oxide layers and of roughness on the films.

3. Results

The oxide thickness of as-deposited film estimated by TEM was about 3 nm, and the surface roughness of grains was within 6 nm. The oxide thickness of the films heated at 400°C was about 6 nm, and Fe atoms were concentrated on the 3 nm thick top layer. The concentration of Fe oxide decreased and Cr oxide layers increased with heating temperature as shown in Figure 1 although the intensity between different elements depends on the detection efficiency of optical emission and does not represent precisely the concentration. The oxygen content seemed to be very low variations on the surface layers. It is considered why there exist

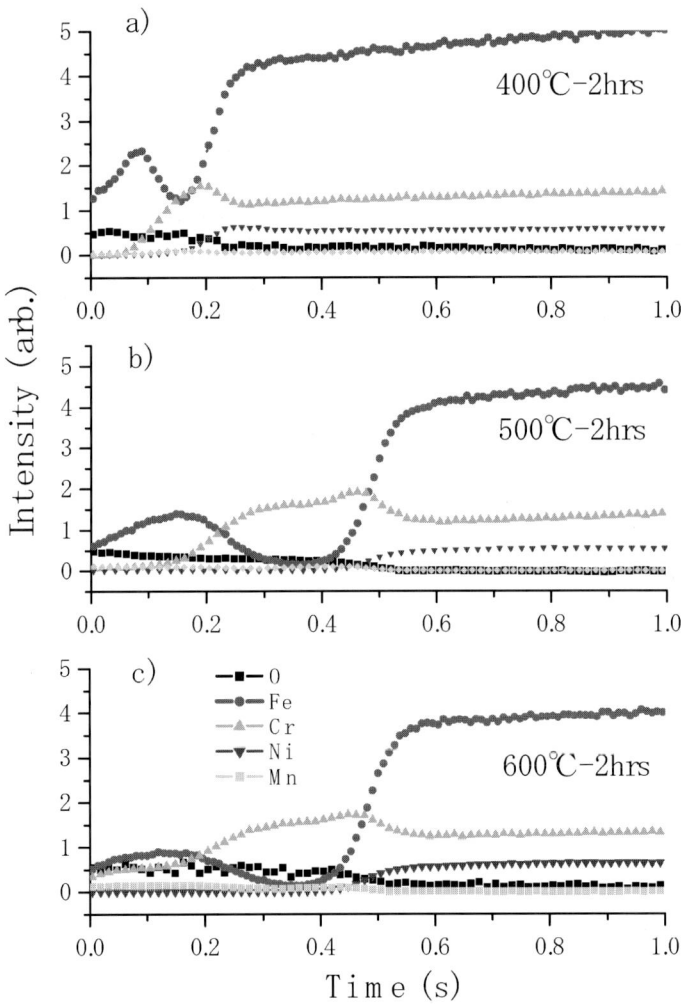

Figure 1. Depth profiles of elements in oxidized stainless steel films on Si oxide/Si wafer by a glow discharge optical emission spectroscopy. Ar sputtering time of 0.5 seconds is corresponding to 12 nm in thickness. Heating temperature: (a) 400°C, (b) 500°C and (c) 600°C.

adsorbed oxygens on the top surface, and absorbed oxygens in the inner layers to form Fe_2O_3 and Cr_2O_3. The oxide thickness of the heated films at 500°C and at 600°C was about 15 nm. Ni atoms were little observed in the inner oxide layers.

DCEMS spectra of as-deposited thin stainless steel films and heated films are shown in Figure 2. A magnetic sextet with broad lines observed in CEMS spectra of as-deposited film is due to a distorted ferritic phase, which is considered to be martensite. When the direction of magnetic moment is parallel in-plane in the most thin surface of ferromagnetic materials, the area intensity ratio of sextet peaks becomes 3 : 4 : 1 : 1 : 4 : 3 for γ-ray incident angle of 90°. The hyperfine field distributions of the substrate films beneath oxide layers were analyzed using this ratio as shown in Figure 3. The internal magnetic field distribution, B_{hf}, obtained from these broad peaks of as deposited film, was average $B_{hf} = 24.8$ T and the isomer shift, $\delta = 0.0$ mm/s. DCEMS could not detect any Fe species else the substrate of as deposited film. The average hyperfine field, B_{hf} was 28.6 T for the film heated at 500°C for 2 hours and 29.0 T for the film heated at 500°C for 4 hours. The magnetic fields became large with heating. It suggests that the heating reduces the structure distortion of as deposited film, and that the ferromagnetic exchange interaction of Fe atoms becomes stronger because the heating results in degassing of Ar in the films and in shortening of distance between Fe atoms. Otherwise, these results most probably reflect that the aging induces solid-state diffusion among the parent elements, which reduces coordination number of iron atoms for the alloying elements. It was confirmed from X-ray diffraction measurements as shown in Figure 4 that the lattice constants become from 0.2864 nm for as deposited film to 0.2858 and 0.2857 nm for the heated films at 400°C and 500°C, respectively.

The paramagnetic singlet of austenite with $\delta = -0.09$ mm/s was observed in CEMS spectra of the samples heated at 500°C. The transformation temperature was lower than thermal equilibrium temperature, where the austenitic phase was normally formed. The magnetic sextet with $B_{hf} = 50.3$ T, quadrupole splitting, $\Delta = -0.15$ mm/s and $\delta = 0.37$ mm/s, was observed in CEMS spectra of the films heated at 500°C. The hyperfine field obtained was a little smaller than that of hematite powder, α-Fe_2O_3 (52 T), and larger than that of maghemite powder, γ-Fe_2O_3 (49 T). X-ray diffraction patterns of the heated films suggested that the iron oxides might be maghemite γ-Fe_2O_3. It is known that maghemite is formed as intermediate layers at the first stage of oxidation of magnetite by heating at 400°C, and that α-Fe_2O_3 is formed at temperatures higher than 500°C. However, on the thin surface, the iron oxides may be assigned to $(Fe, Cr)_2O_3$. The intensity of the peaks of the oxide layers decreased with the increase of oxidation temperature, and there were no clear XRD peaks showing the presence of the other oxides. It suggests that iron oxides become non-crystalline by mixing with large amounts of Cr oxides.

The area intensity (11.5%) of iron oxides produced on the surface at 500°C was smaller than that (18%) of the sample heated at 400°C. It suggests that the concentration of Fe decreased relatively on the surface oxide layers by diffusion

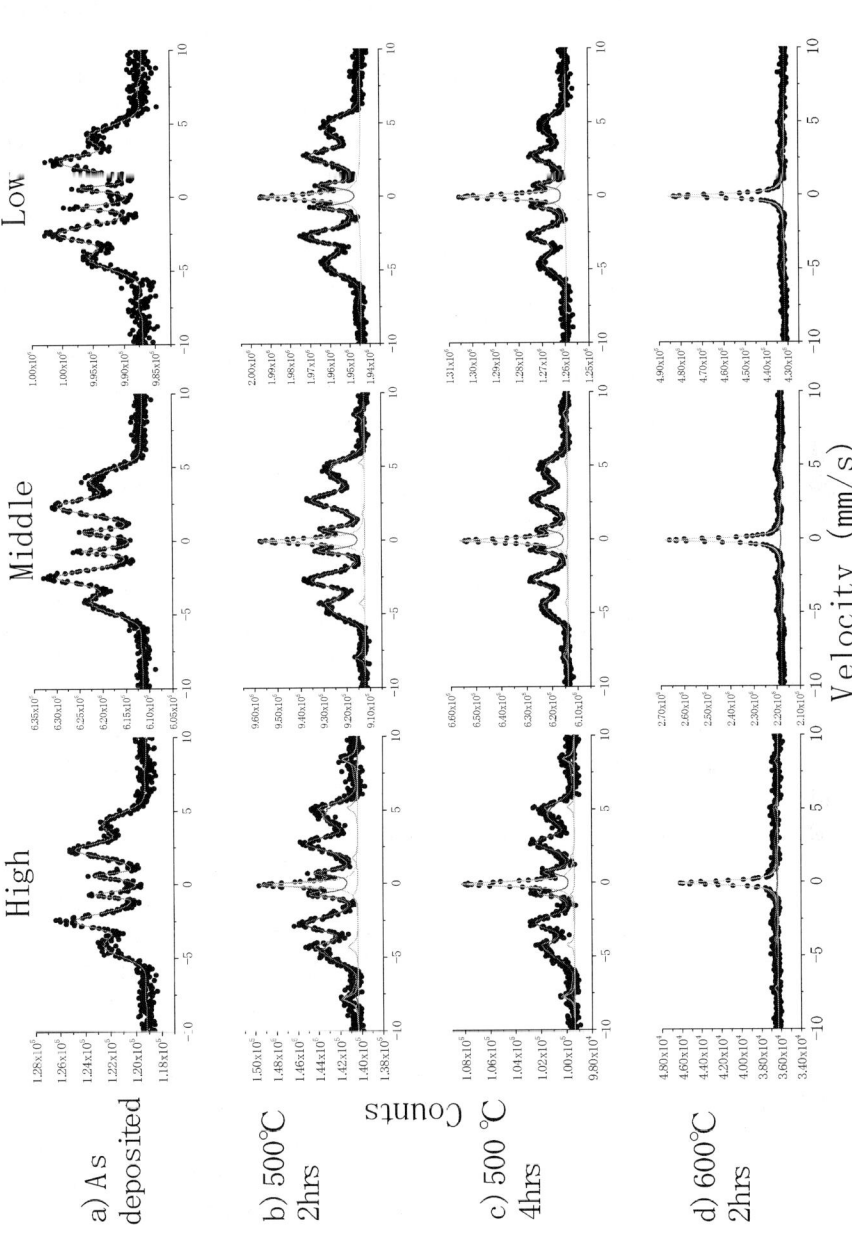

Figure 2. DCEMS spectra of thin stainless steel films; (a) as deposited, (b) after heated at 500°C for 2 hours, (c) at 500°C for 4 hours, and (d) at 600°C for 2 hours. All emitted electrons were divided into three detected parts of high (>6.5 keV), middle (3–6.5 keV) and low (1–3 keV) energy electrons.

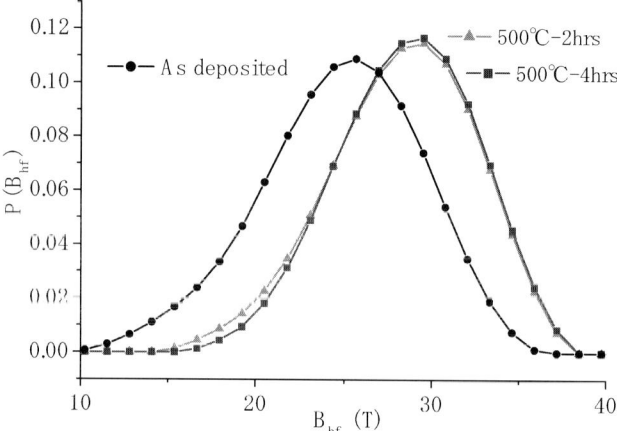

Figure 3. Hyperfine field distributions of thin stainless steel films: (a) as deposited, (b) after heated at 500°C for 2 hours and (c) for 4 hours after magnetic components of iron oxides were removed.

of Cr atoms into the surface at high temperature. The thickness of oxide layers increased to about 15 nm by heating at 500°C, and Cr atoms were more concentrated at the interface layers of the deposited films heated at 500°C than at 400°C. In the CEMS spectra of samples heated at 600°C, the singlet of austenitic phase dominated although the small amounts of sextets of ferritic phase and magnetic oxides were also detected. The relative area of singlet was 63%, 73% and 77% for high, middle, and low energy regions of detected electrons, respectively. It suggests that austenitic phase is easily formed in deep layer, and that ferritic phase remains at the interface. These are also supported by the fact that the composition at the interface of the deposited film was modified by formation of Cr rich oxide layer.

4. Conclusions

Thin stainless steel films deposited by sputtering austenitic AISI304 steel with Ar gas were characterized mainly by DCEMS using a He gas counter. DCEMS gives a simple layer by layer analysis of films within 100 nm in depth. The hyperfine fields of ferritic phase formed in as deposited stainless steel films became large due to degassing of Ar and/or diffusion of alloying elements upon heating so as to form ferritic phase. The magnetic phase can be converted partially into austenitic phase at 500°C, and largely at 600°C. The thin oxide layers on the stainless steel are basically composed of Fe_2O_3 and Cr_2O_3 layers. The intermediate oxide layers of Cr_2O_3 grew up to about 15 nm in thickness with the increase of heating temperature. The ferritic phase remained at the interface of the substrate, and oxide layers and austenitic phase was formed at the deep layer. The layer structures of deposited and oxidized films were clarified. A simple DCEMS is a useful method for surface and interface analysis of thin stainless steel films.

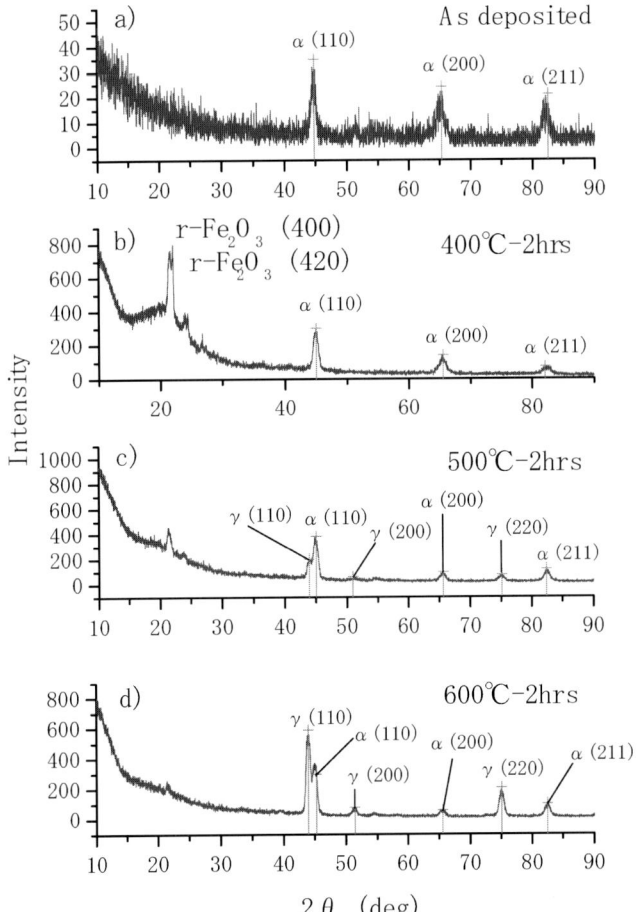

Figure 4. Thin XRD patterns of deposited and heated stainless steel films at various temperatures for 2 hours: (a) as deposited film, (b) at 400°C, (c) at 500°C and (d) at 600°C. The lattice constants of ferritic α-phase and austenitic γ-phase in deposited and heated films are shown in the following table

Lattice constant	As deposited	400°C	500°C	600°C
a_α (nm)	0.2864	0.2858	0.2857	0.2855
a_γ (nm)	–	–	0.3573	0.3571

Acknowledgements

Authors thank Mr. T. Okubo, and Mr. K. Fugawa of University of Tokyo for supporting experiments, and Ministry of Education, Culture, Sports, Science and Technology for science research found. This study was partly supported by the Hungarian Science Foundation (OTKA T043687).

References

1. Nomura, K. and Ujihira, Y., *Analyt. Chem.* **60** (1988), 2564.
2. Nomura, K. and Ujihira, Y., *Analyt. Sci.* **3** (1987), 125.
3. Cho, B., Choi, E., Chung, S., Kim, K., Kang, T., Park, C. and Kim, B., *Surf. Sci.* **439** (1999), L799.
4. Nomura, K., Hosoya, Y., Nishimura, H. and Terai, T., *Czechoslovakia J. Phys.* **51** (2001), 773.
5. Nomura, K., Ujihira, Y. and Vertes, A., *J. Radioanalytical and Nuclear Chemistry, Articles* **202** (1996), 103.
6. Nomura, K., In: M. Miglierini and D. Petridis (eds), *Mössbauer Spectroscopy in Materials Science, NATO sciences Series 3 66,* Kluwer Academic Publishers, 1999, p. 63.
7. Kuzmann, E., Vertes, A. Chisholm, C. U., Watson, A., El-Sharif M. R. and Anderson, A. M. H., *Hyp. Interact.* **54** (1990), 824.
8. Kuzmann, E., Principi, G., Tosello, C., Havancsak, K., Stichleutner, S., Gerocs, I., Homonnay, Z. and Vertes, A., *Nucl. Instr. Methods Phys. Res. B* **183** (2001), 425.
9. Bouberker, B., Eymery, Y., Denanot, M. F. and Sayouty, E. L. H., *J. Magn. Magn. Mater.* **133** (1994), 470.
10. Nomura, K. and Ujihira, Y., *J. Mater. Sci.* **25** (1990), 1745.
11. Hesse, J. and Rubartsch, A., *J. Phys. E* **7** (1974), 526.
12. Klencsar, Z., Kuzmann, E. and Vertes, A., *Hyp. Interact.* **202** (1996), 103.

Hyperfine Interactions **156/157**: 637–641, 2004.
© 2004 *Kluwer Academic Publishers. Printed in the Netherlands.*

CEMS Study on Fe Films Deposited by Laser Ablation

K. NOMURA[1] and Y. YAMADA[2]

[1]*Department of Applied Chemistry, School of Engineering, The University of Tokyo, 7-3-1 Hongo, Bunkyo-ku, Tokyo 113-8656, Japan*
[2]*Department of Chemistry, Science University of Tokyo, 1-3 Kagurazaka, Shinjuku-ku, Tokyo 162-8601, Japan*

Abstract. Iron thin films deposited on Al foil by pulsed laser ablation were characterized by conversion electron Mössbauer spectrometry (CEMS). The hyperfine fields became large and the isomer shift was closed to that of bulk α-Fe with increasing the thickness of deposited films. The small grains are so reactive that small particles of Fe^{III} oxides were produced in thin deposited layers. The magnetic orientation of the deposited films became from parallel to in-plane to at random. The small grains were considered to grow by aggregating with ablation time.

Key words: [57]Fe CEMS, iron film, laser ablation, thin film.

1. Introduction

Pulsed laser deposition has been developed as suitable technique for preparation of epitaxial thin films [1]. Pulsed laser deposition is a simple technique and has possibilities to produce clusters, to evaporate multi-components and to prepare films with the same compositions as the target. Further it has a high deposition rate in dense reactive gas atmosphere. There are some drawbacks such as small area of structural thickness and homogeneity of the deposited film. Pulsed laser deposition with substrate rotation and beam scan is promising in fabrication of uniform films [2]. On the other hand, Mössbauer spectrometry is highly sensitive to detecting the local electron density around resonant nuclei and the magnetic properties. Especially conversion electron Mössbauer spectrometry (CEMS) is useful for characterization of thin films 100 nm thick or less [3]. There are few Mössbauer studies of laser deposited films. Yamada *et al.* have studied cluster species isolated in Ar matrix at low temperature [4]. In this study, pulsed laser deposited iron films are characterized by CEMS.

2. Experimental details

Enriched [57]Fe metal was used as target of pulsed laser ablation. The deposited films were prepared on aluminum foil of 40 μm in thickness with pulsed laser of

Nd:YAG (wave length (532 nm), energy a pulse (200 mJ/pulse), pulse time width (6 ns) and irradiation frequencies (2 Hz)) in vacuum of 10^{-3} Pa. Iron films with various thickness were obtained by varying the laser irradiation time. CEMS spectra were measured using a source of 25 mCi ^{57}Co(Cr) and a homemade proportional counter with 2 π stradian, flowing He + 5%CH$_4$ gas at the rate of 10 cc/min. Both high energy and low energy CEMS spectra were simultaneously obtained by discriminating the detected electrons into two parts of high (>6.5 keV) and low energy (<6.5 keV) regions. Mössbauer spectra were analyzed by Hesse's method [5], varying the ratio of peak intensity of P$_2$ and P$_5$ to P$_3$ and P$_4$, and the better-fitted spectra were chosen.

3. Results

Figure 1 shows CEMS spectra and magnetic hyperfine distributions of iron films deposited by pulsed laser ablation. The spectra changed, depending on the deposited thickness. The hyperfine distributions were calculated for broad magnetic components in CEMS spectra The full disk was obtained from CEMS spectra, measured by detecting conversion electrons with high energy (>6.5 keV) and the empty box by detecting electrons with low energy (<6.5 keV). The former represents relatively the top surface layers and the latter the deep layers. There is not clear difference between high and low energy electron Mössbauer spectra, especially for thinner films. The average hyperfine fields of high-energy electrons detected for 87 nm thick film (B_{hf} = 24.1 T) were a little smaller, and the isomer shits (δ = 0.16 mm/s) were larger than those of low energy electrons (B_{hf} = 24.7 T, and δ = 0.13 mm/s). The surface different layers in several nm thin films were not detected, but the surface layers of several tens nm thick films by DCEMS. It suggested that top surface layers of several nm thicknesses were partially oxidized in air.

CEMS spectra of thin film with 6 nm in thickness were decomposed into the doublet and the magnetic relaxation components. The latter were fitted as hyperfine distribution. The doublet with δ = 0.37 mm/s and Δ = 1.10 mm/s, Area Int. = 35% is considered to be superparamagnetic peaks due to small particles of Fe^{III} oxide species because there is no paramagnetic Fe^{III} oxides except oxyhydroxde Fe^{III} compounds. The broad magnetic components fitted have the hyperfine parameters with average δ = 0.17 mm/s, Δ = −0.11 mm/s, average B_{hf} = 18.4 T. The average δ of average B_{hf} distribution decreased from 0.17 mm/s to 0.13 mm/s with the increase of the thickness. These may be assigned to small particles of metallic iron oxidized partially because of positive δ. It is considered that iron oxides were produced by reaction with the remaining oxygen atoms in 10^{-3} Pa pressure and/or with air after taken out from the vacuum chamber. We used an oil diffusion pump for vacuum chamber, which is generally used for physical vapor deposition using heater. The oxidation is unexpected to be serious in this vacuum level.

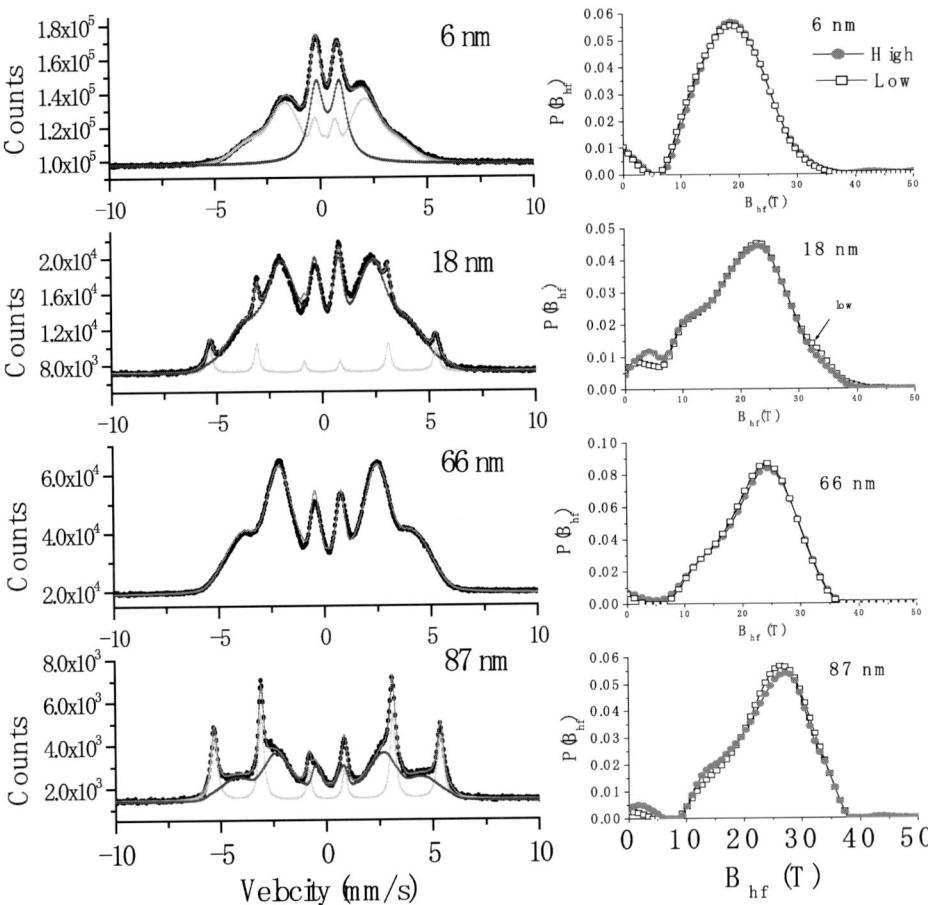

Figure 1. CEMS spectra and hyperfine distributions of iron films deposited by pulsed laser ablation. The hyperfine distributions were calculated from broad magnetic components in CEMS spectra. The full disk dots were obtained from CEMS spectra, measured by detecting the conversion electrons with high energy (>6.5 keV) and the empty box dots by detecting the electrons with low energy (<6.5 keV). A doublet is assigned to small particles of Fe^{III} oxide. Sharp magnetic sextet is assigned to α-Fe with large particles.

With increasing the thickness of the deposited films, the average δ of broad sextet decreases and average B_{hf} increased from 18 to 25 T. For 18 nm thick film, the sharp sextet with area intensity = 6.6% was observed in both relative top and deep layers. The average B_{hf} was 19.6 T for high-energy electrons detected, and 20.2 T for low energy electrons detected. In the case of the other thick films, the average B_{hf} for high-energy electrons detected was a little smaller than that for low energy electrons detected. These suggest that the grains are relatively small in top surface layers. The sharp sextet of bulk Fe was not included in the fitting of CEMS spectrum of 66 nm thick film, but was included for 18 nm and 87 nm thick films. It is considered that thickness uniformity of the deposited film with 18 nm

Table I. Mössbauer parameters of deposited films by pulsed laser ablation

Sample's thickness	δ (mm/s)	Δ (mm/s)	B_{hf} (T)	$I_{2,5}/I_{3,4}$	Area Int. (%)
6 nm					
Surface[*1]	av. 0.17	−0.11	av. 18.4	3.5	65
Fe^{III}	0.37	1.10	–	–	35
Deep[*2]	av. 0.18	−0.14	av. 18.6	3.5	68
Fe^{III}	0.39	1.11	–	–	32
18 nm					
Surface	av. 0.21	−0.05	av. 19.6	3.2	93.4
Fe	0.01	–	33.0		6.6
Deep	0.21	−0.05	20.2	3.2	93.7
Fe	0.02	–	33.0		6.3
66 nm					
Surface	av. 0.16	−0.02	av. 22.1	3.1	100.0
Deep	av. 0.16	−0.02	av. 22.4	3.1	100.0
87 nm					
Surface	av. 0.16	−0.01	av. 24.1	3.0	65
Fe	0.03	–	33.1		35
Deep	av. 0.13	−0.01	av. 24.7	3.0	71
Fe	0.03	–	33.1		29

[*1] Surface and [*2] Deep means the analyzed results of Mössbauer spectra obtained by detecting high energy (>6.5 keV) electrons, and low energy (<6.5 keV) electrons, respectively.

in thickness was by chance affected by deviation of focused laser beams. With increasing further the thickness of the deposited films, the average B_{hf} increased and the sharp sextet appeared strongly in place of broad sextet. Fe grains in the films grow with the ablation time or the thickness. The film deposited by laser ablation is different from that ratios by heater evaporation. The CEMS spectra were fitted with varying the intensity of $I_{2,5}/I_{3,4}$. The peak intensity ratios, which showed the smallest values of least square parameters, became from 3.5 to 3.0 as listed in Table I. It suggests from the ratios that magnetic moments are relatively parallel to film plane at the beginning of deposition, and that the magnetic moments are closed to at random with the increase of thickness of the deposited films.

Laser evaporated particles have high translation energy because the kinetic energy of clusters is in the range of 1 to 100 eV, whereas the kinetic energy of atoms evaporated by heater is around 0.1 eV. The accumulated films by laser ablation are considered to grow with increasing the deposition time because the surface diffusion energy is large. It is found that CEMS is useful for characterization of the deposited films prepared by laser ablation.

4. Conclusions

Fe films deposited by pulsed laser ablation were characterized by CEMS. Small particles produced by laser ablation are so reactive that Fe^{III} oxides can be produced, and the deposited films can be affected by the laser ablation time or the thickness of deposited films. The average δ of hyperfine field distribution decreased with the increase of the thickness. Magnetic moments are parallel to film plane at the beginning of deposition, and the magnetic moments become at random with the increase of the deposited film thickness. A simple depth selective CEMS using a gas proportional counter enables us to characterize the deposition process of pulsed laser ablation. In order to obtain pure metallic film by pulsed laser ablation, high vacuum is necessary, and the sample should be taken out from chamber after waiting for a while in vacuum.

Acknowledgements

The authors thank to Mr. K. Matsumoto, Mr. S. Hirayama, Miss. S. Masukawa, and Mr. A. Chikamatsu, who were students in Science University of Tokyo, for the preparation of samples.

References

1. Dijkamp, D., *Appl. Phys. Lett.* **51** (1987), 619.
2. Nagaishi, T. and Itozaki, H., *J. Vac. Sci. Technol. A* **14** (1996), 1995.
3. Nomura, K., Ujihira, Y. and Vertes, A., *J. Radioanaly. Nucl. Chem. Articles* **202** (1996), 103.
4. Yamada, Y. and Tominaga, T., *Applied Radiation and Isotopes* **52** (2000), 157.
5. Hesse, J. and Rubartsch, A., *J. Phys. E* **7** (1974), 526; Klencsar, Z., Kuzmann, E. and Vertes, A., *Hyp. Interact.* **202** (1996), 103.

Hyperfine Interactions **156/157**: 643–647, 2004.
© 2004 *Kluwer Academic Publishers. Printed in the Netherlands.*

The Investigation of the Magnetic Properties of Metallic Multilayers by Angle Dependent Mössbauer Spectroscopy

V. V. PANCHUCK, V. G. SEMENOV and V. M. UZDIN
Saint-Petersburg State University, Saint-Petersburg, Russia

Abstract. Hyperfine fields (hff) in Fe/V and Fe/Co superlattices have been investigated using angle dependent Mössbauer spectroscopy. Conversion electron Mössbauer (CEM) spectra were measured for different orientation of radiation beam relative to the crystallographic axes of the sample. Hff distributions were extracted from the texture-free spectra in which the ratio between different lines is independent on the hff directions. In Fe/V multilayers we found well-defined magnetic texture in the superlattice plane. To connect hff with magnetic moments we performed microscopic calculations of the electronic structure for Fe/V multilayers with alloyed interface. In Fe/Co multilayers the second and the fifth line of the spectra can be almost fully suppressed for appropriate beam direction in the plane of the sample. It means that all magnetic moments of Fe atoms are ordered in the same direction and, therefore, Fe/Co multilayers have to show strong magnetic anisotropy.

Key words: angle dependent Mössbauer spectroscopy, magnetic texture, metallic multilayers, hyperfine field distributions, magnetic moment.

1. Introduction

Metallic magnetic multilayers present a wide class of low-dimensional magnetic systems, demonstrating a number of new phenomena important for fundamental magnetism and for applications. Transport and magnetic properties of these systems strongly depend on the interface structure at atomic scale. Therefore the investigation of interface roughness and alloying during the sample growth is the problem of large importance. Most experimental techniques give only indirect information about chemical and magnetic roughness of the interfaces. Interpretation of experimental data proves to be very difficult and an ambiguous problem. Moreover it needs a model to connect observed experimental values with the microscopic interface structure.

For the multilayers where iron is used as magnetic material, quantitative information about atomic moments near the interface can be derived from an analysis of magnetic hff that are obtained by Mössbauer spectroscopy. The appearance of satellite lines in the Mössbauer spectra is usually associated with a different local environment of the ^{57}Fe atoms. Often in bulk metals and alloys all hff directions have equal probability. This gives $3 : 2 : 1$ ratio for the lines in the CEMS.

Existence of a preferable direction in the multilayers leads to the breaking of this symmetry. For the normal incidence of radiation on the surface the ratio between spectral lines proves to be $3 : X : 1$, where the parameter X depends on the out-of-plane hff contribution. To avoid the appearance of additional parameters in the fitting procedure and to determine the magnetic texture of the multilayers we have built an original detector with several windows, which allows performing CEMS-measurements for different incidence angles without sample rotation. Below we present results of experiments with Fe/V and Fe/Co superlattices.

2. Results and discussion

Superlattices [Fe(20ML)/V(10ML)]×25, [Fe(10ML)/V(5ML)]×20, [Fc(7ML)/V(10ML)]×29, [Fe(5ML)/V(5ML)]×20 and [Fe(5ML)/Co(5ML)]×35 were prepared by magnetron sputtering on MgO substrates at Uppsala University. Characterization of the sample was reported elsewhere [1, 2]. CEM-spectra were obtained for normal incidence of the γ-radiation as well as for grazing incidence and different in-plane directions. Typical CEM spectra of superlattice Fe(7ML)/V(10ML) are depicted in Figure 1. Other Fe/V superlattices gave similar spectra. Analysis of amplitude of lines in the spectra leads to the conclusion that all hff are in the plane of the superlattice, but the difference between spectra (b) and (c) shows that there is well defined in plane texture with preferable direction of magnetic moments along diagonal of the superlattice plane.

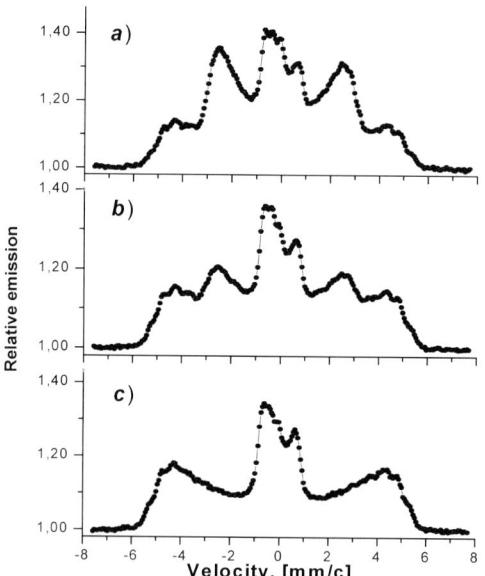

Figure 1. CEM-spectra of the Fe(7ML)/V(10ML) superlattices. (a) Normal incidence of radiation on the superlattice, (b) grazing incidence along side of the superlattice, (c) grazing incidence along diagonal of the superlattice.

To find the distribution of hff we used texture-free spectrum obtained as normalized sum of spectra measured for three perpendicular directions of incident radiation (Figure 2). The ratio between different lines in such a spectrum should be 3 : 2 : 1 independent on the distribution of hff directions [3]. Thus the hff distribution can be found without additional fitting parameters for arbitrary ordering of magnetic moments and hff in the superlattice.

The distribution of hff obtained from the spectrum depicted in Figure 2 is shown in Figure 3. It contains several roughly equidistant peaks between zero field and the field of bulk Fe. To compare hff and magnetic moment distributions we performed microscopic calculations of the electronic structure for Fe/V multilayers with alloyed interface. Intermixing was introduced to the theory by the "floating" algorithm [4, 5]. This algorithm presupposes exchange of adatoms with substrate atoms and floating of adatoms on the upper layers during deposition. Fraction of floated atoms ζ is the only parameter, which determines the intermixing at the interface. Due to the existence of a preferred direction – the growth direction – the chemical profile of the interface proves to be different on both side of the interface. This is the main reason for different alloying on upper and lower interfaces, reported in previous experimental work [6]. Calculations of magnetic moments were performed within the framework of Periodic Anderson Model. Coulomb repulsion on site was taken into account in mean field approximation [5, 7]. Figure 4 presents the distribution of magnetic moments obtained for superlattice Fe(7ML)/V(10ML) with intermixing parameter $\zeta = 0.5$.

Comparison of Figures 3 and 4 shows that in general proportionality between hff and magnetic moments takes place. However all peaks in the hff distribution are wider than peaks in the distribution of magnetic moments. There is also a difference in the relative area of the peaks, especially for the low-field contribution. Two reasons can explain these differences. First, calculations were performed for zero temperature whereas spectra were measured at room temperature. Second, in our theoretical model we neglected all kinds of interface roughness except atomic scale intermixing. In real sample there are many type of large-scale roughnesses such as

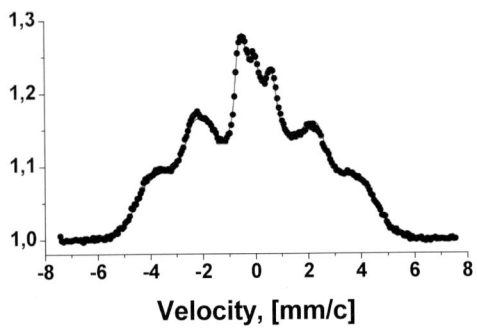

Figure 2. Texture-free spectrum of Fe(7ML)/V(10ML) superlattices.

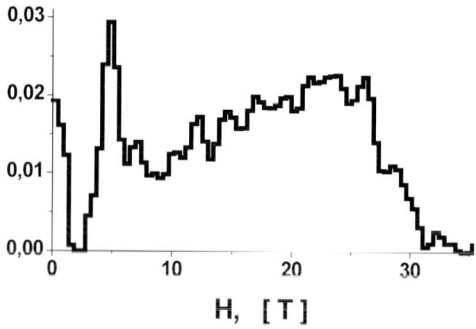

Figure 3. Hff distribution on Fe atoms in the superlattice Fe(7ML)/V(10ML).

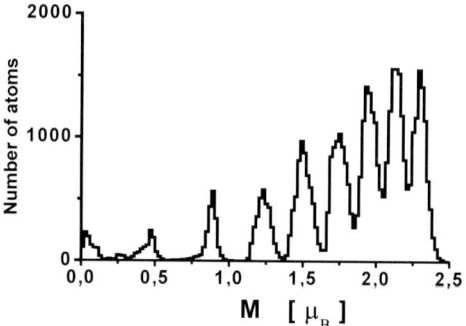

Figure 4. Calculated distribution of magnetic moments on Fe atoms in the superlattice Fe(7ML)/V(10ML).

steps, embedded clusters and islands, which can lead to the wide distribution of the moments and to an increase of low-moment contributions.

CEM-spectra of [Fe(5ML)/Co(5ML)]×35 superlattice demonstrate that all magnetic moments and hff field align along the [110] direction in the plane of the sample [8]. Second and fifth lines of the spectra are fully suppressed, if the radiation beam is incident on the sample along this direction (Figure 5b). For perpendicular direction in the plane and for normal incidence of radiation on the surface we obtain very similar spectra (Figures 5c,d) with large contributions from the second and fifth lines. These results clearly demonstrate that fitting of the single spectrum with additional parameters responsible for the out of plane component of the moment could lead to a wrong conclusion about the perpendicular-to-plane component of magnetic moments. Hff localized on interface Fe atoms are enhanced. There is no oscillation of hff, which were reported in [8]. Ordering of all moments in one direction can be the signature of strong magnetic anisotropy in Fe/Co multilayers. Recent *ab initio* calculations of magnetic anisotropy, however, did not give a large value [9].

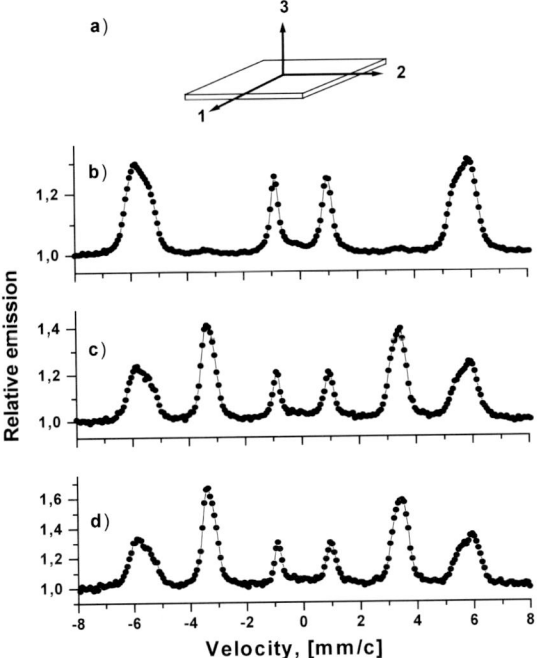

Figure 5. CEM-spectra of the Fe$_5$/Co$_5$ superlattices. (a) The directions of incidence radiation on the sample surface, (b) grazing incidence along direction 1, (c) grazing incidence along direction 2, (d) grazing incidence along direction 3.

Acknowledgements

We would like to thank L. Häggström, B. Lindgren and M. A. Andreeva for providing the samples and for useful discussion. This work was partially supported by RFBR (03-02-17168-a), the Russian Ministry of Higher Education (grant E02-3.4-255), and INTAS project (01-0386).

References

1. Kalska, B., Häggström, L., Blomquist, P. and Wäppling, R., *J. Phys.: Condens. Matter* **13** (2001) 2963.
2. Andersson, G., Nordström, E. and Wäppling, R., *Europhys. Lett.* **60** (2002), 731.
3. Greneche, J. M., Henry, M. H. and Varret, F., *J. Magn. Magn. Mater.* **26** (1982), 153.
4. Uzdin, V., Labergerie, D., Westerholt, K., Zabel, H. and Hjörvarsson, B., *J. Magn. Magn. Mater.* **240** (2002), 481.
5. Uzdin, V. M., *Comp. Mat Sci.* **24** (2002), 186.
6. Shinjo, T. and Keune, W., *J. Magn. Magn. Mater.* **200** (1999), 598.
7. Uzdin, V. M., Keune, W., Schrör, H. and Walterfang, M., *Phys. Rev. B* **63** (2001), 104407.
8. Swinnen, B., Meersschaut, J., Dekoster, J., Langouche, G., Cottenier, S., Demuynck, S. and Rots, M., *Phys. Rev. Lett.* **78** (1997), 362.
9. Eriksson, O., Bergqvist, L., Holmström, E., Bergman, A., LeBacq, O., Frota-Pessoa, S., Hjörvarsson, B. and Nordström, L., *J. Phys.: Condens. Matter* **15** (2003), S599.

Hyperfine Interactions **156/157**: 649–651, 2004.
© 2004 *Kluwer Academic Publishers. Printed in the Netherlands.*

Concluding Remarks

J. R. GANCEDO
Instituto "Rocasolano", CSIC C. Serrano 119, 28006-Madrid, Spain

My most sincere thanks to the organizers for choosing me to address you at the end of the Conference.

An important fact gives brilliance to this Conference: Prof. Mössbauer himself, the father of the creature, is with us.

Our community can be proud too of having a spectrometer, Mimos II, on its way to Mars.

It is a pity that a number of customary participants of previous ICAME's did not attend the present one, because of the fear of being exposed to an unsafe area, which, as you have realized, was not the case. We are flooded by mass media information, and, paradoxically, quite misinformed.

Data concerning topics and participation are summarized in Tables I and II.

Two facts are quite revealing:

- *The decline of presentations in Catalysis.*
- *The increasing proportion of presentations involving Synchrotron radiation was employed.*

The last statement strengthens my idea about the widening gap between "rich" and "poor" groups, not only in relation to the means, but to the ability to incorporate new sophisticated methodologies.

I am convinced that it is my duty to make a few comments on the talk of Prof. John Stevens, who showed a preoccupation for the observed decrease in the number of publications where Mössbauer Spectroscopy is employed. Although certainly this community has to take actions either to enjoy a flourishing revival, or at least to extend its own scientific life, certain remarks should, in my view, be brought into consideration:

(1) "What was obvious to do with the technique, has been obviously done". Forty years is quite a long period of time, and especially talking about an experimental discipline.

(2) Many other techniques have arisen, and the continous advances in the field of electronics have provided us with competitive tools not confined to one, or, in best of the cases, to a few isotopes.

In this sense it is mandatory to remember a comment given many ICAMEs ago by one of the most brilliant former members of this community, Prof. Uli Gonser. He said something similar to the following: *Two common approaches*

Table I. ICAME 2003 summary of contributions (13 Invited lectures, 12 Plenary talks, 30 Oral contributions (paralell), 80 Posters)

Scientific area	Number of Contributions
Lattice Dynamics and Magnetism	42
Chemistry	33
Nano	28
Soils and Minerals	24
Materials	18
Surfaces	17
Experimental and Methodology	13
Biology	10
Archeometry	4
Corrosion	3
Batteries	2
Catalysis	1

Table II. Particitants

Oman	24	Belarus	1
Germany	19	Brazil	1
Japan	17	Denmark	1
France	11	Finland	1
Russia	10	Hungary	1
India	5	Iceland	1
Sweden	5	Israel	1
Czech Republic	4	Italy	1
Poland	4	Kazakhstan	1
South Korea	4	Lebanon	1
Belgium	3	Malaysia	1
Egypt	3	Pakistan	1
Austria	2	Saudi Arabia	1
Colombia	2	Slovak Republik	1
Spain	2	South Africa	1
USA	2	United Kingdom	1
Argentina	1	Ukraine	1

can be given when doing research. "I have a panacea technique and give me any problem". And "I have a problem. What techniques, available to me, can I use to solve it?" Everybody should be aware that the second approach is the correct one. And nowadays availability of techniques is abundant.

(3) If you perform a search using any Data base, just introducing in the search the words "magnetic" and "Mössbauer", a very large number of citations will appear, and all of you know why. The clue is the usefulness of the method.

Well, leaving these remarks rather unconcluded I only want to add a couple of things.

Firstly, to thank, what I think that on the behalf of the majority of all of you, the Organizing Committee of Oman ICAME, who treated us with their so well recognized Arabian hospitality.

And secondly, I am happy to express everybody's wishes that all of you and many more will be at Montpellier in 2005.

Hyperfine Interactions **156/157**: 653–656, 2004.

Author Index to Volumes 156/1–4 – 157/1–4 (2004)